CODES

THE GUIDE TO SECRECY
FROM ANCIENT
TO MODERN TIMES

DISCRETE
MATHEMATICS
AND
ITS APPLICATIONS

Series Editor
Kenneth H. Rosen, Ph.D.

Continued Titles

DISCRETE MATHEMATICS AND ITS APPLICATIONS
Series Editor KENNETH H. ROSEN

CODES

THE GUIDE TO SECRECY FROM ANCIENT TO MODERN TIMES

RICHARD A. MOLLIN

Chapman & Hall/CRC
Taylor & Francis Group

Boca Raton   London   New York   Singapore

Published in 2005 by
Chapman & Hall/CRC
Taylor & Francis Group
6000 Broken Sound Parkway NW, Suite 300
Boca Raton, FL 33487-2742

© 2005 by Taylor & Francis Group, LLC
Chapman & Hall/CRC is an imprint of Taylor & Francis Group

No claim to original U.S. Government works
Printed in the United States of America on acid-free paper
10 9 8 7 6 5 4 3 2 1

International Standard Book Number-10: 1-58488-470-3 (Hardcover)
International Standard Book Number-13: 978-1-58488-470-5 (Hardcover)
Library of Congress Card Number 2005041403

Library of Congress Cataloging-in-Publication Data

Mollin, Richard A., 1947-
    Codes: the guide to secrecy from ancient to modern times / Richard A. Mollin.
        p. cm.
    Includes bibliographical references and index.
    ISBN 1-58488-470-3 (alk. paper)
    1. Computer security. 2. Data encryption (Computer science) I. Title.

QA76.9.A25M67 2005
005.8'2--dc22   2005041403

Taylor & Francis Group
is the Academic Division of T&F Informa plc.

Visit the Taylor & Francis Web site at
http://www.taylorandfrancis.com

and the CRC Press Web site at
http://www.crcpress.com

Preface

This book has been written with a broad spectrum of readers in mind, which includes anyone interested in secrecy and related issues. Thus, this is a tome for the merely curious, as well as history-minded readers, amateur mathematicians, engineers, bankers, academics, students, those practitioners working in cryptography, specialists in the field, and instructors wanting to use the book for a text in a course on a variety of topics related to *codes*. We will look at this topic from all aspects including not only those related to cryptography (the study of methods for sending messages in secret), but also the notion of codes as removal of noise from telephone channels, satellite signals, CDs and the like.

The uninitiated reader may consider the following. Imagine a world where you can send a secret message to someone, and describe to anyone listening in precise detail how you disguised the message. Yet that person could not remove the disguise from that message no matter how much time or how many resources are available. Well, that world exists in the here and now, and the methodology is called *public-key cryptography*. It permeates our lives, from the use of a bank card at an *automated teller machine* ATM to the buying of items or bank transactions over the Internet. You can even purchase items over the Internet and do so anonymously, as you would using hard cash. In this book, you will find out how this is done.

Do you ever wonder how secure your private conversation is over a cell phone? In general, they are not secure at all. In this book, you will find out how they can be made secure. And those transactions over the Internet, just how secure are they? Can these methods be trusted? In this text, you will learn which methodologies are secure and which are not. Here is an excerpt from the end of Chapter 2 that is apt. "What made all of the above not just possible, but rather a *necessity* — that good old *mother of invention* — was the advent of the *Internet*. While information secrecy, as we have seen throughout history, was strictly the purview of governments and their agents, the Internet, and its associated e-mail and e-commerce activities, demanded a mechanism for the ordinary citizen to have *their* privacy concerns addressed. ... Few of us actually understand the mechanisms behind all of these protocols that we use every day (although this book will foster that understanding), yet cryptography has become everybody's business, hence everybody's concern. Therefore it is almost a personal duty that each of us learn as much as possible about the underlying mechanisms that affect our security, our privacy, and therefore our well-being."

What are smart cards and how do they affect your life? This book reveals the answers. What are biometrics and how do they affect you? Several of your identity characteristics such as fingerprints, retinal data, voice prints, and facial geometry, to mention a few, can be embedded in smart cards to identify you to a bank, for instance. Perhaps you have allergies to some medicines, such as penicillin, and this information can be embedded in a medical smart card so that in the event of an accident, appropriate measures can be taken that may save your life. Read this book to find out how this is done.

How did all this begin and where is it headed? Read Chapter 1 to learn about the rumblings of the art of secrecy carved in stone almost four millennia ago and how it evolved to the present where it permeates nearly every aspect of your life.

◆ Features of This Text

• The text is *accessible* to virtually anyone who wishes to learn the issues surrounding secrecy. To this end, Appendix A contains all necessary mathematical facts for the novice, or as a fingertip reference for the initiated. Other appendices, such as Appendix E, contain the requisite probability theory for background needed to understand Information and Coding Theory in Chapter 11, for instance. Moreover, the main text is geared to gently introduce the necessary concepts as they arise. The more difficult or advanced topics are marked with the pointing hand symbol ☞ for the more advanced (or adventurous) reader.

• There are nearly *200 examples, diagrams, figures, and tables* throughout the text to illustrate the history and concepts presented.

• More than *200 footnotes* pepper the text as further routes for information-gathering. Think of these as analogues of *hyperlinks* in the Internet (see page 328), where you can click on a highlighted portion to get further information about a given topic, or ignore it if you already have this knowledge or are not interested. These links provide avenues to pursue information about related topics that might be of separate interest to a wide variety of readers.

• There are more than *80 mini bibliographies* throughout the text of those who helped to develop the concepts surrounding codes, as well as historical data in general to provide the human side of the concepts introduced.

• There are just under *300 references* for further reading in the bibliography. This provides further pointers for the reader interested in pursuing topics of interest related to what is presented herein. Moreover, it provides the foundation for the facts presented.

• The *index has nearly 5000 entries*, and has been devised in such a way to ensure that there is maximum ease in getting information from the text.

• To the *instructor* who wishes to give a course from this text: There are more than *370 exercises in Appendix G* separated according to chapter and even the appendices A–F. (Some are marked with a ✩ symbol for those particularly challenging problems.) The wealth of material in this book allows for more than one course to be given on various aspects of secrecy and even a mini-course in coding and information theory (see Chapter 11). With nearly *50 Theorems, Propositions, and related material*, and more than *60 equations*, the background is amply covered. Moreover, this text is self-contained so that no other reference is needed since the aforementioned appendices have all possible background and advanced material covered in detail (see the Table of Contents for the information covered in each appendix).

• The webpage cited below will contain a file for *updates*. Furthermore, comments via the e-mail address below are also welcome.

◆ **Acknowledgments**: The author is grateful to various people for their time in proofreading various aspects of this project. Thanks go to Professor John Brillhart, who received portions pertaining to his expertise, and as a pioneer in computational number theory with his seminal work in primality testing and factoring, it is an honour to have had him on board. I am grateful to my American colleague Jacek Fabrykowski, a mathematician who devoted his time to looking at the material. A special thanks to my former student (now working cryptographer), Thomas Zaplachinski, whose invaluable expertise in the field helped to keep the material current and accurate. A nonspecialist, Michael Kozielec, assisted greatly in giving me the valuable perspective of the uninitiated for this project which was highly beneficial in setting the proper tone for the book. Thanks go to Ken Rosen, the series editor, who always works diligently to promote the books in the series, and another special thanks to Bob Stern, my senior editor, who makes the transition from copy to finished product a seamless task. For specific information, especially on fine-tuning of details on MULTICS, and related information, thanks go to Brian Kernighan for providing background data.

Richard Mollin, Calgary

website: http://www.math.ucalgary.ca/~ramollin/

e-mail: ramollin@math.ucalgary.ca

About the Author

Richard Anthony Mollin received his Ph.D. in mathematics (1975) from Queen's University, Kingston, Ontario, Canada. He is now a full professor in the Mathematics Department at the University of Calgary, Alberta, Canada. He has to his credit over 170 publications in algebra, number theory, computational mathematics, and cryptology. This book is his eighth, with [164]–[170] being the other seven.

Dedicated to the memory of Pope John Paul II
— God's shepherd of the people.

Contents

List of Figures

Chapter 1

From the Riddles of Ancient Egypt to Cryptography in the Renaissance — 3500 Years in the Making

It was the secrets of heaven and earth that I desired to learn.
Mary Shelly (1797–1851), English novelist
— from *Frankenstein* (1818), Chapter 4

1.1 Antiquity — From Phaistos

Imagine an inscription created some 3600 years ago that nobody, to this day, has been able to decode! It exists and is carved on a clay disk, called the *Phaistos* (pronounced *feye-stos*) disk, roughly 16 centimeters (6.3 inches) in diameter, unearthed from the (old) palace of Phaistos, one of the most important locations of Minoan culture on the island of Crete, now part of Greece.

The Messara Plain is the most sizable and fertile on Crete. Only five kilometers (3.1 miles) from the coast, it ascends to form a chain of hills on the most eastern of which sits Phaistos, which was, according to Greek mythology, the residence of Rhadamanthys, one of Zeus' sons. Another son of Zeus was Minos, from which the name for the Minoan civilization derives. This civilization flourished from approximately 3000 BC to 1100 BC. Crete was the principal location of Bronze Age culture and centre of the eminent civilization in the Aegean Sea.

When this author visited Crete on a lecture tour in August of 2003, the first sight of Phaistos was a phenomenal experience, but perhaps more subdued

1

than that of Henry Miller, the famed American author who spent a few hours there in 1939 during his five-month trip to Greece. He is purported to have said: "God, it's incredible! I turned my eyes away, it was too much to try to accept at once .... I had reached the apogee, I wanted to give, prodigally and indiscriminately of all I possessed .... I wanted to stay forever, turn my back on the world, renounce everything." These anecdotes serve to give the well-deserved impression that Greece, in general, and Crete with the Phaistos site, in particular, are cradles of civilization — deserve to be praised in the highest terms — and a trip there is highly recommended. Now back to the Phaistos Disk itself.

Sometime in the evening of July 3, 1908, an excavator was the first person to unearth and view the the Phaistos Disk. At the center of the (so-called) A side or front side of the disk is an eight-petalled rosette, whereas on the B side there is a helmet sign. On both sides are inscriptions, consisting of a total of 242 symbols, 123 on the front and 119 on the back, and they spiral away from the center on the front and toward it on the back. The problem with finding the meaning of the symbols is that the disk is unique in that there are no other known texts written

Figure 1.1: View of hills and valley to the west from Phaistos.

Figures 1.1–1.4 were photographed by and courtesy of Bridget Mollin.

in the script of the Phaistos Disk, and the shortness of the existing text means that we do not have enough clues to achieve results with statistical methods. (Later, we shall learn more about statistical analysis of disguised texts such as these, called *ciphertexts*, in order to achieve the undisguised text, called *plaintext*.) The uniqueness of the disk means that there are no deductions that can be drawn from other objects in the Minoan culture as a means to begin *deciphering*, meaning the removal of the disguise to achieve the plaintext. Similarly, *enciphering* (also called *encrypting*), means disguising, the turning of plaintext into ciphertext. Later we will learn more about the difficulty of deciphering when there is very little ciphertext available. There are those who believe it is possible to decipher the disk, and several authors have published their versions of what they believe the plaintext to be. These range from a methodology for the execution of sexual rites at the palace of Phaistos to offerings to appease the gods. However, there appears to be no general agreement. No doubt there will be even more interpretations in the future. For the reader interested in more detail on this fascinating story, see Ballister's excellent and very readable, detailed, and entertaining book [12], where he concludes with: "How much longer the charming bearer of secrets and its potential solvers compete with one another, and who in the end will win, only the future will show. Until then, I recommend to everyone to visit the archeological museum in Heraklion to enjoy the beauty and the (as yet) mysterious aura of the Phaistos Disk."

Figure 1.2: Phaistos disk.

(In the above figure, the A side is on the left, and the B side on the right.)

Earlier we made some references to Greek mythology. There are other references in this type of myth to *cryptography*: the study of methods for sending messages in secret, which we now understand to mean the study of methods for transforming of plaintext into ciphertext. (The word "cryptography" comes from the Greek *kryptós* meaning *hidden* and *gráphein*, meaning *to write*.) We will learn a lot more about the cryptographic anecdotes in Greek mythology in Section 1.2. For now, this is a convenient juncture to introduce some terms related to cryptography, and discuss their origins. *Cryptanalysis* is the study of methods for defeating cryptography. The ety-

Figure 1.3: Phaistos royal apartments.

mology of the word is from the Greek *kryptós*, as above, and *analýein, to untie.*
Therefore, to say someone *crypt-analyzed* a text, means they deciphered it. (Later in the text, we will learn a great deal about cryptanalytic techniques.) The term *cryptology* is used to encompass the study of both cryptography and cryptanalysis. The (English) term "cryptography" was coined in 1658 by Thomas Browne, a British physician and writer, whereas the term "cryptology" was coined by James Howell in 1645. Yet, the modern usage of the word "cryptology" is probably due to the advent of David Kahn's encyclopedic book [131], *The Codebreakers*, published in 1967, after which the word became synonymous with the embodiment of the studies of both

Figure 1.4: Phaistos krater, Kamares style.

cryptography and cryptanalysis. Of course, *cryptographers, cryptanalysts*, and *cryptologists* are those practicing cryptography, cryptanalysis, and cryptology, respectively. Lastly, the term *cipher* (which we will use interchangeably with the term *cryptosystem*) is a method for enciphering and deciphering. Later, when we have developed more maturity in our cryptographic travels, we will be more precise, but this will serve us for the current path we are traversing. Now we continue with our discussion of antiquity and carry a new concrete set of terms to help pave our way.

Not only do the Greeks of antiquity have stories about cryptography, but also ancient Egypt has some fascinating history in the cryptographic arena. In fact, the oldest text known to employ a deliberate disguise of writing occurred almost 4000 years ago in Egypt. This is our next story.

Ancient Egypt

A nobleman, Khumhotep II, was responsible for the erection of several monuments for the Pharaoh Amenemhet II. In around 1900 BC, a scribe used hieroglyphic symbol substitution (which, in this case meant the replacing of some ordinary hieroglyphic symbols with some more exceptional ones) in his writing on the tomb of the nobleman to tell stories of his deeds. (The term *hieroglyph* means *secret carving* and is actually a Greek translation of the Egyptian phrase, *the god's words*. Hieroglyphs are actually characters used in a system of *pictorial* writing, usually, but not always, standing for sounds.) The scribe was not actually trying to disguise the inscription, but rather intended to impart some prestige and authority to his writing. Think of this as resembling the use of flowery or legalistic language in a modern-day formal document. (As most of us know, some modern-day legal documents might as well be enciphered since the ordinary individual has a hard time understanding the legalese.)

Today, the primary goal of cryptography is secrecy, which was not the intent of such scribes discussed above. The scribe's method of symbol substitution is one of the elements of cryptography that we recognize today. The use of substitutions *without* the element of secrecy, however, is called *protocryptography*. Other scribes in later years did add the element of secrecy to their hieroglyphic substitutions on various tombs. Yet, even here, the goal seems to provide a *riddle* or puzzle, which would act as an enticement to read the epitaph, which most readers could easily unravel. The obsession with the afterlife and the proliferation of tomb inscriptions resulted in a propensity of the visitors to ignore the inscriptions. When the scribes tried to revive a deteriorating interest in their craft by making these puzzles more *un*intelligible, visitors to the tombs eventually lost all interest, and the technique was abandoned. Thus, although the scribes of ancient Egypt engaged in a sort of game playing involving riddles, included were the basic elements of secrecy and symbol substitution, so we conclude that cryptography was indeed born in ancient Egypt.

These early rumblings of cryptography can be said to have sown the seeds that would develop later in various cultures. The ancient Assyrians, Babylonians, Egyptians, and Hebrews (whose contributions we will discuss in Section 1.2, along with their influence on biblical interpretations from a cryptographic point of view) all used protocryptography for the purpose of magnifying the importance of the revealed writings. For instance, the Babylonian and Assyrian scribes would often use unusual cuneiform symbols to *sign off* the message with a date and signature, called *colophons*. Again, the intent was not to disguise but to display the knowledge of cuneiform held by the individual scribe for future generations to admire. (The etymology of cuneiform is from Latin and Middle French origin meaning *wedge-shaped*.)

Now we turn to some other aspects of cryptographic finds from antiquity. From ancient Mesopotamia, one of the oldest extant examples of cryptography was found in the form of an enciphered cuneiform tablet, containing a formula for making pottery glazes. This tablet, found on the site of Selucia on the banks of the Tigris river, dates back to about 1500 BC. Mesopotamian scribes used cuneiform symbols in these formulas to encrypt their secret recipes. However, later, when the knowledge of the formulas for glaze making they were trying to protect became widespread common knowledge, their cryptographic sleights of hand became unnecessary and so later inscriptions were written in plaintext. The Mesopotamian civilization actually exceeded that of Egypt in its cryptographic evolution after having matched it in its early stages of development.

During the period of Mesopotamia under the Seleucids (312–64 BC), when cuneiform writing was in its final period, some scribes would convert names to numbers. Such cuneiform writing, in colophons, has been found in Urak, which is in modern-day Iraq, and is known to have been written at the end of the Seleucid period. This would be a major advance in cryptographic techniques if it were not for the fact that these "codes" could be easily cryptanalyzed since colophons are well known with only a couple of numbers for many plaintexts. In fact, some tablet pieces from this Mesopotamia period have been found in Susa, in modern-day Iran, consisting of cuneiform numbers in a column next to cuneiform symbols. Now, in modern-day terminology, if we have a column of plaintext symbols next to a column of ciphertext numbers, that is an example of a *code-book*, since you can look up the code and find the plaintext next to it. Hence, *if* this find in Susa is what it purports to be, it is the oldest code book in the known world. There are not enough of these tablet pieces for the experts to make a definitive decision on the matter. It makes great fodder for stories about antiquity, however.

Codes and the Rosetta Stone

We digress here for a moment to discuss the important term "codes". At the outset of the chapter, we cavalierly used the term "decode". However, what we really meant was "decipher" or "decrypt", since ciphers are applied to plaintext independent of their semantic or linguistic meaning. Throughout history the term "code" has become blurred with that of "cipher" and has come to mean (in many people's minds) any kind of disguised secret. However, today the word "code" has a very specific meaning in various contexts. It is usually reserved for the kind of meaning we have given above when we defined a "codebook", a dictionary-like listing of plaintext and corresponding ciphertext. A *cryptographic code* means the replacement of linguistic groups (such as groups of words, or phrases) with numbers, designated words, or phrases, called *codegroups*. This is the meaning that we shall use throughout. Moreover, today there are *error-correcting codes*, which have nothing to do with secrecy, but rather refer to the removal of "noise" from, say, a telephone line or satellite signal; namely, these codes provide a means of fixing portions of a message that were corrupted during transmission. We will look at such codes in Chapter 11. The codes with which we are concerned here are the ones defined above, which are *cryptographic* codes, since they have to do with secrecy. Now we return to our historical narrative.

At the beginning of the second century BC, some stonework was created in Egypt that would prove to be, some 2000 years later, the gateway to an understanding of virtually all Egyptian hieroglyphs that came before it. It was discovered in August 1779 by a Frenchman named Bouchard near the town, known to the Europeans as Rosetta, which is 56 kilometers (35 miles) northeast of Alexandria. It is called the *Rosetta Stone*, an irregularly shaped black basalt stone about 114 centimeters (3 feet 9 inches) long by 72 centimeters (2 feet 4.5 inches) wide, and 28 centimeters (11 inches) thick. It was discovered with three of its corners broken.

When the French surrendered to the British in Egypt in the spring of 1801, it came into British possession and now sits in the British Museum. On it are three different writing systems: Greek letters, hieroglyphics, and demotic script, *the language of the people*, which is a cursive form of writing derived from *hieratic*, a simplified form of Egyptian hieroglyphics. Hence, this provided an opportunity to decipher Egyptian hieroglyphic writing on a scale not seen before. Ostensibly, the inscriptions were written by the priests of Memphis in the ninth year of the reign of Ptolemy V Epiphanes (205–180 BC), in his honour for the prosperity engendered by his reign. To celebrate, they made golden statues of him in Egyptian temples, and made copies of the decree that his birthday be made a "festival day forever". This edict was cut into basalt slabs in the three writings and placed in the temples near the statues. Hence, the presumption by scholars was that the three writings were of the same plaintext — a code book — what a wonderful opportunity!

The first major breakthrough was made by a British physician, Thomas Young, in 1814. For him the sciences were a hobby. Nevertheless, his knowledge of modern and ancient languages served him well. He managed to decipher (correctly, it turns out) several of the hieroglyphs, but stopped there, since he could see no further progress possible with what he knew.

In 1821, Jean-François Champollion (1790–1832) took up where Young left off, and by 1822, this Egyptologist deciphered nearly the entire hieroglyphic list with Greek equivalents. He was the first to discover that the signs fell into three categories: (1) alphabetic; (2) syllabic; and (3) determinative (meaning a mute explanatory sign). A symbol might stand for the object or idea expressed (such as the English verb *hear* represented by the picture of an *ear*, or the verb *whine* depicted by a bottle of *wine*). He also discovered the opposite of what was expected, namely, he proved that the hieroglyphs on the Rosetta Stone were a translation from the Greek, and not the converse. Thus, the work of these two men, Young and Champollion, formed the seminal work upon which all serious future work on deciphering hieroglyphic texts was based. The discovery of the Rosetta Stone opened the door and let in the light to obliterate a darkness that had held force for almost four millennia and unlocked the secrets of the ancients. Even the very thoughts of Ramses II as he fought in battle, inscribed on the walls of Luxor and Thebes, were revealed, theretofore having only been meaningless ciphertext. It is an unfortunate end that young Champollion, the major contributor who truly saw the light, died in 1832, at the age of forty-one. He was a brilliant young man, who at the age of seventeen, was already reading papers on Egyptology. He later studied in Paris, learning Arabic, Coptic, Hebrew, Persian, and Sanskrit, which served him well in his later cryptanalysis of the hieroglyphs. In particular, his knowledge of Coptic allowed him the final breakthrough that saw to the depths of the hieroglyphs with its overlaid complexity of signs, sounds, and meaning. (Coptic is an Afro-Asian language spoken in Egypt from about the second century AD, and is considered to be the final stage of ancient Egyptian language.) He died too young to see the full impact of his work, but lived long enough to appreciate the significance of his breakthrough. As we proceed through the text, we will learn of other contributors to cryptology whose work was of the greatest benefit, yet many died in obscurity, their deeds mostly unnoticed. We will try to enlighten those individuals' lives, contributions, and humanity. For now, we move on to other civilizations from antiquity.

China

One of those great civilizations, China, did not develop any meaningful cryptography. Perhaps the reason is that most messages were memorized and sent in person to be delivered orally. Sometimes, if written, usually on rice paper, the message was concealed by covering it with wax, then either swallowing it, or concealing it elsewhere on the body. These techniques are examples, not of cryptography, but rather of *steganography*, the concealment of the *existence* of the message, sometimes called *covert secret writing*, whereas cryptography is *overt secret writing*. (We will study this practice in detail in Section 1.3.) Due to the ideographic (symbolic writing representing things or ideas) nature of the Chinese language, ciphers are ruled out as unworkable. Furthermore, since most of the populace of that time were illiterate, then the mere act of writing would have been a sufficient form of encryption in itself.

India

The India of antiquity did have numerous forms of cryptographic communications that, ostensibly, were used in practice. We mention two of the outstanding contributions from this civilization. One of them is still used today, namely finger communications (which today would be recognized by hearing- and speech-challenged people as *sign language*, or more commonly used today, *signing*). Ancient India called this kind of communication "nirābhāṣa", where joints of fingers represented vowels and the the other parts used for consonants. The second contribution of Indian civilization of antiquity is that they are responsible for the first reference in recorded history for the use of cryptanalysis for political purposes. A classic book on the craft of statehood, written at the end of the fourth century BC by Kauṭilya, called the *Artha-śāstra*, contained suggestions for diplomatic types to use cryptanalysis for obtaining information necessary to their trade. Although no mechanisms are given for carrying out such suggestions, there is some cryptographic maturity seated in the knowledge that such cryptanalysis could indeed be achieved. Later, in Section 1.4, we will see how the Arabs were the first in recorded history to give a *systematic* explanation of cryptanalysis.

The Spartans and Military Cryptography

The first to use *military* cryptography for correspondence were the Spartans, who used a *transposition cipher* device. Before describing it, let us have a look at this new term, "transposition" cipher. First let us clarify and distinguish it from the earlier use of the term, "substitution" cipher. In the case of a substitution, we replace plaintext symbols with other symbols to produce ciphertext. As a simple example, the plaintext might be *palace*, and the ciphertext might be *QZYZXW* when a,c,e,l,p are replaced by Z,X,W,Y,Q, respectively. (The cryptographic convention is to use *lower-case* letters for *plaintext* and *UPPER-CASE* letters for *CIPHERTEXT*.) However with a transposition cipher, we permute the *places* where the plaintext letters sit. What this means is that we do not change the letters, but rather move them around, transpose them, without introducing any *new* letters. Here is a simple illustration. Suppose that we have thirteen letters in our plaintext, and the following is a permutation that tells us how to move the thirteen positions around. The way to read the following is that the symbol in the position number in the top row gets replaced by the symbol in the position number below it in the second row.

$$
\begin{pmatrix}
1 & 2 & 3 & 4 & 5 & 6 & 7 & 8 & 9 & 10 & 11 & 12 & 13 \\
1 & 2 & 3 & 4 & 10 & 7 & 8 & 9 & 5 & 6 & 11 & 12 & 13
\end{pmatrix}
$$

Now, suppose that our plaintext is *they flung hags*. Then the ciphertext will be *THEY HUNG FLAGS*. Notice that the first four and last three plaintext letters remain in the same position as dictated by the above permutation, but the f in position 5 gets replaced by the H in position 10; the l in position 6 gets replaced by the U in position 7; the u in position 7 gets replaced by the N in position 8; the n in position 8 gets replaced by the G in position 9; the g in position 9 gets replaced by the F in position 5; and the h in position 10 gets replaced by the L in position 6. So this is an easy-to-understand method of depicting transposition ciphers that we will use throughout the book. We can see that transposition ciphers depend upon the permutation given, such as the one above, so often transposition ciphers are called *permutation ciphers*.

Now let us return to the Spartans, the great warriors of the Greek states. The Spartans used a transposition cipher device called a *skytale* (also spelled *scytale* in some sources). This consisted of a tapered wooden staff around which a strip of parchment (leather or papyrus were also used) was spirally wrapped, layer upon layer. The secret message was written on the parchment lengthwise down the staff. Then the parchment was unwrapped and sent. By themselves, the letters on the parchment were disconnected and made no sense until rewrapped around a staff of equal proportions, at which time the letters would realign to once again make sense. One use of the skytale was documented to have occurred around 475 BC with the recalling of General Pausanius, who was a Spartan prince. He was attempting to make alliances with the Persians, an act the Spartans regarded as treasonous. Over one hundred years later, a skytale was used to recall General Lysander to face charges of sedition. Thus, the Greeks have been credited with the first use of a device employing a transposition cipher.

The earliest writings on cryptography, as instructional text, is credited to the Greeks. In the fourth century BC, Aeneas Tacticus wrote a book on military science, called *On the Defense of Fortifications*. In this book, an entire chapter is devoted to cryptography. In this chapter, Tacticus also describes several clever steganographic techniques. One of these techniques is to puncture a tiny hole above or below letters in a document to spell out a secret message. Almost two thousand years later, this method was used (with invisible ink and microdots rather than pin pricks) by the Germans during the world wars in the twentieth century.

More credit goes to the Greeks in terms of development of some of the first substitution ciphers. Polybius who lived approximately from 200 to 118 BC was a Greek historian and statesman. He invented a means of enciphering letters into pairs of numbers as follows.

The Polybius Square

Table 1.1

	1	2	3	4	5
1	a	b	c	d	e
2	f	g	h	ij	k
3	l	m	n	o	p
4	q	r	s	t	u
5	v	w	x	y	z

Label a 5 by 5 square with the numbers 1 through 5 for the rows and columns, and string the English alphabet through the rows, considering "ij" as a single letter, as given in Table 1.1.

Then, look at the intersection of any row and column (with row number listed first and column number listed second) as the representation of the letter in question. For instance, k is 25 and q is 41. Hence, the letters are plaintext and the numbers are ciphertext. This device is called the *Polybius checkerboard* or *Polybius square*. Polybius' intended use of his square was to send messages great distances by means of torches and hilltops. The sender would hold a torch in each hand, then raise the torch in the right hand the number of times to signal the row, and the torch in the left hand the number of times to signal the column. There is no evidence that these were actually used in this fashion or any other in ancient Greece. However, there are many variations of his cipher that have been constructed. The reader may even concoct one by pairing different letters than "ij", and stringing the alphabet in a different way from the straightforward one given in Table 1.1. One such interpretation of Polybius' cipher involved turning the digits into sounds. A known application in the twentieth century was the one developed by Russian prisoners who used *knocks* to convey speech. For instance, using Table 1.1, a prisoner might knock on a wall twice, followed by three knocks for the letter "h", then proceed in this fashion to send a complete message. Hence, this came to be known as the *knock cipher*.

Polybius' substitution cipher has found great acceptance among cryptographers up to modern times, who have used it as the basis for numerous ciphers. We will mention some as we encounter them later in our cryptographic voyage.

Julius Caesar

Although the ancient Greeks made no claim to actually using any of the substitution ciphers that they invented, the first use in both military and domestic affairs of such a cipher is well documented by the Romans. In *The Lives of the Twelve Caesars* [276, page 45], Suetonius writes of Julius Caesar: ".... if there was occasion for secrecy, he wrote in cyphers; that is, he used the alphabet in such a manner, that not a single word could be made out. The way to decipher those epistles was to substitute the fourth for the first letter, as d for a, and so for the other letters respectively." What is being described here is a simple substitution cipher used by Julius Caesar. He not only used them in his

domestic affairs as noted above by Seutonius, but also in his military affairs as
he documented in his own writing of the *Gallic Wars*.

Table 1.2

Plain	a	b	c	d	e	f	g	h	i	j	k	l	m
Cipher	D	E	F	G	H	I	J	K	L	M	N	O	P

Plain	n	o	p	q	r	s	t	u	v	w	x	y	z
Cipher	Q	R	S	T	U	V	W	X	Y	Z	A	B	C

This substitution cipher is even easier to use than that invented by Polybius,
which we discussed above. In this case there is merely a shift to the right of
three places of each plaintext letter to achieve the ciphertext letters. This is
best illustrated by Table 1.2.

Table 1.2 is an example of a *cipher table*, which is defined to be a table
of (ordered) pairs of symbols (p, c), where p is a plaintext symbol and c is its
ciphertext equivalent. For instance, in the Caesar cipher table, (b, E) is the pair
consisting of the plaintext letter b together with its ciphertext equivalent E.
An example of a cryptogram made with the Caesar cipher is: *brutus* becomes
EUXWXV. Also, this simple type of substitution cipher is called a *shift cipher*.
Moreover, the mechanism for enciphering in the Caesar cipher is a shift to the
right of three letters. So the value 3 is an example of a *key*, which we may regard,
in general, as a *shared secret* between the sender and the recipient, which *unlocks*
the cipher. So 3, in this case, is the *enciphering key*. Since shifting 3 units left
unlocks the cipher, then 3 is also the *deciphering key*. This is an example of
a *symmetric-key cryptosystem*, namely, where one can "easily determine" the
deciphering key from the enciphering key and vice versa. (We will formalize this
notion in Chapter 3, when we study symmetric-key cryptosystems in detail,
but for now, this will suffice.) Thus, the key must be kept secret from all
unauthorized parties. (This is distinct from a cryptosystem, about which we
will learn in Chapter 4, where the enciphering key can be made publicly known!
Yet, nobody can determine the deciphering key from it.) There is a method of
employing the Caesar cipher with numbers that simplifies the process. Consider
Table 1.3 that gives numerical values to the English alphabet.

Now, if we take *zebra* as the plaintext, the numerical equivalent is
$25, 4, 1, 17, 0$, and using the Caesar cipher we add 3 to each number to get
the ciphertext. However, notice that when we get to x, y, z, adding 3 will take
us beyond the highest value of 25. The Caesar cipher, Table 1.2, actually loops
these three letters back to A, B, C.

Table 1.3

a	b	c	d	e	f	g	h	i	j	k	l	m
0	1	2	3	4	5	6	7	8	9	10	11	12

n	o	p	q	r	s	t	u	v	w	x	y	z
13	14	15	16	17	18	19	20	21	22	23	24	25

Hence, what we have to do here is to throw away any multiples of 26 and
treat them as zeroes in our addition, and only accept nonnegative numbers (the

positive integers and 0) in our scheme. (This is called *modular arithmetic* in mathematical terms; in this case, modulo 26, and here 26 is called the *modulus*). We perform modular arithmetic in our daily lives when we look at our clocks as mod 24 arithmetic. Once the 24 hours are done, we begin again to count from zero to the midnight hour. This is what we will do here modulo 26. We need a symbol other than = to denote our addition since the outcome will not be strict equality, but rather equality *after* throwing away multiples of 26. Since we might change the value of 26 for some other ciphers, then we need to keep track of it as well. We do this by writing

$$25 + 3 \equiv 2 \,(\text{mod } 26),$$

for instance, in our current example since $25 + 3 = 26 + 2$, which is just 2 when the 26 is discarded. Continuing then, we get that the plaintext numerical equivalents $25, 4, 1, 17, 0$ become $2, 7, 4, 20, 3$, and using Table 1.3, the ciphertext message becomes *CHEUD*. Once sent, the recipient uses the key 3 to decipher by first converting the ciphertext to letters via Table 1.3, then calculating, for instance $2 - 3 \equiv 25 \,(\text{mod } 26)$, since $2 - 3 = -1 = 26 - 1 = 25$, given that multiples of 26 are treated as 0 and no negative numbers are allowed in our arithmetic, described above. (In other words, -1 is the same as 25 modulo 26, and we must choose 25 since only the nonnegative numbers less than 26 are allowed.) Similarly, all other numbers are decrypted to yield $25, 4, 1, 17, 0$, which, via Table 1.2 becomes *zebra*.

The Caesar cipher is a simple example of more general ciphers called *affine ciphers* about which we will learn when we revisit the Caesar cipher in Chapter 3. The introduction of the Caesar cipher is an opportunity to solidify our understanding of ciphers in general. First, we describe it verbally, followed by an illustration. As we have seen, a cipher not only involves a set of plaintext/ciphertext pairs (p, c), but also a key k used to encipher and decipher. Moreover, the key has to satisfy certain properties. We want to ensure that when we encipher a plaintext element using the key, there is *only one* possible ciphertext element, and there is *only one* possible decryption to plaintext possible. (In mathematical terms each key is called a *one-to-one* function.) Thus, we may describe a *cipher* or *cryptosystem* as a set (a collection of distinct objects) of plaintext/ciphertext pairs (p, c) together with (one or more) enciphering keys k, each having a corresponding deciphering key d, called the *inverse* of k, such that $k(p) = c$ and $d(c) = p$. In other words, the action of enciphering using k, denoted by $k(p) = c$ is "unlocked" by d when d is applied to c, denoted by $d(c) = p$. Hence, the action of k followed by d has the *unique* result of doing "nothing" to p, namely,

$$d(k(p)) = d(c) = p.$$

(In mathematical terms, this action is called an *identity function* since it identifies the original object with itself, p in this case.) These properties ensure a well-defined cryptosystem, a definition that we will be using throughout.

Diagram 1.1 A Generic Cryptosystem

(I): Encryption

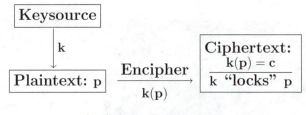

(II): Decryption

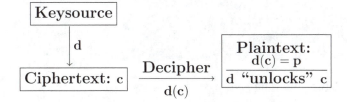

Anglo-Saxon Britain and Scandinavia

Thus far, we have concentrated on the great civilizations of antiquity in Rome, Greece, and Asia. However to the north, in Anglo-Saxon Britain and Scandinavia, cryptographic finds were of high importance as well. We will now look at one of them of note.

In the Rök churchyard in Östergötland, Sweden (dating from the beginning of the Viking era), a ninth-century, thirteen-foot-high slab of granite was discovered. It is known, therefore, as the Rök stone, which has 725 legible texts from the runic language. (See image on the right; courtesy of site owner at *http://www.deathstar.ch/security/encryption/*.)

The *runic alphabet* was used by Germanic people of Britain, northern Europe, Iceland, and Scandinavia from approximately the third to the seventeenth century AD. Although experts are uncertain, it is most probable that runic was developed by the Goths (a Germanic people) from the *Etruscan alphabet* of northern Italy. The inscriptions on the Rök stone are of secret formulas and epic tales. The wealth of letters makes it a treasure chest for the cryptologist.

Figure 1.5: Rök stone.

The Rök stone (see Figure 1.5) is perhaps the best known of the Teutonic runes and Celtic *oghams* (pronounced *oy-hams*). This writing dates somewhere from the first to the fourth century AD, used for (mostly) the Irish language in stone. Runes are abundant in Scandinavia and Anglo-Saxon Britain. (There are approximately 3500 stones with runic inscriptions found in Europe, mostly in Sweden and Norway.) However, there is an older runic stone, containing the oldest extant runic inscription, called the *Kylver Stone* (see Figure 1.6).[1.1]

Figure 1.6: The Kylver stone.

This is a limestone slab dating to the fifth century, which was found in the province of Gotland, Sweden. The inscriptions are of the older runic alphabet, sometimes called *Futhark*, which is is both chronologically and linguistically the oldest testimony to any Teutonic language. (This earliest version of the runic language had 24 letters, divided into three sets, called *ættir*, of 8 letters each. The sounds of the first six letters were *f*, *u*, *th*, *a*, *r*, and *k*, respectively yielding *Futhark*.) The inscriptions on the Kylver stone are facing inside the coffin, most likely to protect the gravesite as some incantation. It contains a *palindrome* (any sequence of symbols that reads the same backward or forward) on it, *sueus*, presumed to be some magical protection, but it has not been deciphered.

These enciphered methods of rune writings are called *Lønnruner* in Norwegian, meaning *secret runes* or *coded runes*. It is not clear that the intention of the carvers was to secrecy, but perhaps, as we saw with the early stages of writings on Egyptian tombs, the rune carver's only purpose was to demonstrate his skills for others to admire (perhaps as puzzles for learning Futhark). Known Ogham writings number nearly 400 in Ireland. These extant examples of Ogham are principally grave and boundary markers. However, there is some evidence of its use by the Druids for documenting stories, poetry, etc. (The *Druids* were the learned class of the ancient Celts, the first historically identifiable inhabitants of Brittany. *Druid* is Celtic for *knowing the oak tree*. Moreover, Julius Caesar, who is perhaps the main source of information about Druids, classified Celts into *druids* as *men of religion and learning*, also *eques* as warriors, and *plebes* as commoners.) It is uncertain if the Druids actually used enciphered oghams for divination or magical purposes. Any carvings in wood have long ago rotted away, leaving only the stone inscriptions. However, in the *Book of Ballymote*, written in 1391 AD, are some fragments of writing, in another system, called *Bricriu's Ogham*, which may be interpreted as an enciphered ogham from an-

---

[1.1]This image from *http://www.runewebvitki.com/index.html*, courtesy of site owner, Rig Svenson.

cient Druid liturgy. (The Book of Ballymote, a collection of Irish sagas, legal texts, and genealogies, along with a guide to the Ogham alphabet — from which much of our present knowledge of Ogham derives — currently sits in the Irish Academy in Dublin.) The main twenty letters of the Ogham alphabet represent the names of twenty trees sacred to the Druids (for instance, *A-Ailim* for *Elm* and *B-Bithe* for *Birch*). The Ogham alphabet was invented, according to the Book of Ballymote, by *Ogma*, the Celtic god of literature and eloquence. In Gaul, he was known as *Ogmios*, ostensibly identified with the Roman hero/god *Hercules*.

In its most rudimentary form, Ogham con- sists of four sets of strokes, which appear like notches in the rock inscriptions, each set con- taining five letters comprised of between one and five strokes, yielding a total of twenty let- ters, mentioned above. These can be seen to be carved into the stone from right to left, or on the edge, in Figure 1.7. In a later development of the language, a fifth set of five symbols were added, called *forfeda*, an Irish term for *extra let- ters*. Ogham is read from top to bottom, left to right.

Ogham markings on standing stones (or *gallán*) have been found as far as Spain and Portugal, in an area once known as Celtiberia, an area of north-central Spain occupied in the third century BC by tribes of Celtic and Iberian peoples. However, some of the inscriptions in Spain date to 800 BC, quite a bit older than the ones in Ireland. The Iberian Peninsula (oc- cupied by Spain and Portugal in southwestern Europe) was colonized by the Celts in 1000 BC.

Figure 1.7: An Ogham stone.

It is part of conjecture that the Celts may have found their way from Celtiberia across the Atlantic to the New World as early as the first century BC. Evidence of this is the discovery of ogham-like carvings in West Virginia in the United States. Readers interested in more detail on Ogham can refer to the relatively recent, easy-to-read, and quite informative book by Robert Graves [115], first published in 1948.

Perhaps one final comment on Druids is in order before we move on. The archeological site in southern England, known as *Stonehenge*, could not have been, as is often claimed, built as a temple for the Druids or Romans since neither was in this location until long after the last stages of Stonehenge were built. The initial stages date back to 3100 BC and were used by Neolithic man who carved the stones with deer antlers, which ostensibly helped to (carbon-14) date them. The final stages of Stonehenge were completed in about 1550 BC. However, there is no cryptography there to interest us.

The Mayan Civilization

Now it is time to leave the Old World and sail across the Atlantic to the New World, where a great people reigned from 2000 BC to 1500 AD, the Mayan civilization. (*Mayan* means *Thrice built*.)

Perhaps some of the most difficult of the languages that have yet to be deciphered, are the *Mayan hieroglyphs*, the only genuine writing system ever devised in the pre-Columbian Americas. This writing system was used by the Mayan Indian peoples of Meso-America from roughly the third to the seventeenth century AD. We use the term *hieroglyph* (see page 4) since the more than 800 symbols are mostly representations of objects, namely, they are pictorial in nature, *pictograms*, and typically we abbreviate this term and refer to them as *glyphs*. Up to the middle of the twentieth century, only minute amounts of mostly numeric data were decrypted. From the middle to the end of the twentieth century progress was made in deciphering numerous Mayan inscriptions, so that by the 1990s a significant number of decipherings were achieved, but much remains to be done. The complexity of the Mayan system is underscored by the fact that a given symbol may represent a complete word. Such glyphs are called *logographs*. A glyph that represents only a sound, syllable, or even just a part of a word is called a *phoneme*. Yet, that is not all. A single logographic symbol might have many meanings. Also, any given glyph could represent a sound, a concept, or both. Hence, there are the interwoven problems of deciphering not only a symbol's logographic meaning — what it represents — but also its phonetic meaning.

Although the reader may find similarities in what we are describing here to what we described in the tackling of the Egyptian hieroglyphs, there are two major differences. First, unlike the Egyptian hieroglyphs, where there were Greek versions, such as on the Rosetta stone, there is no known conversion of Mayan glyphs into another language. Secondly, there are no people alive today who can read or write the glyphs. The Mayan glyphs are unlike the Phaistos disk in that there are a substantial number of sources that have been recovered. Mayan hieroglyphs have been found carved in stone monuments (called *stelae*, meaning *stone trees*), on pottery, jewellery, and to a far lesser extent, in books. The books of the Mayans are called *codices*, most of which were destroyed by Spanish priests, who considered them to be pagan in nature. Four codices are extant. The oldest is the *Paris Codex* dating, it is believed, to the fourth century AD. In Figures 1.8 and 1.9 are representations of two pages of the Mayan zodiac from the Paris Codex, where the constellations are represented by zodiacal animals such as a bird, scorpion, snake, and turtle (there are a total of thirteen zodiacal animals in the Mayan zodiac corresponding to their thirteen constellations). (These digital representations were downloaded from *http://digital.library.northwestern.edu/codex/download.html*, courtesy of Northwestern University Library.)

The most recent codex, the *Grolier Codex*, dating to the thirteenth century, contains exhaustive writings on the orbit of the planet Venus. However, it is estimated that more than half its twenty pages are missing. The other two extant codices are the *Dresden Codex*; and the *Madrid Codex*, dating from about the eleventh and fifteenth centuries, respectively. Of the four codices, the Dresden is the most deciphered. The physical appearance of the codices is quite striking given that they were made of fig bark paper folded into an accordion shape with outside covers of jaguar hide.

Figure 1.8: Paris Codex zodiac 1.

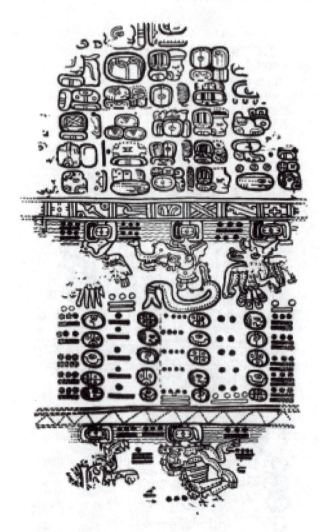

Figure 1.9: Paris Codex zodiac 2.

Epigraphers (those who study ancient inscriptions), working on the Mayan inscriptions are using the Internet and modern-day super-computers to house and dissect the massive body of data gathered over the years. This may be viewed as a task equivalent to trying to crack the code of the Mayans as they would any contemporary cryptosystem. Given the wealth of talent and sophistication of computing and cryptanalytic techniques available today (much of which we will discuss in this book), the day of a complete understanding of the ancient Mayan script and its civilization's secrets may well be at hand.

In Figure 1.10 is a photograph of the *Pyramid of the Magician* in *Uxmal*, Yucatán, Mexico. This was built in *Puuc* style, an architecture used during 600–900

AD, part of the *late classic period*. (The three *periods* of Mayan civilization are *Pre-Classic* (2000 BC–250 AD); *Classic* (250–900 AD); and *Post-Classic* (900–1500 AD).) This pyramid has representations of the rain god *Chac*. In fact, a decryption of glyphs shows that the ruler of Uxmal took the name *Lord Chac* in roughly 900 AD. As with many other cities, Uxmal was abandoned in about 1450 AD. After millennia, the Mayan civilization ceased to be, but nobody knows why, albeit speculation abounds from natural disaster to invasions, one of the great mysteries.

Figure 1.10: Pyramid of the Magician.

Easter Island

To close this section with another fascinating story, we head south, and west to an isolated island 2200 miles west of Chile, now a Chilean dependency, Easter Island. It is the easternmost of the Polynesian islands, famed for its giant stone heads, standing three stories high, called *moais* or *busts*.

In 1722, a Dutch admiral, Jacob Roggeveen, was the first European to visit the island. To commemorate the day of their arrival, the Dutch named it *Paaseiland* or *Easter Island*. However, to its inhabitants, largely of Polynesian descent, it is known as *Rapa Nui* or *Great Rapa*, also *Te Pi te Henua* or *Navel of the World*. Not only were the moais found, but also, tablets inscribed with a language called *rongorongo*. This language still has not been deciphered. Rongorongo is a pictographic language (such as the Egyptian hieroglyphs). Moreover, every other line is written upside down, meaning that the tablet would have to be turned upside down every time a line was read. Experts speculate that the tablets were used by priests for purposes of worship, so they are often called *sacred* tablets. However, nobody really knows. Only some thirty or so tablet fragments remain, so as with the Phaistos disk with which we began this section, there is not enough data to make a definitive analysis of the script. Thus, as with the Mayan mysteries, rongorongo remains one of the few languages left that have not been deciphered. Some images of rongorongo inscriptions are given in Figures 1.12 and 1.13. Figure 1.12 is a portion of a rongorongo tablet. Figure 1.13 is the *Santiago Staff*, a walking stick. It was was obtained, in 1870, from the French colonist Dutrou-Bornier. He maintained that it had belonged to an *ariki* or *king*. It is entirely covered with rongorongo signs, inscribed along its length.

Figure 1.11 is an image of one of the roughly 600 giant stone busts that pepper the island. Although they were initially objects of worship by the inhabitants, when Captain James Cook reached the island in 1774, he found that most of them had been deliberately knocked over.

The population had been reduced from 3000 people to roughly 600 men and little more than a couple dozen women. Ostensibly a civil war had taken its toll on the aborigines there. Although the population again reached 3000 by 1860, a Peruvian-launched slave trade, coupled with smallpox, nearly annihilated the population, so that by 1877, there were only 111 inhabitants left. The population again increased by the end of the nineteenth century. In 1888, Chile annexed Easter Island, and turned it into a sheep-raising community. In 1965, the islanders became Chilean citizens, maintaining their culture and ancestral affiliations. In fact, each February the inhabitants meet for celebrations of the island's past with a revival of old skills and customs.

Figure 1.11: Easter Island Moais.

Given the fact that *antiquity* refers to times up to the Middle Ages and we have covered both the Old and New worlds, this is an appropriate juncture at which to conclude this section. We have only barely scratched the surface of the history of antiquity as it applies to cryptography, but the reader will have a sufficient sense of our past to carry forward.

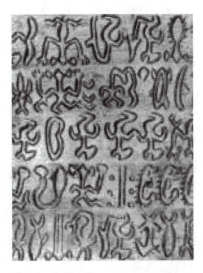

Figure 1.12: Rongorongo tablet.

Figure 1.13: Santiago Staff Segment.
(Figures 1.12–1.13 are courtesy of *http://www.rongorongo.org/*, site owner.)

1.2   Cryptography in Classical Literature

Classical quotation is the parole *of literary men all over the world.*
Samuel Johnson (1709–1784), English poet, critic, and lexicographer

The epic Greek poet Homer, perhaps one of the greatest literary figures of all time, wrote the *Iliad* (among other great stories such as the *Odyssey*). In it, a hero of Greek mythology, Bellerophon, son of Glaucus, and grandson of Sisyphus, fled Corinth after having killed Bellerus, for which he acquired the nickname *Bellerophontes*. He came, as a suppliant, to Proteus, King of Argos, whose wife Anteia fell in love with the handsome hero at first sight. However, he rejected her, and in the role of a "woman scorned", she told the King that Bellerophon had tried to seduce her. Believing his wife, but unwilling to risk the Furies' (goddesses of vengeance) wrath, the king sent Bellerophon to Anteia's father, Iobates, the King of Lycia, with an enciphered message in a folded tablet, the plaintext of which said: "Pray remove the bearer from this world, he has tried to violate my wife, your daughter."

Iobates, for reasons similar to his son-in-law's, was unwilling to directly ill-treat a royal guest. Instead he asked Bellerophon to do him the favour of slaying the *Chimera*, a rather nasty, fire-breathing, she-monster with the head of a lion, body of a goat, and tail of a snake. However, Bellerophon, being no fool, consulted the seer Polyeidus, who advised Bellerophon to first trap and tame the winged horse Pegasus. Bellerophon had been given the gift of a golden bridle by the goddess Athena (after which the city of Athens is named, and why she is considered the *city protectress*, but more commonly, the goddess of war, handicraft, and practical reason in Greek mythology). This gift proved to be timely since Bellerophon, upon finding Pegasus drinking from a well at Periene, on the Acropolis of Corinth, was able to throw the bridle over his head. Then he was able to fly over the Chimera on Pegasus's back, firing a volley of arrows, and finally thrust a spear, which had a clump of lead affixed to it, into the monster's mouth. The Chimera's fiery breath melted the lead, which flowed into her throat, down into her body, searing and killing her.

Iobates was not done. He sent Bellerophon to war against the Solymians and Amazons, but they too were defeated when he flew over them, dropping large rocks on their heads. Before returning to Iobates, he was able to conquer Carian pirates in the Lycian Plain of Xanthus. Iobates sent palace guards to ambush him on his return. However, Bellerophon prayed to Poseidon (god of the sea), to flood the Xanthian plain behind him as he advanced on foot. Poseidon heeded the prayers and sent waves forward as Bellerophon approached the palace where Iobates waited. Since no man or monster could stop him, the Xanthian women offered themselves to him, if he would relent. Being far too modest, Bellerophon turned and ran, the waves retreating along with him. Now, finally, Iobates was convinced that the enciphered message must have been in error. He then demanded the truth from Anteia and upon getting it, begged forgiveness from Bellerophon, offered his daughter Philonoë in marriage, and made him heir to the Lycian throne. Together with the fascinating aspects of

Greek mythology that the above anecdote illustrates, it also shows that even when a message is correctly deciphered, the plaintext message itself may be in error.

Hebrew Literature and the Bible

In the Hebrew literature, the most common technique of letter substitution is called *atbash*, in which the first and last letters of the Hebrew alphabet are interchanged, and the remaining similarly permuted, namely, the penultimate (second to last) letter and the second interchanged, and so on. In fact, the word *atbash* itself is an example of what it denotes. The reason is that it is composed of *aleph*, *taw*, *beth*, and *shin*, the first, last, second, and penultimate letters of the Hebrew alphabet.

Atbash is used in the Bible, in order to add mystery rather than hide meaning. The importance of its use therein is that it inspired the European monks and scribes of the Middle Ages to rediscover and invent new substitution ciphers. Through this development, cryptography was reintroduced into Western culture, and the modern use of ciphers may be said to have grown from this phenomenon. An example of the use of atbash in the Bible is given in *Jeremiah 25:26*: "All the kings of the north, near and far, one after the other; all kingdoms upon the face of the earth [and after them the king of Sheshach shall drink]." *Sheshach* is formed from *babel* by substituting the letters of the Hebrew alphabet in reverse order. The first letter of babel is *beth*, the second letter of the Hebrew alphabet, and this is replaced by *shin*, the penultimate letter. The last letter of babel is *lamed*, the twelfth letter of the Hebrew alphabet, and this is replaced by *kaph*, the twelfth-to-last letter.

In the Bible, there is a well-known *cryptogram* (meaning the final message after encryption, the ciphertext, encapsulated and sent), although this one does not involve atbash. It occurs in the Old Testament in the *Book of Daniel*, which was originally written in Aramaic, a language related to Hebrew, and generally thought to have first appeared among the Arameans (a Semitic people of the second millennium BC in Syria and Mesopotamia) roughly around the late eleventh century BC. The setting is the great banquet given by Belshazzar, the Chaldean king, for a thousand of his lords. As it says in *Daniel 5:5–5:6*, "Suddenly, opposite the lampstand, the fingers of a human hand appeared, writing on the plaster of the wall in the king's palace. When the king saw the wrist and hand that wrote, his face blanched; his thoughts terrified him, his hip joints shook, and his knees knocked." The king sought his wise men to decipher the message. Either they could not or would not do so, since the message was bad news for the king, who was slain that very night. In any case, Daniel was brought before the king and easily interpreted the words for him. "This is the writing that was inscribed: MENE, TEKEL, and PERES. Translation: MENE: God has numbered your kingdom and put an end to it; TEKEL, you have been weighed on the scales and found wanting; PERES, your kingdom has been divided and given to Medes and Persians."(*Daniel 5:25–5:28*)

In the above, Daniel deciphers the three terms via a play on words. *Mene* is associated with the verb meaning *to number*; *Tekel* with the verb *to weigh*;

Peres with the verb *to divide*. Moreover, there is an additional play on the last term with the word for *Persians*. In any case, Daniel's accomplishment made him the first cryptanalyst, for which he became "third in the government of the kingdom." (*Daniel 5:29*)

It is unknown why the king's wise men, for the above-described biblical cryptogram, could not interpret what is essentially a plaintext, so the only answer can be their fear of revealing the bad news to the king. Of course, interest in biblical cryptography has been a topic of interest for many to this day. However, it can be argued that there is *only* the use of protocryptography since the essential element of secrecy is missing.

Troy

We now return to Greek mythology as the backdrop to one of the greatest decryptions in history. Homer's *Iliad* and *Odyssey*, as well as Virgil's *Aeneid* contain accounts of a city named *Troy*, but were they all myths? Let us begin with the account of the *Trojan War*, fought between the Greeks and Trojans. There is no cryptography in this tale, so don't look for any. This fascinating story is presented here both to set the stage for the real-world attempt to find the site of Troy, and the cryptographic secrets to which this quest finally led, as well as to delineate the rich historical and cultural connections with the search for this understanding.

King Priam of Troy and his wife Hecuba had a son Paris, whom Zeus invited to judge a beauty contest between his wife Hera and two of his daughters, Athena and Aphrodite. The scene of the contest is the celebration of the wedding of Peleus, father of Achilles, and Thetis, the water nymph. The need for the contest arose from the fact that Eris, goddess of strife, arrived at the celebration, despite not having been invited. She brought a golden apple upon which was written "For the fairest." Hera, Athena, and Aphrodite all made claim to the apple, appealing to Zeus for a decision; hence the invitation to Paris since Zeus did not want to do it himself for obvious reasons.

Of course, each of the goddesses tried to curry favour with Paris by their offerings. Hera offered power; Athena, military glory; and Aphrodite, a woman as beautiful as herself, Helen of Sparta, for his wife. Paris gave the apple to Aphrodite.

Helen was the only female child of Zeus, and was mortal, but her beauty was world-renowned. She married Menelaus, King of Sparta, and as fate would have it, Paris was sent as Trojan ambassador to Sparta. Paris and Helen instantly fell in love, left Sparta for Troy, taking a great amount of wealth from the city's coffers with them. The Spartans appealed to Troy for return of the treasure, by sending a delegation with Odysseus, King of Ithaca, and Menelaus, the betrayed husband. The Trojans refused, and so the Spartans prepared for war.

The Greeks amassed a fleet of 1000 ships, the largest contingent of which was led by the commander-in-chief, Agamemnon. The Greek army landed on the beaches of Troy and settled down for a siege that lasted more than a decade. Prince Hector, son of King Priam, and leaders of the Trojan army had much success against the Greek forces by breaking through their fortifications and burning their ships. Among the battles was the famous showdown between Achilles and Hector outside the walls of Troy. Achilles won, attached the corpse to his chariot and dragged it away. Later, Priam went to the Greek encampments, pleading for the return of Hector's body and Achilles returned it to Priam for a ransom.

The background of Achilles now comes into play. When he was an infant, his mother bathed him in the waters of the river Styx, which according to myth, resulted in his invulnerability to any weapon. However, his heel, which the waters did not touch since his mother held him by one foot, was his one vulnerable spot. In the tenth year of the war, Paris, with the help of Apollo, killed Achilles with an arrow that pierced his heel. Later, Philoctetes, a leader of a contingent of Greek ships, was able to kill Paris with an arrow shot from the bow of Hercules.

There are many other battle stories, but the ultimate tells of the Greeks contriving the scheme of building a wooden horse that they filled with armed warriors. (In a sense, this was a steganographic technique!) To make it appear that they were abandoning the war, the Greek army withdrew. To celebrate their victory, the Trojans tore down part of their wall and dragged the horse into the city. Later that night, when the Trojans had fallen asleep, the hidden Greek soldiers emerged from within the horse, opened the gates, and signalled the main army, which was in hiding. King Priam was slaughtered at the alter by Achilles's son Neoptolemus; Hector's infant son, Astyanax, was thrown off the walls; and the women, Hecuba, and Cassandra, the daughter of Priam, and Andromache, the wife of Hector, were taken as prisoners.

After the war, the gods considered the sacking of Troy (the best account of which is in Virgil's *Aeneid*) a sacrilege, particularly in view of the desecration of the temples. Thus, they punished many of the Greeks. For example, Menelaus' ships wandered the seas for seven years, while Agamemnon returned to Argos only to be murdered by his wife, Clytaemnestra, and her lover, Aegisthus. Of particular importance (as we will see below) is that Odysseus (known as Ulysses to the Romans) was forced to wander the seas for ten years before returning home to Ithaca, alone. Poseidon had been so angered by Odysseus' putting out the eye of Polyphemus, the cannibal cyclops, and son of Poseidon, that all his ships and all his men were lost on the voyage back to Ithaca. In Ithaca, he disguised himself and killed the princes who were trying to seduce his wife, Penelope, into marrying one of them, and trying to kill his son, Telemachus. After so long an absence, Odysseus had to prove his identity by being able to string the famous *bow of Odysseus*, which was a task no other man had been able to accomplish. Moreover, he was able to tell Penelope the secret tale of their marriage bed, which Odysseus had built around an olive tree. Numerous other tales spring from the Trojan War, and it can legitimately be argued that few stories in our culture have been the inspiration of so many artists, writers, sculptors, and playwrights. It also inspired one particular archaeologist.

One person who believed that Troy was not myth made it his life's goal to prove it, and the story of Troy in the *Iliad* inspired him. Heinrich Schliemann was born in Mecklenburg, North Germany, on January 6, 1822. At the early age of seven, he was given a book, *Jerrer's Universal History*, by his father. The book contained a painting of Troy in flames. Throughout his life he believed that the Homeric versions were more than myth, and he became obsessed with Troy and Homer.

After some early work of little significance, he ultimately managed to acquire a huge fortune using his innate merchant skills, by the time he was thirty. Then he married Katherina, a niece of a business acquaintance. Although the marriage lasted fifteen years, and produced a son and two daughters, it was filled with violent verbal arguments, separations, reconciliations, and ultimately divorce.

He had an aptitude for languages, mastering fifteen by the age of thirty-three, including ancient and modern Greek. In 1851, he first visited America, became a U.S. citizen, and opened a bank in California during the gold rush, which added to his fortune. In 1858, he took an extensive tour of the Middle East, returning to America a second time in 1868, trying to reconcile with his wife, but it was doomed to failure. He then began another extensive voyage of wandering, this time setting foot for the first time on the island of Ithaca. Here he began to dig and excavate, finding what he believed to be the remains of Odysseus and his wife, Penelope. After Ithaca, he travelled to the Peloponnese, Mycenae, the Dardanelles, and the Plain of Troy. Now he was ready to relinquish his business ventures and settle into an extended search for Troy. He was also ready for divorce and finding a new wife. This time, he would not leave it to chance. He wrote a letter in the winter of 1868 to his old friend Vimpos, who had taught him Greek earlier in his life. Now Vimpos was archbishop of Athens. Schliemann appealed to him to find him a Greek wife. After his divorce from Katherina the next year, he arrived in Athens in August and married his new young bride, Sophie.

There had been speculation among scholars and archeologists that a probable site of Troy could be the hill of Hisarlik, in modern-day Turkey. Schliemann had visited the area in 1868, and now was convinced of it. In 1870, with his eighteen-year-old bride by his side, he made some preliminary excavations at Hisarlik. By late 1871, he and numerous workers under his command drove deeply into the northern slope of the hill. Schliemann was a novice at this, and there were little precedents in the archeological world to guide him at that time. The scale and magnitude of the venture was unprecedented. So, believing that Troy was at the lowest levels, when he encountered a building of relatively late date that impeded his progress he demolished it without attempting to record any of it (which would make modern-day archeologists shudder). By 1873, he uncovered the ruins of a city, which he believed to be the Troy of Homer's *Iliad*, and what he thought was Priam's gold.

Not wanting to part with the treasure, he smuggled the gold and jewellery (some of which he believed to have been worn by Helen herself, and with which he adorned Sophie) out of Turkey to Athens. In 1874, after numerous political and legal problems, he was able to offer the Greek government a suggestion that he be able to keep part of the treasure during his lifetime, but that it would revert to Greece after his death. Moreover, the government agreed to let Schliemann dig at Mycenae, his next project, under the supervision of the Archeological Society of Greece, at his own expense, conditional upon his handing over all the finds. They did give him exclusive rights on publishing his discoveries for up to three years.

After two years of legal battles with the Turks, which ended in his having to pay them compensation, he finally began to dig at Mycenae. With Sophie by his side and an entourage of workers, in the summer of 1876, Schliemann began to dig in an area known as the *Lion Gate*, which is a gateway upon which sit two lions carved in stone. This is a gateway to what Schliemann called the *Citadel of the Atridae*, a flat top of a hill crowned with a vast ring of walls. He found upright slabs 87 feet in diameter, beyond which there was a circular stone altar, more gravestones, and a gold ring. At this point, the workers were dismissed, leaving only Heinrich, Sophie, and Stamatkis, the *ephor* who represented the archeological society. They uncovered a total of six graves within the ring of stone slabs, a Grave Circle. Each grave was a shaft of varying depths, containing a total of nineteen bodies of men, women, and children, many laden with gold. Numerous treasures were uncovered, from bronze daggers inlaid with gold designs having various engravings in the men's graves, to engraved golden crowns in the women's graves, treasure of gold masks and crowns. He was certain that he had found the tomb of Agamemnon and Cassandra, among others. Was Schliemann right? Later dating techniques showed that if Agamemnon actually lived, it would have been around 1180 BC, the presumed date of the Trojan War. However, the finds at the Citadel were earlier, around 1600 BC. The excavations Schliemann made at Hisarlik turned out to be the site of Troy but the dating was off by several hundred years. He had dug *past* the level on which Troy itself did reside! He had dug through the very walls of Troy to get to where he thought it was. One of the *upper* levels *was* Priam's Troy. So *if* the treasure found by Schliemann at the lower levels was from a much earlier age, who were the owners? Scholars touted Schliemann for his intuitive acuity, but posed that the objects were older than the period of the Trojan War, older than Homer. There was someone else who shared that belief.

In 1882, a thirty-one year old Englishman, Sir Arthur Evans (1851-1941), came to visit the Schliemanns in Athens, having been introduced by his father whom Schliemann had met in England. He was interested in looking at some of the bead seals and signet rings that Schliemann had found at Mycenae. He believed that they were Aegean, but they fascinated him because he saw elements of ancient Egypt in them. He wanted to unravel the puzzle. Now we continue with the fascinating story that will take us back full circle to Crete, and stories surrounding it, that we discussed in Section 1.1, and a cryptological find that stunned the world.

Evans was born in Nash Mills, England, the son of a paper manufacturer and amateur archeologist of Welsh descent. He was educated principally at the University of Oxford, England, and the University of Göttingen, Germany. He was a recognized scholar who became the curator of the Ashmolean Museum at the University of Oxford from 1884 to 1908, and was appointed as *extraordinary professor of prehistoric archaeology* at Oxford in 1909. He had a long-standing interest in sealstones and ancient coins, one of the reasons he had sought out Schliemann.

Knossos

Meanwhile, Schliemann was seeking out further diggings signalled by the Homeric writings, this time on Crete. There was Idomeneus, leader of the Cretan contingent at the siege of Troy, and many other Cretan stories to inspire him. Schliemann applied to the Turkish government, who then ruled Crete, in 1883 to dig at the site of Knossos. When he had finished his latest diggings three years later, at Tiryns, Schliemann arrived on Crete. He had sought to buy the site on which Knossos sits, but got involved with a shady owner who was trying to cheat him, so he broke off the negotiations and never again considered it. In 1890, a year after he had an operation on his ear, which had been giving him great pain in the last few years, Schliemann was travelling home to Athens across Europe for Christmas. At Naples, his ear troubles returned, so he consulted a doctor. Feeling somewhat better, he visited the ruins of Pompeii, mentioned to him by his father in his youth, but the pain returned with a vengeance. The next day, Christmas day, on his way to see the doctor whom he had visited earlier, he collapsed on the street in a state of paralysis. Eventually, he received medical attention, but it was too late. The inflammation had spread from his ear to his brain and he died the day after Christmas.

Figure 1.14: An artist's rendition of life at Knossos.

This was photographed by Bridget Mollin at the Iráklion Archeological museum, as were Figures 1.15–1.19; see Section 1.1.

However, Evans, who respected his predecessor, and who had been so enchanted eight years earlier when he and his wife visited Schliemann, and viewed the Mycenaean treasures, would indeed carry the torch to unlock many theretofore unsolved mysteries. However, whereas Schliemann had been driven by a belief in the Homeric tales as literal truth, Evans was guided by scientific curiosity. He was drawn, in part, to Crete by the *milkstones*, which are Minoan sealstones, engraved with hieroglyphic symbols. According to folklore, the Minoan women wore these after giving birth with the expectation that this would increase their milk production for nursing. Evans believed that they might be the key to unlocking the language of the Minoan civilization.

In March of 1900, Evans began his excavations at Knossos. The site of Knossos is slightly to the south of Iráklion. It is a quadrangular mound with two steep slopes to the east and south, but roughly level with the surrounding terrain on the other two sides, a mound called *Kephala*. By the end of the month he had the first inscribed clay tablet in hand, and by the end of the first week of April a wealth of them to behold, with the same script as he saw on those sealstones that had lured him there. But he found more than he had sought. There were pieces of art, so refined and beautiful that they could only have been created by a great civilization. Evans hired more workers until those under his command numbered over one hundred. Then on April 5, his first — perhaps greatest — visual find, a picture (see Figure 1.19, page 34) of one of the peoples who had inhabited Knossos on that mound at Kephala, and he named him a Minoan after Minos, the presumed ruler of Crete, and mythological son of Zeus. It was becoming increasingly clear that the Kephala mound held majestic palaces, ranging six acres in magnitude, truly the remains of a magnificent civilization. The first palaces at Knossos were believed to have been built around 2000 BC. He had bored back two millennia BC and saw the opportunity to unlock the mystery of nearly 2000 years of human civilization. The key was in the script.

Figure 1.15: Knossos Linear B Tablet

Evans saw four distinct kinds of script, so he began to classify them. The oldest, the type he had seen on the sealstones, was pictographic, and he called these hieroglyphic script of class A. The stylized form of this that he found on clay tablets, (of which over 3000 were eventually unearthed), which seemed to be an evolutionary spinoff of class A from the Mycenaean period, he called hieroglyphs script of class B. There were two further simplifications of classes A and B, which were more linear than the aforementioned hieroglyphic types. These he dubbed linear script of class A and the most recent, the linear script of class B. It turns out that linear B, as it eventually came to be known, was found only at Knossos, but linear A, again a simplification as it came to be known, could be found all over Crete. It was determined that the two classes of linear A and B did not live together. Linear B had replaced linear A. But the interrelationships were not clear among the four classes. Classification became more refined dividing symbols into sets based on agricultural types, ideographs, phonetic, or numerical. However, all this classification was not deciphering. He still could not read the language.

It should be noted that although Evans used the term "hieroglyphs" with reference to the discovered Minoan scripts, they were not related to the Egyptian hieroglyphs. However, it should also be noted that the stylized pictures of Minoans that they found were the so-called *Keftiu* or people of the islands, found on the walls of Egyptian tombs, of *non*-Egyptian type, with whom the Egyptians both fought and traded. Vases (*rhytons*) were found in Knossos exactly as depicted in the paintings on Egyptian tombs in Thebes. The excitement among Egyptologists could be heard round the world.

Linear B was closely related to the writings discovered on Cyprus dating to 2000 BC, known as *Cyproti-Minoan*. On the other hand, linear A appears on the Phaistos disk, discussed in Section 1.1. However, as we have seen, the Phaistos disk appears to be unique. Linear B consists of ideographs, 87 syllabic letters, numerical symbols, and symbols of weights and measures (the latter two in the decimal system), with the occasional combination of ideographs and syllabic letters. Evans did not decipher linear B. It took about half a century for that to come to fruition, yet linear A remains a mystery.

Evans received numerous honours for his discoveries including a knighthood in 1911. He died in 1941 before seeing the deciphering of the clay tablets that he brought to light. That would be for another young man to do.

In 1936, Michael Ventris was in the audience when Sir Arthur Evans gave a lecture on Minoan writings. Ventris was then only fourteen, but he developed an enthusiasm for the challenge posed by the undeciphered Minoan script, which led him in later life to contact experts, begin reading, learning, and working as a cryptanalyst would.

Figure 1.16: A Knossos symbol: double axe.

By 1952, Ventris, an architect, not an archeologist, had deciphered linear B. Moreover, he verified that the language was indeed an early Greek dialect. This demonstrated that during a disputed time period, 1400–1125 BC, the Greek mainland dominated Crete. Hence, earlier versions of the Late Bronze Age of the Aegean region had to be rewritten. When the clay tablets were deciphered, they were found largely to be bureaucratic trivia concerning insignificant business transactions. However, the value of the tablets is in that to which their existence silently attests. They were 400 years older than Homer, and linear B is the written form of the language spoken at the time of the Trojan War. Furthermore, they help to separate the parts of Homer's writings that are historical from those embellished for the heights of mythological fame. Ventris' cryptanalytic success took us back by more than 700 years from what had previously been deciphered. Until his success, we only had evidence of Greek writing from about 750 BC. Now we could go back to roughly 1500 BC. Moreover, it gives us a glimpse of Greece in the Bronze Age, and clears up some previously unclear areas. Hence, the fact of the decryption and its methodology to obtain it, using only analytic techniques, attests to its greatness. It may not have unlocked secrets as great as the Pharos, as the Rosetta Stone allowed, but it gave us a clearer picture of our shared history farther back than anyone before could accomplish, put the writings of Homer in a clearer light, and gave us the language of ancient Troy.

Figure 1.17: Knossos fresco: blue dolphins.

Ventris, as so many others who contributed to our cryptological heritage, died far too young. He had an accident while driving home late one night near Hatfield, England, at the age of 34. Yet his contribution, another door opened to the past, and the light it shines for us lives on.

For the reader interested in the words of the discoverers themselves, we recommend Evans' own work [78], and Schliemann's works [233]–[234]. For an account of Ventris' cryptanalysis of linear B, see [52].

More From Greek Literature

We now look at three other figures from Greek literature, who lived much later than Homer. Our first figure is Thucydides, who is considered to be one of the greatest Greek historians, primarily for his writing of the *History of the Peloponnesian War* between Athens and Sparta in the fifth century BC. His work was divided into eight books ending in the events of the autumn of 411 BC, almost seven years before the end of the war. Yet his work stands tall as a definitive record, presumed to be the first, of a political and moral analysis of a country's policies on war, (giving his viewpoint as a native Athenian). Our interest in him here is a link with the previous section, since he wrote about how, in 475 BC, the Spartan General Pausanias was recalled from the field using

an enciphered skytale (page 9). In fact, it is only through his writings that we
know of this.

We also mentioned (page 9), that one hundred years later General Lysander
was recalled by a skytale. This, of course, we do not know from Thucydides, but
rather from Plutarch, who was the author of over 220 books, and 60 essays on
topics ranging from ethics and religion to the political and the literary. He can
be said to have had the earliest influence on the western concept of the *essay*,
the *biography*, and *historical writing*.

Our last Greek character for this section is Herodotus, who lived circa (484–
425 BC). He was the author of perhaps the first great historical text, *Histories*,
which dealt with the Greco-Persian Wars. He is believed to have been born in
Halicarnassus, a Greek city in southwest Asia Minor, under Persian rule. His
contribution to our topic was not so much to cryptology as to steganography,
which we have already encountered in our travels, and which we will study in
detail in Section 1.3.

Figure 1.18: Palace ruins at Knossos.

Herodotus' first tale is that of General *Harpagus*, who served under *Astyages*,
king of the Median Empire (in modern-day Iran). (In fact, according to the writings of Herodotus, the creator of the Median kingdom was Deioces who reigned
from 728 to 625 BC, and founded the Median capital of *Ecbanta*, modern-day
Hamadan.) King Astyages sent Harpagus with an army to defeat King Cyrus
II of Persia. However, Harpagus wanted revenge for Astyages' murder of his
son some years earlier and saw this as the golden opportunity. Thus, instead
of confronting Cyrus with his army, Harpagus inserted a message, proposing an

alliance, inside the body of a slain rabbit. Then he had his messenger dress as a hunter and sent him to deliver the missive in the unskinned hare, to Cyrus. The hidden message was well received by Cyrus who immediately agreed to join Harpagus as an ally. Together they deposed Astyages. Under Cyrus, Harpagus was a potent military force, helping, among other feats, to conquer Asia Minor. An example of one of his escapades is the following. Harpagus besieged Xanthus, the main city of Lycia, in 540 BC, killing the Lycians to the last man.

Herodotus tells another tale related to our discussions thus far. The ancient Greek city, *Miletus*, of western Anatolia (30 kilometers (20 miles) south of the modern-day Turkish city of *Söke*), came under Persian rule of king Darius I in the latter part of the sixth century BC, as did the other Greek cities of Anatolia. However, in 499 BC the tyrant, Histiaeus, led a revolt against Persia. This revolt marked the beginning of the Greco-Persian Wars. One anecdote from this period comes from Herodotus.

Histiaeus served Darius by ensuring that tyrants of other cities would not destroy the Danubian bridge over which the Persians were to return from the Scythian campaign (circa 513 BC). For this Darius rewarded Histiaeus with Thracian territory. However, for good reason as it turned out, Darius became suspicious of Histiaeus and recalled him to the Persian court at Susa. There he became a prisoner, in effect, if not in fact. Darius installed Histiaeus' son-in-law Aristagoras as the new ruler of Miletus. Ostensibly, Histiaeus tattooed a message on the shaven scalp of a trusted slave, kept him hidden until a new head of hair grew back, then sent him off to his son-in-law with the message to revolt against Persia. This marks the end of the steganographic part of the story, but it is worth recalling what happened to Histiaeus.

Histiaeus tried to convince Darius that he could stop the revolt. Ultimately he was successful and was allowed to leave Susa. However, when he returned to the Lydian coast, the *satrap* or *provincial governor*, Artaphernes, was suspicious of him, so he was driven out. Histiaeus became a pirate at Byzantium, and after numerous unsuccessful forays to reestablish himself, he was captured. He suffered the ignominious fate of being crucified at Sardis (capital of ancient Lydia, near present Izmir, Turkey) by Artapherenes.

One last story from Herodotus, perhaps the most important from an historical viewpoint, should suffice before we turn our attention to another classical instance of the use of cryptography. Again, it involves steganography. It follows the death, in 486 BC, of Darius I, succeeded by his son, who came to be known as Xerxes the Great (circa 519–465 BC), best known for his massive invasion of Greece.

Herodotus tells us about Demaratus, former king of Sparta, who was dethroned by Cleomenes I, on erroneous charges of illegitimacy, after which he fled to Persia. After the death of Cleomenes, Leonidas became king of Sparta. While in exile, Demaratus learned of Xerxes plans for invasion, and felt obliged to warn Sparta. To do this, he scraped the wax off a pair of wooden folding tablets, wrote on the wood that was thereby revealed, warning of the impending invasion. Then he recovered the wood with wax, sealing the message, giving the appearance of a blank folding tablet that would pass scrutiny, a fine stegano-

graphic technique for that time. When the message was received, only Leonidas'
wife, and Clemenes' daughter, *Gorgo*, discovered it. She told the others to scrape
off the wax to reveal the message on the wood underneath, and so the Greeks
were warned. The message turned out to be a death knell for Leonidas. He led
troops to defend the pass at Thermopylae where he died.

Figure 1.19: Prince of Knossos.

The details of the above battle are worth the telling for their historic importance. The battle took place in August of 480 BC, and the narrow pass (only

6 kilometers (4 miles) in length), in which it was fought, has been immortalized for this and subsequent conflicts. In this particular battle, Leonidas held the Persians for three days in heroic fighting against unbelievable odds. The Persians were eventually led along another mountain pass by the Greek traitor, *Ephialtes*, allowing the Persians to outflank Leonidas's troops. Leonidas sent the majority of his men to safety, leaving only 300 Spartan soldiers and their *helots* (state-owned serfs of the Spartans), and 1100 Boeotian troops (Boeotia was a district in east-central Greece, which allied as the *Beotian League* in 550 BC under the leadership of Thebes). All of the men died in that battle. The Persian victory at Thermopylae cost them a very high price in lives lost. Moreover, the vast majority of the Greek soldiers and their ships escaped to the Isthmus of Corinth where they rejoined the main Greek forces. To commemorate this battle (of great heroism against massive odds) a marble and bronze monument was erected in 1955.

Another epic battle at Thermopylae occurred in 279 BC when the Greeks held and delayed the invading Celts. Although the Celts sacked Delphi in Greece that year, they suffered massive defeat against the Aetolians. The *Aetolian League* was a federal state in ancient Greece, which developed into a leading military power (having allied with Boetia around 300 BC), and can be said to be responsible for the driving out of the invasion of 279. A related battle occurred in the pass many years later. The Aetolians were one of the Greek powers not happy with Rome's growing power in Greece. They asked the Seleucid king *Antiochus III* to be their commander-in-chief of the Aetolian League. With the help of the Aetolians, Antiochus occupied Euboea in 192 BC, but by 191, the Romans, outnumbering him with troops by two to one, cut off his reinforcements in Thrace and outflanked his position at the Thermopylae pass, forcing him to retreat. Later his fleet was wiped out. Eventually in defeat, his kingdom was reduced to Syria, Mesopotamia, and western Iran. In 187, he was murdered near Susa, where he was trying to extract tribute to keep his empire afloat.

The Kāma-sūtra

There are more snippets of cryptography in other classical texts upon which we would like to touch before we close the door on this section. For instance, the *Kāma-sūtra* of *Vātsāyana* lists cryptography as the forty-fourth and forty-fifth of sixty-four arts or *yogas* of which people should not only be aware, but also put into practice, according to the texts. The Kāma-sūtra was written near the end of the fourth century AD or the beginning of the fifth century, but there is no certainty. In fact, Vātsāyana says that his work is a compilation of earlier works, so dating the cryptographic parts becomes even more problematic. A rough translation of the relevant portions of the two aforementioned yogas is given as follows: "The art of understanding writing in cipher, and the writing of words in distinctive fashion. The art of speaking by altering the forms of words. It is of various types. Some speak by altering the start and end of words, others by adding superfluous letters between every syllable of a word, etc."

Around a thousand years after the Kāma-sūtra appeared, cryptography was used to conceal magical spells in a manuscript by Arnaldus de Bruxella. The manuscript dates to around 1473 AD, in Naples, Italy. In it is a spell for making a *philosopher's stone*, but the essential portion of the spell (about five lines) is enciphered. In alchemy, a base metal was considered to be a state of "disease" in a noble metal such as gold. Thus, to "cure" the disease was to turn the base metal into gold. The philosopher's stone was considered to be the vehicle for actually transmuting base metals into gold. Hence, having the spell to create the stone meant great power in the hands of those who held knowledge to do so.

Casanova

An amusing anecdote concerning the breaking of a cipher in the eighteenth century occurred in 1757 involving the famous Casanova, who received a cryptogram for safekeeping from his wealthy friend, Madame d'Urfé. She believed that the cryptogram could never be cryptanalyzed given that she held the keyword in her memory and had never written it down or disclosed it to anyone. Nevertheless, Casanova was able to do just that. He determined the plaintext of the enciphered manuscript, which contained a description for the transmutation of baser metals into gold. He was also able to recover the key via his calculations. She was incredulous at the revelation. Casanova later wrote in his memoirs: "I could have told her the truth — that the same calculation which had served me for deciphering the manuscript had enabled me to learn the word — but on a caprice it struck me to tell her that a genie had revealed it to me." The keyword? NEBUCHADNEZZAR, or in Italian *NABUCODONOSOR*.

Shakespeare

The next story is about perhaps the greatest story-teller of all, *William Shakespeare*, also known as the Bard of Avon. In 1878, Ignatius Donnely, an American, self-styled, pseudo-scientist, began looking for steganographic evidence in the Shakespearean works that the "real" author was Sir Francis Bacon. Others, largely amateurs, followed in his footsteps looking for cryptographic evidence, which of course, they found since in works as vast as Shakespeare's, one can devise schemes to read anything one wants into the works. There is even speculation, and a kind of analysis (by *Baconites*) of passages from the original folio: *William Shakespeare's Comedies, Histories, and Tragedies.* published in 1623, that there is an enciphering of Sir Francis Bacon's name (with various spellings) therein. However, the vast majority of serious scholars see it as manifest that Shakespeare is indeed the author of the works. We will learn more about Sir Francis Bacon in Section 1.5. Shakespeare appears to have been aware of the *need* for ciphers, since he was certainly aware that messages can be intercepted by the unintended. In *Henry V*, a plot is being hatched against the king. Henry uncovers the plot, ostensibly through an interception of letters written by the traitors, proving their guilt, to which they confess, and they are put to death. There is no evidence of cryptography here, but the clear need for it is present, since interception of sensitive documents can lead to dire consequences, such as the aforementioned executions.

Edgar Allen Poe

We conclude this section with an American writer of the nineteenth century. Edgar Allen Poe (1809–1849) gained fame for his tales of the macabre (see Figure 1.20). He initiated the genre of the *detective story* with his 1841 publication of *Murders in the Rue Morgue*. Perhaps his best-known poem, which ranks high in American literature, is *The Raven*, published in 1845.

He showed an early interest in cryptograms starting in 1839 when he wrote articles on ciphers for *Alexander's Weekly Messenger*, a Philadelphia newspaper. All the cryptograms in his articles were simple substitution ciphers, and he really did not have the cryptanalytic skill that would warrant the reputation he developed. Yet he may be responsible for more people becoming interested and learning about cryptography than the most skilled cryptanalyst. This is largely due to turning his attention to literary cryptology in his story, *The Gold Bug*, published in 1843. This story won him a prize of $100 from the *Philadelphia Dollar Newspaper*, contributing to his fame. From a cryptographic viewpoint, the latter is the most outstanding of his works since it revolves around the cryptanalysis of a secret message. It was first published in book form in 1845 in a collection of his *Tales*. *The Gold Bug* may be considered to be one of his detective stories, but it has the element of having added a seductive, bewitching aspect to the cryptography used in the story. The solution of the cryptogram in the story leads to great wealth and the one who breaks the cipher takes on the role of sorcerer of a sort, since divination leads to the buried treasure, all this from a manuscript with occult-like symbols. This helped to popularize the story and thereby aided in increasing interest in the subject of cryptology itself. Other writers followed in his footsteps with cryptographic detective tales, but Poe created the template. Poe died on October 7, 1849 in Baltimore, Maryland, and was buried in the Westminister Presbyterian churchyard there. (See [186] for a collection of his works.)

We have only scratched the surface of the volumes of writers we could cite here, yet what we have covered gives us a sufficient appreciation that allows us to move on to other aspects. We now turn to a look at cryptology in the Europe of the Middle Ages, and some occult associations.

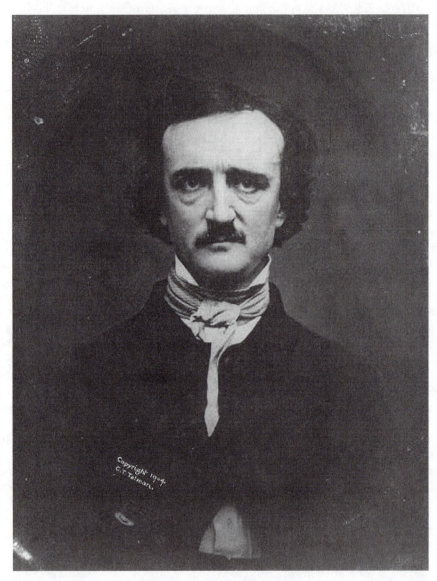

Figure 1.20: Edgar Allan Poe.

This photograph was taken in 1848. Courtesy of the Library of Congress, Prints and Photographs Division, copyright by C.T. Tatman, 1904; reproduction no. LC-USZ62-10610.

1.3 The Middle Ages

All historians have insisted that the soundest education and training for po-litical activity is the study of history, and that the surest and indeed the only way to learn how to bear bravely the vicissitudes of fortune is to recall the disasters of others.

Polybius (ca. 200–118 BC), Greek historian and statesman

The Middle Ages refer to that period in Europe from roughly 500 to 1500 AD. (Historical scholars would put the *end* of the Middle Ages anywhere from the end of the thirteenth to the fifteenth centuries since this is the *beginning* of the Renaissance, but that depends on the area of Europe and other factors.) We have already had an overlap with this period in Sections 1.1 and 1.2, wherein we discussed, for instance, the runic stones dating back to the ninth century AD, and the interest of the monks of the Middle Ages in ciphers inspired by the Bible. Moreover, with the fall of the Roman empire, western Europe fell into the Dark Ages (roughly 500–1000 AD), characterized by rampant illiteracy, frequent warfare, and intellectual darkness, including the lack of any serious development of cryptography. In Section 1.4, we will be able to fill in much of the (non-European) time period with the contributions by the Arabs. For now we begin with a philosopher of the thirteenth century.

Roger Bacon

Roger Bacon (1220–1292 AD) was a philosopher with the Franciscans whose association began when he joined them in 1257. Among his interests were alchemy, astronomy, languages, optics, and mathematics. He was truly a vision-ary as evidenced by the fact that he considered the possibility of "flying ma-chines", "horseless carriages", "motorboats", "microscopes", and "telescopes" centuries before they were invented. Indeed he was one of the first medieval advocates of experimental science. Our interest in him stems from his work, *Epistle on the Secret Works of Art and the Nullity of Magic*, written around 1250. Seven simple ciphers are described therein. For instance, he suggests the use of *only* consonants, or contrived symbols, and even shorthand. In fact he wrote:

"A man is crazy who writes a secret in any other way than one which will conceal it from the vulgar."

Although it is a bit off the topic, it is worth mentioning the influence of Ba-con's ideas, even after his death. Bacon believed in the existence of a habitable land to the west by sea and in the Aristotelian view of a short westward pas-sage to India. These ideas were repeated also, to the word, by Cardinal Pierre d'Ailley, bishop of Cambrai (1350–1420) in his work *Imago mundi*, an ency-clopedic world geography. A copy of this book found its way into the hands of Christopher Columbus, who was highly influenced by it. Columbus' copy is now kept (with several hundred of his marginal comments) at the Biblioteca Columbina, Seville, Spain. Now we turn to a story about the first European text on cryptography.

The Western Schism

When Pope Urban VI (1378–1389) was elected pope in 1378, it began the great *Western Schism* in the Roman Catholic Church, plaguing it for four decades. Although competent, Urban became a tyrannical reformer, which led to the revolt by the thirteen French cardinals led by *Robert de Genève*. They left Rome and four months later declared Urban's election "null", because they felt that it had been made under a cloud of fear. On September 20, 1378, they elected Robert de Genève as *Clement VII* (1342–1394), the first *antipope* (one elected to be pope in opposition to one canonically chosen, in this case, Urban). The new antipope now saw a clear need for a cryptosystem for his new enterprise at his new home in Avignon, France. One of his secretaries was Gabrieli di Lavinde, from Parma, who took up the task. In 1379, he devised a combination of code (book) and cipher. (Recall the definitions of *cipher* [page 4] and *code book* [page 6].) He established not only a simple substitution cipher, but also a list (code book) of plaintext words together with two-letter ciphertext equivalents, which came to be known as a *nomenclator*, a cross-breeding of the code (book) and the cipher. To envision this, think of a telephone book as half of a nomenclator, with the telephone number as the code. The other half is a means to turn a telephone number into the name without having to search the entire book knowing only the phone number. Such mechanisms for *reverse directories* would be the other half, or cipher half of the nomenclator, as a simple illustration. For more than four and a half centuries nomenclators would be used throughout Europe. Although Lavinde's code book consisted of only a few plaintext/ciphertext pairs, it grew to more than several thousand over the centuries.

It is worth mentioning the outcome of the schism. European countries were divided over the pope/antipope dispute; the Aragon, Castile, France, Navarre, Portugal, Savoy, and Scotland sided with Clement, while Bohemia, Flanders, the Holy Roman Empire, Hungary, north/central Italy, and Poland backed Urban. The Papal States fell into anarchy. When Urban died in Rome on October 15, 1389, it was suspected that he was poisoned. When Clement went to his grave on September 16, 1394, in Avignon, he still believed in his legitimacy, and this was echoed by King Charles V, who on that day, proclaimed him "the true Shepherd of the Church." In 1409, with the Roman pope, Gregory XII, and the Avignon antipope, Benedict XIII, both in power, the cardinals met at a council in Pisa, and elected a *third* pope Alexander V, succeeded shortly thereafter by John XXIII. The German King Sigismund wanted an end to the schism, so he convinced John XXIII, in 1414, to hold the Council of Constance. The council deposed him, received the resignation of Gregory XII, and dismissed the claims of Benedict XIII. In November of 1417, pope Martin V was elected, and the schism ended.

Geoffrey Chaucer

Another of the few cryptographic authors of the Middle Ages, but perhaps the most famous, was Geoffrey Chaucer (1342–1400), who was easily the most prominent British poet before Shakespeare. In the 1390s he wrote his best-known work, the unfinished *Canterbury Tales*. Our interest, however, comes from a work written in 1392, namely, *The Equatorie of the Planetis*, which was a supplement to his 1391 *Treatise on the Astrolabe*. In the *Equatorie*, he included six passages written in cipher. The cryptosystem that he used consisted of a substitution alphabet of symbols, where for instance, the letter *h* might be represented by a symbol looking like the Greek letter σ, *sigma*. He also talks of a cipher "using magic figures and spells." This brings us to the association that cryptography has had with the occult and black magic.

Codes and the Occult

One of the best-known enciphered *magic* writings is the *Leiden papyrus*, which was actually written in the third century BC. It has both Greek and demotic symbols as enciphering techniques to hide the "magic recipes", among which are spells for making potions that would give a man an incurable skin disease, and another on how to make a woman desire a man. Of course, none of these work. However, it is a precursor to the kinds of "magical" associations that cryptography had in the Middle Ages.

Part of the reason for the lingering air of the occult attached to cryptography today is due to the association with secret spells and incantations, the history of which we will discuss. It was assumed, in the Middle Ages, that these incantations bestowed power on the sorcerer who voiced them, and that the removal of a disguise from a secret is somehow miraculous or magical. However, the extraction of information by modern cryptographic techniques has become an objective science, whereas the unfortunate association with divination, or insight into the future, is subjective and at best an amusing distraction in our modern world. Through education about cryptology, we can remove this aura of the occult and better understand it as a science with a fascinating history. Now let us learn more about why such an aura lingers.

In the late Middle Ages, February 2, 1462, Johannes Trithemius was born in Trittenheim, Germany. In 1482, subsequent to attending school in Heidelberg, he entered the Benedictine monastery of Saint Martin in Sponheim, Germany. In a very short time he was designated abbot, probably due to the recognition of his clear and outstanding brilliance. He became a prolific writer, known for his biographical dictionaries, and an encyclopedic bibliography, *Liber de scriptoribus ecclesiasticus*, published in 1494, which earned him the title of the *Father of Bibliographies*. This has become a reference work on church writers that is used to this day. However, Trithemius had interests on the darker side.

Trithemius authored books on alchemy, witchcraft, planetary angels, and general topics of the occult. In particular, in his book *Steganographia*, Trithemius describes techniques that we today call aspects of *steganography*, the etymology of which is from the Greek *steganos* meaning *impregnable*, and from a *secrecy* point of view, this means *concealing* the very *existence* of the message itself, rather than the cryptographic goal of disguising the message. We have actually encountered various uses of steganography in earlier sections (pages 8

and 32–34). He began writing *Steganographia* in 1499, intending it to become
an eight-volume series. For over a century the manuscript was circulated before
it was finally printed in 1606. The Roman Catholic Church, in 1609, seeing the
text as sorcery, placed it on its index of *Prohibited Books*, where it remained for
over two centuries. Nevertheless, other printings occurred, including as late as
1721. Yet, this book, above all others, solidified the common belief, especially
among his colleagues, that Trithemius was a "wonder-working magician". This
gave him the reputation as a sorcerer, which lingers to this day.

In modern books, Trithemius is grouped with other famous occultists such
as Paracelsus (1493–1541), a German-Swiss physician and alchemist, who ironically, established the role of chemistry in medicine. Another of his contemporary occultists with whom Trithemius is compared today is Heinrich Cornelius
Agrippa von Nettesheim (1486–1535), who was an acknowledged expert on the
occult, as well as court secretary to Charles V, and among many other activities,
university teacher, and public advocate at Metz, at least until he was condemned
for defending an accused witch. One of his books, *De occulta philosophia*, included numerology and fostered magic as the highest road to knowing God.
Ultimately, he was imprisoned and branded a heretic. From a historical viewpoint and to better understand Trithemius's writings, it is worth discussing
other items in Agrippa's book. His notion of God as magician is called hermeticism (from Hermes Trismegistus, mythical inventor of a magic seal for keeping
vessels airtight, and thus the origin of the modern phrase *hermetically sealed*).
Hence, many hermeticists' goal was to reinterpret the Bible using ciphers. Such
writings were held to include the wisdom of Egyptians at the time of Moses,
and ostensibly written by an ancient Egyptian purported to have received divine
knowledge about the physical world at the time Moses received his knowledge
from God about the moral world. Along with the numerology in Agrippa's book
(one of the components of hermeticism) is also an explanation of the world in
terms of cabalistic[1.2] analysis of Hebrew letters (another aspect of hermeticism).
Hebrew letters were believed to have magical powers when arranged in certain
combinations. Hermeticists believed that breaking the "code of the Bible" would
reveal all the secrets of the universe. In the late Middle Ages, the resurgence
of neo-Platonism provided an acceptance by increased followers of hermetic beliefs. Later, when the beliefs were largely proved to be fraudulent, hermeticism
still had some followers and even influenced thinkers of the Renaissance, and
beyond, including Sir Isaac Newton.

In *Steganographia*, Trithemius describes only some elementary steganⁿo-
graphic techniques, and much of it has nothing at all to do with cryptography,
but rather with magical incantations, thought transference, computation of numerical values of the names of angels, and other concepts from the beliefs of
hermeticism.

Trithemius did turn to more serious cryptography later, but we are going to
save a discussion of these accomplishments until Section 1.5, where we will be

---

[1.2]Cabalism refers to a system of Jewish mysticism and magic using ciphers as a device for
interpretation of scripture.

looking at the Renaissance influences. This is fitting since Trithemius is both a man of the (end of) the Middle Ages, as well as a man of the (beginning of) the Renaissance, given the fluid borders between these two time lines discussed at the outset of this section. We will see that the above work pales in comparison to what he contributed later. Stay tuned for more on this fascinating historical figure in our cryptographic overview.

Figure 1.21: A modern-day steganographic device.

Courtesy of the CIA website. See

http://www.cia.gov/cia/information/artifacts/dollar.htm,

where it is stated: "This hollow container, fashioned to look like an Eisenhower silver dollar, is still used today to hide and send messages or film without being detected. Because it resembles ordinary pocket change, it is virtually undetectable as a concealment device."

1.4   Cryptology and the Arabs

He shall live, and unto him shall be given all the gold of Arabia.
The Book of Common Prayer (1662), *Psalm 72, v. 15*

In this section, we will learn how the flourishing Arab civilization discovered cryptanalytic techniques and published the first systematic analysis of it and of cryptology proper. The earliest known contribution from Arab civilization comes from an author of 725 (unless otherwise specified, we will be talking about dates *AD*, henceforth), Abū 'Abd al-Raḥmān al-Kahalīl ibn Aḥmad ibn 'Amr ibn Tammām al Farāhīdī al-Zadī al Yaḥmadī, who wrote the *Kitāb al-mu'ammā*. This writing was inspired by a cryptogram, written in Greek, sent to him by the Byzantine emperor. It is purported that his reasoning for its solution went along the lines of assuming that the cryptogram began with words similar to "In the name of God..." and he was able to deduce the first few letters based upon this assumption. He worked from there to decrypt the entire message. Ostensibly, this took him one month to solve. Arab cryptanalysis was in its infant stages, but that would change.

The Arabs' invention of cryptanalysis was rooted in religious scholarship where theologians analyzed the Koran, trying to establish the time line of events, by counting the frequencies of words contained in each of Muhammad's revelations. Their reasoning was that if a high frequency of certain more recently evolved words were found in a given revelation, then that would be one to place later in the time line. They also looked at the commonality of letters, among other aspects of cryptanalysis that we consider to be fundamental today. Their earliest known description of such letter frequency analysis was created in the ninth-century by the author Abū YūsūfYa'qūb ibn Is-hāq ibn as-Sabbāh ibn 'omrān ibn Ismaīl al-Kindī (but, we will just call him al-Kindī). His treatise is entitled *A Manuscript on Deciphering Cryptographic Messages*, rediscovered in the Sulaimaniyyah Ottoman Archive in Istanbul in 1987. Although al-Kindī wrote nearly three hundred books on various topics including mathematics and medicine, *our* interest is in the cryptanalytic text since it represents the first recorded instance of a treatise on cryptanalysis involving "letter frequencies".

In order to understand what al-Kindī discovered in the realm of letter frequency analysis and to set up our discussions for later analysis in the text, let us look at the English language from the perspective of most frequently occurring words, and letters.

The statistical data shown in Table 1.4 are taken from this author's book [170, page 203]. The most common words in order of frequency distribution are:

Frequency of Words in English

Table 1.4	THE, OF, ARE, I, AND, YOU, A, CAN, TO, HE, HER, THAT, IN, WAS, IS, HAS, IT, HIM, HIS

Frequency of Letters Ending Words in English

Table 1.5	E, T, S, D, N, R, Y

Table 1.5 is a list of the most common letters to *end* a word, in order of frequency distribution, which is an example of *positional frequency*, wherein the frequency count of the position of a given letter is taken in ratio with the total number of letters occurring in that position over all English texts.

Now to illustrate al-Kindī's idea, suppose that we have the letter *g* occurring most often (in word endings) in a ciphertext for a plaintext known to be in English. Then we would deduce that the letter *e* is the most likely candidate for the plaintext letter. If the second most frequently occurring letter is *k* in ciperhtext, then we would guess, via Table 1.5, that the corresponding plaintext letter is *t*, and so on. Similarly, one could use Table 1.4 on the most frequently occurring words to deduce the plaintext.

Of course, the above tables are not written in stone. There can be the problem of too little ciphertext to make any reasonable statistical inferences (as with the Phaistos disk, see page 2). Moreover, it could be a specialized language about some esoteric subject in which case the frequencies will deviate from the standard. There is no table, or perhaps even set of tables, which can definitively lay claim to being *the* one that is canonical for all situations in a given language. Yet, the above tables will provide us with a general overview and therefore a working template to discuss cryptanalytic matters throughout the text as they arise, and we will bring more to the fore as we need them.

Another contribution from Arab civilization, albeit of less significance than that of al-Kindī, dates to 855. The author Abū Bakr Ahmad ben 'Ali ben Wahhshiyya an-Nabatī published his book, *Kitāb shauq almustahām fī ma'rifat rumūz al-aqlām*, or *Book of the Frenzied Devotee's Desire to Learn About the Riddles of Ancient Scripts*, in which numerous cipher alphabets were included that were typically used for magic spells. Almost five hundred years later, in 1350, 'Abd al-Rahmān Ibn Khaldūn created his work, *The Muqaddimah*, an historical survey detailing how government bureaucrats used symbols including "the names of perfumes, fruits, birds, or flowers" as a code for regular letters in order to encipher correspondence among officials of the tax and army bureaus. The name of this particular kind of cryptography was called *qirmeh*, which sprang up later in sixteenth-century Egypt, and even was used in financial record-keeping of Istanbul and Syria as late as the nineteenth century.

Another major work to come out of the Arab influence on cryptology was completed in Egypt in 1412 by an author named, Shihāb al-Dīn abu 'l-'Abbās Ahmad ben 'Ali ben Ahmad 'Abd Allāh al-Qalqashandi. (We will just call him Qalqashandi.) His work was a prodigious fourteen-volume encyclopedia called *Subh al-a 'sha*. Our interest is in the section on cryptology. Some parts of the section deal with steganographic techniques, such as invisible ink, and the hiding of messages within letters. Qalqashandi claimed that most of his cryptological ideas came from an author of the fourteenth century, none of whose writings are extant, but cites a list of seven cryptosystems deriving from these writings. The list is of historical significance insofar as it marked the first intermingled use of both substitution and transposition ciphers (a common feature of modern-day cryptosystems). Furthermore, it provided the first cipher ever to have more than one ciphertext for a given plaintext symbol. Of greatest note are the tables of letter frequency analysis and other cryptanalytic analyses such as letters that cannot occur together in the same word. Here, again, is a contribution from the Arab civilization to cryptanalysis. Such cryptanalytic techniques are the reason that, for instance, simple substitution ciphers (the monographic ones) are so easy to cryptanalyze, because the frequency of letters in the ciphertext is the same as the frequency of letters in the plaintext. Thus, once the language is known, the plaintext can usually be relatively easily deduced. Hence, from the above, we may conclude that cryptanalysis was therefore born with the Arabs.

Outside cryptography, Arab scholars enlightened Europe's exit from the Dark Ages by preserving much of the mathematics from antiquity. One of the major reasons for this was a vision had by Caliph al-Ma'mun (809–893). In this epiphany, he was visited by Aristotle, after which he was compelled to have all the Greek classics translated into Arabic.

Another story is that of the Persian mathematician and astronomer, Muḥammed ibn Musa al-Khwarizmi, who lived under the caliphate of al-Ma'mun. We owe al-Khwarizmi for the introduction of the Hindu-Arabic number system. Around roughly 825, he completed a book on arithmetic, which was later translated into Latin in the twelfth century under the title *Algorithmi de numero Indorum*. This book was one of the principal means by which the Hindu-Arabic number system was introduced to Europe after being launched in the Arab world. This accounts for the widespread, but mistaken, belief that our numerals are arabic in origin. (Numerals dating from 150 BC have been found inscribed in a cave at Nana Ghat, close to Bombay, India. Moreover, the first documented appearance of a zero,[1.3] as we know it, is an inscription on a birch-bark manuscript dated 400 AD, discovered in 1881 at Bakhashali, a village in northwest India.) Not long after the Latin translations appeared in Europe, readers began contracting al-Khwarizmi's name until it became the norm to associate *algorithm* with these numerals. (Today we use the term "algorithm" to mean any methodology following a set of rules to achieve a goal.) Al-Khwarizmi also wrote a book on elementary algebra, *Hisab al-jabr wa'l-muqābala*. The word *algebra* is derived from *al-jabr* or *restoration*.[1.4] His third major work was *Kitab ṣurat al-arḍ*, which translates best as *Geography*. This assisted in his construction of a world map for al-Ma'mun, including a determination of the circumference of the earth by measuring the length of a degree of a meridian through the Plain of Sinjar in Iraq, amazing achievements!

[1.3]The goose-egg symbol 0 for the zero is *sifr* in Arabic, from which our zero is ultimately derived. In the thirteenth century AD, the term sifra was introduced into the German language as *cifra*, from which we get our present-day word *cipher*.

[1.4]In the Spanish work *Don Quixote* (Part I in 1605, and Part II in 1615), by Miguel de Cervantes, the term *algebrist* is used for *bone-setter* or *restorer*.

1.5 Rise of the West

Oh, East is East, and West is West, and never the twain shall meet,
Till Earth and Sky stand presently at God's great Judgement Seat...
Rudyard (Joseph) Kipling, (1865–1936), English writer and poet
— from *The Ballad of East and West* (1892)

The word *Renaissance* literally means *rebirth*. It was coined by fifteenth-century scholars to separate the fall of ancient Greece and Rome from its rebirth and rediscovery in the middle of their own century. The fall of Constantinople in 1453 may be considered one of the dividing lines since scholars fled to Italy, bringing with them knowledge, irreplaceable books and manuscripts, as well as the classical Greek tradition of scholarship. The earliest sign of the Renaissance was the intellectual movement called *humanism*, perhaps given its biggest surge by the aforementioned influx of scholars. Humanism, born in Italy, had as its subject matter: human nature, unity of truth in philosophy, and the dignity of man. Perhaps most importantly, humanism yearned for the rebirth of lost human spirit and wisdom. While medieval thinkers preferred the idea of "one man, one job", the *Renaissance man* was a versatile thinker, thirsting for an education in all areas of knowledge, and becoming an expert in many. It is one of those men with whom we begin our discussion.

Leon Battista Alberti

If there is to be a holder of the title *Father of Western Cryptography*, it must go to Leon Battista Alberti (1404–1472). He was not only an architect, sculptor, writer, and all round-scholar, but also one of the prime movers in the development of the theory of art in the Renaissance, not to mention his contributions to cryptology, a true Renaissance man.

Alberti was born on February 14, 1404, in Genoa, Italy, the illegitimate son of a wealthy banker, Lorenzo di Benedetto Alberti. Yet, in this time of Florentine Italy, illegitimacy was less of a burden, and more of a reason to succeed. Alberti was raised as Battista in Venice where the family moved shortly after he was born. (He adopted the name Leon later in life.) At the age of 10, he had already learned Latin and his father was teaching him mathematics. His formal education was at the University of Bologna, where he ultimately earned a degree in law. However, he quickly turned his interests to artistic, and ultimately scientific thought. Alberti not only taught himself music, became an expert at playing the organ, and wrote sonnets, but also wrote on art, criminology, sculpture, architecture, and mathematics. In 1432, he went to Rome where he became a secretary in the Papal Chancery, and he remained in the arms of church for the rest of his life. In 1434, he went to Florence as part of the papal court of Eugenius IV. It was in the papal secretariat that he became a cryptographer. In fact, he was a friend of Leonardo Dato, a pontifical secretary who might have instructed Alberti in the state of the art in cryptology.

In order to understand Alberti's contributions, we need to examine some concepts first. A *homophone* is a ciphertext symbol that always represents the same plaintext symbol. For instance, with the Caesar cipher in Table 1.2 (page 11), the letter D is always the ciphertext for the plaintext letter a, so D is a homophone in the *monoalphabetic* cipher known as the Caesar cipher. Here "monoalphabetic" means that there is only one *cipher alphabet*, which means the set of ciphertext equivalents used to transform the plaintext. The row of ciphertext equivalents below the plaintext in Table 1.2, for instance, is the cipher alphabet for the Caesar cipher. A *polyphone* is a ciphertext symbol that always represents the same *set* of plaintext symbols, typically a set consisting of at most 3 plaintext symbols. With homophones or polyphones, there is no option for change since the relationship between plaintext and ciphertext is fixed. However, a cipher is called *polyalphabetic* if it has more than one cipher alphabet. In this type of cipher, the relationship between the ciphertext substitution for plaintext symbols is variable. Thus, since each cipher alphabet (usually) employs the same symbols, a given symbol may represent several plaintexts.

Alberti conceived of a disk with plaintext letters and numbers on the outer ring and ciphertext symbols on an inner movable circle. Alberti divided his ring and corresponding circle into 24 equal segments, called *cells*, each containing a symbol.

A representation of Alberti's disk is pictured in Figure 1.22. We have altered his original presentation since he had ciphertext in lower case and plaintext in upper case, the reverse of what we have as a convention.

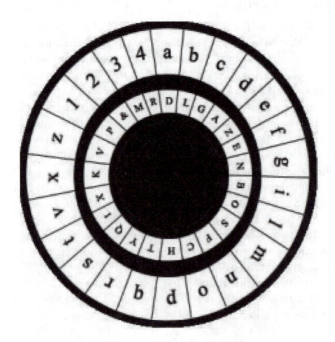

Figure 1.22: Alberti disk.

In Figure 1.22 the plaintext letter z is enciphered as V, so in this setting (one of the 26 possible cipher alphabets), the plaintext *zebra*, for instance, would be enciphered as *VZLYD*. However, there is nothing new at this juncture that is any different from, say, the Caesar cipher with the cipher alphabet having the letter c below the Z. Alberti had an idea, however (which is why he wanted the inner circle to be able to rotate). This idea would revolutionize the forward movement of cryptological development. After a random number of plaintext words had been enciphered, usually three or four, Alberti would move the inner disk to a new setting. Hence, he would now be using a *new* cipher alphabet. Suppose that he moved the inner circle so that z sits over K. Then *zebra* would be enciphered as *KADTR*, a new ciphertext for the same plaintext as above since we have a *new* cipher alphabet. This is polyalphabeticity in action, literally! In fact, with his cipher disk, Alberti invented the first polyalphabetic cipher in history. Yet, he did not stop there.

Alberti had 20 letters, as depicted in Figure 1.22[1.5] and including the numbers 1 through 4 in the outer ring of his original disk. In a book, he used these numbers in two-, three-, and four-digit sets from 11 to 4444 yielding 336 ($= 4^2 + 4^3 + 4^4$) codegroups. Beside each digit he would write a phrase such as "Launch the attack" for the number 21, say. Then, with the setting in Figure 1.22, the code group 21 is enciphered as *&P, enciphered code*. Alberti was the first to discover it, and it is a testimony to his being centuries ahead of his time that enciphered code, when it was rediscovered at the end of the nineteenth century, was simpler than that of Alberti!

Johannes Trithemius

Polyalphabeticity had another ally, and we have already met him in Section 1.3. In early 1508, Trithemius turned himself to the task of writing a book dedicated solely to a serious cryptographic analysis, called *Polygraphia*, with the official title, *Polygraphiae libri sex, Ioaonnis Trihemii abbatis Peopolitani, quondam Spanheimensis, ad Maximilianum Caesarem*, or *Six Books of Polygraphy, by Johannes Trithemius, Abbot at Wurzburg, formerly at Spanheim, for the Emperor Maximilian*. However, Trithemius died on December 15, 1516, in Wurzburg before the book was published. In July of 1518 it finally went to press, and was reprinted (and plagiarized) many times after that. *Polygraphia* can be said to be the first printed book on cryptography. In his book, he invented a cipher where each letter was represented as a word taken from a sequence of columns. The resulting sequence of words turned out to be a legitimate prayer. Perhaps more importantly, from the viewpoint of the advancement of cryptography, he also described a *polyalphabetic cipher*. Another way to think of such a cipher is that there is more than one enciphering key, namely, that a given symbol may be encrypted in different ways depending upon where it sits in the plaintext. An accepted modern form for displaying this type of cipher is a rectangular substitution table, about which we will learn a great deal more as we

---

[1.5]This excludes the letters h, k, and y, deemed to be unnecessary, and since j, u, and w were not part of his alphabet, this left 20 letters. The inner circle consists of the 24 letters of the Latin alphabet, put in the cells at random, including $\&$.

Figure 1.23: Leon Battista Alberti.
(Courtesy of the Archaeological Museum of Bologna, Italy.)

continue our journey. Given below is the Trithemius tableau where all possible shifts (modulo 24) appear as rows below the plaintext, each row representing a distinct cipher alphabet (key), a total of 24 cipher alphabets (keys) in all, polyalphabeticity! Trithemius used 24 letters, excluding the letters j and v.

The Trithimius Tableau

	a	b	c	d	e	f	g	h	i	k	l	m	n	o	p	q	r	s	t	u	w	x	y	z
a	A	B	C	D	E	F	G	H	I	K	L	M	N	O	P	Q	R	S	T	U	W	X	Y	Z
b	B	C	D	E	F	G	H	I	K	L	M	N	O	P	Q	R	S	T	U	W	X	Y	Z	A
c	C	D	E	F	G	H	I	K	L	M	N	O	P	Q	R	S	T	U	W	X	Y	Z	A	B
d	D	E	F	G	H	I	K	L	M	N	O	P	Q	R	S	T	U	W	X	Y	Z	A	B	C
e	E	F	G	H	I	K	L	M	N	O	P	Q	R	S	T	U	W	X	Y	Z	A	B	C	D
f	F	G	H	I	K	L	M	N	O	P	Q	R	S	T	U	W	X	Y	Z	A	B	C	D	E
g	G	H	I	K	L	M	N	O	P	Q	R	S	T	U	W	X	Y	Z	A	B	C	D	E	F
h	H	I	K	L	M	N	O	P	Q	R	S	T	U	W	X	Y	Z	A	B	C	D	E	F	G
i	I	K	L	M	N	O	P	Q	R	S	T	U	W	X	Y	Z	A	B	C	D	E	F	G	H
k	K	L	M	N	O	P	Q	R	S	T	U	W	X	Y	Z	A	B	C	D	E	F	G	H	I
l	L	M	N	O	P	Q	R	S	T	U	W	X	Y	Z	A	B	C	D	E	F	G	H	I	K
m	M	N	O	P	Q	R	S	T	U	W	X	Y	Z	A	B	C	D	E	F	G	H	I	K	L
n	N	O	P	Q	R	S	T	U	W	X	Y	Z	A	B	C	D	E	F	G	H	I	K	L	M
o	O	P	Q	R	S	T	U	W	X	Y	Z	A	B	C	D	E	F	G	H	I	K	L	M	N
p	P	Q	R	S	T	U	W	X	Y	Z	A	B	C	D	E	F	G	H	I	K	L	M	N	O
q	Q	R	S	T	U	W	X	Y	Z	A	B	C	D	E	F	G	H	I	K	L	M	N	O	P
r	R	S	T	U	W	X	Y	Z	A	B	C	D	E	F	G	H	I	K	L	M	N	O	P	Q
s	S	T	U	W	X	Y	Z	A	B	C	D	E	F	G	H	I	K	L	M	N	O	P	Q	R
t	T	U	W	X	Y	Z	A	B	C	D	E	F	G	H	I	K	L	M	N	O	P	Q	R	S
u	U	W	X	Y	Z	A	B	C	D	E	F	G	H	I	K	L	M	N	O	P	Q	R	S	T
w	W	X	Y	Z	A	B	C	D	E	F	G	H	I	K	L	M	N	O	P	Q	R	S	T	U
x	X	Y	Z	A	B	C	D	E	F	G	H	I	K	L	M	N	O	P	Q	R	S	T	U	W
y	Y	Z	A	B	C	D	E	F	G	H	I	K	L	M	N	O	P	Q	R	S	T	U	W	X
z	Z	A	B	C	D	E	F	G	H	I	K	L	M	N	O	P	Q	R	S	T	U	W	X	Y

To illustrate its use, we suppose that the plaintext is *maximilian*, then the ciphertext is achieved by looking at the first row for the first letter under the letter m, which is M, then for the second letter a of the plaintext look at the letter below it in the second row, which is B, for the third letter of the plaintext x, look at the letter below it in the third row, Z, and so on to get the ciphertext *MBZMQORQIX*. If we have plaintext that is longer than 24 letters, then we can start over again in the first row and repeat the process, (mod 24 arithmetic in action). Notice that unlike a simple *mono*alphabetic substitution cipher, such as the Caesar cipher, having only one cipher alphabet — the row below the plaintext — a given plaintext in a polyalphabetic letter does not always go to the same ciphertext letter. For instance, in our plaintext, the letter i goes to M in the first instance, O in the second instance, and Q in the third instance, since i sits in the fourth, sixth, and eighth places of the plaintext corresponding to the fourth, sixth, and eighth row entries of ciphertext (in other words in the corresponding cipher alphabet determined by that row) sitting below i, namely,

M,O, and *Q*, respectively. Later, we will see how another later cryptographer, Blaise de Vigenère (see page 55), was inspired by this tableau to create one that took the idea further.

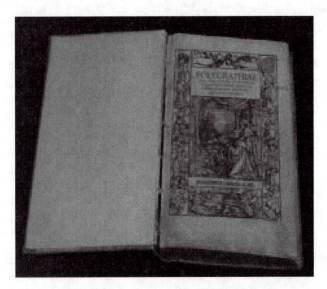

Figure 1.24: Polygraphia.

Image courtesy of the National Cryptologic Museum of The National Security Agency, Rare Books Collections. See *http://www.nsa.gov/museum/books.html*.

The attentive reader will have noticed that the Trithemius tableau (necessarily a square since there are exactly as many rows (cipher alphabets) as there are letters in the alphabet) has an advantage over Alberti's method since the cipher alphabet is changed with *each letter* enciphered, rather than after an arbitrary number of enciphered words as with Alberti's method. Moreover, the ordered table makes a quick look-up possible at a glance for each of the cipher alphabets.

Trithemius also gave examples where he switched alphabets after exhausting 24 letters of plaintext rather than starting over with the first row of the above tableau again. This is a variation of the above simple scheme. Moreover, the aforementioned method is the first cipher to use a *progressive key* where all possible cipher alphabets are exhausted before any are used again. Modern ciphers have used more variations on this theme since we now have computers to employ such key progressions. Moreover, the substitution table that he used is now a standard feature of modern-day cryptography.

Giovanni Battista Belaso

Our next ally and proponent of the advancement of polyalphabeticity is another from Italy, Giovanni Battista Belaso. Neither Alberti nor Trithemius conceived of using a key or key words in their polyalphabetic ciphers. The first in recorded history to do so was Belaso in 1553. His idea was to use a *keyphrase*, which he called a *countersign*, repeated as often as needed, to select the cipher alphabets. (We may think of this as the modern invention of the notion of a *password*.) Here is how his countersign works. First, we are going to be using Trithemius's table (page 51).

BELASO'S KEYPHRASE POLYALPHABETIC CIPHER

To employ Belaso's idea, we do three things to *encipher*.

(a) Put the plaintext letters in a row.

(b) Above each plaintext letter place the keyphrase letters, repeated as often as necessary, to cover all the plaintext.

(c) Replace each letter of the plaintext with the letter at the intersection of the row labelled by the keyphrase letter and column labelled by the plaintext letter in Trithemius's table.

We illustrate these rules with the following.

Example 1.1 *We will suppose the keyphrase, used by Belaso, is* OPTARE ME-LIORA*, and the plaintext is* countersign is key. *Then one places the keyphrase over the plaintext, repeated until the plaintext runs out as illustrated below.*

o	p	t	a	r	e	m	e	l	i	o	r	a	o	p	t
c	o	u	n	t	e	r	s	i	g	n	i	s	k	e	y
Q	D	O	N	L	I	D	X	T	P	B	A	S	Y	T	R

For example, the letter o *labels the row that intersects the column headed by the letter* c *at the ciphertext letter* Q, *and so on.*

To *decipher* using the Trithemius square, we do three things.

(a) Put the ciphertext letters in a row.

(b) Put the keyword letters above the ciphertext letters, repeating them as required, to cover all ciphertext.

(c) Locate the column labelled by each keyword letter, and find the row in which the ciphertext letter sits below it. Then the label of that row is the plaintext.

Applying these rules to Example 1.1, we get the following.

Example 1.2

o	p	t	a	r	e	m	e	l	i	o	r	a	o	p	t
Q	D	O	N	L	I	D	X	T	P	B	A	S	Y	T	R
c	o	u	n	t	e	r	s	i	g	n	i	s	k	e	y

For instance, since the letter o *sits over the ciphertext letter* Q, *the row of which is labelled by* c, *then this is the first letter of plaintext, and so on.*

Employing standard alphabets in his use of a keyphrase, Belaso created a polyalphabetic cipher with much greater flexibility than that of Alberti or Trithemius. With this use of a keyphrase, Belaso ensured that, instead of repeating the enciphering of each letter with 24 standard cipher alphabets, as Trithemius proposed, the key could be changed at will. For example, if the key were compromised in some fashion, it could be discarded and a new one issued to, say, diplomats of the day for their correspondence. Even with keys of length 13, as in the above keyphrase from Belaso, there are 24^{13} possible encipherments of a given plaintext letter, more than a hundred quadrillion choices. Quite an advancement! Nevertheless, however prodigious this contribution seems to be, it would be for another individual to put together all the pieces in order to create the first forerunner of a modern polyalphabetic cipher.

Porta and Cardano

Giovanni Battista Porta (1535–1615) was born in Naples, Italy, in 1535. At the age of 22, he published his first book, *Magia Naturalis*, or *Natural Magic*, a text on "experimental magic". However, in 1563, he published *De Furtivis Literarum Notis*, which contained the cryptographic advances in which we are interested. In this book is the first appearance of a *digraphic cipher*, meaning a cipher in which two signs represent a single symbol.

(Later, we will see how this notion was reinvented in the twentieth century by Lester Hill using only elementary matrices (page 111), and how the first *literal* digraphic cipher was invented much later. Here Porta is using signs rather than letters.) Moreover, he introduced some of the modern fundamentals of cryptography, namely a separation of *transposition ciphers* and *substitution ciphers*, as well as *symbol substitution* (substituting an unusual symbol for a letter). Porta also looked at methods, albeit elementary by modern standards, of cryptanalyzing polyalphabetic ciphers. In fact, in a second edition of his book, published in 1602, Porta added a chapter with these cryptanalytic observations. Although Porta also ultimately did glue together the ideas of Alberti, Belaso, and Trithemius, by adding mixed alphabets and shifts, to produce what we consider to be a basic polyalphabetic substitution cipher, there was work to be done to make polyalphabetic ciphers more secure, the essence of which was in how the key was used.

Figure 1.25: Natural Magic.

Image courtesy of Scott Davis: *http://homepages.tscnet.com /omard1/jportat5.html*.

The first to see how this could be accomplished was Girolamo Cardano (1501–1576). Cardano was born on September 24, 1501, in Pavia, Duchy of

Milan, Italy. In his younger years, he assisted his father who was a lawyer and a mathematics lecturer primarily at the Platti foundation in Milan. Cardano himself came to be known as one of the greatest mathematicians of his time. He wrote more than 130 books in his lifetime. The two that are best known for his mathematical contributions are *Liber de ludo aleae*, or *Book on Games of Chance*, considered to be the first book on probability theory, and *Ars Magna* (1545), considered to be one of the great books in the history of algebra. The ones of interest to us from a cryptographic perspective were his books, *De Subtilitate* (1550), and a follow-up called, *De Rerum Varietate* (1556). In these two books, he introduced the idea of an *autokey*, meaning that the plaintext, itself, is used as its own key. However, Cardano implemented the idea in a flawed manner, which allowed for multiple possible decryptions as well as the fact that, in his implementation, the receiver of the message was in no better position than a cryptanalyst at trying to determine the first plaintext word, from which there would be total decryption. Thus, the idea of an autokey has not been attributed to Cardano. He is remembered for an invention of a steganographic device, which we call the *Cardano grille*. Cardano's idea involved the use of a metal (or other rigid substance) sheet consisting of holes about the height of a written letter and of varying lengths. The sender of a message places the grill on a piece of paper and writes the message through the holes. Then the grille is removed and the message is filled in with some innocuous verbiage. Use of the Cardano grille continued well into the seventeenth century, and has even popped up in various places in modern times. Thus, it is the case that due to his flawed idea for an autokey, he is remembered largely for his steganographic device. He died on September 21, 1576, in Rome with his fame not attached to the greater cryptographic record that he sought. That fame would go to another.

Blaise de Vigenère

Blaise de Vigenère (1523–1596) had his first contact with cryptography at age 26 when he went to Rome on a two-year diplomatic mission. He familiarized himself with the works of his predecessors, Alberti, Belaso, Cardano, and Trithemius. His own work, published in 1585, containing his contributions to cryptography, is called *Traicté des Chiffres*. Vigenère discussed steganographic techniques, and a variety of cryptographic ideas. Among them was the idea for an autokey polyalphabetic substitution cipher.

He employed the idea that Cardano had invented of using the plaintext as its own key. However, he added something new, a *priming key*, which is a single letter (known only to the sender and the legitimate receiver), that is used to decipher the first plaintext letter, which would, in turn, be used to decipher the second plaintext letter, and so on. To understand the details of how this works, we use a Vigenère square, given on page 56, with the full 26-letter alphabet, as opposed to Trithemius' use of 24. It rightfully deserves to be called a *Trithemius* square, as the reader will note, but history has deemed it to have Vigenère's name attached to it.

THE VIGENÈRE TABLEAU

	a	b	c	d	e	f	g	h	i	j	k	l	m	n	o	p	q	r	s	t	u	v	w	x	y	z
a	A	B	C	D	E	F	G	H	I	J	K	L	M	N	O	P	Q	R	S	T	U	V	W	X	Y	Z
b	B	C	D	E	F	G	H	I	J	K	L	M	N	O	P	Q	R	S	T	U	V	W	X	Y	Z	A
c	C	D	E	F	G	H	I	J	K	L	M	N	O	P	Q	R	S	T	U	V	W	X	Y	Z	A	B
d	D	E	F	G	H	I	J	K	L	M	N	O	P	Q	R	S	T	U	V	W	X	Y	Z	A	B	C
e	E	F	G	H	I	J	K	L	M	N	O	P	Q	R	S	T	U	V	W	X	Y	Z	A	B	C	D
f	F	G	H	I	J	K	L	M	N	O	P	Q	R	S	T	U	V	W	X	Y	Z	A	B	C	D	E
g	G	H	I	J	K	L	M	N	O	P	Q	R	S	T	U	V	W	X	Y	Z	A	B	C	D	E	F
h	H	I	J	K	L	M	N	O	P	Q	R	S	T	U	V	W	X	Y	Z	A	B	C	D	E	F	G
i	I	J	K	L	M	N	O	P	Q	R	S	T	U	V	W	X	Y	Z	A	B	C	D	E	F	G	H
j	J	K	L	M	N	O	P	Q	R	S	T	U	V	W	X	Y	Z	A	B	C	D	E	F	G	H	I
k	K	L	M	N	O	P	Q	R	S	T	U	V	W	X	Y	Z	A	B	C	D	E	F	G	H	I	J
l	L	M	N	O	P	Q	R	S	T	U	V	W	X	Y	Z	A	B	C	D	E	F	G	H	I	J	K
m	M	N	O	P	Q	R	S	T	U	V	W	X	Y	Z	A	B	C	D	E	F	G	H	I	J	K	L
n	N	O	P	Q	R	S	T	U	V	W	X	Y	Z	A	B	C	D	E	F	G	H	I	J	K	L	M
o	O	P	Q	R	S	T	U	V	W	X	Y	Z	A	B	C	D	E	F	G	H	I	J	K	L	M	N
p	P	Q	R	S	T	U	V	W	X	Y	Z	A	B	C	D	E	F	G	H	I	J	K	L	M	N	O
q	Q	R	S	T	U	V	W	X	Y	Z	A	B	C	D	E	F	G	H	I	J	K	L	M	N	O	P
r	R	S	T	U	V	W	X	Y	Z	A	B	C	D	E	F	G	H	I	J	K	L	M	N	O	P	Q
s	S	T	U	V	W	X	Y	Z	A	B	C	D	E	F	G	H	I	J	K	L	M	N	O	P	Q	R
t	T	U	V	W	X	Y	Z	A	B	C	D	E	F	G	H	I	J	K	L	M	N	O	P	Q	R	S
u	U	V	W	X	Y	Z	A	B	C	D	E	F	G	H	I	J	K	L	M	N	O	P	Q	R	S	T
v	V	W	X	Y	Z	A	B	C	D	E	F	G	H	I	J	K	L	M	N	O	P	Q	R	S	T	U
w	W	X	Y	Z	A	B	C	D	E	F	G	H	I	J	K	L	M	N	O	P	Q	R	S	T	U	V
x	X	Y	Z	A	B	C	D	E	F	G	H	I	J	K	L	M	N	O	P	Q	R	S	T	U	V	W
y	Y	Z	A	B	C	D	E	F	G	H	I	J	K	L	M	N	O	P	Q	R	S	T	U	V	W	X
z	Z	A	B	C	D	E	F	G	H	I	J	K	L	M	N	O	P	Q	R	S	T	U	V	W	X	Y

THE VIGENÈRE AUTOKEY POLYALPHABETIC CIPHER

(a) Put the plaintext letters in a row.

(b) Place the priming key letter below the first plaintext letter. Then put the first plaintext letter below the second, the second below the third, and so on to the penultimate below the last.

(c) Replace each letter of the plaintext with the letter at the intersection of the row labelled by the plaintext letter and column labelled by the key letter.

Example 1.3 *Let us first choose a priming key, say* x, *and assume that the plaintext is* form secret diction.

f	o	r	m	s	e	c	r	e	t	d	i	c	t	i	o	n
x	f	o	r	m	s	e	c	r	e	t	d	i	c	t	i	o
C	T	F	D	E	W	G	T	V	X	W	L	K	V	B	W	B

For instance, the row labelled f *intersects with the column labelled* x *at the ciphertext letter* C, *and so on. To decipher, the receiver knows the priming key* x, *so this letter is placed above the ciphertext letter* C *and looks in the row labelled* x *to find the letter* C, *then the label of the column in which* C *sits is the plaintext, namely* f, *and so on, as follows.*

x	f	o	r	m	s	e	c	r	e	t	d	i	c	t	i	o
C	T	F	D	E	W	G	T	V	X	W	L	K	V	B	W	B
f	o	r	m	s	e	c	r	e	t	d	i	c	t	i	o	n

Unfortunately, as is the case too often, Vigenère's idea was forgotten and reinvented at the end of the nineteenth century. However, what was rediscovered was a weakened version of his idea. Essentially it amounted to exactly what Belaso has discovered, which we discussed on page 53, applied to the Vigenère square rather than that of Trithemius, so we need not replay it here.

One obvious improvement to the above is to have not a single priming key letter, but rather a priming *keyphrase*. Moreover, in the interests of security, the keyphrase should be as long as possible and feasible. Later we will see a very secure cipher where the key is as long as the plaintext itself, called the *one-time pad* (page 83). For instance, consider the following depiction of the more general idea of extending Vigenère's idea to a keyphrase.

Example 1.4 *Suppose that we want to encipher, again:* form secret diction, *but this time using the priming* keyphrase: *"xanadu". Then we proceed as in steps* (a)–(c) *on page 56, this time with our more general keyphrase sitting below plaintext letters before introducing the plaintext into the key, as follows.*

f	o	r	m	s	e	c	r	e	t	d	i	c	t	i	o	n
x	a	n	a	d	u	f	o	r	m	s	e	c	r	e	t	d
C	O	E	M	V	Y	H	F	V	F	V	M	E	K	M	H	Q

Then to decipher, we proceed as in Example 1.3, but with the full keyphrase this time, rather than the key letter, as follows.

x	a	n	a	d	u	f	o	r	m	s	e	c	r	e	t	d
C	O	E	M	V	Y	H	F	V	F	V	M	E	K	M	H	Q
f	o	r	m	s	e	c	r	e	t	d	i	c	t	i	o	n

The Vatican and Cipher Secretariats

Before we meet our next character, who will help us close the door on the Renaissance and this section, we must backtrack a bit in time to set the stage in another scene populated by the Italian City States, the Vatican, and Cipher Secretariats.

In Pavia, Italy on July 4, 1474, Cicco Simonetta, secretary to the Dukes of Sfoza, oligarchs of Milan, wrote the first known manuscript devoted *solely* to cryptanalysis. He wrote thirteen rules for symbol substitution ciphers. Later, another Italian, Giovani Soro, was appointed *Cipher Secretary* for Venice in 1506. Soro's cryptanalytic prowess gained him two assistants and an office in the Doge's Palace above the Sala di Segret, in 1542. (The *Doge's Palace* was the official residence of the doges in Venice. The *Doge* (from the Latin *dux* or *leader*, or *duke*, in English) was the highest official of the republic of Venice for more than a millennium (roughly 800–1800 AD). They represented the virtual emblem of the sovereignty of the Venetian State.) Soro, and his highly placed assistants, worked on the most elevated level of security, deciphering all messages from foreign powers, intercepted by the Venetians.

Cryptographic assistants were also available at the Vatican. The practice became of such high consequence to the popes that the office of *Cipher Secretary to the Pontiff* was established in 1555. The first of these was Triphon Benicio de Assisi. Assisi assisted Pope Paul IV during warring times with King Philip II of Spain. In 1557, Assisi was adept at deciphering the King's cryptograms. By September 12, 1557, peace was made, due in no small measure to the cryptanalytic skills of Assisi.

In the late 1580s, the Argentis, a family of cryptologists, took over the cipher secretariat. They were the first to institute certain cryptanalytic methods, use of which later became widespread. This included a *mnemonic* or *memory aid* key to mix a cipher alphabet. Of great interest to us is Matteo Argenti, who wrote a 135-page book on cryptology, which many consider to be the height of Renaissance cryptography. The Argentis distributed polyalphabetic ciphers to cardinals for their personal use, but failed to trust them for the bulk of their cryptographic traffic. When they used these ciphers, they employed relatively long keys, for reasons cited below.

It was Matteo Argenti who laid claims to being able to cryptanalyze certain autokey polyalphabetic ciphers. Yet part of this success was due to the use of "weak keys", some of which could be easily guessed. However, by the time Vigenère had developed his ideas and they were refined, the methods of mixing alphabets and using large keys was sufficient to thwart the cryptanalysts of the day. Nevertheless, the nomenclators (discussed earlier, see page 40), held sway for three more centuries over its more powerful cousin, the polyalphabetic autokey cipher. The reasons for this stem from the user more than the cipher. Cryptographers of the day were not enamored with the slowness of polyalphabetic ciphers, of having to always keep track of cipher alphabets, and what they perceived as a lack of precision, too much room for errors, and so on. Although not popular in the main, polyalphabetic ciphers did play a role, often a vital one at the time. We will learn more about this in the next section.

We close this section with Sir Francis Bacon, (1566–1626) whom we already discussed on page 36. He developed a steganographic device where one simply changes the typeface of random text to hide the existence of a message. He also invented a cipher, called *the bilateral cipher* (which today would be known as 5-bit encryption), in which he used a combination of substitution and steganography.

In the Chapter 2, we have another 500 years to put under the microscope.

Chapter 2

From Sixteenth-Century Cryptography to the New Millennium — The Last 500 Years

The age of chivalry is gone. That of sophists, economists, and calculators, has succeeded; and the glory of Europe is extinguished forever.
    Edmund Burke (1729–97), Irish-born whig politician and writer
    — from *Reflections on the Revolution in France* (1790)

2.1   Three Post-Renaissance Centuries

We begin with cryptographic tales surrounding the French, British, and Spanish monarchs in the sixteenth century. In 1556, Philip II of Spain ascended to the throne. In that year, he decided to discard the (deeply compromised) ciphers used by his father Charles V. Philip turned to an idea of Giovani Soro (a cryptographer we discussed in Section 1.5 (see page 58), by dividing his cryptosystems into two sets: *cifra general*, used for communications between the king and his ambassadors; and *cifra particular*, used by an individual messenger and the king. Philip's use of Soro's ideas became the template for Spanish cryptography well into the seventeenth century.

Meanwhile, in France there was a mathematician named François Seigneur De La Bigotiere Viète, (1540–1603). Viète, as shown in Figure 2.1, is known as the father of modern algebraic notation, largely due to his book, *In Artem Analyticem Isagoge*, or *Introduction to the Analytical Art*, published in 1591. This book could actually pass as a modern text in elementary algebra. His ability at cryptanalysis, however, is our chief interest. He was an assistant to King Henry IV (Henry de Bourbon-Navarre) of France, and a Huguenot sympathizer. (The Huguenots were the Protestants in France in the sixteenth and seventeenth centuries.) Henry had come to the throne in 1589, but it would take him almost a decade to secure his kingdom. The problem was *La Sainte Liguea*, or the *Holy League*, a French Catholic faction opposed to the *protestant* king. Philip supported the Holy League for which reason Henry declared war on Philip in January of 1595. Viète was brilliant in his cryptanalysis of Spanish letters from Philip, destined for the Holy League.

When Philip found out about Viète's cryptanalysis of his letters to commanders in France, he was absolutely stunned, having had the firm belief that his ciphers were unbreakable. He looked for other reasons, even going so far as to complain to the pope that "black magic" was being used against him and Spain. He was rebuked, since the pope had Giovanni Batista Argenti in his employ (see page 58), who was a powerful cryptanalyst, so he understood the real nature of the cryptological world, something that Philip sorely lacked. This lack would come back to haunt him.

Figure 2.1: François Viète. From Galérie Française, ou Collection des Portraits (Didot, 1842), I, plate 24.

Ultimately, Henry defeated the Holy League and their Spanish allies at Fontaine-Française in Burgundy in June of 1595, and retook Amiens from Spanish control in September of 1597. On May 2 of 1598, the *Peace of Verins* was reached between France and Spain on May 2 of 1598. In that year, Henry ended more than four decades of persecution against the Huguenots, by putting forth the *Edict of Nantes*, which was their charter of religious and political freedom.

Mary, Queen of Scots

Philip was going to encounter even more trouble, largely due to his inability to understand the powerful cryptanalytic techniques available. Philip had a dream to overthrow Queen Elizabeth I of England, establish a marriage with Mary Queen of Scots, and thereby secure a shared Catholic throne with her. However, letters written between Philip and his half-brother Don Juan of Austria, detailing his invasion plans for England, were intercepted by William of Orange, who was leader of the Dutch and Flemish revolt against Spain. William gave the missive to his cipher secretary, Philip van Marnix van Sint Aldegonde (1540–1598), a brilliant cryptanalyst, whose decryption of the correspondence revealed Philip's planned invasion of England. William then gave the plaintext to an English agent in the employ of Sir Francis Walsingham, Principal

Secretary to Queen Elizabeth. This cryptanalysis of Philip's intentions was accomplished in 1577, but Philip did not invade until 1588, during which time the English were able to fortify their defenses substantially. In fact, of the so-called invincible armada of Spain, numbering about 130 ships and around 19,000 men, only 69 ships returned to Spain and as many as 15,000 men perished, either in battle or on the long voyage home. This was the first major strike at the heart of the most powerful European power of that age, ensured by the actions of a cryptanalyst.

Meanwhile, back in England in 1577, after receiving the deciphered cryptogram revealing Philip's plans for invasion, Walsingham set about to establish a cipher school in London. He employed a man named Thomas Phelippes, as his cipher secretary. Phelippes was destined to be the first eminent cryptanalyst in England's history, and the downfall of Mary Queen of Scots.

After Mary lost her own throne in Scotland, and failed in a final attempt to regain her crown, in 1568, she made the crushing mistake of fleeing south to her cousin Queen Elizabeth hoping for asylum. (Mary Stuart was the only child of King James V of Scotland and his French wife, Mary of Guise.) The Queen merely imprisoned Mary, since she was too much of a threat, given that English Catholics considered Mary to be the rightful Queen of England. Mary spent 18 years in various prisons (actually in castles and manors, where she was under house arrest and heavy guard). Moreover, all of her ingoing and outgoing correspondence did not reach their destinations, that is, until January 6 of 1586, when Gilbert Gifford, a former seminarian, smuggled a pile of correspondence in to her that had accumulated at the French Embassy in London. The embassy had kept the letters fearing that if sent, they would not reach Mary. Gifford fixed that problem. Moreover, he began smuggling out letters from Mary to others.

A young Catholic man named Anthony Babington hated Queen Elizabeth for the atrocities committed against his family and Catholics in general at the time. Anti-Catholic policies included public disemboweling of live victims, accused of being traitors for the mere fact of being a loyal Catholic, or even for just being a priest, loyal to the Vatican. Babington hatched a plot to assassinate Queen Elizabeth and wanted Mary's blessing to do so. Gifford delivered an enciphered letter from Mary to Babington revealing that she had heard about the plot from her friends in France, and that she wanted to hear from him. Babington put together an encrypted letter, outlining the details of the plot, and even added the steganographic benefit of placing the message in a beer barrel to get past the guards. However, all this was for naught since Gifford had been recruited and was working for Walsingham from the outset. All Mary's correspondence had been brought to him for Phelippes to cryptanalyze, then sealed again and sent to their destinations with nobody in Mary's camp being any wiser.

When Gifford brought Babington's letter to Walsingham, and Phelippes deciphered it, the plot was revealed. Yet, Walsingham was biding his time. He let the letter go to Mary and when she responded to Babington, and Phelippes decrypted it, this spelled the end for her. Yet, Walsingham wanted to get everyone involved in the plot, so he had Phelippes forge a *P.S.* asking Babington to name those "best suited" for the assassination. Ultimately, Babington and his cohorts were caught, and put to death by being, among other atrocities, disemboweled alive. Mary went on trial and met the headman's axe (an axe in no small part set in motion by a cryptanalyst) on February 6, 1587.

Antoine Rossignol

Now we turn to cryptographic developments in seventeenth century Europe. France's first recognized full-time cryptologist was Antoine Rossignol who served both Louis XIII and Louis XIV. In fact, on his deathbed, Louis XIII insisted to his Queen that Rossignol be kept at court as a man necessary to the security of the state. Rossignol also used his cryptanalytic skills to assist Cardinal Richelieu (1585–1642). (Perhaps the quote that best epitomizes both the political bent and character of Richelieu is: "If you give me six lines written by the most honest man, I will find something in them to hang him.") Rossignol's initial rise to prominence is due to the following anecdote. In 1626, the French intercepted a cryptogram, carried by a messenger from the city of Réalmont, which was being held by the Huguenots. Rossignol cryptanalyzed it. The plaintext revealed that the Huguenots were on the edge of collapse. Rossignol had the letter sent back to the city together with the plaintext beside it. With their secret plight revealed, the Huguenots surrendered without more fighting, again the result of a cryptanalyst's skills.

Perhaps of greatest technical importance were Rossignol's improvements to the nomenclators of the time (see page 40). These consisted of only one part, meaning a single list of plaintext and code in alphabetical (or numerical) order. In other words, they were listed in parallel, a system that was in place since the start of the Renaissance. Rossignol determined that this parallel order allowed for a means of cryptanalysis just by looking at what numbers stood for which words. For instance, if he were able to find that 64 stood for *launch* and 98 stood for *lethal*, then no number between 64 and 98 could represent *letter* since its code would have to be higher than 98 given the parallel matching of code and plaintext. Also, if he wanted to find the code for *legal*, he knew it had to be between 64 and 98, again because of the parallel matching. This allowed a cryptanalyst too much advantage. He devised a method to thwart such attempts. Rossignol insisted upon two lists, a *tables à chiffer*, consisting of plaintext letters in alphabetical order, and code numbers in random order; and the *second* part, used for decoding, called the *tables à déchiffer*, with plaintext letters randomized and codes symbols in numerical order, the birth of *two-part* nomenclators. (Think of these as similar to a bilingual dictionary.) The revised and improved nomenclators were vital components of cryptology for over four centuries. It is a testimony to Rossignol's cryptanalytic skills that the word "rossignol" has entered the vocabulary of French slang to mean "a tool that picks locks". Rossignol also was a prime mover in the establishment of the *Cabinet Noir*, or *Black Chamber* — a headquarters for cryptanalysis and intelligence gathering — which began France's firm grip on cryptography, reading cryptograms of foreign countries throughout the seventeenth century.

John Wallis and the Black Chambers

The origins of the black chambers in England can be attributed to John Wallis (1616–1703), (see Figure 2.2) who may be considered Rossignol's contemporary. Wallis was cofounder (along with John Wilkins[2.1] (1616–1703)) of the Royal Society. Wallis was primarily a mathematician, perhaps one of the very best England had seen up to that time. For instance, his book, *Arithmetica Infinitorum* (1656), had a profound influence on Sir Isaac Newton's invention of the calculus. Wallis also invented the symbol ∞ for infinity, and numerous other contributions may be cited, but our primary interest is in his cryptological interests, and he had many.

In 1640, Wallis was ordained by the Bishop of Winchester, and in that same year received his Master's Degree. In his twenties, Wallis began looking at ciphers. In fact, in 1642, at the time of the Civil War between the Royalists and the Parliamentarians, he was cryptanalyzing Royalist messages for the Parliamentarians. As a reward, he was given charge of the Church of Saint Gabriel in London in 1643. In that year, his mother died, leaving him an independently wealthy man with a substantial estate in Kent where he was born on November 23, 1616. In 1645, he began meeting with a group (including Wilkins) that would eventually lead to the establishment of the Royal Society in England.

Figure 2.2: John Wallis.

As further reward for using his cryptanalytic skills in support of the parliamentarians, he was appointed to the *Savilian Chair* of geometry at Oxford in 1649, a position he held for over half a century until his death. However, he engaged in many other activities.

His greatest cryptological efforts came late in his life. He was employed in 1689 as a cryptanalyst by King William III (1650–1702), and Mary II (1662–1694), reporting to their Secretary of War, the Earl of Nottingham. (William ruled jointly with Mary from 1689 until her death in 1694, then solely until his death in 1702. He came to the English throne from the house of Orange, in the Netherlands, and thus he is often called William of Orange.)

In the summer of 1689, Wallis cryptanalyzed intercepted cryptograms (largely nomenclators) that had been sent between Louis XIV and his ambassador in Poland. This included Louis' attempts to instigate a war between

---

[2.1]Wilkins, in his book *Mercury, or the Secret and Swift Messenger*, introduced into the English language, the terms *cryptologia*, or *secrecy in speech* and *cryptographia*, or *secrecy in writing*. He also introduced the term *cryptomeneses* as a general term for secret communication.

Poland and Prussia, and promoting a marriage between the Prince of Poland and the Princess of Hanover, which would have been advantageous to Louis. Wallis continued his cryptanalysis including the breaking of important cryptograms for the king, all of which earned him the title of *Father of Cryptology* for England (as had Rossignol earned such a title in France). Wallis died on October 28, 1703, in Oxford, England.

The most iron-clad, efficient, and effective of the black chambers during eighteenth-century Europe was that of Vienna, called the *Geheime Kabinets-Kanzlei.* As an illustration, the cryptograms were usually deciphered, resealed, and sent on to their destinations within three hours of their having been dropped off at the chamber at their usual arrival time of 7 in the morning. The chamber was effectively compartmentalized so that one section might contain language experts, or translators, and another might contain people copying letters or stenographers, all working in concert. Moreover, to reduce stress among their cryptanalysts, they were given staggered working times, one week on and one week off. It is due to the effectiveness of these black chambers that cryptographers, by the end of the century, began turning to polyalphabetic ciphers. The monoalphabetic ones were falling like dominos in the face of the concentrated and talented efforts of these centers of cryptanalysis and intelligence gathering.

In England, certain individuals were appointed as *Decypherers for the English crown.* In 1703, the *Decyphering Branch* was established, to decrypt documents as a means of uncovering plots and schemes against the state. They had no fixed location, but mostly worked at home and submitted their findings. They were indeed more secret than the U.S. Secret Service. They were funded by the secret-service money issued to the Secretary of the Post Office from Parliament. The first to bear the title of Decypherer was Wallis's grandson, William Blencowe. He was also the first Englishman to be paid a regular wage for cryptanalysis. One of his successors was Edward Wiles who was appointed as Decypherer to the crown in 1719. Wiles decrypted a cipher that revealed Sweden's plan to create an uprising in England. For this and other cryptanalytic feats, he ascended to become Canon of Westminister. By 1742, he had been appointed Bishop of St. David's. He brought his two sons, Edward Jr. and William, into the cryptanalysis sector of the decyphering branch in the middle 1750s. Although he died in 1773 (buried in Westminister Abbey), his sons carried on his work and dominated the cryptanalytic sector after his death.

By 1714, the decyphering branch was collaborating with the black chamber at Nienburg, Germany, which was supported by George I (1714–1727) of England. (George I was Georg Ludwig, elector of Hanover (1698–1727), who was the first Hanoverian king of Great Britain.) During the eighteenth century the decyphering branch cryptanalyzed the dispatches of roughly sixteen countries with an average output reaching as high as one per day.

By the middle of the eighteenth century, the decyphering branch was preparing England's diplomatic nomenclators. Typically, these nomenclators had four-figure code groups and various homophones. Despite weaknesses in their system, their use continued well into the end of the century.

2.2 The American Colonies

The sun of Great Britain will set whenever she acknowledges the independence of America ... the independence of America would end in the ruin of England
Lord Shelburne (1737–1805), British Whig politician, Prime Minister
— *spoken in the House of Lords* (October, 1782)

In the eighteenth-century American colonies, cryptology was not as sophisticated as that in Europe. Certainly there were no black chambers, and really no organized effort to do research into intelligence gathering, develop cryptanalytic skills, or anything of the sort. Nevertheless, a development did begin with some early tentative baby steps. We will review them by starting at the end of the eighteenth century.

During the American revolution, there was virtually no cryptanalysis being done until near the end of the war since there were virtually no interceptions of cryptograms. However, most of what was deciphered at the end of the conflict was accomplished by one

Figure 2.3: George Washington.

man, James Lovell, a member of the Continental Congress, who may rightly be considered to be the pioneer of *American* cryptanalysis.

Washington and Jefferson

By 1781, Lovell had already been using a version of the Vigenère cipher (see pages 55 and 56). In that same year, when colonial forces intercepted a British cryptogram, Lovell was given the task of breaking it. It proved to be an easy task for Lovell. However, by the time the information was revealed it was too late to be of any military value. Yet, Lovell decrypted keys, which he kept in anticipation of their being of use later on. Indeed, George Washington (1732–1799), (see figure 2.3)[2.2] the commander-in-chief of the colonial armies (1775–83), and subsequent first president of the United States (1789–97), was able to use them. Washington had been able to surround the British commander in the southern colonies, Lord Cornwallis, (1738–1805) at Yorktown. His forces intercepted a British letter, which he gave to his secretary to decrypt. Washington had received the keys from Lovell's earlier decryption and the letter was easily deciphered using them. This gave Washington important information about the British positioning and strengths. Later, when more British cryptograms were intercepted, Lovell was able to easily break the

---

[2.2]The lithograph above is courtesy of the Library of Congress, reproduction no. LC-USZ62-117116, Prints and Photographs Division, created/published around 1828.

code since the British were still using the same keys. This allowed the colonial forces to prevent reinforcements from reaching Cornwallis, who surrendered five days later on October 19, 1781. This victory at Yorktown ended the fighting and ensured Washington's victory at the end of the American War for Independence.

Figure 2.4: Thomas Jefferson.

One of the founding fathers who sought to improve the means of secret communications was certainly the most forward-thinking of them all, *Thomas Jefferson*[2.3] (1743–1826), (see Figure 2.4) the draftsman of the Declaration of Independence of the United States, the nation's first secretary of state (1789–1794), its second vice-president (1797–1801), and its third president (1801–1809). In 1785, he compiled a nomenclator to aid in his correspondence with Madison and Monroe, a method that he used until 1793. Perhaps Jefferson's most important cryptographic contribution was what he called his *wheel cypher*, (see Figure 2.5) invented in the 1790s.[2.4] This device consisted of 36 concentric wooden disks, each approximately 1/6 of an inch thick, and 2 inches in diameter with a mix of the English alphabet inscribed on the outer edge. Moreover, each disk had its own number, and the key consisted of an agreed-upon sequence of these numbers for correspondents to use. The correspondent would assemble their disks in this (key) sequence on a metal spindle. Here is how it worked.

To encrypt the first 36 letters of the plaintext, the sender found the first letter on the first wheel, second letter on the second wheel, lined up with the first, then the third in the third wheel lined up with the first two, and so forth. The ciphertext consisted of any of the 25 remaining parallel rows of letters on the disk. The sender would select one of them, write it down as the ciphertext for the first 36 letters of plaintext, then repeat the above process for each remaining block of 36 plaintext

Figure 2.5: Wheel cypher.

[2.3]The above lithograph of Thomas Jefferson is courtesy of the Library of Congress, reproduction no. LC-USZ62-117117, Pictures and Photographs Division, created around 1828.

[2.4]The above replica of Jefferson's wheel cypher is courtesy of the National Cryptologic Museum of the National Security Agency. See *http://www.nsa.gov/museum/wheel.html*.

letters until the cryptogram was completed.

The recipient would align the first 36 letters of ciphertext parallel to his spindle as had the sender. Then at one of the other 25 rows would sit the (obvious) plaintext. The process would be repeated for the remaining blocks of 36 letters each until the entire cryptogram was turned into the original plaintext.

Given the above delineation of how it worked, Jefferson's wheel cypher was therefore a polyalphabetic cipher with the plaintext as the key, quite an amazing invention for that time and, as we shall see, for some time to come.

His wheel cypher and his idea were filed away and completely forgotten until 1922 when it was rediscovered in the Library of Congress. It had been reinvented many times and one of the forms was the U.S. Navy Strip Cipher, M-138-A, used in World War II. In fact, many cryptanalysts in U.S. government agencies in the early twentieth century could not cryptanalyze Jefferson's system! Thus, Jefferson truly deserves the title *Father of American Cryptography*.

Wadsworth, Wheatstone, and Playfair

Yet another American invented a cipher disk, this one with gears. In 1812, Colonel Decius Wadsworth was given a position as the first chief of ordnance of the U.S. Army, a post he held until 1821. In 1817, while at this post, he invented a device that was a brass cipher disk in a wooden container 6 and 1/2 inches in diameter and roughly 3 inches high. The outer alphabet had 26 letters together with the integers 2 through 8 inclusive (33 symbols in all); and the inner alphabet had just the original 26 letters. He included a brass plate with two small openings that align to identify the plaintext and ciphertext equivalents. The container itself enclosed two gears, (one with 33 and the other with 26 teeth) to rotate the disks. To set up correspondence, the sender and recipient merely agree on a sequence for the ciphertext and a starting point, which would be a symbol in the brass plate opening for both plaintext and its ciphertext chosen equivalent. For instance, *W* might be in the opening at the outer disk, while *a* is at the opening of the inner disk. This introduced differing numbers of symbols for plaintext and ciphertext resulting in a progressive cipher in which alphabets are used irregularly, depending on the plaintext used. Thus, Wadsworth's device was a progressive system that was polyalphabetic. The reader will recall that Trithemius also introduced a progressive key (see page 52). However, Trithemius' progression was *regular* on 24 cipher alphabets, whereas Wadsworth's progression was *irregular* on 33 cipher alphabets, much more secure. Unfortunately for Wadsworth, his idea died with him, and credit went to someone across the Atlantic.

Charles Wheatstone (1802–1875) worked on many areas from acoustics to inventing the electric telegraph before Morse. His many achievements earned him a knighthood in 1868. He also delved into the cryptographic arena. In 1867, at the Exposition Universelle in Paris, Wheatstone unveiled his *cryptograph*, which was essentially the same as Wadsworth's gear cipher, only a weaker version. Wheatstone's device had an outer ring consisting of 27 plaintext symbols (26 letters and a blank), and an inner ring with mixed ciphertext alphabet consisting of 26 letters. He put two clock-style hands over them, one long and one short, which were employed in unison to choose plaintext and ciphertext equivalents. However, given that there is $27 - 26 = 1$ unit difference between plaintext and ciphertext in his device, while Wadsworth had $33 - 24 = 9$ units difference meant that Wheatstone's apparatus was far less secure than that of Wadsworth. In another, perhaps fitting, misattribution, a cipher that Wheatstone *did* create and which was far superior to the device for which he is known, has been attributed to another. In 1854, Wheatstone invented the first *literal* digraphic cipher in history. (The attentive reader will recall that on page 54 we mentioned that Porta created the first digraphic cipher, but with *signs*, rather than letters.) However, Wheatstone's friend Lord Lyon Playfair, who sponsored it at the British Foreign Office, has his name attached to it. This is how it works.

Consider the Table 2.1 (where the letters *WX* are considered as a single symbol):

The Playfair Cipher

	A	*Z*	*I*	*WX*	*D*
Table 2.1	*E*	*U*	*T*	*G*	*Y*
	O	*N*	*K*	*Q*	*M*
	H	*F*	*J*	*L*	*S*
	V	*R*	*P*	*B*	*C*

Pairs of letters are enciphered according to the following rules.

(a) If two letters are in the same row, then their ciphertext equivalents are immediately to their right. For instance, VC in plaintext is RV in ciphertext. (This means that if one is at the right or bottom edge of the table, then one "wraps around" as indicated in the example.)

(b) If two letters are in the same column, then their cipher equivalents are the letters immediately below them. For example, ZF in plaintext is UR in ciphertext, and XB in plaintext is GW in ciphertext.

(c) If two letters are on the corners of a diagonal of a rectangle formed by them, then their cipher equivalents are the letters in the opposite corners and the same row as the plaintext letter. For instance, UL in plaintext becomes GF in ciphertext and SZ in plaintext is FD in ciphertext.

(d) If the same letter occurs as a pair in plaintext, then we agree by convention to put a Z between them and encipher.

(e) If a single letter remains at the end of the plaintext, then a Z is added to it to complete the digraph.

One merely reverses the rules to decipher.

Example 2.1 *Suppose that we know the following was enciphered using the Playfair cipher.*

EJ DJ DJ EJ GA VO IE JY NK YV TI VO ZU

To illustrate each of the rules for columns, rows, and diagonals, we choose certain groupings. For instance, the first pair EJ *occurs on a diagonal with* TH *as the opposite ends (respectively) of the diagonal. The* VO *in the sixth grouping have letters that are in the same column, so we choose the letters above them* HE *(respectively). The letters in* NK, *the ninth grouping, are in the same row, so we choose the letters to the* left *of them* ON *(respectively), and so on to get:*

THIS IS THE WHEATSTONE CIPHER,

where the last letter Z *is ignored as the filler of the digraph.*

In the end, the Playfair cipher was adopted as the British army's field cipher, so the war office kept it a secret. Although Playfair himself did not claim the cipher to be his own invention, it has come to be known as his. The misattribution has to do with Playfair's unbridled promotion of Wheatstone's cipher, especially to the British Foreign Office, where they began to refer to it as Playfair's cipher.

Now it is time to cross back over the Atlantic and resume our analysis of cryptology in the American colonies.

Samuel Morse and His Code

Samuel Finley Breese Morse (see figure 2.6) was born on April 27, 1791 in Charlestown, Massachusetts, to Reverend Jedediah Morse and Elizabeth Breese. Jedediah was also known as the "father of American geography" and was the author of the first text on the subject, *Geography Made Easy*, published in 1784, followed by 24 editions.

Samuel got his education in Massachusetts, and graduated from Yale College in 1810. After going to England to study painting, he returned in 1815, becoming a wayfaring portrait painter and settling finally in New York in 1825. He founded the National Academy of Design and was its first president, from 1826 to 1845. Although Samuel had no formal training in electricity, he came to the realization that electric current could be used to convey information over wires.

In 1832, he first conceived of a telegraph and had developed a working model by 1837, almost simultaneously with Wheatstone, cited above, and his associate Sir William Cook. Cook and Wheatstone took out a joint patent in 1837 for the first electric telegraph put in use by the British railway system. By 1838, Samuel had invented *Morse Code* (as it has come to be known) of dots and dashes as a convenient method of representing letters for sending telegraphic messages. In 1854 he was granted patent rights by the U.S. Supreme Court. Moreover, Morse's method became more popular than the Cook-Wheatstone system, and eventually by the mid-nineteenth century, Europe had its own version of Morse code.

In 1844, a telegraph line between Baltimore and Washington was completed, and the first message sent on May 24 was "What has God wrought!" By 1861, the United States was linked coast to coast by telegraph. Morse died on April 2, 1872, in New York City.

Figure 2.6: Samuel Morse.

This daguerreotype portrait is courtesy of the Library of Congress, reproduction no. LC-USZ62-110084, created between 1844 and 1860 from the studio of M.B. Brady (circa 1823–1896), photographer. Morse first met Daguerre in Paris, and according to the Library of Congress, this may be the earliest daguerrotype made in America. Louise-Jacques-Mandé Daguerre (1787–1851) was the French painter and physicist who invented the first practical method for photography.

The American Civil War

Since telegraph operators would necessarily have to read the messages they were sending, the original sender would often want to encipher it. Then the telegraph operator would send the cryptogram in Morse Code. During the Civil War, the U.S. Military Telegraph Corps used *route ciphers*, meaning words transposed with codes they called *arbitraries* thrown in to confuse cryptanalysts. They would often use *nulls*, nonsense words, to thwart cryptanalysis. President Abraham Lincoln (see Figure 2.7) had young cipher operators in their early twenties who were adept at cryptanalyzing Confederate correspondence. Some Confederate commanders used a form of Vigenère cipher, but they used it badly. In fact, some officers were given the choice of cipher, and it is known that Confederate General Albert S. Johnson decided to use a Caesar cipher on one occasion!

Lincoln's cryptanalysts had no trouble in deciphering the messages, but the same could not be said of the Confederates. Sometimes they made such bad use of their own ciphers, the recipients could not decipher them, nor could they even come close to decrypting the Union correspondence. Lincoln's assassin, John Wilkes Booth, was known to have used a Vigenère cipher, which was actually introduced at his trial. Although no connection could be made between Booth, his associates on trial with him, and the cipher, they were all put to death. Even at the end of the war, roughly two weeks after General Lee surrendered, a Vigenère cipher was used by Jefferson Davis and the key he used was *COME RETRIBUTION* to send the incredibly defiant message *active operations to be resumed in forty-eight hours.*

Figure 2.7: Abraham Lincoln.

Courtesy of the Library of Congress, reproduction no. LC-USZ61-1938, Pictures and Photographs Division. It was the last sitting, four days before Lincoln's assassination.

We close this section with an interesting anecdote of a noncryptographic nature about Booth. On April 15, 1865, Dr. Samuel A. Mudd treated Booth's broken leg early on the morning after the assassination. He also provided a place for Booth to rest. Mudd was arrested, found guilty of aiding and abetting the assassins, and sentenced to life in prison at Fort Jefferson in the Dry Tortugas. Four years after he went to prison, President Andrew Johnson pardoned him and he was released. However, to this day, his name has gone down in ignominy. The standard modern-day phrase, "Your name is Mudd", means that a person's actions have made him *persona non grata*, Latin for an *unacceptable* or *unwelcome person.*

Figure 2.8: Confederate cipher.

Courtesy of the National Security Agency Public Photo Gallery.
See *http://www.nsa.gov/gallery/photo/photo00020.jpg.*

Figure 2.9: Confederate cipher disk.

Courtesy of the Confederate Secret Service Camp 1710.
See *http://home.earthlink.net/~ cssscv/index.html.*
Original on display at the NSA museum at Fort Meade, MD.

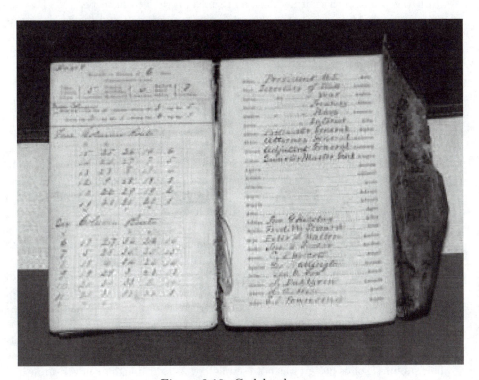

Figure 2.10: Codebook.

2.3   Nineteenth-Century Cryptography

Thought can with difficulty visit the intricate and winding chambers which it
inhabits.   **Percy Bysshe Shelly (1792–1822)**
                                  English poet, husband of Mary Shelly
                           — from Speculations on Metaphysics *(1815)*

In the nineteenth century, one man may be said to have been the vision-
ary pioneer when it comes to foreseeing the modern-day automatic electronic
computer, *Charles Babbage* (1791–1871). Babbage was born in the Walworth,
Surrey area, in London, England on December 26, 1791 to Benjamin Babbage
and Elizabeth Teape. Benjamin was a wealthy banker, who left Charles a size-
able fortune upon his death. This financed his many lifelong interests, from
areas as diverse as archeology, to submarine navigation, mathematics in general, and cryptology in particular.

In 1810, he attended Trinity College, Cambridge, and by 1817 had received
his master's degree. In 1816, Babbage was elected as a fellow of the Royal
Society. He had lived at Devonshire Street in London until 1828 when he took up
a position as Lucasian Professor of Mathematics at the University of Cambridge.
Then he moved to 1 Dorset Street, Manchester Square, London, where he lived
until his death. Babbage held the position at Cambridge until 1839.

It was in the mid-1830s that Babbage envisioned a machine (which he called
the *analytical engine*) executing arithmetical operations via instructions from
punched cards, having memory to store data, and other fundamental aspects
of modern computers that developed more than a century after he conceived of
them. He started on what he called the *Difference Engine Number 1* in 1823,
but abandoned it after a decade of work. By the end of 1834, he had conceived
of his analytical engine, but he began work on a less ambitious project, his
Difference Engine Number 2. However, the government had been funding his
project from the outset, and by this time with no concrete results, they withdrew
their support, so his design was not completed. Yet, although he never published
his notebooks, they were discovered in 1937. By 1991, at the British Science
Museum, the Difference Engine Number 2 was built to original designs in order
to commemorate the bicentennial of Babbage's birth. It is accurate to 31 digits,
as Babbage had envisioned, and it is the first of his machines to be completed.

Another important factor in Babbage's failure to complete the construction
of any of his devices must certainly have been that the technology of the day
was woefully insufficient to make the precision parts that his designs required.
That it took roughly 150 years for one of them to be built is probably testimony
to that statement. Although Babbage never completed any of his machines, his
conception of the analytical engine is the vehicle for his fame as a visionary of
the modern digital computer.

More important for us, Babbage is also known for his penetrating cryptanalytic skills. The Vigenère cipher was considered, up to the mid-nineteenth
century, to be unbreakable, and it achieved the title of the *chiffre indéchiffrable*.
However, in the mid-1850s, Babbage cryptanalyzed the Vigenère cryptosystem.

Yet, as with most of his discoveries, he did not publish this fantastic break-through. There is speculation that since the breaking of the chiffre indéchiffrable occurred after the start of the Crimean War, British intelligence may have wanted Babbage to keep it a secret. (The Crimean War, October 1853 to February 1856, was fought primarily on the Crimean peninsula between the Russians and the British, French, and Ottoman Turks.) The British felt that this secrecy would give them an advantage over the Russians for several years. Thus it is that Babbage died on October 18, 1871 in London, without it being revealed that he broke the Vigenère cipher. That credit would go to another.

Frederich W. Kasiski (1805–1881) was born on November 29, 1805, in Western Prussia. He enlisted in East Prussia's thirty-third infantry at the age of 17, and retired in 1852 as a major. Although interested in cryptography during his military career, he did not publish any of his ideas until after his retirement. In 1863, he published *Die GeheimschRiften und die Dechiffrir - Kunst*, a general solution to cryptanalyzing polyalphabetic cryptosystems with repeating keywords, including the famed Vigenère cipher, a long-sought-after breakthrough.

The central idea behind Kasiski's attack is the keen observation that repeated portions of plaintext enciphered with the same part of a key must result in identical ciphertext patterns. Hence, barring coincidence, one would expect that the same plaintext portions corresponding to repeated ciphertext were enciphered with the same position in the key. Therefore, the number of symbols between the start of repeated ciphertext patterns should be a multiple of the *keylength* (the number of characters in the key). For example, if the repeated ciphertext is *ABC*, called a *trigram*, and if the number of letters between the *C* and the occurrence of *A* in the next trigram *ABC* is, say, 15, and this is not an accident, then 18 is a multiple of the keylength. Since it is possible that some of the repeated ciphertext segments are coincidental, a method of analyzing them, called a *Kasiski examination*, is to compute the greatest common divisor (gcd)[2.5] of the collection of all the distances between the repeated sections. Then choosing the largest factor occurring most often among these *gcd*s is (probably) the keylength. Once a probable keylength ℓ, say, is obtained, a frequency analysis can be performed on a breakdown of the ciphertext into ℓ classes (with an individual class containing every ℓ-th character) to determine the suspected key. The following is an illustration of the Kasiski method for finding the keylength.

Example 2.2 *Suppose that "keys" is the keyphrase and "these are the safest aims" is the plaintext. Then consider the following Vigenère enciphering.*

k	e	y	s	k	e	y	s	k	e	y	s
t	h	e	s	e	a	r	e	t	h	e	s
D	**L**	**C**	**K**	**O**	**E**	**P**	**W**	**D**	**L**	**C**	**K**

[2.5] For the reader unfamiliar with this concept and related notions, see Definition A.11 in Appendix A on page 470.

k	e	y	s	k	e	y	s	k
a	f	e	s	t	a	i	m	s
K	J	C	K	D	E	G	E	C

Notice that DLCK *is a block that occurs twice, at the beginning and end of the first table, and the distance between the occurrence of the first* D *and the second (in the second block) is* 8. *Also, the diagram* CK *occurs at the end of the first table and again* 4 *units away in the second table. Hence, since* gcd(8, 4) = 4, *this is the probable keylength by the Kasiski examination, which is indeed correct.*

Toward the end of the nineteenth century, another book, which may be seen as taking the torch passed by Kasiski, was published. In 1883, *La Cryptographie militaire* was published by Jean-Guillaume-Hubert-Victor-François-Alexandre-Auguste Kerckhoffs von Nieuwenhof (but we will just call him Kerckhoffs).

Kerckhoffs was born on January 19, 1835, of Flemish descent, in Holland. His education involved almost two years of study in England plus degrees obtained at the university in Liège. After some teaching positions and some travelling, Kerckhoffs married and settled down in a town outside Paris. He taught languages there for a number of years. By 1876, he had earned his Ph.D. and by 1881 became a professor of German in Paris. While there, he wrote the aforementioned book, which many consider to be the most succinct text on cryptography ever written.

In his book, Kerckhoffs elucidated several basic tenets. In modern times, one of these has come to be known as Kerckhoffs' Principle and has been incorporated into modern cryptographic methodology.

Kerckhoffs' Principle

In assessing the security of a cryptosystem, one should always assume the enemy knows the method being used.

The telegraph had made possible the introduction to cryptology of a new device, the *field cipher*, a rapid means for the military to send secure, secret messages in a theater of war. Kerckhoffs also instituted several tenets for field ciphers (from which his above principle has been gleaned).

Kerckhoffs' Principles for Field Ciphers

1. The cryptosystem should be practically unbreakable (breakable in theory, perhaps, but not in practice).

2. A compromised cryptosystem should not inconvenience the correspondents. (This is the one from which his aforementioned principle seems to be derived since it says that the enemy may know the cryptosystem, but one should still be able to send messages since the enemy cannot cryptanalyze with this knowledge and *without* the key.)

3. The key should be easy to both remember, and change at will.

4. Cryptosystems must be amenable to being sent by telegraph.

5. The mechanisms of the cryptosystem must be easily portable and entirely operable by a single entity.

6. The cryptosystem must be easy to use, without reference to any manuals, or the need for deep mental effort.

These six tenets are, of course, utopian in nature. Even modern-day ciphers would struggle to achieve all six conditions (where we can replace *telegraph* by *computer* in condition 4). Also, the second condition basically says (and this is implicit in his aforementioned principle) that secrecy lies in the *keys* and not in the cipher itself. Later, when we delve into modern ciphers, we will see that this is as true today as ever. "Key Management", as it has come to be known, is essential since a cryptanalyst who can break a key is better off than one who knows only the cryptosystem itself. Kerckhoffs' book stands tall as one of the great books on cryptology.

We close this section with a story about a French military officer at the end of the nineteenth century, wherein cryptology played a crucial role.

The Dreyfus Affair

On October 15, 1894, Captain Alfred Dreyfus was arrested and charged with high treason by the French government which claimed evidence that he had given military secrets to German and Italian officials. An Italian military attaché, Colonel Alessandro Panizzardi, later sent a cryptogram to Rome, which was intercepted by French cryptanalysts. Part of the deciphered message said: "If Captain Dreyfus has not had relations with you, it would be wise to have the ambassador deny it officially, or avoid press comment." This seemed to indicate that Panizzardi disavowed any contact with Dreyfus. In order to be certain, the French decided to trick Panizzardi into sending a telegram whose contents were known to them, for then they would have the key to decryption of what he had sent. Panizzardi bought the ruse, enciphered the telegram, and sent it to Rome. Subsequently, the French were able to verify the deciphering of the original message. Nevertheless, this failed to exonerate Dreyfus, since certain individuals would rather have had an innocent man go to prison than to admit an error had been made in his arrest. Thus, they prevented the telegram from being admitted at his trial. Hence, Dreyfus was convicted of treason, and sent to Devil's Island. Upon appeal, the telegram was admitted, but it would take several years before he would see justice. When exoneration did come, it included reinstatement in the Legion of Honour. The true criminal in the matter was arrested. Major Ferdinand Walsin Esterhazy had used several cardboard (Cardano) grilles that implicated him in having secret correspondence with a German attaché.

We now turn to the dawn of the twentieth century, with a world war brewing and a major sequence of turning points for the advancement of cryptology. Stay tuned, for the stories get better.

2.4   Two World Wars

Whosoever, in writing a modern history, shall follow truth too near the heels,
it may happily strike out his teeth.   **Walter Raleigh** (1552–1618)
English explorer and courtier
— from The History of the World; Preface (1614)

On December 12, 1901, Guglielmo Marconi (1874–1937) (see Figure 2.11) received signals from atop a hill in St. John's, Newfoundland across the Atlantic from Poldhu in Cornwall, England. This great achievement created a worldwide sensation, the first trans-Atlantic radio message. It marked the beginning of the era of wireless communication. The next several decades would see an explosion of development of radio communication, broadcasting, and navigation applications. For cryptography, however, it presented the problem of ease of interception by unintended recipients. However, despite the lack of security, there was often no alternative to wireless transmission, since it allowed central authorities to communicate with their armed forces. When the "Great War", World War I, broke out in 1914, all the main countries involved in the war were using wireless.

In 1916, the British army suffered losses in the thousands during the battle to take Ovillers-la-Boiselle on the Somme. The British were eventually successful in capturing it, but in the enemy trenches they found a complete plaintext of their orders to take the objective! It seems that a brigade major had read the orders, in plaintext, over a field telephone, despite protests from subordinates. This flagrant disregard for the need for secrecy, and therefore disregard for the lives that would be lost, led to the development of *trench codes* for field armies.

In early 1916, the French began to develop trench codes, which began as telephone codes, due to indiscretions such as the above. General Dubail requested that trench codes be implemented, which dictated that in normal phone conversations, certain words would be spelled out in code rather

Figure 2.11: Guglielmo Marconi.

The photograph of Marconi is courtesy of the Library of Congress, Reproduction no. LC-USZ62-39702, Prints and Photographs Division, Copyright 1908.

than spoken. Originally, these telephone codes had a small collection of two-letter codewords, which eventually grew to three-letter code words. These were subsequently adopted for wireless where early one-part codes grew into two-part codes in later implementations.

By 1917, the Germans were using trench codes, which evolved into enciphered code, as we shall see later. Ultimately, when the Americans joined the fray, they too adopted trench codes. Moreover, they had a brilliant cryptologist at their disposal, named Parker Hitt, who had worked as a Signal Corps instructor. In 1915, Hitt published an influential booklet on cryptology called *Manual for the Solution of Military Ciphers*. This was a practical manifestation of how cryptology should be used in the field. Also, in 1913, Hitt rediscovered the wheel cipher in strip form, and this led to the Navy strip cipher M-138-A (see page 66 for comparison with Jefferson's original invention). In 1916, Major Joseph O. Mauborgne put Hitt's strip cipher back into the intended Jeffersonian cylindrical form, strengthened the alphabet construction, and produced what came to be known as the M-94 device, which remained in service until the early part of World War II.

Britain did not have a formal cryptology bureau. However, with interception of German cryptograms, they quickly saw the need for one. A group was put together by Sir Alfred Ewing, the admiralty's director of naval education. They initially operated out of Ewing's office at the admiralty, but as the group grew and their activities increased, they were moved into Room 40 of the Admiralty Old Building. They became legendary as the *Room 40 Group* for their remarkable cryptanalytic feats. In [131, Chapter 9], Kahn devotes an entire chapter to them.

One of the major intelligence coups of the Room 40 Group occurred in 1917. They had intercepted a telegram sent by the Germans over Swedish and American cables routed through Britain. When deciphered, it indicated that the telegram was from the new German Foreign Minister, Arthur Zimmermann, to the German embassy in Mexico City. It proposed to the Mexican government that they invade the United States to reclaim territories lost in the 1848 war. Germany was offering military assistance. This was an insane move since Mexico could not possibly have the capacity to attack the United States under any circumstances. The Room 40 Group saw the provocative nature of the telegram, but ultimately the *Zimmermann Telegram*, as it went down in infamy, was handed to the American ambassador to Britain on February 23, 1917. Of course, the Americans were outraged, especially when Zimmermann admitted the validity of the telegram. It contributed to the United States' declaration of war against Germany that April.

The breaking of the Vigenère cryptosystem in the nineteenth century, coupled with the advent of radio, and the looming First World War, in the early twentieth century, brought into sharp focus the need for the development of new, strong, and effective cryptosystems. Nothing much happened in this arena until the dying days of World War I. It is worth noting that in the last year of World War I, 1918, the Americans employed eight American Indians from the Choctaw tribe to convey vital messages across insecure communication channels in their native tongue. The native American languages are extremely complex, difficult to learn, and certainly for the Germans, nearly impossible to understand, so it was an effective means of enciphering important data. In World War II, the Americans similarly employed Navajos to transmit important messages in their native language. They became known as the *Windtalkers*, the name of a Hollywood film, released in 2002, celebrating their achievements. The enemy never broke the native codes.

On March 5, 1918, the famous German *ADFGVX* cipher went into service. It was invented by Colonel Fritz Nebel, who was a communications officer in the Kaiser's army. The cipher got its name from the fact that these were the only six letters used in the cipher. These *specific* six letters were chosen since their Morse code equivalents were sufficiently dissimilar so as to minimize errors. To convolute the cryptosystem, the Germans used a combination of substitution and transposition techniques. This it how it worked.

THE GERMAN ADFGVX FIELD CIPHER

The Germans used a table (see Table 2.2) where the twenty-six letters of the alphabet plus the ten digits (with 10 represented by ϕ) populate the six-by-six square, where the coordinates of each letter and digit are uniquely determined by the six letters. For instance, the coordinate of H is FX.

Table 2.2

	A	**D**	**F**	**G**	**V**	**X**
A	B	3	M	R	L	I
D	A	6	F	ϕ	8	2
F	C	7	S	E	U	H
G	Z	9	D	X	K	V
V	1	Q	Y	W	5	P
X	N	J	T	4	G	O

Thus, for instance, *The Germans are there* would be enciphered as:

XF FX FG XV FG AG AF DA XA

FF DA AG FG XF FX FG AG FG

However, this is only the *transitional* ciphertext, which was then placed in another rectangle to be transposed into the *final* ciphertext using a numerical key as follows. We think of the letters of the key *RADIOS* as having numerical equivalents according to the alphabetic order of the letters, namely A corresponds to 1 since it is the letter in RADIOS that appears first in the alphabet, then D corresponds to 2, and so on. Then place the above transitional ciphertext by rows into a matrix as follows in Table 2.3.

R	A	D	I	O	S
5	1	2	3	4	6
X	F	F	X	F	G
X	V	F	G	A	G
A	F	D	A	X	A
F	F	D	A	A	G
F	G	X	F	F	X
F	G	A	G	F	G

Table 2.3

Now the final ciphertext is obtained by "peeling off" the *columns* in the above rectangle according to the order of the numbers as follows and grouping the letters in convenient six-letter pieces.

FVFFGG   FFDDXA   XGAAFG

FAXAFF   XXAFFF   GGAGXG

To decipher, we reverse the process.

Example 2.3 *Suppose that we want to decipher the following, assuming that it was encrypted using the above cipher.*

FAXAAF   FAFAAF   XVGGVG

FAFFXA   XDADFF   GVFGXG

First, we place each group in the **RADIOS** *table according to its position, the first going under column* **A**, *the second under column* **D** *and so on, as in Table 2.4.*

R	A	D	I	O	S
5	1	2	3	4	6
X	F	F	X	F	G
D	A	A	V	A	V
A	X	F	G	F	F
D	A	A	G	F	G
F	A	A	V	X	X
F	F	F	G	A	G

Table 2.4

Then we unravel by taking them out by rows into groups, as follows.

XFFXFG   DAAVAV   AXFGFF

DAAGFG   FAAVXX   FFFGAG

Now, we look up each digraph in Table 2.3, to get the plaintext:

THE ALLIES ARE CLOSER

In the spring of 1918, the Germans were planning a major offensive, presumably to ensure the defeat of the Allies before the arrival of American troops. The ADFGVX cipher turned out to be the toughest field cipher known up to that time, and the Allies could not break the initial cryptograms. Then some of those cryptograms were brought to the attention of the best cryptanalyst in France's Bureau du Chiffre, Georges Jean Painvin (1886–1980) (see Figure 2.12).[2.6] The Bureau du Chiffre was considered to be one of the best black chambers of the day, and Painvin was one of their top stars. The Allies needed to know where the Germans were planning to make their major thrust in the upcoming invasion. They needed Painvin to break the ADFGVX cipher.

Figure 2.12: Georges Painvin.

[2.6]The photograph of Painvin is courtesy of the site owner who cited it as public domain at *http://www.annales.org/archives/x/painvinimages.html.*

On March 21, 1918, the German offensive began with punishing ferocity, forcing the retreat of British and French troops. Painvin was under incredible pressure to find the key. He worked day and night, by his own admission, losing 10 kilograms (22 pounds), but by the evening of June 2, he had broken the cryptosystem. Now the French military knew where the Germans would attack and they did so on June 9. The French held their own and the Germans suffered heavy casualties. Once American forces arrived, the Germans were on the defensive, and ultimately by November, they had to admit defeat.

The One-Time Pad

In 1918 another event occurred that would have massive cryptological consequences. Gilbert S. Vernam, a cryptologist working for the American Telegraph and Telephone Company (AT&T) came to the realization that if the Vigenère cipher were used with a truly random key, with keylength the size of the plaintext, called a *running key*, then the Babbage/Kasiski attack would fail. At this time, AT&T was working closely with the armed forces, so the company reported this to the Army. It came to the attention of Major Mauborgne, head of the Signal Corps' research and engineering division. (When Mauborgne was still just a first lieutenant in 1914, he had published the first solution of the Playfair cipher, see page 68. Also, see page 79 for his refinement of Hitt's rediscovery.) He played with Vernam's idea and saw that if the key were reused, then a cryptanalyst could piece together information and recover the key. Hence, he added the second component to the Vernam idea. The key must be used once, and only once, then destroyed. Now, the idea was complete. Use the Vigenère cipher with a truly random running key that is used *exactly* once, then destroyed. The system is called the *one-time pad*, and sometimes, perhaps inappropriately in view of Mauborgne's contribution, the *Vernam cipher*. Since the key is as long as the plaintext, and the key selected is truly random, and used exactly once, then the ciphertext is completely random as well. Thus, the one-time pad is unbreakable. In other words, it is impossible to crack by any cryptanalytic methods.

It would take until 1949 with Shannon's concept of perfect secrecy (which we will discuss in Chapter 11) that the one-time pad was *proved* to be unbreakable (see page 440). In other words, the one-time pad is not only experimentally and practically unbreakable, but theoretically proven to be so. This was the goal of cryptography, absolute secrecy. Yet, perfection has a price; there are two distinct problems it faces. Finding truly random keys is not as easy as it might seem. Even modern-day computers cannot generate truly random numbers since they are finite-state devices, meaning that eventually they repeat, and so are predictable. The best that one can expect from computers is what is called *pseudorandomness*, which is a computer's simulation of a random number in the sense that, at least in appearance, they have the statistical properties of truly random numbers. This in itself is an entire area of study. (Knuth [138, pages 149–189] spends some 40 pages discussing the very definition of randomness!) Moreover, to generate random sequences in such a fashion that they must be produced for each message is a monumental task, especially since each one has to be the size of the plaintext, which is the second problem. To have keys the size of the plaintext creates unwieldy key management problems. Yet, it is a completely provable secure cryptosystem. Thus, it is used by those who need it for absolute secrecy, such as, for instance, the protection of missile launch codes. Then it is practicable, but for low-level security such as e-mail messages between government officials for the day-to-day running of business, other means must be used.

It is part of the folklore that Soviet spies used one-time pads to send messages, and that they were also used in German diplomatic systems starting in the late 1920s.

In our modern computer age, one can translate all plaintext and ciphertext into numerical data, in particular, into binary.[2.7] Since we have a sequence of zeros and ones now, we can perform addition modulo 2, so we again end up with zeros and ones, ideal for transmission. This is how the one-time pad was used in the legendary *hot line* between Washington and Moscow, inspired by the Cuban missile crisis of the 1960s. They used what was called the *one-time tape*, which was a physical manifestation of the Vernam cipher. At the American end, this took the form of the ETCRRM II or *Electronic Teleprinter Cryptographic Regenerative Repeater Mixer II*. The manner in which the one-time tape worked was that there existed two magnetic tapes, one at the enciphering source, and one at the deciphering end, both having the same running key on them. To encipher, one performs addition modulo 2 with the plaintext and the bits on the tape. To decipher, the receiver performs addition modulo 2 with the ciphertext and the bits on the (identical) tape at the other end.[2.8] Thus, they had instant deciphering and perfect secrecy if they used truly random keys, each used only once, and the tapes were burned after each use. The same keys cannot be used twice since the one-time pad would then be open to an attack since the key k can be computed by addition modulo 2 of the plaintext with the ciphertext.[2.9] Thus, we see that today one-time pads are most practicable for military and diplomatic purposes when unconditional security is of the utmost importance.

Vernam is known for other discoveries. A patent was filed in September of 1918 (and granted with issuance in July 1919) for a cipher that Vernam invented, which was the first polyalphabetic cipher automated using electrical impulses. For this, he has earned the title of the *Father of Automated Cryptography*.

[2.7]Recall that any $n \in \mathbb{N}$ can be represented in the form $n = a_0 + 2a_1 + 2^2 a_2 + \cdots + 2^t a_t$ where $a_j \in \{0, 1\}$ and t is a nonnegative integer. The a_js are called *bits*, which is a contraction of *bi*nary dig*its*. Typically, we will use the notation throughout for this binary representation as: $n_{10} = (a_t a_{t-1} \ldots a_1 a_0)_2$ to denote that our base 10 integer n has a binary representation as given. For instance, the binary representation of 100 is: $(100)_{10} = (1100100)_2$ since $100 = 2^6 \cdot 1 + 2^5 \cdot 1 + 2^4 \cdot 0 + 2^3 \cdot 0 + 2^2 \cdot 1 + 2^1 \cdot 0 + 2^0 \cdot 0$. See Appendix A for more information on basic mathematical facts.

[2.8]This process is often called *XORing* since it is use of *exclusive or*, which we will denote later in the text by $\oplus$ (see page 116).

[2.9]This is an example of what is called a *known-plaintext attack*, which means an attack where a cryptanalyst has both some plaintext and its corresponding ciphertext from an intercepted cryptogram from which to deduce the plaintext in general, or the key.

However, not only was the device a commercial failure, but the stock market crash caused Vernam to lose his job at AT&T. He then went to work for an organization that later merged with Western Union. From that time, he was granted some sixty-five patents, among which was the fully automated telegraph switching system. He was even visionary enough to have invented one of the first versions of a binary digital enciphering of pictures. However, for all these amazing achievements, he died in relative obscurity on February 7, 1960, in Hackensack, New Jersey, after years of battling Parkinson's disease.

The Friedmans

Others who had been on the periphery of the discovery of the one-time pad also had a major impact. William Frederick Friedman[2.10] (1891–1969) (see Figure 2.13) was born in Kishinez, Russia, on September 14, 1891. In 1892, his father fled the antisemitic regulations in Czarist Russia, and his family joined him in Pittsburgh the following year. (Actually, William had been born with the name Wolfe, but his father changed it to William after he became an American citizen.) William obtained his bachelor's degree from Cornell, then joined the Riverside Laboratories (which today would be considered to be a "think tank"), outside Chicago, in 1915. There Friedman met Elizabeth Smith (1892–1980), whom he married in 1917 (see Figure 2.14). One of the projects being researched at Riverside was the contention that hidden messages in Shakespeare's works proved that Bacon was the real author (see page 36).[2.11] However, the Friedmans soon turned their attention to cryptology. William was training cryptologists at Riverbank, and for course material he wrote eight publications (which are collectively known as the *Riverbank Publications*).

Today they are highly regarded as containing the basic essentials of cryptological material. Perhaps his greatest cryptological contribution (and he thought so himself when he was looking back over his career) was his conceiving of the *Index of Coincidence*, which appeared in his monograph no. 22 of the numerous ones that he published. The importance of this discovery was not only that it introduced a statistical methodology for cryptanalyzing polyalphabetic ciphers, but also, it demonstrated the intimate link between cryptology and mathematics, a link that would get more entwined as the twentieth century unfolded. Friedman's Index of Coincidence (for a ciphertext $\mathfrak{C}$) is defined as

Figure 2.13: William Friedman.

[2.10]Figures 2.13–2.14 are courtesy of the National Security Agency *Hall of Honor*. See: *http://www.nsa.gov/honor/index.html*.

[2.11]Friedman and his wife debunked this claim in an excellent book [98], called *Shakespearean Ciphers Examined*, published in 1958.

the probability that two letters selected at random from $\mathcal{C}$ are identical. Below we show how to mathematically demonstrate that the index of coincidence for a monoalphabetic cipher is about 0.065, and the index of coincidence for a polyalphabetic cipher is somewhere between 0.0385 and 0.065. For very long keywords, the index of coincidence for polyalphabetic ciphers will be closer to 0.0385. Hence, by a simple analysis of intercepted ciphertext, a cryptanalyst can relatively easily determine the type of cryptosystem being used. This was quite a breakthrough. Moreover, his idea contained a mechanism for determining the probable keylength, as had Kasiski. Here is how it works.

First we need a table of letter frequencies for the English alphabet. This well-known, standard table (presented here as Table 2.5) augments Tables 1.4 and 1.5, which we presented on pages 44 and 45, when we discussed letter frequencies in Section 1.4.

Now suppose that n stands for the number of letters in a ciphertext, $\mathcal{C}$, and n_j stands for the number of letters in the j-th position of the English alphabet. In other words, n_1 is the number of occurrences of the letter a in $\mathcal{C}$, n_2 is the number of occurrences of the letter b in $\mathcal{C}$, and so on. Without getting into the reasons for it, the Index of Coincidence, $\mathcal{IC}$, is given as approximately the following.

Figure 2.14: Elizabeth S. Friedman.

$$\mathcal{IC} \approx \left(\frac{n_1}{n}\right)^2 + \left(\frac{n_2}{n}\right)^2 + \cdots + \left(\frac{n_{26}}{n}\right)^2 .$$

So if we want to compute $\mathcal{IC}$ for the English language from Table 2.5, and since each of the numbers in the table is a percentage, then we divide each by 100, and get: $\mathcal{IC} \approx (0.8167)^2 + (0.01492)^2 + \cdots + (0.00074)^2 = 0.065$, which explains the aforementioned Index of Coincidence for monoalphabetic ciphers, since the frequency is invariant. (Note that the symbol $\approx$ means "approximately equal to". It is not a strict equality but this is good enough since we are dealing with a statistical analysis wherein approximations are good enough for our investigations.)

Relative Letter Frequencies for English
Table 2.5

a	b	c	d	e	f	g	h	i
8.167	1.492	2.782	4.253	12.702	2.228	2.015	6.094	6.966
j	k	l	m	n	o	p	q	r
0.153	0.772	4.025	2.406	6.749	7.507	1.929	0.095	5.987
s	t	u	v	w	x	y	z	
6.327	9.056	2.758	0.978	2.360	0.150	1.974	0.074	

Now for any language, such as English, with a twenty-six-letter alphabet, in which each letter has the *same* frequency, we get

$$\mathcal{IC} = 26(1/26)^2 = 0.038,$$

which is approximately half of the above Index of Coincidence for English. Hence, the Index of Coincidence helps us in determining if the ciphertext comes from a monoalphabetic or polyalphabetic cipher in the following manner. The closer the $\mathcal{IC}$ is to 0.065, the more likely it is that the message came from a monoalphabetic cipher. If the $\mathcal{IC}$ is much less than 0.065, the cipher is most likely polyalphabetic since frequencies are evened out by polyalphabetic cryptosystems. Hence, the closer the $\mathcal{IC}$ is to 0.038, the greater the chance is that the cipher is polyalphabetic. This was a major contribution by Friedman since he tied a mathematical tool, statistical analysis, to the study of cryptography. Another young professor did the same in another area of mathematics.

Lester S. Hill published a short paper [123] in which he put a cipher, known today as the *Hill Cipher* (which we will study in detail on page 111), into an algebraic framework. This was a reinvention and expansion of Porta's idea (see page 54). Hill obtained his Ph.D. in mathematics from Yale in 1926. He was hired to teach mathematics at Hunter College in New York in 1927, and he remained there until his retirement in 1960. He was the first to successfully use general algebraic concepts to reveal cryptography through mathematics. A.A. Albert (see Footnote 2.15 on page 97) was so impressed with Hill's ideas that he used them in some simple cryptosystems with his own tailoring to suit the situation at hand. Hill's rigorous mathematical approach was certainly one of the pioneering efforts that helped to build today's solid grounding of cryptography in mathematics. Hill died in Lawrence Hospital in Bronxville, New York, after suffering though a lengthy illness.

Now we return to the life of the Friedmans. Soon after his marriage to Elizabeth, William became the director of the Department of Codes and Ciphers, among his other duties, at Riverbank. After the outbreak of World War I, Riverbank offered its services to the government, and since no such federal agency existed at the time, Riverbank became the *de facto* cryptographic center for the American government. One of the first accomplishments the Friedmans achieved was the following. The Germans had been encouraging Hindu radicals to work toward independence from Britain in the hopes of diverting attention and strength from the war effort. Some of these radicals, who lived in the United States, were sending messages about arms shipments. It turns out they were trying to buy arms in the United States and ship them from the West Coast. The Friedman's deduced that the codebook used by these radicals was a German-English dictionary published in 1880. This aided William in his testimony given at the trial of 135 Hindu radicals in San Francisco.

The Friedmans quit Riverbank toward the end of 1920. In 1921, Friedman joined the American Black Chamber, where he eventually headed the Research and Development Division and stayed there until its dissolution in 1929. One man may be said to be chiefly responsible for the creation and (possibly) the dissolution of the American Black Chamber.

After World War II, Friedman continued in government signals intelligence until 1949 when he became head of the code division of the new *Armed Forces Security Agency*, which evolved into the *National Security Agency* (NSA). At NSA he became the chief cryptologist. By the late 1960s his health faded. He died in 1969 in Washington, D.C., and was buried at Arlington National Cemetery. For all his accomplishments and pioneering efforts, he has been dubbed *America's greatest cryptologist*.

Elizabeth Friedman largely worked on the civilian side. She gained some initial fame when she broke the codes and ciphers of "rum runners" in the 1920s during Prohibition. By 1927 she had been hired by the U.S. Coast Guard, and broke thousands of codes for them. During World War II she joined the Office of Strategic Studies (OSS), where she was one of their outstanding cryptologists. After her husband's death in 1969, she retired and lived until 1980. She is buried with her husband at Arlington.

Herbert Yardley[2.12] (see Figure 2.15) was born on April 13, 1889 in Worthington, Indiana. As a young man, he recognized his gift for cryptanalysis when he was hired as a "code clerk" in the State Department at the age of twenty-three.

After the declaration of war in April 1917 by the United States, Yardley was made head of the newly established cryptology section of the Military Intelligence Division, MI-8. In May of 1919, he submitted his idea with a plan for a permanent cryptology establishment, which came to be known as the *American Black Chamber*. Its operation was exceptional, cryptanalyzing more than 45,000 enciphered telegrams from various countries. By 1929, however, the Black Chamber was shut down by the Secretary of State, Henry L. Stimson, who disapproved of the Chamber saying "Gentlemen do not read each other's mail". In 1931, Yardley published a book entitled *The American Black Chamber*, which was an exposé of the United States' weak, if not defenseless, status in the arena of cryptol-

Figure 2.15: Herbert Yardley.

ogy. It caused a furor in many circles. In fact, when he tried to publish a second book, *Japanese Diplomatic Secrets*, it was suppressed by the U.S. government. He involved himself in real estate speculations in the late 1930s, and served as enforcement agent in the Office of Price Administration during World War II. He died of a stroke on August 7, 1958, in Silver Spring, Maryland.

After his service in the American Black Chamber, Friedman moved to the Army's Signals Intelligence Service (SIS), and by 1935 was replaced by Major Haskell Allison as head of SIS. Meanwhile, Elizabeth was employed as a

---

[2.12]Figure 2.15 is courtesy of the National Security Agency *Hall of Honor*. See: *http://www.nsa.gov/honor/index.html*.

cryptologist for the Treasury Department. Her picture now sits along with her husband's in the N.S.A. Hall of Fame for her cryptological contributions.

By 1938, Joseph Mauborgne, now a two-star general, was heading a group at SIS to look at Japanese cipher systems, since it was beginning to look like a new war was brewing. He asked Friedman to head up the new division at SIS, and Friedman agreed. The Japanese cipher machine was called *Purple*, given its name from the Japanese cipher of the same name in which their correspondence was written. It proved to be incredibly difficult, but by August of 1940, Friedman's team had constructed an exact replica of the Purple machine,[2.13] (see Figure 2.16) allowing them to decipher an increasing amount of Japanese traffic. (Copies of the machine were also given to the British to decrypt correspondence between the Japanese and the Germans.) However, Friedman suffered a nervous breakdown and was hospitalized on January 4, 1941, after which his work schedule was severely cut back.

Figure 2.16: Purple machine replica.

The information that was obtained from breaking Purple, the Americans called *MAGIC*. This name was given by Rear Admiral Walter S. Anderson, probably for reasons surrounding the associations with the occult that we discussed in Section 1.3. MAGIC has come to be known as the code name for the joint Army and Navy operation, first set up in 1939 to break Japanese codes.

Pearl Harbour, Midway, and Post–World War II

On December 7, 1941, a message to the Japanese Embassy in Washington was intercepted. The decryption showed that the Embassy was being ordered to end all negotiations with the United States. The implication of impending war was crystal clear, and this message was to have been delivered to the American State Department only hours before the attack on Pearl Harbour. However, the ruse — to come as close to the attack before giving formal notice — failed since the embassy's first secretary Katzuso Okumura, was still typing the formal notification for the State Department when the bombs began raining down on Pearl Harbour. They had started a war without formal declaration, a failure that would be part of the charges against Japanese war criminals on trial after

---

[2.13]Figure 2.16 is a representation of the 1941 Purple Machine Replica, courtesy of the CIA website *http://www.cia.gov/cia/publications/facttell/intel_overview.html*.

the war. Later a Joint Congressional Committee met for an investigation of the Pearl Harbour attack and concluded that the war efforts of America's cryptanalysts had shortened the war, and saved thousands of lives. We will now have a look at how some of that was accomplished.

American cryptanalysts were able to decipher a highly secret cryptogram detailing the itinerary of the Japanese Navy Admiral Isoruko Yamamoto's plane tour of the Solomon Islands. Thus, the Americans were able to pinpoint his whereabouts and shoot down his plane. American cryptanalysts also helped to ensure that Japan's lifeline was rapidly cut, and the German U-boats were defeated. Perhaps the best-known and most vital success was the Battle of Midway. The cryptanalysts were able to give complete information on the size and location of the Japanese forces advancing on Midway. This enabled the Navy to concentrate a numerically inferior force in precisely the right place at the right time that turned the tide of the Pacific War. This was a stunning victory for American cryptanalysts.

Another outcome of World War II was an outstanding advance in cryptanalysis by the Americans. To discuss it, we must go back to the invention of the electric typewriter, which opened the way for electromechanical enciphering. The first electric contact rotor machine was invented by Edward Hugh Hebern in 1915. He used two electric typewriters randomly connected by twenty-six wires. Hence a plaintext letter key hit on one typewriter would yield a ciphertext letter to be printed on the other machine. These wire connections were the seminal idea for the idea of a rotor, namely, a way of varying the monoalphabetic enciphering. By 1918, Hebern had a device that embodied the rotor principle. He filed a patent in 1921, but did not receive it until 1924. In 1919, patents were also filed for rotor enciphering machines by Alexander Koch and Avrid Gerhard Damn, the latter for half rotors. Damn owned a company called Aktiebolaget Cryptograph or Cryptograph Incorporated. In 1922, Emanuel Nobel, nephew of the famed Alfred Nobel, put Boris Caesar Wilhelm Hagelin to work in Damn's company. Hagelin simplified and improved one of Damn's machines. This was such a success that the Swedish army placed a large order with Damn's firm. When Damn died in 1927, Hagelin took over the operation of the firm. Later, he developed the rotor-based cipher machine, called the Converter M-209 by the American military; this was so successful that in the early 1940s more than 140,000 were manufactured. Hagelin's M-209 used a version of the self-decrypting *Beaufort cipher*. (The Beaufort cipher was a variant of the Vienère cipher, and was published by Admiral Beaufort's brother after his death in 1857, in the form of a four by five inch card. Admiral Sir Francis Beaufort (Royal Navy), was also the creator of the *Beaufort scale*, an instrument used by meteorologists to indicate wind velocities on a scale from 0 to 12, where 0 is calm and 12 is a hurricane.) Royalties from the sales of Hagelin's cipher machine made him the first millionaire of cryptography. Perhaps, Hagelin had Thomas Jefferson to thank since his wheel cypher inspired the development of rotor machines (see page 66).

In 1918, the German Arthur Scherbius applied for a patent on a rotor enciphering machine using multiple rotors. In 1923, a corporation was formed to manufacture and sell his machine, which he called *Enigma*. In 1934, the Japanese Navy bought the Enigma for their own use, and developed it into the Japanese cryptosystem called Purple, which we discussed above. However, the Japanese version was unlike Enigma in that it used stepper switches, similar to those used in telephone exchanges. When the SIS built a machine to replicate Purple, they made the unwitting decision to use exactly the same telephone stepper switch used by the Japanese designer! This accounts for Friedman's group at SIS being able to duplicate a machine they had never seen.

In 1932 Hebern designed a machine with five rotors, the HCM. In 1936, a rotor machine, based on the Hebern machine, called the *SIGABA* (see page 93), or M-134-C was developed and used with great success by the U.S. military in World War II. (It was also called the CSP-889, or ECM Mark II, by the Navy.) It was so well designed that all the efforts by the Army's cryptanalysts to break it failed. As it would be learned later, the Germans also could not break the Americans' cryptograms enciphered with the *ABA*s, as they were nicknamed. The fundamental idea of electronically controlled rotors was created by William Friedman, and he implemented it in the original M-134 device, which had five rotors that encrypted plaintext, the motion of which was controlled by a paper tape.

Then Frank Rowlett[2.14] (1908–1998) (see Figure 2.17) created the vital concept of the SIGABA, namely, the idea of using rotors to control the rotors that enciphered the plaintext. Rowlett was one of Friedman's earliest assistants, since 1929, and was part of the team that broke Purple at SIS. The SIGABA had fifteen rotors, ten of which were conventional 26-contact rotors, and five of which had smaller rotors with only ten contacts on each side. Moreover, the rotors were divided into three sets. Five of the 26-contact rotors, called cipher rotors, encrypted or decrypted a message in the same fashion as the Hebern rotor machine. Another five 26-contact rotors were called control rotors, and the five 10-contact rotors were called index rotors. In the

Figure 2.17: Frank Rowlett.

1940s the SIGABA would prove to be the securest of the machines developed in the West, and it never fell into enemy hands.

When Hitler came to the stage, the cryptographers of the Wehrmacht made the decision to use the Enigma, upon which they made improvements for their security purposes. However, the German Enigma cryptosystem was cryptanalyzed by British researchers at Bletchley Park, which is a Victorian country mansion in Buckinghamshire, halfway between Oxford and Cambridge. In August of 1939, the Government Code and Cypher School was seconded there. Perhaps one of the most important among these researchers was Alan Mathison

---

[2.14]Figure 2.17 is courtesy of the National Security Agency *Hall of Honor*. See: *http://www.nsa.gov/honor/index.html.*

Turing (1912–1954).

Turing was born on June 23, 1912 in London. He studied under Alonzo Church (1903–1995) at Princeton and obtained his doctorate in 1938. During World War II, he was employed in the British Foreign Office, where he got involved in cryptanalyzing the Enigma cryptosystem. Toward this end, he conceived of a machine called the *BOMBE*, the first prototype of which arrived at Bletchley on March 14, 1940. However, it was not as successful as they had hoped and to compound the problem, the Germans had changed the method of how they managed keys, deleting repetitions, so decryptions dropped dramatically. They needed a new improved BOMBE, which was not delivered until later that year, on August 8. In less than two years there were eighteen working versions of the BOMBE at Bletchley Park. By September of 1941, Field Marshal Rommel's Enigma cryptograms to Berlin were being cryptanalyzed. In fact, William Friedman visited Bletchley Park in 1941, exchanging information on techniques for attacking Purple for British information on breaking Enigma. By 1942, they had dug deeply into cryptanalyzing Enigma, which played a major role in the Allied victory.

Figure 2.18: Midway exhibit.

(Courtesy of the National Security Agency Public Photo Gallery.
See *http://www.nsa.gov/gallery/photo/photo00010.jpg*.)

Figure 2.19: SIGABA.

(Courtesy of the National Cryptologic Museum of the National Security Agency. See *http://www.nsa.gov/museum/big.html.*)

Figure 2.20: Purple cipher switch.

(Courtesy of the National Security Agency Public Photo Gallery.
See *http://www.nsa.gov/gallery/photo/photo00016.jpg.*)

Figure 2.21: BOMBE.

(Courtesy of the National Security Agency Public Photo Gallery.
See *http://www.nsa.gov/gallery/photo/photo00013.jpg.*)

Figure 2.22: Enigma.

(Courtesy of the National Cryptologic Museum of the National Security Agency. See *http://www.nsa.gov/museum/enigma.html.*)

2.5 The Postwar Era and the Future

We shall see that cryptography is more than a subject permitting mathematical formulation, for indeed it would not be an exaggeration to state that abstract cryptography is identical with abstract mathematics.

Abraham Adrian Albert[2.15] (1905–1972)

Host Feistel may be considered to be one of the early pioneers in the drive to secure privacy for the public at large using cryptography. Born in Germany in 1914, he emigrated to the United States in 1934, but would not obtain a U.S. citizenship for another decade. In fact, in 1941, with Germany having declared war on America, he was placed on a (sort of) house arrest, where his movements were restricted to the Boston area where he lived. Yet, surprisingly, on January 31, 1944, the house arrest was lifted, he was granted U.S. citizenship, and the very next day he was given security clearance that allowed him to work at the Air Force Cambridge Research Center (AFCRC).[2.16] There he set up a cryptography research group that developed some outstanding cryptographic algorithms. In particular, they developed the *MARK XII*, which is widely used in American aircraft. It is known that the NSA had an ambivalent attitude toward Feistel's group. On the one hand, they exerted pressure to steer his work, while at the same time they considered his group to be a threat. Consequently his group was dissolved in the late 1950s. Then Feistel moved to MIT's Lincoln Laboratory, followed by a move to MITRE Corporation, a spinoff of the MIT lab. When he tried to form a cryptography group there, again NSA exerted pressure on MITRE, so his efforts failed, and his group did not materialize.

A.A. Albert, a friend of Feistel, advised him to go to IBM, since they were hiring the brightest scientists to do their own innovative work, a kind of think tank. Feistel began work at their Watson Laboratory in Yorktown Heights, New York. There he created a cryptosystem used in the IBM2984 banking system, known today as the *Alternative Encryption Technique*, but then it was called *Lucifer*.[2.17] This cryptosystem was the predecessor of the first commercially

---

[2.15] Albert was born in Chicago, Illinois, on November 9, 1905. He studied under L.E. Dickson at the University of Chicago, receiving his Ph.D. in 1928. His elegant work on the classification of division algebras (see Appendix A, page 484) earned him a National Research Council Fellowship. This provided him with the opportunity to secure a postdoctoral position at Princeton, after which he spent a couple of years at Columbia University, then returned to Chicago in 1931. His book, *Structure of Algebras*, published in 1939, remains a classic today. The events of World War II induced Albert to take an interest in cryptography. The above quote is taken from his lecture on mathematical aspects of cryptography at the American Mathematical Society meeting held in Manhattan, Kansas, on November 22, 1941. His numerous achievements would take several pages to describe. Suffice it to say he has had a lasting influence. He died on June 6, 1972, in Chicago.

[2.16] There is speculation that something may have been going on behind the scenes between Feistel and the U.S. government (see Levy's excellent book *Crypto* [151] for an account of some of these possible scenarios as well as with other related cryptographic activities).

[2.17] Years later, Feistel said that if it had not been for the Watergate scandal that rocked Washington, the NSA would probably have shut down the Lucifer project, as they had so many of his earlier efforts. In fact, in the early 1970s, patent secrecy orders were placed on some of Feistel's inventions by the U.S. government.

available algorithm (namely for use with unclassified computer data) officially announced in 1977 as the *Data Encryption Standard* (DES).[2.18]

DES is an example of a *block cipher*, about which we will learn the details in Chapter 3 (as well as an entire class of ciphers, called *Feistel ciphers*, in honour of the groundbreaking work he did in those early years). Basically, block ciphers encipher fixed size blocks of data. For DES this is a block size of 56 bits, which is too small for modern-day data transfer. Its key size, at 56 bits, is also inadequate for modern usage, as we shall demonstrate below.

Lucifer was modified by the NSA, before it became the Data Encryption Standard. There was, and in some circles still is, controversy that the NSA had slipped in a "back door" into the standard, which would allow them an easy method for deciphering messages encrypted with DES. This suspicion was even investigated in 1978 by the U.S. Senate Select Committee on Intelligence, the findings of which are, of course, classified. However, an unclassified summary of their investigation stated that the NSA had no improper involvement in the design of DES. Yet, many remain skeptical since the details of the investigation were not made public. Despite such concerns, DES was used by banking, commerce, and industry until the end of the twentieth century, when it reached the end of its tenure as a secure cryptosystem.

At the *CRYPTO*[2.19] conference, in 1993, M.J. Weiner presented an efficient key-search design that would have taken 3.5 hours (at that time) on a machine costing one million U.S. dollars to do an *exhaustive search of the keyspace*, also called a *brute force attack*, which means that all possible keys are tried to see which one is being used by the communicating entities. We will come back to this issue when we look at the replacement for DES, the new AES (see Footnote 3.10 on page 150). By 1998, the 56-bit keylength used by DES was becoming increasingly under attack by modern methods. In that year, a group led by Paul Kocher (about whom we will learn more later when we talk about security issues, see page 176), custom-built a computer for about a quarter of a million U.S. dollars, which they used to find a DES key in roughly fifty-six hours. The plaintext read: "It's time for those 128-, 192-, and 256-bit keys." Six months later, in January 1999, the same team did this in less than twenty-four hours. This and other developments spelled the end for DES since the keylength was just too small to withstand cryptanalytic advances. By August of 2000, DES was replaced with a *non*-Feistel cryptosystem called the *Advanced Encryption Standard* (AES), which allowed for 128-, 192-, and 256-bit keys. We will discuss it in detail in Section 3.5.

The 1970s also saw a revolutionary change in the manner in which keys were handled. Cryptography was about to go public. In a paper [69], published in

---

[2.18]A complete description of DES is given in the U.S. *Federal Information Processing Standards* Publication number 46 (or FIPS-46) Springfield, Virginia, April 1977. It was updated to FIPS-1 in 1988, then again to FIPS-2 in 1993 — see the FIPS homepage: *http://www.itl.nist.gov/fipspubs/*. The *American National Standards Institute* (ANSI) approved DES as a private sector standard in 1981 — see the ANSI homepage at: *http://www.ansi.org/*.

[2.19]CRYPTO is a conference on cryptology held annually in late August at the University of California at Santa Barbara.

1976, Whit Diffie and Martin Hellman conceived of a method for two entities,[2.20] who have never met in advance or exchanged keys, to establish a shared secret key by exchanging messages over an open (unsecured) channel.[2.21] We will learn the mathematical means for how this works in Chapter 4. Up to the time of this idea, all cryptosystems, including DES, were looking for mechanisms to securely distribute secret keys. This is because once a symmetric enciphering key is known, an entity can easily deduce the deciphering key from it. Now, with the introduction of the Diffie-Hellman idea, which has come to be known as the *Diffie-Hellman Key-Exchange*,[2.22] entities could exchange keys in the open and ensure privacy. It seems contrary to the very notion of secrecy. However, that is the brilliance of the scheme, use two essentially *different* keys, one for enciphering that can be made public, and one for deciphering that can be kept private, a *key pair*. No longer would the key be *symmetric* (the deciphering key easily determined from the enciphering key and vice versa). Now there would be an *asymmetric* key pair, the advent of *public-key cryptography* (PKC). How could this possibly work?

Public-Key Cryptography (PKC)

Before giving an introduction to the Diffie-Hellman idea, let us look at an analogy, a standard one, for PKC, which will provide an easy-to-understand scenario to give the reader an understanding of how a public key can work. First we will introduce the first two characters (entities) in our cryptographic cast, *Alice* and *Bob*. Suppose that Bob has a *public* wall safe with a *private* combination known only to him. Moreover, the safe is left open and made available to passers-by. Then, anyone, including Alice, can put messages in the safe and lock it. However, only Bob can retrieve the message, since, even Alice, who left the message in the safe has no way of retrieving it.

In order to give a general overview of the basic Diffie-Hellman idea, we need the notion of a *one-way function*, which we may view, at this juncture, as a method of enciphering that cannot be reversed. For instance, if you write a message on a piece of paper, then burn it, that is an example of a one-way function since retrieving the message is impossible. One says, in mathematical terms, that this is a function whose values are easy (computationally feasible) to compute, but calculating that inverse is *computationally infeasible*, meaning

---

[2.20]Henceforth, by an *entity* we will mean any person or thing, such as a computer terminal, which sends, receives, or manipulates information.

[2.21]From now on, by a *channel* we will mean any means of communicating information from one entity to another. A *secure* channel is one that is not physically accessible to an adversary, whereas an *unsecured* channel is one from which entities, other than those for whom the information was intended, can delete, insert, read, or reorder data.

[2.22]In some parts of the literature, this is called the *Merkle-Diffie-Hellman Key-Exchange* since R.C. Merkle was working on these same ideas at that time. Merkle was a graduate student at the University of California at Berkeley, and was working on an idea for a one-way function involving certain puzzles. This would evolve later into what we now call the *knapsack ciphers*, none of which have survived cryptanalysis today. We will come back to this topic in later chapters. Merkle actually proposed joint work in a letter he wrote to Hellman in February 1976. However, it turned out that the Diffie-Hellman idea was both more efficient and more secure than Merkle's idea.

that the task cannot be carried out in reasonable computational time. As Diffie and Hellman put it in [69], a computationally infeasible task is one whose "cost as measured by either the amount of memory used or the runtime is finite but impossibly large." (Typically, this means that it would take hundreds, if not millions, of years on the fastest computer known.) However, if you burn the paper, how does the intended recipient read the message? You need additional information built into your one-way function so that the intended recipient can recover the message. This additional information is called a *trapdoor*. Mathematically speaking, a *trapdoor* in a one-way function is additional information that makes the finding of the inverse a feasible task, but without the trapdoor information, the task is computationally infeasible (see Chapter 4). For now, think of a trapdoor as information that allows you to invert the function (decrypt the message), but if you do not know it, you cannot invert the function. It is easy enough, as our paper-burning example indicated, to find one-way functions, but getting those with trapdoors requires a bit more effort. So now let us see how the Diffie-Hellman idea works.

Alice and Bob have never met, but want to establish a secret means of communicating with one another. Bob and Alice both have unique public keys, which we may envision as long strings of bits, published in some public data base of keys that anyone can look up. Both Alice and Bob also have private keys that they keep secure and known only to themselves, namely, *only* Bob knows his private key[2.23] and *only* Alice knows her private key. Now, Alice takes a message and uses Bob's public key via a one-way function to encipher the message in a manner that *only* Bob's private key can decipher. So when Alice sends the cryptogram, the only person in the world who can decipher it is Bob, with his private key. Now suppose that another of our cast of characters, eavesdropping adversary *Eve*, intercepts the message. Without Bob's private key, she has only trial and error at her disposal to try to cryptanalyze it, probably taking millions of years, so her interception is useless. Thus, since Bob is the only person who has *both* elements of the key pair, he can decipher the message instantly. The message might contain the symmetric-key k, say, and a reference to the symmetric-key algorithm, such as DES, say. Similarly, Bob uses Alice's public key and a one way function to encrypt a response, which would say that he agrees to use DES with symmetric-key k for their correspondence, and sends this to Alice, who uses her private key to decrypt, and she is the *only* one who can do so. In the Diffie-Hellman scheme, k is the shared secret key independently generated by both Alice and Bob. The key exchange is complete since Alice and Bob are in agreement on k. Hence, over an unsecured channel, they have established a secure means of communicating.

The observant reader may wonder why they do not just use this key pair for

---

[2.23]We use the convention that the term *private* key is reserved for use in association with public-key cryptography, also called *asymmetric-key* cryptography, whereas the term *secret* key is reserved for symmetric-key cryptography. The cryptographic community has adopted this convention since it takes two or more entities to share a secret (such as the *symmetric* secret key), whereas it is truly *private* when only *one* entity knows about it (such as with the asymmetric private key).

all of their correspondence rather than using it to set up a key exchange for use with a symmetric-key cryptosystem. The reason has to do with efficiency, as we will see in detail in Chapter 4. Public-key methods are extremely slow compared to symmetric-key methods. In later discussions, we will see how both the public-key and symmetric-key cryptosystems come to be used, in concert, to provide the best of both worlds combining the efficiency of symmetric-key ciphers with the increased security of public-key ciphers, called *hybrid cryptosystems*.

The Diffie-Hellman paper [69] was the "door-opener" to *public-key cryptography* in that it was the landmark, since it had the first cryptographic protocol[2.24] with public-key properties including the idea of a trapdoor one-way function, a partial solution to the public-key cryptosystem, and digital signatures (see Chapter 4). At the end of their paper Diffie and Hellman state: "Skill in production cryptanalysis has always been heavily on the side of the professionals, but innovation, particularly in the design of new types of cryptographic systems, has come primarily from amateurs." They even go on to mention the "cryptographic amateur", Thomas Jefferson, and his wheel cipher and the fact that it was used two centuries after its invention (see pages 66 and 67). Also, they talk about the amateurs responsible for the rotor ciphers (see page 90).

In summary, the Diffie-Hellman key exchange allowed two entities to set up a shared secret symmetric key, but they did not provide any method for enciphering messages, or any way to extend to digital signatures, digital data strings that associate a given message with its sender. As Diffie and Hellman put it at the outset of their paper, "We propose new techniques for developing public key cryptosystems, but the problem is still largely open." This would take a couple more years.

RSA and PKC

In 1978, a paper [230] was published by R. Rivest, A. Shamir, and L. Adleman. In this paper they describe a public-key cryptosystem, including key generation and a public-key cipher, whose security rests upon the presumed difficulty of factoring integers into their prime factors.[2.25] This cryptosystem, which has come to be known by the acronym from the authors' names, the *RSA cryptosystem* has stood the test of time to this day, where it is used in cryptographic applications from banking, and e-mail security to e-commerce on the Internet. We will be discussing all these applications as we progress through the text, and we will provide the details of the RSA algorithm in Chapter 4. The astonishing aspect of the RSA cipher is that it rests upon mathematical developments from the eighteenth century, merely updated to our modern-day information-based computer world. In the RSA paper [230], Alice and Bob

---

[2.24]By a *protocol*, in general human terms, we will mean *prearranged etiquette* such as understood behavior at a formal dinner party. On the other hand, a *cryptographic protocol* means an algorithm, involving two or more entities, using cryptography to achieve a security goal, which might involve issues of authentication, privacy, and secrecy, all of which we will discuss in detail later in the text.

[2.25]See theorem A.1 on page 469 in Appendix A.

make their first appearance as sender and recipient of messages. These characters were quickly adopted by the cryptographic community, and were expanded to include a family of characters, such as Eve, and a host of others whom we will meet as our horizons broaden in our travels.

As the following diagram illustrates, if Alice wants to send a message to Bob, she looks up his public key e_B in a public data base and encrypts her message m with it to get $e_B(m) = c$, as ciphertext. If Eve is listening in, she has only question marks in her head since she does not have access to Bob's securely protected private key d_B, which is required to decipher the cryptogram. Of course, for this to work, $d_B(e_B(m)) = m$ must hold for all messages m, and it must be impossible (or computationally infeasible) for anyone to decipher m from e_B without knowledge of d_B, to which only Bob has access. (Think of d_B as Bob's trapdoor information (his unique key) for unlocking the encrypting (one-way) function e_B, to recover m. Using the analogy described on page 99, $e_B(m)$ is his wall safe, which Alice locked with the message m inside, and to which only he has the combination (key).) Hence, unlike a symmetric key cryptosytem, an asymmetric key cryptosystem or PKC, has two distinct keys for each person, a public one, such as Bob's e_B, which everyone can access, and a truly private one, such as Bob's d_B, which he and only he knows, and keeps secure. Hence, we make the distinction between asymmetric-key encryption or PKC, and secret-key encryption or SKC, as illustrated on page 13 where both the enciphering and deciphering keys must be kept secret.

Diagram 2.1 A Generic Public-Key Cryptosystem

(I): Encryption

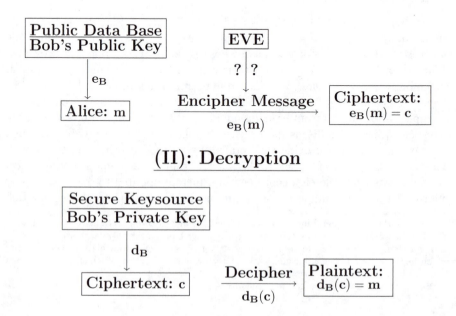

(II): Decryption

The Secret Development of PKC

Now that we have learned about the pioneering efforts of Diffie, Hellman, and Merkle in establishing PKC, it is time to remind ourselves that there is always activity behind the shroud of government agencies involved in cryptology. It is now *public* knowledge that the notion of PKC had already been discovered years earlier by British cryptographers, but not *officially* released until relatively recently. In December of 1997, five papers, [58], [76], [77], [281], [292], were released by the Communications-Electronics Security Group (CESG), which is the technical authority on official cryptographic applications for the British Government Communications Headquarters (GCHQ), whose duty is to ensure information security for their government.

Public-key methodologies were first discovered by the CESG in the early 1970s. We begin by talking about the author of two of the aforementioned papers, who may be seen as the prime motivator. James H. Ellis (1924–1997) was born in Australia, but his parents returned to London when he was still a baby. After graduating from the University of London, he was employed at the Post Office Research Station at Dollis Hill, whose cryptography section moved to join the (newly formed) CESG in GCHQ in 1965. There Ellis became a leading figure in British cryptography. The British government asked Ellis to investigate the key-distribution problem since management of large amounts of key material needed for secure communication was problematic for the military. In January of 1970, Ellis established the fundamental ideas behind public-key cryptography in [76]. He called his method *nonsecret encryption* (NSE). Hence, the discovery of the idea of public-key cryptography predated Diffie, Hellman, and Merkle by more than a half dozen years. In [77], published (internally) in CESG in 1987, Ellis describes the history of NSE. In this paper, he says: "The task of writing this paper has devolved to me because NSE was my idea and I can therefore describe their developments from personal experience." Also, in this paper Ellis cites the 1944 publication [282] (by an unknown author for Bell Laboratories), which he describes as an ingenious idea for secure conversation over a telephone. This was his inspiration for NSE. Ellis states, in the aforementioned paper, that this is how the idea was born, that secure communication was possible if the recipient took part in the encryption process. At the end of his paper Ellis concludes that the Diffie-Hellman idea "was the start of public awareness on this type of cryptography and subsequent rediscovery of the NSE techniques I have described." Shortly after his death in 1997, GCHQ/CESG released the five publications cited above. According to a spokesman for the British government, the release of the papers was a "pan-governmental drive for openness" by their Labour party.

Another author of one of the aforementioned papers is the second on the scene. Clifford Cocks joined CESG in September of 1973, where he became acquainted with Ellis's ideas for NSE. He naturally moved to the idea of a one-way function since he had studied number theory at the University of Cambridge as a student. Cocks claimed that it took him only a half hour to invent the notion in [58], dated November 20, 1973, wherein he essentially describes what we now call the RSA cryptosystem, with any differences being entirely superficial.

Our last character authored two of the papers under discussion. Malcolm Williamson joined CESG in September of 1974. He learned from Cocks about the NSE idea, but found it difficult to believe. By trying to disprove the existence of NSE, he discovered a notion equivalent to the Diffie-Hellman key-exchange protocol. This means that the discovery of (a notion equivalent to) RSA preceded that of (a notion equivalent to) Diffie-Hellman, which is the opposite of what occurred in the public domain. In [281], dated January 24, 1974, Williamson describes what we now call the Diffie-Hellman key-exchange protocol, and in [292], dated August 10, 1976, Williamson improved upon the ideas [281] he put forth in 1974.

In an interview in the the *New York Times* in December of 1997, Williamson said that he felt bad knowing that others were taking credit for solutions found at CESG. However, he concluded that this was just one of the restrictions to which you agree and accept when you work for a government agency on secrecy projects. On the other hand, Hellman has said that these things are like stubbing your toe on a gold nugget left in the forest: "If I'm walking in the forest and stub my toe on it, who's to say I deserve credit for discovering it?" Hellman also stated that he, Diffie, and Merkle were all "working in a vacuum". He claimed that if they had had access to the classified documents over the previous three decades, it would have been a great advantage. Diffie commented that the history of ideas is hard to write because people find solutions to different problems and later find out that they have discovered the same thing as someone else. In fact, Diffie did have meetings with Ellis in 1982, but Ellis never once disclosed his discoveries. It is up to historians to sort out the details and the claims, but it is certain that the ideas for public key cryptography were known (in the classified domain) well in advance of the (publicly acknowledged) efforts of Diffie, Hellman, and Merkle.

Perhaps the big difference between the CESG discoveries and those in the public sector is that the individuals at CESG were "government-tied". In other words, they were extremely reluctant to develop their ideas since, first it went against established practice, and second, even though they verified the validity of public key, they knew it was far too slow compared to symmetric-key methods. Thus, they never considered the use of hybrid cryptosystems that evolved in the public domain, since the "cryptographic amateurs" were willing to take their ideas to the limit, and they did so with amazing success.

After the introduction of the RSA public-key cipher, numerous other PKC schemes came into being, which we will discuss in later chapters, along with associated digital signature schemes, and other related schemes that we will discover in due course. Some of these schemes had false starts and some had weaknesses that it took years to discover and for attacks to be developed to which they finally succumbed. One such type of cryptosytem is the *knapsack cryptosystem*. In the late 1970s, these cryptosystems came into being with the work of Merkle and Hellman (see [160]), but this was broken by Shamir (see [247] and [248]) in the early 1980s. Also, in 1982, a new knapsack public-key cipher, the *Chor-Rivest knapsack cryptosystem*, was introduced (see [56] and

[57]). It took nearly two decades for this one to be broken, but it was cracked in 2001 by S. Vaudenay (see [285]). It sometimes takes many years for attacks to be developed so that a given cryptosystem will succumb, and the knapsack family is a good example. Chor-Rivest was the last-standing secure knapsack cryptosystem, so now they are chiefly of theoretical interest.

What made all of the above not just possible, but rather a *necessity* — that good old *mother of invention* — was the advent of the *Internet*. While information secrecy, as we have seen throughout history, was strictly the purview of governments and their agents, the Internet, and its associated e-mail, and e-commerce activities, demanded a mechanism for the ordinary citizen to have *their* privacy concerns addressed. We now have *personal identification numbers* (PIN)s to identify ourselves to *automated teller machines* (ATM)s as well as to engage in banking on secure Web pages, all of which use public-key cryptography to guarantee that credit cards, banking, and other sensitive personal data travels securely to the intended target. Few of us actually understand the mechanisms behind all of these protocols that we use every day (although this book will foster that understanding), yet cryptography has become everybody's business, hence everybody's concern. Therefore it is almost a personal duty that each of us learn as much as possible about the underlying mechanisms that affect our security, our privacy, and therefore our well-being.

Wireless telephones and e-mail traffic are notoriously insecure. Anyone with minimal technology can "listen in" on electronic and voice conversations. There are ways to ensure privacy in these matters and we are going to learn about them.

By the mid-1990s, we had the standardization of digital signature algorithms such as the *Digital Signature Standard* (DSS), and Internet cryptographic algorithms for protecting e-commerce, such as Rivest's *RC4* algorithm, and others about which we will learn the details later. There is much information for us to process.

The Future

By the time the light of the twenty-first century shone upon us, we had the new AES, and a promise of outstanding, if not, incredible possibilities for the future. We will learn all about *smart cards*, including methods for storing your medical data in such a fashion that if you were in an accident, the information on the card could save your life. We will learn about secure methods of protecting medical data bases, as well. There are numerous levels and types of cards, all of which will be our tools and we will understand them. As well, there is the area of *biometrics* such as fingerprints, eye retina scans, voice patterns, and facial geometry, that can be used for identification, including fighting terrorism. All will be ours to understand and appreciate. Then there is the realm of the fantastic, what has not yet come to pass, but which has the potential to do so.

Quantum cryptography means the *possibility* of using quantum mechanical properties of subatomic particles to give us encryption on a scale that, if it ever came into being, would eliminate all classical symmetric and asymmetric cryptosystems as if they were an extinct species of dinosaur. The reason is that the existence of a true *quantum computer* could not only outperform classical computers (for instance, breaking RSA since it could factor integers efficiently!), but also do things that classical computers *cannot* do, such as generating truly random numbers! We would even be on the brink of such phenomena as *teleportation*! We will understand how this has been done already on a subatomic level. But we will leave this topic for the last chapter of the book. We have a lot to learn before we get there.

We now have an overview of our history under our belts, a bird's-eye view of how we got here crytoplogically. In the next chapter, we begin to learn the details of all the mechanisms we have talked about to this point. Let us get started.

Figure 2.23: The Cray XMP

The above image is courtesy of the National Cryptologic Museum. See *http://www.nsa.gov/museum/cray.html* where the following caption is given: "This Cray XMP was donated to the museum by Cray Research, Inc. It denotes the newest era of partnership between NSA and the American computer industry in the employment of computers for cryptologic processes."

Chapter 3

Symmetric-Key Cryptography

Nature is a temple, where, from living pillars
confused words are sometimes allowed to escape;
here man passes, through forests of symbols, which
watch him with looks of recognition.

Charles Baudelaire (1821–67), French poet and critic
— translated from *Les fleurs du mal* (1857), correspondence no. 4

3.1 Block Ciphers and DES

In Chapters 1 and 2 we saw, through the vehicle of historical discourse, a fair amount of cryptological terminology. We now review and summarize parts of this as a background to our discussion of block ciphers in general, and DES in particular.

On page 3, we began to learn about the very basic notions of cryptography, and we developed more detailed knowledge as we learned about various cryptosystems that were developed throughout history. However, the notions in Chapters 1 and 2 were introductory and informal. Although those concepts were sufficient to give us a vehicle for understanding the evolution of the subject of cryptology, we need more. Now we strive for a clearer mathematical picture, since in the final analysis, as we saw in these chapters, cryptology evolved, and its modern-day manifestation is deeply rooted in a mathematical framework. If we wish to have any concrete understanding of cryptology today, we are obliged to understand its inner mathematical structures. We begin our quest for more precision by formalizing the definition of cryptosystem given informally on page 4.

The reader unfamiliar with some of the background required may review Appendix A, which has all the material required to understand what follows. In particular, the reader unfamiliar with the notions of congruences, modular inverses, greatest common divisors, classes, prime factoring and related ideas, should digress, at this juncture, to review the material presented for them in order to ease their introduction to the following topics.

First we introduce some preliminary terminology. Both plaintext and ciphertext are written symbols from some finite set $\mathcal{A}$, called the *alphabet of definition*, which may consist of letters from an alphabet such as English, Greek, Hebrew, or Russian, and may include symbols such as ✂, ☞, ☎, ©, ®, or any symbols we choose to use to send messages. The message space, which we will denote by $\mathcal{M}$ from now on, is a finite set of strings of symbols from the alphabet of definition, which we will denote by $\mathcal{A}$ henceforth. If $m \in \mathcal{M}$, then m is called a *plaintext message unit*. The ciphertext space, which we will denote by $\mathcal{C}$ in what follows, consists of strings of symbols from $\mathcal{A}$ for the ciphertext. If $c \in \mathcal{C}$, then c is called a *ciphertext message unit*. The symbol, $\mathcal{K}$ called the *keyspace*, will be used to denote the set of parameters from which we choose our keys for a given cryptosystem.

On pages 12 and 13, we introduced the notion of a generic cipher and illustrated the process of encryption and decryption. We also gave an informal verbal description of the enciphering and deciphering transformations. The reader needing a refresher of these concepts should review those pages before proceeding to the following formal definition.

◆ Ciphers/Cryptosystems

An *enciphering transformation* (also called an *enciphering function* or *encryption function*) is a bijective function

$$E_e : \mathcal{M} \to \mathcal{C}$$

where the key $e \in \mathcal{K}$ uniquely determines E_e acting upon plaintext message units $m \in \mathcal{M}$ to get ciphertext message units $E_e(m) = c \in \mathcal{C}$. A *deciphering transformation* (also called a *deciphering function* or *decryption function*) is a bijective function determined by a given key $d \in \mathcal{K}$, acting upon ciphertext message units $c \in \mathcal{C}$ to get plaintext message units $D_d(c) = m$. The application of E_e to m, namely, the operation $E_e(m)$, is called *enciphering* $m \in \mathcal{M}$. The application of D_d to c is called *deciphering* $c \in \mathcal{C}$.

A *cryptosystem* or *cipher* consists of a set of enciphering transformations

$$\{E_e : e \in \mathcal{K}\}$$

and the corresponding set of deciphering transformations

$$\{D_d : d \in \mathcal{K}\} = \{E_e^{-1} : e \in \mathcal{K}\}.$$

In other words, for each $e \in \mathcal{K}$, there exists a unique $d \in \mathcal{K}$ such that $D_d = E_e^{-1}$, with

$$D_d(E_e(m)) = m \text{ for all } m \in \mathcal{M}.$$

The keys (e, d) are called a *key pair*. The pairs of plaintext symbols and their ciphertext equivalents:

$$\{(m, E_e(m)) = (m, c) : m \in \mathcal{M}\} \text{ is called a *cipher table*.}$$

A cryptosystem is called *symmetric-key* (also called *single key*, *one key*, and *conventional*) if, for each key pair (e, d), it is computationally easy[3.1] to determine d knowing only e and to determine e knowing only d. (Often $e = d$ in symmetric-key ciphers, adding more justification for the term "symmetric-key".)

Remark 3.1 *We caution the reader that the term "cipher" is not used uniformly throughout the literature. We have clarified and set our meaning here so there is no confusion. Moreover, as discussed on page 6, the term "codes" was historically blurred with the notion of "cipher". However, we maintain the definition of "cryptographic codes" given on page 6 to distinguish them from "ciphers" and non-cryptographic codes described therein. Lastly, the term "cipher table" is sometimes used in the literature and throughout history to mean what we have defined to be a "cryptosystem". However, the more precise meaning we have given to the term "cipher" here makes the context clear, and the term "cipher table" is also well defined. For instance, cipher Table 1.2 on page 11 for the Caesar cipher is such an example, whereas the description of the enciphering and deciphering transformations that make up the Caesar cipher itself are components of the cryptosystem established in this definition.*

In Chapter 1, we encountered the notion of monoalphabetic ciphers (those with a single-cipher alphabet/key, as we determined therein), with the Caesar cipher as a worked example. Then we witnessed the evolution of the polyalphabetic cipher (those with more than one cipher alphabet), with Trithemius's tableau as a vivid example.

With monoalphabetic ciphers, an alteration of one letter in plaintext alters exactly one letter in ciphertext. This makes the finding of the key by a frequency analysis of the ciphertext a relatively easy task. In polyalphabetic ciphers, such as the Vigenère cryptosystem, which we also studied in detail in Chapter 1, for instance, the use of blocks of letters corresponding to the keylength makes this more difficult, but still feasible since there is no interaction among the characters in each block. The following more general type of cryptosystem avoids these failings by enciphering blocks of many characters simultaneously, so that changing a symbol in one plaintext block, should (potentially) result in a corresponding change in all symbols in the corresponding ciphertext block.

We have already encountered an example of a "block cipher" on page 98. Here we present a formal definition for the sake of completeness and for easy reference.

◆ Block Ciphers

A *block cipher* is cryptosystem that separates the plaintext into strings, called blocks, of fixed length $n \in \mathbb{N}$, called its *blocklength*, and enciphers one block at a time.

An example of one of the most basic kinds of block ciphers is the Caesar cipher discussed on pages 11 and 12. We now look at a class of ciphers of which

---

[3.1]A *computationally easy* problem means one that can be solved in expected polynomial time (see the section on complexity in Appendix A, especially page 501.)

the Caesar cipher is a simple example. Before giving the formal definition, we may think of the following type of cipher as a combination of an *additive modular cipher*, such as the Caesar cipher where we add a fixed amount, say, b, to the plaintext numerical equivalent, m, to get $m + b \pmod{n}$, for a fixed modulus $n \in \mathbb{N}$; and a *multiplicative modular cipher*, where we take a plaintext message unit and multiply it by a fixed amount, say, a, to get ciphertext given by $am \pmod{n}$. Each of these values a, b, may be regarded as keys, so that used in combination we get that the pairs (a, b) make up the *keyspace*. When we combine all of this into one cryptosystem, we get the following type of cipher.

◆ **Affine Ciphers**

An *affine cryptosystem* is a symmetric-key block cipher, defined as follows. Suppose that both the message space $\mathcal{M}$ and ciphertext space are both $\mathbb{Z}/n\mathbb{Z}$ for some integer $n > 1$, and let the keyspace be given by

$$\mathcal{K} = \{(a, b) : a, b \in \mathbb{Z}/n\mathbb{Z} \text{ and } \gcd(a, n) = 1\}.$$

Then for $e, d \in \mathcal{K}$, and $m, c \in \mathbb{Z}/n\mathbb{Z}$, the enciphering and deciphering transformations are given, respectively, by the following:

$$E_e(m) \equiv am + b \pmod{n}, \text{ and } D_d(c) \equiv a^{-1}(c - b) \pmod{n}.$$

The Caesar cipher is the simple application of an affine cipher, where $n = 26$, $a = 1$, and $b = 3$. Here is a slightly more complicated example.

Example 3.1 *Let $n = 26$, and $\mathcal{M} = \mathcal{C} = \mathbb{Z}/26\mathbb{Z}$. Define an affine cipher via,*

$$E_e(m) \equiv 5m + 11 \equiv c \pmod{26}, \text{ and } D_d(c) \equiv 21(c - 11) \pmod{26},$$

since $21 \equiv 5^{-1} \pmod{26}$. Now suppose that we are aware of the following cipher-text having been enciphered with the above cryptosystem (Kerckhoff's Principle in action, see page 76).

$$c = (11, 10, 10, 25, 24, 5, 21, 25, 8, 20, 5, 18, 23, 11, 18, 5, 5, 11, 23, 1).$$

Then to decipher, we use D_d on each ciphertext message unit. For instance,

$$D_d(11) \equiv 21(11 - 11) \equiv 0 \pmod{26}, \quad D_d(10) \equiv 21(10 - 11) \equiv 5 \pmod{26},$$

and so on, (where the reader can now fill in the blanks), to achieve the plaintext,

$$m = (0, 5, 5, 8, 13, 4, 2, 8, 15, 7, 4, 17, 18, 0, 17, 4, 4, 0, 18, 24).$$

Now, if we want the plaintext in English, we go to Table 1.3 on page 11, to get

affine ciphers are easy.

Affine ciphers are just special cases of the following, which we informally encountered on page 8. This can now be put into a well-defined mathematical notion. The reader unfamiliar with, or in need of a reminder of at least the notation for, the notion of permutations should review page 9 where we introduced the informal notion therein.

◆ Substitution Ciphers

A substitution cipher is defined as follows. Let $\mathcal{A}$ be an alphabet of definition consisting of n symbols, and $\mathcal{M}$ be the set of all blocks of length $r \in \mathbb{N}$ over $\mathcal{A}$. The keyspace, $\mathcal{K}$, consists of all ordered r-tuples $e = (\sigma_1, \sigma_2, \ldots, \sigma_r)$ of permutations σ_j on $\mathcal{A}$. The enciphering and deciphering transformations are defined by the actions below, respectively. If $e \in \mathcal{K}$ and $m = (m_1 m_2 \ldots m_r) \in \mathcal{M}$, then

$$E_e(m) = (\sigma_1(m_1), \sigma_2(m_2), \ldots, \sigma_r(m_r)) = (c_1, c_2, \ldots, c_r) = c \in \mathcal{C},$$

and for $d = (d_1, d_2, \ldots, d_r) = (\sigma_1^{-1}, \sigma_2^{-1}, \ldots, \sigma_r^{-1}) = \sigma^{-1}$,

$$D_d(c) = (d_1(c_1), d_2(c_2), \ldots, d_r(c_r)) = (\sigma_1^{-1}(c_1), \sigma_2^{-1}(c_2), \ldots, \sigma_r^{-1}(c_r)) = m.$$

If all the keys are the same, namely, $\sigma_1 = \sigma_2 = \cdots = \sigma_r$, then this cryptosystem is called a *simple substitution cipher* or *monoalphabetic substitution cipher*. If the keys differ, we call this cryptosystem a *polyalphabetic substitution cipher*.

Thus, we know that the Caesar cipher is an example of a block cipher that is a monoalphabetic substitution, whereas the Vigenère cipher, studied earlier (see page 56), is an example of a polyalphabetic substitution.

As we have seen in Chapters 1 and 2, simple substitution ciphers suffer from the inherent weakness that a frequency analysis can be done on the ciphertext, whereas polyalphabetic substitution ciphers are more secure than the monoalphabetic ones.

When a substitution block cipher replaces one or more symbols by groups of ciphertext symbols, we call this a *polygram substitution cipher*. We encountered one of these already on page 68, namely, the Playfair cipher, which is an example of a digraphic cipher. In fact, as we saw therein, this was the *first* literal digraphic cipher. Another polygram substitution cipher we have already mentioned in our historical travels is the Hill cipher. We now describe this cipher in detail.

The reader will need a tiny bit of elementary matrix theory, all of which is supplied in Appendix A (see pages 491–494).

The Hill Cipher

Choose fixed $r, n \in \mathbb{N}$ and let the keyspace

$$\mathcal{K} = \{e \in \mathcal{M}_{r \times r}(\mathbb{Z}/n\mathbb{Z}) : e \text{ is invertible}\},$$

and let the message space, $\mathcal{M}$ and ciphertext space $\mathcal{C}$, both be $(\mathbb{Z}/n\mathbb{Z})^r$. This stands for r copies of the integers $\mathbb{Z}/n\mathbb{Z}$, meaning ordered r-tuples of integers modulo n (see Definition A.20 in Appendix A on page 477). Then for $m \in \mathcal{M}$, $e \in \mathcal{K}$, $c \in \mathcal{C}$, the enciphering transformation is given by

$$E_e(m) = me = c,$$

and the deciphering transformation is given by

$$D_d(c) = cd = ce^{-1}.$$

Note that

e is invertible, namely, $d = e^{-1}$ exists if and only if $\gcd(\det(e), n) = 1$.

(See Appendix A, page 493.)

The above definition tells us that changing one character of plaintext will usually change r letters of ciphertext. Hence, frequency analysis of ciphertext is less effective, especially for large r. However, the Hill cipher succumbs to known plaintext attacks (see Footnote 2.9 on page 84). Now we illustrate it with a simple example, which is intended for the uninitiated reader. Merely revisit pages 491–493 in Appendix A to see the simple methods for two-by-two matrices illustrated there.

Example 3.2 *Let $n = 26$ and $r = 2$, so*

$$\mathcal{A} = \mathbb{Z}/26\mathbb{Z}, \qquad \mathcal{M} = \mathcal{C} = (\mathbb{Z}/26\mathbb{Z})^2,$$

and $\mathcal{K}$ consists of invertible 2×2 matrices with entries from $\mathbb{Z}/26\mathbb{Z}$. Thus, if $e \in \mathcal{K}$, then $\gcd(\det(e), 26) = 1$ (see part (b) of Theorem A.25 on page 493). For instance, take

$$e = \begin{pmatrix} 2 & 5 \\ 3 & 4 \end{pmatrix}$$

for which $\det(e) = -7$. Suppose that we want to encipher the plaintext:

message by matrix.

First we get the numerical equivalents from Table 1.3 on page 11:

$$12, \quad 4, \quad 18, \quad 18, \quad 0, \quad 6, \quad 4, \quad 1, \quad 24, \quad 12, \quad 0, \quad 19, \quad 17, \quad 8, \quad 23. \qquad (3.1)$$

Thus, we may set

$$m_1 = (12, 4), \quad m_2 = (18, 18), \quad m_3 = (0, 6), \quad m_4 = (4, 1), \quad m_5 = (24, 12),$$

$$m_6 = (0, 19), \quad m_7 = (17, 8), \quad and \quad m_8 = (23, 25),$$

where we have used a "z" with numerical value 25 to make up the last ordered pair m_8. Now use the enciphering transformation defined in the Hill Cipher. (Remember that once we get the entries in the final ciphertext matrix, we have *to reduce modulo* 26, *namely, throw away all multiples of* 26 *until we are left with a nonnegative numerical value less than* 26.)

$$E_e(m_1) = (12, 4) \begin{pmatrix} 2 & 5 \\ 3 & 4 \end{pmatrix} = (10, 24) = c_1,$$

$$E_e(m_2) = (18, 18) \begin{pmatrix} 2 & 5 \\ 3 & 4 \end{pmatrix} = (12, 6) = c_2,$$

$$E_e(m_3) = (0, 6) \begin{pmatrix} 2 & 5 \\ 3 & 4 \end{pmatrix} = (18, 24) = c_3,$$

$$E_e(m_4) = (4, 1) \begin{pmatrix} 2 & 5 \\ 3 & 4 \end{pmatrix} = (11, 24) = c_4,$$

$$E_e(m_5) = (24, 12) \begin{pmatrix} 2 & 5 \\ 3 & 4 \end{pmatrix} = (6, 12) = c_5,$$

$$E_e(m_6) = (0, 19) \begin{pmatrix} 2 & 5 \\ 3 & 4 \end{pmatrix} = (5, 24) = c_6,$$

$$E_e(m_7) = (17, 8) \begin{pmatrix} 2 & 5 \\ 3 & 4 \end{pmatrix} = (6, 13) = c_7,$$

and

$$E_e(m_8) = (23, 25) \begin{pmatrix} 2 & 5 \\ 3 & 4 \end{pmatrix} = (17, 7) = c_8.$$

Now we use Table 1.3 to get the ciphertext letter equivalents and send

KYMGSYLYGMFYGNRH.

as the cryptogram. Now we show how decryption works. Once the cryptogram is received, we must calculate the inverse of e, *which is*

$$e^{-1} = d = \begin{pmatrix} 18 & 23 \\ 19 & 22 \end{pmatrix}.$$

To see why this is the case, see Example A.12 on page 493 in Appendix A, and note that the multiplicative inverse of

$$\det(e) = -7$$

modulo 26 *is given by*

$$(-7)^{-1} \equiv 11 \pmod{26},$$

from Example A.5 on page 478 in Appendix A. Now apply the deciphering transformation to the numerical equivalents of the ciphertext as follows:

$$D_d(c_1) = D_{e^{-1}}(10, 24) = (10, 24) \begin{pmatrix} 18 & 23 \\ 19 & 22 \end{pmatrix} = (12, 4) = m_1,$$

$$D_d(c_2) = D_{e^{-1}}(12, 6) = (12, 6) \begin{pmatrix} 18 & 23 \\ 19 & 22 \end{pmatrix} = (18, 18) = m_2,$$

and so on until we achieve the original plaintext numerical equivalents. The letter equivalents now give us back the original plaintext message via Table 1.3.

On page 9, we looked at the concept of transposition/permutation ciphers. We can now formalize this definition as well. First, as an introductory mechanism, think of the following type of cipher as one in which the characters of the plaintext are reordered according to some agreed-upon procedure by, say, Alice and Bob, who are corresponding with one another (securely). For instance, as we saw on page 9, the skytale is an historical example of such a cipher. A modern-day example is in most local newspapers, namely, an *anagram puzzle*, which is a meaningful rearrangement of an already meaningful plaintext. The clear conclusion, from the fact that these are solved on a regular basis by ordinary readers, is that these kinds of ciphers are easily cryptanalyzed.

To informally introduce the notion of a "permutation" f on a set $\mathcal{S} = \{1, 2, \ldots, r\}$, we may think of f as a bijective function from $\mathcal{S}$ to itself. A naive question that arises is: How many such permutations are there? We see that for the first element 1, of $\mathcal{S}$ there are r choices to which it can be mapped, and for the second choice, 2, there remain $r - 1$ choices to which it can be mapped (since we cannot map the first element to *more* than one element given that f is a function), and similarly for the third element, 3, there are $r - 2$ elements to which we can map it, and so on. Hence, the total number of such permutations is $r!$. This explains the cardinality of the keyspace below.

◆ **Permutation/Transposition Ciphers**

A *simple transposition cipher* or *simple permutation cipher* is a symmetric-key block cryptosystem having blocklength $r \in \mathbb{N}$, with keyspace $\mathcal{K}$ being the set of permutations on $\{1, 2, \ldots, r\}$. The enciphering and deciphering transformations are given as follows, respectively:

For each $m = (m_1, m_2, \ldots, m_r) \in \mathcal{M}$, and $e \in \mathcal{K}$,

$$E_e(m) = (m_{e(1)}, m_{e(2)}, \ldots, m_{e(r)}),$$

and for each $c = (c_1, c_2, \ldots, c_r) \in \mathcal{C}$,

$$D_d(c) = D_{e^{-1}}(c) = (c_{d(1)}, c_{d(2)}, \ldots, c_{d(r)}).$$

In the above, notice that e implicitly defines r since e is a permutation on r symbols. Moreover, in such cryptosystems, the cardinality $|\mathcal{K}| = r!$. The following is a simple, perhaps amusing, illustration of how such permutation ciphers work in transposing the *places* where the plaintext letters sit.

Example 3.3 *Suppose that $r = 13$ and $\mathcal{M} = \mathcal{C} = \mathbb{Z}/26\mathbb{Z}$. We apply the key*

$$e = \begin{pmatrix} 1 & 2 & 3 & 4 & 5 & 6 & 7 & 8 & 9 & 10 & 11 & 12 & 13 \\ 9 & 12 & 6 & 13 & 1 & 7 & 4 & 10 & 2 & 3 & 11 & 5 & 8 \end{pmatrix}$$

to the plaintext,

$$m = (m_1, m_2, m_3, m_4, m_5, m_6, m_7, m_8, m_9, m_{10}, m_{11}, m_{12}, m_{13}) =$$

$$(b, r, i, t, n, e, y, s, p, e, a, r, s),$$

to get

$$E_e(m) = (m_{e(1)}, m_{e(2)}, m_{e(3)}, m_{e(4)}, m_{e(5)}, m_{e(6)}, m_{e(7)}, m_{e(8)}, m_{e(9)}, m_{e(10)},$$

$$m_{e(11)}, m_{e(12)}, m_{e(13)}) =$$

$$(m_9, m_{12}, m_6, m_{13}, m_1, m_7, m_4, m_{10}, m_2, m_3, m_{11}, m_5, m_8) =$$

$$(P, R, E, S, B, Y, T, E, R, I, A, N, S).$$

On pages 45 and 46, we learned that Qalqashandi was the first to introduce the intermingled use of substitution and transposition in a single cipher. In a more modern-day setting, we saw how such a combination of substitution and transposition was used in the World War II ADFGVX field cipher, whose use is illustrated on page 80. Even closer to the modern day is the DES cipher, which employs some of the best combinations of transposition and substitution. Although we mentioned DES on page 98, we gave no indication of how this block cipher works. It is now time to learn about this cryptosystem in detail. Although this cryptosystem is no longer used as the standard, the fact that this block cipher ruled the roost for about a quarter century makes it deserving of a closer look, if for no other reason than historical, in keeping with Chapters 1 and 2. Moreover, certain stronger ciphers derived from it are still valid and in use.

DES is a symmetric-key block cipher, encrypting *octograms of bytes*[3.2] with a key based on a permutation, then sixteen substitutions followed by another permutation. This is another way of saying that DES enciphers 64-bit blocks of plaintext to produce 64-bit blocks of ciphertext, using the same key for encryption and decryption, a key that is based on a combination of substitution and permutation techniques. However, in the interests of ease of presentation, we are going to illustrate DES in a simplified form introduced by Ed Schaefer [232] in 1996. This version uses only 8-bit plaintext and 10-bit keys to produce 8-bit ciphertext, not secure, but pedagogically more satisfying for our purposes. (For a complete description of the entire DES algorithm, see [169, pages 86–99].) We look at each component of DES and build the edifice until the final construction of the cryptosystem, dubbed *S-DES*.

[3.2]A *byte* is an 8-bit binary integer, see Footnote 2.7 on page 84. Therefore, an octogram of bytes is a collection of eight bytes, or a block consisting of a 64-bit integer. In what follows, we will suppress base-integer subscripts. For example, $(110)_2$ will be written as (110) with the context understood.

3.2 S-DES and DES

Pass then through this moment of time in harmony with nature, and end your journey in content, as an olive falls when it is ripe, blessing nature who produced it, and thanking the tree on which it grew.

Marcus Aurelius (121–180 AD), *Roman philosopher and emperor* — from *Marcus Aurelius and His Times* (see [11, page 43]).

We begin with an informal overview to describe the mechanisms behind S-DES, which is a simplified version of DES, presented for pedagogical purposes. Although DES reached the end of its cryptographic usefulness by the beginning of the twenty-first century, it is valuable to look at its design and implementation in order to understand how it stood the test for roughly a quarter century before the cryptanalytic onslaught brought on by new mathematical and computational power caused it to fall from grace.

◆ Overview of S-DES

As with any cipher, the encryption function takes the plaintext m and the key k as input. For S-DES, m has bitlength 8 and k has bitlength 10.

First, m is put through what is called an initial permutation **IP**, followed by two rounds of the same function (described below), which uses both permutation and substitution in its execution, the first round followed by a swap of the left and right 4 bits of the output. The 8-bit output of round two is put through the inverse permutation $\mathbf{IP}^{-1}$ to form the 8-bit ciphertext.

Each round of S-DES is described as follows. The 8-bit input is split into left and right 4-bit blocks, L and R. Then there is an expansion of R to 8 bits via an expansion permutation **E**, to get $\mathbf{E}(R)$. The 8-bit result, $\mathbf{E}(R)$, is added modulo 2, denoted by $\oplus$, to an 8-bit subkey SK, generated from k in a separate S-DES key generation stage; see Footnote 2.8 on page 84 for a motivation of the use of $\oplus$ with the one-time pad.[3.3] The resulting 8-bit output $E(R) \oplus SK$ is separated into left and right 4-bit strings, L_1 and R_1, which are fed into two separate substitution boxes, S_1 and S_2, respectively, called *S-boxes* (described in detail below), which are publicly known lookup tables that take 4-bit inputs and output 2-bit strings, L'_1 and L'_2. The resulting 4-bit string, (L'_1, L'_2), is put through a permutation **P**, to produce a 4-bit output Z. Last, Z is added modulo 2 with L to form L', and (L', R) is the output of the round.

All of the above is illustrated in Diagram 3.1, which is a single round of the S-DES cipher. Then, before giving a detailed description of S-DES that will extrapolate the above to a full explanation of all the detailed features of the (simplified) cipher, we look at the motivations behind the design of DES itself.

[3.3]For the reader needing a reminder, see the detailed treatment of modular arithmetic on pages 475–478 in Appendix A. We also had a brief elementary introduction to modular arithmetic on page 12. As mentioned in Footnote 2.8, addition modulo 2 is often called *XORing* in the computer science community since addition modulo 2 is bit by bit *exclusive or* addition.

Diagram 3.1 An S-DES Round

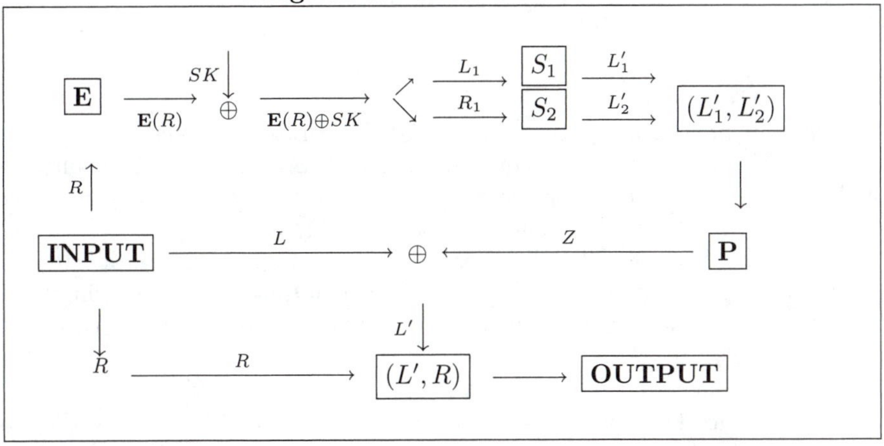

◆ DES Design Principles

In a 1994 publication, [60], Coppersmith described the criteria used in the design of DES. The focus is principally upon the design of the S-Boxes and the permutation function that processes their outputs. There is an interesting story behind this publication and what led up to it. Almost twenty years before Coppersmith decided (or rather was allowed), to publish this knowledge, it was known that IBM researchers had discovered an attack on DES, later known as differential cryptanalysis (see Footnote 3.4 on page 127). This was, let us say, not met with great joy by the NSA, since they had known about it for some time and it was classified information. Moreover, added to this lack of joy at NSA was the fact that IBM researchers had discovered methods for thwarting the attack. Hence, the NSA went out of its way to sanction IBM and classify the IBM discoveries. Not only was this attack a powerful tool against DES, but also many other ciphers, and the NSA did not want this information to be leaked. Coppersmith was one of the IBM researchers who worked on the methods for combating the attack. The compliance by IBM to the NSA demands for secrecy only contributed to the speculation about potential secret back doors through which NSA could cryptanalyze the DES cryptosystem. This is part of the background to the controversy we discussed on page 98, which led to investigations mentioned therein. Ultimately, the information became public through independent discoveries, and as we have seen, the governmental agencies, such as NSA, could no longer control the flow of information. The advent of the Internet, public-key cryptography, and all the interrelated activities in the public domain saw to that.

One important aspect of block ciphers, especially DES, that requires elucidation is the notion of *linearity*. A *linear cipher* is one for which each output bit is a linear combination of the input bits. An example of such a cipher is the Hill cipher discussed on pages 111–113. The Hill cipher is easily broken with a known-plaintext attack. The reason is that since a key matrix e acts upon a plaintext matrix m to produce a ciphertext matrix c via $c = me$, then this can be analyzed in such a fashion that ultimately an inverse matrix m^{-1} can be found, so that

$$e = m^{-1}c$$

and the key is recovered.

The only *non*-linear aspect of DES are the S-Boxes. Hence, an inherent design of DES stipulates that no output bit of an S-Box can be a linear function of the input bits. If they were, then the entire cryptosystem would be linear and could be broken with a known-plaintext attack.

Now we list the principles that were revealed by Coppersmith in his article, which concentrated upon the S-Boxes and their output. Thus, ensuring non-linearity was the key to ensuring that the cryptosystem could not easily be broken.

1. Linearity in the S-Box construction must be avoided. In other words, no bit output by an S-Box is allowed to be anywhere near a linear function of the input bits.

2. Each row of an S-Box should include all possible output bit combinations.

3. If two inputs to an S-Box differ in precisely one bit, or by exactly two middle bits, then the outputs must differ in a minimum of two bits.

4. If two inputs to an S-Box differ in their first two bits, but have identical last two bits, the two outputs must be distinct.

5. There are other criteria such as 2–4, which were designed to thwart differential cryptanalysis, and pertain primarily to the permutations that take the outputs of the S-Boxes. Since these criteria are very technical, we do not go into the details for the sake of efficiency. The reader may consult Coppersmith's paper [60] directly for the specifics, if necessary.

Now, we are ready for a detailed description of S-DES. First, recall our discussion and notation for permutations given on page 8, and the follow-up given in the preceding section. The enciphering and deciphering in S-DES requires several basic components. We begin with two of them that are permutations.

◆ Initial Permutation

Let $m = (m_1m_2m_3m_4m_5m_6m_7m_8)$ be the byte of plaintext input. Then the initial permutation **IP** acts according to the following transposition of places where the plaintext sits, namely, **IP** retains all the plaintext bits, but merely permutes them according to the rule given below.

IP								
j	1	2	3	4	5	6	7	8
IP(j)	2	6	3	1	4	8	5	7

Therefore, the action of **IP** on x is given in the following.

IP								
j	1	2	3	4	5	6	7	8
$m_{\textbf{IP}(j)}$	m_2	m_6	m_3	m_1	m_4	m_8	m_5	m_7

Hence, $\textbf{IP}(m) = (m_2 m_6 m_3 m_1 m_4 m_8 m_5 m_7)$. For instance, if $m = (10010111)$, then $\textbf{IP}(m) = (01011101)$.

The next component is also a permutation used at various stages of S-DES.

◆ Expansion Permutation

This permutation, denoted by **EP**, takes a *bitstring* (binary number) of length 4 (its *bitlength*), and expands it into a byte according to the following.

EP								
j	1	2	3	4	5	6	7	8
$\textbf{EP}(j)$	4	1	2	3	2	3	4	1

For instance, if $x = (x_1 x_2 x_3 x_4)$ is the input, then the following table gives us the action of **EP** on it.

EP								
j	1	2	3	4	5	6	7	8
$x_{\textbf{EP}(j)}$	x_4	x_1	x_2	x_3	x_2	x_3	x_4	x_1

Hence, $\textbf{EP}(x) = (x_4 x_1 x_2 x_3 x_2 x_3 x_4 x_1)$. For example, if $x = (1001)$, then $\textbf{EP}(x) = (11000011)$.

A very important aspect of S-DES is the key schedule. In other words, we need to understand how the keys are used and generated in the cipher.

◆ S-DES Key Generation

S-DES uses a 10-bit secret (shared) symmetric key

$$k = (e_1 e_2 e_3 e_4 e_5 e_6 e_7 e_8 e_9 e_{10}),$$

say, and employs k to generate two 8-bit (sub)keys for deployment at various stages of the encryption and decryption process. Here is how that is accomplished.

First, a permutation $\mathbf{P}_{10}$ is applied to k according to the following.

$\mathbf{P_{10}}$										
j	1	2	3	4	5	6	7	8	9	10
$\mathbf{P}_{10}(j)$	3	5	2	7	4	10	1	9	8	6

Thus, $\mathbf{P_{10}}(k) = (e_3e_5e_2e_7e_4e_{10}e_1e_9e_8e_6)$.

Secondly, there is a circular left shift of 1 place, denoted by **LS1**, on each of the left five bits and the right five bits, as follows. $\mathbf{LS1}(e_3e_5e_2e_7e_4) = (e_5e_2e_7e_4e_3)$, and $\mathbf{LS1}(e_{10}e_1e_9e_8e_6) = (e_1e_9e_8e_6e_{10})$. Hence, under this shifting process, $(e_3e_5e_2e_7e_4e_{10}e_1e_9e_8e_6)$ becomes

$$(e_5e_2e_7e_4e_3e_1e_9e_8e_6e_{10}). \tag{3.2}$$

Then we apply yet another permutation called $\mathbf{P_8}$, which selects 8 of the 10 bits and permutes them as follows.

$\mathbf{P_8}$								
j	1	2	3	4	5	6	7	8
$\mathbf{P_8}(j)$	6	3	7	4	8	5	10	9

Applying $\mathbf{P_8}$ to (3.2) yields

$$\mathbf{P_8}(e_5e_2e_7e_4e_3e_1e_9e_8e_6e_{10}) = (e_1e_7e_9e_4e_8e_3e_{10}e_6) = k_1,$$

where k_1 is now our first subkey for use later.

Now, we return to (3.2) and perform a left shift of two places, denoted by **LS2**, on both the left and right 5-bit pieces to get

$$\mathbf{LS2}(e_5e_2e_7e_4e_3) = (e_7e_4e_3e_5e_2), \text{ and } \mathbf{LS2}(e_1e_9e_8e_6e_{10}) = (e_8e_6e_{10}e_1e_9),$$

yielding $(e_7e_4e_3e_5e_2e_8e_6e_{10}e_1e_9)$ to which we apply $\mathbf{P_8}$ to get

$$\mathbf{P_8}(e_7e_4e_3e_5e_2e_8e_6e_{10}e_1e_9) = (e_8e_3e_6e_5e_{10}e_2e_9e_1) = k_2,$$

where k_2 is our second subkey for use in the S-DES cipher.

The next essential component of the S-DES cryptosystem is an important method of substitution, and an innovation of Feistel in his development of the original DES. (However, it is believed, in some quarters, that Feistel got the idea from the NSA; see [151, page 42]).

◆ **S-Boxes**

An S-Box or *substitution box* for S-DES is a 4×4 matrix with entries from $\mathbb{Z}/4\mathbb{Z}$ (put into binary) with rows and columns labelled from 0 to 3 (put into binary), that takes a 4-bit input and outputs a 2-bit string as follows.

If $(x_1x_2x_3x_4)$ is the input, then the output is given by one of the two S-Boxes used in S-DES, defined as follows.

S_0		x_2	0	0	1	1
		x_3	0	1	0	1
x_1	x_4					
0	0		01	00	11	10
0	1		11	10	01	00
1	0		00	10	01	11
1	1		11	01	11	10

and

S_1		x_2	0	0	1	1
		x_3	0	1	0	1
x_1	x_4					
0	0		00	01	10	11
0	1		10	00	01	11
1	0		11	00	01	00
1	1		10	01	00	11

Thus, for example if $x = (x_1x_2x_3x_4) = (1101)$ is our input bitstring of length 4, then if we wish to employ the first S-Box, we get $\mathbf{S_0}(1101) = (11)$, since $(x_1x_4) = (11)$ represents the fourth row, and $(x_2x_3) = (10)$ represents the third column, the entry at the intersection of which is 11. Similarly, if we want to use the S-Box $\mathbf{S_1}$, then $\mathbf{S_1}(1101) = (00)$.

Perhaps the most complicated part of S-DES is the function that does the combining of permutation and substitution.

◆ The S-DES Round Function

First, we need to describe a mapping F that takes bitstrings of length 4 using a subkey SK, and outputs bitstrings of length 4.

Let $x = (x_1x_2x_3x_4)$ be the input. Then F first uses the expansion $\mathbf{EP}$ to produce $\mathbf{EP}(x)$, as described on page 119. Then this 8-bit result is added to the subkey SK, modulo 2. Recall that addition modulo 2 is denoted by $\oplus$. Thus, this result is denoted by $\mathbf{EP}(x) \oplus SK = (y_1y_2y_3y_4y_5y_6y_7y_8) = y$. For the sake of convenience, we will denote the left four bits of a given byte, such as y, by

$$\mathbf{L}(y) = (y_1y_2y_3y_4) \text{ and the right four bits by } \mathbf{R}(y) = (y_5y_6y_7y_8).$$

The next action of F is to feed $\mathbf{L}(y)$ into $\mathbf{S_0}$ to produce $\mathbf{S_0}(\mathbf{L}(y)) = (z_1z_2)$, and feed $\mathbf{R}(y)$ into $\mathbf{S_1}$ to get $\mathbf{S_1}(\mathbf{R}(y)) = (z_3z_4)$. Thus, under this action y gets sent to $(\mathbf{S_0}(\mathbf{L}(y))\mathbf{S_1}(\mathbf{R}(y))) = (z_1z_2z_3z_4) = z$. Next, we apply the following permutation to z.

$\mathbf{P_4}$				
j	1	2	3	4
$\mathbf{P_4}(j)$	2	4	3	1

Therefore, we get $\mathbf{P_4}(z) = (z_2z_4z_3z_1) = Z$, which is the final outcome for F, namely, $F(x, SK) = Z$.

Now, the definition of the round function, denoted by f_{SK}, which takes an 8-bit plaintext t and a subkey SK, is given as follows.

$$f_{SK}(t) = (\mathbf{L}(t) \oplus F(\mathbf{R}(t), SK), \mathbf{R}(t)).$$

Thus, the round function only alters $\mathbf{L}(t)$, the left four bits of t, leaving $\mathbf{R}(t)$ unaltered. However, there is a reason for f_{SK} being called a round function, since there are two rounds. The next mechanism, the penultimate one, is a means of swapping left and right bits.

◆ The Switch/Swap Function

The switch function, denoted by $\mathbf{SW}$, merely exchanges the left and right four bits of an input m. Hence, if $m = (\mathbf{L}(m), \mathbf{R}(m))$ is an 8-bit input, then

$$\mathbf{SW}(m) = (\mathbf{R}(m), \mathbf{L}(m)).$$

The last aspect of S-DES is the inverse of the initial permutation.

◆ **The Inverse of IP**

The inverse of **IP**, naturally denoted by $\mathbf{IP}^{-1}$, is given by the following.

$\mathbf{IP}^{-1}$								
j	1	2	3	4	5	6	7	8
$\mathbf{IP}^{-1}(j)$	4	1	3	5	7	2	8	6

An easy means for finding the inverse of any permutation is as follows. Take the table for **IP** on page 118, for instance. To find the inverse, just read off in numeric order (determined by the second row), the terms in the first row. For instance, the term in the first row sitting above 1, in the aforementioned table for **IP**, is 4, so 4 is the first term in the table for $\mathbf{IP}^{-1}$. The term in the first row sitting above 2 is 1, so 1 is the second entry in the table for $\mathbf{IP}^{-1}$, and so on, to construct the above. Note that the reason for the above to work is that $\mathbf{IP}(\mathbf{IP}^{-1}(j)) = j$ for all j under consideration. So since **IP** takes 1 to 2, then $\mathbf{IP}^{-1}$ must take 2 to 1, and so forth. (See Definition A.5 on page 467.)

Now, we are in a position to describe the totality of the S-DES cipher.

◆ **The S-DES Cryptosystem**

Given a 10-bit key k and an 8-bit plaintext m, to encipher, we execute the following.

◆ **S-DES Encryption**

1. Apply **IP** to m.

2. Apply f_{k_1} to the output from step 1. (This is round 1.)

3. Apply **SW** to the output of step 2.

4. Apply f_{k_2} to the output of step 3. (This is round 2.)

5. Apply $\mathbf{IP}^{-1}$ to the output of step 4.

Hence, the plaintext 8-bit message unit m gets sent to the 8-bit ciphertext message unit c, the output of step 5, under this sequence of steps of the S-DES cipher. To decrypt, we perform the following.

◆ **S-DES Decryption**

1. Apply **IP** to c.

2. Apply f_{k_2} to the output from step 1. (This is round 1.)

3. Apply **SW** to the output of step 2.

4. Apply f_{k_1} to the output of step 3. (This is round 2.)

5. Apply $\mathbf{IP}^{-1}$ to the output of step 4.

The following is derived from Ed Schaefer, the creator of S-DES, [232].

Example 3.4 *Suppose we are given plaintext bitstring* $m = (10100101)$ *and key bitstring* $k = (0010010111)$. *First we generate our subkeys as follows.*

1. $\mathbf{P_{10}}(k) = 1000010111$.

2. $\mathbf{LS1}(10000) = (00001)$ *and* $\mathbf{LS1}(10111) = (01111)$.

3. $\mathbf{P_8}(0000101111) = (00101111) = k_1$.

4. $\mathbf{LS2}(00001) = (00100)$ *and* $\mathbf{LS2}(01111) = (11101)$, *(applying* $\mathbf{LS2}$ *to the output of step 2.)*

5. $\mathbf{P_8}(0010011101) = (11101010) = k_2$, *(applying* $\mathbf{P_8}$ *to the output of step 4.).*

Now we encrypt as follows. First we calculate $\mathbf{IP}(m) = (01110100)$. *Then we need to calculate the round function for the first round* $f_{k_1}(01110100) = (\mathbf{L}(01110100) \oplus F(\mathbf{R}(01110100), k_1), \mathbf{R}(01110100))$. *We do this as follows.*

1. $\mathbf{EP}(0100) = (00101000)$.

2. $\mathbf{EP}(0100) \oplus k_1 = (00101000) \oplus (00101111) = (00000111)$.

3. $\mathbf{S_0}(0000) = (01)$ *and* $\mathbf{S_1}(0111) = (11)$.

4. $\mathbf{P_4}(0111) = (1110) = F(\mathbf{R}(01110100), k_1)$.

5. $\mathbf{L}(01110100) \oplus F(\mathbf{R}(01110100), k_1) = (0111) \oplus (1110) = (1001)$.

6. $f_{k_1}(01110100) = (10010100)$.

Now we apply the switch function, $\mathbf{SW}(10010100) = (01001001)$. *The reader may now verify the second round, namely,*

$$f_{k_2}(01001001) = (\mathbf{L}(01001001) \oplus F(\mathbf{R}(01001001), k_2), \mathbf{R}(01001001)) = (01101001).$$

Last, we apply the inverse of the initial permutation, $\mathbf{IP}^{-1}(01101001) = (00110110)$, *which is the ciphertext.*

To decrypt, we reverse the process. First feed c into $\mathbf{IP}$ *to get*

$$\mathbf{IP}(c) = (01101001),$$

then apply f_{k_2} *to get (with the reader filling in the details),*

$$f_{k_2}(0110 \oplus F(1001, k_2), 1001) = (01001001).$$

Then $\mathbf{SW}(01001001) = (10010100)$. *Next,*

$$f_{k_1}(1001 \oplus F(0100, k_1), 0100) = (01110100),$$

then the final application yields the original plaintext, $\mathbf{IP}^{-1}(01110100) = (10100101) = m$.

Diagrams 3.2 and 3.3 give a succinct presentation of S-DES.

Diagram 3.2 The S-DES Encryption Flow Chart

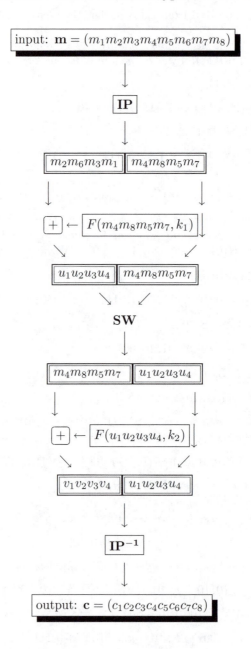

The action between **IP** and **SW** is round 1, namely, the execution of f_{k_1}, and the action between **SW** and **IP**$^{-1}$ is round 2, the action of f_{k_2}.

Diagram 3.3 The S-DES Decryption Flow Chart

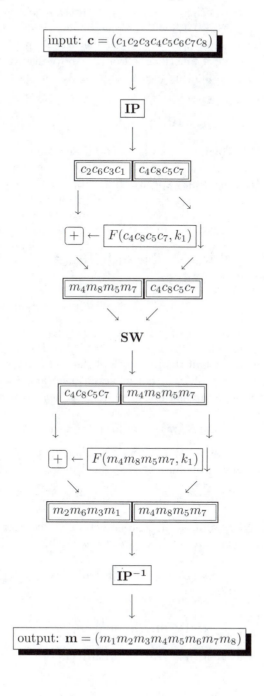

◆ **Analysis of S-DES and Comparison to DES**

Schaefer relabeled S-DES as *baby DES* since it is a much simpler block cipher than the full-blown DES. S-DES will encipher one block at a time and there are 2^8 possible plaintext blocks since we are dealing with 8-bit plaintext bitstrings. In terms of composition of functions, all of the above discussion of S-DES can be encapsulated in the following (see Definition A.5 on page 467 in Appendix A).

$$(\mathbf{IP}^{-1} \circ f_{k_2} \circ \mathbf{SW} \circ f_{k_1} \circ \mathbf{IP})(m) = \mathbf{IP}^{-1}(f_{k_2}(\mathbf{SW}(f_{k_1}(\mathbf{IP}(m))))) = c.$$

Full DES takes 64-bit plaintext blocks, a 56-bit key, from which sixteen 48-bit subkeys are generated, and sixteen round functions, which we will label f_{k_j} for $j = 1, 2, \ldots, 16$. Hence, we may specify (full) DES now as a single composition of functions.

$$(\mathbf{IP}^{-1} \circ f_{k_{16}} \circ \mathbf{SW} \circ f_{k_{15}} \circ \mathbf{SW} \circ \cdots \circ f_{k_1} \circ \mathbf{IP})(m) = c.$$

Moreover, in DES, we have eight S-Boxes $\mathbf{S}_j$ for $j = 1, 2, \ldots, 8$, each having 4 rows and 16 columns, where

$$\mathbf{S_j}(m_1 m_2 m_3 m_4 m_5 m_6)$$

picks out the entry in row $(m_1 m_6)$ and column $(m_2 m_3 m_4 m_5)$, which represents 16 possible entries, in binary, for each such row. Also, $\mathbf{P_4}$ in S-DES, is replaced by $\mathbf{P_{32}}$ in DES, which is half the bitlength of the input in either case.

One of the weaknesses of DES that makes it unsuitable for use, and ranks it as below standard for the modern day are its *weak keys*, which are keys k such that

$$E_k(E_k(m)) = m \text{ for all } m \in \mathcal{M}.$$

DES has four of these as follows, where an exponent means the repetition of that bitstring the number of times the exponent dictates.

$$k \in \{(0^{28}, 0^{28}), (1^{28}, 1^{28}), (0^{28}, 1^{28}), (1^{28}, 0^{28}) \in \mathbb{Z}^{28} \times \mathbb{Z}^{28}\}.$$

With these keys, encryption is the same function as decryption, so these keys must be avoided. There are also *semiweak keys*, which are key pairs (k_1, k_2) such that

$$E_{k_1}(E_{k_2}(m)) = m \text{ for all } \in \mathcal{M}.$$

There are six of them. They are listed as follows:

$$((01)^{14}(01)^{14}, (10)^{14}(10)^{14}),$$

$$((01)^{14}(10)^{14}, (10)^{14}(01)^{14}),$$

$$((01)^{14}(0)^{28}, (10)^{14}(0)^{28}),$$

$$((01)^{14}(1)^{28}, (10)^{14}(1)^{28}),$$

$$((0)^{28}(01)^{14}, (0)^{28}(10)^{14}),$$

and

$$((1)^{28}(01)^{14}, (1)^{28}(10)^{14}).$$

Each of these 56-bit key pairs will encipher plaintext to identical ciphertext. In other words, one key in the pair can decipher messages enciphered with the other key in the pair. Hence, these key pairs generate only two different subkeys, each of which is used eight times in the DES algorithm. They have to be avoided.

Another weakness of DES is the *complementation property*, described as follows. Let $c(k)$ denote the bitwise complementation of an input key k in DES. In other words, replace all 0's with 1's and all 1's with 0's. DES satisfies the following, which the reader may verify by trying this complementation on Diagram 3.1 on page 117.

DES Complementation Property

$$E_{c(k)}(c(m)) = c(E_k(m)).$$

In plain words, if one enciphers the complement of the plaintext with the complement of the key (the left side of the equation), then one gets the complement of the original ciphertext (the right side of the equation).

This says that complementation of the plaintext yields complementation in the ciphertext, and this means that a chosen-plaintext attack[3.4] against DES only has to test half of the keyspace of 2^{56} keys, namely, 2^{55} of them.

As mentioned in Chapter 2 (see page 98), DES reached the end of its ability to deliver as a secure cryptosystem by the end of the twentieth century, and, of course, S-DES is a a weaker version intended only to display the basic principles behind its construction. In Section 3.5, we study its successor, the Advanced Encryption Standard (AES). For now, we need to look more deeply into the design principles underlying DES since they are important from several perspectives for an understanding of symmetric-key block ciphers.

◆ Feistel Ciphers

A *Feistel cipher* is a block cipher that inputs a plaintext pair (L_0, R_0), where both halves L_0 and R_0 have bitlength $b \in \mathbb{N}$ and outputs a ciphertext pair (R_r, L_r), where R_r and L_r have bitlength $b \in \mathbb{N}$ for each $r \in \mathbb{N}$, according to an iterative process, making it what is called an *iterated block cipher*. A

[3.4] A *chosen-plaintext attack* means that a cryptanalyst chooses plaintext, is then given the corresponding ciphertext, and analyzes the data to determine the encryption key. One of the best-known chosen plaintext attacks against iterated block ciphers is *differential cryptanalysis* (DC). The original idea was developed by Murphy [175] in 1990, as an attack on another block cipher. It was improved and perfected by Biham and Shamir [23] and [24] in 1993, who used it to attack DES. DC involves the comparisons of pairs of plaintext with pairs of ciphertext, the task being to concentrate on ciphertext pairs whose plaintext pairs have certain "differences". Some of these differences have a high probability of reappearing in the ciphertext pairs. Those which do are called "characteristics", which DC uses to assign probabilities to the possible keys, with an end-goal being the location of the most probable key.

key k is input and subkeys k_j for $j = 1, 2, \ldots, r$ are generated from it via a specified key schedule. Generally, $k_j \neq k_i$ for $j \neq i$, and $k \neq k_j$ for any j. A function F, called a *round function* (iterated over r rounds, all of which have the same construction, described below), acts on plaintext pairs (R_{j-1}, k_j) for $j = 1, 2, \ldots, r$ in a prescribed fashion, in concert with a switching function. The ciphertext output is (L_j, R_j), where $L_j = R_{j-1}$ and

$$R_j = L_{j-1} \oplus f(R_{j-1}, k_j). \tag{3.3}$$

In other words, if

$$(L_0, R_0) = (R_{-1}, R_0)$$

is the initial input plaintext, then for rounds $j = 1, 2, \ldots, r+1$, $(L_{j-1}, R_{j-1}) = (R_{j-2}, R_{j-1})$ is input and

$$(L_j, R_j) = (R_{j-1}, L_{j-1} \oplus f(R_{j-1}, k_j)) \tag{3.4}$$

is output.

The methodology prescribed for each round is that a substitution is executed on the left-hand data, from the previous round, via the action in the right-hand side of (3.3), to yield

$$(L_{j-1} \oplus f(R_{j-1}, k_j), R_{j-1}). \tag{3.5}$$

This is followed by a permutation yielding (3.4), which essentially results from a swap of the two halves of the data in (3.5). This process turns out to be a configuration of a methodology called the *substitution-permutation network* (SPN) put forth by Shannon, [250], about whom we will say more below.

The above Feistel encryption is essentially the same algorithm as the deciphering scheme. To decipher, one inputs the ciphertext, with the use of the subkeys in the reverse order. Hence, we have a nice feature for implementation in that essentially the same algorithm is used for both encryption and decryption.

We now look at some design features of Feistel ciphers. We outline only the barest of statements about each principle, which we will expand in the section immediately following this list.

◆ **Feistel Design Principles**

1. **Block Size**: A large blocklength is chosen for increased security, with a 64-bit blocklength having been common, but blocklengths of 128 bits or more, becoming standard due to modern demands stemming from increased cryptanalytic developments.

2. **Keylength**: When first developed, a 64-bit keylength was used, but, as we have seen, this has not survived the cryptanalytic onslaught. Now typically 128-bit keylengths are becoming standard.

3. **Rounds and Round Functions**: More rounds mean more security, with typically sixteen rounds being most common. A round function with increased complexity adds to the security.

4. **Subkeys**: Generation of subkeys from an input key during the operation of the algorithm aids in thwarting cryptanalysis.

S-DES and DES are examples of Feistel ciphers (with the only deviation from the above being that DES and S-DES begin and end with permutations). S-DES has a round function given above with $r = 2$, and subkey generation described in the above key schedule. DES is a Feistel cipher with $r = 16$.

Now we are in a position to explain the intimate details of just how the substitutions and permutations are used in Feistel ciphers in general, and DES in particular.

◆ **Confusion and Diffusion**

DES is basically a block cipher combining fundamental cryptographic techniques, *confusion* and *diffusion*. *Confusion* obscures the relationship between the plaintext and the ciphertext, which thwarts a cryptanalyst's attempts to study the ciphertext by looking for redundancies and statistical patterns. The best way to cause confusion is through the use of a complex substitution algorithm. (Note that a simple linear substitution such as some we have studied earlier, would add negligible confusion. It is *necessary* to have a deeply complex substitution algorithm in order to cause confusion.)

Diffusion dissipates the redundancy of the plaintext by spreading it over the ciphertext, which frustrates a cryptanalyst's attempts to search for redundancies in the plaintext through observations of the ciphertext. The simplest manner in which we can cause diffusion in a binary block cipher is through repeatedly performing a permutation on the data followed by the application of a function to that permutation. This results in bits from different positions in the plaintext contributing to the same position in the ciphertext. Since DES involves an initial permutation followed by sixteen rounds of substitution, then a final permutation, DES essentially employs a sequence of confusion and diffusion techniques.

In 1949, Shannon published [250] in which the terms "confusion" and "diffusion" were introduced. His idea was to thwart frequency analysis by cryptanalysts, such as those we have studied in Chapters 1 and 2. We will learn more about Shannon later; see Section 11.1, pages 425–426.

The plaintext block size in DES of a 64-bit key input (reduced to 56-bit in the algorithm, since eight of the bits are parity check bits that are discarded) proved to be insecure for modern purposes. The new AES, which we will study in Section 3.5, has a 128-bit keylength, which is common in much of modern-day cryptosystems. (Many of us will see at the bottom of our browsers, when logging into a secure Web site such as online banking, something akin to "*connection secure — RC4: 128-bit encrypted.*" This is referring to Rivest's secure 128-bit *RC*4 cipher, a "stream" cipher, which we will study in Section 3.7.) The greater the number of rounds in a Feistel cipher, the greater the security. Today, sixteen rounds is typical. Of course, the greater the complexity of the round function, the greater the difficulty for a cryptanalyst à la Shannon [250]. In fact, Shannon laid down commandments in the 1940s for secure symmetric-key cryptosystems. He first echoed Kerchoff's Principle (see page 76). Furthermore, he stipulated that any secure cipher must include *both* confusion and diffusion techniques, as does DES, for instance.

Let us now review classical ciphers in this light. Monolaphabetic substitution ciphers fail Shannon's criterion on both counts since no confusion or diffusion exists, given that all plaintext symbols are sent to the same ciphertext symbols and there is no transposition. With polyalphabetic substitution ciphers such as Vigenère, there is the use of confusion, since plaintext letters do not go to the same ciphertext letters, but they fail at diffusion since there is no transposition. Transposition ciphers use diffusion by definition but confusion is not necessarily employed, certainly not often effectively if it is. Now, we return to to DES.

◆ Double DES

One may strengthen DES by multiple encryptions (which means the application of the encryption algorithm several times, in the same fashion as we would compose functions numerous times). For instance, there is *double DES* wherein we have two keys k_1 and k_2 so that encryption is given by

$$E_{k_2} \circ E_{k_1}(m) = E_{k_2}(E_{k_1}(m)) = c \text{ for any } m \in \mathcal{M},$$

and decrypt via

$$m = D_{k_1}(D_{k_2}(c)) = D_{k_1} \circ D_{k_2}(c).$$

On the surface, it would seem that the ostensible keylength in the double DES scheme involves $2 \times 56 = 112$ bits, which would be a significant increase in security over DES. However, reality has a way of interfering with expectations. Double DES has only a 56-bit keylength security level (which makes it only negligibly better in use than the original DES, which has 55-bit keylength security due to the complementation property described on page 127). This weakness of double DES was proved by Merkle and Hellman [161] in 1981. They show that the security is reduced from 112 bits to 56 bits by making use of the *meet-in-the-middle attack*, which we now describe in the interest of completeness. Moreover, this form of attack is closely related to another attack (called the "birthday attack", which we will study in Section 7.1).

The meet-in-the-middle attack was introduced in 1977 by Diffie and Hellman [70]. It is based upon the following simple observation. Since

$$E_{k_2}(E_{k_1}(m)) = c, \text{ then } D_{k_2}(c) = E_{k_1}(m),$$

given that $D_{k_2} \circ E_{k_2}$ is the identity function, by definition. The way the attack works is that we are given a known plaintext/ciphertext pair (m_1, c_1), and we set up a table, which we will call T_1, of (sorted) values consisting of all 2^{56} possible values of $E_{k_1}(m)$. Now we start calculating another table consisting of all possible values of $D_{k_2}(c)$, one at a time, checking each one against the values in table T_1. If there is a match, say (K_1, K_2), then we take another known plaintext/ciphertext pair (m_2, c_2), and check for the equality:

$$E_{K_1}(m_2) = D_{K_2}(c_2).$$

If so, we accept this key pair as the legitimate keys.

To see why this works, and to set up our discussion of the so-called birthday attack later, consider the following. Suppose that we have an N-element set of values and we want to find a match of two of them. We split the values into two sets of n_1 and n_2 values, say. There are $n_1 n_2$ pairs of elements and each pair has a chance of $1/N$ in matching up. Hence, the match will likely occur when $(n_1 n_2)/N$ is close to 1. Thus, if we choose

$$n_1 \approx n_2 \approx \sqrt{N},$$

we achieve maximum efficiency in this search. Now, go back to the specific situation with double DES. Since $N = 2^{112}$ and $\sqrt{N} = 2^{56}$, we see why the effective keylength security of double DES is 2^{56}. This level of multiple encryption is therefore insufficient. We need more.

At the end of the twentieth century when DES had reached the end of its reign, and before the AES came into effect, the *National Institute of Standards and Technology* (NIST)[3.5] proposed an interim standard as follows; see [94].

◆ Triple DES

Let E_e and D_d denote the DES enciphering and deciphering transformations, respectively, and let k denote a DES key. We employ three keys k_j for $j = 1, 2, 3$. Then enciphering of plaintext is achieved via

$$E_{k_3}(D_{k_2}(E_{k_1}(m))) = c,$$

and deciphering occurs via

$$D_{k_1}(E_{k_2}(D_{k_3}(c))) = m.$$

Multiple encryptions strengthen the cipher so long as we do not have $k_1 = k_2$ or $k_2 = k_3$, since then,

either $D_{k_2} \circ E_{k_1}$ or $E_{k_2} \circ D_{k_3}$ is the identity function

so we are back at square one with single DES. It is allowed that $k_1 = k_3$, or that all are distinct.

It turns out that multiple encryption of DES would be rendered useless if it were the case that for any given keys k_1 and k_2, there existed a key k_3 such that

$$E_{k_3}(m) = E_{k_2}(E_{k_1}(m))$$

for all plaintext inputs m. (This property, if it held, would be tantamount to DES permutations being closed under composition, and this would happen if DES satisfied the property that the set of permutations is closed as a group under composition.) Then multiple encryptions would be reduced to single encryptions and again we would be back to square one. However, in 1992,

---

[3.5]See the NIST homepage: *http://www.nist.gov/*.

Campbell and Weiner saved the day by proving, in [51], that DES is not a group. In fact, they showed that a lower bound on the size of the group generated by composing the set of permutations is 10^{2499}. Thus, since we are safe on these issues, then with the proper choice of three keys triple DES has the effective keylength of 168 bits, making it a reasonable alternative, and triple DES is resistant to the meet-in-the-middle attack. That being said, triple DES still inherits the disadvantages of DES, such as weak keys, semiweak keys, and the complementation property mentioned earlier. (It should be pointed out, in anticipation of the next section, that part of the ANSI X59.52 Triple DES Modes of Operation Standard, involving the CBC mode described on the next page, was cryptanalyzed in 2002 (see [22]). As a result, ANSI removed this mode from the proposed standard.)

There are other strengthenings of DES possible. Rivest developed a provably strong improvement to DES, called *DESX*. It simply does the following. Choose three keys k_1, k_2, k_3, and encipher by executing

$$k_1 \oplus E_{k_2}(k_3 \oplus m).$$

In other words, we add a 64-bit key k_3 modulo 2 to the input plaintext m before encryption, then we encipher the result with key k_2, and lastly add the 64-bit key k_1, modulo 2, to the ciphertext. In 1996, both Killian and Rogaway [136] and Rogaway [231] demonstrated the improved security of DESX over DES. The security of DESX against the DC attack (see Footnote 3.4 on page 127) is roughly equivalent to that of DES.

An attack developed more recently than DC is one by Matsui [156] in 1994, called *linear cryptanalysis* (LC). This is one of the most prominent known-plaintext attacks[3.6] against block ciphers. (See [122] for a nice tutorial treatment of both LC and DC.) LC uses linear approximations to describe the behavior of the block cipher under attack. Matsui successfully used LC against DES to obtain a key with 2^{43} known plaintexts (see [157]).

In general, block ciphers with larger S-boxes are less susceptible to DC and LC attacks. The next block cipher that we describe is therefore stronger than DES since it has larger S-boxes. First, we look at "modes of operation" for block ciphers, which allows us to apply them to a variety of situations.

[3.6] A *known-plaintext attack* occurs when a cryptanalyst has both ciphertext and plaintext from intercepted cryptograms as data from which to deduce the plaintext in general, or the key. In the case of a simple cipher such as the Caesar cipher, for instance, only *one* plaintext-ciphertext pair needs to be known to determine the key, which is instantly known to be the distance the enciphered symbol is shifted from the plaintext symbol, namely 3 units.

3.3   Modes of Operation

In architecture as in all other operative arts, the end must direct the operation. The end is to build well. Well building hath three conditions. Commodity, firmness, and delight.

Henry Wotton (1568–1639), *English poet and diplomat*
— from *Elements of architecture* (1624, page 1)

We need to examine how block ciphers, such as DES, may be applied to a variety of situations, called *modes of operation*. Symmetric-key block ciphers have five *modes of operation* recommended by NIST, and defined in FIPS 81, December 2, 1980, as well as in ANSI $X3.106 - 1983$, with the number of modes expanded from four to five in Special Publication $800 - 38A$, December 2001. These modes (initially intended for DES) are meant to address every conceivable application for cryptology to which block ciphers can be applied.

Before describing the formal details of each mode, we present a brief verbal introduction.

◆ **Block Cipher Modes — Overview**

1. **Electronic Codebook (ECB):** Each 64-bit block of plaintext is enciphered with the same key, albeit independently. This mode is typically used to send small amounts of data such as a symmetric key.

2. **Cipher Block Chaining (CBC):** The input is the addition, modulo 2, of the previous 64 bits of ciphertext with the succeeding 64 bits of plaintext. Normally, this mode is used as a general-purpose block-transport mechanism, but also may be employed for authentication purposes.

3. **Cipher Feedback Mode (CFB):** This mode employs a chaining mechanism similar to CBC. It uses prior ciphertext as input and outputs pseudorandom strings that are added, modulo 2, with plaintext to produce the next quantity of ciphertext. This mode is employed as a stream-cipher-oriented means for general-purpose messaging since it processes $n \in \mathbb{N}$ bits at a time.

4. **Output Feedback (OFB):** This is comparable to CFB mode with the exception that its input is the prior block cipher's output. This mode is usually employed for stream-cipher-oriented communications, especially those requiring message authentication, such as a MAC (see Chapter 7).

5. **Counter Mode (CTR):** The ciphertext is formed via a modulo 2 addition of a plaintext block with an enciphered counter, which is updated for each succeeding block. This mode is remarkably easy to use, and is typically utilized for high-speed transmission. In fact, this is the least-known of the modes, but is rapidly gaining ground with working cryptographers in the field as an excellent means of using block ciphers in a variety of situations.

◆ Block Cipher Modes — Details

In what follows E_k is the enciphering function for the block cipher E using the key k, whereas $D_k = E_k^{-1}$ denotes the decryption function.

◆ Electronic Code Book (ECB)

We begin with the simplest of the modes. In ECB mode, we input a sequence m_j for $j \geq 1$, of 64-bit plaintext blocks, each of which is enciphered with the same key, producing a string of ciphertext blocks c_j. In other words,

$$\text{enciphering is } E_k(m_j) = c_j \text{ and deciphering is } E_k^{-1}(c_j) = m_j.$$

The problem with this is that two identical plaintext blocks get sent to identical ciphertext blocks, which can be exploited by a cryptanalyst. Some experts feel that this weakness is sufficient to render it insecure for any use, while others feel that it is ideal for sending small amounts of data such as the sending of a DES key. It certainly should not be used for sending large amounts of data in any case. The aforementioned weakness of ECB is overcome in the next mode.

◆ Cipher Block Chaining (CBC)

In CBC mode, we first let IV be an initialization vector (meaning a 64-bit input bitstring), set $c_0 = IV$, and let k be the 64-bit input key. Given a sequence m_j of 64-bit plaintext blocks, for $j \geq 1$, we recursively define

$$\text{encryption by } c_j = E_k(c_{j-1} \oplus m_j), \text{ and decryption by } m_j = E_k^{-1}(c_j) \oplus c_{j-1}.$$

Thus, the weakness of ECB mode is eliminated by the modulo 2 addition of plaintext blocks with previous ciphertext blocks, thereby randomizing the plaintext with the previous ciphertext. Essentially, this means that we have "chained together" the sequence of enciphering plaintext blocks. This obscures the relationship between the plaintext and ciphertext, substantially reducing the data for a cryptanalyst to use effectively.

Next is the not-so-obvious problem of how to choose IV. Most texts recognize the problems with leaking information about IV, and therefore suggest keeping it as secure as the key, since a cryptanalyst can derive information from it by posing as a sender using the *man-in-the-middle attack*.[3.7] However, few cite the best solution to this problem. We should *not* have a *fixed IV* or even a randomized IV since there remains the problem (the one for which it is deemed necessary to keep IV a secret), namely, either method requires that the recipient of the message has to know this IV. In the case of a fixed IV we return to the ECB problem in encryption of the first block of each message. With the randomized IV, we require a secure randomizer at hand, for each message, which adds more effort in the use of the cipher, since as we will discover later,

---

[3.7]To describe this attack, we introduce another of our cryptographic cast of characters, *Mallory, the malicious active attacker*. (This is as opposed to Eve, our *passive* eavesdropper.) The principal idea in the man-in-the-middle attack is that Mallory assumes a position between Alice and Bob. Mallory can stop all or parts of messages being sent between them and substitute his own data. In this way, he impersonates Alice and/or Bob who believe they are communicating with each other, while they are really talking to Mallory.

obtaining secure randomizers is a difficult task. There is a better method, which essentially uses the idea behind the one-time-pad (see page 83).

First, a *nonce* is a unique number used *exactly once* in a given protocol. (This is derived from **n**umber used **once**.) As with the one-time-pad, a nonce should never be used *more* than once. In this fashion, we eliminate the need to keep the nonce secret. A *nonce-generated IV* is one where the IV is enciphered with the block cipher in CBC mode as follows.

1. Using a counter that starts at 0, assign a number to the message and use this number to generate a (unique) nonce.

2. Encipher the nonce with the block cipher, such as DES, to generate the IV.

3. Encipher the message in CBC mode using the IV.

4. Instead of sending $c_0 = IV$ as above, add the message number appended to the front of the ciphertext.

5. To ensure that there is a safeguard built in to guarantee the nonce is never accepted more than once by a recipient, the receiver will not accept messages with an assigned number less than or equal to the previously assigned message numbers.

If there were a popularity contest among the modes, CBC would probably win as the most utilized of them all. It certainly is an excellent all-purpose application for sending block data. However, others are gaining ground.

◆ Cipher Feedback Mode (CFB)

In CFB mode, again we input IV, m_j as above, and set $c_0 = IV$. Then we produce subkeys by enciphering the previous ciphertext block. In other words, for $j \geq 1$,

$$E_k(c_{j-1}) = k_j, \text{ then produce ciphertext: } c_j = m_j \oplus k_j.$$

CFB encryption is similar to CBC encryption in that the chaining mechanism causes ciphertext block c_j to depend on m_k for $k \leq j$. Moreover, the same issues with the IV remain.

◆ Output Feedback Mode (OFB)

In OFB mode, we input IV, k, m_j for $j \geq 1$ as above, and set $k_0 = IV$. Then subkeys are computed by repeatedly encrypting the initialization vector, in a mechanism described by the following.

OFB Feedback Mechanism

$$k_j = E_k(k_{j-1})$$

Then m_j is enciphered via

$$c_j = m_j \oplus k_j \text{ for } j \geq 1.$$

In ECB and OFB modes, changing one input block m_j causes *exactly one* ciphertext block c_j to be changed. This is valuable in such applications as the encryption of satellite transmissions. In CBC and CFB modes, a change to input block m_j changes $c_j, c_{j+1}, \ldots$. This turns out to be useful in applications involving message authentication. In other words, these latter two modes can be used to produce a *message authentication code* (MAC). What this means is that the MAC can be used as an *electronic signature* (or *digital signature*, which we will study in Section 4.3), that will convince the receiving party of the authenticity of the message.

In OFB mode, the block cipher is used to generate a pseudorandom stream of keys. This is an example of a *keystream* about which we will learn much more in Section 3.6, when we study *stream ciphers*. The IV has to be random, so it can either be chosen randomly or generated as a nonce as in CBC mode. Moreover, only the enciphering function is needed since enciphering is exactly the same method as deciphering. Also, since the keystream is generated in the above fashion, then there is no padding[3.8] required. In other words, one needs only send a ciphertext as long as the plaintext (and not have to pad to fill in the blocklength).

A major weakness of OFB mode is that if the same IV is ever used for two different messages, then a cryptanalyst, Eve say, can add ciphertext modulo 2 to recover plaintext. To see why, assume that c_i and c_j were enciphered using the same keystream, k_i. Then

$$c_i \oplus c_j = m_i \oplus k_i \oplus m_j \oplus k_i = m_i \oplus m_j,$$

and now Eve has a means of computing the difference between two plaintexts. This is a disaster if Eve knows one of the plaintexts already since then she readily gets the other. Moreover, even if she does not know either one, there are means of recovering both from information about the differences between them (see [130], as well as [286] for active attacks on OFB). We will return to OFB when we study stream ciphers in Section 3.6.

◆ Counter Mode (CTR)

Counter mode (CTR) has been around since 1980 or so, but was not standardized until December of 2001 by NIST, as mentioned at the outset of this section (see [73]). Thus, it has not appeared in most textbooks as a mode of operation. However, it has recently been gaining in popularity and many consider it to be the best mode. As with OFB, it is a stream cipher, the methodology for which we now describe.

A nonce n is concatenated with the counter i and enciphered to form a single block of key for $i = 1, 2, \ldots$,

$$k_i = E_k(n, i);$$

[3.8] *Padding* means appending a randomly generated bitstring of suitable length to the plaintext prior to encryption, a practice also called *salting*, since we change the "taste" of the message, so the result is called a *salt*. Moreover, the random bitstring must be independently generated for each separate encryption.

and ciphertext is obtained via

$$c_i = m_i \oplus k_i \text{ for given plaintext blocks } m_i.$$

Therefore, the counter and nonce must fit into a single block (for instance, a 128-bit block in most modern-day ciphers would not present a problem). For reasons discussed in above modes, the nonce must be used exactly once for each plaintext block encrypted.

To decipher, the same set of nonce/counter concatenated values are used as follows to recover plaintext:

For each $i = 1, 2, \ldots$, execute $k_i = E_k(n, i)$, then $m_i = c_i \oplus k_i$.

CTR does not suffer the problems cited for other modes because all the k_i are distinct since they are encipherings of a concatenation of nonce and counter, used only once. Then all plaintext m_i get enciphered via k_i to distinct ciphertext values, so two keyblocks (formed by the ciphertext values) are never the same.

CTR is an all-purpose block-oriented method that is highly useful for high speed transmissions, the reason being that the keystream can be paralleled to any desired level. The structure of CTR, moreover, ensures that its use is as secure as that of the underlying block structure.

CTR, as with OFB, does not require padding, whereas CBC does. CTR may, in fact, be considered to be a simplification of OFB, which solves one of the problems inherent in the latter. The counter replaces the feedback mechanism in OFB, discussed earlier, and this provides a formidable feature of CTR.

CTR Random Access Property

A ciphertext block c_j need not be deciphered in order to decipher c_{j+1}.

With the chaining modes such as CBC, one must decipher c_j in order to decipher c_{j+1}.

CTR, due to its high speed configurations, is used in network security applications, such as *IPSec*, or *IP security*, which we will study in detail in Section 8.3. Another palatable feature of CTR is its simple structure in that, unlike ECB and CBC, CTR requires only the implementation of the enciphering scheme, not the deciphering algorithm. For instance, if the underlying block cipher were AES, this matters a lot since the encryption and decryption transformations differ so greatly, as we will see later in the chapter (see also page 308, where we discuss the use of AES-CTR in IPSec). This simplifies matters since key scheduling for deciphering is not needed in the CTR implementation. Perhaps, from a security viewpoint, the greatest selling feature of CTR is that it is *provably* secure. For all these reasons, it appears that CTR is on its way to dominance as the mode of choice.

3.4   Blowfish

If plans related to secret operations are prematurely divulged the agent and all those to whom he spoke of them shall be put to death.

Sun Tzu (ca. 400 B.C.), *Chinese warrior and philosopher*
— from *The Art of War* ([279, page 147, no. 15])

In 1994, Bruce Schneier developed the formidable Blowfish cryptosystem. Schneier obtained his bachelor's degree in physics from the University of Rochester, and his master's degree in computer science from American University. Currently he is president of *Counterpane Systems*, a consulting firm in Minneapolis, which essentially deals with computer security, and specializes in cryptography. He is the author of several books [238]–[240], as well as numerous publications on cryptological issues. He has written a number of articles for magazines and is a contributing editor to *Dr. Dobb's Journal*, editing the *Algorithms Alley* column. Additionally, he serves on the boards of directors of the *International Association of Cryptologic Research* and the *Voter's Telcom Watch*. He is considered to be a leading influence in today's cryptographic community, with sought-after opinions on cryptological matters.

Below, we will describe the Blowfish cipher briefly without details about, for instance, the specific initialization strings used and the like. We gave an exhaustive description of S-DES and its analysis in the previous section to give us sufficient background to appreciate this symmetric-key block cipher. For Schneier's comments see, [235]–[237].

◆ The Blowfish Cipher

The Blowfish cipher encrypts 8-byte blocks of plaintext into 8-byte blocks of ciphertext. It has a key k, with keylength variable from 32 to 448 bits, namely, from one to fourteen 32-bit strings, stored in a K-array, $K_1, K_2, \ldots, K_{14}$. The key k is used to generate eighteen 32-bit subkeys (precomputed before any encryption or decryption occurs), and stored in a P-array, $P_1, P_2, \ldots, P_{18}$. There are four 8×32 S-boxes with 256 entries each, denoted by $S_{j,0}, S_{j,1}, \ldots, S_{j,255}$, for $j = 1, 2, 3, 4$. These make up the S-array. We let $E_{P,S}(m)$ denote the ciphertext that results from using Blowfish to encipher m with arrays P and S.

▼ Subkey Generation

1. Initialize P_1 and the four S-boxes with a fixed string (in a fashion that utilizes the fractional part of π.)

2. Perform $P_j \oplus K_j$ for $j = 1, 2, \ldots$, as often as needed to exhaust the P-array, reusing the elements of the K-array, if necessary. For example, if the keylength is 448 bits, then the full fourteen units of the K-array will be utilized as follows. $P_j \oplus K_j$ for $j = 1, 2, \ldots, 14$, Then reuse the first four to get, $P_{15} \oplus K_1$, $P_{16} \oplus K_2$, $P_{17} \oplus K_3$, and $P_{18} \oplus K_4$.

3. Using the subkeys in steps 1 and 2, encrypt the 64-bit block consisting of all zeros.

4. Replace P_1 and P_2 with the output of step 3, namely, $E_{P,S}(\{0\}^{64}) = P_1 P_2$.

5. Encrypt the output of step 3 with the modified subkeys, namely, the current P and S arrays.

6. Replace P_3 and P_4 with the output of step 5, namely, $E_{P,S}(P_1 P_2) = P_3 P_4$.

7. Continue the above process until all entries of the P-array have been replaced, namely, $E_{P,S}(P_{j-1} S_j) = P_{j+1} P_{j+2}$ for $j = 4, 5, \ldots 16$. Then replace all four S-boxes in order, starting with $E_{P,S}(P_{17} P_{18}) = S_{1,0} S_{1,1}$, $E_{P,S}(S_{1,0} S_{1,1}) = S_{1,2} S_{1,3}$, and continue until $E_{P,S}(S_{4,252} S_{4,253}) = S_{4,254} S_{4,255}$.

In total, there are 521 iterations required to generate all the subkeys. Hence, to test for a single key there would be a total of 522 executions of the encryption algorithm to test for a single key, making a brute-force attack much more difficult. In fact, with the use of a 448-bit keylength, the cipher is virtually unbreakable in the face of brute-force attacks.

Next, as with DES, there is a complicated function to iterate over sixteen rounds.

▼ **Round Function**

The round function F takes a 32-bit input m that is divided into 4 bytes, which we will label a, b, c, d. Then F acts on them as follows:

$$F(m) = ((S_{1,a} + S_{2,b} \,(\mathrm{mod}\ 2^{32})) \oplus S_{3,c}) + S_{4,d} \,(\mathrm{mod}\ 2^{32}).$$

▼ **Encryption and Decryption**

To encipher, we first separate the 64-bit plaintext into 32-bit left and right blocks $L_0^{(e)}$ and $R_0^{(e)}$, respectively. Let $R_j^{(e)}$ and $L_j^{(e)}$ be the right and left halves after round j and execute, for $j = 1, 2, \ldots, 16$,

1. $R_j^{(e)} = L_{j-1}^{(e)} \oplus P_j$.

2. $L_j^{(e)} = F(R_j^{(e)}) \oplus R_{j-1}^{(e)}$.

3. $\mathbf{SW}(R_j^{(e)}, L_j^{(e)})$.

After the sixteenth round is completed, perform a switch on $R_{16}^{(e)}$ and $L_{16}^{(e)}$ to undo the last swap. Then execute, $L_{17}^{(e)} = R_{16}^{(e)} \oplus P_{18}$ and $R_{17}^{(e)} = L_{16}^{(e)} \oplus P_{17}$.

To decipher, we do the same as we did for enciphering, with the exception that the P_j are used in reverse order. In other words, to decipher, execute the following (where $R_j^{(d)}$ and $L_j^{(d)}$ are the right and left halves after round j), for $j = 1, 2, \ldots, 16$,

1. $R_j^{(d)} = L_{j-1}^{(d)} \oplus P_{19-j}$.

2. $L_j^{(d)} = F(R_j^{(d)}) \oplus R_j^{(d)}$.

3. $\mathbf{SW}(R_j^{(e)}, L_j^{(e)})$.

Lastly, unswap the last pair, then execute, $L_{17}^{(d)} = R_{16}^{(d)} \oplus P_1$ and $R_{17}^{(d)} = L_{16}^{(d)} \oplus P_2$. Diagrams 3.4 and 3.5 illustrate Blowfish encryption/decryption.

Diagram 3.4 Blowfish Encryption

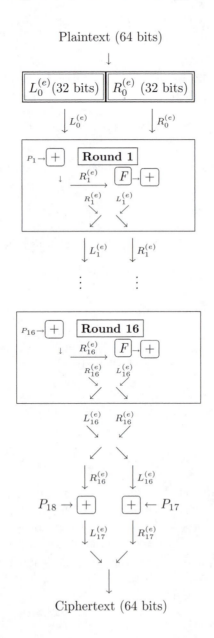

Plaintext (64 bits)

Ciphertext (64 bits)

Diagram 3.5 Blowfish Decryption

Ciphertext (64 bits)

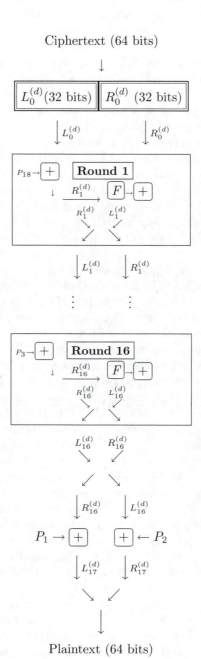

Plaintext (64 bits)

▼ Analysis and Summary

Blowfish differs in its decryption from most block ciphers, such as DES, in that it does not reverse the order of encryption but follows it. The subkeys are, as is usual with most block ciphers, used in reverse order. Also, unlike DES, the S-boxes in Blowfish are key-dependent, and subkeys as well as S-boxes are created by repeated execution of the Blowfish enciphering transformation itself. This has proven to be a strength against cryptanalysis. In comparison with how most Feistel ciphers work on only *one* half of the blocks in each round, Blowfish works on *both* halves in each round, again adding to strength against cryptanalysis. This would explain why the cipher is being used in over 150 products thus far. (This is according to Schneier's Web site *http://www.schneier.com/blowfish.html*. The reader may also download a free source copy of Blowfish from the site, since it is unpatented, royalty-free, and no license is required, all of which were part of Schneier's intentions in the design of Blowfish.) Additionally, Blowfish is extremely fast to execute, much faster than DES, or triple DES. (To see timings and verify this, see Schneier's Web site referenced above.)

There is a notion shared by DES and other block ciphers such as Blowfish, called the *avalanche effect*, which means that the change of a single input bit amplifies into a change in about half the bits of ciphertext. The way this works in a block cipher with many rounds is that a change in a single bit of input generally results in many bit changes after one round, even more bit changes after another round, until, eventually, about half of the block will change. The analogy from which the name is derived is to an avalanche involving snow, where a tiny preliminary snowslide can result in a dramatic deluge of snow. Perhaps Feistel said it best in the 1973 *Scientific American* article [81, page 22]: "As the input moves through successive layers the pattern of 1's generated is amplified and results in an unpredictable avalanche. In the end the final output will have, on average, half 0's and half 1's...". Blowfish has a very strong avalanche effect since, in the jth round, every bit of the left side of the data affects every bit of the right, and every subkey bit is affected by every key bit. The result is that the function F has the best possible avalanche effect between the key P_j and the right half of the data after each round. Moreover, Schneier deliberately made F independent of the rounds since the P-array already is round-dependent.

Since the Blowfish S-boxes are key-dependent, every bit of the input to F is used as input to only one S-box. DES, on the other hand, has several inputs to two S-boxes, but the DES S-boxes are not key-dependent.

From Blowfish evolved a cipher designed by Schneier, and a team of others. They called it *Twofish*. It became one of the five finalists for the successor to AES, announced on August 9, 1999 by NIST (in round two of their competition). In [85], Schneier and Ferguson (a member of the Twofish design team) make arguments for the advantages of Twofish. Moreover, the entire team wrote a book on the cipher [241] to which the reader is referred for details. Yet, it was not chosen as the AES. That distinction went to a *non*-Feistel cipher, which we will present in the next section.

3.5 ☞ The Advanced Encryption Standard

Not to go back, is somewhat to advance,
And men must walk at least before they dance.

Alexander Pope (1688–1744), English poet
— from *Imitations of Horace* (1738, Bk. 1, Epistle 1)[3.9]

On November 26, 2001, NIST announced, in FIPS 197 (see [93]), that the *Advanced Encryption Standard* (AES) would be *Rijndael*, and that it would take effect on May 26, 2002. The name "Rijndael" (just call it "Rain Doll") was derived from the names of Rijndael's Belgian designers, Vincent **Rij**men and Joan **Dae**men.

Joan Daemen was born in Belgium in 1965. In 1988 as a member of the research group **CO**mputer **S**ecurity and **I**ndustrial **C**ryptography (COSIC), he began a Ph.D. in cryptography, which he completed in 1995. By the spring of 1998, he joined the newly formed *Proton World International*, a Brussels-based company whose focus is on high-level banking security applications, and he remains there to this day. He is currently designing protocols for smart cards, and related applications. Moreover, he continues, on occasion, to collaborate with his former COSIC colleague, Vincent Rijmen.

Vincent Rijmen was born in Belgium in 1970. After obtaining a degree in electrical engineering at the Katholieke Universiteit Leuven, he joined COSIC, where he, too, was working on his Ph.D., which he obtained in 1997. His preferred area of research has always been cryptanalysis of block ciphers. In fact, the title of his Ph.D. thesis is *Cryptanalysis and design of iterated block ciphers*. Among his pursuits has been the evolution of computer security systems.

The Rijndael cipher is based upon the 128-bit block cipher, called *Square*, which Rijmen and Daemen originally designed with a concentration on resistance against LC (see page 132). Later Lars Knudsen engaged in more cryptanalysis of the Square cipher. A paper by these three authors, describing the details of Square, was presented at the workshop for *Fast Software Encryption* in the spring of 1997 in Haifa, Israel. (Consequently, Rijndael has been called *Son of Square* and alternatively Square has been called *Mother of Rijndael* by their creators.) In that spring of 1997, Daemen and Rijmen began working on a variant of the Square cipher that would allow for key and block lengths of 128, 192, and 256 bits. They called their new cipher design "Rijndael" and submitted it to NIST by the June 1998 deadline. The rest, as noted above, is history.

The first item of importance is that Rijndael is *not* a Feistel cipher. Yet, the reader will recognize similarities to DES. We will see that the modulo 2 additions, $\oplus$, will add key material to the data. As with DES, the S-boxes will add nonlinearity. However, the S-boxes were designed so that the complementation property suffered by DES is avoided (see page 127).

We begin by providing a preliminary verbal introduction to the AES cipher, before presenting the details of the Rijndael mechanisms in action.

[3.9]The symbol ☞ will denote advanced material henceforth.

◆ AES — Preliminary Overview

1. **Non-Feistel Structure**: As noted in the leadup to this overview, AES is not of a Feistel construction. Instead, the entire data block is processed in parallel, during each round, using a combination of substitution and permutation.

2. **Keys**: The input key, which may be variable in length as we shall see, will be assumed, for the purposes of this introductory discussion, to be of keylength 128 bits. This key is expanded into a matrix of forty-four 4-byte words, wherein four distinct words play the role of the round key for the succeeding round.

3. **Rounds**: For both encryption and decryption (see Diagrams 3.6 and 3.7 on pages 148 and 149, with **Nr** = 10), the AES cipher begins with an add round key stage, followed by nine rounds, each round having four stages, which in addition to the add round key stage, are called, bytesub, shift rows, and mix columns (all described in the detailed delineation of AES below). This is followed by a tenth round having three stages (with the mix columns eliminated for this round, since its inclusion would unnecessarily slow the algorithm).

4. **Round Stages**: Only the add round key stage uses the key. The other three stages provide confusion (bytesub), diffusion (mix columns), and nonlinearity (S-boxes). These three stages do not add security by themselves since they do not use the key. Moreover all stages are reversible.

5. **Decryption**: The decryption uses the expanded key in reverse order. However, the decryption algorithm is not the same as the encryption algorithm. One needs the inverse lookup table of the S-box, and the inverse mix columns, which is distinct from the enciphering mix column operation. It is this fact that causes decryption to be slower than encryption, namely, the inverse mix columns operation is a more complex operation that can take a third longer than encryption on 8-bit processors. Yet, this is not seen as a disadvantage since many implementations do not need deciphering, such as CFB mode (see page 135).

6. **S-boxes**: The S-boxes are all identical, and map bytes to bytes. The AES S-box was designed to be highly resistant to cryptanalytic attacks. In particular, the designers ensured that there is a low correlation between the input bits and the output bits. This is the reason, cited on page 143, that AES does not have the DES complementation property. The S-boxes are also invertible, but not self-inverses.

7. **Last Add Round Key Stage**: Since only the add round key stage uses the key, the cipher begins and ends with this stage. By itself, the add round key stage (a virtual one-time-pad) will not add enough security, but its interaction with the other three stages provides a highly efficient and secure cryptosystem.

◆ AES — Detailed Description

Much of the following description is taken from this author's book [169], since we maintain that it remains the best explanation at this level. The mathematics required for this section is contained in Appendix A.

◆ The Advanced Encryption Standard (AES) — Rijndael

In order to give even a brief description of Rijndael, we need to describe the essential components of it.

▼ The State

The *state* is the intermediate cipher resulting from application of the round function. The state can be depicted as a $4 \times \mathbf{Nb}$ matrix, with bytes as entries, where $\mathbf{Nb}$ is the block length divided by 32. For instance, if the input block has 256 bits, then $\mathbf{Nb} = 8 = 256/32$, and the state would appear as a matrix

$$(a_{i,j}) \in \mathcal{M}_{4 \times 8}((\mathbb{Z}/2\mathbb{Z})^8)$$

of bytes. In this case, the state has 32 bytes. For an input block of 192 bits, the state would have 24 bytes as a $4 \times \mathbf{Nb} = 4 \times 6$ matrix, and for a block of length 128, it would have 16 bytes as a $4 \times \mathbf{Nb} = 4 \times 4$ matrix. Thus, we have variable state size.

Note that the input block (or *plaintext* if the mode of operation is ECB) is put into the state (matrix) by column: $a_{0,0}, a_{1,0}, a_{2,0}, a_{3,0}, a_{0,1}, a_{1,1} \ldots$, and at the end of the execution of the cipher the bytes are taken from the state in the same order.

▼ The Cipher Key

As with the state, the *cipher key* is portrayed as a $4 \times \mathbf{Nk}$ matrix of bytes, where $\mathbf{Nk}$ is the keylength divided by 32. For instance, if the key length is 128 bits, then the cipher key is $(k_{i,j}) \in \mathcal{M}_{4 \times 4}((\mathbb{Z}/2\mathbb{Z})^8)$. Hence, we have variable key size 16, 24, or 32 bytes, depending on key length 128, 192, or 256 bits.

▼ Key Schedule and Round Keys

The *round keys* can be derived from the cipher key by means of the following *key schedule*. There are two parts.

(1) The total number of round key bits equals $B \cdot (\mathbf{Nr}+1)$, where B is the block length and $\mathbf{Nr}$ is the number of rounds defined for each case in Table 3.1 on page 146. For instance, if the block length is 128 bits and $\mathbf{Nr} = \mathbf{12}$, then 1664 round key bits are required.

(2) The cipher key is expanded into the *expanded key* in the following fashion. The expanded key is a linear array of 4-byte *words* (i.e. columns of the key matrix), where the first $\mathbf{Nk}$ words contain the cipher key. All other words are defined recursively in terms of previously defined words.

▼ The AES S-Box

For the sake of convenience, ease of presentation, and due to the highly technical nature of the S-box in AES, the details are in Appendix D on page 527.

▼ Round Function

Most block ciphers employ the Feistel structure in the round function. However, the round function used by *Rijndael* does *not* have the Feistel structure. Instead, the round function in Rijndael is comprised of three distinct invertible functions, the details of which we will learn in what follows.

First, we note that the *number of rounds*, denoted by **Nr**, is defined via Table 3.1.

	Nr	**Nb = 4**	**Nb = 6**	**Nb = 8**
Table 3.1	**Nk = 4**	10	12	14
	Nk = 6	12	12	14
	Nk = 8	14	14	14

In Table 3.1, we are including the final round, (described below), which slightly differs from the other rounds in that step (3) below is eliminated.

The round function consists of four steps, each with its own name and its own particular function.

(1) **Bytesub (BSB)**: In this step, bytes are mapped by an invertible S-box, and there is only one single S-box for the complete cipher. Thus, for instance, the state (position) matrix,

$$(a_{i,j}) = (8i + j - 9) \text{ (for } 1 \leq i \leq 32, \, 1 \leq j \leq 8)$$

would be mapped, elementwise, by the S-box to the state matrix $(b_{i,j})$ via

$$a_{i,j} \longrightarrow \boxed{\text{S-box}} \longrightarrow b_{i,j}.$$

This guarantees a high degree of nonlinearity by operating on each of the state bytes $a_{i,j}$ independently.

(2) **Shift Row (SR)**: In this step, depending upon the value of **Nb**, row j for $j = 2, 3, 4$ of the state matrix is shifted x_j units to the right, where x_j is defined by Table 3.2.

	Nb	x_2	x_3	x_4
Table 3.2	4	1	2	3
	6	1	2	3
	8	1	3	4

For instance, if **Nb** $= 4$, then

$$
\begin{pmatrix}
a_{0,0} & a_{0,1} & a_{0,2} & a_{0,3} \\
a_{1,0} & a_{1,1} & a_{1,2} & a_{1,3} \\
a_{2,0} & a_{2,1} & a_{2,2} & a_{2,3} \\
a_{3,0} & a_{3,1} & a_{3,2} & a_{3,3}
\end{pmatrix}
\xrightarrow{\boxed{\text{SR}}}
\begin{pmatrix}
a_{0,0} & a_{0,1} & a_{0,2} & a_{0,3} \\
a_{1,3} & a_{1,0} & a_{1,1} & a_{1,2} \\
a_{2,2} & a_{2,3} & a_{2,0} & a_{2,1} \\
a_{3,1} & a_{3,2} & a_{3,3} & a_{3,0}
\end{pmatrix}
$$

The **SR** step introduces high diffusion over multiple rounds and interacts with the next step.

(3) Mix Column (MC): As with the S-box description, the MC description is given in Appendix D on page 529 for the sake of simplicity of presentation. All one needs to know about this step, at this juncture, is that it linearly combines bytes in the columns, and creates high intracolumn diffusion.

(4) Round Key Addition (RKA): In this step, a round key is added modulo 2 to the state. For example,

$$(a_{i,j}) \oplus (k_{i,j}) = (b_{i,j}),$$

where $\oplus$ is addition modulo 2, $(a_{i,j})$ is the state matrix, $(k_{i,j})$ is the round key matrix, and $(b_{i,j})$ is the resulting state matrix. Thus, this step makes the round function key dependent.

There is significant parallelism in the round function. All four steps of a given round operate in parallel on bytes, rows, or columns of the state.

Then round keys are extracted from the expanded key as follows. The first round key consists of the first **Nb** words, the second round key consists of the following **Nb** words, and so on.

◆ **Stepwise Description of the Rijndael Cipher**

Step 1 (Initial Addition Round) There is an initial RKA step.

Step 2 (Rounds) There are **Nr** $- 1$ rounds executed.

Step 3 (Final Round) A final round is executed (omitting the **MC** step).

Hence, the detailed sequence of steps for Rijndael is an initial round key addition, then **Nr** $- 1$ rounds of **BSB, SR, MC, RKA** each, followed by a final round consisting of **BSB, SR, RKA**. Unlike DES, Rijndael does not require a "swapping step" in its rounds since the **MC** step causes every byte in a column to alter every other byte in the column.

Deciphering Rijndael is executed by reversing the steps using inverses and a modified key schedule. Encryption and decryption diagrams are given in Diagrams 3.6 and 3.7.

Diagram 3.6 AES Encryption

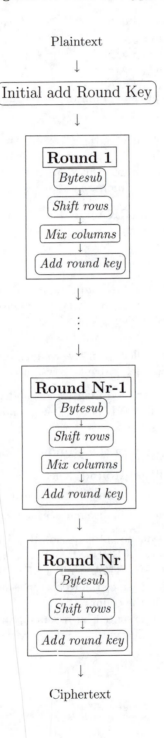

Diagram 3.7 AES Decryption

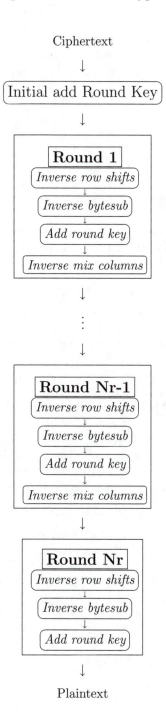

Ciphertext

↓

Initial add Round Key

↓

Round 1
Inverse row shifts
Inverse bytesub
Add round key
Inverse mix columns

⋮

Round Nr-1
Inverse row shifts
Inverse bytesub
Add round key
Inverse mix columns

Round Nr
Inverse row shifts
Inverse bytesub
Add round key

Plaintext

▼ Security of Rijndael

The design of Rijndael practically eliminates the possibility of weak or semi-weak keys, which exist for DES. Moreover, the design of the key schedule virtually eliminates the possibility of equivalent keys. Although the mechanisms of LC and DC can be adjusted to present attacks on Rijndael, it appears that Rijndael's design is sufficient to withstand these cryptanalytic onslaughts, since its S-box is nearly perfect for resistance to DC and the $\mathbb{F}_{2^8}$ equivalent of LC.

A chosen plaintext attack, called the *square attack*, which is a dedicated attack on the Square cipher, can be used as well, since Rijndael inherited many features from Square. However, for seven or more rounds in Rijndael, no such attack, faster than exhaustive key search,[3.10] has been found. Other attacks, such as Biham's *related-key attack*, or the *interpolation attacks* introduced by Jakobsen and Knudsen have little chance of success against Rijndael due to the diffusion and nonlinearity of Rijndael's key schedule and the complicated construction of the S-box.

The S-box was designed to avoid any suspicions of a trapdoor being built into the cryptosystem. (Recall that this was a problem with DES; see page 98).

▼ Concluding Comments

Unlike the Feistel structure of the round function, such as in DES, where some of the bits of the intermediate state are simply put into a different position unchanged, the Rijndael round function is comprised of three different invertible transformations, called *layers*, through which every bit of the state is treated in a similar fashion, called *uniformity*. The **BSB** step in each round is a *nonlinear mixing layer* (confusion). **SR** is a *linear mixing layer* (intercolumn diffusion), and **MC** is also a linear mixing layer (interbyte diffusion within columns). Then there is the *key addition layer*. These layers ensure that the Rijndael round does *not* have a Feistel structure. The layers are predominantly based upon the application of what the designers call the *Wide Trail Strategy*, which is a devised system for providing resistance against LC and DC, discussed in Daemen's doctoral dissertation of March 1995. Essentially this strategy means that **MC** makes it impossible to find LC and DC attacks that involve "few" active S-boxes.

For further information on Rijndael, such as attacks on reduced rounds and alternative mathematical methods for describing AES, see [83] and [84]. Also, for further, relatively recent research on security of AES against LC, see [132].

Rijndael is well tailored to modern processors (Pentium, RISC, and parallel processors). It is also ideally suited for ATM, HDTV, Voice, and Satellite. Uses for Rijndael include MAC by employing it in a CBC-MAC algorithm. It is also possible to use it as a *synchronous stream cipher*, a *pseudorandom number generator*, or a *self-synchronizing stream cipher* (the latter, by using it in CFB mode), and we will learn about all of these concepts in the next section.

[3.10]An *exhaustive search of the keyspace* or *brute force attack*, means that all possible keys are tried to see which one is being used by communicating parties.

3.6 Stream Ciphers

Some people seem to think that stream ciphers are bad in some way. Not at all! Stream ciphers are extremely useful, and do their work very well.

Neils Ferguson and Bruce Schneier
— from *Practical Cryptography* (see [85, page 73])

Up to this juncture, we have studied the first kind of symmetric-key cryptosystem, the *block cipher*. This section is devoted to the second kind, *stream ciphers*. First, we need the following.

◆ **Keystreams, Seeds, and Generators**

If $\mathcal{K}$ is the keyspace for a set of enciphering transformations, then a sequence $k_1 k_2 \cdots \in \mathcal{K}$ is called a *keystream*. A keystream is either randomly chosen, or is generated by an algorithm, called a *keystream generator*, which generates the keystream from an initial small input keystream called a *seed*. Keystream generators that eventually repeat their output are called *periodic*.

Finding sources of truly random numbers is a difficult task at best. In fact, it is impossible to generate arbitrarily long bitstrings and prove they are random. Hence, we settle for what computers can give us. Pseudorandom (recall the discussion on page 83) number generators (PRNG)s are a topic for an entire text. However, we will not be concerned with the intricacies of such investigations. We will assume that we have at our disposal a *cryptographically secure pseudorandom number generator* (CSPRNG) ; see Appendix B on page 506.

◆ **Cryptographically Secure Pseudorandom Number Generators**

A bit-producing number generator algorithm is a CSPRNG if the sequences of bits produced satisfy the following properties.

1. The sequence of bits must be statistically random. One way of stating this mathematically is that no polynomial-time algorithm can distinguish the output of this number generator from that of a truly random number generator with probability greater than $1/2$. This makes the number generator a PRNG.

2. For every given output bit, the next output bit must be computationally infeasible (see page 99) to predict, even given knowledge of all previous bits, knowledge of the algorithm being used, and knowledge of the hardware. This is the property that make the PRNG *cryptographically secure*, so this turns it into a CSPRNG.

With *truly random* sequences of numbers, each number is statistically independent of other numbers in the sequence, so they are unpredictable. However, with PRNGs, care must be taken to ensure that they are *cryptographically secure*, namely, that they are unpredictable in the sense of property 2 above. We will assume henceforth, that this has been done. Hence all of our keystreams will be assumed to be generated by keystream generators that are CSPRNGs.

Now we are ready to look at the topic of this section. Before the formal definition, think of a stream cipher as a cryptosystem where plaintext messages are encrypted character by character, just before sending the cryptogram.

◆ Stream Ciphers

Let $\mathcal{K}$ be a keyspace for a cryptosystem and let $k_1 k_2 \cdots \in \mathcal{K}$ be a keystream. A cryptosystem is called a *stream cipher* if encryption upon plaintext strings $m_1 m_2 \cdots$ is achieved by repeated application of the enciphering transformation for $j \geq 1$,

$$E_{k_j}(m_j) = c_j,$$

and deciphering occurs as repeated application of the deciphering transformation for $j \geq 1$,

$$D_{k_j^{-1}}(c_j) = m_j.$$

If there exists an $\ell \in \mathbb{N}$ such that $k_{j+\ell} = k_j$ for all $j \in \mathbb{N}$, then we say that the stream cipher is *periodic* with *period* ℓ.

Generally speaking, stream ciphers are faster than block ciphers, and are easier to describe, since stream ciphers encrypt individual plaintext message units, usually one bit at a time.

All keystream generators are periodic except for *one-time pads*. This is the cryptosystem where the (randomly generated) key (used only once) is the size of the plaintext. (Recall the discussion on pages 83 and 84).

◆ The One-Time-Pad

The one-time-pad is a stream cipher with alphabet of definition $\mathcal{A} = \{0, 1\}$ that enciphers in the following fashion.

Given a bitstring $m_1 m_2 \cdots m_\ell \in \mathcal{M}$, and a keystream $k_1 k_2 \cdots k_\ell \in \mathcal{K}$,

the enciphering transformation is given by

$$E_{k_j}(m_j) = m_j \oplus k_j = c_j \in \mathcal{C},$$

and the deciphering transformation is given by

$$D_{k_j}(c_j) = c_j \oplus k_j = m_j,$$

for all $j = 1, 2, \ldots, \ell \in \mathbb{N}$. The keystream is randomly chosen and never used again.

The one-time-pad is a stream cipher wherein the keystream does not repeat, so it is not periodic. Moreover, it is an example of another notion we encountered on page 83, namely, that of a running-key, a notion shared by a cipher we described on page 56, namely, the Vigènere cipher (translated into arithmetic modulo n). Recall that this is the cryptosystem where there is a keyphrase used as a priming key with which to encipher the plaintext. Formally, this is given as follows.

◆ Vigenère Ciphers

Fix $r, n \in \mathbb{N}$, and let $\mathcal{M} = \mathcal{C} = (\mathbb{Z}/n\mathbb{Z})^s$, the elements of which are ordered s-tuples from $\mathbb{Z}/n\mathbb{Z}$, and $\mathcal{K} = (\mathbb{Z}/n\mathbb{Z})^r$ where $s \geq r$. For $e = (e_1, e_2, \ldots, e_r) \in \mathcal{K}$, and $m = (m_1, m_2, \ldots, m_s) \in \mathcal{M}$, let

$$E_{e_j}(m_j) = m_j + e_{j \,(\mathrm{mod}\ r)} \,(\mathrm{mod}\ n) \text{ for all } j = 1, 2, \ldots, s,$$

and for $c = (c_1, c_2, \ldots, c_s) \in \mathcal{C}$, let

$$D_{d_j}(c_j) = c_j - e_{j \,(\mathrm{mod}\ r)} \,(\mathrm{mod}\ n) \text{ for all } j = 1, 2, \ldots, s.$$

This cryptosystem is called the *Vigenère cipher* with period r, which is why the subscript on the key is taken modulo r (where we choose r rather than 0 in order to keep all subscripts positive). If $r = s$, then this cipher is an example of a *running-key cipher*.

Thus, the one-time-pad is an simple example of a running-key cipher. (Note that this also says that the Vignère cipher becomes a Vernam cipher if we assume that the keystream is truly randomly generated and never repeated.) The Vigenère cipher is an example of a stream cipher with period length r, where the key $e = (e_1, e_2, \ldots, e_r)$ provides the first r elements of the keystream $k_j = e_j$ for $1 \leq j \leq r$, after which the keystream repeats itself, until the plaintext is exhausted. Now, we turn to a general discussion of stream ciphers, but will revisit our two examples later. Typically, stream ciphers are classified as follows.

◆ Synchronous and Asynchronous Ciphers

A stream cipher is said to be *synchronous* if the keystream is generated without use of the plaintext or of the ciphertext. This is called keystream generation *independent* of the plaintext and ciphertext. A stream cipher is called *self-synchronizing* (or *asynchronous*) if the keystream is generated as a function of the key and a fixed number of previous ciphertext units. If the stream cipher utilizes plaintext in the keystream generation, then it is called *nonsynchronous*.

The distinctions between block and stream ciphers are more readily seen in practice than in theory. Stream ciphers encrypting one bit at a time are not suitable for software implementation since bit manipulation is time-consuming. Where stream ciphers win out is in the arena of error propagation. Obviously, with a block cipher, a single error will corrupt at least a block's worth of data, whereas implementation of a synchronous stream cipher can guarantee that a single bit error will result in only a single bit of corrupted plaintext. Thus, synchronous stream ciphers would be useful where lack of error propagation is critical. However, use of self-synchronizing stream ciphers can result in error propagation. If the keystream is acting on the nth ciphertext digit and an error occurs, then the deciphering of up to n subsequent ciphertext digits may be incorrect.

An example of a synchronous stream cipher is DES operating in OFB mode, whereas an example of an asynchronous stream cipher is DES in CFB mode. An example of a nonsynchronous cipher is given by reinterpreting an idea of Vigenère.

◆ The Autokey Vigenère Cipher

Let $n = |\mathcal{A}|$ where $\mathcal{A}$ is the alphabet of definition. We call $k_1 k_2 \cdots k_r$ for $1 \leq r \leq n$ a *priming key*. Then given a plaintext message unit $m = (m_1, m_2, \ldots, m_s)$ where $s \geq r$, we generate a keystream as follows:

$$k = k_1 k_2 \cdots k_r m_1 m_2 \cdots m_{s-r}.$$

Then we encipher via

$$E_{k_j}(m_j) = m_j + k_j \pmod{n} = c_j \text{ for } j = 1, 2, \ldots, r,$$

and

$$E_{k_j}(m_j) = m_j + m_{j-r} \pmod{n} = c_j \text{ for } j > r,$$

and decipher via

$$D_{k_j}(c_j) = c_j - k_j \pmod{n} = m_j \text{ for } j = 1, 2, \ldots, r,$$

and

$$D_{k_j}(c_j) = c_j - m_{j-r} \pmod{n} = m_j \text{ for } j > r.$$

This cipher is nonsynchronous since the plaintext serves as the key from the $(r+1)$th position onward, with the simplest case being $r = 1$. Here is a simple example.

Example 3.5 *Given a priming key $k = k_1 k_2 = 72$ and $n = 26$ in the autokey Vigenère cipher, suppose we want to decrypt the Vigenère ciphertext*

LPXEHGM,

using Table 1.3 on page 11. Converting ciphertext to numerical equivalents, we have

$$11, 15, 23, 4, 7, 6, 12.$$

Thus, we compute the following:

$$m_1 = c_1 - k_1 = 11 - 7 = 4, \quad m_2 = c_2 - k_2 = 15 - 2 = 13,$$

$$m_3 = c_3 - m_1 = 23 - 4 = 19, \quad m_4 = c_4 - m_2 = 4 - 13 \equiv 17 \pmod{26},$$

$$m_5 = c_5 - m_3 = 7 - 19 \equiv 14 \pmod{26}, \quad m_6 = c_6 - m_4 = 6 - 17 \equiv 15 \pmod{26},$$

and

$$m_7 = c_7 - m_5 = 12 - 14 \equiv 24 \pmod{26}.$$

Via Table 1.3, the letter equivalents give us

ENTROPY

This now gives us the opportunity to see the formulation of the notion of an autokey cipher that we first discussed when we met Cardano on page 55.

◆ Autokey Ciphers

An *autokey cipher* is a cryptosystem wherein the plaintext itself (in whole or in part) serves as the key (usually after the use of an initial priming key).

As is the case with the autokey Vigenère cipher, the plaintext is introduced into the key generation after the priming key has been exhausted.

Perhaps the most common of the stream ciphers is the following type.

◆ Binary Additive Stream Ciphers

A *binary additive stream cipher* is a synchronous stream cipher for which all of the digits in the ciphertext, keystream, and plaintext are binary, and output is achieved by addition modulo 2.

Many, if not most, keystream generators have the following as their basic component. We will discuss the uses for the following after we have described and illustrated the concept. The following notion is a mechanism for sender and receiver to agree upon an easy way to generate long bitstrings.

◆ Linear Feedback Shift Registers

A *linear feedback shift register* (LFSR) is a mechanism for providing fast number generation, but is not cryptographically secure. We provide a brief description here for completeness and to illustrate why it is both fast and insecure, yet is a building block for more secure schemes. An LFSR is comprised of three parts.

1. A *shift register* of length $\ell \in \mathbb{N}$, consists of a sequence of ℓ registers (memory cells) labelled $0, 1, 2, \ldots, \ell - 1$, each capable of holding one bit and each having one input and one output.

2. A *tap sequence* is an ℓ-tuple of bits:

$$(c_1, c_2, \ldots, c_\ell),$$

with $c_\ell = 1$.

3. A *state* s_j, of the LFSR is the bitstring describing the contents of the registers for states numbered, $j \in \{0, 1, \ldots, \ell - 1\}$, given by

$$s_j = (k_{(\ell-1,j)} k_{(\ell-2,j)} \cdots k_{(0,j)}),$$

namely, register i has bit $k_{(i,j)}$ in state j, so the *first coordinate* of the subscript denotes the *register*, and the *second coordinate* determines the *state* of the bit. For instance, the *initial state* is given by the bitstring,

$$s_0 = (k_{(\ell-1,0)} k_{(\ell-2,0)} \cdots k_{(0,0)}),$$

called the *seed*.

▼ LFSR Operation

0. Input the seed, and set $j = 1$.

1. The bit, $k_{(0,j-1)}$, in register 0 is *output* as the next bit in the keystream. In other words, $k_{(0,j-1)}$ is *tapped* as the next keystream bit, and becomes part of the output sequence, and we store it with the label K_{j-1}.

2. The bit in register i, for $i = 0, 1, \ldots, \ell - 2$, is shifted one register to the right, namely, the contents of register i becomes $k_{(i-1,j-1)}$.

3. Register $\ell - 1$ is given as the following input,

$$k_{(\ell-1,j)} = c_1 k_{(\ell-1,j-1)} \oplus c_2 k_{(\ell-2,j-1)} \oplus \cdots \oplus c_\ell k_{(0,j-1)}. \qquad (3.6)$$

   This step is called the *linear feedback*.

4. If $s_0 \neq s_i$, set $i = i+1$ and go to step 1. Otherwise, set $i = L$, and terminate the algorithm, with output keystream given by

$$k = (K_{L-1} K_{L-2} \ldots K_0),$$

   which is said to have period length L.

Diagram 3.8 shows the result of the first bit iteration.

Diagram 3.8 A Linear Feedback Shift Register

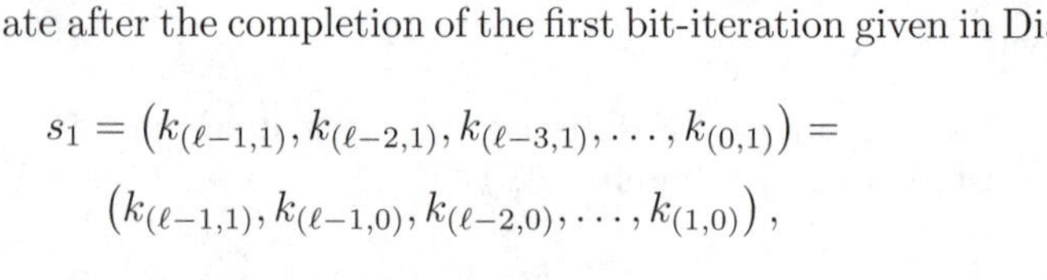

Thus, the state after the completion of the first bit-iteration given in Diagram 3.8 is

$$s_1 = \left(k_{(\ell-1,1)}, k_{(\ell-2,1)}, k_{(\ell-3,1)}, \ldots, k_{(0,1)} \right) =$$
$$\left(k_{(\ell-1,1)}, k_{(\ell-1,0)}, k_{(\ell-2,0)}, \ldots, k_{(1,0)} \right),$$

where,

$$k_{(\ell-1,1)} = c_1 k_{(\ell-1,0)} \oplus c_2 k_{(\ell-2,0)} \oplus \cdots \oplus c_\ell k_{(0,0)}.$$

A very simple illustrating instance of the LFSR is given in the following.

Example 3.6 *Suppose that we have an LFSR with $\ell = 4$, and tap sequence*

$$(c_1 c_2 c_3 c_4) = (0101)$$

and initial state

$$s_0 = (k_{(3,0)} k_{(2,0)} k_{(1,0)} k_{(0,0)}) = (1101).$$

Then we calculate the following.

j	$k_{(3,j)}$	$k_{(2,j)}$	$k_{(1,j)}$	$k_{(0,j)}$
0	1	1	0	1
1	0	1	1	0
2	1	0	1	1
3	1	1	0	1

For instance, for $j = 1$, the state after the first bit iteration is given by

$$s_1 = (k_{(3,1)} k_{(2,1)} k_{(1,1)} k_{(0,1)}) = (0110),$$

where from (3.6),

$$k_{(3,1)} = c_1 k_{(3,0)} \oplus c_2 k_{(2,0)} \oplus c_3 k_{(1,0)} \oplus c_4 k_{(0,0)} =$$

$$0 \cdot 1 \oplus 1 \cdot 1 \oplus 0 \cdot 0 \oplus 1 \cdot 1 = 0,$$

$$k_{(2,1)} = k_{(3,0)} = 1, \quad k_{(1,1)} = k_{(2,0)} = 1,$$

and

$$k_{(0,1)} = k_{(1,0)} = 0.$$

The LFSR has period length $L = 3$, since

$$s_3 = s_L = (1101) = s_0.$$

The output bitstring consists of the rightmost entry in each of the above table's rows for each (distinct) bit iteration $j = 0, 1, 2$, namely,

$$k = (K_2 K_1 K_0) = (101).$$

There is a very palatable, simple, easy-to-understand matrix method of describing the above. Consider the following *tap matrix* derived from the tap sequence, and *state matrix* derived from the states.

$$C = \begin{pmatrix} c_1 & c_2 & c_3 & \cdots & c_{\ell-1} & c_\ell \\ 1 & 0 & 0 & \cdots & 0 & 0 \\ 0 & 1 & 0 & \cdots & 0 & 0 \\ \vdots & \vdots & \vdots & & \vdots & \vdots \\ 0 & 0 & 0 & \cdots & 1 & 0 \end{pmatrix} \quad \text{and} \quad S_i = \begin{pmatrix} k_{(\ell-1,i)} \\ k_{(\ell-2,i)} \\ k_{(\ell-3,i)} \\ \vdots \\ k_{(0,i)} \end{pmatrix},$$

so,

$$CS_i = S_{i+1} \text{ for } i = 0, 1, \ldots, L - 1.$$

For instance, take the case in Example 3.6.

$$C = \begin{pmatrix} 0 & 1 & 0 & 1 \\ 1 & 0 & 0 & 0 \\ 0 & 1 & 0 & 0 \\ 0 & 0 & 1 & 0 \end{pmatrix} \quad \text{and} \quad S_2 = \begin{pmatrix} 1 \\ 0 \\ 1 \\ 1 \end{pmatrix},$$

so

$$CS_2 = \begin{pmatrix} 1 \\ 1 \\ 0 \\ 1 \end{pmatrix} = S_3 = S_L = S_0.$$

LFSRs are amenable to hardware implementations, the costs of which are low, and LFSRs are extremely fast. The choice of the bits to be tapped can ensure a statistically random appearance of output bits. Thus, an LFSR can be used as a PRNG, but not a CSPRNG, since LFSRs are very susceptible to known-plaintext attacks. All a cryptanalyst needs is a few bits of consecutive plaintext and corresponding ciphertext, then by adding modulo 2, eventually the bits of the key are determined. Bitstrings output by a single LFSR are not secure since sequential bits are linear, so it only takes 2ℓ output bits of the LFSR to determine it, even if the feedback scheme is unknown to the cryptanalyst. Thus, LFSRs become a trade-off between speed and security. If one is not concerned with security, but rather speed, such as in cable television transmission, then LFSRs are a good bet. For systems of communications requiring high security, they are not. However, they can be built into *non*-linear PRNGs. For examples of this and a deeper insight into LFSRs, see [111], [169, pages 119–126], and [283], as well as [287] for a cryptanalytic perspective.

In the concluding section of this chapter, we will look at the most popular and highly secure stream cipher.

Diagram 3.9 A Generic Linear Feedback Shift Register

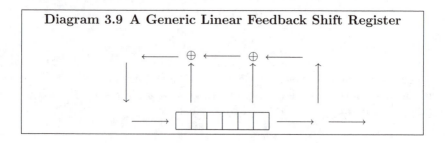

3.7 RC4

If we could find the answer to that [why it is that we and the universe exist], it would be the ultimate triumph of human reason — for then we would know the mind of God.

Stephen Hawking
— from *A Brief History of Time* (see [121, page 175])

RC4 is a stream cipher invented by Rivest in 1987 for Data Security. Ronald L. Rivest received a B.A. in mathematics from Yale University in 1969 and a Ph.D. in computer science from Stanford university in 1974. He is a co-inventor of the RSA public-key cryptosystem (nongovernmental version, see pages 101–104) and founder of RSA Data Security (now called RSA Security after having been bought by Security Dynamics). Among his numerous, outstanding honours and positions are Fellow of the American Academy of Arts and Science; Fellow of the Association for Computing Machinery; member of the National Academy of Engineering; Director of the the Financial Cryptographic Association; Director of the International Association for Cryptologic Research; Fellow of the World Technology Network; member of MIT's Laboratory for Computer Science; member of MIT's laboratory's Theory of Computing Group; a leader of the MIT Cryptography and Information Security Group; and currently the Andrew and Erna Viterbi Professor of Electrical Engineering and Computer Science at MIT. He, together with Adleman and Shamir, was awarded the 2000 IEEE Koji Kobayashi Computers and Communications Award, as well as the Secure Computing Lifetime Achievement Award. Moreover, he founded Peppercoin^{TM}, a company that provides a digital payment service for merchants, which ostensibly allows merchants to process small digital transactions for only pennies. He is widely respected as an expert in cryptographic design and cryptanalysis.

RC4 was kept a secret, but in 1994 it was somehow leaked to the *cyberpunks remailers list* on the Internet. Then it quickly spread to the sci.crypt newsgroup, then to numerous other sites on the Internet. It is probably safe to say that RC4 is the most widely used stream cipher in existence today (see page 129).

◆ The RC4 Cryptosystem

As usual, we describe the cipher via each of its components, which in this case is minimal, and easy to understand.

▼ The State Vector

The *state vector* (row matrix) is a 256-byte vector labelled

$$S[j] \text{ for } j = 0, 1, 2, \ldots, 255.$$

We *initialize* S by setting

$$S[j] = j \text{ for each } j = 0, 1, 2, \ldots, 255.$$

The keylength k of the input (unique) nonce, or key, K, may vary from 1 to 256 bytes.

▼ Initial Permutation of the State Vector

Set $i = j = 0$. Then execute the following.

1. Set $j \equiv j + S[i] + K[i \pmod{k}] \pmod{256}$.

2. **SW**$(S[i], S[j])$. (Swap the components.)

3. Set $i = i + 1$. If $i \not\equiv 0 \pmod{256}$, go to step 1. Otherwise, terminate the initial permutation.

After the initial permutation, the key K is discarded/destroyed since it is a unique key to be used only once. Next, we describe the actual streaming.

▼ Keystream Generation

Set $i = j = 0$. Then execute the following.

1. Set $i \equiv i + 1 \pmod{256}$.

2. Set $j \equiv j + S[i] \pmod{256}$.

3. **SW**$(S[i], S[j])$.

4. If $i \not\equiv 0 \pmod{256}$, go to step 1. Otherwise go to step 5.

5. Set $KS \equiv S[S[i] + S[j]] \pmod{256}]$.

Use KS to encipher/decipher the next byte of plaintext/ciphertext by addition modulo 2. Then go to the keystream generation to get the next key to encipher/decipher the next plaintext/ciphertext byte, and continue until the entire message is encrypted/decrypted.

▼ Analysis and Summary

As we shall see in Chapter 10, RC4 is used for secure wireless transmission (badly as it turned out, at first, but only because the keys for RC4 were generated and used improperly, so the RC4 cipher is not at fault, but rather the implementation). In fact, RC4, as with all stream ciphers, is easily cryptanalyzed if the same input key is used more than once. Sometimes this problem can be eradicated by "hashing" (which we will study in Chapter 4) the key with a nonce each time it is used and sending the nonce along with the message. RC4 is used in the SSL/TLS (See Section 5.7) standard for communication between Web browsers and servers. When RC4 is compared, in terms of speed, to block ciphers, it wins hands down.

On a Pentium II, RC4 operates at 45 Mbps[3.11] whereas DES operates at 9 Mbps and triple DES at a mere 3 Mbps. Thus, using a CSPRNG will ensure that a stream cipher is as secure as a block cipher, and, as with RC4, typically much faster. This may explain the release in the mid-1990s, by Netscape, of a browser with RC4 as its encryption function utilizing a 128-bit key, which is typically employed today for much of the Internet traffic in electronic commerce, especially banking.

[3.11]This means *megabits per second* where a *megabit* is 10^6 bits.

Chapter 4

Public-Key Cryptography

A golden key can open any door.

Late sixteenth-century proverb

4.1 The Ideas behind PKC

We were first introduced to Alice and Bob on page 99, where we described the idea behind *public-key cryptography* (PKC) by using the "open safe" analogy. It is now time to give this a mathematical context.

◆ **Public-Key Cryptosystems (PKC)**

A cryptosystem consisting of a set of enciphering transformations $\{E_e\}$ and a set of deciphering transformations $\{D_d\}$ is called a *public-key cryptosystem* or an *asymmetric cryptosystem* if, for each key pair (e, d), the enciphering key e, called the *public key*, is made publicly available, while the deciphering key d, called the *private key* (*see page 100*), is kept secret. The cryptosystem must satisfy the property that it is computationally infeasible to compute d from e.

In order to motivate the study of PKCs, rather than leave the issue to the end, we begin by comparing the two types of cryptosystems, SKCs and PKCs, and their mutual interdependence in the modern world, with some illustrations before we begin to describe the various types of PKCs and their uses.

◆ **PKCs and SKCs — A Comparison**

1. **Security**: With a PKC only the private key needs to be kept a secret, concealed by one entity, and public keys may be distributed freely. With an SKC there must be a shared secret key known by at least two entities. No PKC has been proven secure, yet except for the one-time-pad, this is also true for SKCs.

2. **Longevity**: With PKCs, key pairs may be used without change in most cases over long periods of time, years in some situations. With SKCs, there may have to be a change of keys for each session.

161

3. **Key Management**: If a multiuser large network is being used (without a key server) then fewer private keys will be required with a PKC than with an SKC. For instance, if $n \in \mathbb{N}$ entities are communicating, using DES, say (see page 98), then the number of keys required to allow any two entities to communicate is $n(n-1)/2$. Also, every user on the system has to store $n-1$ keys. This is called *key predistribution*. With a public-key cryptosystem, only n keys are required for any two entities to communicate since only one (public) key for each entity has to be stored. Hence, SKC, by itself, on the Internet is completely unworkable. Internet e-commerce cannot be supported by SKCs alone.

4. **Key Exchange**: In a PKC, no (private) key exchange between communicating entities is necessary. (Note that this tells us that the Diffie-Hellman key-exchange protocol, discussed on page 166, is not a public-key cryptosystem, although it contained the basic original ideas for it.) With an SKC, it is difficult and risky to exchange a secret key. In fact, one of the principal uses of PKC is for the exchange of a secret symmetric key, an important point to which we will return before we conclude this chapter.

5. **Digital Signatures and General Authentication**: Another of the principal roles played by PKC is that of providing digital signatures since they offer virtually the only means for securely doing so. On the other hand, the principal use of SKCs is bulk data enciphering.

6. **Efficiency**: PKCs are slower than SKCs. For instance, the RSA cryptosystem is roughly a thousand times slower than DES.

7. **Key sizes**: The key sizes for a PKC are significantly larger than that required for an SKC. For instance, the private key in the RSA cryptosystem should be 1024 bits, whereas with an SKC, generally 128 bits will suffice. Usually, private keys are ten times larger than secret keys. PKC key sizes, for such ciphers as RSA for instance, are getting so huge that some PKC implementations are switching to ECC (see page 190). An example is that Motorola uses ECC in its wireless phones; see Section 9.2.

8. **Nonrepudiation**: This means that the sender of a message cannot deny having sent it. With PKCs we can ensure nonrepudiation with digital signatures, whereas with SKCs, we need Trent as a trusted third party.

We may summarize one salient point derived from the above: PKC is *not* meant to *replace* SKC, but rather to *supplement* it for the goal of achieving maximum security and efficiency. This is done as follows. The general motivation behind modern cryptographic usage, especially on the Internet for e-commerce, is to employ PKC to obtain symmetric keys, which are then used in an SKC. Such cryptosystems are called *hybrid cryptosystems* or *digital envelopes*, which have the advantages of both types of cryptosystems. Here is how they work in practice.

Alice and Bob have access to an SKC, which we will call S. Also, Bob has a public-private key pair (e, d). In order to send a message m to Bob, Alice first generates a symmetric key, called a *session key* or *data encryption key*, k to be used only once. (The property of producing a new session key each time a pair of users wants to communicate is called *key freshness*.) Alice enciphers m using k and S obtaining ciphertext $E_k(m) = c$. Using Bob's public key e, Alice encrypts k to get $E_e(k) = k'$. Both of these encryptions are fast since S is efficient in the first enciphering, and the session key is small in the second enciphering. Then Alice sends c and k' to Bob, who deciphers k with his private key d, via $D_d(k') = k$. Then Bob easily deduces the symmetric deciphering key k^{-1}, which he uses to decipher, $D_{k^{-1}}(c) = D_{k^{-1}}(E_k(m)) = m$.

Hence, the PKC is used only for the sending of the session key, which provides a digital envelope that is both secure and efficient, a very nice and elegant resolution of the above problems.

Diagram 4.1 (Digital Envelope — Hybrid Cryptosystem)

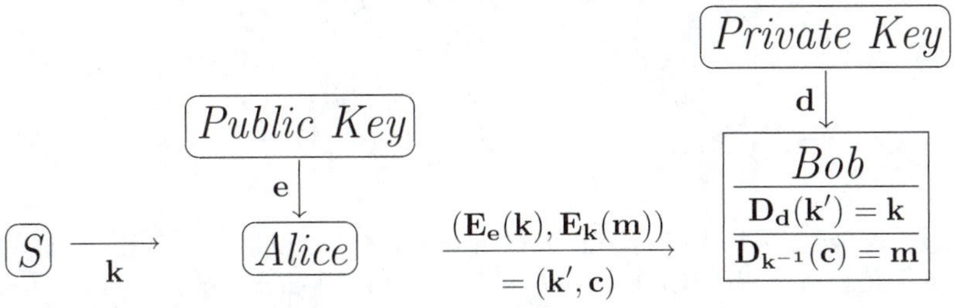

Example 4.1 *Suppose that the symmetric-key cryptosystem, S, that Alice and Bob agree to use is a permutation cipher (see page 114) with parameters $r = 7$, $\mathcal{M} = \mathcal{C} = \mathbb{Z}/26\mathbb{Z}$, and key*

$$k = \begin{pmatrix} 1 & 2 & 3 & 4 & 5 & 6 & 7 \\ 3 & 4 & 7 & 6 & 5 & 1 & 2 \end{pmatrix}.$$

Alice wants to send

$$m = \textbf{travels}$$

to Bob. Alice converts m to numerical equivalents via Table 1.3 on page 11 to get $m = (19, 17, 0, 21, 4, 11, 18)$ to which she applies k to get

$$c = E_k(m) = (0, 21, 18, 11, 4, 19, 17).$$

She then proceeds to encipher k using Bob's public key as follows.

Since m has seven letters, then we may encipher the key k (second row) as a 7-digit, base 10 integer:

$$k = 3 \cdot 10^6 + 4 \cdot 10^5 + 7 \cdot 10^4 + 6 \cdot 10^3 + 5 \cdot 10^2 + 1 \cdot 10 + 2 = 3476512. \quad (4.1)$$

Using a modulus,

$$n = pq = 8179 \cdot 9547 = 78084913,$$

with public key $e = 7$, and private key $d = 22304911$, Alice then enciphers k as

$$k' = E_e(k) = k^e = 3476512^7 \equiv 62221916 \,(\text{mod } 78084913),$$

and sends the pair $(k', c) = (k^e, c)$ to Bob. Bob receives the pair and makes the following calculations.[4.1]
He computes

$$D_d(k') = (k')^d \equiv 62221916^{22304911} \equiv 3476512 \equiv k \,(\text{mod } n).$$

Bob then converts this back to its original format via (4.1), and is able to easily deduce the deciphering key

$$k^{-1} = \left(\begin{array}{ccccccc} 1 & 2 & 3 & 4 & 5 & 6 & 7 \\ 6 & 7 & 1 & 2 & 5 & 4 & 3 \end{array} \right),$$

which he applies to c to get

$$D_{k^{-1}}(E_k(m)) = D_{k^{-1}}(0, 21, 18, 11, 4, 19, 17) = (19, 17, 0, 21, 4, 11, 18) = \textbf{travels}.$$

Now we are ready to look at the various PKCs. The ideas behind the most famous PKC, namely, RSA, about which we will learn the details in the next section, is based upon the simple mathematical idea of exponentiation and related matters. The first of the related matters is a notion that we need to set up the first exponentiation cipher. The security of many cryptosystems depends upon the difficulty of solving certain problems such as the following.

If we are dealing with real numbers then finding e from α^e is called the logarithm function. In $\mathbb{F}_p^*$ (or more generally in any finite group) this is called the *discrete logarithm problem* (DLP). Thus, we present the problem formally as follows.

Discrete Log Problem (DLP):

Given a prime p, a generator m of $\mathbb{F}_p^*$, and an element $c \in \mathbb{F}_p^*$, find the unique integer e with $0 \le e \le p - 2$ such that

$$c \equiv m^e \,(\text{mod } p). \tag{4.2}$$

[4.1]What Bob does is to employ Euler's theorem (see Theorem A.14 on page 479). This motivates what we will study in the next section.

The DLP is often called simply *discrete log*. Here $e \equiv \log_m(c) \pmod{p-1}$. If p is "properly chosen", this is a very difficult problem to solve. One of the ways that p has to be properly chosen is to insist upon $p-1$ having at least one large prime factor. This is due to the Silver-Pohlig-Hellman algorithm, which allows for efficient calculation of discrete logs when $p-1$ has only small prime factors. Due to the technical nature of the algorithm, we have placed a description of it in Appendix D on page 530.

It can be shown that the complexity of finding e in (4.2) when p has n digits is roughly the same as factoring an n-digit number (for instance, see [183]). Therefore, computing discrete logs is virtually of the same degree of difficulty as factoring, and since there are no known tractable factoring algorithms, we assume that the *integer factoring problem* (IFP) is intrinsically difficult (see page 509). Hence, cryptosystems based upon the discrete log problem are assumed to be secure. Yet, there is no verification of this abstractly in the sense that no nontrivial lower bounds have been found for the complexity of integer factorization. (See the discussion of complexity on pages 500–505 in Appendix A.)

A symmetric-key cipher whose security depends upon the discrete log problem is our next topic. It involves the name of a contributor whom we will see often; a brief discussion of his achievements follows.

Martin E. Hellman was born on October 2, 1945. He obtained all his academic degrees in electrical engineering: his bachelor's degree from New York University in 1966; his master's degree in 1967; and his Ph.D. in 1969, the latter two from Stanford. He was employed at IBM and at MIT, but returned to Stanford in 1971. He remained there until 1996, when he received his Professor Emeritus status. We already learned on page 99 that he was one of the pioneers of PKC. He has been involved in computer privacy issues going back to the debate over the DES keylength in 1975 (see Levy's book [151] for some background to this fascinating story). He has not only demonstrated his scholarship with numerous publications, but also has excelled in teaching. He was recognized with four teaching awards; three of these were from minority-student organizations. He is now retired from research and teaching. He and Dorothie, his wife of some 35 years, live on campus at Stanford.

Now it is time for us to go back to symmetric-key cryptography (SKC) and learn about an exponentiation cipher that will help us to set the stage for PKC in general and RSA in particular.

◆ The Pohlig-Hellman Symmetric-Key Exponentiation Cipher

(a) A secret prime p is chosen and a secret enciphering key $e \in \mathbb{N}$ with $e \leq p-2$.

(b) A secret deciphering key d is computed via $ed \equiv 1 \pmod{p-1}$.

(c) Encryption of plaintext message units m is: $c \equiv m^e \pmod{p}$.

(d) Decryption is achieved via $m \equiv c^d \pmod{p}$.

Example 4.2 *Let $p = 167$, and set $e = 69$, with plaintext $12, 4, 18, 0, 6$. Then we encipher each by exponentiating as follows, where all congruences are modulo 167.*

$$12^{69} \equiv 85; \quad 4^{69} \equiv 50; \quad 18^{69} \equiv 96; \quad 0^{69} \equiv 0; \quad 6^{69} \equiv 27.$$

Then we send off the ciphertext. To decipher, we need the inverse of e modulo $166 = p - 1$, and this is achieved by using the Euclidean algorithm (see Theorem A.3 on page 470) to solve

$$69d + 166x = 1,$$

which has a solution $d = 77$ for $x = -32$, and this is the least positive such value of d. So we may decipher via, $85^{77} \equiv 12$ and so on to retrieve the plaintext.

▼ Analysis

Since knowledge of e and p would allow a cryptanalyst to obtain d, then both p and e must be kept secret. The security of this cipher is based on the difficulty of solving the DLP, namely, an adversary, without knowledge of e or d, would have to compute $e \equiv \log_m(c) \pmod{p - 1}$.

The Pohlig-Hellman cipher is an example of the use of *fixed-exponent exponentiation* where the base may vary but the exponent is fixed. The next algorithm, which we mentioned briefly on page 99, is an example of the use of *fixed-base exponentiation* where the exponent may vary but the base is fixed. This algorithm is a prime motivator for PKC, and its security depends upon the DLP.

◆ The Diffie-Hellman Key Exchange Protocol

Suppose that Alice and Bob have not yet met nor exchanged keys, but they want to establish a shared secret key k by exchanging messages over an unsecured channel. First Alice and Bob agree on a large prime p and a generator α of $\mathbb{F}_p^*$ ($2 \leq \alpha \leq p - 2$). These need not be kept secret, so Alice and Bob can agree over an unsecured channel. Then the protocol proceeds as follows.

(1)   Alice chooses a random (large) $x \in \mathbb{N}$ and computes the least positive residue X of α^x modulo p, then sends X to Bob (and keeps x secret).

(2)   Bob chooses a random (large) $y \in \mathbb{N}$ and computes the least positive residue Y of α^y modulo p, then sends Y to Alice (and keeps y secret).

(3)   Alice computes the least positive residue of Y^x modulo p, and Bob computes the least positive residue of X^y modulo p. Since

$$Y^x \equiv \alpha^{yx} \equiv \alpha^{xy} \equiv X^y \equiv k \pmod{p},$$

they have a shared secret key k.

Example 4.3 *Suppose that the parameters are $p = 1187$, $\alpha = 2$, $x = 285$, and $y = 781$. Then $X \equiv 2^{285} \equiv 1013\,(\mathrm{mod}\ p)$; $Y \equiv 2^{781} \equiv 7\,(\mathrm{mod}\ p)$; $Y^x \equiv 7^{285} \equiv 870\,(\mathrm{mod}\ p)$; and $X^y \equiv 1013^{781} \equiv 870\,(\mathrm{mod}\ p)$. Hence, $k = 870$ is the shared secret key.*

▼ **Analysis**

In the Diffie-Hellman protocol, k is the shared secret key independently generated by both Alice and Bob. The key exchange is complete, since Alice and Bob are in agreement on k. The Diffie-Hellman Protocol differs from the Pohlig-Hellman cipher in that the latter requires that both p and e be kept secret since d could be deduced from them, whereas in the former p and α may be made public due to the intractability of the DLP. However, there is a subtler problem here that we need to discuss, not only in reference to the above, but also for later use (see page 186).

A cryptanalyst, Eve, listening to the channel would know p, α, X, and Y, but neither x nor y. Thus, Eve faces what is called the

---

Diffie-Hellman Problem (DHP):

find $\alpha^{xy}\,(\mathrm{mod}\ p)$ given $\alpha, \alpha^x\,(\mathrm{mod}\ p)$ and,

$\alpha^y\,(\mathrm{mod}\ p)$ (but not x or y).

If Eve can solve the DLP, then she can clearly solve the DHP. Whether the converse is true or not is unknown. In other words, it is not known if it is possible for a cryptanalyst to solve the DHP without solving the DLP. Nevertheless, the consensus is that the two problems are equivalent. Thus, for practical purposes, one may assume that the Diffie-Hellman Key Exchange Protocol is secure as long as the DLP is intractable.

Given that we have discussed the achievements of one of the developers of this algorithm, it is now time to talk about the other.

Bailey Whitfield Diffie was born on June 5, 1944. Diffie, by his own admission, was not a good student in high school, but it was not for lack of ability, rather lack of focus. His ability did shine brightly when his less-than-stellar high school performance was overshadowed by his strikingly high marks on entrance examinations for MIT. He entered MIT in 1961 and graduated in 1965, later accepting a job at Mitre Corporation. There he worked under the tutelage of Ronald Silver, (one of the authors of the algorithm we describe on page 530). He worked on development of the mathematical symbolic manipulation package *Mathlab*, which later developed into the powerful symbolic mathematical software package called *MACSYMA*. Silver taught Diffie a great deal and inspired him to look further into cryptographic issues. In 1969, Diffie left Mitre and joined John McCarthy's Artificial Intelligence Lab at Stanford. By 1975, as described on page 99, the collaboration with Hellman, with input from Merkle, created the breakthrough. For his involvement, along with Hellman and Merkle,

in the discovery of the notion of PKC, he was awarded a Doctorate in Technical Sciences (*Honoris Causa*) by the Swiss Federal Institute of Technology in
1992. His current position is Chief Security Officer at Sun Microsystems, in Palo
Alto, California, where he has been since 1991. He has numerous awards from
the Association of Computing Machinery (ACM), IEEE, NIST, NSA, and the
Franklin Institute. The reader wanting more details of his involvement in public
policy concerning cryptography, and his opposition to limitations on the use of
cryptography by individuals and corporations, should consult Levy's book [151],
wherein Diffie is a central character.

There are two more ingredients for the RSA recipe that we need before we
close this section. The first we encountered briefly on page 99, and we now
formalize the notion.

◆ One-Way Functions

A one-to-one function f from a set $\mathcal{M}$ to a set $\mathcal{C}$ is called *one-way* if $f(m)$ is
"easy" to compute for all $m \in \mathcal{M}$, but for a randomly selected c in the image of
f, finding an $m \in \mathcal{M}$ such that $c = f(m)$ is computationally infeasible. In other
words, we can easily compute f, but it is computationally infeasible to compute
f^{-1}.

Diagram 4.2 One-Way Function

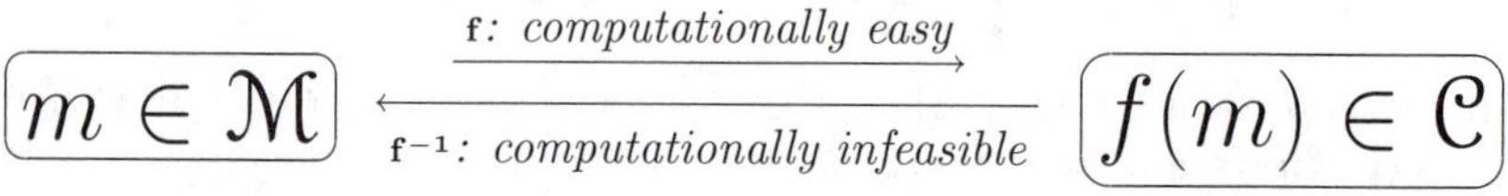

$$\boxed{m \in \mathcal{M}} \quad \begin{array}{c} \xrightarrow{\text{f: } computationally\ easy} \\ \xleftarrow{\text{f}^{-1}\text{: } computationally\ infeasible} \end{array} \quad \boxed{f(m) \in \mathcal{C}}$$

One-way functions have a plethora of cryptographic uses. For instance, we
talked about PRNG earlier and mechanisms for making CSPRNGs (see page
151). One means of creating a CSPRNG using a one-way function is as follows.
Let f be a one-way function and assume a nonce n is known to us. Then define

$$f(n + j) = r_j \text{ for } j = 1, 2, \ldots.$$

If b_j is the least significant bit[4.2] of r_j, then the sequence $b_0 b_1 \ldots$ will be a
CSPRNG.

Another use of one-way functions, about which we will see more later, is
password security. Suppose that Alice has a password p and f is a one-way
function. Then p can be stored as $f(p)$ on a computer. When Alice logs in to
her account, the computer takes p, calculates $f(p)$, and checks that it matches
the stored value. A cryptanalyst who gets hold of the password file will have
only the $f(p)$ value for each user, and obtaining p is computationally infeasible.

The above being said, there is no rigorous mathematical proof that one-way
functions actually exist. Yet, as we saw on page 99, we have working definitions,

---

[4.2]In a given base-b representation of $a \in \mathbb{N}$, $a = (a_n, a_{n-1}, \ldots, a_1, a_0)_b$, the digit a_0 is called
the *least significant digit*, and a_n is called the *most significant digit*.

pragmatic ones, that serve us well. Moreover, we now have "candidate" one-way functions such as the DLP and the IFP, discussed on page 165 — see also Appendix C on page 509. The reader interested in a deeper analysis of this issue may consult books on complexity theory that go well beyond the basics covered in Appendix A (see [101] for instance).

The reader may wonder at this point how it is that we could devise a cryptosystem using one-way functions. The recipient of a message enciphered with a one-way function would ostensibly be no better off than a cryptanalyst at finding the plaintext since computing the inverse is computationally infeasible. This is correct, so the recipient needs more information, the idea for which is contained in the next notion.

◆ **Trapdoor One-Way Functions**

A *trapdoor one-way function* or *public-key enciphering function* is a one-way function,

$$f : \mathcal{M} \mapsto \mathcal{C},$$

satisfying the additional property that there exists information, called *trapdoor information*, or simply *trapdoor*, that makes it feasible to find $m \in \mathcal{M}$ for a given $c \in \text{img}(f)$ such that $f(m) = c$, but without the trapdoor this task becomes infeasible.

Diagram 4.3 Trapdoor One-Way Function

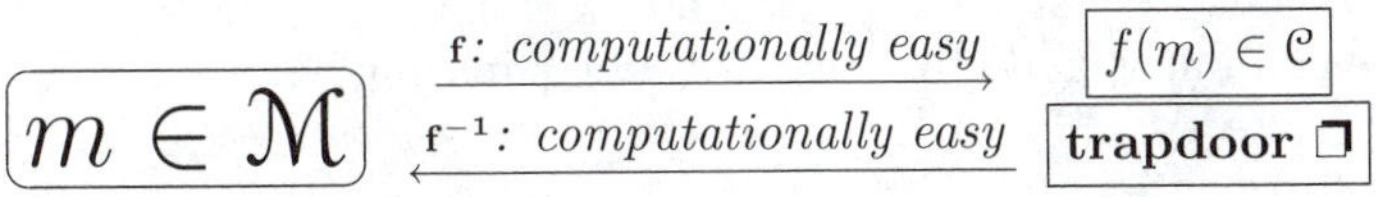

The essential idea behind the Diffie-Hellman key exchange is the use of trapdoor one-way functions. The Diffie-Hellman Protocol, discussed earlier in this section, allows for entities who have never met or exchanged information to establish a shared secret key by exchanging messages over an unsecured channel. Since exponentiation modulo p is polynomial time, then enciphering is easy. However, finding the inverse, solving the DLP, is computationally infeasible, without the trapdoor, namely, one of the secret pair (x, y). This can be made to work in a more general context, using the IFP, that is most germane to our discussion of RSA in the next section, and will provide a nice motivator.

Example 4.4 *Let* $f(x) \equiv x^e \,(\text{mod } n)$ *where* $n = pq$ *with* $p \neq q$ *primes, and suppose that*

$$de \equiv 1 \,(\text{mod } (p-1)(q-1)).$$

Then applying f *is computationally, easy but finding*

$$f^{-1}(x^e) \equiv f^{-1}(f(x)) \equiv x^{ed} \equiv x \,(\text{mod } n) \tag{4.3}$$

is computationally infeasible without the trapdoor d.

▼ Analysis

First, we note that $(p-1)(q-1)$ in Example 4.4 is just the Euler function (see page 479) applied to pq, since $\phi(n) = \phi(pq) = (p-1)(q-1)$, and the application of f^{-1} in Equation (4.3) is just Euler's generalization of Fermat's Little Theorem (see Theorem A.14 on page 479). This little bit of elementary number theory is really all that is behind the RSA cipher, so understanding this is sufficient to understand the entire cryptosystem.

In order to show that the finding of the trapdoor d in Example 4.4 is based upon the IFP, in this case factoring n, we need to show that computing d is "as hard as" factoring n. Determining what *as hard as* means will involve some discussion of probabilities.

If we can factor n, obviously we can compute $(p-1)(q-1)$. Then we can use the Euclidean algorithm to find d from e (in computationally feasible time), since we need merely solve $ed + x(p-1)(q-1) = 1$ for x and d. A simple example is $e = 7$, $p = 101$ and $q = 167$. Then solving $7d + 100 \cdot 166 \cdot x = 1$ is easily achieved, with $x = -2$, and $d = 4743$. In fact, it can be shown that being able to compute d can be converted (with an arbitrarily high probability) into an algorithm for factoring n (see [170, page 65] for instance). In other words, knowledge of d can be converted into an algorithm for factoring n (with an arbitrarily small probability of failure to do so). Thus, to say that finding d is as hard as factoring the modulus is not a proven fact, rather a conjecture based upon some (rather solid) evidence. We will formalize this conjecture in the next section.

Note, as well, that if we have $\phi(n)$ and n, then we can factor n. The reason is that we can find p and q by successively computing

$$p + q = n - (p-1)(q-1) + 1 \text{ and } p - q = \sqrt{(p+q)^2 - 4n}, \qquad (4.4)$$

so we get

$$p = \frac{1}{2}\left[(p+q) + (p-q)\right] \text{ and } q = \frac{1}{2}\left[(p+q) - (p-q)\right].$$

Hence, finding d or finding $\phi(n)$ means we can factor n.

The next topic is a mechanism for creating a digital "fingerprint" of data.

◆ Hash Functions

A *hash function* is a computationally efficient function that maps bitstrings of arbitrary length to bitstrings of fixed length, called *hash values*. A *one-way hash function* is a hash function that is a one-way function as described above. The process of using a hash function on a message is called *hashing the message*.

In order for a hash function to be cryptographically viable, it must be a one-way hash function to prevent easy unauthorized retrieval of the original bitstring. Thus, one-way hash functions are called *cryptographic hash functions*. Hash values produced by such functions are used as a concentrated representative of the original bitstring, so they can be used as a unique identifier of it. Thus, this type of hash value is called a *message digest*, *imprint*, or *digital fingerprint.* We will mean a cryptographic hash function when we use the term *hash function* henceforth.

There is a special terminology for one-way hash functions. A hash function h is called *weakly collision resistant* if it is computationally infeasible to find for a given x_1, *a value* $x_2 \neq x_1$ such that $h(x_1) = h(x_2)$. A hash function h is called *strongly collision resistant* if it is computationally infeasible to find *any pair of values* (x_1, x_2) such that $x_1 \neq x_2$ and $h(x_1) = h(x_2)$. The reader is cautioned that these collision-related terms are not used consistently throughout the literature.

We will be looking at hash functions from numerous perspectives, some of the most important of which are in message authentication, as one might expect from a device that mimics a "fingerprint". We study hash function in depth later (see page 252).

We conclude this section with an important algorithm that is essential to the efficiency of many cryptosystems including RSA. It is our last example of yet another kind of exponentiation algorithm, called *basic exponentiation*, which can be used with any base b and any exponent r.

Our basic goal is to calculate the least positive residue of x^d modulo n for any given $x, d \in \mathbb{N}$. To do this with a single exponentiation would overflow the memory of most computers for sufficiently large d. We could reduce memory requirements by starting with x and multiplying by x, $d - 1$ times, reducing modulo n at each step, but even here, for large d, the methodology is still too slow. Fortunately, there is an efficient algorithm of squaring and reducing modulo n in successive steps, that is quite efficient.

The Repeated Squaring Method

Given $d, n \in \mathbb{N}$, $d > 1$, and

$$d = \sum_{j=0}^{k} d_j 2^j, \, d_j \in \{0, 1\},$$

the goal is to find $x^d \pmod{n}$.

First, we initialize by setting $c_0 = x$ if $d_0 = 1$, and set $c_0 = 1$ if $d_0 = 0$. Also, set $x_0 = x$, $j = 1$, and execute the following steps:

(1) Compute $x_j \equiv x_{j-1}^2 \pmod{n}$.

(2) If $d_j = 1$, set $c_j = x_j \cdot c_{j-1} \pmod{n}$.

(3) If $d_j = 0$, then set $c_j \equiv c_{j-1} \pmod{n}$.

(3) Reset j to $j + 1$. If $j = k + 1$, output $c_k \equiv x^d \pmod{n}$, and terminate the algorithm. Otherwise, go to step (1).

4.2   RSA

*If in other sciences we should arrive at certainty without doubt and truth
without error, it behooves us to place the foundations of knowledge in mathematics*

Roger Bacon (ca. 1214–1292), English philosopher and scientist
— from *Opus Majus* (bk. 1, ch. 4)

We have given some biographical information for one of the three gentlemen who make up the acronym for this section (see page 159). It is time to
give some information about the the other two. Adi Shamir is an Israeli cryptographer, who is currently the Borman Professor in the Applied Mathematics
Department of the Weizman Institute of Science in Israel. He obtained his Ph.D.
from Stanford in 1977 after which he did postdoctoral work at Warwick University in England. Shamir's name is attached to a wide variety of cryptographic
schemes, many of which we will study in this text, including the Fiat-Shamir
identification protocol, RSA, DC (see Footnote 3.4 on page 127), and his polynomial secret-sharing scheme, to mention only a few. On April 14, 2003, the
ACM formally announced that the A.M. Turing Award (essentially the "Nobel
Prize of computer science") would go to Adleman, Shamir, and Rivest for their
developmental work on PKC.

Leonard Adleman was born on December 31, 1945, in San Francisco, California. He received his B.Sc. in mathematics from the University of California
at Berkeley in 1972 and his Ph.D. there in 1976. His doctoral thesis was done
under the guidance of Manuel Blum, and was titled, *Number Theoretic Aspects of Computational Complexity.* He is married with three children, and
is currently Henry Salvatori Professor of Computer Science and Professor of
Molecular Biology at the University of Southern California Los Angeles, California, where he has been since 1980. His professional interests are algorithms;
computational complexity; computer viruses; cryptography; DNA computing;
immunology; molecular biology; number theory; and quantum computing. His
most recent activity is the building of a DNA computer, which has the potential
for a vastly faster computation for the future. He noticed that a protein, called
polymerase, which produces complementary strands of DNA, resembles the operation of a Turing machine (see page 503). Adleman reached the conclusion
that DNA formation essentially functions in a fashion similar to a computer, so
he is interested in constructing a viable DNA computer.

Although the Diffie-Hellman key-exchange protocol, discussed on page 166,
was the genesis of a profound investigation into the notion of PKC, their scheme
did not provide a complete solution to the establishment of a complete PKC.
They only provided a mechanism for the exchange of keys, and by the authors'
own admission, left open the problem of establishing a working secure PKC (see
page 99). The first to (publicly) do this, as we know, have their names attached
to the acronym that did provide such a solution.

◆ **The RSA Public-Key Cryptosystem**

We break the algorithm into two parts with the underlying assumption that Alice wants to send a message to Bob.

(I) RSA Key Generation

1. Bob generates two large, random primes $p \neq q$ of roughly the same size, and computes both $n = pq$ and

$$\phi(n) = (p-1)(q-1).$$

The integer n is called his (*RSA*) *modulus*.

2. He selects a random $e \in \mathbb{N}$ such that $1 < e < \phi(n)$ and $\gcd(e, \phi(n)) = 1$. The integer e is called his (*RSA*) *enciphering exponent*. Then using the extended Euclidean algorithm (see page 471), he computes the unique $d \in \mathbb{N}$ with $1 < d < \phi(n)$ such that

$$ed \equiv 1 \,(\text{mod } \phi(n)).$$

3. Bob publishes (n, e) in some public database and keeps d, p, q, and $\phi(n)$ private. Thus, Bob's (*RSA*) *public-key* is (n, e) and his (*RSA*) *private key* is d. The integer d is called his (*RSA*) *deciphering exponent*.

(II) RSA Public-Key Cipher

enciphering stage:

In order to simplify this stage, we assume that the plaintext message $m \in \mathcal{M}$ is in numerical form with $m < n$. Also, $\mathcal{M} = \mathcal{C} = \mathbb{Z}/n\mathbb{Z}$, and we assume that $\gcd(m, n) = 1$.

1. Alice obtains Bob's public key (n, e) from the database.

2. She enciphers m by computing $c \equiv m^e \,(\text{mod } n)$ using the repeated squaring method given on page 171, and sends $c \in \mathcal{C}$ to Bob.

deciphering stage:

Once Bob receives c, he uses d to compute $m \equiv c^d \,(\text{mod } n)$.

Example 4.5 *Suppose that Bob chooses* $(p, q) = (9221, 7489)$. *Then* $n = 69056069$ *and* $\phi(n) = 69039360$. *If Bob selects* $e = 7$, *then solving* $1 = 7d + \phi(n)x$ *(for* $x = -4$), *he gets* $d = 39451063$, *his private key. Also,* $(69056069, 7)$ *is his public key. Alice obtains Bob's public key and wishes to send the message* $m = 7289258$. *She enciphers using Bob's public key to get*

$$c \equiv m^7 \equiv 19407420 \,(\text{mod } n),$$

which she sends to Bob. He uses his private key d *to decipher via*

$$c^d \equiv 19407420^{39451063} \equiv 7289258 \equiv m \,(\text{mod } n).$$

▼ Block Size

We cannot properly encipher the plaintext message unit if it is a numerical value $m \geq n$. (The reader may try an example, say, $m = 72892588$, in Example 4.5, and see that information is lost (under modular reduction) and the system fails.) When $m \geq n$, we must subdivide the plaintext numerical equivalents into blocks of equal size, a process called *message blocking*. If we are dealing with numerical equivalents of the plaintext in base N integers for some fixed $N > 1$, then message blocking is accomplished by choosing that unique integer ℓ such that $N^{\ell} < n < N^{\ell+1}$. Then we write the message as blocks of ℓ-digit, base N integers (with zeros packed to the right in the last block if necessary), and encipher each separately. Since $N^{\ell} < n$, each block of plaintext corresponds to an element of $\mathbb{Z}/n\mathbb{Z}$. Therefore, since $n < N^{\ell+1}$, then each ciphertext message unit can be uniquely written as an $(\ell+1)$-digit, base N integer in $\mathcal{C} = \mathbb{Z}/n\mathbb{Z} = \mathcal{M}$.

▼ Modulus Size

For the modern day and the near future, an RSA modulus of 1024 to 2048 bits would be considered secure. Certain RSA moduli of n digits that are a product of two primes of approximately the same size are denoted by RSA-n, called an *RSA challenge number*. These are published on the Internet and the reader may request the list from *challenge-rsa-list@rsa.com*. These are numbers for which rewards are offered to factor them. We will return to some concrete examples of these numbers shortly.

▼ Security

In the RSA cipher, the four items $d, p, q, \phi(n)$ form the trapdoor and knowledge of any one of them reveals the remaining three items. In other words, they are not independent items. Also, to ensure a secure cryptosystem, there must be "preprocessing" of plaintext message units before the enciphering stage. In [37], it is shown that implementing the RSA cryptosystem as described on page 173, which we will call *plain* RSA (namely, *without* any preprocessing of plaintext message units) is insecure in the following sense. The attack against plain RSA given in [37] shows that even though an m-bit key is used in plain RSA, the *effective security* is $m/2$ bits. Hence, it is essential that, before encryption, a preprocessing step be implemented that uses the *Optimal Asymmetric Encryption Padding* (OAEP) (introduced in [15]) such as [185], a recent standard from RSA Labs. There are methods for adding *randomness* to the enciphering stage in RSA (see [15] for instance). In order to obtain a *secure* RSA cryptosystem from a *plain* RSA cryptosystem, there should be an application of a preprocessing function to the plaintext *before enciphering*. In [15], there are new standards for "padding" plain RSA so that it is secure against certain chosen ciphertext attacks (see Footnote 4.3 on page 176).

One example that is pertinent at this juncture is the following. Suppose that Alice and Bob use the RSA cryptosystem but they choose the same RSA modulus n, and enciphering (public) keys e_A and e_B, respectively, with $\gcd(e_A, e_B) = 1$. Suppose that Eve intercepts two cryptograms m^{e_A} and m^{e_B}, enciphering the same message m, sent by a third entity to Alice and Bob, respectively. Given that $\gcd(e_A, e_B) = 1$, the extended Euclidean algorithm allows Eve to solve $e_A x + e_B y = 1$ for some $x, y \in \mathbb{Z}$. Then, Eve calculates:

$$c_A^x c_B^y \equiv (m^{e_A})^x (m^{e_B})^y \equiv m^{e_A x + e_B y} \equiv m \pmod{n},$$

and this is done without knowledge of a factorization of n or of knowledge of either private keys. This is called *common modulus protocol failure* (CMPF). This is not a failure of the RSA cryptosystem, rather a very bad implementation of it. In fact, the CMPF shows, in no uncertain terms, that an RSA modulus should *never* be used by more than one entity. The CMPF illustrates the fact that no matter how strong a cipher might be, a bad implementation of it will render the scheme to be insecure, and useless. The true security of RSA requires a proper implementation. For instance, even the RSA modulus size of 2048 bits suggested above, is useless in the face of a bad implementation such as the CMPF. We will come back to implementation issues below.

On pages 169 and 170, we discussed an instance that is tantamount to the encryption and decryption of the RSA cipher. Therein, we talked about a notion that we can now name.

The RSA Conjecture
Cryptanalyzing RSA must be as difficult as factoring.

Although there is no proof of this conjecture, the aforementioned discussion in the previous section tells us that the evidence is strong and the general consensus is that the conjecture is valid. A good reason for believing this is that the only known method for finding d given e is the extended Euclidean algorithm applied to e and $\phi(n)$. Yet, to compute $\phi(n)$, we need to know p and q, namely, we need to know how to factor n.

Given the above statement, it is worth a few more words on the extended Euclidean algorithm. This algorithm calculates the $\gcd(e, \phi(n))$, and when $\gcd(e, \phi(n)) = 1$, it calculates the $e^{-1} \pmod{\phi(n)}$. This is accomplished relatively quickly.

Although factoring is indeed an intrinsically difficult problem, it is nowhere nearly as hard as it was at the inception of RSA. In Martin Gardiner's *Scientific American* article [100], in 1977, it was trumpeted that it would take millions of years to factor the RSA-129 challenge number. A reward of $100 (U.S.) was offered at that time (see page 174). In 1994, however, reality got in the way of that perception. After only eight months of trying, the authors of [10] factored it. They used a variation of a factoring algorithm called the *Multipolynomial Quadratic Sieve* (MPQS) (the details of which we discuss, along with other factoring issues, in Appendix C. In fact, the reader desiring an in-depth look at factoring and its consequences should consult this appendix for the details.) The authors used over 600 researchers by distributing their quadratic sieve operations to hundreds of physically separated computers all over the world. The term for this is *factoring by electronic mail* coined by Lenstra and Manasse in [148]. The lesson here is that we should never underestimate the potential breakthroughs in mathematical factoring techniques such as CFRAC (1970) and

MPQS (1985). Ironically, this possibility was addressed in the aforementioned article: "Rivest and his associates have no proof that at some future time no one will discover a fast algorithm for factoring composites as large as .... [but] they consider the possibility extremely remote." What was a problem that could take millions of years in 1977; was reduced to mere months by 1994 owing to such breakthroughs.

This is an appropriate juncture to introduce the notion of a *mips year* which is defined to being equivalent to the computational power of a computer rated at one million instructions per second (mips) and used for one year, which is tantamount to approximately $3 \cdot 10^{13}$ instructions. The RSA-129 challenge number took 5000 mips years.

There are numerous cryptosystems that are called *equivalent to the difficulty of factoring*. For instance, there are RSA-like cryptosystems whose difficulty to break is as hard as factoring the modulus. It can be shown that any cryptosystem for which there is a constructive proof of equivalence to the difficulty of factoring, is vulnerable to a chosen-ciphertext attack.[4.3] We have already seen that factoring an RSA modulus allows the breaking of the cryptosystem, but the converse is not known. In other words, it is not known if there are other methods of breaking RSA, but some new attacks presented concerns.

◆ Attacks on RSA

In what follows, the attacks on RSA must really be seen as attacks on particular *implementations* of RSA. Hence, taken together, the following present a cogent argument and criteria for secure implementations of RSA.

In 1995, Paul Kocher, then a Stanford undergraduate, (see [139]) discovered that RSA could be cryptanalyzed by recovering the decryption exponent through a careful timing of the computation times for a sequence of decryptions. This weakness was a surprising and unexpected discovery,[4.4] and although there are means of thwarting the attack, it was another wake-up call. This is another lesson for cryptographers: never to be overconfident, and always be alert to the unexpected.

In order to understand this new attack, we need some statistical notions and probabilistic notions; see Appendix E. Let R be a randomized algorithm (see page 500) that produces $t \in \mathbb{R}$ as output where t is the amount of time it takes for the computer to complete a calculation for a given input. We record the outputs $t_1, t_2, \ldots, t_r$ for given inputs and compute the *mean* $m = (t_1 + t_2 + \cdots t_r)/r$.

[4.3]In a *chosen-ciphertext* attack, the cryptanalyst chooses the ciphertext and is given the corresponding plaintext. This attack is most effective against public-key cryptosystems, but sometimes is effective against symmetric-key ciphers as well. One way of mounting a chosen-ciphertext attack is to obtain access to the machinery used to do the encryption. This was accomplished prior to World War II when the Americans were able to replicate the Japanese cipher machine, *Purple* (see page 89).

[4.4]Two of the major reasons this attack was so stunning were that it came from a completely unexpected area, and it is a *ciphertext-only* attack, which means the cryptanalyst has access only to the ciphertext, obtained through interception of some cryptograms, from which to deduce the plaintext, without any knowledge whatsoever of the plaintext.

The *variance* of the $\{t_j\}$ is defined to be

$$\text{var}\left(\{t_j\}_{j=1}^r\right) = \frac{(t_1 - m)^2 + (t_2 - m)^2 + \cdots (t_r - m)^2}{r},$$

and the *standard deviation* is just the square root of the variance. It can be shown that for another set of outputs $s_1, s_2, \ldots, s_r$,

$$\text{var}\left(\{s_j\}_{j=1}^r\right) + \text{var}\left(\{t_j\}_{j=1}^r\right) \approx \text{var}\left(\{s_j\}_{j=1}^r + \{t_j\}_{j=1}^r\right). \tag{4.5}$$

An analogue for the following attack is for a thief to watch someone turning the lock on a safe and measuring the time it takes to go to each combination number in order to guess the combination, quite clever.

▼ Timing Attack

Here is what Eve knows: the hardware, such as a smart card or computer, that is being used; the RSA modulus n; and prior to her attack, Eve has measured the time values that it takes the hardware to compute $x_i^d \equiv c_i \pmod{n}$ in the repeated squaring method (given on page 171) for some large number r of ciphertexts x_i. Eve wants to obtain the RSA decryption exponent $d = \sum_{j=0}^k d_j 2^j$, which she knows is odd, so she already has $d_0 = 1$. Suppose that she has obtained $d_0, d_1, \ldots, d_{\ell-1}$ for some $\ell \in \mathbb{N}$. Since Eve knows the hardware, she therefore knows the time t_i (for each x_i) that it takes the hardware to compute $c_\ell, \ldots, c_k$ in the repeated squaring method. She now wants to determine d_ℓ. If $d_\ell = 1$, then the multiplication $x_\ell \cdot x \pmod{n}$ is calculated for each ciphertext x_i, and Eve knows that this takes time q_i, say. However, if $d_\ell = 0$, this multiplication does not take place. Now suppose it takes time s_i for the computer to complete the calculation *after* the multiplication. Then by Equation (4.5), if $d_\ell = 1$,

$$\text{var}\left(\{t_i\}_{i=1}^r\right) \approx \text{var}\left(\{q_i\}_{i=1}^r\right) + \text{var}\left(\{s_i\}_{i=1}^r\right) > \text{var}\left(\{s_i\}_{i=1}^r\right), \tag{4.6}$$

and if $d_\ell = 0$, then this fails to hold. Hence, Eve can determine d_ℓ, and similarly d_j for each $j > \ell$. This simple observation allows for Eve to find d without having to factor n.

To summarize: the attack essentially consists of simulating the computation to some point, then building a decision criterion (4.6), with only one correct interpretation possible, depending on the selected value, and finally deciding the bit value by observing whether (4.6) holds. This attack is most effective against smart cards and such devices where timing measurements can be obtained precisely.

So how do we defend against Eve's (passive) attack? There are two basic methods for defense against Kocher's timing attack. The simplest is to ensure that the modular exponentiation being used always takes a *fixed* amount of time. This may be accomplished by adding a suitable delay factor in each operation. The second method of defense is attributable to Rivest. The method is called *blinding*. Suppose that b is the ciphertext and e is the RSA encryption exponent. Then prior to deciphering b, a random $r \in (\mathbb{Z}/n\mathbb{Z})^*$ is chosen and $b' \equiv b \cdot r^e \pmod{n}$ is computed, followed by $c' \equiv (b')^d \pmod{n}$. Then the computer sets $c = c' \cdot r^{-1} \pmod{n}$ thereby exponentiating a random b' with d that is totally unknown to Eve, thereby thwarting her attack.

▼ Power Cryptanalysis Attack

A section, albeit small, is deserving for another outstanding idea from Kocher, called *power cryptanalysis*. By a very careful measurement of the computer's *power consumption* during decryption, Eve could recover the secret key. This works since during multiprecision multiplications the computer's power consumption is necessarily higher than it would normally be. Hence, if Eve measures the length of these high consumption episodes she can easily decide when the computer is performing one or two multiplications, and this gives away the bits of d. The only protection is some kind of physical shielding of the power output. See *http://www.cryptography.com/resources/whitepapers/DPA.html*.

▼ Low Public RSA Exponent Attacks

For the sake of efficiency, one would like to use small public RSA exponents. However, one has to be careful not to compromise security in so doing. One typically used public exponent is, surprisingly, $e = 3$. However, if the same message m, in a single block, is sent to three different entities, having pairwise relatively prime RSA moduli n_j, with $m < n_j$ for $j = 1, 2, 3$, this allows recovery of the plaintext. Here is how it is done. Using the Chinese remainder theorem, there is a solution to $x \equiv c_i \equiv m^3 \pmod{n_i}$ for each $i = 1, 2, 3$. Since $m^3 < n_1 n_2 n_3$, then $x = m^3$. By computing the cube root of the integer x, we retrieve m. Furthermore, this attack can be generalized to show that a plaintext m can be recovered if e is the RSA enciphering exponent and m is sent to $k \geq e$ recipients with pairwise relatively prime RSA moduli n_i such that $m < n_i$ for $i = 1, 2, \ldots, k$. Recognizing these issues, certain experts have suggestions.

The authors of [149] suggest: "Values such as 3 and 17 can no longer be recommended, but commonly used values such as $2^{16} + 1 = 65,537$ still seem to be fine. If one prefers to stay on the safe side one may select an odd 32-bit or 64-bit public exponent at random." One of the reasons they suggest $2^{16} + 1$ is that $(65537)_{10} = (10000000000000001)_2$, so encryption using the repeated squaring method needs only 16 modular squarings and 1 modular multiplication; and here to recover the plaintext by the generalized method of the above paragraph would require sending the same message to $2^{16} + 1$ entities, not likely to occur. In any case, this generalized method of attack can be thwarted by padding a randomly generated bitstring of suitable length to the plaintext message prior to encryption. In fact, this is something discussed in the security section on page 174, namely, the necessity of preprocessing plaintext before the RSA cipher is employed. However, one has to be careful that the padding itself is secure. For instance, the attack in the paragraph above works because the messages are linearly related, allowing use of the Chinese Remainder Theorem. In fact, a generalization of results by Coppersmith (see [61]), were given by Hastad in [120], called the *Strong Hastad Broadcast Attack.*. He proved that *any* fixed polynomial applied as padding is insecure. Therefore, to defend against his attack is to pad with a *randomized* polynomial, not a fixed one.

There are also attacks against *naive* padding algorithms. Let us enlist Alice, Bob, and Mallory to describe one such method. Alice pads a message m that she wants to send to Bob, then enciphers it, and transmits it. However, malicious Mallory, applying a classic case of the man-in-the-middle attack, intercepts the message and prevents it from reaching Bob. When Bob does not respond to her message, Alice assumes he did not receive it, so she decides to randomly pad m again, encrypts it, and sends it to him. Now Mallory intercepts the second message and has two different encipherments of m using different random pads. In [61], Coppersmith describes a method for Mallory to retrieve m. For instance, if $e = 3$, then when the pad has maximum bitlength less than a ninth that of the message length, Mallory can efficiently recover m (see [170, page 117] for illustrations and a deeper discussion).

Perhaps the most powerful attack developed by Coppersmith is his *partial key exposure attack* (see [61]), which can be described as follows. Given an RSA modulus $n = pq$ of bitlength ℓ, the bitlength of either p or q is about $\ell/2$. Knowledge of either the $\ell/4$ most significant bits of p or the $\ell/4$ least significant bits of p, can be shown to allow one to efficiently factor n. There is another related attack (see [36]), which says that knowledge of the $\ell/4$ least significant bits of the decryption exponent d, allows one to find d in $O(e\log_2(e))$ steps. Thus, if e is small, a cryptanalyst can deduce d from just a few bits. Thus, we have another lesson for the cryptographer: ensure that *all bits* of d are secure.

▼ Low Secret RSA Exponent Attacks

Again, as with the reason for choosing small public exponents, we want increased efficiency in the decryption process, so we choose small secret exponents. For instance, given a 1024-bit RSA modulus, the decryption process can have efficiency increased ten-fold with the choice of small d. However, Weiner [280] developed an attack that yields a total break of the RSA cryptosystem, (by a *total break*, we mean that a cryptanalyst can recover d, hence retrieve all plaintext from ciphertext).

Weiner's Attack

If $n = pq$ where p and q are primes such that $q < p < 2q$ and $d < n^{1/4}/3$, then given a public key e with $ed \equiv 1 \pmod{\phi(n)}$, d can be efficiently calculated.

We conclude that use of small decryption exponents, in the sense given by Weiner above, lead to a total loss of security in the RSA cryptosystem. Weiner's method was improved in [35] by Boneh and Durfee who showed that RSA is insecure if $d < n^{0.292}$. Hence, the gain in efficiency for using such small decryption keys is a cryptographically lethal exercise.

4.3 Digital Signatures

> *Language is only the instrument of science, and words are but the signs of ideas; I wish, however, that the instrument might be less apt to decay, and the signs might be permanent, like the things which they denote.*
>
> **Samuel Johnson** (1709–84), English poet, critic, and lexicographer
> — from preface of *A Dictionary of the English Language* (1755)

We had a brief introduction to the notion of a digital signature when we mentioned applications of OFB mode on page 136. Now it is time to dig deeper into this significant application. First, we ask: Why do we want a digital signature in cryptography? This is best answered by bringing in Alice, Bob, and Mallory to give us an illustration. They will demonstrate the issues surrounding *entity authentication*, meaning verification of the identity and data origin of a legitimate entity in a protocol by another legitimate entity; and *impersonation*, meaning the assumption of the identity of a legitimate entity by an adversary.

Suppose that Alice wishes to send a message m to Bob, whose public key is e, using a PKC. Suppose further that Mallory, impersonating Bob, sends Alice his public key e' and Alice assumes this is Bob's public key. She sends $m^{e'}$, which Mallory intercepts, and using his private key d' computes $(m^{e'})^{d'} = m$. Then he encrypts m with Bob's public key and sends m^e to Bob. Neither Alice nor Bob knows that they have been duped by Mallory. This is illustrated in Diagram 4.4.

Diagram 4.4 (Impersonation Attack on PKCs)

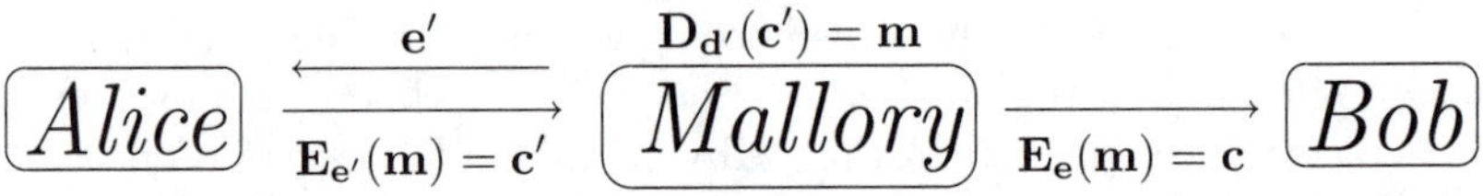

$$\boxed{Alice} \quad \overset{e'}{\underset{\mathbf{E_{e'}(m)=c'}}{\longleftrightarrow}} \quad \boxed{Mallory} \quad \overset{\mathbf{D_{d'}(c')=m}}{\underset{\mathbf{E_e(m)=c}}{\longrightarrow}} \quad \boxed{Bob}$$

This provides an answer to our question above. We need a mechanism for authentication to thwart impersonation; we need digital signatures, which are formally defined below.

◆ **Digital Signature Schemes (DSS)**

Let $\mathcal{M}$ be a *message space*, $\mathcal{K}$ a *keyspace*, and $\mathcal{S}$ a set of bitstrings of fixed length, called a *signature space*. For $k \in \mathcal{K}$, to produce a digital signature, we have a *digital signature algorithm*,

$$\text{sig}_k : \mathcal{M} \mapsto \mathcal{S}.$$

To verify the signature, we have a *digital verification algorithm*,

$$\text{ver}_k : \mathcal{M} \times \mathcal{S} \mapsto \{0,1\} = \mathbb{F}_2,$$

where $ver_k(m,c) = 1$, when $\text{sig}_k(m) = c$ is authentic, and $ver_k(m,c) = 0$ when it is not. A *digital signature scheme* (DSS) is comprised of a digital signature algorithm and a digital verification algorithm.

▼ **Criteria for a Secure DSS**

1. If Alice signs a message m with $\text{sig}_k(m)$, it must be computationally infeasible for an adversary to retrieve the pair $(m, \text{sig}_k(m))$, called the *unforgeable property*.

2. If Bob receives $\text{sig}_k(m) = c$ from Alice, then Bob must be able to verify that this is Alice's signature using $\text{ver}_k(c)$, called the *authentic property*.

3. After being transmitted, neither Bob nor Mallory can alter m, called the *not alterable property*;

4. Bob must be able to instantly detect if an m is being resent, called the *not reusable property*.

▼ DSS Types

1. A DSS *with message recovery* means that the message being sent is not required as input to the verification algorithm.

2. A DSS *with appendix* means that the message is required as input for the verification algorithm.

A significant benefit of RSA is that it can be used for both enciphering plaintext and signing messages. Thus, our first DSS is naturally derived from the RSA cipher, and it is an example of the first type of scheme, one with message recovery (see [206]). Before explaining it in formal terms, we give a brief introductory explanation.

The executions for encryption and signing involve the same RSA calculations modulo the given RSA modulus n. For instance, if Alice wants to sign a message m, she easily computes $c \equiv m^{d_A} \pmod{n}$ using her private RSA key, d_A, and sends (c, m) to Bob. Bob, or anyone in possession of her public key e_A, uses it to verify the signature via $c^{e_A} \equiv m \pmod{n}$. Formally, this is presented as follows.

◆ RSA Signature Scheme

Setup Stage: Alice wishes to send a message $m \in \mathcal{M} = \mathbb{Z}/n\mathbb{Z} = \mathcal{C}$ to Bob. She selects an RSA modulus $n = pq$ and an RSA key pair (e, d) obtained via the RSA key generation algorithm given on page 173. The keyspace is $\mathcal{K} = \{k = (n, p, q, e, d) : ed \equiv 1 \pmod{\phi(n)}\}$, where n, e are public and p, q, d are private.

Signing Stage: Alice's private digital signature sig_k is given by

$$\text{sig}_k(m) \equiv m^d \equiv c \pmod{n},$$

and $\text{ver}_k \equiv c^e \pmod{n}$ is her public verification algorithm. She sends (m, c) to Bob.

Verification Stage: Bob obtains Alice's public (e, ver_k), and computes $\text{ver}_k(m, c)$ which is 1 precisely when $m \equiv c^e \pmod{n}$, in which case he accepts the signature, and rejects it otherwise.

Example 4.6 *If Alice generates* $n = pq = 3023 \cdot 3359 = 10154257$, $e = 7$, $d = 7248483$, *then* $\phi(n) = 10147876$. *If she wishes to send* $m = 1111101$ *to Bob, she computes* $\mathrm{sig}_k(m) \equiv 1111101^d \equiv 5134234 \,(\mathrm{mod}\ n)$ *and sends* $(m, c) = (1111101, 5134234)$ *to Bob. After getting Alice's public data* (e, ver_k), *he computes* $\mathrm{ver}_k(m, c) = 1$ *since* $c^e \equiv 5134234^7 \equiv 1111101 \equiv m \,(\mathrm{mod}\ 10154257)$.

A real-world analogue of the above RSA DSS is Alice's signing a postcard and sending it to Bob. Alternatively, Alice could write a letter on paper, sign it, and put it in an envelope, which gets sent to Bob. There is a variant of the RSA DSS given above, which has this as its analogue, namely, to digitally sign the message, then encrypt it. This variant of the RSA DSS is an example of the second kind of DSS, namely, one with an appendix.

Thus, after the above signing stage, she would add an *encryption stage*, where she enciphers with Bob's public exponent e_B, so $(m, c)^{e_B}$ is sent. Then Bob uses his private RSA exponent d_B to calculate $((m, c)^{e_B})^{d_B} \equiv (m, c) \,(\mathrm{mod}\ n)$, and he uses Alice's public RSA exponent to compute $c^e \equiv m \,(\mathrm{mod}\ n)$. This further encryption of the entire message with Bob's public key ensures confidentiality, as does the analogue of sending a sealed letter, rather than a postcard. This variant of the RSA signature scheme can be applied to any DSS with message recovery, namely, by hashing the message and signing the hash, thereby turning it into a DSS with an appendix.

▼ Analysis

As with the RSA cipher itself, we must ensure that the above DSSs are properly set up and the private data is kept secure. We assume this has been done.

The first thing that we observe in the RSA DSS with message recovery is that anyone can verify Alice's signature since e is made public, but only Alice can sign the message since $\mathrm{sig}_k = d$ is private. This also ensures that Alice cannot deny later that she sent the message, since nobody else could have computed m^d. This is an example of *nonrepudiation* (see page 162). Another safeguard is to ensure that a digital signature is not reused, which can be ensured by appending a *timestamp*. For instance, instead of just sending the message m, Alice would have a message with a timestamp t, so the original message would be $M = (m, t)$.

If we choose a small public exponent (see page 178) the verification is considerably faster than the signing. Thus, the RSA DSS is well suited to circumstances where signature verification is the primary operation used. In order to make it even more efficient, we must introduce another in our cast of cryptographic characters, *Trent*, the *trusted third party* (TTP). If we enlist Trent to create a certificate of identification for Alice, which he has to do only once, then verification may take place numerous times by Bob and other entities with whom Alice has communication. It can be shown that for messages no longer than half the RSA modulus, the RSA DSS with message recovery is most efficient, whereas if message blocking is required, then the most bandwidth-efficient[4.5] method is

---

[4.5] *Bandwidth* is the width of the range of frequencies that an electronic signal occupies on

the RSA signature scheme with appendix.

We close this section with a description and discussion of the first DSS recognized by any government. In August of 1991, NIST proposed the *Digital Signature Standard* (DSS) and in May of 1994, it became FIPS 186, (see [91]). Although this evolved into a new standard in the twenty-first century, for simplicity, we present the original standard here, (see [92] for the current DSS, which uses key sizes of 1024 bits or *more*).

◆ Digital Signature Algorithm (DSA — the DSS)

Setup Stage:

1. Alice selects a prime q with 160 bits. Then she selects a prime p with bitlength a multiple of 64 between 512 and 1024, satisfying the property that q divides $p - 1$.

2. She chooses an $\alpha \in \mathbb{F}_p^*$ of order q modulo p. This can be done, for instance, by selecting a primitive root a modulo p and setting $\alpha \equiv a^{(p-1)/q} \pmod{p}$.

3. A cryptographic hash function $h : \mathbb{F}_q^* \mapsto \mathcal{B}_{160}$ (bitstrings of length 160) is selected. She chooses a private key $e \in \mathbb{N}$ such that $e < q$ and computes $\beta \equiv \alpha^e \pmod{p}$.

4. She publishes (p, q, α, β) and keeps private her key e.

Signing Stage: Alice performs the following in order to sign a message $m \in \mathbb{F}_q^*$. In what follows, we will assume that any powers of α or β have been reduced modulo p before being used in any congruence modulo q:

1. Select a random $r \in \mathbb{N}$ such that $r \leq q - 1$.

2. Compute $\gamma \equiv \alpha^r \pmod{q}$.

3. Compute $\sigma \equiv r^{-1}(h(m) + e\gamma) \pmod{q}$.

4. Alice sends m and $\text{sig}_k(m, r) = (\gamma, \sigma)$ to Bob.

Verification Stage: Bob executes the following steps:

1. Obtain Alice's public data (p, q, α, β).

2. Compute $\delta_1 \equiv \sigma^{-1}h(m) \pmod{q}$ and $\delta_2 \equiv \sigma^{-1}\gamma \pmod{q}$.

3. Compute $\delta \equiv \alpha^{\delta_1}\beta^{\delta_2} \pmod{q}$.

4. $\text{ver}_k(m, (\gamma, \sigma)) = 1$ if and only if $\delta \equiv \gamma \pmod{q}$, in which case Bob accepts, and rejects otherwise.

---

a given transmission medium. In other words, it is the speed of data on a given transmission path, usually measured in Mbps (see Footnote 3.11, page 160).

▼ Analysis

First we show why, in step 4 of the verification stage, the criterion actually verifies Alice's signature. It does so since, first of all,

$$\delta_1 + e\delta_2 \equiv \sigma^{-1}h(m) + e\sigma^{-1}\gamma \equiv \sigma^{-1}(h(m) + e\gamma) \equiv r \,(\text{mod } q),$$

then

$$\gamma \equiv (\alpha^r \,(\text{mod } p)) \equiv (\alpha^{\sigma^{-1}h(m)+e\sigma^{-1}\gamma} \,(\text{mod } p)) \equiv (\alpha^{\delta_1}\beta^{\delta_2} \,(\text{mod } p)) \equiv \delta \,(\text{mod } q).$$

Of course, the key e must be kept private or the scheme can be broken, since anyone in possession of e can sign any data and thereby impersonate Alice. Moreover, if r is used more than once, e can be recovered by a cryptanalyst (easily verified, given our many previous related discussions on such matters).

In order to see why the DSA depends upon the DLP for its security, we look at step 2 of the setup stage. Since the Silver-Pohlig-Hellman attack (discussed on page 530) is useless against large prime factors of $p-1$, then this is sufficient to thwart such attacks, and computing r from knowledge of the public γ is deemed to be computationally infeasible. This is the DLP. Moreover, the reader may wonder why we did not just choose a primitive root a modulo p rather than $\alpha \equiv a^{(p-1)/q} \,(\text{mod } p)$. The reason is that it is a generally held opinion that many pieces of information about divisors of $p-1$ can collectively add up to something useful, so DSA avoids this potential problem by keeping all congruences as modulo q data in the signing and verification stages.

An advantage of DSA is that in a precomputation stage, the exponentiation of α can be done offline and need not be part of the signature generation. Another positive feature is that DSA has relatively short signatures of 320 bits so the signing can be done efficiently. Some disadvantages of DSA include the fact that it cannot be used for key exchange. Moreover, the modulus at a mere 512 bits can be a drawback for security, so the prime p should actually be chosen such that $2^{1023} < p < 2^{1024}$ for long-term security. There is another potential problem that one would not imagine and is difficult to detect, namely, the building of a *subliminal channel* into DSA. This is a method of signing an innocuous message with subliminal bits hidden in it. This could be as little as one bit per message or as much as two bytes per message. For the reader interested in how this is done in detail see [287, pages 300 and 301].

DSA evolved into the new *Digital Signature Standard* in FIPS 186-1 announced by NIST on December 15, 1998, and this included the RSA DSS. On February 15, 2000, NIST announced the approval of FIPS 186-2, and this included the upgraded DSS, the RSA DSS, and the *Elliptic Curve Digital Signature Algorithm* (ECDSA), about which we will learn later in the text.

The governmental plans for DSA are akin to that of the role played by DES. They include applications such as cash transactions, data exchange, data storage, electronic mail, and software distribution, to mention a few.

In the next section, we learn about the DSS upon which the DSA was based.

4.4 ElGamal

The best laid schemes o' mice and men Gang aft a-gley.

Robert Burns (1759–96), Scottish poet
— from *Death and Dr. Hornbrook* (1787, st. 3)

The title for this section is the name of a major contributor to several cryptographic schemes. Taher ElGamal was born in Cairo, Egypt, on August 18, 1955. He obtained his bachelor's degree in electrical engineering from Cairo University, in 1977. Both his master's degree and his Ph.D. were obtained from Stanford University in 1981 and 1984, respectively. His doctorate was done under the supervision of Martin Hellman (see page 165). While at Stanford, he helped to pioneer digital signatures and PKC. He founded Security Inc. in 1988, which later became the Kroll-O'Gara Information Security Group, where he became president of its Information Security Group. From 1991 to 1993, ElGamal was the Director of Engineering at RSA Security, Inc., where he produced the RSA cryptographic toolkits and the initial VeriSign certificate issuance products. From 1993 to 1995, he was Vice President of Advanced Technologies at OKI Electric. From 1995 to 1998, he held the position of Chief Scientist of Netscape Communications where he pioneered Internet security technology such as Secure Sockets Layer (SSL), the standard for Web security, to be discussed in Section 5.7. Other accomplishments include development of Internet credit card payment schemes. He also serves on the boards of directors of Phoenix Technologies; RSA Security, Inc.; hi/fn, Inc.; Security Dynamics; ValiCert Inc.; and Register.com, and is a member of the technical staff at Hewlett-Packard Laboratories since 1984. ElGamal is a respected leader in the worldwide information security industry.

The following cryptographic scheme bases its security upon the DLP (see (4.2), page 164). The cryptosystem was first published in [74] in 1985.

The following is performed assuming that Alice wants to send a message m to Bob, and $m \in \{0, 1, \ldots, p-1\}$ (equivalent to the actual plaintext).

(I) ElGamal Key Generation

1. Bob chooses a large random prime p and a primitive root α modulo p.

2. Bob then chooses a random integer a with $2 \leq a < p - 1$ and computes α^a (mod p).

3. Bob's public key is (p, α, α^a) and his private (session) key is a.

(II) ElGamal Public-Key Cipher

Enciphering stage:

1. Alice obtains Bob's public key (p, α, α^a).

2. She chooses a random natural number $b < p - 1$.

3. She computes α^b (mod p) and $m\alpha^{ab}$ (mod p).

4. Alice then sends the ciphertext $c = (\alpha^b, m\alpha^{ab})$ to Bob.

Deciphering stage:

1. Bob uses his private key to compute $(\alpha^b)^{-a} \equiv (\alpha^b)^{p-1-a} \pmod{p}$.

2. Then he deciphers m by computing $(\alpha^b)^{-a}m\alpha^{ab} \pmod{p}$.

Example 4.7 *Suppose that Alice wants to send the message $m = 1010$ to Bob using the ElGamal cipher. Bob chooses $p = 1481$, $\alpha = 3$, and $a = 7$, his private key. He computes $\alpha^a \equiv 3^7 \equiv 706 \pmod{p}$. Bob's public key is therefore $(p, \alpha, \alpha^a) = (1481, 3, 706)$, which Alice downloads from some public database. She chooses $b = 96$ and computes both $\alpha^b \equiv 3^{96} \equiv 737 \pmod{p}$ and $m\alpha^{ab} \equiv 1010 \cdot 706^{96} \equiv 521 \pmod{p}$. The ciphertext is $c = (737, 521)$, which Alice sends to Bob. He uses his private key to compute*

$$(\alpha^b)^{p-1-a} \equiv 737^{1473} \equiv 940 \pmod{p},$$

and

$$(\alpha^b)^{-a}m\alpha^{ab} \equiv 940 \cdot 521 \equiv 1010 \pmod{p},$$

thereby recovering m.

▼ Analysis

Key Generation Options: Although it is preferable in step 1 of key generation, one need not choose a primitive root, provided one chooses an element $\alpha \in (\mathbb{Z}/p\mathbb{Z})^*$ whose order is close to the size of p. In other words, the smallest $r \in \mathbb{N}$ such that $\alpha^r \equiv 1 \pmod{p}$ must be nearly as large as p. Such α are called *near-primitive roots*. In the case of a primitive root, $r = p - 1$.

Security Issues: The random number b generated by Alice in step 2 of the enciphering stage must be kept secret since one can recover $m = m\alpha^{ab}(\alpha^a)^{-b}$ from knowledge of it, given that $m\alpha^{ab}$ and α^a are made public. Furthermore, b should never be used twice. Suppose that Alice uses b for two different messages m_1 and m_2, and Eve knows m_1. Then this is how Eve can obtain m_2. The two ciphertexts are $c_1 = (\alpha^b, m_1\alpha^{ab})$ and $c_2 = (\alpha^b, m_2\alpha^{ab})$. Then she calculates that

$$m_2\alpha^{ab}m_1^{-1}m_1\alpha^{-ab} = m_2.$$

We conclude the discussion of security issues with a detailed argument to show that indeed the security of the ElGamal cipher is based upon the DLP. To see this, we demonstrate first that the ElGamal cipher is equivalent to the Diffie-Hellman key-exchange protocol. Assume Eve can solve the DHP (see page 167), and she desires to get m from $c = (\alpha^b, m\alpha^{ab})$. Since she can solve the DHP, she can determine $\beta \equiv \alpha^{ab} \pmod{p}$ from α^a and α^b. Therefore, she can reconstruct the message $m \equiv \beta^{-1}m\alpha^{ab} \pmod{p}$, In other words, if Eve can break the Diffie-Hellman cipher, she can break the ElGamal.

Now assume that Eve can cryptanalyze the ElGamal cipher above. Then she can obtain any message m from knowledge of $p, \alpha, \alpha^a, \alpha^b$, and $m\alpha^{ab}$. If Eve wants to get α^{ab} from $p, \alpha, \alpha^a, \alpha^b$, she computes $(m\alpha^{ab})m^{-1} \equiv \alpha^{ab} \pmod{p}$. In other words, we have shown that cryptanalyzing ElGamal is tantamount to cryptanalyzing Diffie-Hellman. In fact, the ElGamal cipher may be viewed as a Diffie-Hellman key exchange on $k = \alpha^{ab}$, which is used to encrypt m in step 3 of the enciphering stage. Thus, we have demonstrated that although Diffie-Hellman is not itself a public-key cryptosystem, it is the basis for the ElGamal public-key cryptosystem. Furthermore, ElGamal's cipher has difficulty equivalent to the Diffie-Hellman key-exchange. Moreover, as noted on page 167, if Eve can solve the DLP, she can solve the DHP. The converse is not known, but the consensus is that it is true. Hence, we assume that the security of the ElGamal cipher is based upon the DLP. Last, as with RSA, a modulus of 1024 to 2048 bits is recommended for long-term security.

Deciphering Verification: The reason Bob's deciphering stage works is due to the fact that

$$(\alpha^b)^{-a} m\alpha^{ab} \equiv m\alpha^{ab-ab} \equiv m \pmod{p}.$$

In 1985, ElGamal developed a DSS (see [74] and [75]). It turns out that these publications were perhaps a little hasty since he had not applied for patent rights, thereby forfeiting his rights to those patents. Variations of ElGamal's scheme did get patented by others such as Schnorr (see [151, pages 180 and 181]), whose identification scheme we will study in Chapter 5. The RSA DSS studied on pages 181–183 is deterministic, whereas the following is a randomized algorithm. Also, RSA is a DSS with message recovery, whereas ElGamal's is a DSS with appendix.

◆ ElGamal Signature Scheme

The goal is for Alice to sign and send a message to Bob for verification. The message should be hashed before signing, but for the sake of simplicity, we will not do this and leave the issue for a discussion in the analysis after the description of the DSS.

Key Generation Stage: First Alice engages in ElGamal key generation as described on page 185 for Bob. Thus, Alice's public key is (p, α, y), α being a primitive root modulo a large random prime p (with intractable DLP in $\mathbb{F}_p$) and her private key is a, where $y \equiv \alpha^a \pmod{p}$. The message to be signed is $m \in \mathbb{F}_p^*$.

Signing Stage: Alice performs each of the following:

1. Select a random $r \in (\mathbb{Z}/(p-1)\mathbb{Z})^*$.

2. Compute $\beta \equiv \alpha^r \pmod{p}$ and $\gamma \equiv (m - a\beta)r^{-1} \pmod{p-1}$.

3. For $k = (p, \alpha, a, y)$ the signed message $\text{sig}_k(m, r) = (\beta, \gamma)$ is sent, along with m, to Bob.

Verification Stage: Bob does each of the following:

1. Using Alice's public key (p, α, y) verify that $\beta \in \mathbb{F}_p^*$ and reject if not.

2. Compute $\delta \equiv y^\beta \beta^\gamma \pmod{p}$, and $\sigma \equiv \alpha^m \pmod{p}$.

3. $\text{ver}_k(m, (\beta, \gamma)) = 1$ if and only if $\sigma \equiv \delta \pmod{p}$. Otherwise reject.

Example 4.8 *Let $p = 5531$, with primitive root $\alpha = 10$. Alice selects $a = 351$ as her private key and computes $\alpha^a \equiv 10^{351} \equiv 3122 \equiv y \pmod{5531}$. Thus, her public key is $(p, \alpha, y) = (5531, 10, 3122)$. If $m = 1129$, and she chooses $r = 151$, then she computes $\beta \equiv 10^{151} \equiv 1257 \pmod{5531}$. Then she computes*

$$\gamma \equiv (m - a\beta)r^{-1} \equiv (1129 - 351 \cdot 1257) \cdot 151^{-1} \equiv 52 \pmod{5530},$$

and sends $\text{sig}_k(1129, 151) = (\beta, \gamma) = (1257, 52)$ to Bob. First Bob verifies that $\beta \in (\mathbb{Z}/p\mathbb{Z})^*$, then computes,

$$\delta \equiv y^\beta \beta^\gamma \equiv 3122^{1257} 1257^{52} \equiv 3865 \equiv 10^{1129} \equiv \alpha^m \pmod{5531},$$

so Bob accepts the signature as valid.

▼ **Analysis**

Suppose that Mallory tries to forge Alice's signature on m by choosing a random $r_1 \in (\mathbb{Z}/(p-1)\mathbb{Z})^*$ and computing $\beta' \equiv \alpha^{r_1} \pmod{p}$. Mallory is now in the position of having to compute

$$\gamma' \equiv (m - a\beta')r_1^{-1} \pmod{p-1}.$$

However, if the DLP in $\mathbb{F}_p$ is intractable, then this computation is infeasible so only a guess at the value of γ' is possible with a probability of success being $1/p$. For large p, this is insignificant.

As noted prior to the description of the ElGamal scheme, we purposely did not hash the message. However, one *must* hash the message or else Mallory can forge a signature on a random message. Here is how he does it.

Suppose that Mallory selects $r_1, r_2 \in (\mathbb{Z}/(p-1)\mathbb{Z})^*$. He then computes

$$\beta_1 \equiv \alpha^{r_1} y^{r_2} \pmod{p} \quad \text{and} \quad \gamma_1 \equiv -\beta_1 r_2^{-1} \pmod{p-1}.$$

Now we show that (β_1, γ_1) is a valid signature for the message

$$m_1 \equiv \gamma_1 r_1 \pmod{p-1}.$$

We have, $y^{\beta_1} \beta_1^{\gamma_1} \equiv \alpha^{a\beta_1} \alpha^{(r_1 + ar_2)\gamma_1} \equiv \alpha^{a\beta_1} \alpha^{(r_1 + ar_2)(-\beta_1 r_2^{-1})} \equiv$

$$\alpha^{a\beta_1} \alpha^{-\beta_1 r_1 r_2^{-1} - \beta_1 a} \equiv \alpha^{-\beta_1 r_1 r_2^{-1}} \equiv \alpha^{\gamma_1 r_1} \equiv \alpha^{m_1} \pmod{p}.$$

In any case, a hash function h must be applied to the orig inal message, and the hash is signed. Thus, Mallory would have to find a me ssage m' such that $h(m') = m$, which he has a very low probability of doing if h is strongly collision-resistant. However, if step 1 of the verification stage i s not enforced, then Mallory can forge certain signatures of his own choosing if b e has a previous legitimate message signed by Alice, as demonstrated in the fol! owing.

Suppose that a previous legitimate signature by Alice for a message m is (β, γ). Furthermore, suppose that Mallory is lucky and m^{-1} (mod $p-1$) exists, and Mallory chooses a message m_1 to forge. Mallory computes both congruences $t \equiv m_1 m^{-1} \pmod{p-1}$ and $\gamma_1 \equiv t\gamma \pmod{p-1}$. By the Chinese remainder theorem, he can also compute a solution $x = \beta_1$ to the cong ruences:

$$x \equiv \beta t \pmod{p-1} \text{ and } x \equiv \beta \pmod{p}.$$

Thus,

$$y^{\beta_1} \beta_1^{\gamma_1} \equiv \alpha^{a\beta_1} \beta^{t\gamma} \equiv \alpha^{a\beta t} \alpha^{rt\gamma} \equiv \alpha^{t(a\beta+r\gamma)} \equiv \alpha^{tm} \equiv \alpha^{m_1 n \, im^{-1}} \equiv \alpha^{m_1} \pmod{p}.$$

Hence, (β_1, γ_1) is accepted as a valid signature by the ve rification stage for m_1, if step 1 in that stage is ignored. The essential nature of step 1 in the verification stage was first observed in [26].

The value r chosen by Alice in the signing stage has to be kept secret, or there is a total break of the system since Mallory can g et a from knowledge of r. Since β, γ, and m are known, then knowledge of r me ans that he may compute $a \equiv (m - r\gamma)\beta^{-1} \pmod{p-1}$.

Also, if Alice is careless and uses r for the signing of two different messages, then Mallory can get r and break the system as abc ve. Here is how he gets r.

If $\text{sig}_k(m_1, r) = (\beta, \gamma_1)$ and $\text{sig}_k(m_2, r) = (\beta, \gamma_2)$, then

$$y^{\beta} \beta^{\gamma_1} \equiv \alpha^{a\beta + r\gamma_1} \equiv \alpha^{m_1} \pmod{p},$$

and

$$y^{\beta} \beta^{\gamma_2} \equiv \alpha^{a\beta + r\gamma_2} \equiv \alpha^{m_2} \pmod{p}.$$

Therefore,

$$\alpha^{m_2 - m_1} \equiv \beta^{\gamma_2 - \gamma_1} \equiv \alpha^{r(\gamma_2 - \gamma_1)} \pmod{p}.$$

Hence,

$$m_2 - m_1 \equiv r(\gamma_2 - \gamma_1) \pmod{p-1}.$$

If $\gcd(p-1, \gamma_2 - \gamma_1) = g$, then

$$\frac{m_2 - m_1}{g} \equiv \frac{r(\gamma_2 - \gamma_1)}{g} \pmod{(p-1)/g}.$$

Thus,

$$r \equiv \left(\frac{m_2 - m_1}{g}\right) \left(\frac{\gamma_2 - \gamma_1}{g}\right)^{-1} \pmod{(p-1)/g},$$

since $\gcd((\gamma_2 - \gamma_1)/g, (p-1)/g) = 1$, and once we have r we can get a as above.

We close this section with some advanced material for which the reader will need some knowledge of elliptic curves. We have presented all necessary material in Appendix A (see page 498).

☞ **ElGamal Public-Key Elliptic Curve Cryptosystem (ECC)**

We assume that E is an elliptic curve over $\mathbb{F}_p$ where p is prime and H is a cyclic subgroup of $E(\mathbb{F}_p)$ generated by a point $P \in E(\mathbb{F}_p)$. Alice wants to send a message to Bob whose public key is (E, P, aP) and whose private key is the natural number $a < p - 1$. Alice executes the following.

Enciphering stage:

1. Choose a random natural number $b < p - 1$.

2. Consider the plaintext message units embedded as points m on E.

3. Compute $\beta = bP$ and $\gamma = m + b(aP)$.

4. Send the ciphertext $E_e(m) = c = (\beta, \gamma)$ to Bob.

Deciphering stage:

Once Bob receives the ciphertext, the plaintext m is recovered via the private key as
$$D_d(c) = m = \gamma - a\beta.$$

Example 4.9 *Consider the elliptic curve group E given by $y^2 = x^3 + 4x + 4$ over $\mathbb{F}_{13}$. It can be shown that $|E(\mathbb{F}_p)| = 15$, which is necessarily cyclic. Also, $P = (1,3)$ is a generator of E. Assuming that Bob's public key is $(E, P, 4P)$ where $a = 4$ is the private key and $m = (10,2)$ is the message that Alice wants to send to Bob, then Alice performs the following. Alice chooses $b = 7$ at random. Then she calculates*

$$E_e(m) = E_e((10,2)) = (bP, m + b(aP)) = (7P, (10,2) + 7(4P)) =$$
$$((0,2), (10,2) + 7(6,6)) = ((0, 2), (10,2) + (12,5)) = ((0,2), (3,2)) = (\beta, \gamma) = c.$$

Then Alice sends c to Bob who uses the private key to recover m via

$$D_d(c) = (3,2) - 4(0,2) = (3,2) - (12,5) = (3,2) + (12,8) = (10,2) = m.$$

The ElGamal cipher given on page 185 has what is called a *message expansion factor* of 2 over $\mathbb{F}_p$, meaning that the ciphertext is roughly twice the size of the plaintext. However, an elliptic curve implementation of ElGamal has a message expansion factor of approximately four, because there are p plaintexts, but each ciphertext is comprised of four field elements. Moreover, and perhaps more seriously, plaintext message units m lie on E and there does not exist an appropriate (both theoretically and practically) method of deterministically generating such points. A more workable scheme is the *Menezes-Vanstone ECC*, which is described in [169, pages 245 and 246] from which the above example was adapted. A big advantage of PKC ECCs is that key sizes are much smaller than, say, RSA.

Chapter 5

Cryptographic Protocols

The shadow cloaked from head to foot,
Who keeps the keys of all the creeds.
Alfred Lord Tennyson (1809–1892), English poet
— from *In Memoriam A.H.H.* (1850), canto 23

5.1 Introduction

The term *cryptographic protocol* was briefly introduced to us in footnote 2.24 on page 101. This chapter is devoted to studying several particular such protocols including: key establishment; key agreement; identification; commitment; secret sharing; electronic voting; protocol layer analysis with SSL as an Internet protocol providing authentication and secrecy for session-based communications; and we conclude with digital cash schemes and e-commerce.

Before presenting the wealth of information listed above, we need an overview. In any interaction among people, there is a certain level of risk, trust, and expected behavior implicit in the interchange. This may be inspired to be conducted properly for any reason ranging from fear of prosecution under the law, to the desire to act in an ethical manner due to societal influences. However, in a cryptographic protocol, trust has to be kept to the lowest possible levels. In any such protocol, if there is an absence of a mechanism for verifying, say, authenticity, one must assume that other participants are dishonest, if for no other reason than self-preservation (or in the case of e-commerce, the preservation of a positive balance in one's bank account). One must take this approach as a default, unless there is a clearly specified secure interface to deal with authentication of the entities with whom one is communicating.

In Section 5.7, we will discuss the various layers that make up the hierarchy of communications within a given protocol. These layers allow each level to speak to the next level up. Moreover, any alteration on one protocol layer does not affect other layers. This setup greatly eases the burden of work in creating and maintaining communications networks. We leave the actual details of these mechanisms for Section 5.7. We concentrate on some other features to occupy this section, including some types of protocols that we will not cover in the succeeding sections, but rather highlight them here as an introductory feature of this chapter.

One aspect, not explicitly described in the following sections of this chapter, but deserving of some preliminary commentary, is the notion of a *subliminal channel*, meaning covert methods for an adversary to send missives hidden within a legitimate message. Perhaps the most widely-known such example is the hiding of bits in a digitized photograph, or commercial digital television message. Much of this type of subliminal message protocoling is accomplished via steganography. The reader is referred to the book [137], which is devoted entirely to steganographic methodologies, for such information. Some modern schemes use nonces, such as the DSA (discussed on page 184), and the ElGamal signature scheme. We will encounter subliminal channels again when we describe the topic of nuclear test ban treaty compliance in Section 9.6.

Two other intertwined types of protocols with which we end this section are of value in many situations. For instance, suppose that Bob is a CIA agent and he wants to buy secrets from Alice who is a Russian double agent. Bob does not want to reveal what he knows and what he does not, since the consequences could range from the merely embarrassing to the downright dangerous. Hence, he would like to buy secrets from Alice, but not have her know, in advance, which ones. In other words, Alice transfers information, containing one or more secrets, to Bob in such a way that upon completion of the protocol, Alice does not know (is "oblivious" as to) which of the secrets Bob received. This mechanism has a name.

◆ Oblivious Transfer Cryptographic Protocol

We deal with the simplest scenario where only one secret is transferred, but this is easily extrapolated to numerous secrets. Alice sends Bob two messages, only one of which he receives, and Alice does not know which one. The PKC and SCK used are RSA and Rijndael, respectively.

1. Alice generates two RSA key pairs and sends the two public keys to Bob.

2. Bob selects a Rijndael key, k, and chooses, at random, one of Alice's public keys to encrypt it, and sends this to Alice.

3. Alice uses her two private keys to attempt to decipher the cryptogram, only one of which is successful, but she does not know which one, since she does not know the secret Rijndael key. So one is binary gibberish and the other is the legitimate Rijndael key, k.

4. Alice enciphers her two secret messages, one with k, and the other with the binary gibberish key, and sends both to Bob. (For the sake of simplicity, we will assume that the two secret messages are indeed distinct, that is, that Alice is not trying to cheat Bob.)

5. Bob uses his secret Rijndael key on both messages, but only one is successful in yielding one of the secrets, and Alice does not know which one.

To see how this works with symbols, suppose that Alice's RSA key pairs are (e_1, d_1) and (e_2, d_2), and Bob's Rijndael key is k. Assume he chooses e_1 to encipher k and sends $E_{e_1}(k)$ to Alice, who forms $D_{d_1}(E_{e_1}(k)) = k$ and $D_{d_2}(e_1(k)) = k_2$, but she does not know which of k or k_2 is the legitimate key. She forms $E_k(m_1)$ and $E_{k_2}(m_2)$, which she sends to Bob. Once Bob receives these, he uses k^{-1} to get $D_{k^{-1}}(E_k(m_1)) = m_1$, and $D_{k^{-1}}(E_{k_2}(m_2)) = m'$, where the latter is gibberish and the former is the legitimate secret. Alice, however, does not know which one he received. (If Bob needs to verify that Alice did not cheat in step 4, then he asks Alice for her private keys so he can verify the outcome of both possible transfers.)

This is only one of numerous oblivious transfer protocols, some of which are noninteractive. This, however, gives the reader a flavour of the methodology involved. Now we turn to the use of oblivious transfer within the scope of yet another type of protocol.

◆ Cryptographically Secure Contract Signing Protocol

Alice and Bob want to sign an important contract, and they are using Rijndael keys.

1. Alice and Bob, independently, select a set of $n \in \mathbb{N}$ key pairs,

$$\mathcal{S} = \{(\ell_j, r_j) : 1 \le j \le n;\ \ell_j, r_j \in \mathcal{K}\},$$

   where ℓ_j and r_j are from the keyspace $\mathcal{K}$. (These pairs are randomly selected, so there is no special relationship between the left and right sides of any given pair.)

2. Alice and Bob, independently, generate n pairs of signatures,

$$\mathcal{M} = \{S_j = (L_j, R_j) : 1 \le j \le n\},$$

   where L_j and R_j are the left and right halves of their respective signatures. Also, each S_j, for the sake of simplicity, will be assumed to be accompanied by a time stamp and a digital signature of the contract itself. The contract will be considered to be signed if both L_j and R_j for a given message pair, can be produced by each of them.

3. Alice and Bob, independently, sign each message as follows:

$$\mathcal{C} = \{(\ell_j(L_j), r_j(R_j)) :\ 1 \le j \le n\},$$

   then they send each other their respective pairs of encrypted messages, namely, $2n$ keys in the form of n pairs sent to each other, ensuring that they tell each other which is left and which is right for each pair.

4. Using the oblivious transfer protocol, Alice and Bob send each other exactly one half of each key pair, namely, either ℓ_j or r_j, so neither of them knows which half they have.

5. Alice and Bob independently decrypt what messages they can, ensuring as they do so that they do indeed have a legitimate message in each case.

6. Alice and Bob alternate in sending the bit for each of their $2n$ keys until all verifying bits have been received by both of them.

7. Once step 6 is complete, they can each decrypt the other half of each message and the contract is signed.

If there is a question of cheating, Alice and Bob can exchange private RSA keys at the end of the contract signing as mentioned in the oblivious transfer protocol to provide a verification step. However, cheating would likely be detected at step 5 since for large enough n, each has only a 1 in 2^n chance of escaping detection. Hence, both have incentive to complete the protocol fairly. There is an additional problem if either Alice or Bob has significant resources over the other. For instance, if after sending a sufficient number of bits in step 6, Alice has the computing power to get the rest of the bits, and Bob does not, she is at an advantage since she can stop sending bits, and claim the contract to be signed since she can produce the signed portions of both halves. Hence, this protocol should not be implemented unless both Alice and Bob have roughly equal resources in computing power.

In general, the building of cryptographic protocols relies upon the building bricks, called *primitives*, by which we mean cryptographic tools used to ensure information security. For instance, SKC primitives consist of symmetric-key ciphers (both block and stream ciphers); MACs; digital signatures; pseudorandom sequences; and identification tools. PKC primitives consist of public-key ciphers; digital signatures; and identification tools. Primitives not involving keys are hash functions (unkeyed); one-way permutations; and random sequences. When building a protocol, it is essential that all possible hypotheses used in the design are explicitly identified, and an analysis is made of what effect a breach of any of those hypotheses might have upon the security of the protocol. A *protocol failure* occurs when it is possible for an adversary, such as Mallory, to manipulate the protocol to his advantage without breaking any of the underlying primitives such as the encryption scheme. In this case, the protocol fails to meet the goals for which it was intended. Typically, a protocol failure occurs when there is a weakness in one of the underlying primitives that is magnified by the protocol; or there is an oversight in the implementation of the protocol that allows manipulation without the breaking of the primitive itself. For instance, if the one-time pad is used in a careless fashion, then there could be protocol failure, even though the one-time pad is itself secure. As always, a bad implementation or improper use of a secure mechanism can compromise the entire enterprise.

5.2 Keys

You know everyone is ignorant, only on different subjects.
Will Rogers (1879–1935), American actor and humorist

The term *key agreement* refers to a protocol where two entities, acting in concert, contribute to the generation of a symmetric key. We have seen one of the most famous of these in the Diffie-Hellman key-exchange protocol discussed on page 166. This brings up the topic of the less accurate term *key-exchange*, which is often used in reference to key agreement, the reason being that two entities perform an *exchange* of information resulting in their *agreement* on a shared key. *Key establishment* is a means of using cryptography to establish a shared secret (symmetric) key.

Key *distribution*, also called *key transfer* or *key transport*, is a protocol where one entity generates a symmetric key and sends it to other entities, usually over a network. On page 162, we discussed key predistribution and the issues surrounding it from both perspectives of SKCs and PKCs. With an SKC, there is the so-called n^2-*problem*, on an n-user network, where the number of keys required is $n(n-1)/2$. To avoid this problem in an SKC, we can employ Trent who needs only a single key shared with each user. Also, we want protocols to do more than just establish shared keys. We need to avoid impersonation, since for instance, the Diffie-Hellman protocol is open to an impersonation attack (see page 180), so we require protocols that play both roles of establishing keys for users and mutual entity authentication.

The following is a key authentication and establishment protocol using only SKC, and the employment of Trent, plus the introduction of some new characters in our cryptographic play.

Kerberos is the three-headed dog of Greek mythology that stood guard at the gates of Hades. It is also the name of an authentication protocol developed at MIT in 1989. According to MIT, the initial intent was to have not only authentication, but also accounting and auditing features. However, the last two "heads" were never added. The Kerberos project originated from a larger endeavor at MIT, called *Project Athena*, the purpose of which was secure communication across a public network for student access of their files. Kerberos is the authentication protocol aspect of Project Athena, and is based upon a client-server-verifier model described as follows. (In Section 8.5, we look at the client-server model as a general architecture, but for now, we will be content with the informal description given below.)

We require the introduction of more in our cast of cryptographic characters. First, a *client*, Carol, is a user (which might in reality be software or a person) with some goal to achieve, which could be as simple as sending e-mail or as complex as installation of a system's software. A *server* (and verifier), Victor, provides services to clients, which might involve anything from e-commerce to accessing personal files.

◆ Kerberos

Kerberos was designed with a goal of authenticating clients who desire access to servers in a network. We prepare for the formal, but simplified, description by giving a brief introductory overview. We will be describing a basic version of Kerberos that provides both entity authentication and key establishment using an SKC, denoted by E, that we assume has built-in data integrity features, and Trent, as our trusted third party.

▼ Preliminary Description

Carol sends a request to Trent to establish a session key (see page 163), which she can use to communicate with Victor, and includes her identity data in the request. Using the key he shares (only) with Victor, Trent generates an E-encrypted *ticket*, t, for Victor, which includes Carol's identity data, certain time constraints, and a copy of the session key. He also uses the key he shares (only) with Carol to encipher a message, m, containing a copy of the session key, Victor's identity data, and some time constraints. He sends t and m to Carol.

She cannot alter t, without being detected, since it is enciphered with a key known only to Trent and Victor. Carol verifies time constraints, and decrypts m using the key she shares with Trent. Then she uses the copy of the session key to E-encrypt an *authenticator*, a, which contains her identity data, and some fresh time constraints. She sends a and t to Victor.

He decrypts t with the key that he shares with Trent. This gives him the copy of the session key, which he now uses to decrypt a, where he checks both time constraints and the identity data in a and t. If they match, and the time constraints are valid, Carol is verified to Victor. He uses the session key to encipher the time data she sent in the authenticator and sends it to her.

Carol decrypts using the session key and if the time data matches, Victor is authenticated to her. Hence, they have a established a session key with which to securely communicate.

▼ Kerberos Authentication Protocol — Simplified

Basic Assumptions: Trent selects a random key k, a timestamp t, and a validity period L, called a *lifetime*. Carol and Trent share a secret symmetric key $k_{C,T}$, and Victor and Trent share one, $k_{V,T}$. Also, I_C, I_V, and I_T are identity strings for Carol, Victor, and Trent, respectively. Moreover, at the outset, Carol and Victor share no secrets.

Protocol Activities:

1. Carol sends her request for a session key to use with Victor, together with her identity string I_C to Trent, who computes $m_C = E_{k_{C,T}}(k, I_V, t, L)$. He also computes

$$m_V = E_{k_{V,T}}(k, I_C, t, L),$$

   called a *ticket* for Victor, and sends both m_C and m_V to Carol.

2. Carol uses $E_{k_{C,T}}^{-1}$ to retrieve k, I_V, t, and L from m_C. She verifies that t and L are valid, and that I_V is the identity of Victor. She then creates a fresh

timestamp t_C, and computes

$$m'_V = E_k(I_C, t_C),$$

called the *authenticator*, which she sends to Victor along with m_V.

3. Victor uses $E_{k_{V,T}}^{-1}$ to get k from the ticket, m_V. Then he uses E_k^{-1} to decrypt m'_V. He checks that the two copies of I_C from m_V and m'_V match. He checks that t_C is valid. Then he checks that his current time is within the lifetime L specified by m_V. If these three facts hold, he declares Carol to be authentic, and he computes $m'_C = E_k(t_C)$, which he sends to her.

4. Carol applies E_k^{-1} to m'_C, and checks that t_C matches the value she created in step 2. If it does, she declares Victor to be authentic and now has a session key k to communicate with him.

▼ **Analysis**

Any timestamp in the protocol must be within the *expiration window*, which can be any agreed fixed amount. Also, checking that a given time t is within the expiration window can be accomplished by subtracting t from the current time, which must be within some mutually accepted fixed time interval. The role of the timestamp t and the lifetime L is to thwart Mallory from storing old messages for retransmission at a later time (a replay attack).[5.1] If any of the checks against t in the above protocol fail, then the protocol terminates since a *stale* timestamp has been discovered. The lifetime L also has the advantage of allowing Carol to reuse Victor's ticket without contacting Trent, so step 1 can be eliminated over the lifetime of the ticket. However, each time Carol reuses the ticket, she must create a new authenticator with a fresh timestamp, but the same session key k. The use of timestamps means that there must be synchronized clocks in the network. Cryptanalysts must be prevented from modifying clocks to guarantee the security of the scheme.

In the full version of Kerberos, there is another entity who grants the tickets, and Trent's role is merely to authenticate. Thus, in the full Kerberos model, Trent is a trusted authority, called the *Kerberos authentication server*. The Kerberos protocol is based upon predistribution protocols of Needham and Schroeder (see [177] and [178]), full descriptions and analysis of which can be found in [170, pages 167–169]).

In the above scheme, we used only an SKC and Trent to establish a shared secret key as well as mutual authentication. However, the scheme heavily depends upon synchronized clocks, which is difficult to achieve. When synchronized clocks are not available, the following scheme is required.

The next protocol uses a PKC and ensures both key agreement and mutual entity authentication. The scheme involves what is known as a *challenge-response* protocol, which we now describe.

[5.1] A *replay attack* (also called a *playback attack*) on a protocol involves the use of information gathered from a previous execution of the protocol in an attempt to deceive.

On page 180, we discussed and illustrated an impersonation attack on PKCs. One mechanism for thwarting such attacks is for Alice to get a random number r from Bob, sign it with her private key d, which Bob can then verify. This is illustrated in Diagram 5.1.

Diagram 5.1 (Challenge-Response Protocol)

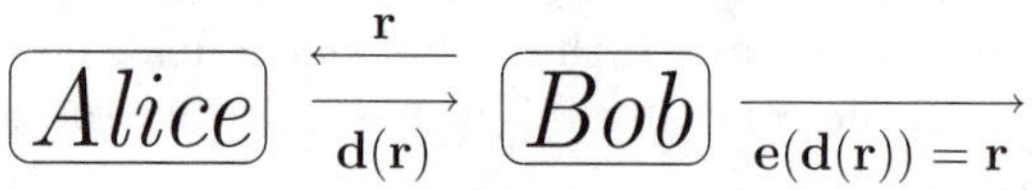

$$\boxed{Alice} \begin{array}{c} \xleftarrow{\ \ r\ \ } \\ \xrightarrow{\ d(r)\ } \end{array} \boxed{Bob} \xrightarrow{\ e(d(r))\ =\ r\ }$$

The random number r is the *challenge* from Bob to which Alice returns the response $d(r)$.

◆ **Three-Way Authentication and Key Agreement**

Basic Assumptions: In order to describe the scheme we need a PKC, whose encryption we will denote by E and whose decryption we will denote by D. We also need a DSS with signature and verification pair (sig, ver). Moreover, Alice and Bob have two key pairs, their PKC key pairs (e_A, d_A), (e_B, d_B), respectively, and their DSS key pairs (s_A, v_A) and (s_B, v_B), respectively. Also, I_A and I_B are their respective identity data strings. The goal is for Alice and Bob to agree on a session key and mutual authentication.

Protocol Steps:

1. Alice selects a nonce r_A and sets $t = (I_B, r_A)$, signs it, $\text{sig}_{s_A}(t)$, and sends $m_1 = (t, \text{sig}_{s_A}(t))$ to Bob.

2. Bob verifies Alice's signature, and chooses a nonce r_B and a random session key k. He enciphers k with Alice's public key, $E_{e_A}(k) = c$, sets

    $$t_1 = (I_A, r_A, r_B, c),$$

    and signs it, $\text{sig}_{s_B}(t_1)$. Then he sends $m_2 = (t_1, \text{sig}_{s_B}(t_1))$ to Alice.

3. Alice verifies Bob's signature, and checks that her r_A matches the one she generated in step 1. Once verified, she now is convinced that she is communicating with Bob. She gets k via

    $$D_{d_A}(c) = D_{d_A}(E_{e_A}(k)) = k,$$

    sets $t_2 = (I_B, r_B)$ and signs it, $\text{sig}_{s_A}(t_2)$. Then she sends $m_3 = (t_2, \text{sig}_{s_A}(t_2))$ to Bob.

4. Bob verifies Alice's signature and checks that r_B matches his choice in step 2. If both verifications pass muster, Alice and Bob have mutually authenticated each other's identities and have agreed upon session key k.

▼ Analysis

The checks by Alice in step 3 and Bob in step 4, that their unique random number choices match, are intended to thwart replay attacks. The built-in challenge-response protocol is given as follows. Alice's challenge number r_A is sent to Bob, who responds with his signature affixed to it, thereby verifying himself to her. Bob then sends his challenge r_B to Alice, who responds with her signature affixed, thereby verifying her identity to him. The protocol gets its name from the fact that three messages m_1, m_2, m_3 are exchanged, sometimes called the *three-pass authentication and key-agreement protocol*. This is similar to what is called the X-509 strong three-way authentication protocol (see Section 7.4).

We have seen key establishment, key agreement, and entity authentication schemes based strictly on SKC and strictly on PKC. An example of a hybrid key agreement and authentication protocol, called *Encrypted Key Exchange* (EKE), was introduced by Bellovin and Merritt [17] in 1992, with a patent granted to the inventors in 1993 (see [18]). For a full description and analysis of EKE, see [170, pages 169–171]. Since the inception of EKE, it has evolved into a family of protocols most of which are stronger than the original. For instance, in 1996, the *Simple Password Exponential Key Exchange* (SPEKE) was developed (see [129]). Both EKE and SPEKE allow use of a small password to provide authentication and key agreement over an unsecured channel. However, password-based protocols are subject to *password sniffing*, which is an attack in which an adversary listens to data traffic that includes secret passwords in order to capture and use them at a later time. To give an example from the Internet, we need to define *TCP/IP*, which is the acronym for *Transmission Control Protocol/Internet Protocol*, the set of communications protocols used to connect hosts on the Internet. *Hosts* are those computers that provide services to other computers and to users on a network (such as the Internet). The Internet itself is the globally interconnected network of computers using, mainly, the set of Internet protocols. TCP/IP uses several protocols, the two main ones being TCP and IP. TCP/IP is used by the Internet, and is considered to be the *de facto* standard for transmitting data over networks. We will discuss these protocols in Section 5.7. Now we return to the issue of password sniffing.

Eavesdropping on a TCP/IP network can easily be accomplished against protocols that transmit passwords in the clear. In addition, if password protocols require the passwords to be stored in the host, usually hashed, then for it to be revealed would compromise security. The biggest problem with the EKE family is that they require what is called *plaintext-equivalence* meaning that both the client and the server/host are required to have access to the same secret password or hash thereof. There are versions such as *Augmented* EKE *A-EKE* (see [19]) making EKE a *verifier-based* protocol. We will use the password/verifier terminology to mean the same as the private/public key pairs in PKC *with the modification* that the verifier is stored and kept secret by the server/host. The verifier is similar to the public key in that it can easily be computed from the password, but it is computationally infeasible to compute the password from the verifier. Thus, the verifier-based protocols are those that only require the verifier to be stored (not the password). However, in rectifying the password-storage problem inherent in EKE, A-EKE destroys a desirable property that is possessed by EKE, namely *forward secrecy*, which means that revealing the password to a cryptanalyst does not help in obtaining session keys of past sessions (see [272]). To address these problems and create a protocol that has all the positive and none of the negative features of the EKE family, a new protocol was developed at Stanford University in 1997, called *Secure Remote Protocol* (SRP). SRP differs from the EKE family in that instead of relying on shared secrets such as passwords or their equivalents stored by a server, SRP mandates that the server store a salt value (see Footnote 3.8 on page 136) and a verifier. Without password storage, SRP is more secure than password schemes, performing a secure key exchange in the authentication process. We now describe this protocol as a closing feature of this section.

◆ Secure Remote Protocol (SRP-6) — Latest Version

Carol will interact with Victor to establish a password k, and upon mutual authentication, a session key S will be used to establish a key K to be used to encrypt all future traffic.

Background Assumptions: All computations are carried out modulo a preselected large prime p, and α is a primitive root modulo p, also preselected. The prime p must be a *safe prime*, which means that $(p-1)/2$ must be a prime. These are public values. H is a hash function (typically SHA-1 is used with SRP-6; see page 255). To establish a password k with Victor, Carol picks a random salt s and computes $v \equiv \alpha^d \pmod{p}$, where $d = H(s, k)$, her private key. Victor stores v and s as Carol's password verifier and salt. All equalities below are assumed to be reductions modulo p for convenience.

Protocol Steps:

1. Carol sends I_C and $A = \alpha^a$ (where a is a nonce) to Victor.

2. Victor looks up her password entry, retrieves s and v from the database, and sends both s and $B = 3v + \alpha^b$ (where b is a nonce) to Carol.

3. Both Carol and Victor, independently, compute $u = H(A, B)$.

5. Carol computes her private key $d = H(s, k)$, then she calculates $S = (B - 3\alpha^d)^{(a+ud)}$. Victor independently computes $S = (Av^u)^b$.

6. Both Carol and Victor apply the hash to get $K = H(S)$, the session key.

7. To verify that she has the correct key, Carol sends

$$h_1 = H(H(p \oplus H(\alpha)), H(I_C), s, A, B, K)$$

to Victor, where $\oplus$ is addition modulo 2.

8. Victor computes h_1 and verifies that it matches the value of h_1 sent by Carol, then he sends $h_2 = H(A, h_1, K)$ to her.

9. Upon receipt of h_2 Carol verifies that K is the correct key. If all is valid, then they have a shared session key K.

▼ **Analysis**:

The exchange may be seen as a type of Diffie-Hellman exchange, since the private values a and b correspond roughly to the private values in the Diffie-Hellman key-exchange, and they have similar properties. In fact, the exponentials used in the protocol have been modified over the SRP-3 version to counter dictionary attacks[5.2] as well as casual password sniffing. In SRP-6, this is accomplished by introducing the coefficient 3 of v in step 2 (which was 1 in SRP-3), as well as the addition of sending A in step 1 (whereas only I_C was sent in SRP-3). Adding the coefficient of 3 to v removes a symmetry property in SRP-3 that made it easier to launch a dictionary attack. Moreover, the computation of u as a hash in step 3 (whereas the related variable was sent unhashed in SRP-3), thwarts impersonation attacks. In SRP-3, the order of sending messages and revelation of the related u parameter, before certain steps were executed, opened the protocol to such impersonation attacks. This introduction of the hash eliminates this problem. If Mallory wants to find a value of u for which $u = H(\alpha^a v^{-u}, B)$, then it is infeasible (with a hash function such as SHA-1, for instance), for him to pick a value of u and work back to find an appropriate value of a.

From the above, we see that SRP-6 is designed to thwart dictionary attacks since even if Victor's password database is publicly disclosed, Mallory, for instance, would need an exponential computation to validate a guess, which is more time consuming than he can afford. In any case, Victor uses SRP-6 to store passwords in a form not directly attainable by Mallory.

SRP is relatively immune to the man-in-the-middle attack (see Footnote 3.7 on page 134) because, without Carol's password, Mallory cannot deceive both Carol and Victor. Without Carol's private key, Mallory cannot deceive Victor into thinking he is communicating with Carol. Without v as well, Mallory has no hope of masquerading as Victor to fool Carol. Hence, properly implemented, SRP is perhaps the most secure of authentication schemes with password entry. It is part of a new family of verifier-based protocols, called *Asymmetric Key Exchange* (AKE), where password and verifier are integrated into a single key exchange round. One protocol wherein the use of SRP-6 would be particularly useful is SSL/TLS (see Section 5.7 on page 218), since the server can send its (initial) messages in one pass, rather than than two (for both client and server) as was the case with SRP-3.

SRP solves the long-standing problem of having both ease-of-use and security without sacrificing performance. Also, it has the advantage of forward secrecy. SRP is ideal for a number of applications for which secure password authorization is required. For more technical information, see Internet draft: *draft-ietf-tls-srp-08* at the IETF website (see Footnote 5.3 on page 219):

http://www.ietf.org/internet-drafts/draft-ietf-tls-srp-08.txt.

[5.2] A *dictionary attack* occurs when an adversary takes a list of probable passwords, hashes all the entries on the list, and compares this list to the list of actual enciphered passwords in an effort to find a match.

5.3   Identification

Two things are identical if one can be substituted for the other without af-
fecting the truth.

Gottfried Wilhelm Leibniz (1646–1716), German Philosopher
— translated from *Table de définitions* (1704) in L. Coutourat (ed.)
Opuscules et fragments inédits de Leibniz (1903)

In order to describe our first identification scheme, we need to discuss how
the participants interact. For instance, Alice and Bob will interact in such a
way that Alice is able to "prove" her identity to Bob by convincing him that
she knows a secret without revealing anything about that secret. This type of
identity authentication is valuable in such instances as Alice having to convince
a merchant that she is the owner of a credit card without revealing the password
or PIN for that card. This is important since revealing the PIN, for instance,
would allow someone to impersonate Alice. (In the modern day, this has come
to be known as *identity theft*, where a criminal obtains a person's financial
information and uses it to withdraw funds from bank accounts or uses credit
cards to fraudulently acquire goods and services.) In other words, we need
mechanisms for authentication that will *not* give the authenticator the ability
to impersonate you.

A challenge-response protocol is an *interactive proof of knowledge* involving
two participants, Alice and Bob, say, where Alice is the *prover*, and Bob is
the *verifier*. Alice knows a secret S and must convince Bob of knowledge of
this secret. If she reveals nothing about the secret in so doing, it is called a
zero-knowledge proof of knowledge. However, there are rigorous mathematical
constraints in defining such systems into which we will not delve here. For our
needs, the above and what follows are sufficient. For a mathematical analysis
of the theory of zero-knowledge concepts, see [169, pages 252–261].

We are ready for our first protocol introduced in 1987 by Fiat and Shamir as
an authentication and digital signature scheme in [87]. Later, it was modified
by Feige, Fiat, and Shamir to an identification protocol (see [79] and [80]). For
ease of presentation, we supply a simplified version and we employ Trent to set
the stage for us.

◆ **(Simplified) Feige-Fiat-Shamir Identification Protocol**

Background Assumptions:

First, Trent chooses an RSA modulus $n = pq$, where p and q are large primes
of roughly the same size to be kept secret. Also, a parameter $a \in \mathbb{N}$ is chosen.
Next, Alice and Bob, respectively, randomly select secret $s_A, s_B \in (\mathbb{Z}/n\mathbb{Z})^*$.
Then they compute, respectively, the least residues t_A and t_B modulo n where

$$t_A \equiv s_A^2 \pmod{n} \text{ and } t_B \equiv s_B^2 \pmod{n}.$$

They register their secrets s_A and s_B with Trent, whereas t_A and t_B do not need
to be kept secret. The goal is for Alice to *prove* her identity, via demonstration

of knowledge of s_A to Bob, without revealing anything about s_A, called a zero-knowledge proof of a modular square root of t_A.

Protocol Steps

1. Alice selects an $m \in (\mathbb{Z}/n\mathbb{Z})^*$ and sends $w \equiv m^2 \pmod{n}$ to Bob. (The value m is called a *commitment* and w is called a *witness*.)

2. Bob chooses $c \in \{0,1\}$ and sends it to Alice. (The value c is called a *challenge*.)

3. Alice computes $r \equiv ms_A^c \pmod{n}$ and sends it to Bob. (The value r is called the *response*.)

4. Bob computes r^2 modulo n. If $r^2 \not\equiv wt_A^c \pmod{n}$, then terminate the protocol with Bob rejecting the proof. Otherwise go to step 5.

5. Reset a's value to $a - 1$ and go to step 1 if $a > 0$. If $a = 0$, then terminate the protocol with Bob accepting the proof.

▼ **Analysis**

The above protocol is a three-pass protocol in the sense that three messages are exchanged, w, c, and r. (Compare with the analysis of the three-pass protocol discussed on page 199.) The process in steps 1 and 2 is an instance of the cut-and-choose protocol, whereas the process in steps 2 and 3 (given a witness from step 1), is an instance of a challenge-response protocol. Moreover, each iteration of these rounds (namely, for each value of a used), are sequential and independent. In other words, there is a variation from one round to the other, with the initial randomness giving the guarantee that they will be so. The protocol is designed to ensure that only the prover, Alice, with knowledge of the secret s_A is capable of answering all of the challenges with correct responses, and none of these responses give away any information about the secret.

Interactive proof (of knowledge) systems have to satisfy certain properties in order to be valid. One of them is called (knowledge) *completeness*, which means that if Alice actually knows the secret, s_A in this case, then Bob will always accept Alice's proof. We now demonstrate that the above protocol satisfies this property. If Alice knows s_A, then the response $r \equiv ms_A^c \pmod{n}$ is a square root of $wt_A^c \pmod{n}$ (for any c) so Bob will accept.

Another property that interactive proof (of knowledge) systems must satisfy is that of (knowledge) *soundness*, which means that if Alice can convince Bob with reasonable probability, then she must know the secret. We now show that the above protocol is sound. It is easily seen, from the choice of the challenge c, that Eve can convince Bob, with probability of 50%, that she is Alice if $a = 1$. However, this is the *highest* probability that a *cheating prover* such as Eve can achieve in this protocol. To see this, suppose that Eve can actually convince Bob that she is actually Alice, with probability greater than $1/2$. What this means in the protocol is that Eve knows a value of w for which she can answer both challenges $c = 1$ and $c = 0$. Hence, Eve can find r_1 and r_2 such that

$$r_1^2 \equiv w \pmod{n} \text{ and } r_2^2 \equiv wt_A \pmod{n}.$$

Therefore, Eve can find a square root of t_A, namely, $r_2 r_1^{-1} \pmod{n}$. In other words, if she can successfully impersonate Alice, Eve must know the prover's secret. Hence, the protocol is in possession of the soundness property.

The above being said, the probability of $1/2$ is still too high for a cheater such as Eve. Thus, the parameter a, if set sufficiently high, can reduce this probability to negligible levels. In other words, for $a > 1$, the probability is reduced to 2^{-a}, which for sufficiently high a means that Eve has near zero chance of success. Also, in order to maintain security in the protocol, Alice must respond to at most one challenge for a given witness; she should never reuse a given witness.

Since Alice has communicated *only* that she has *knowledge* of a square root of t_A, then the protocol has the *zero-knowledge property*, which means that the verifier, Bob, learns nothing from the prover, Alice, that could not have been learned *without* Alice's participation. The zero-knowledge property ensures that interacting with Alice, as described in the protocol, does not leak information that can be used to impersonate her.

We conclude the analysis with a discussion of types of protocols to show how the above protocol fits in. *Arbitrated protocols* are those protocols relying on a trusted third party, such as Trent, who will not render preferential treatment to any of the participants. Trent has no allegiances to any of the participants and no particular reason to complete the protocol. Thus, Trent may be considered to be playing the role of a disinterested lawyer. Hence, all participating entities are assured that what is done in the protocol is correct, and that their particular portion of the protocol is complete. The Feige-Fiat-Shamir protocol is an arbitrated protocol.

The above discussion motivates us to complete the discussion of protocol types, of which there are two. A variation of the arbitrated protocol is the *adjudicated protocol*. This requires the introduction of our next character in the cryptographic play, *Judy the adjudicator*. Judy is brought into the protocol only if cheating by participants is suspected. In that case, she comes into the play and analyzes the dispute, rendering a ruling to determine who is right and determining the punishment for the entity who is in the wrong. An example is a scenario where Bob agrees to sell his house to Alice, who gives him a cheque for it. If the cheque is fraudulent, or the keys are fake, they go before Judy to present their case. Judy rules on the evidence presented and the entity who cheated is fined or imprisoned. There is, however, a third kind of protocol involving no third party. A *self-enforcing protocol* is designed to make cheating a virtual impossibility. Cheaters gain no advantage by *not* following the protocol. In Section 5.4, we will encounter an example of such a protocol, *coin flipping by telephone*.

We close this section with an alternative to the Feige-Fiat-Shamir scheme. The following is based upon the intractability of the DLP (see page 164, especially Equation (4.2)). We will require the notion of a *certificate*, which is a quantity of information that has been signed by a trusted authority such as Trent. One type of certificate pertinent to the following, and protocols to be considered later in the text, is an *identification certificate*, which contains identifying information such as a birth certificate or passport. The following was introduced in 1991 (see [242]). Again, we need Trent for this (zero-knowledge) interactive proof of knowledge.

◆ **Schnorr Identification Protocol**

Trent's Actions: Trent selects each of the following parameters.

1. A large prime p such that the DLP in $\mathbb{F}_p^*$ is intractable (say, $p \geq 2^{1024}$).

2. A large prime divisor q of $p - 1$ (say, $q \geq 2^{160}$).

3. $\alpha \in \mathbb{F}_p^*$ such that $\mathrm{ord}_p(\alpha) = q$ (say, $\alpha = \beta^{(p-1)/q}$ where β is a primitive root modulo p).

4. A parameter t such that $q > 2^t$ (usually $t \geq 40$).

5. A secure signature scheme embodying a secret digital-signing algorithm $\mathrm{sig}_{T(k)}$ and a public digital-verifying algorithm $\mathrm{ver}_{T(k)}$ for verification of Trent's signatures. (Typically $\mathrm{sig}_{T(k)}$ involves a cryptographic hash function for security (see page 170), but we will omit this here for increased clarity of presentation.)

Then Trent creates a certificate for Alice as follows:

6. Trent establishes a bitstring containing information I_A that identifies Alice. Then Alice selects a private random nonnegative exponent $e \leq q - 1$ and she computes $v \equiv \alpha^{-e} \pmod{p}$, which she sends to Trent. Upon receipt, Trent generates a signature $s = \mathrm{sig}_{T(k)}(I_A, v)$, thereby blinding I_A with v. Then he sends the certificate $C(A) = (I_A, v, s)$ to Alice.

Three-Pass Identification Protocol: Alice wishes to identify herself to Bob, who verifies her identity by proving knowledge of e (without revealing e).

1. Alice selects a random $k \in \mathbb{Z}/q\mathbb{Z}$, called a *commitment*, and computes

$$\gamma \equiv \alpha^k \pmod{p}.$$

Then she sends her certificate $C(A)$ and γ, called the *witness*, to Bob.

2. Bob computes $\mathrm{ver}_{T(k)}((I_A, v, s)) = 1$, thereby verifying Trent's signature. Then Bob selects a random natural number $r \leq 2^t$, called the *challenge*, which he sends to Alice.

3. Alice computes $y \equiv k + er \pmod{q}$, called the *response*, which she sends to Bob.

4. Bob computes $\delta \equiv \alpha^y v^r \pmod{p}$, and if $\delta \equiv \gamma \pmod{p}$, called the *verification*, he accepts Alice's identity. Otherwise, he rejects it.

Example 5.1 (*With artificially small parameters for illustration only.*) *We dispense with the issue of the certificate and assume it has been handled properly. We proceed with the rest of the protocol. Suppose that we have* $p = 8699$, $q = 4389$ *and* $t = 10$. *We know that* $\alpha = 4$ *has order* 4389 *in* $\mathbb{F}_p^*$, *since* 2 *is a primitive root modulo* p *and* $\alpha \equiv 2^{(p-1)/q} \pmod{p}$. *If Alice's private exponent is* $e = 11$, *she computes* $v \equiv \alpha^{-e} \equiv 4^{-11} \equiv 4111 \pmod{8699}$. *This completes the interaction of Trent and Alice. Now if Alice selects* $k = 110$, *she computes* $\gamma \equiv \alpha^k \equiv 4^{110} \equiv 4572 \pmod{8699}$. *If Bob chooses* $r = 233$ *and sends it to Alice, she computes* $y \equiv k + er \equiv 110 + 11 \cdot 233 \equiv 2673 \pmod{4389}$, *which she sends to Bob who computes*

$$\delta \equiv \alpha^y v^r \equiv 4^{2673} \cdot 4111^{233} \equiv 4572 \equiv \gamma \pmod{4937},$$

so Bob accepts Alice's identity as valid.

▼ **Analysis**
Security: We first demonstrate that the protocol has the soundness property discussed on page 203. Suppose that Mallory has knowledge of γ, and that he has a nonnegligible probability of successfully impersonating Alice. We now show this means that Mallory can actually compute e, which will demonstrate soundness. Mallory can compute a response y that will be accepted by Bob's verification in step 4 of the protocol, so Mallory can compute integers y_1, y_2, r_1, r_2 such that both

$$y_1 \not\equiv y_2 \pmod{p} \text{ and } \gamma \equiv \alpha^{y_1} v^{r_1} \equiv \alpha^{y_2} v^{r_2} \pmod{p}.$$

Hence,
$$\alpha^{y_1 - y_2} \equiv v^{r_2 - r_2} \pmod{p}.$$

Since $v \equiv \alpha^{-e} \pmod{p}$, then

$$y_1 - y_2 \equiv e(r_1 - r_2) \pmod{q}.$$

Since
$$0 < |r_2 - r_1| < 2^t \text{ and } q > 2^t \text{ is prime,}$$

then $\gcd(r_2 - r_1, q) = 1$. Hence, $(r_2 - r_1)$ has a multiplicative inverse modulo q. It follows that Mallory can compute

$$e = (y_1 - y_2)(r_1 - r_2)^{-1} \pmod{q}.$$

What we have shown is that if Mallory has a reasonable (nonnegligible) probability of successfully executing Schnorr's protocol, then he must (essentially) "know" Alice's private exponent e. This is soundness.

Once Alice proves her identity in the fashion prescribed in the protocol, Bob accepts her proof in step 4, so the protocol has the completeness property, also discussed on the aforementioned page. However, soundness and completeness are insufficient to guarantee security. For example, Alice could just reveal her private exponent e to Mallory, who could then impersonate her, and the protocol would still have the soundness and completeness properties. Thus, Alice must ensure that no information about e is leaked, the zero-knowledge property. Alice proves *knowledge of* e (without revealing e) by her response y to the challenge r in step 3 of the protocol. If Mallory does not have knowledge of e, and he tries to impersonate Alice, then he is in the position in step 3 of having to compute y, which is a function of e, in response to Bob's challenge r. However, computing e from v involves solving an instance of the DLP, which is assumed to be intractable. Yet it has not been proved that Schnorr's protocol is secure.

Attacks: There are three main types of attacks on identification protocols in general. We have already met the *replay attack* in Footnote 5.1 on page 197. Defense against such attacks can involve the use of challenge-response methods, or the use of nonces. Another type of attack is the *chosen-text attack*, which is an attack on a challenge-response protocol where Mallory chooses challenges according to some design to recover information about Alice's private key. For example, in Schnorr's protocol, Alice encrypts the challenge r with y, so this attack involves chosen-plaintext (see Footnote 3.4 on page 127). Methods for thwarting such attacks include the embedding of a nonce in each challenge-response. Last, there is the *forced delay attack*, which involves Mallory intercepting a message and relaying it later. This is a type of man-in-the-middle attack (see Footnote 3.7 on page 134). Defense against it may include the use of nonces tied in with short response time outs.

Comparisons: There are variations of Schnorr's protocol that have been proved to be secure under the assumption of a particular discrete log. One such is Okamoto's protocol (see [170, pages 131–133] for a description, analysis, and comparison). However, Okamoto's protocol and other variations that are provably more secure, sacrifice speed. Moreover, even without a proof of security (which is scarce in any case), Schnorr's protocol still has not been cryptanalyzed. In other words, no weaknesses have been found, and with its efficiency and suitability for use in smart cards (to which we will return later in the text) it is an excellent pragmatic choice. When compared with the Feige-Fiat-Shamir protocol, Schnorr's protocol is also much more efficient. The reason for this is that the most computationally intensive operation is the modular exponentiation in step 1 of the protocol, which by design, may be computed offline. In step 3 there is one modular addition and one modular multiplication, so the online computations are very moderate. The computations for Feige-Fiat-Shamir protocol are significantly greater. The Schnorr algorithm was designed with this computational efficiency in mind for such applications as smart cards with low computing power. In general, the Schnorr protocol is quite suitable when Alice has restricted computing power. Notice as well that more computational efficiency is gained by using a subgroup of order $q \mid (p-1)$, which lowers the number of bits needed for transmission. The three-pass protocol involved in steps 1–3 was a built-in design of the protocol to reduce bandwidth (see Footnote 4.5 on page 183), especially in comparison to the Feige-Fiat-Shamir protocol.

5.4   Commitment

He that but looketh on a plate of ham and eggs to lust after it, hath already committed breakfast with it in his heart.

> **C.S. Lewis** (1898–1963), *English literary scholar*
> — from letter (March 10, 1954)

Commitment schemes are valuable in setting up practical cryptographic applications such as electronic voting, which we will study in Section 5.6. For simplicity, we will concentrate upon *bit commitment*, meaning the commitment to either a 0 or a 1. The term "bit commitment" was coined by Blum in 1982. He devised the following example to illustrate this type of protocol.

Suppose that Alice and Bob are getting a divorce and are living in different cities. They need to decide how to split their possessions. They have agreed upon everything except who gets the car. Flipping a coin is an option, but they do not trust each other enough to do this since they are physically separated. The following is a version of Blum's solution (see [28]).

◆ Coin Flipping by Telephone (Using a One-Way Function)

Alice and Bob know a one-way function f, but not its inverse f^{-1}. Moreover, f reliably produces even and odd numbers with equal probability (as would a fair coin in a coin flip).

Protocol Steps

1. Bob selects an integer x at random and sends the value $f(x) = y$ to Alice.

2. Alice makes a guess as to whether the number x is odd or even and sends the guess to Bob.

3. Bob tells Alice whether the guess is correct.

4. Bob sends Alice the value x.

5. Alice confirms that $f(x) = y$ (verification step).

The above protocol is an example of *flipping coins into a well*, which is a metaphor for the following scenario. Bob is next to a well and Alice is physically removed from it. Bob throws a coin into the well, and can see it clearly at the bottom, but cannot reach it. Alice cannot see the result of the coin toss until Bob allows Alice to come to the well to have a look.

Commitment schemes must satisfy certain properties. There is the *sender*, Bob, and the *receiver*, Alice. Bob *commits* to a bit b and sends it to Alice in encrypted form. Bob sends additional information to Alice enabling her to recover b. There are requirements. Once Bob sends b, Alice does not learn anything about b. This is called the *concealing property*. Second, Bob cannot change the value of b after he commits, called the *binding property*. Binding must be a satisfied property even if Bob tries to cheat. Third, if both Alice and

Bob follow the protocol, Alice will always receive the committed value b, called the *viability property*.

Thus, bit commitment schemes are really the digital analogues of the use of opaque sealed envelopes, wherein the sealing of a message in an envelope commits the sender to the message, while keeping it secret.

The above concepts are now illustrated using an SKC to describe a bit-commitment scheme.

♦ Bit-Commitment Protocol Using SKC

We assume that E is a symmetric-key cryptosystem available to Alice and Bob.

1. Alice generates a random bitstring R and sends it to Bob.

2. Bob creates a message consisting of Alice's random bitstring R and the bit to which he wants to commit, producing (R, b) (binding). Bob uses a random (secret symmetric) key e to encipher and sends the cryptogram $E_e(R, b)$ to Alice (concealing).

3. When ready to reveal his commitment, Bob sends the key e to Alice.

4. Alice deciphers the message via $E_e^{-1}(E_e(R, b)) = (R, b)$ to reveal the bit and the random bitstring to verify the bit's validity.

The next scheme involves the DLP.

♦ Coin Flipping by Telephone Using Discrete Logs

Suppose that the problem with which Alice and Bob are faced is the determination of who gets the car after the divorce, discussed on the previous page. The problem is to be solved using a coin flip in the following manner.

Protocol Steps:

1. Alice and Bob agree upon a large prime p such that the factorization of $p - 1$ is known.

2. Alice selects two generators $\alpha, \beta \in \mathbb{F}_p^*$ and sends both to Bob.

3. Bob chooses a random integer $x \in (\mathbb{Z}/(p-1)\mathbb{Z})^*$, then he computes exactly one of $y \equiv \alpha^x \pmod{p}$ or $y \equiv \beta^x \pmod{p}$, which he sends to Alice.

4. Alice guesses whether y is a function of α or β and sends the guess to Bob.

5. If Alice's guess is correct, the result of the flip is deemed to be *heads*, and if it is incorrect, it is deemed to be *tails*. Bob sends the result of the coin flip to Alice.

6. Bob reveals x to Alice. Then Alice computes $\alpha^x \pmod{p}$ and $\beta^x \pmod{p}$ to verify both outcomes of the coin tosses.

▼ **Analysis**: If Bob wants to cheat in step 3, he must know two integers u and v such that $\alpha^u \equiv \beta^v \pmod{p}$. To compute u given v, Bob must calculate $\log_\alpha(\beta^v) \pmod{p}$, but for this he must know $\log_\alpha(\beta)$. Yet, it is Alice who chooses α and β in step 2, so Bob is in the position of having to compute a discrete log. This means that the coin flipping relies on the DLP. Moreover, Bob cannot cheat at step 3 by choosing an x such that $\gcd(x, p-1) > 1$, since Alice checks $\gcd(x, p-1)$ in step 6. Bob knows the result of the coin flip in step 3, but cannot change it (binding), and Alice does not know the choice when y is sent to her (concealing). Also, Alice knows the value of the commitment in step 5, and is able to verify it in step 6 (viability), so this is another example of flipping coins into a well.

The next scheme is a variation of the above for playing poker using RSA, where nobody can influence the cards dealt.

◆ **Poker Playing by Telephone**

The following protocol is a mechanism for Alice to deal Bob a five-card hand.

1. Alice and Bob agree upon two large primes p and q, and form an RSA modulus $n = pq$. They also agree upon a set of random numbers c_j for $j = 1, 2, \ldots, 52$, as names for the cards. Both Alice and Bob generate public/private RSA key pairs (e_A, d_A) and (e_B, d_B), respectively.

2. Alice computes $f_j \equiv c_j^{e_A} \pmod{n}$, for each $j = 1, 2, \ldots, 52$, shuffles the numbers, and sends them to Bob.

3. For $j = 1, 2, \ldots, 52$, Bob computes $g_j \equiv f_j^{e_B} \equiv c_j^{e_A e_B} \pmod{n}$, shuffles the numbers, and sends sends them back to Alice.

4. Alice takes each g_j and computes $h_j \equiv g_j^{d_A} \equiv c_j^{e_B} \pmod{n}$. She selects five cards at random and sends them to Bob.

5. Bob computes each $h_j^{d_B} \equiv c_j \pmod{n}$ to get his five-card hand.

▼ **Analysis**: In step 3, Bob cannot determine which of the random numbers represents which card since Alice encrypted and shuffled them. To prevent Alice from cheating, Bob encrypts and shuffles the cards as well. Thus, in step 4, Alice cannot determine which of the g_j represent which cards, so she has to send a random hand to him. The process can be played with reversed roles to deal Alice's hand. However, as we saw with the RSA cryptosystem in Section 4.2, we must ensure that no information whatsoever is leaked about the plaintext. There is one attack that is subtle. RSA preserves Jacobi symbols (see page 482), for instance, $\left(\frac{g_j}{n}\right) = \left(\frac{c_j^{e_A e_B}}{n}\right) = \left(\frac{c_j}{n}\right)$. Thus, in step 4, Alice can determine whether the cards with Jacobi symbol 1 or -1 are better for her and draws only these from the number deck. This means that Alice can get a bit of information from every g_j giving her an advantage over Bob (see [96]).

However, the advantage can be eliminated by merely insisting in step 1 that only numbers c_j with Jacobi symbol 1 be chosen.

We conclude this section with another commitment scheme based upon the DLP, which will allow us to set up a notion required for Section 5.6.

◆ Commitment Scheme Based on the DLP

This scheme allows Alice to commit to a message $m \in \mathbb{Z}/q\mathbb{Z}$ where q is prime.

1. Bob randomly selects primes p and q as in step 1 of the setup stage of the DSA described on page 183. Then he randomly selects distinct generators α, β of $\mathbb{Z}/q\mathbb{Z}$, as in steps 2 and 3 of the DSA setup. He sends (p, q, α, β) to Alice.

2. Alice randomly selects $r \in \mathbb{Z}/q\mathbb{Z}$ and computes

$$c \equiv \alpha^r \beta^m \pmod{p},$$

   her commitment, which she sends to Bob.

3. When it is time to reveal her commitment, she sends r and m to Bob, who verifies that $c \equiv \alpha^r \beta^m \pmod{p}$.

▼ **Analysis**: Since the selection of p, q, α, β is as in the DSA, the above scheme is based upon the DLP. To see why, let us suppose that Alice tries to cheat by selecting $m' \neq m$ as her commitment. Then $\alpha^r \beta^m \equiv \alpha^{r'} \beta^{m'} \pmod{p}$, so

$$\log_\alpha(\beta) \equiv (r - r')(m' - m)^{-1} \pmod{q},$$

which she must compute. Yet for randomly chosen $\alpha, \beta \in \mathbb{Z}/q\mathbb{Z}$, this is deemed to be computationally infeasible. Thus, by selecting α, β randomly in step 1, $\alpha^r \beta^m$ is a means of blinding m, which depends upon the DLP.

The above scheme opens the door to a notion for commitments that we will need later.

◆ Homomorphic Property

If $E(x)$ and $E(y)$ are ciphertext in a given scheme and

$$E(x)E(y) = E(x * y), \qquad (5.1)$$

where $*$ is the operation used on plaintext, the scheme is said to have the *homomorphic property*.

For instance if we set $E(r, m) = \alpha^r \beta^m$ in the above DLP scheme, then for $r, r', m, m' \in \mathbb{Z}/q\mathbb{Z}$, $E(r, m) \cdot E(r', m') \equiv E(r + r', m + m') \pmod{p}$, which satisfies (5.1), so it is an example of a *homomorphic commitment scheme*.

▼ **Analysis**: Homomorphic commitment schemes allow sums of integers to be calculated without revealing either of the summands. We will see a real-world application of this when we discuss electronic voting.

5.5 Secret Sharing

I share no one's ideas. I have my own.
                              Ivan Turgenev (1818–1883), *Russian novelist*
                              — from *Fathers and Sons* (1862), Chapter 9

When we discuss e-commerce on page 228, look at splitting secrets for the digital cash schemes discussed therein. That application is the simplest form, namely, the sharing of two pieces of a secret by two entities who add their respective pieces modulo 2 to recover the secret. This notion is easily generalized to any finite number of entities who may piece together the information to retrieve the secret information. Herein we look at two secret-sharing schemes, which will serve us well, especially in Section 5.6 when we talk about electronic voting.

The two schemes that we describe in this section are ideas from the two preeminent pioneers in the area, who independently worked out the schemes that have found numerous applications. We begin with Shamir's idea from 1979 (see [246]). We begin by generalizing and formalizing the notion of secret splitting that we have already encountered.

◆ Threshold Schemes

Let $t, w \in \mathbb{N}$ such that $t \leq w$, and m is the secret. A (t, w)-threshold scheme invovles Trent computing the pieces m_j, for $j = 1, 2, \ldots$, called *shares* (sometimes called *shadows* of m), among a set of w entities. Each t of the entities can recover m from their shares. It is not possible for $t - 1$ or fewer of the entities to recover m using their shares.

The first question that may arise is that of the existence of such schemes. A mere definition does not mean they will always exist. But in fact, they always do since it can be shown that for any $t \leq w$ and $t > 1$, there exists a (t, w)-threshold scheme (see [141, Theorem 1.7, page 8]). The secret splitting discussed on page 228 is an example of a $(2, 2)$ threshold scheme. The following is also known as *Lagrange Interpolation Scheme*. In fact, the reader must be familiar with the Lagrange interpolation formula, Theorem A.20, given in Appendix A on page 486, as well as the notions surrounding Definition A.39 on page 490. We require good old trusted Trent again.

◆ Shamir's Threshold Scheme

Trent distributes shares of m to $w \in \mathbb{N}$ participants of whom any $t \leq w$ of them will be able to recover m.

Trent's Actions:

1. Choose a prime $p > \max(m, w)$, where p is public, and set $m_0 = m \in \mathbb{Z}/p\mathbb{Z}$.

2. Select $t - 1$ random integers c_j for $j = 1, 2, \ldots, t - 1$ and set

$$p(x) \equiv m + \sum_{j=1}^{t-1} c_j x^j \equiv \sum_{j=0}^{t-1} c_j x^j \pmod{p},$$

where $c_0 = m$.

3. Compute $p(x_k) \equiv m_k \pmod{p}$ for distinct integers $x_k \leq p - 1$ and securely distribute the share (x_k, m_k) to participant $\mathcal{P}_k$ for $1 \leq k \leq w$.

Pooling Shares: Without loss of generality, suppose a group of t participants $\mathcal{P}_k$ for $1 \leq k \leq t$ get together and plug their shares into the Lagrange interpolation formula:

$$f(x) = \sum_{k=1}^{t} m_k \prod_{\substack{1 \leq \ell \leq t \\ \ell \neq k}} \frac{x - x_\ell}{x_k - x_\ell} = \sum_{k=1}^{t} m_k K_k(x),$$

where

$$K_k(x) = \prod_{\substack{1 \leq \ell \leq t \\ \ell \neq k}} \frac{x - x_\ell}{x_k - x_\ell}.$$

In the analysis following, we will show that the next equation must hold:

$$f(x_i) \equiv m_i \pmod{p}, \text{ for } 1 \leq i \leq t \tag{5.2}$$

and from it the following crucial equation must hold:

$$p(0) \equiv f(0) \equiv \sum_{k=1}^{t} m_k K_k(0) \equiv m \pmod{p}, \tag{5.3}$$

so the shares have been pooled to retrieve the secret.

▼ Analysis

Verification of (5.2) and (5.3): To show that (5.2) is valid, we observe that

$$K_k(x_i) = \prod_{\substack{1 \leq \ell \leq t \\ \ell \neq k}} \frac{x_i - x_\ell}{x_k - x_\ell} \equiv 0 \pmod{p}$$

if $i \neq k$ since $K_k(x_i)$ has a factor $(x_i - x_i)/(x_k - x_i)$. Also,

$$K_k(x_k) \equiv 1 \pmod{p}$$

since all factors are of the form $(x_k - x_\ell)/(x_k - x_\ell) = 1$. We note that

$$1/(x_k - x_\ell) \equiv (x_k - x_\ell)^{-1} \pmod{p},$$

so as long as $k \neq \ell$, such inverses exist. Therefore,

$$f(x_i) \equiv \sum_{k=1}^{t} m_k K_k(x_i) \equiv m_i \pmod{p}$$

for $i = 1, 2, \ldots, t$, which is (5.2). Therefore,

$$p(x) \equiv f(x) \, (\mathrm{mod} \; p)$$

since the Lagrange interpolation formula, given on page 486, tells us that f is the unique polynomial with $f(x) \in \mathbb{F}_p[x]$ such that

$$f(x_i) \equiv m_i \, (\mathrm{mod} \; p) \text{ for } 1 \le i \le t.$$

In particular, Equation (5.3) holds with $x = 0$.

Security: In Shamir's scheme, no fewer than t participants can recover m, a property that makes it an example of what is sometimes called a *perfect threshold scheme*. The security of Shamir's scheme does not rely upon the assumed intractability of such problems as the DLP or the IFP. Hence, in practice, the scheme is as secure as a one-time pad in the sense that an exhaustive search of all possible shares will reveal to an adversary that *any* message m could be the secret.

Variations: Let us assume that a bank president wants to control the majority of the shares in a scheme for secret-sharing the combination to the bank's vault. Suppose that $t = 9$ shares are required and the president has 7 shares while other participants have only 1 share. Therefore, the president gets together with two underlings to recover the combination, but without participation by the president, it takes 9 participants to recover it.

Another variation on Shamir's scheme is depicted by the next setting. Assume that two banks A and B hold their securities in the same vault. They wish to create a scheme where 2 participants from bank A and 3 participants from bank B hold shares. Here is how they accomplish the task. Form the product of a linear polynomial p_1 and a quadratic polynomial p_2. Then give $w_1 \ge 2$ employees of bank A a share $p_1(x_i)$ for $1 \le i \le w_1$, and give $w_2 \ge 3$ employees from bank B a share $p_2(y_i)$ for $1 \le i \le w_2$. Then any two participants from bank A can get together and recover p_1 but not p_2, and any 3 participants from bank B can get together and recover p_2 but not p_1. Participants from both banks A and B must work together to recover the full combination determined by the product $p_1 p_2$ acting on the individual shares.

The last secret-sharing scheme is due to the second pioneer who developed his idea in the same year as Shamir. In 1979, Blakely came up with a scheme (see [25]) based upon vectors and matrices (see pages 490–494 in Appendix A). It is not a (t, w)-secret-sharing scheme.

◆ **Blakely's Secret-Sharing Vector Scheme**

The secret message is m_1 to be reconstructed by $t > 2$ participants.

Setup Stage: The following are executed.

1. Choose a large prime $p > m_1$, where p is made public, and select $m_2, m_3, \ldots, m_t \in \mathbb{F}_p$ at random. Then $m = (m_1, m_2, \ldots, m_t)$ is a point in the t-dimensional vector space $\mathbb{F}_p^t$.

2. For each $j = 1, 2, \ldots, t$, select $n_1^{(j)}, \ldots, n_{t-1}^{(j)} \in \mathbb{F}_p$ at random and set

$$c_j \equiv m_t - \sum_{i=1}^{t-1} n_i^{(j)} m_i \pmod{p}.$$

3. Each of the t participants is given the equation for a hyperplane in $\mathbb{F}_p^t$ as follows:

$$\sum_{i=1}^{t-1} n_i^{(j)} x_i - x_t \equiv -c_j \pmod{p},$$

for $j = 1, 2, \ldots, t$, where the intersection of the t hyperplanes must be the point m.

Pooling Stage: The participants convene to recover the secret message as follows. In matrix terminology, the pooling of their equations translates into the following:

$$AX \equiv \begin{pmatrix} n_1^{(1)} & n_2^{(1)} & \cdots n_{t-1}^{(1)} & -1 \\ n_1^{(2)} & n_2^{(2)} & \cdots n_{t-1}^{(2)} & -1 \\ \vdots & \vdots & \vdots & \vdots \\ n_1^{(t)} & n_2^{(t)} & \cdots n_{t-1}^{(t)} & -1 \end{pmatrix} \begin{pmatrix} x_1 \\ x_2 \\ \vdots \\ x_t \end{pmatrix} \equiv \begin{pmatrix} -c_1 \\ -c_2 \\ \vdots \\ -c_t \end{pmatrix} \pmod{p}. \quad (5.4)$$

It follows from Theorem A.26 on page 494 that, if $\det(A) \neq 0$, then there is the unique solution,

$$X = (m_1, \ldots, m_t),$$

so the secret m_1 is recovered.

▼ **Analysis:** Although we cannot be certain that $\det(A) \neq 0$ in (5.4), if we choose p large enough, then it is highly probable that A is indeed invertible. Shamir's method is essentially a special case of the Blakely method since Shamir's method effectively deals with a Vandermonde matrix (see page 494) for A, the determinant of which is zero if and only if some $x_k \equiv x_i \pmod{p}$, but we chose these values to be distinct in step 3 of the algorithm. This gives Shamir's method an advantage over Blakely's method. Moreover, Shamir's method clearly requires each participant to have less information in their respective shares.

When we look to applications involving electronic elections in the next section, secret-sharing schemes will play a role. It is one of the highly important modern-day applications of cryptographic protocols to ensure a secure and legitimate voting process in a free society.

5.6   Electronic Voting

Democracy substitutes election by the incompetent many for appointment by the corrupt few.

George Bernard Shaw (1856–1950), Irish dramatist
— from *Man and Superman* (1903), Maxims: Democracy

Cryptographic protocols can be employed to create scenarios where voters can electronically cast their ballots. Of course, secrecy is of paramount importance when casting such ballots over a network. Furthermore, most often we require authentication, for many reasons, not only including the need to protect legitimate voters, but also to prevent an entity from voting more than once. Another requirement is that each legitimate voter should be able to verify that their vote has been cast. One method of starting such a process is to ensure that each voter has a unique secret bitstring identification number V_j for $j = 1, 2, \ldots, u$. Furthermore, there needs to be a set of *voting authorities* who tally the votes. We will call them $A_1, A_2, \ldots, A_w$. There will also be a *central voting authority* who is trusted by all parties. This will be Trent.

A version of the following was first published in 1997 (see [62]).

♦ A Multiauthority Election Protocol

Background Assumptions: We will assume that the voters are in California and voting on whether to recall their governor, a "yes or no" vote. The voters will communicate through what is called a *bulletin board* that may be viewed as a public data base. Each voter has a section of the bulletin board on which to post messages, and can read the entire board, but no voter can erase anything. A PKC DSS is assumed to have authenticated the origins of any postings to the bulletin board (see Section 4.3 on page 180).

Setup Stage: Trent chooses primes p and q, with $p \equiv 1 \pmod{q}$, and α a generator of $\mathbb{Z}/q\mathbb{Z}$, as in the DSA setup stage (see page 183). Then he randomly selects a secret key $a \in \mathbb{Z}/q\mathbb{Z}$ and makes his public key $k \equiv \alpha^a \pmod{p}$ known. Trent uses the ElGamal encryption scheme described on page 185 to encipher a message m via the choice of a random $b \in \mathbb{Z}/q\mathbb{Z}$ and computing (α^b, mk^b). This is then an example of a scheme with the homomorphic property (see page 211), since if (α^b, mk^b) and $(\alpha^{b'}, m'k^{b'})$ are the encipherings of m and m', then

$$(\alpha^b, mk^b)(\alpha^{b'}, m'k^{b'}) \equiv (\alpha^b \alpha^{b'}, mk^b mk^{b'}) \equiv (\alpha^{b+b'}, mm'k^{b+b'}) \pmod{p}.$$

Trent selects a (t, w)-threshold scheme (see page 212) to split the secret key a among the above-defined w authorities in the following fashion. Authority A_j has secret share (j, a_j) where the a_j for $j = 1, 2, \ldots, w$ are the pieces that make up the secret key a. Trent publishes $\alpha^{a_j} \pmod{p}$ for $j = 1, 2, \ldots, w$ on the bulletin board.

Protocol Steps:

1. **Casting Votes**: Each voter V_j chooses a vote that is one of $v_j \in \{-1, 1\}$ (say, with -1 meaning no, and 1 meaning yes). Then the ElGamal cipher is

used to encrypt the vote by choosing a random $b_j \in \mathbb{Z}/q\mathbb{Z}$ and computing $c_j = (\alpha^{b_j}, \alpha^{v_j} k^{b_j})$, posting it to the bulletin board. We assume that each voter has signed c_j and correctly performed a proof of knowledge (see Section 5.3 on page 202) to verify that the protocol has been validly followed. Failure to meet any of these criteria means that the vote is discarded as invalid.

2. **Tallying the Votes**: Since all the c_j are posted on the bulletin board, then anyone can compute the following:

$$(V_1, V_2) \equiv \left(\prod_{j=1}^{u} \alpha^{b_j}, \prod_{j=1}^{u} \alpha^{v_j} k^{b_j} \right) \equiv \left(\alpha^{\sum_{j=1}^{u} b_j}, \alpha^{\sum_{j=1}^{u} v_j} k^{\sum_{j=1}^{u} b_j} \right) \pmod{p}.$$

Each authority A_j posts $\alpha^{a_j b_j}$ to the bulletin board. Once t honest authorities, say $A_1, A_2, \ldots, A_t$ for simplicity, have posted their data to the bulletin board, then anyone can use their data to compute the following:

$$g \equiv V_1^{\sum_{j=1}^{t} a_j K_j} \equiv V_1^a \equiv \alpha^{a \sum_{j=1}^{u} b_j} \equiv k^{\sum_{j=1}^{u} b_j} \pmod{p},$$

where

$$K_j = \prod_{\substack{1 \le \ell \le t \\ \ell \ne j}} \frac{\ell}{\ell - j}$$

(see page 213).

Thus,

$$V_2 g^{-1} \equiv \alpha^{\sum_{j=1}^{u} v_j} \equiv \alpha^D \pmod{p},$$

where $D = \sum_{j=1}^{u} v_j$ is the difference between the number of yes and no votes. Hence, we see above that (V_1, V_2) is actually the cryptogram enciphering α^D and we used the homomorphic property to get it. Now the tally is accomplished by computing $\alpha^D \alpha^{-u+i} \pmod{p}$ for $i = 0, 1, 2, 3 \ldots$, until we have that $\alpha^D \alpha^{-u+i} \equiv 1 \pmod{p}$. Then that value of $u - i$ is D, the tally.

▼ **Analysis**: Privacy is provided by the in-built ElGamal cipher, which rests upon the intractability of the DLP (see pages 186 and 187). Given a coalition of no fewer than $t \le w$ honest authorities, anyone can verify the tally, a property called *universal verifiability*, something currently *not* available to voters. Also, the message sent by each voter is simple and concise, so the time complexity of the scheme can be shown to be exceptionally low. Our version of the original scheme is necessarily a simplified one for pedagogical reasons.

There is a mechanism for eliminating the role of Trent. To do so, Trent's actions need to be performed by the authorities. Since this is a minor modification, we will not discuss the details.

5.7   Protocol Layers and SSL

Don't express your ideas too clearly. Most people think little of what they understand, and venerate what they do not.
                      Baltasar Gracián (1601–1658), Spanish philosopher

To become acquainted with the notion of a "protocol layer", we must understand its (formal) inception, which began with the following organization. The *International Organization for Standardization* (ISO), embodying members from 148 countries, is a world federation of national standards organizations. ISO is a nongovernmental body, created in 1947 to promote the development of standardization and related activities. The reader already will have noticed that ISO is *not* an acronym. Its roots are from the Greek *isos* meaning *equal*, which will be recognized as the prefix *iso-*, such as in *isometric*. It happened that *equal* devolved to *standard*, and the ISO name was adopted. Additionally, this provides the feature of not requiring translation in each country, as would an acronym. ISO develops precise criteria for such applications as the development of a framework of international standards in computer networks, for instance. (A *network* is a hardware and software communications system.) In 1978, ISO developed a model of network protocols, called a *protocol stack*, which is a layered set of protocols working together to render a set of network functions. The ISO model divides the architecture among seven layers, where we understand a *layer* to be the environment of two or more communications devices in which a particular network protocol operates. The ISO model is called the *Open Systems Interconnection Reference Model* (OSI-RM). OSI is the umbrella name for a set of nonproprietary protocols and specifications, which includes the OSI-RM, having the following seven layers, from the bottom to the top.

◆ **OSI-RM Seven Layer Protocol Stack**

1. **Physical Layer**: This bottom layer deals with electrical and mechanical connections to the network.

2. **Data Link Layer**: This layer splits data into *frames*, which are *data packets* containing the header and trailer information required by the physical layer. The data link layer executes error checking and retransmits correct frames for any corrupted frames it receives, thereby providing an error-free connection to the next layer up to which it sends the frames.

3. **The Network Layer**: This is the communications subnet layer, which decides the routing of packets received from the data link layer to be used by the next layer up. Most commonly, IP is used (see page 199).

4. **The Transport Layer**: This middle layer is essentially the communications system component of a given protocol. For instance, the TCP protocol discussed on page 199 is one such communications system. Although TCP itself is not cryptographically secure, mechanisms can be used to make it

so. For instance, in 1996 the *Internet Engineering Task Force* (IETF),[5.3] the agency that develops protocol standards for the Internet, formed a committee, the *Transport Layer Security* (TLS) working group. Their mandate was to to develop a standard for *Secure Sockets Layer* (SSL) a protocol that originated at Netscape in 1994, and which we will describe in detail in this section. In January 1999, the TLS working group published the TLS protocol. However, TLS is essentially a version of SSL, so we will not describe it here, but rather wait to get the full description of SSL, which is not, in itself, a single protocol, but rather two layers of protocols using TCP to provide a secure connection with *WWW browsers*.

This is an appropriate juncture to explain the Internet terminology that we will be using. For instance, WWW is the acronym for the *World Wide Web*, which is the information network using *HTTP* and *HTML* on Internet host computers. HTTP is the acronym for *HyperText Transfer Protocol*, which is the protocol used to transfer files from an Internet server onto a browser in order to view that page on the Internet. HTTP is a one-way system in the sense that the contents of a page from the server are downloaded to the computer's browser for viewing, but files cannot be transferred to the computer's memory. HTML is the acronym for *HyperText Markup Language*, which is the text format for WWW pages. *Browser* is a short form for WWW browser, which is a software application used to locate and display WWW pages. The two most popular browsers are *Netscape Navigator* and *Microsoft Internet Explorer*. Both of these are *graphical* browsers, meaning they can display graphics as well as text. *Plug-ins* are usually required for presentation of multimedia information.

To summarize, and expand the role of this layer, essentially the transport layer decides how to utilize the network layer to render a virtual error-free connection between hosts. Thus, it both initiates and terminates connections between hosts.

5. **Session Layer**: This layer uses the transport layer to establish a connection between hosts for certain processes. It essentially handles the security side and the creation of the session itself.

6. **Presentation Layer**: This layer executes such functions as text compression and format conversions. This is the mechanism for ironing out differences between two hosts. If there are incompatible processes in the next layer up, the presentation layer allows the process to communicate via the session layer.

7. **Application Layer**: The top layer essentially handles the user's needs. For instance, it deals with such issues as allowing a user to access a remote

---

[5.3]The IETF is indirectly overseen by the *Internet Society*, a nonprofit organization that acts as a conscience and guide for the Internet. The Internet Society supports the *Internet Architecture Board* (IAB), which oversees the technical development of the Internet. In particular, IAB supervises IETF. See *http://ietf.org/*.

resource through a network without having to know if the resource is remote or local, a feature called *network transparency*. It will style itself after the user's particular desires such as email message formatting. This layer also deals with resource allocation and problem partitioning. The presentation layer provides the top layer with familiar local representation of data, which is independent of the format used on the network.

▼ **Analysis**: *Network Connections* embody a set of independent protocols, each in a different layer. The top layer, the applications layer, consisting of user applications programs, is the only variable layer. Each layer uses the layer one step below it and provides a service to the layer one step above it. Each of the network's components on a given host uses protocols applicable to its layer to communicate with its analogous component in another host. Such layered protocols are sometimes known as *peer-to-peer protocols*.

One large advantage of layered protocols is that the mechanism for delivering information from one layer to another is specified clearly as part of the protocol's definition. Also, changes within a protocol layer are prevented from affecting the other layers. This vastly simplifies the task of designing and maintaining network communication systems.

◆ **SSL Protocol — Simplified**

Now we describe SSL, mentioned earlier in the section. SSL is an Internet protocol that provides authenticity and secrecy for session-based communication. It provides a secure channel on the client/server model using a secret sharing scheme. The security model of SSL is that it encrypts the channel by enciphering the bits that go through that channel. As mentioned earlier, SSL began with Netscape who originated it and in 1996, they handed over the specifications of SSL to IETF who worked to standardize the SSL version 3 model, which had been released in 1995. In 1999, the TLS working group released TLS version 1, which has now become the IETF standards-track variant of the SSL version 3 protocol (see [68]). The cryptographic power of SSL/TLS is that it operates at the transport level so HTTP runs on top of SSL, called HTTPS.

To understand the layers of SSL, we must introduce the names of the two main subprotocols to be discussed in detail below: (1) the handshake protocol; which operates above the (2) record protocol. This is illustrated below.

HTTP
SSL Handshake Protocol
SSL Record Protocol
TCP
IP
Data Link Layer
Physical Layer

We begin by describing the lower level of SSL.

◆ **SSL Record Protocol**

This protocol defines the format used to transmit data, and is used by the handshake protocol to exchange messages between client and server. First the message to be transmitted is *fragmented*, which means it breaks the message down into manageable blocks. Then it compresses the data (but this is an optional exercise in SSL). It then applies a MAC (see page 136), enciphers the data, adds a header, and transmits the cryptogram as a TCP unit. This is illustrated in Diagram 5.2.

Diagram 5.2 SSL Record Protocol Actions

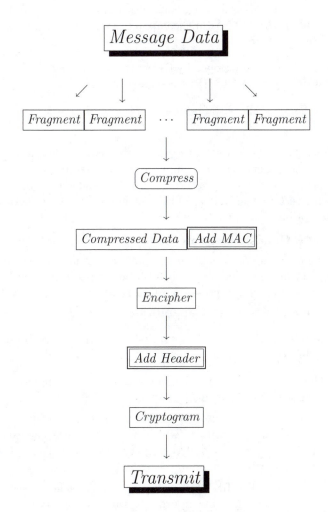

Upon receipt of the transmitted data, it is deciphered, authenticated, decompressed, reassembled, and delivered to users at higher levels.

▼ **Analysis**: The message data is typically fragmented into blocks of 2^{14} bytes, after which data compression is optional. We do not describe the details of the application of the MAC here since it is very similar to the HMAC described in Section 7.2. The encryption is done with an SKC cipher, which can be any of the *suite of ciphers* (or sets of ciphers) supported by SSL, listed below in order of cryptographic strength. These are the cipher suites for SSL implementations that use the RSA key-exchange algorithm.

1. Triple DES (see Section 3.2, page 131), using 168-bit encryption and SHA-1 message authentication (see page 255).

2. The only stream cipher, RC4 (see Section 3.7, page 159), using 128-bit encryption and MD5 message authentication (see page 255). When the RC4 is used, the MAC is first computed, then the MAC and compressed data are enciphered.

3. RC2 (a block cipher developed by Rivest for RSA Data Security), employing 128-bit encryption and MD5 message authentication.

4. DES (see Section 3.2) with 56-bit encryption and SHA-1 message authentication.

SSL supports the above variety of cipher suites since clients and servers may support different ciphers depending upon numerous factors.

The following protocol shows how the server and client authenticate one another, send certificates, and establish session keys. (See Section 8.5 for a general description of the client-server model.)

◆ **The SSL Handshake Protocol**

Below there are actions that are: mandatory, situation-dependent, or optional. We will call those that are either situation-dependent or optional, merely *optional* for simplification of presentation in Diagram 5.3 on page 225.

I Contact and Establish Capabilities:

1. The client sends the server a *client-hello* message, which contains the following fields:

    (i) The client's SSL version number (usually the highest SSL version supported by the client).

    (ii) Cipher suite (usually listed in decreasing order of preference), each element (cipher suite) of which includes both a key-exchange algorithm and the details of the cipher proposed. The following is the *SSL key-exchange suite of algorithms*:

        (a) **RSA**: The RSA public key of the recipient is used to encipher the secret key, but in order to validate the process, a public-key certificate for the recipient must be accessible.

(b) **Authenticated Diffie-Hellman**: In this type of key exchange, it is mandated that the server certificate contains the Diffie-Hellman public-key parameters, authenticated (signed) by Trent as a CA. If the client is required to send a certificate (see step 3 of stage II, below), then the public-key parameters as so included (see step 2 of stage III). Hence, the Diffie-Hellman-generated secret key is fixed in this case.

(c) **Anonymous Diffie-Hellman**: Essentially the Diffie-Hellman key exchange as given on page 166, is used with no authentication.[5.4]

(d) **Fortezza**: The *Key-Exchange Algorithm* (KEA) is the key exchange algorithm used with Fortezza. KEA was declassified by the U.S. Department of Defense on June 23, 1998. KEA requires a 1024-bit prime modulus, generated via the DSS specifications in [91]. Moreover KEA is based on a Diffie-Hellman protocol that uses SKIPJACK for the purpose of reduction of final values to an 80-bit key (see [90]).

(iii) Some randomly generated data consisting of a 32-bit timestamp and 28 bits generated by a CSPRNG (see page 151), both of which are treated as nonces to prevent replay attacks (see page 197).

(iv) List of compression methods supported by the client.

(v) A variable length session ID.

2. The server sends a *server-hello*, which consists of the same parameters as the client-hello. For instance, the server selects a cipher suite from the list proposed by the client, and the server chooses a compression method from the client-proposed list. However, the random field is generated by the server independent of the client-generated random field.

II Key Exchange and Server Authentication:

1. The server sends an identification certificate to the client (required for all key exchanges except anonymous Diffie-Hellman). (If RSA is used, we assume that the server's public key was sent with the certificate.)

2. The server sends a server-key-exchange message (*not* required only if either the server has sent a certificate with authenticated Diffie-Hellman parameters in step 1, or if RSA key exchange is used). If exercised, this contains the server's public keying material.

[5.4]This means that the SSL handshake protocol supports a totally anonymous operation in which neither the client nor the server is authenticated. As we saw with the Diffie-Hellman protocol in particular, and with PKC in general in Section 4.3, impersonation is possible since the entities are not authenticated, leaving the scheme open to the man-in-the-middle attacks. We will study a remote login protocol in Chapter 9 (see page 334), called the SSH protocol that does mandate server authentication.

3. If the server is *not* using anonymous Diffie-Hellman, it may send a request for the client's certificate. Contained in the client certificate request is a certificate type that dictates the PKC to be employed. For instance, if either RSA or DSS is used with authenticated Diffie-Hellman, then authentication (only) is accomplished via an RSA or DSS signature on the certificate.

4. The server sends a *server-hello-done* message.

III Key Exchange and Client Authentication:

1. After receiving the server-hello-done message, the client verifies the server's certificate if sent, and other server-hello parameters. If all is valid, the client responds.

2. If requested, the client sends a certificate. If authenticated Diffie-Hellman is being used, then the client's public-key parameters are included.

3. The client-key-exchange message must now be sent. The key-exchange mode dictates the content as follows:

   (i) If RSA is used, then the client generates a 48-byte *premaster secret*, which is encrypted with the server's public key (sent with certificate in Stage I).

   (ii) If anonymous Diffie-Hellman is employed, then the client's public Diffie-Hellman parameters are sent.

   (iii) If authenticated Diffie-Hellman is used, then the parameters were already sent in step 1 of stage II, so this is a null action.

   (iv) If Fortezza is used, then the client's Fortezza parameters are sent.

3. If a certificate has been requested, the client signs a piece of data that is unique to the handshake and known by both client and server, along with the encrypted premaster secret.

IV Finish Protocol:
To simplify the final stage, we assume that RSA is being used.

1. If the server verifies the client's identity, then the server uses its private key to decipher the premaster secret. Then the server performs a sequence of steps to create the *master secret* from the premaster secret, a one-time 48-byte generated for this session. These same steps are followed by the client to recover the master secret.

2. Both the client and the server use the master secret to generate session keys, which are symmetric keys used to encipher and decipher information exchanged over the course of this SSL session, and to verify its *integrity*, meaning the detection of changes that might have occurred in the time period from transmission to reception.

3. The client sends a *client-finished message* saying that all future messages will be encrypted with the session key, the first message encrypted with the secret session-key independently generated by the client and server.

4. The server sends a similar encrypted *server-finished message*, which assures the client it is communicating with the server since the client sent the premaster secret encrypted with the server's public RSA key, which only the server could have deciphered to calculate the session key.

5. The handshake is now completed and the client and server may exchange application layer information with a secure connection.[5.5]

Diagram 5.3 SSL Handshake Protocol Actions

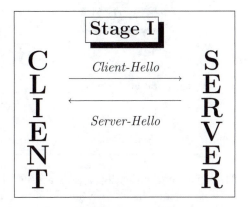

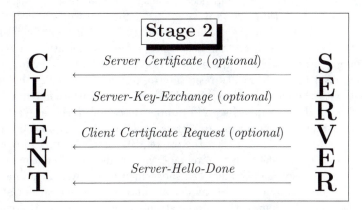

[5.5]Caution must be exercised in certain generic implementations of SSL. See [142], for instance.

Diagram 5.3 SSL Handshake Protocol Actions (continued)

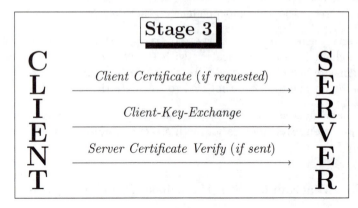

Stage 3

C L I E N T

Client Certificate (if requested) →

Client-Key-Exchange →

Server Certificate Verify (if sent) →

S E R V E R

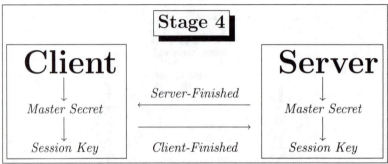

Stage 4

Client
↓
Master Secret
↓
Session Key

Server-Finished ←

Client-Finished →

Server
↓
Master Secret
↓
Session Key

▼ **Analysis**: SSL server authentication allows a user to confirm a server's identity, which is quite important if, for instance, the user is sending a credit card number over a network, and needs to check the receiving server's identity. SSL client authentication allows a server to confirm a user's identity. This is very important if, for example, a bank is sending confidential financial information to a customer, and needs to check the recipient's identity. An enciphered SSL session is protected by a tamper-detection mechanism, which automatically checks to see if information has been altered in transit, a secure hybrid cryptosystem, with the handshake allowing independent creation of symmetric keys for fast enciphering, deciphering, and tamper-detection during the session.

If authentication of client and/or server is chosen, then X.509V3 certificates are used; see page 238. This makes the use of SSL somewhat unwieldy given the necessity of an established PKI to manage the certificates. Yet, the certificates render a scalable key-management scheme, which is a powerful mechanism. Of course, a totally anonymous SSL mode provides no authentication, and opens the scheme up to the man-in-the-middle attack, as noted earlier. However, when users want to take advantage of SSL on their Web site without being associated with their host, then anonymous SSL is the way to go. Hence, the anonymous SSL server has its place, and there are numerous vendors available to sell such packages to the willing.

5.8 Digital Cash Schemes

Some rhyme a neebor's name to lash;
Some rhyme (vain thought!) for needfu' cash;
Some rhyme to court the countra clash,
An' raise a din;
For me, an aim I never flash;
I rhyme for fun.

Robert Burns (1759–1796), Scottish poet
from *To J. S*[mith] (1786), st. 5

Anyone living in the modern world, having access to a computer and a source of funds, is aware of *e-commerce*, that is, basically doing business by computer. This may include the purchasing of goods or services from an Internet site, or conducting banking transactions, for instance. All of this requires the use of *digital cash schemes* that comprise a collection of protocols for transferring money. These are modeled on "hard cash" and its use, as well as credit card or debit card use. Naturally, we want secure transactions, and this involves many facets.

In our discussion below, we will simplify the general scenario by assuming one customer, one vendor, one bank, but potentially many adversaries seeking to steal your money. What makes digital cash and e-commerce possible are hybrid cryptosystems, PKC and its digital signatures, for security and authentication, combined with SKC for efficient transfer of funds. For instance, an example of the use of digital cash is the bank affixing its digital signature to an electronic money order via its private key, and the vendor verifying this by using the bank's public key. In this fashion, the customer may withdraw funds from the bank to pay the vendor both of whom are assured that the proper amounts are securely withdrawn as payment for the goods or services.

Before describing our digital cash scheme, we must be familiar with some terminology from e-commerce. In particular, we are going to describe a scheme called *ECash* as a relatively simple template for Internet purchases, which involves Bob as the customer, the bank, and the vendor.

◆ E-commerce Terminology

▼ **Bank** The *bank* represents any electronic bank that is connected to the Internet and plays an essential but unobtrusive role. The bank has an RSA modulus n, a private key d that it keeps secure, and a public key e.

▼ **Coins** A coin is a pair of integers $(m, m^d) \pmod{n}$ where m is the unique identification number of the coin. The value $m^d \pmod{n}$ is the *bank's signature* or *special digital stamp* on the coin. Coins will typically have different denominations, but for simplicity we will assume that the denomination is the same for each of them, e.g., \$100. To ensure a coin is used only once, the bank records all identification numbers m in its *used coin database*. If m has been so recorded and someone tries to "spend" the coin, called *double spending*, the bank would inform that the coin is a worthless copy.

▼ **Blinding** This is the technique first mentioned on page 177. We repeat it here to put it in the context of e-commerce. Suppose that Bob has an account at the bank, and that m is the identification number of a coin. Bob's selects a random integer $z \in (\mathbb{Z}/n\mathbb{Z})^*$ (called the *blinding factor*), computes

$$s \equiv z^e m \,(\text{mod } n), \qquad (\text{called } blinding).$$

He sends the *blinded message* s to the bank, which computes $t \equiv s^d \,(\text{mod } n)$, and sends t to Bob, who computes $tz^{-1} \,(\text{mod } n)$, called *unblinding*. Since

$$tz^{-1} \equiv s^d z^{-1} \equiv z^{ed} m^d z^{-1} \equiv z m^d z^{-1} \equiv m^d \,(\text{mod } n),$$

given that $ed \equiv 1 \,(\text{mod } \phi(n))$, the bank has blindly signed the identification number, validated with its signature. Of course, some validity checks must be done on Bob before this signing.

▼ **Money Order** This consists of digital data, which contains Bob's identifying data, together with the identification number m of a coin, and its denomination. For instance, a \$100 money order is given by $(\$100, m \,(\text{mod } n), I_B)$ where I_B is a digital data string uniquely identifying Bob. Thus, m is a "blank" coin awaiting the banks signature m^d to validate it.

▼ **Cut-and-Choose Protocol** The classic cut-and-choose protocol, for dividing anything equitably, is described as follows. Alice cuts the thing in half, Bob chooses one half for himself, and leaves the other half for Alice. For instance, if they both want a piece of an apple, this ensures that Alice will be as fair as possible in her cutting, since Bob chooses first. Below, we will show how the bank uses this notion on the money orders generated by Bob.

The next topic has already been covered in detail in Section 5.5. However, for the convenience of the reader, we present the basics for what is needed herein.

▼ **Secret Splitting** This is any protocol that takes a message, and divides it into pieces each of which is meaningless in itself, but when pieced back together yields the original message. For instance, given the assistance of Trent, Alice and Bob can split a message m via the following protocol.

1. Trent generates a random bitstring b with bitlength equal to that of m and creates $b \oplus m = r$, where $\oplus$ is addition modulo 2.

2. Trent gives b to Alice and r to Bob, with b and r having no meaning unto themselves individually.

3. Alice and Bob can piece together the information to retrieve the original message via $b \oplus r = m$.

Before describing the ECash scheme in detail, we will begin with an outline of how it works in verbiage only, then proceed to add the mathematical description.

◆ ECash Scheme – Preliminary Description

Suppose that Bob wants to withdraw $100 from his account at the bank and use it to spend at a vendor to purchase merchandise electronically. The following protocol is enacted.

▼ ECash Withdrawal

1. Bob generates 100 sets of identity strings, and uses a secret-splitting protocol to split each one into two pieces.

2. Bob prepares 100 money orders for $100 each, uses a blinding protocol, and sends the blinded orders to the bank (including his split-identity strings).

3. The bank, after verification of validity using a cut-and-choose protocol, encrypts the identification number of one of the 100 blank coins with its signature to form a valid money order, debits Bob's account, and sends the money order to Bob. If verification fails the bank reports to the authorities.

4. Bob unblinds the signed coin in the money order, which he can now use to spend with a vendor via the money order.

▼ ECash Spending

1. Bob sends the money order to the vendor, who verifies the bank's signature.

2. The vendor uses the secret-splitting generated by Bob in step 1 of the withdrawal stage and sends the money order to the bank for verification.

3. The bank checks the coin against its used coin database. If it is not there, the bank credits the vendor's account with $100 and the vendor sends the goods to Bob, and the transaction is complete. If the coin *is* there, the bank executes step 4.

4. The bank rejects the money order and uses the identity strings to identify Bob.

Now we get the details of how the above is accomplished.

◆ ECash™ Scheme — Full Description

▼ ECash Withdrawal Details

1. Bob generates 100 sets of unique digital data strings $\mathcal{S}_j = \{I_{j_k}\}_{k=1}^{100}$ for $1 \leq j \leq 100$ such that *each* I_{j_k} uniquely identifies him. He then engages in a secret-splitting protocol so that, for each $j = 1, \ldots, 100$, the digital data string is split into two pieces $I_{j_k} = \{L_{j_k}, R_{j_k}\}$ with $I_\ell \oplus I_r$ identifying Bob if and only if $\ell = r$, where $\ell, r \in \{j_k\}_{k=1}^{100}$.

2. Bob prepares 100 money orders for \$100 each:

$$M_j = (\$100, m_j, \{L_{j_k}, R_{j_k}\}_{k=1}^{100}) \quad (1 \le j \le 100),$$

where m_j is a randomly generated number (by Bob's computer) that is an ECash coin's identification number with $m_j \ne m_i$ for $j \ne i$. Bob executes a blinding protocol for each of the money orders by selecting a random $z_j \pmod{n}$ for $1 \le j \le 100$ and sends the 100 blinded money orders $(\$100, z_j^e m_j, \{L_{j_k}, R_{j_k}\}_{k=1}^{100})$ to the bank.

3. Using a cut-and-choose protocol, the bank opens 99 of the money orders and checks that the amounts are all the same, \$100, that $m_j \ne m_i$ for $j \ne i$, and that each $L_{j_k} \oplus R_{j_k}$ is a valid identity string. If the bank sees no evidence of fraud, it (blindly) signs the remaining money order, M_{100}, say, and sends the validated money order,

$$(\$100, z_{100}^e m_{100}, (z_{100}^e m_{100})^d, \{L_{100_k}, R_{100_k}\}_{k=1}^{100}),$$

to Bob, withdrawing \$100 from his account. Otherwise, the bank does not, and discloses the problem, and the transaction is terminated.

4. Bob unblinds to get the ECash coin (m_{100}, m_{100}^d), which he can now spend using the money order M_{100}.

▼ **ECash Spending Details**

1. The vendor verifies the bank's signature by computing $(m_{100}^d)^e = m_{100}$.

2. The vendor gives Bob a random 100-bit binary string $(b_1 b_2 \ldots b_{100})$, and requests that Bob reveal L_{100_k} if $b_k = 1$, and R_{100_k} if $b_k = 0$ for each of $k = 1, 2, \ldots, 100$, which Bob does. The vendor sends the money order to the bank for verification from its database.

3. The bank checks its used coin database to ensure that m_{100} is not there. If it is not, then the bank deposits \$100 into the vendor's account, and records m_{100} in its used coin database along with the identity string selected by Bob via the binary string in step 1. The vendor then sends the goods to Bob along with a receipt, and the transaction is completed.

4. If m_{100} is in the used coin database, the bank rejects the money order. Then it compares the identity string on the bogus money order with the stored identity string attached to m_{100}. If they are the same, then the bank knows the vendor duplicated the money order. If they differ, then the bank knows that the entity who gave it to the vendor must have copied it. Given that the coin (m_{100}, m_{100}^d) was spent with another vendor, then that vendor gave Bob a different binary string. The bank compares the differing strings until it finds a position where the bits differ, say the ith position. This is where one vendor asked Bob to open L_i and the other asked Bob to open R_i. Thus, then bank forms $L_i \oplus R_i$, revealing Bob's identity, which can be reported to the authorities.

▼ Analysis

Untraceability: The above scheme ensures anonymity for Bob, as a legitimate user. In other words, his identity is untraceable. When he spends the coin, the bank must honour it since the bank's signature is on it. However, since it is unable to recognize the specific coin, given that it was blinded when signed, the bank does not know who made the payment. However, if Bob is not a legitimate user and tries to spend the coin twice, the bank can detect him in step 3 of the spending stage. The attentive reader will have noticed that we assumed that the binary strings were different in step 4, if Bob is illegitimate. This is not 100% certain but the probability that they *are* the same is 1 in 2^{100}, which is *extremely* unlikely. This legitimate use of digital cash ensures that it is *anonymous digital cash*. This mimics the use of real paper cash where the use is anonymous and untraceable. The other type of digital cash is *identified digital cash*, which mimics the use of a credit card, allowing a bank to track the transaction as it moves through the system. This is not used often since users want the untraceable property.

Security: This is a property guaranteed by both step 4 of spending, which tells us that the coins cannot be copied and reused, and the fact that the bank keeps its signature d secure, along with identity data.

Integrity: This is a property satisfied by ECash since the scheme is based upon the security of RSA, which we have seen to be valid when properly implemented.

Authenticity This is a property guaranteed by step 4 of spending since Bob, as a legitimate user, is protected from impersonation.

Offline: Since an illegitimate user can be identified in step 4 of spending, then it is not necessary to check the coins immediately since a cheater would be identified later. Thus, the offline property exists for the scheme, since the vendor does not have to check at the time of payment (online), but rather can do so later (offline).[5.6]

Recovery: ECash has a special built-in recovery protocol executed between Bob and the bank that allows all the coins that have been withdrawn by Bob to be reconstructed. Thus, if there is a system crash, or computer crash, in the middle of a payment attempt over the Internet, these reconstructed coins can be redeemed at the bank (but only those coins not already in its used coin database). Recovery of ECash coins can be accomplished over the Internet with the click of a button.

Coin Denominations: Our simplified version of the ECash scheme did not address the issue of different coin denominations. The ECash scheme uses a different RSA public exponent for each denomination, but the same RSA modulus n for each of them. Then the above ECash scheme is executed in parallel for as many iterations necessary to withdraw the required amount.

[5.6] All of the above characteristics thus far in this analysis, are aspects of what Okamoto and Ohta described as the *ideal* digital cash scheme. See [182].

Background: David Chaum is *the* pioneer in digital cash. He invented the notion of digital coins and the basic protocols for digital cash. In fact, the seeds of ECash can be found in Chaum's works, [53] and [54]. Chaum was also a pioneer in ensuring that the (now world-renowned) CRYPTO meetings at U.C. Santa Barbara, and the EUROCRYPT meetings in Europe, would become an annual affair under a single organization, the *International Association for Cryptologic Research*. Chaum was a pioneer of protocols for using PKC to ensure anonymity of electronic users. In fact, it is not even certain when he was born, since his desire for anonymity runs deep. It is known that he was raised in Los Angeles, and ultimately got his undergraduate degree at U.C. San Diego, and did his graduate work at U.C. Berkeley. By 1979 he was already creating ideas for using PKC for authentication via digital signatures. Ultimately this led him to anonymous untraceable schemes embodied in ECash.

Chaum's idea of assigning a unique number to each coin guaranteed the authenticity of the "virtual" money. His idea for "blinding" a digital signature was the key notion to protect a user's anonymity, since even a bank which issues the cash, does not know who has it, and so cannot trace it. He is also responsible for the above notion of preventing double spending while, at the same time, providing anonymity for legitimate users.

In those pioneering days, Chaum did not find a lot of support among colleagues, so he eventually opened his own company to spread his ideas. Thus in 1990, he founded Digicash in Amsterdam, Holland, while working for the Centre for Mathematics and Computer Science (CWI). His company worked on smart card applications, including the world's first automated road toll collection scheme, and held patents on his anonymous digital cash schemes, including ECash, which Digicash invented in 1990. However, as Visa and other credit card companies failed to strike deals with Digicash for use of those patented ideas, they eventually developed their own. By May of 1997 Visa and Master Card completed their own e-commerce standard, SET^{TM} (Secure Electronic Transmission) that allows for security, privacy, integrity, as well as authenticity in the protection of cards used in Internet e-commerce. We will revisit SET in Section 6.3.

Even the so-called "cyberpunks" (see page 159) used an idea of Chaum's for "remailers". These are information launderers, so to speak. One sends a message to some Internet site (having what is called an *anonymous server* or *remailer*), maintained by the cyberpunks who remove all identifying characteristics from the message, and send it on, ultimately, to its final destination without any return address or identifying information.

While Digicash was still trying to get its patents to be used by mainstream business, a former student of Chaum's, Stefan Brands, came up with a complicated but nearly ideal digital cash scheme, and although Chaum claimed that Brands' ideas were derivative of his own, Brands obtained his own patents. (See page 536 in Appendix D for a complete description of Brands' scheme.) Although Digicash did have some limited success, it finally came crashing down in 1998, when Chaum filed for bankruptcy and lost his patents, yet another pioneer who came up short in terms of recognition for his (real-world) efforts.

Chapter 6

Key Management

Every time I make an appointment, I create a hundred malcontents and one ingrate.

King Louis XIV (1638–1715)
— from Voltaire *Siècle de Louis XIV*, 1768 edition

6.1 Authentication, Exchange, and Distribution

Since any (properly implemented) cryptosystem is only as strong as its keys, we need to be concerned about *key management*, the secure generation, distribution, and storage of keys. Generation of cryptographic keys is vital in any cryptosystem. A real-world example, illustrating what can go wrong, is given by SSL discussed in Section 5.7. In the early days of SSL, implementations released by Netscape failed due to *weak keys*. We saw how this was a problem with DES, and related ciphers, in Section 3.2. We encountered numerous secure key-generation schemes such as Blowfish in Section 3.4; AES in Section 3.5; RC4 in Section 3.7; RSA in Section 4.2; and ElGamal in Section 4.4. Thus, we have sufficient illustrations of the mechanisms for doing so.

As for key agreement we saw, on page 180, how an impersonation attack can be launched, and how Diffie-Hellman key exchange is particularly vulnerable to this type of fraud. Also, on page 199, we saw how three-pass protocols can be effective means of authentication. Now we look at a scheme that is considered to be a three-pass variant of the Diffie-Hellman scheme, which solves the problem with the original scheme. The following appeared in [71] in 1992. This is an example of an *authenticated key-agreement protocol*, which means that the key-agreement protocol itself, authenticates the parties, in this case, Alice and Bob, with Trent's help.

◆ **Station-to-Station Protocol (STS)**

Background Assumptions: In the following, it is assumed that $(\text{sig}_A, \text{ver}_A)$ and $(\text{sig}_B, \text{ver}_B)$ are Alice and Bob's respective signature and verification algorithms (see Section 4.3); and that Trent has compiled and made public ver$_A$, ver$_B$, and certified that these are indeed their respective verification algorithms, and not, say, for Eve or Mallory. Alice and Bob want to establish a key k for use with a symmetric encryption scheme E_k, and they develop k in the following fashion, which is Diffie-Hellman with digital signatures added.

Protocol Steps:

1. Alice and Bob agree upon a large prime p and a primitive root α modulo p.

2. Alice selects a random (secret) integer $e_A \in (\mathbb{F}_p)^*$, and Bob selects a random (secret) integer $e_B \in (\mathbb{F}_p)^*$.

3. Alice computes $m_1 \equiv \alpha^{e_A} \pmod{p}$, which she sends to Bob.

5. Bob computes $k \equiv (\alpha^{e_A})^{e_B} \pmod{p}$, and sends

$$m_2 = (\alpha^{e_B}, E_k(\text{sig}_B(\alpha^{e_B}, \alpha^{e_A})))$$

to Alice.

5. Alice computes $k \equiv (\alpha^{e_B})^{e_A} \pmod{p}$ and obtains $\text{sig}_B(\alpha^{e_B}, \alpha^{e_A})$ via E_k^{-1} acting on $E_k(\text{sig}_B(\alpha^{e_B}, \alpha^{e_A}))$.

6. Alice requests that Trent certify that ver$_B$ is indeed Bob's verification algorithm, and if it is so certified, she uses it to verify Bob's signature. Then she sends

$$m_3 = E_k(\text{sig}_A(\alpha^{e_A}, \alpha^{e_B}))$$

to Bob.

7. Bob deciphers via E_k^{-1}, asks Trent to certify that ver$_A$ is Alice's verification algorithm. If so, he uses it to verify Alice's signature.

▼ **Analysis**: With the three messages, m_1, m_2, m_3, this is a three-pass version of Diffie-Hellman, using digital signatures to do the authentication in conjunction with Trent. The STS protocol establishes a key k, mutually confirmed by Alice and Bob, whose identities have been verified to each other, but not to Eve or Mallory. Thus, we indeed have an authenticated key-agreement protocol, so now Alice and Bob can use k to encrypt all subsequent messages between them.

We have seen schemes for distributing keys over large networks (see the Kerberos protocol on page 196 for instance). However, we might want to decide upon keys in advance and *pre*-distribute them. We briefly met this concept on page 162, and the following scheme is designed to deal with the problems discussed therein. The scheme below was introduced by Blom in 1985 (see [27]). However, we present a simplified version given in [32] several years later.

◆ **Blom's Key Predistribution Scheme — Simplified**

Basic Assumptions: We suppose that there is a network of $m \in \mathbb{N}$ users, and that keys are taken from $\mathbb{F}_p$ where $p \geq m$ is a public prime. Each user on the network has a unique identification number such as u_A for Alice. Trent will distribute to each user, over a secure channel, a mechanism for communicating with any other user on the network.

Protocol Steps:

1. Trent chooses three random values $r_1, r_2, r_3 \in \mathbb{F}_p$, and for each user, such as Alice, computes

$$x_A \equiv r_1 + r_2 u_A \pmod{p} \text{ and } y_A \equiv r_2 + r_3 u_A \pmod{p}.$$

Then for each user, such as Alice, he computes the polynomial $f_A(x) = x_A + y_A x$, which is sent over the secure channel to her.

2. If Alice wants to communicate with user u_B for Bob, say, then Alice computes $k_{AB} = f_A(u_B)$ and Bob computes $k_{BA} = f_B(u_A)$. In the analysis below, we show that $k_{AB} = k_{BA} = k$, say, so Alice and Bob now have a means of communicating with k via some chosen SKC.

Example 6.1 *For simplicity, we will assume a network of three users, Alice, Bob, and Carol. If $p = 31$, $u_A = 7$ for Alice, $u_B = 11$, for Bob, $u_C = 17$ for Carol, and Trent selects $r_1 = 2$, $r_2 = 10$, $r_3 = 29$, then*

$$x_A = 2 + 10 \cdot 7 = 72; \quad y_A = 10 + 29 \cdot 7 = 213; \quad so, \quad f_A = 72 + 213x;$$

$$x_B = 2 + 10 \cdot 11 = 112; \quad y_B = 10 + 29 \cdot 11 = 329; \quad so, \quad f_B = 112 + 329x;$$

$$x_C = 2 + 10 \cdot 17 = 172; \quad y_C = 10 + 29 \cdot 17 = 503; \quad so, \quad f_C = 172 + 503x$$

are the respective polynomials for Alice, Bob, and Carol. Thus,

$$k_{AB} = f_A(u_B) = 72 + 213 \cdot 11 = 2415 = k_{BA} = f_B(u_A) = 112 + 329 \cdot 7,$$

$$k_{AC} = f_A(u_C) = 72 + 213 \cdot 17 = 3693 = k_{CA} = f_C(u_A) = 172 + 503 \cdot 7,$$

$$k_{BC} = f_B(u_C) = 112 + 329 \cdot 17 = 5705 = k_{CB} = f_C(u_B) = 172 + 503 \cdot 11.$$

▼ **Analysis**: First we show that $k_{AB} = k_{BA}$, where we assume the equalities are congruences modulo p for convenience.

$$k_{AB} = x_A + y_A u_B = r_1 + r_2 u_A + (r_2 + r_3 u_A)u_B = r_1 + r_2 u_A + (r_2 + r_3 u_A)u_B =$$

$$r_1 + r_2 u_A + r_2 u_B + r_3 u_A u_B = r_1 + r_2 u_B + (r_2 + r_3 u_B)u_A = f_B(u_A) = k_{BA}.$$

Now we show that Blom's scheme is unconditionally secure against an attack by a user, Mallory. In other words, we will show that with the knowledge Mallory has, namely,

$$f_M(x) \equiv r_1 + r_2 u_M + (r_2 + r_3 u_M)x \pmod{p}$$

sent by Trent, all values of $z \in \mathbb{F}_p$ are possible for k_{AB}, which he is trying to cryptanalyze. Since Mallory knows $f_M(x)$, then he knows the coefficients

$$r_1 + r_2 u_M \equiv x_M \pmod{p}, \tag{6.1}$$

and
$$r_2 + r_3 u_M \equiv y_M \,(\mathrm{mod}\ p), \tag{6.2}$$

but not the unknown value $r_1 + r_2(u_A + u_B) + r_3 u_A u_B \equiv z \,(\mathrm{mod}\ p)$. Putting this into a matrix equation we get

$$AX = \begin{pmatrix} 1 & u_M & 0 \\ 0 & 1 & u_M \\ 1 & u_A + u_B & u_A u_B \end{pmatrix} \begin{pmatrix} r_1 \\ r_2 \\ r_3 \end{pmatrix} \equiv \begin{pmatrix} x_M \\ y_M \\ z \end{pmatrix} \,(\mathrm{mod}\ p),$$

where

$$\det(A) \equiv u_M^2 + u_A u_B - (u_A + u_B)u_M \equiv (u_M - u_A)(u_M - u_B) \not\equiv 0 \,(\mathrm{mod}\ p),$$

since $u_M \not\equiv u_A \,(\mathrm{mod}\ p)$ and $u_M \not\equiv u_B \,(\mathrm{mod}\ p)$. Hence, there exists a solution $(r_1, r_2, r_3) \in \mathbb{F}_p^3$, for any possible value $z \in \mathbb{F}_p$ of k_{AB} given the information Mallory has at his disposal. Hence, Mallory obtains no information about k_{AB}.

Although the scheme is unconditionally secure against an attack by any individual user, it is vulnerable to a total break by more than one user acting in concert. For instance, suppose that Mallory conspires with Eve.

Since Mallory has Equations (6.1) and (6.2) and Eve has her two similar equations, then they have the four modular equations

$$x_M \equiv r_1 + r_2 u_M \,(\mathrm{mod}\ p), \ y_M \equiv r_2 + r_3 u_M \,(\mathrm{mod}\ p),$$

$$x_E \equiv r_1 + r_2 u_E \,(\mathrm{mod}\ p), \ \text{and} \ y_E \equiv r_2 + r_3 u_E \,(\mathrm{mod}\ p).$$

Hence, they have four equations in three unknowns from which elementary algebra will yield a unique solution for r_1, r_2, r_3.

However, the scheme can easily be made secure against any $n \in \mathbb{N}$ users acting in concert by altering the choice by Trent in step 1. Trent replaces the polynomial,
$$p(x, y) = r_1 + r_2(x + y) + r_3 xy,$$
by

$$f_x(y) \equiv \sum_{i=0}^{n} \sum_{j=0}^{n} r_{i,j} x^i y^j \,(\mathrm{mod}\ p), \tag{6.3}$$

for randomly chosen $r_{i,j} \in \mathbb{F}_p$ with $r_{i,j} \equiv r_{j,i} \,(\mathrm{mod}\ p)$ for all such i, j. The general setup (6.3) is an aspect of the full Blom protocol (see [27]). It will however, succumb to a conspiracy by $n + 1$ users acting in concert in the same fashion as above. Thus, the above polynomial can be chosen for an appropriate, arbitrarily high, value of n.

6.2 Public-Key Infrastructure (PKI)

Who says that fictions only and false hair
Become a verse? Is there in truth no beauty?
Is all good structure in a winding stair?

George Herbert (1593–1633), English Poet and Clergyman

A *public-key infrastructure*, or PKI, consists of a set of protocols and standards, which support and enable the secure and transparent use of public-key cryptography. PKI is particularly important in *applications* requiring the use of public-key cryptography. For instance, in Section 6.3, we will look at one such application to a secure e-commerce scheme developed for credit card payments over the Internet, which will use the concepts we develop herein. As we shall see, PKI may be used as a tool for authentication, key distribution, and nonrepudiation. Section 6.1 dealt with several issues surrounding authenticity, and key distribution. We discussed nonrepudiation briefly on page 162 when we were comparing SKCs and PKCs; and on page 182 when we looked at the DSS. In this section, we will formalize the notions surrounding PKI so we can use it as a framework for such discussions as that surrounding the credit card payment scheme in the next section.

PKI provides protocols for certification of public keys and verification of certificates. The reason is that if Alice wants to be sure she is communicating with Bob and not Eve, say, then she must have assurance that Bob's public key actually belongs to Bob. This is where the role of a certificate comes into play. We now discuss PKI with the role of providing key management through the use of a *certification authority* (CA) and a *registration authority* (RA).

◆ Role of the CA

The CA is an entity responsible for issuing *public-key certificates*, which are tamperproof data blocks. A certificate contains (at least) the following: entity identification; CA identifier; and a public key. These are used to bind the individual name to the corresponding public key. The CA accomplishes this by affixing its private key as a digital signature, thereby performing *key registration* via the issuing of a certificate. Think of the certificate as being the analogue of a driver's license.

◆ Role of the RA

The RA typically plays the role of assisting the CA by establishing and verifying the identity of entities, called *end users*, who wish to register on a network, for instance. Other functions of the RA may include:

(1) Key predistribution for later online verification.

(2) Initiation of the certification process with the CA for end-users.

(3) Performance of certificate-management functions such as *certificate revocation* (meaning the cancellation of a previously issued certificate).

(4) Key generation (in the absence of a CA to do so).

◆ PKI Services

1. Certificate creation, distribution, management, and revocation.

2. PKI-enabled services, that are *not* part of the PKI, can be built upon the core PKI, and these include secure communications and timestamping.

3. A PKI, in itself, may *not* involve any cryptographic operations with the keys that it is managing. A common feature of all PKIs is a set of certification and validation protocols, since the fundamental core predicate of PKI is the secure management of public keys, as well as nonrepudiation.

◆ Certificates

By a *certificate* we will mean the ISO/ITU-T X.509 Version 3 public-key certificate format. The ITU is the International Telecommunication Union, which was established on May 17, 1865 (as the International Telegraph Union) to manage the first international telegraph networks. The name change came in 1906 to properly reflect the new scope of the Union's mandate. The ITU-T is the ITU Telecommunication Standardization Section, one of three sections of ITU, established on March 1, 1993. In conjunction, ISO (see page 218) and ITU-T form world standards such as the X.509, which is a public-key certificate. Version 3 (as specified in [128]) was developed to correct deficiencies in earlier versions, and has become the accepted standard so that often the term *certificate* is used to mean this version of X.509. Version 3, denoted by X.509V3, contains each of the following fields: (1) version number; (2) certificate serial number; (3) signature-algorithm identifier; (4) issuer name; (5) validity period; (6) entity name; (7) entity public-key information; (8) issuer unique identifier; (9) entity unique identifier; (10) extensions; (11) signature; (12) In addition, the extensions field can contain numerous types such as authority key identifier, extended key usage, and private-key usage period.

◆ PKI Trust Models

In PKIs, the *trust models* are used to describe the relationships of CAs with end users and others. We describe only two of them.

1. *User-Centric Trust.* In this model, each user makes the decision as to which certificates to accept or reject. There is an implementation, used by *Pretty Good Privacy* (PGP), about which we will learn in Chapter 8, when we discuss e-mail security. In this implementation, a user, such as Alice, exchanges certificates which are public keys of those other users with whom she wants to communicate. She protects her certificate from alteration by signing it with her private key. Upon receipt of Bob's certificate, say, Alice acts as a CA by assigning it one of the following levels:

    (1) *Complete trust*, meaning that she trusts Bob and anyone whose certificate is signed with Bob's key.

(2) *Partial trust*, meaning that Alice does not completely trust Bob, so certificates signed by Bob must also be signed by other users (whom she does trust) before she accepts it.

(3) *No trust*, meaning that Alice does not trust Bob and will not trust any certificate signed by Bob.

(4) In some implementations, there is a fourth level of *uncertain*, but this essentially amounts to no trust.

In this way she builds a *web of trust* with other users, but this model is not acceptable for such applications as e-commerce. A more generally secure trust model is described in what follows, where we will need the term *security domain*, which means a system governed by a trusted authority.

2. In the PKI trust model called *cross-certification*, the CAs (in their respective security domains) are required to form a *trust path* between themselves. There are various processes within the framework of this trust model.

(1) The process called *mutual cross-certification* involves CA_1 signing the certificate of CA_2, and CA_2 signing the certificate of CA_1.

(2) If the domains are different, called *interdomain cross-certification*, then *relying parties* (those entities who verify the authenticity of an end user's certificate) are able to trust end users in the other domain. This trust model is clearly suited to e-commerce, such as that engaged by two distinct business organizations.

(3) If two CAs are part of the *same* domain, called *intradomain cross-certification*, then this model can be varied to accommodate a hierarchy of CAs where CA_1 can sign the certificate of CA_2 who is at a lower level, without having CA_2 sign CA_1's certificate, called *unilateral cross-certification*. An advantage of unilateral cross-certification is that it allows relying parties to trust only the top-level root CA, having their certificates issued by the authority closest to them.

Clearly, the trust model is an indispensable part of any PKI. We have described only two of many such models, which is sufficient for our purposes. The reader interested in seeing more of them in greater detail may consult [4], which is a book dedicated entirely to the topic of PKI.

In the following we will need the term, *certificate-revocation list* (CRL), which is a signed data structure embodying a timestamped inventory of revoked certificates.

◆ Certificate Storage

Once generated, a certificate must be stored for use at a later time. For this, CAs require what is called a public *certificate directory*, which is a public database or server accessible for read-access by end users that the CA manages and to which it supplies certificates. This directory is a central storage location that provides an individual, public, central location for the administration and distribution of certificates. As with PKI itself, there is no single standard. Perhaps the most popular is the X.500 series, which is the ISO/ITU-T array of standards with specifications in [127], and is, in fact, the underlying structure in which the X.509 certificate originated. Proprietary directories based on X.500 include *Microsoft Exchange*, for instance. The X.500 series has standardized protocols for obtaining data structures, thus allowing any PKI to have access via a mechanism called a *schema* for the storage of certificates and CRL data structures in a given entity's directory entry.

◆ **Certificate Revocation**

Suppose that Alice's private key has been compromised, which means the corresponding public key can no longer be used for Alice. The process for alerting the rest of the network of users is *certificate-revocation checking*. We can now invoke the earlier analogy in terms of a driver's license. A police officer, upon checking a driver's license, not only verifies the date on the license, but also calls some central police authority to confirm that the license has not been revoked. *Certificate revocation* means marking the certificate as revoked by the CA and placing it in a CRL. CAs issue periodic CRLs to ensure relying parties that the most recent CRL is current, so even if there are no changes, a CRL is issued on time according to the schedule. Also, some certificates are cross-certified between the CAs themselves. To revoke these certificates, we need a separate *authority revocation list* (ARL), which plays the role of CRLs. However, revoking the PKC of a CA is rare and usually occurs when the CAs private key is compromised.

The X.509 Version 2 standard for CRLs, as with the Version 3 certificates, discussed earlier, has extension fields to make the CA's job of revocation easier. They are:

1. *Reason code*, namely, a specification of the reason for the revocation.

2. *Hold instruction code*, which is a mechanism to temporarily suspend a certificate, and contains an *object identifier* (OID), which stipulates the action to be taken if this field is filled.

3. *Certificate issuers*, which has the identity of the certificate issuer.

4. *Invalidity date*, which contains the date and time of the known or suspected compromise.

There is an alternative online mechanism for certificate revocation, the most popular being the *Online Certificate Status Protocol* (OCSP), documented in [176] with HTTP being the most common practical mechanism (see page 219). This is a challenge-response protocol offering a mechanism for online revocation of data from a trusted authority, called an *OCSP responder*. However, as a mere protocol, it does not have the capacity to store revocation data, so the OCSP responder must obtain information from some other source. Thus, latency is involved with its use. Moreover, it is limited to the supplying of information about the revocation status of a given list of certificates, and nothing else. Hence, there is still the need for CRLs.

◆ **Key Backup and Recovery Server**

This gives the CA a mechanism for backing up private keys together with a means of recovering them later should end users lose their private keys. Key recovery is implemented in an individual PKI by its authorities to provide key recovery for its end users. The key recovery server is an automated process to relieve the burden on PKI authorities. To prevent an adversary from accessing an entity's private key and launching an impersonation attack, a CA may support not one, but two key pairs: one for enciphering and deciphering and the other for signature and verification. For instance, in the DSA, discussed on page 183, the key pair cannot be used for encryption and decryption, whereas the Diffie-Hellman key pair, discussed on page 166, cannot be used for signing and verification. The management of key pairs is paramount in any PKI, and dual key pairs has become a central feature of any in-depth PKI.

First, keys must be generated. The best method is for a CA or RA to generate the key pair. Once multiple key pairs for individual entities have been generated, there is a need for multiple certificates, since X.509V3, for example, does not support multiple key pairs in a single certificate. A private key used for signing and verification requires secure storage throughout its lifetime. In this case, we should not back up the key pair, since the compromise of the pair necessitates the generation of a new key pair, and it makes verification of all signatures associated with that key pair impossible. Such key pairs must always be secured, since knowledge of the private key needed for nonrepudiation will allow the owner of the key to claim the adversary engaged in the nonrepudiable act, which would defeat the goal of having the key pair for nonrepudiation. A private key used for decryption must be backed up to enable recovery of enciphered data, and it should not be destroyed once expired since it may be needed for later decryptions. It should be placed in a *key archive*, which is a long-term storage of keying data including certificates. Typically, archives are appended with timestamp and notarization data in order to resolve any future disputes, as well as for audit purposes.

If private keys are lost by end users (and they will be) there should also be an optimal automatic process of key recovery in the PKI. Note that this means the recovery of private decryption keys only, not private signature keys, for the reasons cited above. An alternative method to the CAs storing public keys and certificates for digital signature purposes is the RSA digital envelope (see page 163). Alice can use a secret symmetric session key to encipher, but also she encrypted it, using an RSA public recovery key, when it was generated. Thus, if Alice loses her key, the CA who owns the private RSA recovery key can open the digital envelope and recover Alice's session key. Key recovery can also be accomplished using secret-sharing schemes such as those we discussed in Section 5.5. These key recovery threshold schemes are also very common since they have a nice checks and balances feature. Splitting a private key among shares thwarts attempts by any one entity from surreptitiously capturing private keys, and it allows reconstruction of the key shares without one or more of the trusted entities being present to pool the shares.

♦ Key Updating and Key History

Key pairs must be updated at regular intervals, if for no other reason than to thwart compromise threatened by cryptanalytic attacks. Once the key pair expires, the CA can reissue a new certificate based on the new key pair, or a new certificate for the old key pair can be generated. This gives rise to what is called a *key history*, consisting principally of old private keys. A key history must be maintained by the PKI for such purposes as later decryptions of old data. Ideally, a key history is stored with a CA who has an automated process available to retrieve the data from the key history as it is needed. This is different from key archiving, which meets the need for storing public keys and certificates for digital signature purposes.

Typically, to free the end user from responsibility, there is automatic verification of a certificate each time it is used on the network. Once expiration approaches, the automated system will request a key update from a suitable CA or more likely, an RA. Once the new certificate is created by a CA, it is automatically replaced and requirements on end users are eliminated.

♦ The Future of PKI

The future of PKI is an open book. It is developing, with new standards emerging, at a vigorous pace. For further information, the reader is referred to any of the following:

1. *PKI Forum* at

http://www.pkiforum.org

2. Recall that we have already mentioned the IETF's working group on page 219, see

http://www.ietf.cnri.reston.va.us/html.charters/pkix-charter.html.

3. The Government of Canada:

http://www.cse-cst.gc.ca/en/services/pki/pki.html

4. NIST has a *Federal PKI Technical Working Group* (PKI-TWG) studying PKI infrastructures for use by government agencies:

http://csrc.nist.gov/pki/twg/

4. *The Open Group*, an international vendor and technology-neutral consortium, is developing PKI standards:

http://www.opengroup.org/public/tech/security/pki/cki/

to mention only a few.

6.3 Secure Electronic Transaction (SET)

To travel hopefully is better than to arrive, and the true success is to labour.
Robert Louis Stevenson (1850–1894), Scottish novelist

On page 232, we mentioned the circumstances surrounding how Visa and MasterCard developed SET in 1997. Now we have the tools to describe this scheme in detail. To do so, we need some terminology, and due to the complexity of the scheme, a brief overview before we give the complete description.

First of all, a *payment gateway* is an interface between SET and the existing e-commerce network for authorization and payment. We will use Trent for this role. Alice will be our cardholder, and Bob will be our merchant. Diagram 6.1 is an illustration of the SET mechanism after which we will explain the operation in detail. However, it will be a simplified version to ease the explanation without significantly altering the themes and function of the SET protocols. For instance, in the following we assume that Trent is internally contacting the *acquirer* (a financial institution that processes credit card payments and authorizations), and an *issuer* (a bank that provides Alice with her card), without mentioning them, except parenthetically.

Diagram 6.1 SET Protocol Actions

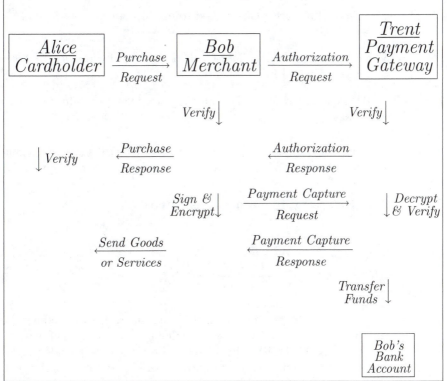

SET is presented via its components, the first being vital and innovative.

◆ SET Dual Signature Protocol

Background Assumptions: This concept of "dual signatures" is new to the DSS arena. Dual signatures permit two data blocks to be intimately linked, yet sent to two different entities for handling. Two mechanisms allowing this to occur within SET are the following.

1. The *order information* (OI) message (which contains order reference generated in the exchange between Alice and Bob in her shopping phase before the first SET message is sent, which we will detail below, but contains no explicit order information such as the cost of Alice's chosen items). The OI is sent to Bob for processing.

2. At the same time, a *payment information* (PI) message is sent to Trent who requires it (typically for authorization with the issuer and acquirer).

▼ Protocol Steps:

1. Given a public cryptographic hash function H, a message digest (see page 170) is created for the OI, called the *OI message digest* $H(OI) = (OIMD)$; and for the PI, called the *PI message digest* $H(PI) = (PIMD)$.

2. The OIMD and PIMD are concatenated to produce a new data block $C = (OIMD, PIMD)$.

3. C is hashed to produce a new data block $H(C)$, called the *payment order message digest* denoted by $(POMD)$.

4. The POMD is encrypted using the signer's private key to produce a (dual) digital signature.

In the following diagram, k is the signer's private key, **DS** is the dual signature, and all other acronyms are as given in the above protocol description.

Diagram 6.2 (SET Dual Signature Illustrated)

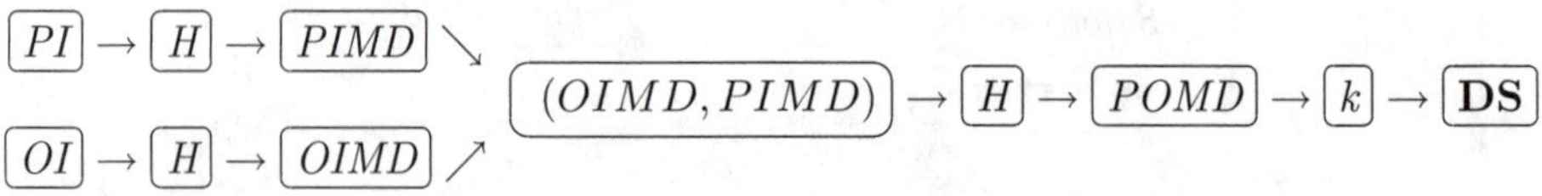

$$\boxed{PI} \to \boxed{H} \to \boxed{PIMD} \searrow$$
$$\boxed{(OIMD, PIMD)} \to \boxed{H} \to \boxed{POMD} \to \boxed{k} \to \boxed{\textbf{DS}}$$
$$\boxed{OI} \to \boxed{H} \to \boxed{OIMD} \nearrow$$

Now we describe the components of the SET scheme, which will employ the dual signature protocol for security. Later we will discuss the details of how the security needs are met by the use of dual signature to provide confidentiality, integrity, and authentication. We begin with the first stage in Diagram 6.2, initiated by Alice, who wants to do business with Bob over the Internet.

◆ **Purchase Request and Response**

Background Assumptions: Alice is assumed to have already shopped online, chosen her goods or services, and placed an order with Bob. Bob is assumed to have sent a completed order to Alice. This is all done outside the purview of SET. It accounts for the order reference mentioned in mechanism 1 of background assumptions in the SET dual signature scheme described above. The CA, in what follows, is assumed to be trusted to issue X.509V3 certificates to all participants. We assume a secure PKI is in place (typically with interdomain cross-certification; see Section 6.2).

▼ **Protocol Messages**

1. **Initiate Request Message**: In this message, Alice includes a request for certificates, and information provided in this request are: (1) the type of credit card; (2) Alice's ID; (3) a nonce, used as a form of timestamp.

2. **Initiate Response Message**: Bob responds with: (1) Alice's nonce that she sent to him; (2) another nonce for her to return in the next message; (3) a transaction ID for this particular purchase, denoted by TID; (4) Bob's signature certificate; (5) Trent's key exchange certificate.

   Then Bob signs the above response with his private signature key and sends it to Alice.

3. **Purchase Request Message**: Alice verifies both Bob and Trent's certificates via their respective CA signatures. Then she creates the OI and the PI, and to both of these she affixes the TID. For the sole purpose of sending the purchase request information, Alice generates a one-time symmetric encryption key k_A. The block of data that Alice forwards to Bob so that he can pass it on to Trent consists of the following:

   (1) The PI.

   (2) The dual signature calculated over the OI and the PI as defined in step 4 of the dual signature protocol steps on page 244.

   (3) The OIMD as defined in step 1 of the dual signature protocol steps on page 244. This is needed by Trent to verify the dual signature, as we will describe in detail below.

   All of the items in (1)–(3) are encrypted with k_A to form the *Trent-encrypted* part of the message, denoted by **TE**. Then, the following is added:

   (4) The *digital-envelope*, which is formed by encrypting k_A with Trent's public key-exchange key, e_T. (The term "digital-envelope" is used here in a different sense than that given on page 163, where we talked about hybrid cryptosystems. Here we have an "envelope" digitally locked by Trent's key, which must be unlocked, or decrypted, before any data in the envelope can be viewed.) k_A is not made available to Bob, who therefore cannot read any of this part of the message. He merely passes it to Trent.

In the message, Alice includes data for Bob, which consists of:

(5) The OI.

(6) The dual signature as in (2) above.

(7) The PIMD as defined in step 1 of the dual signature protocol steps on page 244. (This is needed by Bob to verify the dual signature, as we will detail below).

(8) Alice's certificate C_A, containing her public signature key, s_A. (This is needed by both Bob and Trent.)

The purchase request message is sent to Bob.

4. **Purchase Response Message**: Upon receipt of Alice's message, Bob executes the following steps:

(1) He verifies Alice's certificates via the CA signatures.

(2) He decrypts the dual signature using Alice's public signature key, thereby verifying that the order has not been altered by any entity tampering in transit and that it was indeed signed using Alice's private signature key.

(3) He processes the OI and forwards the aforementioned data to Trent.

(4) Then Bob sends a purchase response to Alice, which includes acknowledgement of the order appended with the TID; signature of the block with Bob's private signature key; and Bob's signature certificate.

Upon receipt of the purchase response message, Alice verifies Bob's certificate, then verifies his signature on the response block. If everything is valid, her database is updated with this fact.

In the Diagram 6.3, **DS** stands for dual signature, and the balance of the acronyms are given above.

Diagram 6.3 Alice's Purchase Request

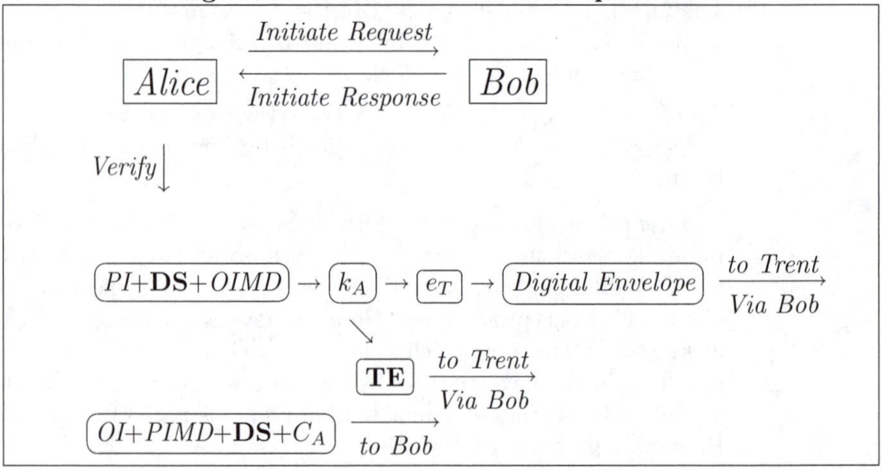

Diagram 6.4 shows the balance of the purchase request/response protocol with Bob's actions, after which Alice verifies and updates as described above.

Diagram 6.4 Bob's Purchase Verification/Response

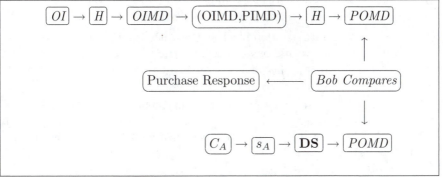

◆ **Authorization Request and Response**

Background Assumptions: Bob, during the processing of Alice's order, must authorize the transaction with Trent via an authorization request, which results in a guarantee that Bob will be paid or the transaction is rejected. This is included in Trent's authorization response.

▼ **Protocol Messages**

1. **Authorization Request Message**: Bob forwards the message from Alice contained in the **TE** and digital envelope as outlined in part 3 (Purchase Request Message) of Protocol Messages on page 245 and as illustrated in Diagram 6.3 on page 246.

   The data generated by Bob, which is sent with the above, includes the following items:

   (i)   TID signed with Bob's private signature key and encrypted with a one-time symmetric key k_B, generated by Bob.

   (ii) k_B is enciphered by Bob using e_T to create another digital envelope **BE**.

   (iii)   Alice's signature key certificate, needed to verify the dual signature.

   (iv)   Bob's signature key certificate, required to verify Bob's signature.

   (v)   Bob's key-exchange certificate, required for Trent's response.

2. **Authorization Response Message**: Trent, upon receipt of Bob's request, executes the following:

   (1)   Verifies all certificates.

   (2)   Deciphers **BE** to obtain k_B, which he uses to decrypt the TID and verify Bob's signature.

(3) Decrypts **TE** to obtain k_A, which is used to verify the PI by comparing the TID in it with the TID obtained in step (2) above.

(4) If the above are all valid, he sends an authorization response to Bob (requested and received from actions of an issuer and acquirer).

Included in the message sent by Trent are:

   (i)   An authorization block signed with Trent's private key and encrypted with a one-time symmetric key k_T, generated by Trent.

   (ii) A digital envelope **AE**, created by Trent via enciphering k_T with e_B, Bob's public key-exchange key.

   (iii) Some information for later payment capture, namely a digital envelope called a capture token, denoted by **CT**, not to be opened by Bob, rather returned with Bob's payment request later.

   (iv) Trent's signature key certificate.

Upon receipt, Bob decrypts **AE** with d_B, his private key-exchange key to get k_T which he uses to decrypt the authorization block. If the payment is authorized, he can provide Alice with the goods and/or services.

In Diagram 6.5, d_T is Trent's private key-exchange key. All other acronyms are as in the above protocol descriptions. We assume that Trent has verified all certificates and signatures in the illustration as well.

Diagram 6.5 Trent's Authorization/Response

$$\boxed{\textbf{BE}} \rightarrow \boxed{d_T} \rightarrow \boxed{k_B} \rightarrow \boxed{\text{TID}}$$

$$\uparrow$$

$$\boxed{\text{Authorization Response}} \longleftarrow \boxed{\textit{Trent Compares}}$$

$$\downarrow$$

$$\boxed{\textbf{TE}} \rightarrow \boxed{d_T} \rightarrow \boxed{k_A} \rightarrow \boxed{\text{TID}}$$

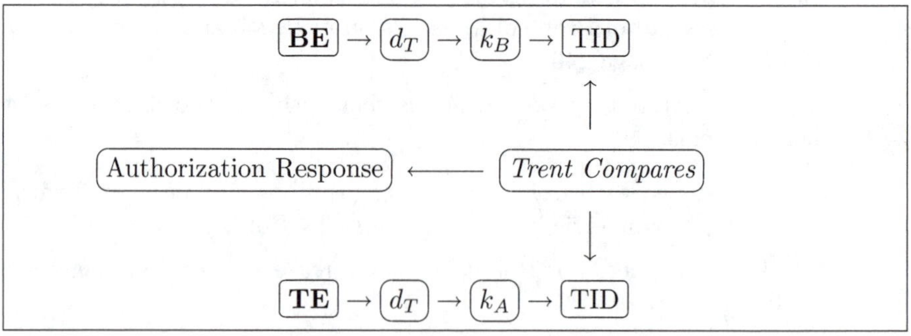

Diagram 6.6 Bob's Verification

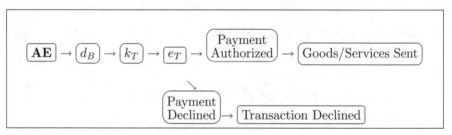

In the event the transaction is declined, this is the end of the SET protocol. However, if all is validated, then we proceed as follows.

◆ **Payment Capture and Response**

Once Alice's order is completed, Bob requests payment from Trent.

▼ **Protocol Messages**

1. **Capture Request Message**: Bob generates, signs, and encrypts a message block that contains the TID; the **CT**; his signature key; and his key-exchange key certificates.

    Upon receipt of the capture request, Trent decrypts and verifies that the **CT** matches what he sent in part (4) of the Authorization Response Message above. (If all is valid, Trent interacts with the issuer over a private network to request payment to Bob's bank account. Otherwise the payment is declined.)

2. **Capture Response Message**: Once the data is verified, Trent sends a message block containing payment details for the transaction and includes his signature-key certificate. Trent then forms a digital envelope and sends it to Bob.

    Upon receipt, Bob decrypts the envelope, verifies the message, and data, then stores the message (for any future reconciliations with the acquirer).

The SET protocol is now complete. Our description was necessarily stripped down since the original SET document, released in 1997 is nearly 1000 pages. We have given all the necessary details to have a reasonable overview of the SET scheme. If nothing else, the reader who has even a passing understanding of the SET mechanisms, must be convinced of the security of this e-commerce scheme. Since shopping on the Internet will almost certainly involve SET-enabled software, the reader will now be convinced of the security of such transactions. We now look at the scheme in detail from several perspectives pertaining to the key features of SET.

◆ **Analysis**

Certificates and PKI: the secure PKI with a trusted CA guarantees that the public keys are actually keys used by the legitimate entities to whom they belong. This is an essential role of PKI. As we have seen, this was a verification for both the key-exchange keys and the signature keys. Hence, the PKI provides trust through the use of X.509V3 digital certificates.

Confidentiality: Alice's account and payment information is secure as it travels over a network. For instance, even Bob does not know Alice's credit card number, and Trent does not know the details of Alice's order. This is guaranteed by the mechanisms in the dual signature. Moreover, since k_A is not made available to Bob in her purchase request, Bob cannot read any of the payment-related details.

Integrity: The data sent by Alice to Bob is guaranteed not to be altered in transit. RSA digital signatures (see page 181) are used and SHA-1 hash schemes (see Chapter 7) are used to guarantee this.

Authentication: The use of not only the aforementioned certificates, but also the use of RSA signatures ensures Bob that Alice is a legitimate card holder of a legitimate account. Similarly, Alice can verify that Bob is a legitimate merchant.

Issuer: We have kept the issuer in the background for the sake of simplicity of presentation. The issuer is typically a bank or some other financial institution that provides Alice with her card, such as a MasterCard or Visa. She interacts with the issuer to open an account. Moreover, the ultimate responsibility for the payment of all authorized transactions put on the card is the issuer. The issuer must have certificates to be processed by a CA if they process any SET messages. In this case, they receive them from the credit card organization, such as Visa. Otherwise, the issuer may have the credit card organization process certificates on their behalf, in which case they are not processing SET messages, and do not require certificates.

Acquirer: We have also kept the acquirer in the background. This is a financial institution that supports merchants such as Bob by providing the service of processing payment of credit cards. Thus, the acquirer pays Bob, and the issuer repays the acquirer. The acquirer must also have certificates that can be processed by a CA. These certificates are obtained from the credit card organization as above.

The Payment Gateway: We have put the task of all three of the issuer, acquirer, and payment gateway on Trent's shoulders, but in reality, Trent must interface with the acquirer at some juncture, since Trent processes Bob's payment messages which must, at some point go through the acquirer.

Certificate Status Inquiry: Again, for simplicity, we did not describe some of the other transaction types in SET. In this type of inquiry either a CA can send a message to Bob or Alice saying that more processing time is needed, or Alice or Bob can send such a message to the CA to check the status of a certificate request, for instance.

Purchase Inquiry: Alice can send this message to check the status of the processing of her order, for instance.

Authorization Reversal: Bob can send this message to reverse an authorization or part of it.

Capture Reversal: Bob may use this to correct errors in earlier capture requests.

Credit or Credit Reversal: These types of messages may be used by Bob to issue a credit or reverse a credit due to a previous error, for example.

Chapter 7

Message Authentication

The true test of a first-rate mind is the ability to hold two contradictory ideas at the same time.

F. Scott Fitzgerald (1896–1940), American Novelist

7.1 Authentication Functions

Issues of authentication were discussed at various points previous to this chapter, such as the presentation of Kerberos in Section 5.2 page 195; in Section 5.7, page 220, in the presentation of SSL and issues surrounding it; in Section 6.1, page 233, when we described authentication issues surrounding key management; and in Section 6.3, page 243, when we delved into the details of SET. We also looked at attacks on authenticity, such as the impersonation attack presented on page 180, and methods for thwarting it. This impersonation attack is essentially a man-in-the-middle attack, which we introduced in Footnote 3.7, page 134. This was in reference to authentication issues in the use of the various modes of operation about which we learned in Section 3.3. Thus, we are fairly well versed in authentication issues to date. Now we want to look at authentication functions, such as MACs, which we briefly mentioned on page 136.

We are concerned in Chapter 7 with message authentication as opposed merely to say, entity authentication, which we addressed on page 180 in the discussion of digital signatures. A *message authentication scheme* is any algorithm for ensuring that messages come from the legitimate source and have not been altered. What is implicit in message authentication is the verification of the message's content; nonrepudiation by sender; origin; receipt; timing; and sequence (of messages) if there is more than one.

As with protocol layers studied in Section 5.7, there are layers to authentication schemes, albeit in the latter, only two basic ones. At the bottom layer, there must exist a function, which produces an *authenticator*, or value affixed to a message as its means of being authenticated. (For instance, recall the Kerberos authenticator on page 197.) This bottom layer function is then used by the upper layer (authentication protocol) ensuring that the receiver can verify the authenticity of the message.

Although there is no hierarchy written in stone as to the bottom layer functions for authentication, the following is generally accepted by the cryptographic community.

◆ Types of Authentication Functions

1. Hash functions, where a publically known cryptographic hash function is the authenticator.

2. Message authentication codes (MAC)s, where a publicly known MAC of the message coupled with a secret key, outputs a fixed-length value, which is the authenticator.

3. Message encryption, wherein the ciphertext is the authenticator itself.

We are going to devote the balance of this section to the first of these, and cover MACs in Section 7.2. Section 7.3 will be a description of the third and a comparison of the three. The concluding Section 7.4 will deal with authentication applications.

◆ Hash functions

We had a brief introduction to hash functions and message digests on pages 170 and 171, where we learned that a (*cryptographic*) *hash function* is a one-way function that is a computationally efficient function mapping bitstrings of arbitrary length to bitstrings of fixed length. The hash value (or *message digest*) is then affixed to the message by the sender, and the message is authenticated by the receiver, who recomputes the message digest. As noted on page 171, the message digest is a "fingerprint" of sorts, and this is the purpose of the hash function: to authenticate the data by fingerprinting it. Moreover, as noted on page 171, it is desirable for the cryptographic hash function to be *strongly collision resistant*. The reason for this latter requirement is to guard against a class of attacks for which we have prepared the reader on page 130, when we discussed the meet-in-the-middle attack.

▼ The Birthday Attack

We are given a hash function $H : \mathcal{S} \mapsto \mathcal{T}$, with $|\mathcal{T}| = n$ and $|\mathcal{S}| > n$, so there is at least one collision. (If $|\mathcal{S}| \geq 2n$, there are at least n collisions. In fact, it can be shown that when $\infty > |\mathcal{S}| \geq 2|\mathcal{T}|$, there exists a probabilistic algorithm that finds a collision for h with probability bigger than $1/2$ (see [159, Fact 9.33]).) We now look at how to find such collisions. First, we describe the analogue of the above that gives this attack its name.

The Birthday Paradox: Suppose there are $n > 1$ balls in a container numbered from 1 to n inclusive. Also, let us assume that $m > 1$ balls are drawn one at a time, listed, and replaced each time (where $m < n$). What is the probability that one of the balls is drawn at least twice?

Let $P_j(n, m)$ be the probability that one ball is drawn at least j times. Then, we are seeking

$$P_2(n, m) = 1 - P_1(n, m).$$

To find $P_1(n, m)$, note that the probability that the second drawn ball is different from the first is $1 - 1/n$, the probability that the third ball is different from the first two is $1 - 2/n$, and so the probability that the first three balls are all different is $(1 - 1/n)(1 - 2/n)$. Continuing in this fashion, we see that the probability that all of the m balls drawn are different is,

$$P_1(n, m) = \prod_{j=1}^{m-1}\left(1 - \frac{j}{n}\right) = \frac{1}{n^{m-1}}\prod_{j=1}^{m-1}(n - j) =$$

$$\frac{(n-1)(n-2)\cdots(n-m+1)}{n^{m-1}}.$$

Thus,

$$P_2(n, m) = 1 - \frac{(n-1)(n-2)\cdots(n-m+1)}{n^{m-1}}.$$

In particular, suppose we want to prove that in any room of 23 people, the probability that at least two of them have the same birthday is greater than 50%. From the above this is a fact since

$$P_2(365, 23) \approx 0.5072972343.$$

This phenomenon is called the *birthday paradox*.

The birthday paradox is a special case of the *occupancy problem*, which is given as follows. Suppose that a container has n balls numbered 1 through n inclusive. Again assume that m balls are drawn one at a time, listed, and replaced each time. Then the probability that exactly ℓ of the m balls are different for $1 \le \ell \le m$ is given by

$$\frac{1}{\ell!}\sum_{j=0}^{\ell}(-1)^{\ell-j}\binom{\ell}{j}j^m P_1(n, m),$$

where $\binom{\ell}{j}$ is the binomial coefficient (see Definition A.14 on page 473 and the discussion following it, as well as Appendix E on probability theory.).

Now we return to the birthday attack that began this discussion. We initially asked how we can find a collision. From the above, the probability that there do not exist any collisions is

$$\prod_{j=1}^{m-1}(1 - j/n),$$

so

$$1 - x \approx e^{-x}$$

for small x values (such as ours). Hence, the probability of no collisions is

$$\prod_{j=1}^{m-1}(1-\frac{j}{n}) \approx \prod_{j=1}^{m-1} e^{-j/n} = e^{-m(m-1)/(2n)}.$$

Therefore, the probability of at least one collision occurring is

$$p_c \approx 1 - e^{-m(m-1)/(2n)}. \tag{7.1}$$

Since

$$e^{-m(m-1)/(2n)} \approx 1 - p_c,$$

then

$$-m(m-1)/(2n) \approx \ln(1 - p_c),$$

(where $\ln(x)$ is the *natural logarithm*, meaning the log to the base e, with e^x being the *natural exponential function*.) Hence,

$$m^2 - m \approx -2n\ln(1 - p_c),$$

and so

$$m^2 \approx -2n\ln(1 - p_c) \approx 2n\ln(1/(1 - p_c)),$$

since we can safely ignore the smaller factor of $-m$ in an approximation, so

$$m \approx \sqrt{2n\ln(1/(1 - p_c))}.$$

If $p_c = 1/2$, then $m \approx 1.17\sqrt{n}$. Clearly then, by hashing over little more than $\sqrt{n}$ random elements of S, we have a greater than 50% chance of finding a collision. This is the birthday attack.

The birthday attack places a lower bound on the number of bits a hash function should have in order to be secure. The reason is that the birthday attack can find a collision in $O(2^{k/2})$ hashings on an k-bit function. Thus, if $k = 64$, then it is not secure against the birthday attack since only 2^{32} hashings are required.

The following illustration of the birthday attack was first presented by Yuval in 1979 (see [297]), and we re-presented it in [170].

● Alice Cheats Bob Using the Birthday Attack

The hash function has 64 bits. Alice wants Bob to sign a contract that he thinks will benefit him, and later she wants to "prove" that he signed a contract that actually robs him of his life savings.

1. Alice prepares two contracts, one that is "good" for Bob, C_G, and one, C_B, which will sign away his savings.

(2) Alice makes very minor changes in each of C_G and C_B. Then she hashes 2^{32} modified versions of C_G and 2^{32} modified versions of C_B.

(3) She compares the two sets of hash values until she finds a collision $h(C_G) = h(C_B)$ and recovers the corresponding preimages.

(4) Alice has Bob sign C_G via the hash of its value.

(5) Later Alice substitutes C_B for C_G whose hash value is the same as that signed by Bob, who has now lost all his money.

From the discussion preceding the Yuval attack above, we see that a birthday attack requires an effort of only the order of 2^{32}. Thus, simple hash functions based on a 64-bit message digest are insecure, so from a cryptographic perspective, they are not worth discussing. (As a counterpoint to the above contract signing scam, we saw on page 193, how to make contract signing secure.)

Modern cryptography requires custom-built hash functions to meet current standards for security. In 1990 (see [228]), Rivest developed a hash scheme, called MD4, designed for software implementation on a 32-bit processor. However, very early after its inception, attacks on it proved it to be insecure under modern cryptanalysis. MD4 was updated to MD5 in 1992 (see [229]), but in 1996 hash collisions of the underlying compression function proved it to be insecure. In fact, the birthday attack can be used on MD5 to find a collision in about 2^{64} iterations, which is quite insufficient for modern cryptosystems. In 1995, the *Secure Hash Algorithm* (SHA-1) was developed for the NSA and standardized by NIST (see [88]). SHA-1 employs a 160-bit hash function. In 2002, NIST updated SHA-1 (see [89]) in what they called the *Secure Hash Standard* (SHS) containing specifications for 256-, 384-, and 512-bit message digests, called (respectively) SHA-256, SHA-384, and SHA-512. Naturally, these upgraded hash standards are much slower than SHA-1, yet the increased security level makes them excellent choices for modern cryptosystems.[7.1] In terms of speed combined with modern-day security requirements the SHA-256 is perhaps the best choice, since the security level is 2^{128}, based upon the above-established fact that the birthday attack on a message digest of size 256 bits produces an effort of about 2^{128} iterations of workload. Similarly, the SHA-1 scheme requires on 2^{80} iterations, and some cryptographers feel that this is insufficient for modern standards. Yet, from our perspective, it embodies the fundamentals of the SHA algorithms, and so deserves to be studied in detail, since it provides a simple method for describing the underlying mechanisms.

◆ SHA-1

Background Assumptions

The algorithm inputs messages of maximum bitlength 2^{64} and outputs 160-bit message digests. The input is divided into blocks of 512 bits.

[7.1] At the August 2004 CRYPTO meeting (see page 232), theoretical attacks against MD5 and the original SHA algorithm show that they are not as secure as originally believed. This may have some consequences for the security of SHA-1 as well, since the attacks had partial success in finding collisions for the latter. Thus, use of SHA-256 seems even more prudent.

▼ **Algorithm Steps**

1. **Padding**: The input message, denoted by m, is padded so that its bitlength $\ell \equiv 448 \,(\text{mod } 512)$. If ℓ is already 448 modulo 512, *before* padding, then we still pad, in this case with 512 bits. The padded message is denoted by M.

2. **Appending**: A block of 64 bits is appended to M.

3. **Buffering**: A 160-bit buffer is employed to hold the intermediate and final outputs of the algorithm. We represent the buffer having five 32-bit registers, labelled $ABCDE$. (The buffer is initialized with specific hexadecimal values that we will not cite here for the sake of simplicity.) We will denote the five initialization values by

$$(H_1 H_2 H_3 H_4 H_5) \to (ABCDE).$$

4. **Processing**: A module consisting of four rounds of 20 steps each employs three different primitive logic functions. We will, for the sake of simplicity, not describe their individual specific functions, rather we will call them f_1, f_2, and f_3. Each of these function inputs three 32-bit data strings or *words* and outputs 32-bit words. The notation is as follows.

   We will assume that there is only one 512-bit block. The procedure can be iterated to accommodate as many such blocks as necessary. M is divided into sixteen 32-bit words, denoted by m_j for $j = 0, 1, \ldots, 15$. Then each m_j is put into temporary storage $m_j \to X_j$. Then we expand the sixteen 32-bit words into eighty 32-bit words as follows.

   First, we need some notation. Let $\oplus$ be addition modulo 2, and let $\mathbf{LS_k}$ be a circular shift left of k places, (for instance, see page 120, where we used a slightly different notation for the $k = 2$ case in our description of S-DES). For $j = 16, 17, \ldots, 79$, assign the following storage:

   $$\mathbf{LS_1}(X_{j-16} \oplus X_{j-14} \oplus X_{j-8} \oplus X_{j-3}) \to X_j.$$

5. **Rounds**: We need to employ four constants c_i for $i = 1, 2, 3, 4$. (These have a certain hexadecimal representation that we need not cite here, again forthe sake of simplicity.) Then each round operates on (the already initialized) buffer's so-called *chaining variables ABCDE*, of 160 bits segmented into five 32-bit words, by updating the contents of the buffer in each step as follows, (where $+$ denotes is addition modulo 2^{32}):

   Round 1: For $j = 0, 1, \ldots, 19$, set,

   $$(\mathbf{LS_5}A + f_1(B, C, D) + E + X_j + c_1, A, \mathbf{LS_{30}}(B), C, D) \to (A, B, C, D, E).$$

   Round 2: For $j = 20, 1, \ldots, 39$, set,

   $$(\mathbf{LS_5}A + f_2(B, C, D) + E + X_j + c_2, A, \mathbf{LS_{30}}(B), C, D) \to (A, B, C, D, E).$$

Round 3: For $j = 40, 1, \ldots, 59$, set,

$$(\mathbf{LS_5}A + f_3(B,C,D) + E + X_j + c_3, A, \mathbf{LS_{30}}(B), C, D) \to (A, B, C, D, E).$$

Round 4: For $j = 60, 1, \ldots, 79$, set,

$$(\mathbf{LS_5}A + f_2(B,C,D) + E + X_j + c_4, A, \mathbf{LS_{30}}(B), C, D) \to (A, B, C, D, E).$$

6. After completion of the fourth round (or 80th step), we assign

$$(H_1 + A, H_2 + B, H_3 + C, H_4 + D, H_5 + E) \to (A, B, C, D, E).$$

which is the output message digest of 160 bits.

One could simplify the rounds as a single set of iterations as follows.
For each $i = 1, 2, 3, 4$, set the following storage for each of the values:

$$j = 20(i - 1), 20(i - 1) + 1, \ldots, 20(i - 1) + 19,$$

$$(\mathbf{LS_5}A + f_i(B,C,D) + E + X_j + c_i, A, \mathbf{LS_{30}}(B), C, D) \qquad (7.2)$$
$$\longrightarrow (A, B, C, D, E),$$

where

$$f_4 = f_2.$$

Diagram 7.1 illustrates a single step in a single round, which is actually one
iteration of (7.2). Diagram 7.2 gives the complete processing of a 512-bit block
(assuming step 4 above is completed).

Diagram 7.1 SHA-1 Single Step

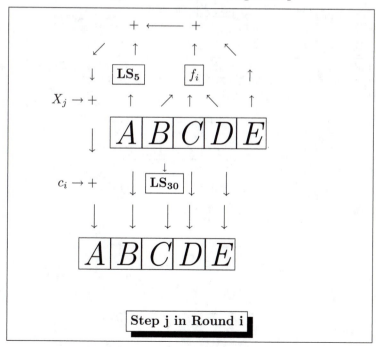

Step j in Round i

Diagram 7.2 SHA-1 Processing of 512-Bit Block

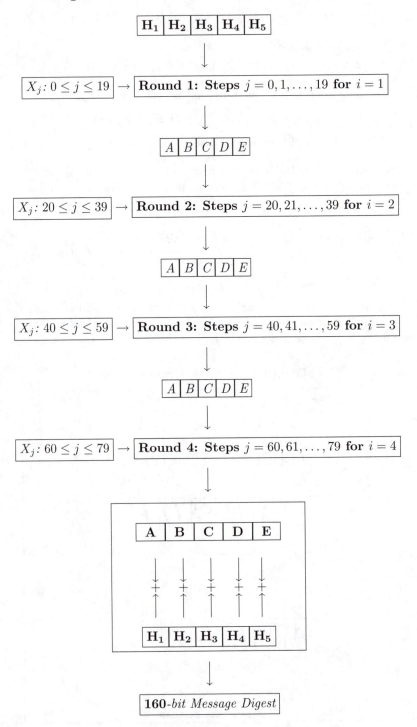

We conclude this section with a brief overview of a hash algorithm developed under the European *Race Integrity Primitives Evaluation* (RIPE) project by a research group that had successfully cryptanalyzed MD4 and MD5 (see [41] and [72]). The following version is the evolution of earlier attempts, so represents the best upgrade, to date, of the group's efforts. We give only a brief description since the details are similar to SHA-1, which we have described in detail and use for comparison.

◆ **RIPEMD-160 Message Digest Algorithm**

The algorithm takes binary inputs of arbitrary length and outputs 160-bit message digests. The input is separated into 512-bit blocks.

▼ **Algorithm Steps**

1. **Padding**: The input m is padded so that its bitlength is congruent to 448 modulo 512.

2. **Appending**: A block of 64 bits is appended to m.

3. **Buffering**: A 160-bit buffer is used to house intermediate and final outputs, and is separated into five 32-bit registers.

4. **Processing**: This is the primary mechanism of the algorithm, consisting of ten rounds of sixteen steps each of processing. The ten rounds are separated into two parallel sequences of five rounds each. The ten rounds have the same structure with five different primitive functions and additive constants, but the order of application differs (to which we will return in the analysis after the algorithm description). Each round inputs the current 512-bit block being processed and updates the contents of the buffer. The output of the fifth round (80th step) is (independently) added to the chaining variable input to the first round (compare to the same idea used in SHA-1; see Diagram 7.2 on page 258).

Analysis and Comparison: RIPEMD-160 uses two parallel sequences of five rounds each to increase the complexity inherent in finding collisions between rounds, since this could be the starting point for finding collisions of the compression function (notably a problem the designers found in cryptanalyzing MD4 and MD5). Since the two sequences are virtually the same in logic design, there were differences introduced: the additive constants differ; the order of applications of the five primitive functions is reversed; and the processing of 32-bit blocks is different.

In comparison with SHA-1, RIPEMD-160 is equally resistant to weak collision attacks. Moreover, the parallel processing in RIPEMD-160 makes it stronger against cryptanalysis than SHA-1. In terms of speed, SHA-1 has 80 (four rounds of 20) steps, whereas RIPEMD-160 has 160 (five paired rounds of sixteen) steps. Thus, SHA-1 is slightly faster than RIPEMD-160, but again it is a trade-off between security and efficiency. There are also MACs based upon RIPEMD-160 (see [42]), but we will not discuss them in the next section where a wealth of MAC information is otherwise available.

7.2   Message Authentication Codes

Success is to be measured not so much by the position that one has reached in life as by the obstacles that one has overcome while trying to succeed.
Booker T. Washington (1856–1915), Educator

In Section 7.1 we looked at the types of authentication functions of which the MAC is the second that we study. A MAC consists of a secret key, a (for authentication), shared by two communicating entities, such as Alice and Bob, and a MAC function h, so that given any message m, a MAC is formed via $h_a(m)$. Then the MAC together with m is sent, by say, Alice to Bob, who computes the MAC on m, via h and a, and compares it to what Alice sent. If they match, Bob can assume the message is authentic.

We have already studied hash functions in Section 7.1 and although hash functions are typically considered to be *non*-keyed, there are types that a *are* keyed. In particular, MACs, whose particular purpose is message authentication, are often called *keyed hash functions*. Thus, the message digest (see page 170) from a MAC is a keyed message digest (message together with key sent), whereas a *nonkeyed* message digest (message alone sent), is called a *Modification Detection Code* (MDC), sometimes called a *Message Integrity Code* (MIC). These are the two types of message digests possible. Recall, however, that a message digest is a *digital fingerprint*, so saying that a MAC is a *keyed* hash function does not mean that the message digest is signed (private key enciphered, in the sense of a digital signature; see page 180), rather it means that the message digest is formed with a secret key.

Diagram 7.3 Generic MAC

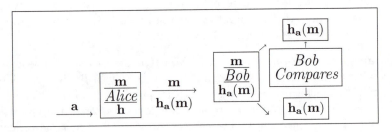

In the above scenario, there is no confidentiality since m is sent in the clear. Eve can intercept the message m and modify it to m', say. However, given that Eve does not know the secret key a, she cannot compute $h_a(m')$, so Bob receives m' and $h_a(m)$. When he computes $h_a(m')$ it will not equal $h_a(m)$, so he discards the message m' as invalid. Yet, to ensure confidentiality of m, one technique is to encipher $(m, h_a(m))$ with an SKC using a shared secret enciphering key k to get $E_k(m, h_a(m))$.

Diagram 7.4 MAC with Confidentiality

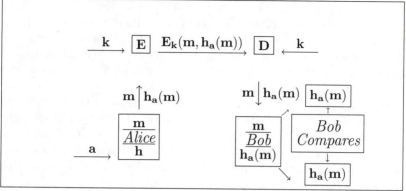

Eve can also intercept the message and send it later in order to deceive (an instance of a replay attack; see Footnote 5.1, page 197). However, this attack can be thwarted by numbering the messages m_1, m_2, m_3, ..., so that Bob accepts message m_j if and only if $j > k$, where m_k is the last message he accepted.

On page 136, we talked about the use of CBC and CFB modes of operation for using block ciphers as MACs. One of the most popular of these MACs is the CBC-MAC used with DES.

◆ CBC-MAC

Background Assumptions: Let m be the n-bit input block, and let E denote the DES cipher, with secret enciphering key k.

▼ **Algorithm Steps**

1. **Blocking and Padding**: Separate m into 64-bit blocks m_1, m_2, m_3, ..., m_n where, if necessary, m_n is padded to the right with zeros to form the last 64-bit block.

2. **CBC Processing**: Block B_j for $j = 1, 2, \ldots, n$ is computed as follows, where $\oplus$ denotes addition modulo 2:

$$B_j = E_k(m_j \oplus B_{j-1}),$$

where B_0 denotes the 64-bit block of zeros.

3. **Final Triple Enciphering**: Using a secret key $k' \neq k$, compute

$$E_k(E_{k'}^{-1}(B_n)) = B_{n'}.$$

4. **The Completed MAC**: $B_{n'}$ is the CBC-MAC.

▼ **Analysis**: The triple encryption in step 3 of the CBC-MAC ensures that brute-force attacks are made much more difficult, and it helps to thwart message forgery. CBC-MACs have the advantage of speed of operation since only addition modulo 2 and DES encryption are applied to the blocks, and this also makes implementation much simpler. In some implementations of the CBC-MAC, an initialization vector different from B_0 is used, but this typically does not strengthen the algorithm. The triple encryption in step 3 does strengthen it, although in some descriptions of CBC-MAC, this is considered to be an optional step. We deem it to be necessary.

On pages 136 and 137, we discussed the importance of the CTR mode over the other modes discussed in Section 3.3. In 2002, D. Whiting, R. Housley, and N. Ferguson developed a combination of CTR mode with CBC-MAC, called CCM, which was proposed as RFC[7.2] 3610 to NIST in June of that year (see [226]). CCM is a generic authenticated encryption block cipher mode, designed for use with 128-bit block ciphers, such as AES.

The CBC-MAC based on DES, discussed above, is FIPS-113 (NIST 1985), and ANSI standard X9.17, sometimes called the *Data Authentication Algorithm* (DAA). The algorithm is illustrated below, where $\boxed{+}$ denotes addition modulo 2 in Diagram 7.5.

Diagram 7.5 CBC-MAC with DES

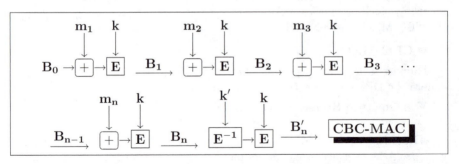

▼ **Cryptanalysis**: There are two brute-force attacks against MACs, either with the secret key as the target, or the MAC itself. By its very nature, a MAC employs a many-to-one function, so if messages are not encrypted, then a cryptanalyst has access to plaintext messages and their associated MACs. Suppose that the keylength ℓ is larger than the MAC bitsize n. Then if Eve knows m and $h_a(m)$, then she can try $h_a(m_i)$ for all i until she gets $h_a(m) = h_a(m_i)$. Such a match will occur since Eve produces 2^ℓ MACs, but $2^n < 2^\ell$ so $2^{\ell-n}$ values will produce a match.

Given the above, we now formulate properties that a MAC function should possess to be secure. First, it should be computationally infeasible for Eve to find an m' such that $h_a(m') = h_a(m)$, given that she has m and $h_a(m)$. Second, given m and m', the probability that $h_a(m) = h_a(m')$ should be 2^{-n} where n is the number of bits in the MAC. This second property thwarts brute-force

---

[7.2]Documents called RFC's, *Requests For Comments*, are the official working notes of the Internet research and development community. See *http://www.rfc-editor.org/rfcxx00.html*.

attacks based upon chosen plaintext (see Footnote 3.4 on page 127). Third, it should be computationally infeasible to compute a valid MAC pair $(m', h_a(m'))$ from a known MAC pair $(m, h_a(m))$, for any new input $m \neq m'$. This property thwarts a cryptanalyst's attempts to create a valid MAC for a given message m', called *existential forgery*, whereas if the secret key itself is obtained, Eve can manufacture a MAC for any selected plaintext, called *selective forgery*.

The security of MACs employing secret-key block ciphers, such as DES or AES, depends upon the security of the underlying block cipher and the secret key. In particular, the new AES-XCBC-MAC-96 algorithm is a variant of the basic CBC-MAC using AES with a *minimum* 128-bit key, although variable key lengths are possible. The AES-XCBC-MAC-96 algorithm and its use with IPsec (which we will study in Section 8.3), released in 2003, is document RFC 3566 (see Footnote 7.2 and *http://www.faqs.org/rfcs/rfc3566.html*). For any CBC-MAC variant, the major computational effort is expended in computing the underlying block cipher, and for AES-XCBC-MAC-96, there is a minimal number of AES operations used, resulting in performance roughly equivalent to the above-described CBC-MAC. The use of a MAC with underlying block cipher such as AES, is suggested since we achieve a desirable level of security. As noted in the aforementioned RFC document: "The security provided by AES-XCBC-MAC-96 is based upon the strength of AES. At the time of this writing there are no practical cryptographic attacks against AES or AES-XCBC-MAC-96."

Ideally, the most desirable MAC function is one that maps randomly from bitstrings of arbitrary length to bitstring of length n. Hence, it seems most reasonable to seek a hash function as the MAC function to do the job. Moreover, hash functions, such as SHA-1 (see page 255), typically execute much faster in software implementations than block ciphers, such as DES. The most successful of the developments in the direction of such a MAC is the HMAC, which is a new standard as a keyed-hash MAC, which is FIPS-198, updated April 8, 2002 (see *http://www.faqs.org/rfcs/rfc2104.html*). HMAC is a generalization of Internet RFC 2104, and ANSI X9.71, and it can be used with any iterative[7.3] hash function, in combination with a shared secret key. The security of HMAC depends on the properties of the underlying hash function.

◆ HMAC Algorithm

Background Assumptions: We assume the HMAC has been prepared as a module wherein the hash function H, (such as SHA-256, for instance, to ensure optimal security, or SHA-1, if speed is a concern, even MD5 if speed is more of a concern than rigorous security) is embedded as a separate module. In this fashion, we have a prepared HMAC that may be installed where needed with no further alterations. Moreover, if H has to be replaced, say, for security reasons, then one merely removes the H-module and replaces it with another

---

[7.3] *Iterative* hash functions, such as SHA-1, split the input into a sequence of fixed blocks $m_1, m_2, \ldots, m_n$ with padding to fill in the nth block, typically of blocklength 512 bits. Then the blocks are processed in ascending order, using a compression function and a fixed size buffer, or intermediate state, again as with SHA-1, for instance, the final value being the output of the hash function.

one.

The message M input to the HMAC is assumed to be separated into blocks M_j of bitlength ℓ each for $0 \leq j \leq \ell - 1$. There are also two padding constants (which we will not explicitly specify here), of bitlength ℓ each, denoted by p_1 and p_2. The secret key is denoted by k, which we will assume to have bitlength ℓ. (Typically, k will have to be padded, but we assume for the sake of simplicity, that this has already been done.) Last, the output of the HMAC is of bitlength $n \leq \ell$.

▼ **Algorithm Steps**

1. Compute $k' = k \oplus p_1$, where $\oplus$ is addition modulo 2.

2. Form the concatenation $M' = (M, k')$.

3. Compute $H' = H(M')$, and pad to bitlength ℓ.

4. Compute $k'' = k \oplus p_2$.

5. Form the concatenation $M'' = (k'', H')$.

6. Compute $H'' = H(M'')$, as the HMAC output.

The above algorithm steps may be succinctly stated as a single equation:

$$H(k \oplus p_2, H(M, k \oplus p_1)).$$

This is illustrated as follows, where $\boxed{+}$ denotes addition modulo 2 in Diagram 7.6.

Diagram 7.6 HMAC

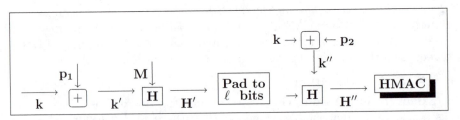

▼ **Analysis**: Step 1 flips one half of the bits of k, whereas step 4 flips the other half. Thus, once H compresses k' and k'', we have essentially produced two pseudorandomly generated keys from the original key k. The designers of HMAC employed this feature with an eye to both offline and online attacks (the underlying assumption being that offline attacks are easier to mount). Use of the key k in step 4 is used to thwart offline attacks.

Typically HMAC is used with MD5 or SHA-1, but as noted earlier, if one desires the very best possible long-term security, SHA-256 is the premium choice since HMAC is already efficient and easy to implement, so the time cost is offset by the security profit.

7.3 Encryption Functions

Minds are like parachutes. They only function when they are open.
James Dewar (1842–1923), Scottish physicist

We have seen the first two types of functions that may be employed to produce a message authenticator in Sections 7.1 and 7.2. Now we turn to the last of these to close the chapter. Since we have SKCs and PKCs, we have different methods for using message encryption for authentication. With an SKC, E, and a secret key, k is sufficient to provide confidentiality and (a degree of) authentication (provided the secret key is kept secure). It should be noted, however, that although Mallory may not know k, he may still alter message content. Yet, not knowing k, he does not know how to alter bits in the ciphertext to produce desired changes in the plaintext. Thus, although the cryptographic community is aware that the mere fact of encryption does not guarantee message integrity, the maintenance of a secure secret key can ensure that tampering is detected. A mechanism for ensuring both confidentiality and integrity along with the encryption process is to use an MDC (see page 260) in a fashion that is essentially equivalent to the use of a MAC depicted in Diagram 7.4 on page 261.

Diagram 7.7 SKC Encryption: Confidentiality & Some Authentication

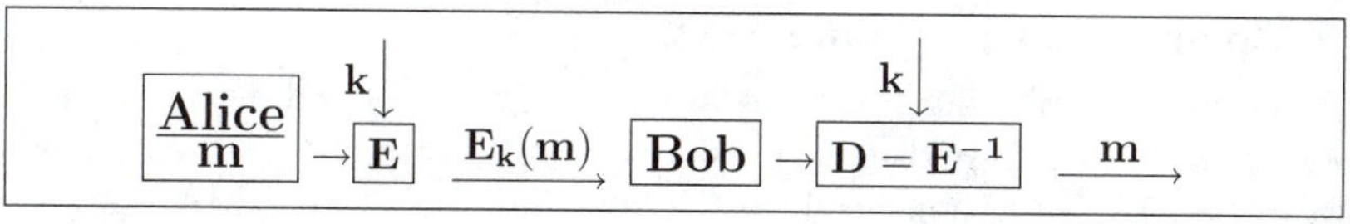

With PKC, straight encryption provides confidentiality, but may not provide authentication as we saw on page 180, since Mallory can mount an impersonation attack. In order to provide authentication with a PKC, Alice must use her private key, d_A, to encipher a message to Bob, who uses Alice's public key, e_A, to decrypt. Hence, in this fashion, Alice is *essentially* providing a "digital signature", although as we saw in Section 4.3, this is not exactly how formal digital signatures are formed. Yet, by using her private key, to which *only* she has access, she is virtually signing the message.

Diagram 7.8 PKC Encryption: Authentication and Signature

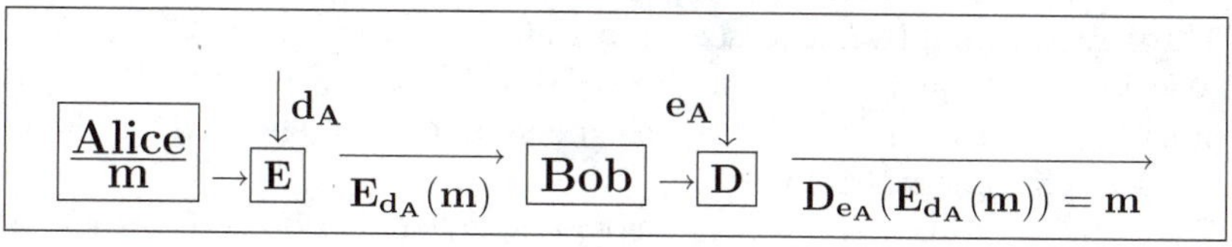

In the above, it is clear that there is no confidentiality since anyone can get Alice's public key to decipher the cryptogram. To provide both authenticity and confidentiality, Alice can first "sign" her message with her private key, then use Bob's public key to ensure confidentiality.

Diagram 7.9   PKC Encryption: Authentication, Confidentiality, and Signature

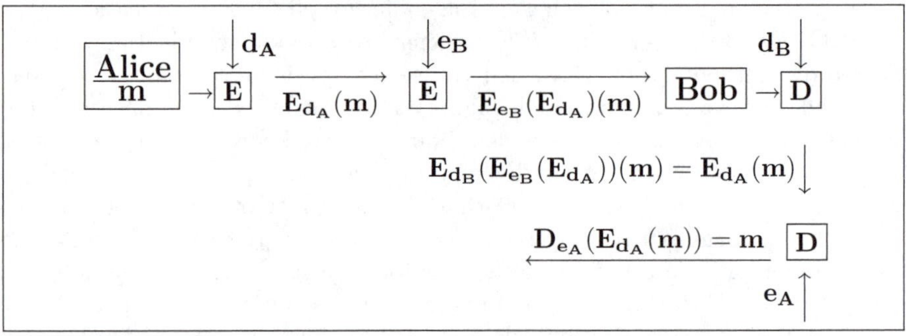

$$\boxed{\begin{matrix}\text{Alice}\\ \hline m\end{matrix}} \to \boxed{\begin{matrix}\downarrow d_A\\ E\end{matrix}} \xrightarrow{\ E_{d_A}(m)\ } \boxed{\begin{matrix}\downarrow e_B\\ E\end{matrix}} \xrightarrow{\ E_{e_B}(E_{d_A})(m)\ } \boxed{\text{Bob}} \to \boxed{\begin{matrix}\downarrow d_B\\ D\end{matrix}}$$

$$E_{d_B}(E_{e_B}(E_{d_A}))(m) = E_{d_A}(m) \Big\downarrow$$

$$\xleftarrow{\ D_{e_A}(E_{d_A}(m)) = m\ } \boxed{\begin{matrix}D\\ \uparrow e_A\end{matrix}}$$

The main disadvantage to the last scheme is that an already slow PKC process must be executed four rather than two times, so should be used only when the highest possible security is needed.

▼ Comparisons and Summary

MAC and enciphering functions are very closely related as we noted above in one instance, when we discussed use of an MDC with encryption on page 265. However, a MAC function need not have an inverse, since a MAC is typically a many-to-one function. Also, PKC encryption can provide a virtual digital signature, but a MAC cannot because, with a keyed MAC, both Alice and Bob share the same secret key.

As for cryptanalysis, a hash function is only as secure as the bitlength of the message digest. For instance, SHA-1 outputs 160-bit message digests, and SHA-256 outputs 256-bit message digests, the latter being a *very secure* modern-day option. Cryptanalyzing a MAC with a brute-force attack is more problematic than that for a hash function since Mallory must know message-MAC pairs to do so, as we have seen in Section 7.2. Effectively, if the (keyed) MAC outputs n-bit message digests and the key has bitlength ℓ, then the effort required to launch a brute-force attack on the MAC is $M = \min(2^\ell, 2^n)$. A similar comment may be voiced for SKC enciphering algorithms. Hence, for an optimally secure MAC today, we would want to see $M \geq 256$ bits.

One issue we have not yet addressed is the *order* of encryption and authentication in general. Should we encipher first, then encrypt, or should we authenticate first, then encipher?

If we encipher first, this has the advantage that if Bob receives a message that is invalid, he discovers this when he attempts to authenticate it. Thus, he can discard the illegitimate message without having to decipher it, a time-saver, especially if a lot of fake messages are in the traffic. Hence, this option is more efficient. If Alice authenticates first, say, with a MAC for instance, then when Alice sends Bob a message, Mallory only gets to see the ciphertext and enciphered MAC. Therefore, Mallory will find it more difficult to launch an attack than if Alice enciphered first since the plaintext and original MAC value are disguised.

So which do we perform first? The question boils down to whether we should fear an active or a passive attack. If we encipher first, Mallory will attack the authentication function first and be able to launch an active attack if successful, where data can be modified. If we authenticate first, Mallory gets to attack the encryption function first and read data, a passive attack. More damage is done in the former than in the latter. Of course, Mallory may well be unsuccessful if a truly strong scheme is used. Hence, as it often does, the issue comes down to sacrificing security for efficiency or sacrificing efficiency for security.[7.4]

We conclude this section with a summary of the key elements of our study of authentication functions.

▼ **Summary of Encryption, MACs, and Hash Functions**

1. **MACs and Hash Functions**: a is a shared secret authentication key, E is an SKC, k is a shared secret key, and h is a hash function.

   (i)

   $$\boxed{\text{Alice}} \xrightarrow{(\mathbf{m}, \mathbf{h_a(m)})} \boxed{\text{Bob}}$$

   provides authentication

   (ii)

   $$\boxed{\text{Alice}} \xrightarrow{\mathbf{E_k(m, h_a(m))}} \boxed{\text{Bob}}$$

   provides authentication and confidentiality

2. **Encryption functions**: d_A is Alice's public PKC key and e_B is Bob's public PKC key.

   (i)

   $$\boxed{\text{Alice}} \xrightarrow{\mathbf{E_k(m)}} \boxed{\text{Bob}}$$

   provides confidentiality and some authentication

   (ii)

   $$\boxed{\text{Alice}} \xrightarrow{\mathbf{E_{d_A}(m)}} \boxed{\text{Bob}}$$

   provides authentication and signature

   (iii)

   $$\boxed{\text{Alice}} \xrightarrow{\mathbf{E_{e_B}(E_{d_A})(m)}} \boxed{\text{Bob}}$$

   provides authentication and signature via d_A, and confidentiality via e_B

[7.4]However, either method is insecure if improperly implemented (see [142], for instance).

7.4   Authentication Applications

There are no such things as applied sciences, only applications of science.
Louis Pasteur (1822–1895), French chemist and bacteriologist

On page 238, we discussed the ISO/ITU-T X.509v3 public-key certificates, which are part of the X.500 series, discussed on page 240. In this section, we look at the X.509 authentication protocols, employing public-key transport, which uses the signing of encrypted keys (see [126]). The X.509 standard recommends the use of RSA as a PKC, and the digital signature scheme (which could be RSA or another DSS) is assumed to use a hash function. The X.509 standard is important since it is used in many of the schemes we have discussed and will study: iPSEC (see Section 8.3 on page 294); SET (see Section 6.3 on page 243); S/MIME (see Section 8.2 on page 287); and SSL (see Section 5.7), to mention a few.

On page 238, we presented the twelve possible fields in a given X.509 certificate, which the reader may want to review before proceeding. Furthermore, since it is most pertinent to this section, the reader must be familiar with Section 6.2 on page 237, concerning PKI issues.

◆ X.509 Strong Authentication Protocols

Background Assumptions: Alice and Bob have PKC pairs for encryption and signatures, (e_A, d_A) and (e_B, d_B), respectively. Moreover, Alice and Bob are assumed (prior to the protocol) to have verified each other's respective public keys, by obtaining those public keys from other certificates $C(A)$ and $C(B)$, from the X.500 directory.

▼ Strong One-Way Authentication

Protocol Steps

1. Alice obtains a timestamp t_A, generates a nonce r_A, obtains a secret key SKC k, and she may (optionally) include a message m. She computes $M_A = (t_A, r_A, C(B), m, e_B(k))$ and sends $d_A(M_A)$ to Bob.

2. Bob uses e_A to get M_A, then he checks that the timestamp t_A has not expired, and that $C(B)$ is his valid certificate. Then he uses his private key, d_B, to get k.

The one-way authentication ensures that Alice is authenticated since only she has d_A. It verifies, via $C(B)$, that the message was indeed intended for Bob. The integrity and originality of m are guaranteed via r_A, since this nonce is a sequential component, which Bob can check for uniqueness within the validity time frame dictated by t_A. This prevents replay and impersonation attacks. In other words, Bob can store the nonce until it expires and reject any new messages that arrive with the same nonce. In any case, Bob now has k as the shared secret SKC key.

Diagram 7.10 X.509 Strong One-Way Authentication

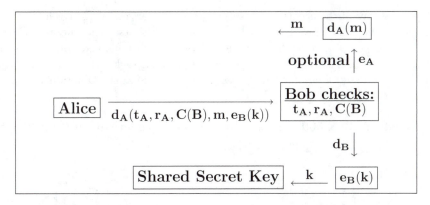

▼ **Strong Two-Way Authentication**

The one-way protocol is executed, then the following steps.

3. Bob obtains a new timestamp t_B, generates a nonce r_B, and (optionally) may send a message m'. He obtains a secret SKC key k' which he encrypts with e_A. Then he computes, $M_B = (t_B, r_B, C(A), m', e_A(k'))$ and sends $d_B(M_B)$ to Alice.

4. Alice now executes analogous actions to those Bob took in step 2 of the one-way protocol. If all is valid, and the option has been exercised, she decrypts k' with her private key d_A. Now she can store k' as another shared key for future use.

Two-way authentication adds to the outcome of the one-way authentication by authenticating Bob, since he is the only one with d_B. Since $C(A)$ is valid. As in the one-way authentication, the integrity and originality of m' is validated.

Digaram 7.11 is a simplified version of the strong two-way authentication (with the actions in Diagram 7.10 by Bob understood as well as the corresponding actions by Alice).

Diagram 7.11 X.509 Strong Two-Way Authentication (Simplified)

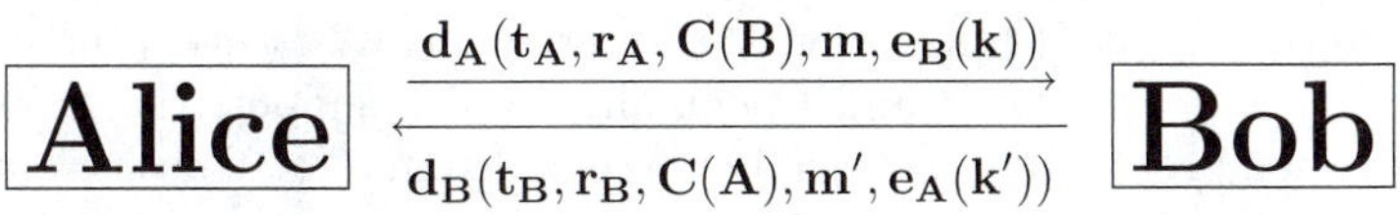

$$\text{Alice} \quad \xrightarrow{\quad d_A(t_A, r_A, C(B), m, e_B(k)) \quad}_{\quad d_B(t_B, r_B, C(A), m', e_A(k')) \quad} \quad \text{Bob}$$

▼ **Strong Three-Way Authentication**

The two-way protocol is executed, then the following step.

5. Alice sends Bob the message $d_A(r_B)$.

Three-way authentication is used (without timestamps) as a vehicle to be employed when synchronized clocks are not available (see the analysis of Kerberos on page 197). Since both Alice and Bob have exchanged nonces, both of them can check the nonce received to detect replay attacks. Recall that on page 199, we discussed a similar three-pass protocol by Shamir. Diagram 7.12 adds step 5 to Diagram 7.11.

Diagram 7.12 X.509 Strong Three-Way Authentication (Simplified)

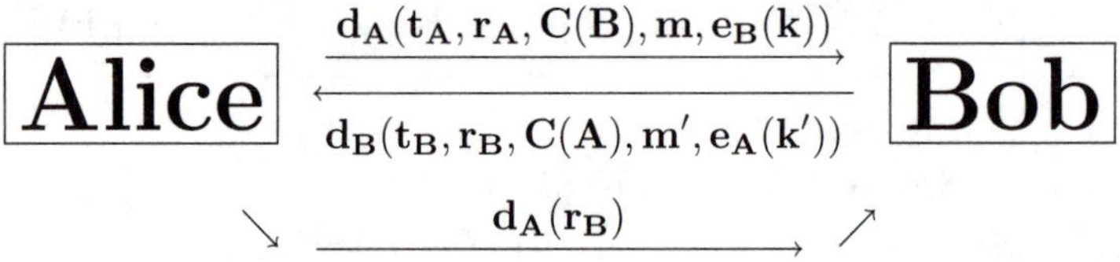

$$
\boxed{\text{Alice}} \quad
\begin{array}{c}
\xrightarrow{\quad d_A(t_A, r_A, C(B), m, e_B(k)) \quad} \\
\xleftarrow[\quad d_B(t_B, r_B, C(A), m', e_A(k')) \quad]{} \\
\searrow \xrightarrow{\quad d_A(r_B) \quad} \nearrow
\end{array}
\quad \boxed{\text{Bob}}
$$

◆ **Authentication and the Internet**

The most common use of X.509 certificates, and the associated strong authentication protocols, is for Internet transactions. (We are now talking about both message and entity authentication.) Typically how this works is that a server, Victor, say, needs to authenticate a user, Alice, say, as follows.

1. Alice sends her X.509 certificate, containing e_A, to Victor.

2. Victor sends a challenge, in the form of a nonce, n_V, back to Alice's browser.

3. Alice's browser encrypts n_V with d_A, and sends $d_A(n_V)$ back to Victor.

4. If Victor can recover n_V using her public key e_A, then he is convinced that Alice is in possession of d_A, and is indeed the person to whom the certificate was issued.

The reason that this X.509 certificate-based authentication is called "strong" is that no password or other secret information is sent over the network. Since the private keys are secure, and since a nonce is used, Mallory cannot gather any data that can be used to recover d_A or to launch a replay attack. All of this, of course, is predicated upon the absolute and unequivocal security of private keys. In this fashion, the X.509 standard for strong authentication is a method superior to simple password-based protocols.

The most popular browsers on the Internet are *Netscape Communicator*, *Microsoft Internet Explorer*, and *Opera*, all of which support the X.509 strong authentication protocols. Server support is enabled in most Web servers, as long as there is an embedded module with SSL/TLS (see Section 5.7). Embedded X.509 certificate support is available, for instance, in *Microsoft Outlook* and *Outlook Express, Netscape Communicator*, and *Mozilla*. In Chapter 8, we will learn about e-mail security in depth.

Chapter 8

Electronic Mail and Internet Security

The new electronic interdependence recreates the world in the image of a global village.

Marshall McLuhan (1911–1980), Canadian communications scholar.
— from *The Gutenberg Galaxy* (1962)

8.1 Pretty Good Privacy (PGP)

Phil Zimmermann (Figure 8.1) was born in 1954 and raised in Florida. His interest in codes began at an early age. For instance, in the fourth grade, he was deciphering some minor codes broadcast on children's television shows. He began reading about codes, and steganography, even creating his own invisible ink out of lemon juice, as well as some of his own original ciphers. This interest continued through his youth, so that by the time he entered Florida Atlantic University in 1972, he turned to computers as a tool for the cryptographic skills that he, independently, had honed over the years.

By the time he was ready to graduate in 1977, he came across Martin Gardiner's article in *Scientific American* [100] (see page 175), about RSA. This merely increased his interest in learning more about cryptography. He even considered implementing RSA on a computer. However, he felt that he had neither the computing machinery available to him, nor the mathematical background to make it work, so he abandoned the idea. This would eventually change.

By 1980, he was already married and raising a family. He moved to New Zealand, largely for political reasons involving his disenchantment with American politics, especially as it pertained to privacy issues at the time. Yet, he found little there (in the sense of a lack of any computer business infrastructure), so after a couple of years he returned home.

With some friends, Zimmermann created a company called Metamorphic Systems. He received a phone call at the company one day, perhaps one that changed the direction of his thinking for good, from a man named Charlie Merritt, who had accomplished what Zimmermann failed to do years ago: the implementation of the RSA PKC on a microcomputer.

Figure 8.1: Phil Zimmermann.

Courtesy of Phil Zimmermann.

NSA had effectively shut down Merritt's company by threatening action if they did not stop exporting their software program outside the United States. Since this was the heart of their enterprise, they had to find another way, calling companies such as Metamorphic Systems to see if their software might be incorporated in the company's hardware for export. The idea excited Zimmermann, and it inspired him to begin writing his own program for e-mail encryption using PKC.

It took a while for the ideas to develop and the relationship to evolve, but by November of 1986, Merritt and Zimmermann had a project for using RSA. Nevertheless, RSA Data Security Inc. had patents on the protocols they wanted to use. Attempts were made to strike a deal with the patent holders, but nothing substantive came out of those discussions.

Zimmermann, undeterred, continued to work on his ideas to produce a cipher without the explicit use of RSA protocols. By 1990, he had developed a communications program, which he called *Pretty Good Privacy*, (PGP) a name derived from a fictitious entity on a radio show, *Ralph's Pretty Good Grocery.*

By 1991, Zimmermann became concerned that some impending legislation by the government might make it illegal for him to launch PGP 1.0, so he turned to the Internet. He uploaded copies of PGP 1.0 to the Internet for anyone to use, that is, *freeware*. His intention was not to profit, but to make encryption available to the masses for privacy considerations. Almost overnight, the program became a hit, and Zimmermann was delighted, but version 1.0 had its failings. He plugged the holes and killed the bugs in 1.0 to produce a vastly superior version 2.0. One particularly important improvement was the addition of certificates. Yet as we saw in Section 6.2, the proper handling of certificates requires a CA, but Zimmermann had no access to a PKI for this independently-generated program, so he had to come up with a new idea. That idea was to make the *users* of PGP, themselves, the CA. To do this, he had the idea of signed keys, as a symbol of "trust", for the communicating parties, something he developed into what he called a *web of trust*, (which we discussed on pages 238 and 239). This web of trust became the users' self-enforcing CA.

In September of 1992, Zimmermann posted PGP 2.0 on the Internet as freeware, and as the light of 1992 faded into memory, Zimmermann was becoming a very famous man indeed. However, fame sometimes engenders costs. In 1993, he was put under criminal investigation, since the government charged that PGP was available to criminals, and they were also concerned about export regulations. The exportation of strong cryptography programs, they maintained, was deemed to be equivalent to illegally exporting munitions!

Fortunately, perhaps because the government finally realized the futility of this war with the Internet as the battleground, they officially dropped the investigation on January 11, 1996.

Zimmermann launched a new company called, *Pretty Good Privacy Inc.* to market the software to commercial enterprises, but due to his lack of business acumen, it was going nowhere fast, so he turned over the reigns to some business types. However, the company eventually went to the brink of bankruptcy before it was sold to *Network Associates Inc.*, (NAI) an established computer firm, where Zimmermann remained as its figurative head, as well as special advisor, and consultant. It is worth

Figure 8.2: Phil Zimmermann, after the charges.

Helen Davis, Denver Post, courtesy of Phil Zimmermann, whose photo was taken right after the Justice Department dropped their case against him in 1996. (He said he was "feeling pretty good".)

ending this anecdote with an ironic note about Zimmermann and the commercial version of PGP. During a party held by NAI at a conference in 2000, Zimmermann staged a demonstration of launching a commercial version of his product over a computer to a market abroad, an act for which he was, years earlier, put under criminal investigation. The new millennium has arrived, and privacy is no longer in the hands of private enterprise or governments.

Zimmermann has received numerous awards for his achievements. Among them are: the Chrysler Award for Innovation in Design in 1995 (see *http://www.chrysler.com/design/design_influences/design_awards/1995/*); the 1995 Pioneer award from the Electronic Frontier Foundation; the Norbert Wiener Award from Computer Professionals for Social Responsibility, for promoting the responsible use of technology, in 1996; a Lifetime Achievement Award from Secure Computing magazine in 1998; the Louis Brandeis Award from Privacy International in 1999; and in 2001, he was inducted into the *CRN Industry Hall of Fame* (see *http://www.crn.com/sections/special/hof/industryHOF_Main.asp*). It is certain that there will be many more such recognitions of his achievements in the future.

PGP has enjoyed remarkable success and is now widely used over the globe as a mechanism for secure e-mail transmission and file storage. It is time to see the details of the algorithm in action.

◆ **Pretty Good Privacy (PGP)**

PGP embodies five protocols for the secure transmission of e-mail messages.

▼ **PGP Protocols**

1. Authentication.

2. Compression.

3. Confidentiality.

4. E-mail compatibility.

5. Segmentation.

Now we look at each of these in detail. We assume that Alice is communicating with Bob.

▼ **Authentication (Digital Signature)**

Protocol Steps

1. Alice creates a message, m, to be used for the purpose of authenticating herself to Bob.

2. SHA-1 (see page 255) is used on m to create a 160-bit message digest, $h(m)$.

3. Alice enciphers $h(m)$ with her private RSA key d_A. She sends $D_A = (d_A(h(m)), m)$ to Bob. On the network, D_A passes through a ZIP compression operation, denoted by Z. (We will learn more about ZIP later.)

4. After decompression, denoted by Z^{-1}, Bob uses Alice's public RSA key e_A to decipher and recover $h(m)$.

5. Bob applies h to the value of m sent by Alice and compares the result to the value of $h(m)$ he deciphered in step 4.

Diagram 8.1 PGP Authentication

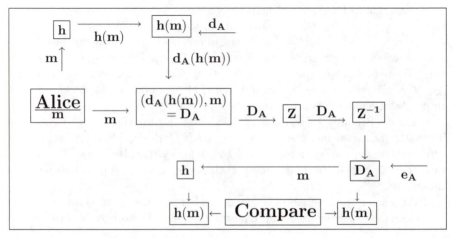

▼ **Confidentiality**

Several mechanisms using SKC may be used for ensuring PGP confidentiality. Among them is Triple DES (3DES) (see page 131), and this is the one we assume will be used in what follows, denoted by E herein. Other options include IDEA (see [159]), and CAST-128 (see [2] and [3]). Moreover, we will assume that 64-bit CFB mode is also used in what follows (see Section 3.3). We will use RSA as our PKC, but ElGamal is also an option (see Section 4.4), as well as Diffie-Hellman/DSS (see pages 166 and 180).

Protocol Steps

Alice wants to send an enciphered message m to Bob.

1. Alice generates a 128-bit nonce k to be used as a one-time-only (session) key for this message, and uses it (after compression of m using Z) via 3DES to get $E_k(Z(m))$.

2. Alice enciphers k with Bob's public RSA key e_B to get $e_B(k)$ and sends $(e_B(k), E_k(Z(m)))$ to Bob.

3. Bob deciphers k with his private RSA key d_B and recovers m with k (after decompression with Z^{-1}).

Diagram 8.2 PGP Confidentiality (Without Authentication)

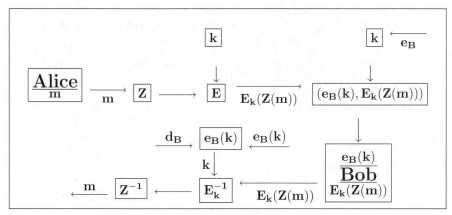

▼ **Authentication and Confidentiality**

This is illustrated in Diagram 8.3 on page 276, with the amalgamation of the previous two protocols as follows.

Steps 1–3 of the authentication protocol are executed, followed by steps 1 and 2 of the confidentiality protocol (acting on $Z(D_A)$ rather than $Z(m)$). Then Bob recovers k with his private RSA key d_B, and uses k to recover the compressed version of D_A via E. Then steps 4 and 5 of the authentication protocol are executed.

Diagram 8.3 PGP Authentication and Confidentiality

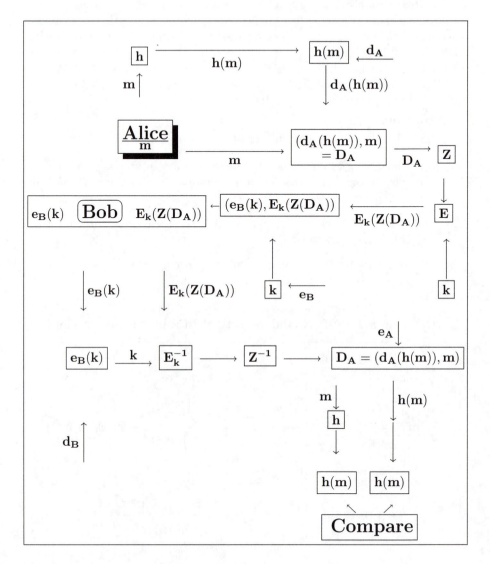

▼ Compression Analysis

For the purposes of efficient e-mail transmission and file storage, PGP has a built-in default mechanism that compresses m after signing but before enciphering. As with our description of the pros and cons of the order of enciphering versus authentication on pages 266 and 267, the order of signing vs. compression deserves some elucidation.

If Alice were to compress m, forming $Z(m)$, then sign it to form, $d_A(Z(m))$, it would be necessary to either store $Z(m)$ for the purposes of later verification by Bob, or once Bob obtains $Z(m)$ via e_A, then it would be necessary to form $Z(m)$

from m for comparison. Both of the latter options entail additional workload over merely storing $(d_A(m), m)$. Furthermore, Z is a randomized operation in the sense that the same input may produce different compressed outputs at different times, say, $Z(m) = x$ at time t_1, and $Z(m) = y$ at time t_2, with $x \neq y$. However, any version can decompress to get the correct version of compression by any other version; in other words,

$$Z^{-1}(x) = Z^{-1}(y) = m.$$

Yet, forming, say, $d_A(m)$ at time t_1 would restrict the PGP scheme to the version of Z applied at time t_1, since we would have to verify Alice's application of that version of the compression at time t_1, which is an unacceptable shackle to put on the security mechanism. Last, speaking of security, enciphering is applied after compression for increased cryptographic security since $Z(m)$ has less redundancy that does m, so cryptanalytic attacks are made much tougher on Mallory.

ZIP compression is a freeware/shareware package that is perhaps the most frequently used compression mechanism for virtually any computing platform. It is based upon an algorithm, called LZ77 (see [299]), developed in 1977. Since a version of LZ77 is used in all versions of ZIP, we will describe the basic features that make up the algorithm. The source data is input to the algorithm as 9-bit words[8.1] (a binary 1 followed by 8-bit ASCII[8.2] representation of the word) to be processed from left to right. The algorithm uses two buffers, called the *sliding-history buffer* and the *look-forward buffer*. The former contains $W \in \mathbb{N}$ already processed words, and the latter contains $W \in \mathbb{N}$ to-be-processed words. The buffers interact as follows. The algorithm tries to match $n \geq 2$ words from the initial part of the look-ahead buffer to a bitstring in the sliding-history buffer. If no match is found, the first word in the look-ahead buffer is output as a 9-bit word, and it is also input to the sliding-history buffer, which discards its last 9-bit word. If a match is found; the longest match bitlength, ℓ, is calculated; the matched word is output as a three-tuple consisting of the bitlength of the word, its indicator value, and a pointer to the prior word of the same value; the ℓ-bit word is input to the sliding-history buffer; and the last ℓ-bit word is discarded from that buffer.

Decompression of the compressed data in the algorithm uses the pointers, bitlength, and value fields to replace the compressed strings with the original text.

[8.1]Typically, the term *word* refers to a fixed-size integer of given bitlength in the main memory of a given computer. Usually, the bitlengths are one of 8, 16, 32, 64, or 128. A word can then be represented by its binary representation as a single word in the computer, such as a 16-bit word having a representation in computer memory as one of the values between 0 and $65535 = 2^{16} - 1$.

[8.2]ASCII is the acronym for *American Standard Code for Information Interchange*. Each symbol is represented as a 7-bit word, and allows for 128 possible symbols to be so represented. Typically, a bit is appended to the 7-bit word as either a parity-check bit or an error-check bit to see if an error occurred in transmission. The mechanism for ASCII conversion is radix-64 transformation, wherein binary blocks of three bytes each are converted into four ASCII symbols, each of which is appended with an error check in the form of a *cyclic redundancy check*; see pages 541 and 542. Note that the term "radix" is a synonym for "base".

Diagram 8.4 LZ77 Compression

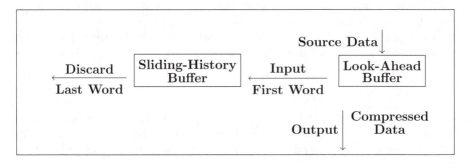

▼ **E-Mail Compatibility**

Typically, PGP sends a stream of bytes of data. However, there are certain e-mail networks allowing only ASCII data to be transmitted. PGP satisfies this requirement by transforming the stream of bytes into a stream of printable ASCII characters, using an encoding technique called *radix* 64, which we describe in Appendix D; see page 541. This inflates the message by 33%, but the aforementioned compression stage offsets this message expansion. In fact, a standard analysis of the PGP mechanism shows that, even with the message expansion, the net compression is approximately one-third.

From a security viewpoint, the aforementioned conversion to ASCII provides a camouflage of the data since it blindly converts to ASCII (even in the case that the original data is already ASCII text). Therefore, it will be unreadable to Eve, even if the message is not enciphered. Optionally, PGP may be formatted so that only the signature part is converted to ASCII, which permits Bob to read the message without conversion, (albeit, PGP should still be used to verify Alice's signature).

In summary, once $h(m)$ is formed, the concatenation, $(m, h(m))$, is signed by Alice to get $d_A(m, h(m)) = D_A$, which is compressed via Z. Then she enciphers via k using E to get, $E_k(Z(D_A))$, then $e_B(k)$ is appended to get $(e_B(k), E_k(Z(D_A)))$, which is converted to ASCII. Upon receipt, Bob reconverts to binary; recovers k via d_B, which he uses to get $Z(D_A)$. This is passed through Z^{-1}, after which he uses e_A to get $(m, h(m))$. He applies h to m; and compares his $h(m)$ to the version sent by Alice.

▼ **Segmentation**

Anyone who has tried to send a very large e-mail attachment knows that certain e-mail sites will "bounce back" the message stating that it exceeds the maximum message length allowable (which typically over the Internet is $5 \cdot 10^4$ bytes). Hence, segmentation, the splitting up of the message into smaller pieces or segments, is necessary. PGP meets this requirement by segmenting a message into manageable, acceptable blocks for easy transmission. Segmentation is done after all of the above processing is completed.

Now that we have described the basics of the fundamental protocols underlying PGP, we look in detail at the various aspects of the message transfer and reassembly, starting with the components of the message itself.

▼ Message Components

There are three basic components in a message m to be sent by Alice.

1. **Session Key**: This component has two facets. First there is Bob's identifier I_{e_B} for his public RSA key e_B, defined by $I_{e_B} \equiv e_B \pmod{2^{64}}$, namely, the least significant 64 bits of e_B. The identifier I_{e_B} is the most efficient means to transfer the key verifier to Bob that does not involve the use of too much space or too much workload to do the verifying. (Note that this identifier is essentially a probabilistic identifier in the sense that it is possible for two different public keys to have the same least significant 64 bits, but the probability is very low given the bitlength involved.)

   The second facet of the session key component is the session key k, itself.

2. **Signature**: This component has four facets. There is the timestamp t_A, which corresponds to the creation time of Alice's signature. Then there is the identifier I_{e_A} for Alice's public key, via $e_A \equiv I_{e_A} \pmod{2^{64}}$ (see the description of this device, presented for Bob's key, in part 1 above). Third, there is the message digest, $h(t_A, m)$, which is formed (with t_A appended to thwart replay attacks). Last, there are the two leading bytes L_1 and L_2, of $h(t_A, m)$, which allows Bob to ensure that the correct public key, e_A, was used to decipher the message for authentication. He does this by comparing the plaintext copy of these bytes with the first two bytes of the deciphered message digest. (Note that in the previous discussion and diagrams, we did not mention, explicitly, the timestamp in order to simplify the presentation. Thus, we are assuming, tacitly, that it is present and handled in the aforementioned fashion.)

3. **Message**: This is the component consisting of the message data, m, itself, accompanied by a timestamp, t_m, specifying the creation time of m, as well as a filename F_m.

Both the message and signature components are ZIP compressed, then enciphered with the session key. The session component together with the compressed components are then converted to ASCII.

In Diagram 8.5, we are assuming that the (otherwise optional) operations of: ensuring confidentiality by forming $e_B(k)$; ensuring authentication by forming $d_A(m, h(t_A, m))$; ZIP compression of the signature and message components is carried out; and ASCII conversion is executed on all components. Each of the symbols in the diagram are defined in the discussion preceding the diagram. Each double box contains a set of operations to be carried out, and the nesting of the boxes dictates the order of the operations from inner to outer.

Diagram 8.5 PGP Message Components

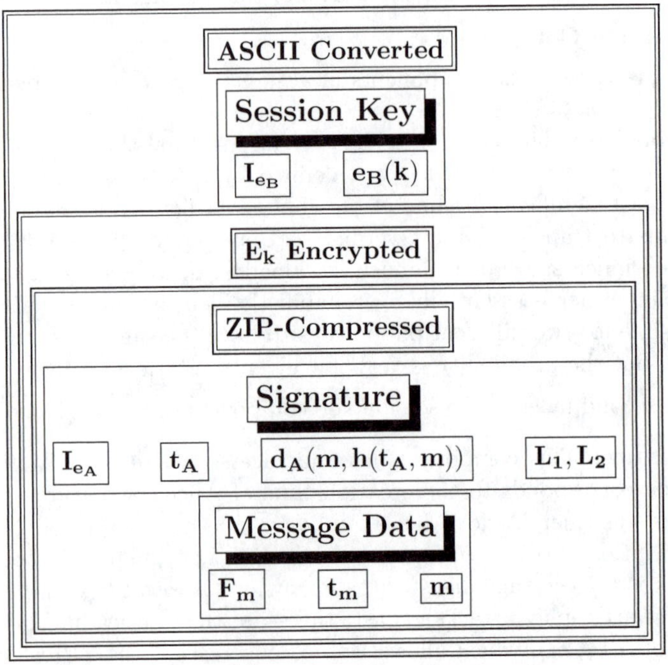

$$
\boxed{
\begin{array}{c}
\boxed{\text{ASCII Converted}} \\[4pt]
\boxed{\textbf{Session Key}} \\[2pt]
\boxed{I_{e_B}} \quad \boxed{e_B(k)} \\[6pt]
\boxed{
\begin{array}{c}
\boxed{\textbf{E}_k \text{ Encrypted}} \\[4pt]
\boxed{\text{ZIP-Compressed}} \\[4pt]
\boxed{\textbf{Signature}} \\[2pt]
\boxed{I_{e_A}} \; \boxed{t_A} \; \boxed{d_A(m, h(t_A, m))} \; \boxed{L_1, L_2} \\[6pt]
\boxed{\textbf{Message Data}} \\[2pt]
\boxed{F_m} \; \boxed{t_m} \; \boxed{m}
\end{array}
}
\end{array}
}
$$

The next topic is a fundamental feature of PGP and is a mechanism for an individual user to communicate with entities it knows, securely, and efficiently.

From the above, it can be seen that the key identities, I_{e_A} and I_{e_B}, for Alice and Bob, respectively, provide authentication and confidentiality. Bob's public and private keys are stored securely at his computer along with public keys of others, such as Alice, with whom he communicates. PGP uses data structures to store them, called *public key rings* and *private key rings*. We now describe each of these in turn, and delineate the schemes by which private keys are securely maintained.

▼ Key Rings

The private key ring is stored only on Alice's computer, which stores the RSA key pairs owned by her, and is accessible only to Alice. In the private key ring, each entry for an entity has the following fields (but typically she will only have one entry, namely, her own public/private key pair).

▼ Private Key Ring Individual Field Entry

1. **Timestamp:** t_A, the creation time of (e_A, d_A).

2. **Key ID:** $I_{e_A} \pmod{2^{64}}$.

3. **Public Key**: e_A.

4. **Private Key**: d_A (enciphered using CAST-128, 3DES, or IDEA). The actual key d_A is not stored on Alice's computer, only the encrypted version. Here is the actual mechanism by which Alice accesses the private key, when needed, in order to achieve maximum security.

Private Key Storage and Access Steps

(i) Alice chooses a passphrase that she will use for enciphering private keys. (It is paramount that she keep this secure, never write it down, or disclose it to anyone.)

(ii) When the PGP program generates a new RSA key pair, such as (e_A, d_A), it will prompt Alice for her passphrase, P, and using SHA-1, a 160-bit hash $h(P)$ is formed, and the passphrase is discarded.

(iii) The program enciphers d_A, using an SKC, E (which is one of 3DES, IDEA, or CAST-128), with $h(P)$ as the key, namely, to form $E_{h(P)}(d_A)$, and discards $h(P)$. Then $E_{h(P)}(d_A)$ is stored on Alice's private key ring.

(iv) Whenever Alice wants to access d_A, she must provide the passphrase. The PGP program provides her with $E_{h(P)}(d_A)$, generates $h(P)$, and deciphers d_A using E with $h(P)$, namely, via

$$E_{h(P)}^{-1}(E_{h(P)}(d_A)) = d_A.$$

5. **User ID**: ID_A, which could be, for instance: *Alice@PGPprivateRing.com*.

▼ **Public Key Ring Individual Entry**

This ring is used to store the public keys of other users, such as Bob, with whom Alice communicates. The following are the fields in Bob's entry, which may be viewed as a public-key certificate (see Section 6.2 on PKI). Items 4, 6, and 8 are under a framework, called a *trust-flag-byte*, the contents of which are described individually in each field entry, and refer to the web-of-trust model described on pages 238 and 239.

1. **Timestamp**: t_B, which is the creation time of the entry.

2. **Key ID**: $I_{e_B} \pmod{2^{64}}$.

3. **Public Key**: e_B.

4. **Owner Trust**: $trust - flag - byte$, which is the trust, assigned by Alice, that indicates the degree to which e_B can be trusted to sign other public-key certificates. When a new public key is to be added to the public-key ring, the PGP program prompts Alice to assign a level of trust to the key owner, Bob in this case. When the level of trust is *complete trust*, then the

public key is also put on Alice's private-key ring. In the case where Bob's key appears on Alice's private-key ring, there is a *buckstop bit*, which is set to 1 in that instance.

5. **User ID**: ID_B, which is Bob's identifier, such as *Bob@PGPpublicRing.ca*.

6. **Key legitimacy**: $trust - flag - byte$, which is the level of trust that the PGP program (which computes this field), imparts to the binding of Bob's user ID to e_B. The means by which this is determined by the PGP program is on a weighted basis, whereby the PGP program bases the weighting upon the signature trust fields present in item 8. There is also a *warnonly bit*, which is set to 1 if Alice only wants to be warned that e_B is only used for enciphering, but is not fully validated.

7. **Signature**: When a new public key, Bob's in this case, is added, one or more signatures could be appended to it, and more may be added later.

8. **Signature Trust**: $trust - flag - byte$, which is the degree of trust that Alice assigns Bob to certify public keys, so is essentially a cached version of field 4 (owner trust), in the following sense. Upon addition of a signature, the PGP program looks through the public-key ring to determine if Bob's signature is among the public-key owners therein. If so, the trust value given in field 4 is assigned, and if not, an *unknown* value is assigned to this field. This field is periodically updated by the PGP program, which scans the public-key ring for all signatures owned by Bob and updates this field to be the same as the owner trust field.

Now that we have the notion of PGP rings, we can give a more detailed and informed description of PGP message generation, processing, and reception.

▼ PGP Message Processing Protocol Via Key Rings

This protocol description, and accompanying diagrams on pages 284 and 285, depict the PGP message generation, and processing upon reception using key rings. Since we fully described the mechanism for ASCII conversion and ZIP compression above, we eliminate those stages for the sake of simplicity. Moreover, we are assuming that both signing and encryption are required.

Protocol Steps

We assume, as above, that Alice is sending a message to Bob.

1. The PGP program obtains Alice's encrypted private RSA key d_A from her private-key ring using ID_A (for instance, *Alice@PGPprivateRing.com*) as an index for so doing.

2. The PGP program requests Alice's keyphrase in order to provide her with this enciphered version, which she provides and d_A is obtained as in part (iv) of private-key storage and access on page 281.

3. Alice generates the message m, and the digital signature $d_A(h(m))$ is formed as in the authentication protocol described on page 274. However, the public-key identifier, I_{e_A}, her public-key identifier from the signature component of the message (see part 2 on page 279), must be appended to the signature since Bob must know which public key is intended for use given that Alice could have many private keys.

4. The PGP program uses a random-number generator to create a session key k, as above, and forms $E_k(m)$.

5. The PGP program gets e_B, Bob's public key from Alice's public-key ring using ID_B (for example, *BOB@PGPpublicRing.ca*) as an index.

6. Then the PGP program forms $e_B(m)$, and $(e_B(k), E_k(m))$ is sent to Bob.

7. Upon reception, the PGP program obtains Bob's encrypted private key, d_B, from his own private key ring using I_{e_B}, from the session key component of the message (see part 1 on page 279), as an index.

8. The PGP program requests Bob's passphrase, which he delivers, and decrypts to get the session key, k, which is used to recover the message $(d_A(h(m)), m)$.

9. The PGP program gets e_A from Bob's public key ring, using I_{e_A} from the signature component of the message (see part 2 on page 279), as an index. This is used to recover the $h(m)$ sent by Alice.

10. The PGP program computes $h(m)$ from Alice's sent message m, and compares it to the $h(m)$, sent by Alice for authentication.

In step 4 above, we mentioned the PGP random-number generation (PG-PRNG). We will not describe the algorithm here since it is based upon the ANSI X9.17 algorithm, which is described in detail in Appendix B (see page 506). However, before we turn to diagrams illustrating the details of the PGP message scheme in action, there are some features of the PGPRNG that deserve to be elucidated. PGPRNG generates random numbers from the content and timing of keystrokes. This provides an intricate and formidable scheme for generating both random and pseudorandom numbers. The PGP mechanism uses the random-number generation for initial seed inputs to PRNGs (see page 151); an alternative input during the actual operation of a PRNG; and the generation of RSA key pairs. The PGPRNG generates pseudorandom numbers for session key generation (see page 275); and to generate an IV for CFB mode (see pages 135 and 275). The PGPRNG employs a 256-byte buffer of random bits in the following fashion. When the PGPRNG anticipates a keystroke, it records the time in 4-byte configuration, then it waits. When it receives the keystroke, it records the time the key was pressed and the byte value of the keystroke. This information is used to generate a key, and this key is used to encipher the current value of the random-bit buffer.

Diagram 8.6 PGP Message Generation and Encryption Via Rings

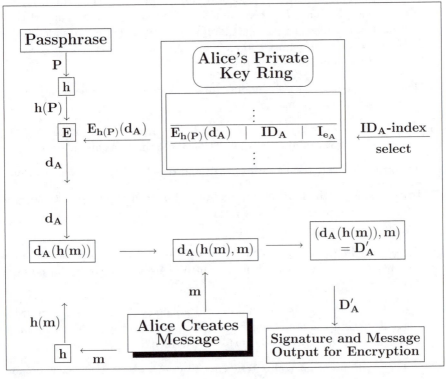

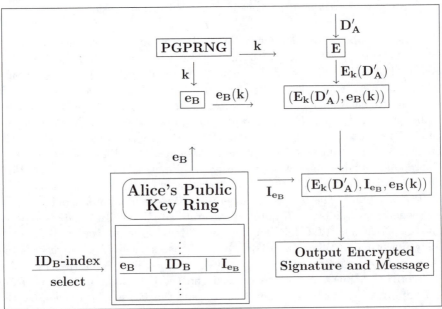

Diagram 8.7 PGP Message Reception, Decryption, and Authentication

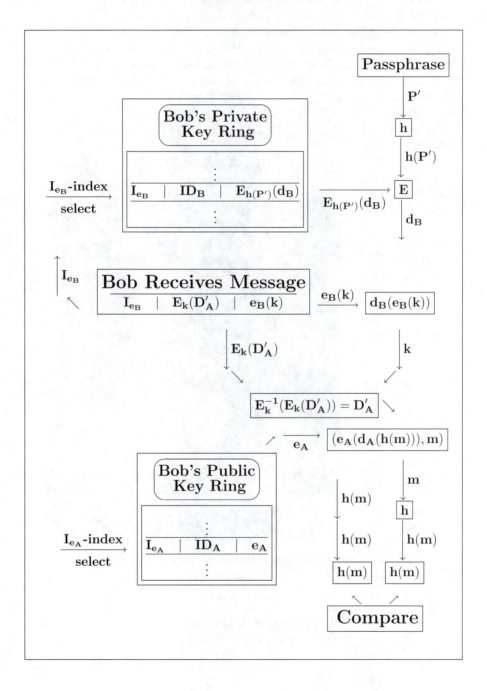

▼ Analysis and Summary

PGP utilizes a package of algorithms, in a general-purpose application, which is operating system and machine independent, embodying only a few simple operations. It is freeware for individuals and of moderate cost to commercial enterprises who enjoy vendor support. Moreover, the scheme is independent of government control (although it has become Internet standard track RFC3156, MIME security with OpenPGP, which we will study in the following section).

The trust model used by PGP does not include a PKI specification, but its web-of-trust approach (see page 238) does provide a convenient trust-use mechanism for the purpose of linking trust with public keys, as depicted by our discussion of public and private-key rings on pages 280–285). This is a particularly clever and innovative means of dealing with one of the principle weaknesses of PKC, namely, the protection of public keys from being compromised.

In conclusion, PGP embodies an interwoven collection of protocols (including public-key management) in an efficient, yet secure manner to ensure authentication and confidentiality of e-mail services, as well as file storage.

Figure 8.3: Phil Zimmermann in Red Square.

From *Computerworld Russia.*

8.2 S/MIME and PGP

Pictures are for entertainment, messages should be sent by Western Union
Sam Goldwyn (born Samuel Goldfish) (1882–1974),
American film producer

A quarter of a century ago, Internet e-mail involved little more then elementary ASCII message exchange, typically among researchers based in universities or government centers. For these clients, security was not much of an issue. Today, e-mail is used by tens of millions of people worldwide for sending a cornucopia of digital information including not only text-based data, but also sophisticated graphics, movies, music, and much more. A substantial amount of this traffic requires security, and much of this has been provided by a scheme called *Secure Multipart Internet Mail Extension* (S/MIME), the initial version of which was developed by a private consortium of vendors. This was an evolution of the original MIME e-mail scheme, developed by IETF, which had no security attached to it. The latest version in this evolution is S/MIMEV3, or S/MIME, version 3, which was made an IETF standard in July 1999. S/MIMEV3 is described in [216]–[220], which contain the following parts: (RFC2630), cryptographic message syntax; (RFC2631), Diffie-Hellman key-exchange method; (RFC2632), certificate handling; (RFC2633), signature/encryption protocols; and (RFC2634), some enhanced security service extensions: signed receipts; security labels; secure mailing lists; and signing certificates. S/MIMEV3 includes PKI attributes such as CRLs, and X.500 certificates used to bind an entity's identification and public key for the secure operation of S/MIME and other PKI-enabled functions. Indeed, S/MIME uses PKI to employ mechanisms for authenticating S/MIME users, to provide digital signatures, ensure confidentiality, nonrepudiation, and more.

Not only has S/MIME been proposed for providing e-mail security services, but also for its use with PGP. There are two proposed standards, OpenPGP and PGP/MIME, both of which are based on PGP, and the latter of which was developed by individuals, some of whom now form PGP Inc. In 1997, the OpenPGP Working Group was formed in IETF to define a standard. OpenPGP is now an IETF proposed standard RFC2440. It appears clear at this point in time that S/MIME will become the industry standard for commerce, while PGP will prevail as the choice for individuals seeking security in their e-mail transactions.

Section 8.1 looked at PGP in depth, and we saw that the scheme provides for mechanisms involving the signing and encryption of data. The same is true of S/MIME. We now look at enhancements in functionality built into S/MIMEV3.

◆ S/MIME Functionality

We assume that Alice is sending a message to Bob.

1. **Enveloped Data**: This function provides for SKC encryption, with a symmetric key k, say, of S/MIME data, D, to form $k(D)$, followed by

enciphering of k with Bob's public key e_B to form $e_B(k)$. Then $e_B(k)$, $k(D)$, identifying data, and specifiers for the cryptographic algorithms being used are sent to Bob.

The various cryptographic algorithms that MUST[8.3] be used by S/MIME are contained in the following *cryptographic suites*, or sets of cryptographic algorithms.

[CS1]: RSA for digital signatures, with RSA key size a minimum of 1024 bits being a MUST (see page 181) and SHA-1 for hashing (see page 255); RSA for key transport; 3DES (see page 131) for content encryption, and at least two independent keys MUST be supported using CBC mode (see page 134), called DES EDES3 CBC.

[CS2]: DSA (see page 183) for digital signatures, with DSA key size of 1024 bits being a MUST, with SHA-1; RSA for key transport; and 3DES for content encryption.

In addition to the above, the following cryptographic algorithms SHOULD be supported for implementation.

[CS3] RSA for digital signatures with SHA-1 for hashing, RSA for key transport, and AES (see Section 3.5) for content encryption.

[CS4] DSA for digital signatures with SHA-1 for hashing, Diffie-Hellman (see page 166) for key agreement, and 3DES for content encryption.

[CS5] DSA for digital signatures, SHA-256 (see page 255) for hashing, Diffie-Hellman for key agreement, and AES for content encryption.

2. **Signed Data**: This function renders a data integrity resource. First, a message digest is formed, h(m), then encrypted with Alice's private key to get $d_A(h(m))$, which is radix-64-encoded, the latter being called *transfer encoding*. This means that the digital equipment being used encodes the data in base-64 to enable the binary data to be transferred, unaltered, through a variety of systems. (This is essential since, for instance, if an 8-bit message is sent through an e-mail portal, which is, say, a 7-bit device, then it could strip the message of important symbols, and any digitally signed message that is altered or stripped of characters is rejected as invalid.) Once so encoded, only S/MIME-enabled users can read the signed data.

3. **Clear-Signed Data**: This is a function allowing *non*-S/MIME-enabled users to view the message content, but not verify the signature. This is

---

[8.3]The terms "MUST", "SHOULD", and "MAY" are precisely defined in [204]. Essentially a "MUST" means that what is referenced is an absolute requirement of the S/MIME protocol, and must be implemented in order to be in compliance with the specification. "SHOULD" means that the referenced feature may be ignored for sound reasons, but it is recommended that the feature be implemented. "MAY", sometimes replaced by the adjective "OPTIONAL", means that an item is truly optional.

accomplished via the digital signature *only* being base-64 encoded. This, therefore, ensures the same data integrity as part 2, but allows for more flexibility in the "read-only" format.

4. **Signed and Encrypted Data**: This is a nesting function allowing for both confidentiality and integrity via either the signing of encrypted data, or the enciphering of signed data.

There are also recommendations for key sizes depending on the cryptographic suite used.

▼ **Public Key Sizes**

1. If the implementation of S/MIME is employing cryptographic suite CS3, then it SHOULD also support RSA key sizes *greater* than 1024 bits.

2. If the S/MIME implementation uses cryptographic suite CS5, then it SHOULD support Diffie-Hellman key sizes *greater* than 1024 bits.

3. If the S/MIME implementation supports key sizes greater than 1024 bits when employing either DSA or RSA for digital signatures, then it SHOULD also support SHA-256.

Deciding upon a cryptographic suite to use may depend on the capabilities of the intended receiver. S/MIME, therefore, makes decision criteria available to the sender for making such a determination.

▼ **S/MIME Decision Criteria for Selecting Cryptographic Suites**

We assume Alice is sending Bob an S/MIME message.

1. **Known Capabilities**: If Alice has knowledge of Bob's cryptographic capabilities from some previous correspondence, then she should choose the item on Bob's list that most closely corresponds to the most preferred S/MIME capability.

2. **Unknown Capabilities**: If Alice has no knowledge of Bob's cryptographic capabilities, but has received at least one message from him in the past, then she should use the algorithm employed by Bob for sending him the message.

3. **Unknown Capabilities and Unknown S/MIME Version**: If Alice has neither any knowledge of Bob's cryptographic capabilities, nor has she had any previous correspondence with him, she should use 3DES, which is required in cryptographic suites CS1–CS2.

▼ **S/MIME Messages**

S/MIME messages embody *cryptographic message syntax objects* as defined in [216] and MIME bodies. In order to process an S/MIME message, one must first prepare what is called a *MIME entity*, which may be a subpart of a message or the whole message, including all its subparts. (A MIME message[8.4] consists of: (1) one of five header fields, which provide information about the body of the message; and (2) a variety of content formats, supporting multimedia e-mail.)

Once the MIME entity is created, it is converted to *canonical form*, which is a format, suitable to the content type, standardized for use between various systems. Then the appropriate transfer encoding is applied to the message content. Then the MIME entity is sent to security services, where it is enveloped, signed, or both.

▼ S/MIME Content Types

1. **Enveloped-Only Data**: The S/MIME content type, called *enveloped data*, consists of enciphered content of any kind, together with encrypted content-enciphering keys for one or more recipients. For each such recipient, a digital envelope is manufactured. This envelope contains the enciphered content itself, together with an attendant encrypted content-enciphering key. This guarantees confidentiality of the message while in transit. The methodology for creating enveloped-content data is given in the following steps.

   We will assume that there is one recipient, Bob, for simplicity, but there may be numerous recipients for whom each of these steps must be carried out.

   [a] Choose an SKC (3DES, for instance), and generate a pseudorandom content-enciphering key k.

   [b] Encipher k with Bob's public key e_B, to get $e_B(k)$.

   [c] Create a block of data for Bob consisting of $e_B(k)$, an identifier, $C(B)$, for Bob's X.509V3 certificate, and an identifier of the algorithm used to encrypt the session key, k, say I_{RSA}, for instance.

   [d] Encipher the message m, with k to get $E_k(m)$.

   [e] Form $ED = (e_B(k), C(B), I_{RSA}, E_k(m))$, the enveloped data, which is base-64 encoded to produce the enveloped data value.

   When Bob receives the message, he strips off the base-64 encoding, uses d_B to get k, which is used to recover m.

   Enveloped-only data in S/MIME provides secrecy without authentication.

2. **Signed-Only Data**: Although it is possible to have more than one signer, we will assume that there is only Alice, for the sake of simplicity. There are also two methods for signing S/MIME messages: (1) *SignedData-MIME* with signed-only data (but this is readable only by S/MIME-enabled

---

[8.4]MIME message specifications are provided in RFC 2045–2049; see [198]–[202].

users); (2) *multipart signed*, which is also called *clear signing* (and this is viewable by all users).

SignedData MIME: The following steps provide the means for constructing a SignedData MIME entity. Again, we assume there is only Alice doing the signing of a given message m.

[i]   Select a message digest algorithm, h, such as SHA-1.

[ii] Hash m, to get $h(m)$.

[iii] Encrypt the message digest with Alice's private key to form $d_A(h(m))$.

[iv] Create a message block for Alice, consisting of: an identifier of her public-key certificate, I_{e_A}; her X.509V3 certificate identifier $C(A)$; an identifier of the hash algorithm being used, I_{SHA-1}, say; an identifier of the algorithm used to encipher $h(m)$, I_{RSA}, say; m itself; and the encrypted message digest. This produces the SignedData MIME,

$$SDM = (I_{e_A}, C(A), I_{SHA-1}, I_{RSA}, m, d_A(h(m))).$$

[v]   Then SDM is base-64 encoded to produce the SignedData MIME value.

Upon reception, Bob, strips off the base-64 encoding, then uses e_A to get $h(m)$. Then he independently computes $h(m)$ from m and compares this with the deciphered value of $h(m)$ to verify Alice's signature.

Clear Signing: This structure allows Alice to communicate with Bob if he is not an S/MIME-enabled user. The body of the multipart/signed MIME is comprised of two parts, the first of which can be of any MIME type, is left to be broadcast in the clear. The second part is actually a special case of the SignedData MIME type, called a *detached signature*, wherein the plaintext of the message is omitted.

Here are the basic steps in producing a clear signed S/MIME entity:

[A] The message m is signed with Alice's private key to form $d_A(m)$.

[B] She forms a data block consisting of

$$CSM = (I_{e_A}, C(A), I_{RSA}, d_A(m)).$$

[C] Then CSM is base-64 encoded to form CSM_{64} and the two-part message (m, CSM_{64}) is sent to Bob.

Bob receives the message, strips the base-64 encoding from the second part of the message, CSM_{64}, to get CSM, uses Alice's public key e_A to recover m and compares it with the message m sent in the clear in the first part of the message.

Signed-only data in S/MIME provides authentication without secrecy.

There is a means in S/MIME to provide both authentication and secrecy.

▼ S/MIME Message Authentication and Secrecy

This involves the *nesting of protocols*. In other words, Alice nests the enveloped-only data with the signed-only data, so that either she signs the message first or envelopes it first. The pros and cons of enciphering first versus authentication first were discussed on pages 266 and 267.

Diagram 8.8 Enveloped-Only S/MIME Message

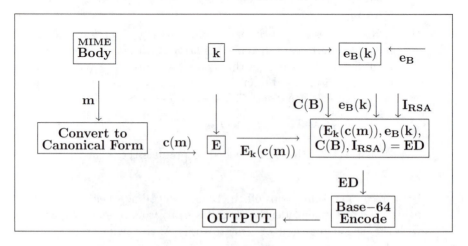

In Diagram 8.9, for the sake of simplicity of presentation, we do not explicitly give the conversion to canonical form, but assume this has tacitly been done.

Diagram 8.9 Signed-Only S/MIME Message

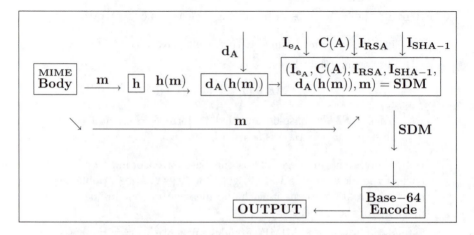

In Diagrams 8.10 and 8.11, which conclude the discussions for this section, we illustrate both clear signing and the combination of enveloping and signing with S/MIME. We have not discussed the actual certificate processing with S/MIME since this is essentially a facet of the PKI structures discussed in Section 6.2, to which we refer the reader for a reminder of this mechanism.

Diagram 8.10 Clear Signed S/MIME Message

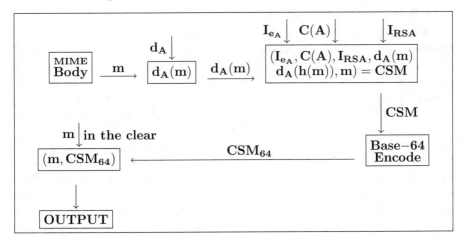

Diagram 8.11 Signed and Enveloped S/MIME Message

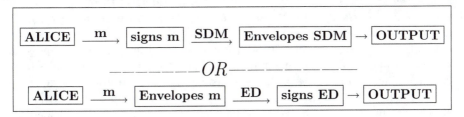

In the next section, we look at IP-level security since there are security issues that cross the protocol layers (see Section 5.7). Such applications-specific security schemes such as PGP and S/MIME, for example, do not address issues where a particular network might want to restrict ingress and egress of data to and from its site. We will see that IPSec (initially mentioned on page 137 in connection with CTR mode), provides security at the network layer, that layer between the bottom physical layer and the next layer up, the transport layer. Recall that we already discussed SSL in depth in Section 5.7, which deals with security at the transport layer. Thus, the next section deals with yet another aspect of protocol level security for our Internet activities, such as remote logins, file transfer, WWW access, as well as e-mail and more.

8.3 ☞ IPSec

*Knowledge is of two kinds. we know a subject ourselves, or we know where
we can find information upon it.*

Samuel Johnson [8.5] (1709–1784)
— from *Life of Samuel Johnson* (1791), 7 April 1775, by James Boswell[8.6]

◆ What IS IPSec?

Internet Protocol Security (IPSec), is a foundation of open standards for
establishing end-to-end security in network architecture. The IETF developed
standards for IPSec under which it ensures cryptographically enforced authenticity, confidentiality, and integrity of data transfer over a public IP network.
Perhaps the most important attribute of IPSec is that it provides security to *all*
traffic at the IP level, including those distributed applications mentioned at the
end of the previous section on page 293.

IPSec is a very complex mechanism and is still evolving. It is not as preva-
lent as SSl/TLS (studied in Section 5.7), but is more secure, as we shall learn in
this section. IPSec's complexity has drawn criticism from some cryptographers,
but even they agree that IPSec is the best there is for secure Internet communications, at this point in time. Succinctly, an IPSec-enabled computer is one
that can authenticate any data packet it receives and encipher any data packet
it sends.

The majority of IPSec security measures are provided by the use of two
traffic security protocols, called the *Authentication Header* (AH), and the *Encapsulating Security Payload* (ESP), although (optionally) through the use of
Internet Key Exchange (IKE), for exchanging keys and negotiating security.
(All of this will be described in detail later in this section.) AH provides only
authentication, meaning verification of the origin and integrity of the message
sent, according to IPSec documentation. ESP, on the other hand, provides both
confidentiality and integrity, albeit AH ensures that more of the message is authenticated. IPSec does not specify the cryptographic suites to be used, since
its security protocols are designed to be cryptographic algorithm-independent.
However, there is a default suite of cryptographic algorithms for use with AH
and ESP if required.

◆ Why Use IPSec?

Everyone from corporations to individuals are increasingly seeking more security for their communications. For business organizations, leasing lines dedicated to their companies provides the security, but the cost can be prohibitive,
and is relatively inflexible when compared to the Internet. Thus, an increasingly
more common choice is being exercised, namely, the *Virtual Private Network*

[8.5]See page 180.
[8.6]Boswell (1740–1795) was a Scottish lawyer, who was Johnson's biographer.

(VPN), based upon IPSec, for secure Internet data transmissions. A description of a simple VPN is as follows. If Alice works for company A and wishes to communicate with Bob who works for company B, both behind their respective security gateways, then her gateway automatically negotiates security with his gateway. In this case, all IPSec processing is done behind these respective gateways so no adversary can determine anything other than the fact that the gates are communicating. Below, we will describe the details of such setups.

Remote logins for workers in large companies, as well as individuals away from home, is becoming commonplace. To do so securely is also becoming a necessary part of this fact. Whether it is for an individual's online banking while on vacation, or a company employee who needs to access sensitive corporate files while at a business meeting away from the workplace, Internet security is becoming a daily fact of life.

An advantage for end users having IPSec-security-enabled software is that they can make local calls to an Internet Service Provider (ISP), and acquire access to a corporate network, for instance. For employees of this corporation, this reduces access costs when travelling or commuting. Moreover, when these employees are at their workstations, they can achieve secure communications with other corporate entities with whom they do business. Even when the network used by employees has its own built-in security mechanism, IPSec complements and intensifies that security. Moreover, unlike SSL/TLS, studied in Section 5.7, the choice of cryptographic algorithm to be used, can be negotiated in secret. With SSL/TLS, the negotiations are done in plaintext. Also, with SSL/TLS, applications, such as e-mail, require that such cryptographic services be requested, whereas IPSec-enabled computers automatically protect e-mail, Web browsing, file transfers, and generally any data communications between itself and any other IPSec-enabled computer. Even if the other computer is not IPSec-enabled, the IPSec-enabled one can allow or disallow messages in a way that is transparent to the user. (*Transparent*, in this context, means hardware or software that works without user interference.)

◆ How Does IPSec Work?

When IPSec is implemented as a boundary between unprotected and protected perimeters (such as in a firewall or router),[8.7] for a host or network, it controls whether data crosses the boundary unrestricted, are subject to AH or ESP security processing, or are discarded (say, if a replay data packet is detected). Paths into an organization are protected against all bypass traffic if it is specified that all outside traffic must pass through IP, say, in a firewall. Moreover, given that IPSec is implemented at the network level, there is no need to alter software or access to servers (see page 218). This also clearly has the advantage that IPSec can be made transparent to end users. Yet, IPSec

---

[8.7]In Section 8.4, we will learn about firewalls in depth. For now, think of them as network gateway-server programs, that shield data of a network site from users situated in other networks. Firewalls provide security in concert with what are called *router programs*, which are mechanisms for directing data, via the best route possible, to the next network site enroute toward the target site. Together they screen all data to decide action.

can provide end users with security when necessary, say, for those employees of a corporation working on highly classified material.

▼ IPSec Services

IPSec security, provided at the IP level, enables a system by using AH and ESP in concert to provide the following services:

1. **Access Control**: using AH and ESP.

2. **Confidentiality**: via enciphering of data or limited traffic-flow security, using ESP for both encryption and authentication.

3. **Connectionless Integrity**: via an in-built IP detection mechanism.

4. **Data Origin Authentication**: using AH.

5. **Rejection of Replay Data Packets**: using AH and ESP.

Now we are ready to look at the various components that make up the IPSec structure. At the time of this writing the RFC 2401 overview of IPSec security architecture has been rendered obsolete by a document currently being updated; see *http://www.ietf.org/internet-drafts/draft-ietf-ipsec-rfc2401bis-01.txt*.

First, we examine how keys are used to set up the IPSec mechanism.

▼ IPSec Key Management

IPSec provides another essential feature of any security protocol, namely, the management of keys for use in data exchange, encryption, and for such, the negotiation of keys with other entities. IPSec further mandates that a record of such key negotiations be kept. There exist two kinds of IPSec support for this key service.

Key Management Techniques

The following provides minimal requirements of IPSec key (and SA) management (see page 302).

Manual: The manual key and SA management is the simplest type. In this case an entity, typically a systems administrator, manually configures each network with keying material and SA management information pertinent to communications with other systems. This is really only practical for small, relatively static communications environments. For instance, in a VPN, with a small number of sites in a single administrative domain, this would be feasible. Manual techniques might also work in larger environments where only a small number of gateways need to be secured. However, in larger networks, in general, this method is not practical.

Automated: With an automated system, on-demand keys may be created for SAs, and is scalable for ever-changing and growing larger networks. Moreover, this type of management enables options not available in manual mode such as antireplay protection, as well as on-demand creation of SAs for the purposes of, say, session-key creation. Although IPSec supports many standards, IKE is the default IPSec key-exchange protocol.

IPSec is divided into two major parts: part one consists of user authentication and key exchange using IKE, and part two consists of bulk data confidentiality and integrity for message and file transfer.

◆ **IPSec Part I: IKE Identity Authentication and Key Exchange**

▼ **Internet Key Exchange (IKE)**

What is IKE?: IKE is an IPsec standard used to ensure security for VPN negotiations and access to networks or remote hosts. IKE is specified in RFC 2409 [212], which specifies an automatic mechanism for establishing security, and does so without the preconfiguration necessary for manual mode, which we discussed above. IKE is a hybrid protocol that evolved from two older protocols called *Oakley* and *SKEME* with an *ISAKMP* (Internet Security Association and Key Management Protocol) TCP/IP-based configuration. The Oakley protocol defines a sequence of key exchanges and specifies their services, typically authentication and identity protection. SKEME is a protocol that defines the methodology for negotiating key exchange.

Why Use IKE?: Although it is not specified that IPSec use IKE, its employment ensures automatic authentication for antireplay security; certification authority services; and on-demand change of IPSec session-based encryption keys (among other built-in services).

How Does IKE Work?: There are two phases to IPSec IKE. In phase one, two IKE peers establish a secure authentication communication channel via an IKE SA and establish a shared secret key. In phase two, the secret key and secure IKE SA, established in phase, are used to send encrypted messages. In these messages, they agree upon secret keys for bulk encryption, cryptographic methods for using them, and other parameters. Different modes for IKE are available for accomplishing the above.

▼ **IKE Modes**

Main Mode: In this mode, there is a three-pronged approach for creating the first phase of an IKE SA, which is used for later transactions. This is similar to the initial phase of SSL/TLS where negotiation to determine cryptographic parameters is done largely in plaintext.

In main mode, there is a six-step message exchange between, say, Alice and Bob, consisting of three two-way passes.

(1) They agree on cryptographic algorithms for use as the IKE SA.

(2) They exchange public keys to be used for Diffie-Hellman exchange, and exchange nonces.

(3) They verify identities via signed nonces.

Upon completion of this phase, authentication and key exchange are completed, and the IKE SA is established.

In what follows, we give a description of the six steps involved in the three-pass IKE main mode, followed by an illustration. What we describe is a simplified version of the scheme, assumed to take place between Alice and Bob. In fact, IKE itself, is a slim-down version of ISAKMP/Oakley.

Background Assumptions: For the Diffie-Hellman part of the exchange, we need the following notation. We use (p_A, s_A), and (p_B, s_B), respectively for Alice's, and Bob's, respectively public/secret Diffie-Hellman keys. Recall from page 166 that $(p_A, s_A) = (\alpha^x, x)$ and $(p_B, s_B) = (\alpha^y, y)$, so

$$p_A^{s_B} = p_B^{s_A} = k.$$

This notational assumption will be made below. Moreover, I_A and I_B are identifying data strings for Alice and Bob, respectively, and we assume that Alice and Bob have RSA public/private key pairs (e_A, d_A) and (e_B, d_B), respectively, where they have exchanged e_A and e_B in advance of the following.

IKE Phase I Using Main Mode

1. **SA Negotiation Initialization**: Alice sends Bob her list of proposed parameters, $\mathbf{S_A}$, such as proposed encryption algorithms, hash functions, pseudo-random generators for hashing messages to be signed, and so on. These will be used to establish an IKE SA. Also, contained in the message is a *header*, $\mathbf{H_A}$, containing a cookie, $\mathbf{C_A}$ (see pages 323–325) for Alice (in order to keep the session state information for her).

2. **SA Agreement**: Bob selects one of each of the parameters from Alice's lists in $\mathbf{S_A}$, such as a single choice of hash function, sole choice of SKC, and so forth. He sends back his list of choices, $\mathbf{S_B}$, together with a header, $\mathbf{H_B}$, containing a cookie, $\mathbf{C_B}$, for his session state data.

3. **Key Negotiation Initialization**: Alice sends Bob her Diffie-Hellman public key $\mathbf{p_A}$, a nonce $\mathbf{N_A}$, and $\mathbf{H_A}$.

4. **Key Generation Completion**: Bob sends his Diffie-Hellman public key, $\mathbf{p_B}$, his nonce $\mathbf{N_B}$, and his header $\mathbf{H_B}$.

   Alice and Bob independently compute $p_A^{s_B} = k$ and $p_B^{s_A} = k$, respectively.

5. **Alice's Identity Verified**: Alice sends

$$(\mathbf{H_A}, k(\mathbf{I_A}, d_A(\mathbf{N_A}, \mathbf{N_B}, k, \mathbf{p_A}, \mathbf{p_B}, \mathbf{C_A}, \mathbf{C_B}, \mathbf{S_A}))),$$

   to Bob who now is able to use k^{-1} and e_A, to verify Alice's identity.

6. **Bob's Identity Verified/SA Established**: Bob sends to Alice

$$(\mathbf{H_B}, k(\mathbf{I_B}, d_B(\mathbf{N_A}, \mathbf{N_B}, k, \mathbf{p_A}, \mathbf{p_B}, \mathbf{C_A}, \mathbf{C_B}, \mathbf{S_B}))),$$

   and Alice may similarly verify Bob's identity.

Diagram 8.12 IKE Main Mode

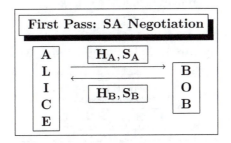

First Pass: SA Negotiation

ALICE $\quad$ $\mathbf{H_A, S_A}$ $\longrightarrow$ $\quad$ BOB

$\mathbf{H_B, S_B}$

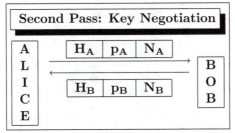

Second Pass: Key Negotiation

ALICE $\quad$ $\mathbf{H_A} \mid \mathbf{p_A} \mid \mathbf{N_A}$ $\longrightarrow$ BOB

$\mathbf{H_B} \mid \mathbf{p_B} \mid \mathbf{N_B}$

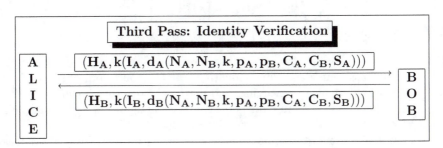

Third Pass: Identity Verification

ALICE $\quad (\mathbf{H_A, k(I_A, d_A(N_A, N_B, k, p_A, p_B, C_A, C_B, S_A)))}$ BOB

$(\mathbf{H_B, k(I_B, d_B(N_A, N_B, k, p_A, p_B, C_A, C_B, S_B)))}$

Aggressive Mode: The essential difference between this mode and main mode is the manner in which messages are configured, which reduces the exchanges depicted in Diagram 8.12 to two. This naturally increases the speed of communication. We make the same background assumptions as for main mode given on page 298.

IKE Phase I Using Aggressive Mode

1. **SA and Key Negotiation Initialization**: Alice sends to Bob $(\mathbf{H_A, S_A, p_A, N_A, I_A})$, where the notation is as above.

2. **SA Agreement and Verification of Bob's Identity**: Bob sends

$$(\mathbf{H_B, S_B, p_B, N_B, I_B}, k(\mathbf{I_B})),$$

so Alice may now compute k using her s_A, then use k^{-1} to verify Bob.

3. **Verification of Alice's Identity and SA Establishment**: Alice sends $(\mathbf{H_A}, k(\mathbf{I_A}))$, so Bob may verify Alice via k^{-1}.

Diagram 8.13 IKE Aggressive Mode

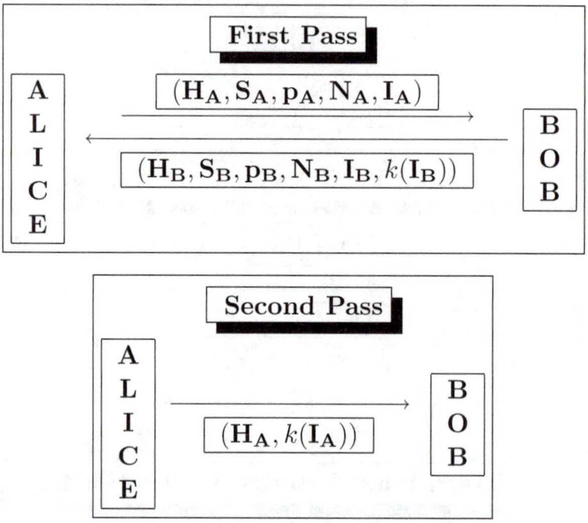

Since aggressive mode sacrifices identity protection in favour of speed, Alice and Bob may exchange identifying information before this exchange. Also, aggressive mode, unlike main mode, does not prevent against a *denial-of-service attack*,[8.8] so there is a sacrifice in security as well.

One of main or aggressive modes is used to create an IKE SA as above with a shared secret key k. Using k, and a keyed hash function plus other values from phase I, they derive three more secret shared keys, k_e to encrypt all phase-II messages, k_a, used as an HMAC to authenticate all phase-II messages, and k_d, which is used to derive the second set of shared secret keys.

For phase II, there is only one mode, whose sole purpose is to transact IP security and keying material, wherein Alice and Bob now have an authenticated secure channel, so every packet is encrypted. Now, they wish to negotiate a new SA called IPSec SA, and a secret key for bulk data encryption in IPSec part II, as well as other parameters including protocol and mode. (Note that IPSec SA, the commonly used term, should not cause confusion even though the entire process is under the IPSec umbrella. The term IPSec SA is merely used to distinguish it from the SA established in phase I under the tutelage of IKE.)

[8.8] A denial-of-service attack (DOS) is one that impedes the normal functioning of communications sites. This may involve anything from disruption of the entire network to *suppressing* all messages to a particular target site, the antithesis of which accomplishes the former, namely, by *overloading* the network with messages.

IKE Phase II — Quick Mode

1. Alice sends the following, where $\mathbf{S_{ID}}$ is a 32-bit session ID to differentiate the phase-II session setup; $\mathbf{SA_2}$ is a list of parameters for IPSec SA; and $\mathbf{N'_A}$ is a new nonce uniquely identifying Alice's message to thwart replay attacks:

$$k_e(\mathbf{C_A}, \mathbf{C_B}, \mathbf{S_{ID}}, \mathbf{SA_2}, \mathbf{N'_A}).$$

2. Bob responds with the following where $\mathbf{SB_2}$ is his list of choices from $\mathbf{SA_2}$; $\mathbf{SPI_B}$ is his security parameters index (see page 303) authorization; and $\mathbf{N'_B}$ is his unique nonce to identify his message:

$$k_e(\mathbf{C_A}, \mathbf{C_B}, \mathbf{SPI_B}, \mathbf{SB_2}, \mathbf{N'_B}, \mathbf{k_a}(\mathbf{N'_A})).$$

3. Alice acknowledges receipt by sending $k_e(\mathbf{N'_B})$.

4. Using k_d, in part, Alice and Bob independently and simultaneously generate a secret key K that they will use in part II for bulk data encryption.

In phase II there is an option for establishing a Diffie-Hellman shared secret key k' in order to calculate K. Since k' would not be based upon any previously shared secrets, it would have what is called *perfect forward secrecy*[8.9] (PFS). This means that if this particular secret, k', is compromised, not all of the encrypted data is compromised. Without PFS, if k is compromised in phase I, then all secrets derived from it, including k_d, are compromised. However, this D-H option is not an automatic default since negotiating k' is costly in terms of time.

Part I Summary: In phase I, Alice and Bob established an IKE SA and shared secret keys for use in phase II. This makes phase II an efficient mechanism since only SKC keys are used rather than PKC devices (without the D-H option). In phase II, using the phase-I keys, they establish an IPSec SA and thereby a shared secret key for bulk data encryption as well as other parameters not explicitly specified for IPSec SA in the above (such as protocol and mode). There is a lifetime associated with the IPSec SA, and once it expires, Alice and Bob's computers automatically reestablish a new IPSec SA, without Alice and Bob being involved, namely, a transparent process to them.

Diagram 8.14 IPSec Part I — Summary

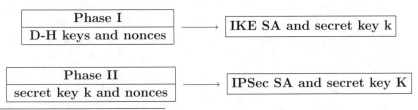

[8.9]This is not related to Shannon's notion of perfect secrecy that we will study in Chapter 11, nor is the forward secrecy we discussed on page 200.

◆ IPSec Part II: Bulk Data Confidentiality and Integrity

Now that we know how IPSec SAs are created, we now explore how they are used in detail for part II bulk encryption as well as integrity for message or file transport. In fact, it is possible for a single IKE SA to create several IPSec SAs, which may be employed for varying tasks. For instance, the established SA from phase I above, can be used to establish, in phase II, say, SA_m and SA_f, where SA_m and its associated keys and parameters are used for encrypted e-mail, and SA_f and its (different) associated keys and parameters are used for transport of encrypted database files.

Earlier in this section, we learned a bit about the two types of IPSec protocols, AH and ESP. The function of these protocols is to protect the confidentiality and/or message integrity of data packets.

The modes, which we study below, control how much of the data packets are protected by these protocols. The details will now be presented for the individual IPSec SA, its modes, parameters, security databases, and interoperability.

▼ Security Association (SA)

This is a one-way "connection" or relationship — *unicast traffic* — which supplies security to the traffic it carries, allowing only one of AH or ESP to be used. If *both* AH and ESP are required, then *two* SAs must be created *and* coordinated — *multicast traffic* — to ensure a security shield via this application. Typically, there is a two-way IPSec-enabled transmission between two SAs (one in either direction). Since this is such common usage, IKE is set up to explicitly create SA pairs, which must be of the same mode, defined as follows.

SA Modes

1. **Tunnel Mode**: SA tunnel mode is essentially SA applied to an IP tunnel, which means that an entire packet is protected as it travels from one site of an IP network to another without being screened by any routers (a "tunnel"), to examine any *inner* IP header. What this means, in practice, is that when a packet $\mathcal{P}$ leaves the original host, and gets to the boundary of its hosts's firewall, there is a determination of whether $\mathcal{P}$ needs IPSec processing. If so, it encases $\mathcal{P}$ with an *outer* IP header, and is sent to the target site.

   When $\mathcal{P}$ is enroute, intervening routers screen only the outer IP header, and upon reaching the target site, this outer header is stripped off by the target's firewall and the inner packet is delivered to the target.

   Hosts, shielded by firewalls, may communicate via tunnel mode, without invoking IPSec. This is accomplished via communications where the above-described "unprotected" data packets are sent by SAs in tunnel mode set up by IPSec software in the firewall.

2. **Transport Mode**: SA transport mode is typically used between a pair of hosts for protection of upper-layer protocols and selected IP header

fields. There are two current IPSec environments, IPv4 and IPv6. When a host uses IPv4 with AH or ESP, the IP header is followed by the payload data. With IPv6, the IP headers and the IPv6 extensions are followed by the payload. In transport mode, AH authenticates the IP payload, and selected parts of the IP header, whereas EPS in transport mode encrypts the IP payload, and optionally authenticates it, but not the IP header.

When we referred to the packet $\mathcal{P}$ in part 1, we meant, and will mean throughout our discussion, one of IPv4 or IPv6 data packets. These versions are specified in the document that obsoletes RFC 2401, cited on page 296; as well as in documents being updated, which make obsolete RFC 2402, [207] and RFC 2406, [211]; see *http://www.ietf.org/internet-drafts/draft-ietf-bmwg-ipsec-term-04.txt*, dated August 2004.

SA Parameters

1. Security Parameters Index (SPI)

This bitstring uniquely identifies an SA relative to a security protocol such as AH or ESP. The SPI is located within the AH and ESP headers so that the target site can select the type of SA under which to process the packet.

If the SA is employed for unicast traffic only, then a locally assigned bitstring is sufficient to specify an SA. If multicast traffic is supported by the IPSec implementation, then it MUST[8.10] support multicast SAs. However, in this instance, a sender SHOULD put traffic into different packets to avoid the improper discarding of low-priority packets, which may occur due to the in-built reject-replay mechanism.

2. IP Destination Address

This parameter dictates the target IP address for the SA, and is allowed to be only a unicast address. The target may be an end user, but it may also be a firewall or network system router. Note that an IP address (also known as an *Internet address*) is a unique 32-bit string allotted to a host and used for all communication with that host.

3. Security Protocol Identifier

This parameter stipulates the SA, namely, whether it is an AH or ESP.

In any IP packet,the SA is uniquely identified by the Destination Address in the IPv4 or IPv6 header and by the SPI in the enclosed extension header, one of AH or ESP. In Diagrams 8.15–8.19, we will illustrate only the IPv6 version since it is more extensive than the IPv4 model.

[8.10]Note that, as in Footnote 8.3 on page 288, MUST, MUST NOT, REQUIRES, SHALL, SHALL NOT, SHOULD, SHOULD NOT, MAY, and OPTIONAL are to be interpreted by the document specifications given in [204].

Diagram 8.15 Standard IPv6 Packet

Original IP Header (possibly options)	Extension Headers (if present)	TCP	DATA

Diagram 8.16 IPv6 AH Packet in Transport Mode

Original IP Header	Site-by-Site Destination Routing Info.	AH	Destination Options	TCP	DATA

Diagram 8.17 IPv6 AH Packet in Tunnel Mode

New IP Header	Extension Headers (Optional)	AH	Original IP Header	Extension Headers	TCP	DATA

Diagram 8.18 IPv6 ESP Packet in Transport Mode

Original IP Header (possibly options)	Site-by-Site Destination Routing Info.	ESP Header	Destination Options	$\cdots$

$\cdots$	TCP	DATA	ESP Trailer	ESP Authentication

Diagram 8.19 IPv6 ESP Packet in Tunnel Mode

New IP Header	New Extension Headers	ESP Header	Original IP Header	$\cdots$

$\cdots$	Original Extension Header	TCP	DATA	ESP Trailer	ESP Authentication

An SA in AH mode MUST have associated AH information containing the authentication algorithm; keys; key lifetimes; and any related data necessary for the interoperability of the IPSec implementation. Similarly, in ESP mode an SA MUST have the encryption and authentication algorithm; keys; initialization values; key lifetimes; and any other data essential to the implementation.

There are also means of using layered security protocols via IP tunneling, called *iterated tunneling*. In these cases, the options involve tunnels, each of which can begin and end at any given IPSec site along the route. Both parts of the illustrated configurations of Diagram 8.20 involve the host-to-host tunneling described in the discussion of tunnel mode on page 302.

Diagram 8.20 Iterated Tunneling

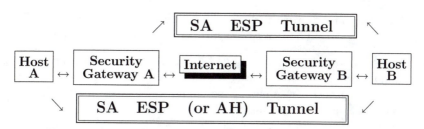

Configuration I: Host-to-Host and Host-to-Gateway

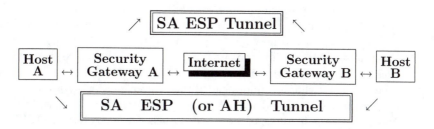

Configuration II: Host-to-Host and Gateway-to-Gateway

In configuration I, the host-to-gateway tunnel allows Host B to reach Host A's security gateway, after which it may gain access to a server behind the gateway. Both Host A and Host B are IPSec enabled to communicate via an SA, via the tunnel through which they are connected.

In configuration II, the gateway-to-gateway tunnel may provide both authentication and confidentiality for all traffic between the two networks. Moreover, since the tunnel is in ESP mode, it also contributes a certain qualified amount of traffic confidentiality. With the SA tunnel between hosts, we have end-to-end security in either configuration.

The next protocol to be discussed in detail is a major component of IPSec, which we mentioned at the outset of this section, the AH. It renders services that protect against attacks levelled at networks, such as *spoofing*, where an adversary creates packets with some other entity's IP address, then exploits those software applications that are based upon IP authentication; as well as replay attacks; and *packet sniffing*, where an attacker reads login and database information.

▼ Authentication Header Protocol

These fields are presented top-down in order (see Diagram 8.21 on page 307).

1. **Next header**: This 8-bit field identifies the header for the higher-level protocol immediately following AH (such as ESP or TCP, for instance).

2. **Payload length**: This 8-bit field contains the length of the AH contents.

3. **Reserved**: This 16-bit field is reserved for future use.

4. **Security parameters index**: This 32-bit field identifies the SA for this packet by specifying a set of security parameters for use herein.

5. **Sequence number**: This 32-bit field supplies a monotonically increasing counter value for each packet sent with a given SPI, used to track the order of packets. Moreover, the counter establishes protection against replay attacks. The means by which this is achieved is as follows.

   When a new SA is created, the sender initializes the sequence to zero, and each time a packet is sent using this SA, the sender increments the counter and the result is put into the sequence number field. Since the default mechanism is enabling *antireplay*, the sender must ensure that the counter has an upper bound of $B = 2^{32} - 1$, beyond which it is forbidden to increment (since otherwise, there will be more than one packet with the same sequence number given that 2^{32} cycles back to zero). Once B is reached, a new SA with a new key is negotiated. On the other hand, the receiver ensures that a window, called the *antireplay window*, is created for received packets, checked via a MAC, which discards unauthorized ones. Moreover, in the instance of a discarded packet, the receiver SHOULD be capable of sending a message with reasons for the dropping of the packet along with date, time, and sequence number of the packet. (See page 311 where details on the antireplay window and associated notions are detailed.)

6. **Authentication data**: This variable length (modulo 32, which might necessitate padding), field consists of the *Integrity Check Value* (ICV) for this packet.

   The ICV is essentially the output of a truncated MAC defined as follows. The compliant implementations are HMAC with SHA-1 or MD5, but also AES with CBC (see [208] for a description of HMAC-MD5-96; [209] for a description of HMAC-SHA-1-96; [224] for a description of AES-XCBC-MAC-96; and see pages 263–264 for a reminder of our description of HMAC in general). Whichever is used, the complete HMAC value is calculated, then truncated using the authentication field default value of 96 bits.

   The ICV is calculated using, (1) certain IP header fields, namely, those whose values are predictable when they arrive at the AH SA, or those that do not change in transit, the latter process called *immutable*; (2) the entire contents of the AH header except for the authentication data field, which is set to zero for calculation purposes at both origin and target; (3) the complete upper-level protocol data, assumed to be immutable in transit, such as an IP packet in tunnel mode (see Diagram 8.20).

Diagram 8.21 Authentication Header Fields

Next Header ← 8-bit →	Payload Length ← 8-bit →	Reserved ← 16-bit →
Security Parameters Index (SPI) ← 32-bit →		
Sequence Number ← 32-bit →		
Authentication Data ← Variable →		

Transport and Tunnel Mode AH

We discussed transport and tunnel modes in general terms earlier, and illustrated AH transport mode (IPv6 implemented) in Diagram 8.16. This case is considered to be an *end-to-end payload* — it is immutable — and remains untouched by routers between its origin and target sites. For this reason, AH appears after the original IP header and routing information. Authentication, in this case, extends to the entire packet, excluding only mutable fields that are set to zero for MAC calculations.

In AH tunnel mode, the entire original IP packet is authenticated, again, except for the mutable fields (see Diagram 8.17). Unlike transport mode, the AH can be used by either hosts or security gateways. In fact, when AH is used in a security gateway, tunnel mode must be employed. Thus, the new IP header may contain addresses for firewalls or other security gateways.

▼ Encapsulating Security Payload (ESP)

ESP Fields

As with AH, we provide a top-down description of the ESP fields followed by a diagram illustrating the same.

1. **Security Parameters Index (SPI)**: This 32-bit field names the specific SA to be used.

2. **Sequence Number**: This 32-bit field is similar to the corresponding AH field, with a monotonically increasing counter that guards against replay attacks. (See the explanation for AH above.)

3. **Payload Data**: This variable-length field consists of the enciphered data for the packet being transmitted, which may be at the transport level (thus, in transport mode), or IP packet (in tunnel mode).

4. **Padding**: This field provides space for up to 255 bytes, which might be necessitated by the enciphering algorithm being used. The suite of compliant encryption algorithms are 3DES-CBC (see [214]), which the

IPSec implementation MUST have, but SHOULD NOT have DES-CBC as specified in [210]; AES-CBC (see [225]), which it SHOULD have; and AES-CTR which it SHOULD have (see page 137). The Padding Field is used, for instance, to expand the plaintext (consisting of the Payload Data, Padding, Pad Length, and Next Header Fields described below), to the desired length when an encryption algorithm specifies, say, some fixed number of bytes.

5. **Pad Length**: This 8-bit field identifies how much of the encrypted payload is padding.

6. **Next Header**: This 8-bit field identifies the type of data carried in the payload data field by identifying the first header in that payload.

   The ESP format dictates that the Pad Length and Next Header Fields be right justified within a 32-bit word (see Diagram 8.22).

7. **Authentication Data**: This variable-length field (modulo 32) contains the value representing the ICV computed over the ESP packet minus the Authentication Data Field.

Diagram 8.22 ESP Header Fields

Security Parameters Index (SPI) ← 32-bit →		
Sequence Number ← 32-bit →		
Payload Data ← Variable →		
Padding ← 0 – 255 Bytes→		
Padding Field (Continued)	Pad Length ← 8-bit →	Next Header ← 8-bit →
Authentication Data ← Variable →		

Transport and Tunnel Mode ESP

We illustrated the general discussion of transport mode with ESP (IPv6 implemented) in Diagram 8.18. In this case, all parts of the packet, except the original IP header and routing information are encrypted. Authentication (which is optional) covers ciphertext and ESP header. The IP header and plaintext IP extension headers are not encrypted since they have to be examined by intermediate routers.

As with AH, either hosts or security gateways may employ tunnel mode ESP. See Diagram 8.19 for an illustrated ESP IPv6 implemented tunnel mode. When ESP is used at a security gateway, again as with AH, tunnel mode must be employed (see Diagram 8.20). In this case, encryption occurs only between external hosts and a security gateway, or between security gateways. In this fashion, hosts do not need to execute any enciphering, so key distribution is made easier, since fewer keys are required. From a security standpoint, this mode is valuable since it thwarts *traffic analysis*.[8.11] Last, as with AH, authentication is optional and essentially covers the same features as with AH.

▼ Combining Security Associations

As observed in our earlier examination of SAs, they are one-way relationships. Yet, we may wish to employ more than one of them, in which case we must set up new SAs for each instance. We already considered and illustrated one mechanism for combining them, namely, iterated tunneling (see Diagram 8.20). Now we look at the other method of combining SAs.

Transport Adjacency

Transport adjacency (see Diagram 8.23) pertains to the mechanism where multiple transport SAs are applied to the same IP packet (without using tunneling SAs). Both AH and ESP IP packets may be combined by this methodology. In this case, the IP packet is processed *only* at its target destination.

The use of either of the methods: iterated tunneling or transport adjacency, is called *security association bundling* (SA bundling).

Diagram 8.23 Transport Adjacency

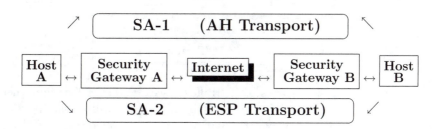

Transport adjacency may also be used to bundle SAs with, for instance, the inner one being an ESP SA and the outer one being an AH SA, thereby applying authentication after encryption. (We refer the reader to pages 266 and 267 for our discussion of the pros and cons concerning the order of encryption vs. authentication.) Enciphering, in this case, is applied to the IP payload, then AH is applied in transport mode so that authentication is the umbrella for ESP and the original IP header, extensions included, but mutable fields excluded. One could simply use a single ESP SA and invoke the authentication option to

---

[8.11]Traffic analysis refers to the scrutiny of frequencies and lengths of enciphered messages, by an adversary, in an effort to guess the nature of the communication being observed. From this an opponent could discover the location and identity of communicating entities.

avoid the costs of using two SAs, but the nesting of two SAs ensures that more fields are authenticated.

If one wants to authenticate first, in a bundling of SAs, then an inner AH transport SA may be formed with an outer ESP tunnel SA. Thus, authentication is applied to the IP payload and the IP header, plus extensions, but minus mutable fields. Then the IP packet is processed by ESP in tunnel mode. The outcome is that the complete (authenticated) inner packet is encrypted and new outer IP header and extensions are added.

Diagram 8.24 Nesting SAs: Authentication After Encryption

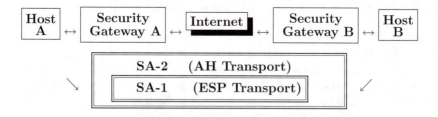

Diagram 8.25 Nesting SAs: Encryption After Authentication

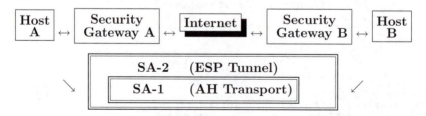

Although numerous aspects of IP traffic processing and IPSec implementation are local matters, and thus not subject to standardization, there are external features of the process that require such systematization in order to guarantee interoperability and render a lower bound on the management capacity that is crucial for effective use of IPSec. In order to accomplish this, we need the following.

▼ IPSec Security Databases

IPSec possesses two formal databases: the *Security Policy Database* (SPD), and the Security Association Database (SAD). SPD prescribes the guidelines that govern the configuration of all incoming and outgoing IP traffic. SAD embodies parameters that are associated with each keyed SA.

Security Association Database (SAD) Parameters

The following parameters are used to define an SA and each SA has an entry in the SAD.

1. **Security Parameter Index (SPI)**: This 32-bit value is the unique identifier of the SA, chosen by the receiver of the SA. If the SAD entry is for an outbound SA, the SPI is used to build the packet's AH or ESP header. In the case of a SAD entry for an inbound SA, the SPI is employed to assign traffic to the suitable SA.

2. **Sequence Number Counter (SNC)**: This is either a 64-bit or a 32-bit value that is employed to generate the Sequence Number Field in either AH or ESP headers. The default is a 64-bit but a 32-bit may be negotiated.

3. **Sequence Counter Overflow**: This is a flag signifying whether overflow of the SNC should generate an auditable event, thereby preventing the sending of any more packets on the SA (but rollovers may be permitted).

4. **Antireplay Window**: This is a 64-bit counter and a bit map used to indicate whether an inbound (AH or ESP) packet is a replay. Accommodation for 32-bit numbers are made, but the default is 64-bit. It is possible for the receiver, in certain situations, to disable antireplay. If so, the Antireplay Window is ignored for this SA.

5. **AH Authentication Algorithm**: These are parameters associated with the use of AH, which include keys and their lifetimes. Of course, this parameter is required only if AH is supported.

6. **ESP Encryption Algorithm**: These are parameters related to the use of ESP such as keys and mode.

7. **ESP Integrity Algorithm**: This involves the keys and other parameters involved in ESP integrity, but if this service is not chosen, these will be null fields.

8. **ESP Combined Mode Algorithm**: In this case, the keys, and so on, as are chosen above, but only if both encryption and integrity are selected to be used with ESP.

9. **Lifetime of this SA**: Typically this parameter is a byte count that specifies the life span of the SA, and upon completion of that duration of use, the SA either (1) must be replaced by a new SA with a new SPI, or (2) terminated. This parameter includes an indication as to which of (1) or (2) should occur. This parameter may also be expressed as a time count, and any compliant implementation MUST support both types of lifetimes, as well as simultaneous use of both. Furthermore, if the packet does not get delivered during the lifetime of the SA, the packet SHOULD be discarded.

10. **IPSec Protocol Mode**: This is a choice of tunnel or transport, which indicates which mode of AH or ESP is applied to traffic on this SA.

There may be other minor parameters added to the above, but the bulk of what is required is contained in this list.

Security Policy Database (SPD) Parameters

IPSec documentation refers to SAs as management constructs used to enforce security policies for traffic crossing an IPSec boundary. Therefore, since the SPD is responsible for the screening of all inbound and outbound traffic (IPSec and *non*-IPSec), it is necessary to have a clear indication of what services are offered and in what manner. To do so SPD needs what are called *selectors*, which are top-level protocol field values. These are defined as follows.

1. **Destination IP Address**: Typically this is a list of IP addresses of those systems sharing the same SA, especially if the IPSec implementation is operating behind a gateway. Yet, this selector will support a single IP address.

2. **Source IP Address**: As with the Destination IP Address, this is usually a list, if there is sharing of the same SA, but this could be a single address in the case of a simple configuration.

3. **Next Layer Protocol (NLP)**: This is obtained from one of the IPv4 Protocol field, or IPv6 Next Header field. Other selectors depend on the NLP value. For instance, if a port such as TCP is used, there are selectors for Source and Destination Ports, each of which is a list of values. If the NLP is a mobile header, there is a selector for IPv6 Mobility Header Message Type, which is an 8-bit value that identifies a specific mobility message. There may be others, depending on the message type.

4. **Name**: This is a symbolic identifier for an IPSEC origin or target address, which may be an X.500 distinguished name or an operating system identifier.

5. **Data Sensitivity Level**: This is an indicator of the security level of the information being transferred, such as classified or unclassified.

There is a third, not often mentioned, database — the *Peer Authorization Database* (PAD) — which is also needed within a secure IPSec architecture. The reason that this database is often ignored in descriptions of IPSec is that the PAD may already be integrated within the SA management protocol itself. Nevertheless, it is important to understand the PAD functions. The PAD establishes a connection between an SA management protocol (such as IKE), and the SPD. Among the PAD duties are defining the range of identities that a peer (one of a set of entities that are in the same protocol layer or the equivalent layer of another system), is authorized to represent when SAs are negotiated with a peer; defining how to authenticate a peer (such as via a certificate); and verification of the authorization of SPD traffic selectors relative to the authorized peer of the SA management protocol. PADs may also be needed to locate secure gateways.

8.4 Internetworking and Security — Firewalls

Things won are done; joy's soul lies in the doing.
William Shakespeare (1564–1616)
— from *Troilus and Cressida* (1602)

IP, or Internet Protocols, provide services for connecting hosts over various disparate networks, as we have seen. To accomplish this, however, each IP must be embedded, not only at each host site and its associated network, but also in routers. This presents challenges for these routers since they connect such dissimilar systems. Here are some of the differences routers face.

◆ Network Dissimilarities

Address Labels: The various schemes for networks to allocate a target address to data in an Internet mechanism may range from 48-bit assignments to encoded decimal representations. Therefore, some kind of universal standardization is needed together with a central archive for record keeping.

Fragmentation: On page 221, we already met the concept of message fragmentation. Fragmentation is required because of network disparities in maximum packet sizes permitted.

Interfaces: A router must be designed to execute its duties irrespective of the disparate hardware and software interfaces among networks.

Network Dependability: A router must be independent of the differences in network reliability, which may range from unreliable to end-to-end dependability.

◆ Firewalls

All the above being said, the primary concern is with local security, so we need firewalls (see Footnote 8.7 on page 295). The term "firewall" is taken from the firefighting profession, wherein a firewall is a barrier constructed to prevent the spread of fire. In the computer world, it means keeping the flames of disaster, ubiquitous on the Internet, away from your local network, and preventing entities from inside the local network from opening a "door" that will let those flames in. A firewall may be defined as a combination of hardware and software, located at the interface between two networks, that enforces an access control security policy between them. For instance, these security gateways[8.12] may screen IP addresses, or ports requested on incoming connections, to decide what traffic is permitted into the local network.

[8.12] A *gateway* is an access point on a network that plays the role of an entrance to another network. For instance, when we discussed SET in Section 6.3, we looked at *payment gateways*. More generally, a *node* on the Internet is a connection point, typically with the capacity to read, process, and forward data to other nodes. Thus, a node may be a computer or other device. For a user at home, an ISP (see page 295), is a gateway. For a business enterprise, a gateway node may play the role of both a proxy server (see Footnote 8.14 on page 317), and firewall.

Origins of Firewalls: Development of firewall architecture has been contemporaneous with the evolution of the Internet. Not surprisingly, initial funding for firewall research was the domain of the U.S. Department of Defense. The origins of the first commercial firewall architecture may be traced to the mid-to-late 1980s with Cisco Systems, who introduced (static) *packet filters*. In the late 1980s and early 1990s the next generation of firewalls, called *circuit level firewalls* came out of research at AT&T Bell Labs. Then the third generation of firewalls came to attention in the early 1990s, out of work from Bell Labs and others, with *application layer firewalls*. A fourth generation, called *dynamic packet filtering firewalls*, sometimes called *stateful inspection*, was epitomized by *Firewall-1*, the first user-friendly firewall architecture, released by Check Point Technologies in 1994. This essentially replaced static packet filtering as a standard. Today there is the fifth generation of firewall, called *Kernel Proxy Architecture*, the first commercial incarnation being Cisco System's *Centri Firewall*, released in 1997. All of the aforementioned types will be discussed below.

▼ **Firewall Design Principals**: If the security goal of a local network that has its own local security policy is to explicitly deny all transmission that fail those criteria, then the following firewall design goals should be sought: (1) all data traffic into and out of the local network must physically be directed through the firewall; and (2) the firewall must be impenetrable.

The local security policy will dictate the level of monitoring, and what traffic will be permitted or denied access. Typically a local network will want a balance between protection of that local system from threats, and access to the Internet.

What A Firewall Can Do

▶ First of all, in general terms, firewalls guard against unauthorized access from outside the protected local network, but allow access from within the local network to the outside. A more intricate firewall scheme will ensure that certain entities within the local network are prevented from accessing certain sensitive documents inside, as well as prevent users from within the local network from sending confidential, sensitive, or vulnerable data outside the firewall.

▶ A firewall provides a single *choke point* where security, audit, tracking (of logins, Internet usage, etc.), and other management functions may be concentrated into a single system. Security alarms can also be set.

▶ A firewall may be employed as a foundation upon which to implement IPSec (see Section 8.3). Some opinions in the cryptographic community even suggest that IPSec usage will replace firewalls altogether, but that remains to be seen. What can be done is to use a firewall to establish VPNs via IPSec employed in firewall-to-firewall tunnel mode (see Diagram 8.20).

▶ Firewalls may also serve the function of *Network Address Translator* (NAT), by which it can alter data in packets to change the network address, which means one set of IP addresses is used for local network traffic and another is used for external traffic. The firewall would have a *NAT box* installed to make all the requisite IP address translations. In this fashion, the firewall hides all local network IP addresses. Moreover, behind the firewall in the local network, the use of a distinct set of IP addresses means there is no conflicting intersection with IP addresses from outside.

What Firewalls Cannot Do

◗ A firewall cannot thwart attacks that go around it. There might be a dial-out server behind the firewall, for instance, that circumvents it by dialing directly to an ISP.

◗ If there are hackers within the local network, the firewall will not detect them. Possibly, an employee of a corporation operates in concert with Mallory outside the corporation to steal vital data by giving him needed passwords. No firewall can prevent this.

◗ A firewall is not an antivirus program. Thus, infected files or programs may get through. A firewall is not the place for virus-control software, since there are simply too many ways for viruses to be sent. It would be virtually impossible for a firewall to filter every piece of data for a possible virus. Furthermore, even if it could be implemented, it would still only guard against viruses from the Internet. There are viruses that come in CDs, via modems, as well as the Internet. A better mechanism is to have antivirus software installed in every individual computer in the local network.

◗ A firewall is only as secure as the operating system (OS) in which it sits. If there are weaknesses in the OS, a firewall cannot protect against them.

▼ **Basic Kinds of Network Firewalls**

1. **Packet Filters — Screening Routers**: A simple firewall configuration is called a *packet filter*, which records the permitted origins and target IP addresses, as well as port number[8.13]. If a packet has an address that is not on its list, it is discarded. Given its simplicity, this type of firewall is both efficient and is transparent to users, as well as being inexpensive to implement. However, this very simplicity makes it vulnerable to such attacks as *network layer address spoofing*.

   Spoofing: In general (not necessarily computer-related) terms, spoofing means assuming another entity's identity. In a computer context, IP spoofing, faking the origin of a message, was an idea tossed around the cryptographic community in the 1980s. It first appeared in reality when there was a problem discovered with the TCP protocol, called *sequence prediction* (see [116] for a discussion of a story related to spoofing and the introduction of the first Internet worm, called the Morris Worm; see also pages 407–409 where we discuss worms in general). Later Bellovin [16], wrote an article discussing the TCP/IP problems. Unfortunately, IP spoofing is a problem intrinsic to the TCP/IP model. Yet there are

---

[8.13]Port numbers are integers ranging from 0 to 65,000, which allow data to be sent directly to a specific device that is "tuned in" to the designated port on a target computer. Port numbers less than 1024 are for use and assignment only by a systems administrator. Typically, a port on a computer is specified by the IP address (of the computer on which the port is active), followed by a colon, and the number of the port, such as 123.214.2.7:60.

measures to be taken as we will see below. First, we look at some spoofing attacks.

In the case of Mallory, say, trying to breach a firewall, he might use a (source) IP address of a local network host in the hope of his packet being delivered by a system that "trusts" the IP addresses of internal hosts. Some examples of IP spoofing are man-in-the-middle attacks (see Footnote 3.7 on page 134). For instance, there is the *routing redirect attack*, where data is redirected from the original host to Mallory's host, say. There is also the *source routing attack* where Mallory redirects individual packets. IP spoofing is used almost always in denial-of-service attacks (see Footnote 8.8 on page 300), wherein Mallory might spoof a source IP address to thwart tracing his steps, and thus stopping the attack is made that much more difficult. These are but a few of many attacks involving spoofing. Incidentally, one misconception about spoofing is that it involves anonymous Internet access, which is not the case.

Diagram 8.26 Simple Firewall: Packet Filter

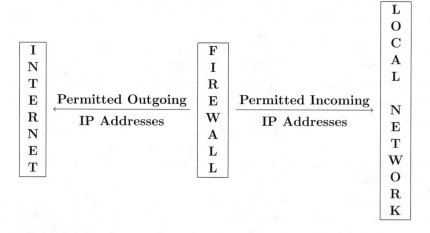

2. **Stateful Inspection Packet Filters — Dynamic Filtering**: Since the aforementioned packet filter firewall bases its decisions on whether the IP address or port number correspond to those listed in the packet filter's configuration, the filtering process is *static*. However, there is a methodology wherein it is possible to incorporate the notion of the *state* of a connection into a packet filter. This is accomplished by using a *state table* and some data in the TCP headers to record those packets previously given access within a connection. In other words, stateful inspection keeps track of an IP packet over a period of time; that is, "remembers" the interaction between the local network and the Internet, say. This makes it possible to thwart unauthorized incoming traffic. This implementation of a packet filter is called *stateful inspection packet filtering*. Packets leaving the local

network that require a particular kind of incoming packet are recorded. Any packet coming into the local network are allowed only if they embody an appropriate response. Whereas static packet filtering essentially only checks headers, dynamic filtering of packets looks at the packet in context, namely, all the way to the application layer. With dynamic filtering, a network administrator is allowed to define the guidelines to satisfy the requirements of the local network.

3. **Application-Level Gateways — Proxy Servers:**[8.14] These types of firewalls are also called *application proxies*, since they require two ingredients, a *proxy server* and a *proxy client*. Suppose that a user, Alice say, in the local network wants to connect to a service on the Internet. Her request, together with her authentication ID, is first sent to the proxy server at the gateway/firewall using a TCP/IP application such as HTTP or FTP. The proxy server, acting in the role of the Internet server, assesses the request, and based upon the local network security policy, allows or denies Alice's wish. If approved the proxy server sends the data, as TCP pieces, to the proxy client, which contacts the actual Internet server. Then connections are established between the Internet server and the proxy client, which relays them to the proxy server for transfer to Alice. Hence Alice's outbound connections are always made to the proxy server, and the Internet's connections are always made with the proxy client. There is never a direct connection between Alice and the Internet server.

Application gateways execute intricate record keeping and audit of traffic passing through them, as well as the traditional access restrictions required of any firewall. These firewalls may be used as NATs (see page 314). The reason is that the data exits the firewall after having been processed by an application, which usually conceals the source address of the data. Thus, the complexity of this type of firewall slows performance and reduces transparency. On the other hand, they are more secure than packet filters, and render thorough audit records. Moreover, since they do not operate at the TCP/IP level, rather at the applications level, they need to screen only a small number of permissible applications.

There are several more advantages to the use of application gateways. They recognize and administer high-level protocols such as HTTP and FTP. At the same time, application gateways present the semblance that they are connecting directly with external servers. They can also be employed *within* the local network to route services to other servers therein.

[8.14]A *server* may be viewed as a program, or computer, that provides services to other programs, or computers. A *proxy server* is a server that acts as a go-between for a user in a business enterprise, say, and the Internet so that enterprise can ensure security and control, as well as possibly caching. A *cache* is a memory location that stores data for quick access. For example, if a user requests a WWW page and the proxy server has a cache with that page already in it, downloaded previously for another user, say, then that page can be forwarded immediately to the next user on request. This saves a great deal of time over the server having to actually request the WWW page from where it really sits on the Internet.

Lastly, these gateways can be used for caching (as described in Footnote 8.14), and may be employed for user authentication.

There are some disadvantages to application gateways such as the fact that the local network cannot run a network server on the firewall server. Also, if a new protocol has to pass through the gateway, a new proxy has to be implemented, which causes inefficiencies. Moreover the complication of the process further reduces efficiency since modifications to configurations often have to be made.

Diagram 8.27 Application Level Gateway/Firewall

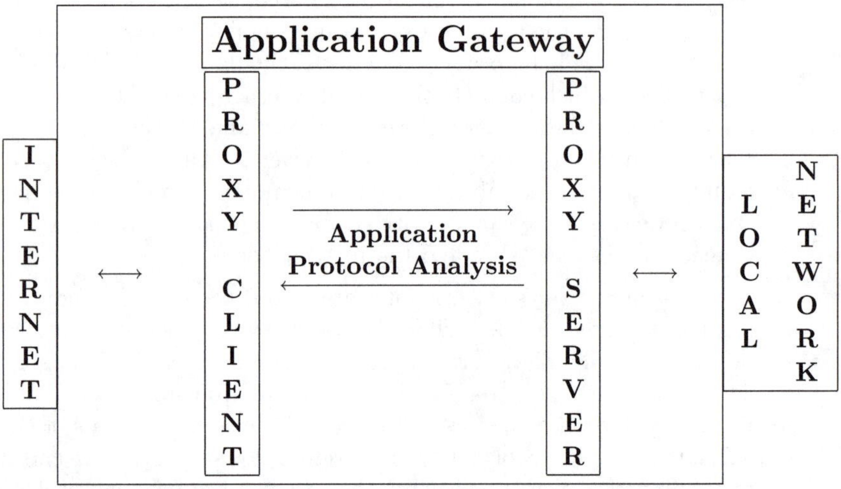

4. **Circuit-Level Gateway**: These firewalls are very fast, but have limited security checks. They are a type of proxy server where a virtual "circuit" is established between the local network and the proxy server, which receives requests, via the circuit, from Alice in the local network; and after changing the IP address, delivers data to the Internet host. Any user outside the local network sees only the IP address of the proxy server, and when it receives a response, it is relayed back through the circuit to Alice.

The security checks are restricted to the firewall's checking of permissions for Alice to send her message to the Internet, based on local security policy, and whether the target Internet host has permission to receive Alice's data. If a connection is established, no further checks are done. Hence, circuit-level gateways are best used when Alice is a trusted local network user.

These gateways transmit TCP connections, such as TELNET, wherein once the connection is established, the firewall forwards data unrestricted.

This makes circuit-level gateways more secure than static packet filters, but less so than application gateways, since there is no applications-level checking. The circuit-level firewall security is essentially the decision pertaining to which connections will be permitted. Whereas the applications-level gateway operates (necessarily) at the applications level, the circuit-level gateway functions at the session level, which explains the means by which the proxy sets a virtual circuit between Alice and the Internet host on a session-by-session basis.

Disadvantages to the circuit-level gateway are that they are restricted to TCP protocol access, they have limited ability to audit events, and they cannot interpret the application protocol being employed.

Now we turn to a circuit-level gateway implementation, which is considered to be an Internet standard firewall. First, we need to expand our understanding of several notions. On page 199, we (informally) defined the term (computer) host to mean those computers that provide services to other computers and to users on a network (such as the Internet). There is more to it. A host has associated with it a *host number*, and coupled with its *network number*, forms its unique IP address (see page 303). The host number is that part of the IP address that determines which computer on the subnetwork[8.15] is being addressed. The Network number is that part of the IP address that designates the specific network to which the host belongs.

The term *IP reachability* is often used synonymously with *Internetworking*, which means any technology and associated mechanisms allowing communications across disparate computer networks. The following firewall has a basic function which is to provide hosts on either side of it to communicate *without* direct IP reachability.

◆ The Socks Firewall/Proxy

SOCKSv5 is an IETF standard (see [194]) known as the *Authenticated Firewall Traversal* (AFT). SOCKS (derived from *SOCKetS*), is a networking proxy protocol allowing hosts on one side of the SOCKS server to access hosts on the other side of the SOCKS server without direct IP reachability. When used as a firewall, SOCKS redirects requests for connections from both sides of the SOCKS server; so acts as a proxy server. The SOCKS protocol makes connection requests, establishes proxy circuits, relays data, and authenticates clients. This is accomplished as follows.

First an application client, Bob say, sends the SOCKSv5 server a request for connection. If the request succeeds, Bob sends a list of authentication

---

[8.15]A *subnetwork* is a set of computer systems under the control of a single administrative domain that uses a specific network-access protocol. Forming subnets, *subnetting*, allows a network supervisor to segment the host part of an IP address into more than one *subnet*, which is interconnected, but independent portion of a network.

schemes that he can support. Then the SOCKSv5 server selects one, or
if none of the methods intersects nontrivially with the network administrator's security policy, no connection is made with Bob. If a method
is available, the SOCKSv5 server sends the choice to Bob, after which
authentication is set up between Bob and the server.[8.16]

Once authenticated, Bob sends his request to the SOCKSv5 server, and
that request must contain the IP address of the application server with
which Bob wishes to connect. Then the SOCKSv5 server evaluates Bob's
request and either rejects it or accepts it. If it is accepted, then using
the address sent by Bob, the server connects to the specified application
server, and establishes a circuit between Bob and it, notifying Bob in the
process. Once established, the circuit conveys data between Bob and the
external server with the SOCKSv5 server screening each fragment of data
and relaying it between the two.

There is an earlier version of the protocol, SOCKSv4, but it had some
issues that were not fully considered or were omitted altogether such as
authentication, which SOCKv5 addresses completely. SOCKSv5 is used
as a firewall, proxy server in VPNs, as well as a single communications protocol that authenticates users and establishes a communications channel.
SOCKSv5 uses the same channel for both authentication and communication establishment, which has a higher degree of integrity guarantees built
into the process. Moreover, it does so without direct IP reachability.

Socks may be configured to work with virtually any application, and it
can set up not only TCP connections, but also UDP connections via a
proxy.[8.17] UDP capacity is another improvement of SOCKSv5 over its
former version. This is a valuable addition since UDP provides a couple
of services not available with TCP. One is an (optional) capacity, called
a *checksum*, meaning a value related to the contents of a packet, sent
with the packet, or stored to detect if the data has been altered during
transmission. The other UDP feature (over TCP) is that it provides port
numbers to help differentiate user requests. SOCKS uses *sockets*[8.18] to
record and track a given connection.

5. **Kernel Proxy Firewall**: The fifth and latest generation of firewall is the

---

[8.16]There are, in fact, two support authentication mechanisms for SOCKv5. They are RFC
1929 [195], and RFC 1961 [196].

[8.17]A *UDP* is a *User Datagram Protocol*, which is a communications protocol providing
service for network communications that use IP. In fact, UDP is an alternative to TCP. UDP
actually transfers what is called a "datagram" from one computer to another. A datagram is
an independent data unit not requiring preprocessing in order to be transported from origin
to target site on the network. Datagram is a term that has been replaced by the word *packet*,
and either term is simply meant to refer to any message unit that the IP handles and the
Internet transfers from one site to another. UDP differs from TCP in that it does not keep
track of the order in which packets arrive at the target site. Thus, since UDP does not process
the sequence of packets, time is saved, so UDP is used over TCP when there is only a small
amount of data to process. Both TCP and UDP are transport layer mechanisms.

[8.18]Think of a socket as one endpoint of an interprocess communication link between two
entities on a network, and each entity establishes their own socket.

most intelligent. It has the capacity to do stateful inspection of network packets at *every* protocol layer of the network stack. It does so via the existence of a proxy within the kernel (core of the firewall), and relays packets on a session-by-session basis using a custom TCP/IP stack. In this fashion, each packet is screened at every layer from the physical to the application and back. Yet, despite this complexity, the filtering can be done efficiently. It accomplishes this via the kernel embodying the full set of available proxies. The kernel stands ready to proxy *any* protocol layer and execute full security checks.

The proxy server examines each incoming packet against a network security policy. If the packet passes this security point, it is checked against existing sessions. If the packet belongs to one, it is relayed to the proxy stack for that session. Each such proxy stack is dynamically built for each session. If the packet does not belong to an existing session, a new proxy stack is created and the packet is relayed to that stack for analysis.

Each of the dynamically created stacks analyze the network packet for those protocols determined by the specific session. Each packet may be discarded at a given layer if it does not meet security standards, or it may be modified at the pertinent protocol proxy. Furthermore, each proxy layer records state information for a given session.

If there are particular requested services, the proxy establishes an application layer extension. This renders the specific services, such as caching, without sacrificing efficiency. If no such additional services are needed, the packet does not go to the applications level.

There is also a *native network stack*, which stands alone without changes and has its own separate security policy allocated to it. Packets may be passed to the native stack after passing security checks/modifications, or the packets may be delivered to other computers, if so destined.

The new firewall architecture marries the need for some of the best possible security with exceptional performance. It still suffers from the failing of all firewalls as outlined on page 315, of course, but is a fantastic stride forward for network security.

There are hybrid systems employing combinations of the above firewalls using what is called a *bastion host*, which is a host that a local network designates as the *only* computer allowed to be accessed directly from the Internet, and used to shield the local network from security breaches. Usually, bastion hosts are stages for either application-level or circuit-level gateways. An example would be what is called a *screened subnet firewall* wherein a packet filter firewall is positioned on either side of the bastion host, thereby creating an isolated subnetwork. Another example is one configured to have both the packet filter and application gateway firewalls positioned on either side of the bastion host. Numerous such configurations are possible. The endgoal is maximum security with minimum time.

8.5   Client–Server Model and Cookies

*Any general statement is like a cheque drawn on a bank. Its value depends
on what there is to meet it.*

                                        Ezra Pound (1885–1972), American poet
                                        — from *The ABC of Reading* (1934)

We have informally discussed client-server model applications throughout
the text such as Kerberos in Section 5.2 and SSL in Section 5.7. Now we look
at the general nature of such models, largely from the perspective of "cookies",
which we will define and study in detail below. The so-called client-server model
is one of the central features of Internetworking. It is time to settle on a general
definition of these terms.

◆ Client-Server Model

A *client*, when considered as part of software, is a computer program (although we may use Alice and/or Bob in these roles), which relies on a server
to perform some operations. (In the client-server model, the term "program"
may be replaced by the term "computer" on which the program runs, sometimes
called a "host computer", but this computer is typically employed for more tasks
than just the client-server architecture). Think of a client as a "requester of services". A *server* in this context, is a computer program that provides access
(for the client) to WWW formats and protocols (or to where HTML documents
are stored). Think of a server as a "provider of services". The *client-server
model* is a relationship between two programs in which one program, the client,
makes a request of the other program, the server, which fulfills the request.

Client-Server Origin and Role: The client-server model was introduced
in the 1980s as message-based modular software, intended for use over a network.
The motivation was to improve functionality, versatility, interoperability, and
scalability over a single mainframe computer with time sharing. It is possible to
configure the client-server architecture so that it operates on a single computer;
in other words, the same machine serves the role of both client and server.
However, the intention for, and full value of, the client-server model is realized
over a network with physically separated client and server machines. This is
because the client-server model was introduced largely to address the limitations
of file-sharing architecture where the server downloaded files from the shared
location to the desktop environment. This type of architecture was strained by
a large number of online users, and large volumes of data. The client-server
architecture, in contrast, was a means by which the file server was replaced by
the database server. By employing a *database management system* (DBMS),
user enquiries could be answered directly, thereby reducing network traffic via
an enquiry and response rather than total file transfer. The term *Intranet* means
the employment of Internet technology for a given organization to implement
client-server applications. To do this, a corporation for example, would merely
have to change its code on an HTTP server, as opposed to updating, the client
code on numerous desktop computers in its organization.

Among the simplest forms of servers are the file servers, whereas among the more advanced servers are the database servers. As mentioned earlier, use of a file server to transfer data over a network slows the process considerably. In the client-server model, the client sends a request to a server, which processes the request on its own power to find the requested data rather than transferring all the information back to the client to find its own data.

Client-Server and HTTP: On page 219, we were introduced to HTTP via protocol layers studied in Section 5.7. Here is how HTTP fits into the client-server model. When Alice opens her WWW *browser* (a client software program used for locating and viewing different types of Internet resources such as data on a WWW site), she indirectly makes use of HTTP. Each WWW server contains an HTTP *daemon*[8.19] (pronounced *dee-muhn*), denoted by *HTTPD*, which is a continuously running program (by itself under the operating system), whose sole purpose is to (wait for and) handle requests that a given computer system receives periodically. Alice's browser is an HTTP client that makes requests to a server by, say, opening a WWW file via the typing in of a *Uniform Resource Locator*[8.20] (URL). By so doing, her browser formulates an HTTP request and sends it to the IP address indicated by the URL. The HTTP daemon at the server site receives the request and sends back a response in the form of requested files. Unfortunately, HTTP is what is known as a *stateless protocol*, which means that each time Alice visits a WWW site (or even when she just clicks to another location from that site), the server sees this as her first visit. In other words, the server forgets all that has transpired after each request unless there is a means to somehow "stamp" Alice so that the server will remember the details of her last visit. The following is a mechanism for accomplishing this task.

◆ **Cookies**[8.21]

What is a cookie and how does it fit into the client-server model? In simplest terms, a cookie is data (for future use) that is stored by a server on the client side of a client-server model. For instance, a cookie might record Alice's preferences when visiting, say, QQQ.com. The cookie is a means by which the server can store its own data about Alice on Alice's own computer.

Analogy: An analogy is a voucher Alice gets when she brings her shoes to a cobbler, Corbett, say. If she returns for her shoes without that voucher, Corbett will not be able to locate her shoes. To him, she could be a new customer. Alice's

---

[8.19]The etymology is from the Greek meaning an attendant supernatural being, on a hierarchy between gods and humans.

[8.20]A *URL* is the acronym for *Uniform Resource Locator*, which is the global address associated with given data. The first part of the URL specifies which protocol to use, and the second part indicates the domain name. For example, *http://www.math.ucalgary.ca/~ramollin/* indicates that this is a WWW page and the HTTP protocol should be used. The second part is the domain name where my homepage is located.

[8.21]The origin of the term "cookie" is uncertain, although its inventor, Netscape, claims it was a name chosen at random. Some claim that it was derived from a similar Unix operating system transaction called a "token". On MAC computers, the cookies are kept in a list called "magic cookie", whereas on IBM CPUs, they are in a file called "cookies.txt".

voucher is necessary for Corbett to maintain record keeping, and it establishes a formal relationship (which we will call a *state*), between him and Alice.

Cookies and HTTP: In Internet terms, a server, when returning an HTTP object to Alice, includes a cookie that has a description of the range of URLs for which that cookie is valid. Any future HTTP requests made by Alice that fall in that range will include the current value of the cookie from Alice sent back to the server. This means that she can shop online and store information about the currently selected items, and it frees Alice from retyping her user ID for each visit. The sites at which she shops can store preferences on her computer, and have Alice supply those preferences every time she visits that site. For instance, the QQQ.com server provides the cookie to Alice's browser, which stores it in its memory as a text file. Each time her browser sends a request to QQQ.com (when she types in its URL for example), the cookie is sent back to the server.

Types of Cookies: There are different types of cookies. For instance, a *session cookie* (or *transient cookie*), is one that is erased when Alice closes her browser, because the session cookie is stored in temporary memory and discarded after the browser is closed. These transient cookies do not obtain information from Alice's computer. Rather they store data in a session ID format, which does not explicitly identify Alice. Another type of cookie is the *persistent cookie* (also called, *permanent* or *stored* cookie), which is a cookie set with an expiration date and is stored on Alice's hard drive[8.22] until it expires (or else Alice, herself, deletes it). Persistent cookies gather information about Alice, including her WWW surfing behaviour or her preferences at, say QQQ.com. The QQQ.com server may use this information to present Alice with a customized welcome page with, say "Hello Alice", the next time she visits.

Alice's browser automatically updates her cookies every time she revisits a site, since once the browser is closed, the cookies are resaved to disk.

Effect of Cookies: In the final analysis, a cookie is simply a piece of text, not a program, and only Alice's browser can store cookies on her hard drive, if it is a persistent cookie. The data is stored in a special file called a cookie list, and is done without the knowledge or consent of Alice. However, it cannot be used for, say, a virus, so it is harmless in that regard. Moreover, the number of cookies allowed for storage on Alice's hard drive is also restricted. Most browsers conform to RFC 2109 (see [203]), which puts a limitation of 300 cookies that may be stored on a given hard drive (with a 4096 byte-per-cookie maximum). This involves a limit of 20 cookies per WWW site, so if 15 sites maximize the cookies on Alice's hard drive, then the next time a cookie is to be set, Alice's browser will discard her least used cookie to free space for the new cookie.

When Alice returns to QQQ.com, her browser will automatically and again, without her knowledge or consent, transmit the cookie containing her personal data to QQQ.com's server.

[8.22] A *hard disk*, also called a *disk drive*, is part of a unit that stores (and provides efficient access to) large blocks of data on one or more electromagnetically charged surfaces.

Cookie Ingredients: Cookies transport between server and client as an HTTP header, and the formal specifics of this header as defined in RFC 2109. There are six parameters that can be assigned to a cookie. The first two are mandatory and are set by pairing them together. The others (set optionally), configured manually or automatically, typically separated by semicolons.

1. **Name**: This is any alphanumeric value (excluding semicolons, commas, and white space), used to identify the cookie.

2. **Value**: This cookie value may be any scalar.

3. **Expiration Date**: This determines the valid lifetime of the cookie and if not explicitly set, defaults to the end of the session, as long as Alice's browser is open.

4. **Path**: This sets the subset of URL paths on a domain for which the cookie is valid. If a path is not specified, the default is the path of the document that created the cookie.

5. **Domain**: This is the textual equivalent of a numerical IP address. When searching a cookie list, a comparison is made between the *tail* of the valid host domain name (such as QQQ.com), and the tail of the cookies on the list. For instance, it might be shopping.QQQ.com, which indeed satisfies the *tail matching* for the domain QQQ.com. Because of this tailmatching, no domain is allowed to set a cookie with fewer than two dots, in order to distinguish among tails such as those containing *.com*, *.ca*, *.gov*, and so on. Thus, for instance, QQQ.com would not be an allowed cookie on the list. Moreover, the server setting the cookie must be a member of that domain. For instance, WWW.QQQ.com cannot set a cookie for the domain WWW.RRR.com, since the security breaches would be severe.

6. **Secure Label**: If this label is set to TRUE, then the cookie may only be sent over a secure channel, typically HTTPS (see page 220). The default is FALSE, since most WWW sites do not need secure connections.

Basically cookies are pieces of textual data generated by a WWW server for storage on a client's computer for future access. Cookies are embedded in HTML information that flows between the client browser and the server. Most often both the storage of, and access to, cookies goes unnoticed by the client. However, any client, concerned about privacy issues can set their computer to notify of any attempt to set a cookie, and will ask permission. Of course, this may become a headache since there will be a lot of "alerts". The crucial issue is for the client to be "aware" of the issues, which this section addresses. Cookies cannot damage your computer or give out private data on you without your giving it out at a WWW site in the first place. The bottom line is that cookies were meant as a mechanism to make it easier for you to access your favorite WWW sites by storing information, so you do not have to login each time you visit, a process impossible before the advent of cookies due to the stateless nature of HTTP.

8.6   History of the Internet and the WWW

Networks aren't made of printed circuits, but of people

Cliff Stoll

— from *The Cuckoo's Egg* (see [273, page 392])

It is appropriate to close this chapter with a look at the origins of the stage upon which the applications (which we have been describing), perform, the Internet and World-Wide Web (WWW), the principal information retrieval scheme for the Internet. The Internet is the system architecture underpinning the globally interconnected network of computers, the set of all the computer networks connected, via routers, all over the globe.[8.23] Now we look at where they began, how they developed, and how the infrastructure evolved.

Computer networks existed as far back as the late 1950s in the form of special-purpose systems (such as the inception of the airline registration system called SABRE). By the early 1960s, time-sharing systems were in use in many leading-edge corporations. These systems allowed multiple employees of the firm to access the computer virtually simultaneously. Such computers came to be known as *hosts*, and with a vision toward a host-to-host network. By 1969, the first implementation of a host-to-host, general-purpose network was put into service, called ARPANET, for the U.S. Department of Defence's *Advanced Research Projects Agency* (ARPA).[8.24] ARPANET supported host-to-host, time-sharing connections in the United States, mainly at government-supported research sites such as universities.[8.25] ARPANET also contained one of the first e-mail manifestations in the form of a protocol called *simple mail transfer protocol* (SMTP).[8.26] From SMTP evolved the *file transfer protocol* (FTP) needed for use with much larger data packages than those typically encountered in an e-mail transmission. In order to process these bigger blocks of data, ARPANET used a new segmenting mechanism called *packet switching*, which broke down large blocks of data into manageable packets for independent dispatch, and later reconstruction at the target site. This new notion for processing of packets via segmentation and reconstruction was one of the earliest means of communication *without* a dedicated channel.[8.27] Although packet networks were created in the private sector in the 1970s, such as Telnet in the U.S.,

---

[8.23]There exist isolated *internets* that are not connected to *the* Internet, but still follow Internet Standards (see RFC 1602, [193]).

[8.24]The RFC series (see Footnote 7.2 on page 262), began in 1969 as part of ARPANET.

[8.25]However, the military people wanted separate communications, so they created *Milnet*, which nevertheless, remained connected and accessible by ARPANET users.

[8.26]The other modern-day e-mail standard is *Post Office Protocol* (POP). When e-mail arrives at an SMTP server, it is forwarded to a POP server where it is stored until accessed by the user who logs on to the POP server with username and password. Then the POP server retrieves the mail and sends it. The newest version, POP3, can be used with or without SMTP.

[8.27]A *dedicated channel* is a channel reserved exclusively for one type of communication. The term is often used to mean a *leased or private line*. On the other hand, a *dedicated server* is a particular computer in a network reserved specifically for the purpose of fulfilling the needs of the network. Typically, however, most servers are not dedicated in today's world, since the computer may be employed to be a server in addition to performing other duties. The antithesis of dedicated is *general purpose*.

these were not host-to-host connections. Instead, they were *virtual* circuits over packet networks. In the 1970s and 1980s, the host-to-host networks remained in government control.

ARPA was replaced by DARPA, the Defence Advanced Research Project Agency, which may be seen as having played a seminal role in the establishing of a mini-version of the Internet via its researchers employing a network for their communications. Essentially, DARPA employed a combination of ground and satellite-based packet networks. This allowed a combination of ground-based radio system transportable access to computing facilities, coupled with a satellite-based connection between the United States and Europe. However, there was no interconnection among the ground-based net, the satellite-based net, and other networks; i.e., the modern-day Internet was not yet born.

By the mid-1970s, the notion of data packets evolved into a scheme called *Transmission Control Protocol* (TCP), allowing interconnected networks all over the globe to transmit and receive data. This new protocol contained a world-wide addressing scheme permitting routers to deliver data packets to their target sites. This new addressing method was called the *Internet Protocol* (IP). By the mid-1980s, the TCP/IP scheme was effectively adopted worldwide.

The National Science Foundation (NSF) in the U.S. played a significant role in establishing TCP/IP as a universal standard. In the mid-1980s, they funded the first five supercomputing centers, and the development of NSFNET, a network to connect these centers. By the late 1980s, a commercial distribution of networks was developed in the private sector called the Commercial Internet Exchange (CIX), since private enterprise was not allowed to use NSFNET for their transactions. However, by 1993, federal legislation allowed NSF to open NSFNET to commerce. As a consequence, in 1995, NSF dropped its support of NSFNET, since they saw the willingness of the private sector to support a communications network on their own. This, as with cryptography discussed earlier, marked the end of government control of the Internet, and permitted the proliferation of private sources to carry the torch. For instance, at the grassroots level, the IETF has developed and maintained standards (see page 219). By the late 1990s, the number of *Internet Service Providers* (ISPs) had mushroomed, and we now have tens of millions of ISP subscribers, with no end in sight.

In 1988, the Corporation for National Research Initiatives provided the first commercial Internet connection linking e-mail, called *MCI* mail. Following this inception, other e-mail providers entered the fray and Internet traffic has never been the same. In September of 1993, the National Center for Supercomputing Applications at the University of Illinois introduced *Mosaic*, which was the first of a new breed of computer programs called a *browser*, which made it easier to access, obtain, and display Internet files. Embedded in Mosaic was a collection of protocols, developed at Centre Européen de Recherche Nucléaire (CERN), for an Internet application called the *World-Wide Web*. From Mosaic Communications Corporation evolved Netscape Communications Corporation, established in April 1994, to develop Mosaic for commercial use. Mosaic was released officially in December of 1994, after which it swiftly became the pre-dominant browser. Later, Microsoft Corporation developed *Internet Explorer*, which was derived from the Mosaic idea. In fact, Mosaic was the first program to produce a multimedia graphical user interface (GUI).[8.28]

In the 1980s, CERN saw a clear and increasing need for researchers, students, and visiting scientists to quickly become conversant with the latest developments in physics and information processing. CERN's project included the use of their hardware and software to implement some elementary browsers for individual users, at their workstations, who incorporated their ideas into the framework.

In March of 1989, the WWW was initiated as an information retrieval system based upon the client-server model. To operate the scheme, the researchers at CERN created a protocol named *HyperText Transfer Protocol* (HTTP) a measure initiated to standardize server-client communications. The WWW browser was officially released in January of 1992, and the acceptance of the WWW was accelerated by the aforementioned creation of Mosaic. The WWW swiftly ascended into the stratosphere in terms of the number of users.

The WWW allows users to access the universe of data all connected to each other via *hypertext* (also called *hypermedia links*) or simply *hyperlinks*, which are electronic interconnections that tie together blocks of data permitting easy access by users. The way this works is that hypertext is essentially an aspect of a computer program permitting a user to choose a word or phrase and obtain more data on it — a definition or related commentary within the text — for example. Mosaic introduced this notion to the WWW to allow users to employ the *point-and-click* option they had on their personal computers for some time. For instance, point at the text "hypertext" at a WWW site, and one might be taken to a document with comments on "hyperlinks". This provides users with instant access, cross-referencing to a large array of linked relevant data pertaining to their target idea. It allows users to access small pieces of data at any given time, digest it, and move on to more data through more links.

A hypertext document and its associated hyperlinks are written in *HyperText Markup Language*, which comes with an assigned URL. The user may contribute to the documents on the WWW by creating their own *homepage* written in HTML, which is a simple, easy-to-learn language. The user merely dictates the structure and content they want on their site, and the detailed presentation and extraction of information is left to the user's browser.

The future of the Internet is unbounded in terms of the features it may offer us, as is the potential for hypertext in the WWW scheme. We are, in this new millennium, on the verge of an explosion of information technology that will rival the changes that the advent of the twentieth century produced. The dawn of the twentieth century did not see the air and ground travel, which we now take for granted, nor was it a computer-dominated world, without which our own would collapse. What will the next century bring for us?

[8.28] A GUI refers to the use of pictures, as well as text, to display the output of a program. This may be presented in the form of icons or buttons, for instance, which a user can control via a mouse-controlled pointer. Although the concept of a GUI was conceived at Xerox's PARC laboratory in the late 1970s, it was Apple with its Macintosh operating system that first employed it in a computer for general use. The term *multimedia* refers to the interaction between computer and user including graphics, text, video, speech, and often hypertext.

Chapter 9

Applications and the Future

The future ain't what it used to be.

Yogi Berra (1925–) American baseball player

9.1 Login and Network Security

When we *login* (sometimes called *signing in*), to a computer, we must provide a passphrase, which may be as simple as a single word (typically called a *password*), or a sequence of words used to identify us uniquely for secure access to the system. The encrypted passphrase will be accompanied by our plaintext *username ID*. A user ID, authenticated by its associated passphrase, determines the privileges allotted to the user, which may vary from personal e-mail access to *superuser* status, where actions may be executed that are protected by the operating system.

If we are trying to login from home, or a hotel when on a trip, to gain access to a computer at work, for instance, this is called *remote login*. In this case, passwords may travel over unsecured channels, making them susceptible to eavesdropping by Eve or interception by Mallory. Mechanisms exist for dealing with these situations. One strong method, IPSec, was studied in Section 8.3. Of course, while workers are at their workplace, firewalls would likely be in place to prevent attacks, and IPSec deals with communications between such security gateways, as we have seen. A secure PKI indirectly assists here since the X.509V3 certificates are part of the IPSec protocol, including user transparency on certain issues. Moreover, we have the strong X.509 authentication protocols studied in Section 7.4, which also employ X.509 directories and other PKI structures. We have methods for secure authentication in e-commerce, such as SET studied in Section 6.3. Secure session-based communication via SSL was explored in Section 5.7. E-Mail security via PGP and S/MIME were described in Chapter 8, and message authentication itself was discussed in Chapter 7. Now we delve further into password protection.

On page 168, we described the use of one-way functions in the role of password security. Also, we have already been introduced to the concept of a "salted" message (see Footnote 3.8 on page 136). Passphrases may be salted, or as we saw with the explanation of SRP (see page 200), a salt and a verifier can be used to eliminate the need for direct password-based schemes.

▼ Why Use Salt?

The purposes behind salting a passphrase are threefold.

1. Eliminating the visibility of duplicate passphrases on a user's file.

2. Increasing the bitlength of the passphrase, to thwart password-guessing.

3. Helping to thwart attacks such as the dictionary attack (see Footnote 5.2 on page 201).

▼ Proactive Password Selection

Since human beings are notoriously lazy about choosing proper passwords, instead selecting easy-to-remember words, and neglecting security, there needs to be a means for ensuring that user-chosen passwords are acceptable. This is where a proactive password checker comes into play. This built-in checker will determine if a user-selected password is acceptable, and reject it if not, prompting the user to try again. System enforcement may contain some of the following criteria.

Passphrase Selection Criteria

Parts 1–4 below refer to the criteria for a proactive checker itself, whereas the remainder are more for a given user to consider when choosing a passphrase.

1. All passphrases must have at least ten symbols.

2. There must be at least three of: lower case letters; upper case letters; numeric; and characters such as !,#,),&,*,❋, and so on.

3. No symbols should be repeated.

4. No actual words should be used.

5. No personal data such as birthdays, or telephone numbers should be used.

6. Memorize the passphrase. Never write it down and do not store it on your computer as a file.

Of course, the above criteria are also known by Mallory, so he knows which passphrases he should *not* try, but if properly implemented, a brute-force attack is made less likely to succeed.

There exist methods for creating effective and efficient passphrase checkers, which do not require lots of space and time as would, say, a list of stored "unacceptable" phrases. There is a Markov model (see [64]), and a Bloom filter model, both of which are probabilistic methods (for details, the interested reader may consult: [267] and [268]).

Attacks on Passwords

We have met some attacks on passwords already in our travels, such as *password sniffing* (see page 199); the *birthday attack* (see page 252); *spoofing* (see page 305); and the *dictionary attack* (see Footnote 5.2 on page 201). Moreover, there is also password-cracking software available. Some of these we saw only in passing so we expand our discussion here.

There are attacks based upon the aforementioned human laziness in choosing passwords. Mallory understands these human weaknesses and exploits them in a group of attack methods known collectively as *social engineering attacks* (see page 394).

There are several attacks Mallory may employ to gain access to sensitive information as follows.

▼ Packet Sniffers

A *packet sniffer* is a program that monitors, captures, and analyzes network traffic, or databases (legitimately or illegitimately). For instance, a database might be (illegitimately) scrutinized by Mallory to detect passwords. If he is successful at gaining access to a system-level password, Mallory can create a new account that can be used at will as a *back door* to get into the network and its resources, including the altering of core system files, such as the password for the system administrator account, the list of server services and permissions, and the login information for other machines, containing critically confidential information. This could create chaos since the daily workings of the network are up for grabs, and Mallory's network packet sniffer can be modified to include his information or change system information in a network packet, forcing network connections to behave erratically, at best.

Packet sniffers can also be used legitimately as follows. A *snoop server* is a server that uses a packet sniffer to capture network traffic for analysis. For example, an employer might want to use a snoop server to monitor the WWW sites visited by their employees.

Snoop servers typically operate in *promiscuous mode*, which is a networking mode allowing a network device (a unit of removable hardware), to access all packets, irrespective of their target addresses. In this manner, a snoop server for instance, can seize any data packet, copy, and store it to a file for later analysis and reporting. For example, the Sun operating system, *Solaris*, has a feature called the *snoop command* permitting administrators to capture packets with an attendant packet description or summary. However, this also permits intruders (running the Solaris OS), to scrutinize the traffic over the network.

In general, a promiscuous mode is used for legitimate monitoring of network activity. This might involve the performance of diagnostic testing to try to resolve such problems as bottlenecks in the flow of traffic, or general troubleshooting to identify a variety of performance problems. Modern sniffers can be configured to automatically alert administrators when a performance problem is triggered by some preset standard, which they set as a local bound.

A packet sniffer can be configured to store copies of packets in memory or hard drive.[9.1] This might be done via temporary storage in a buffer for later analysis. Employers might want to monitor any number of employee activities such as who visits the employee's site; what an employee downloads, including streaming audio and video; contents of incoming and outgoing e-mail messages; which sites the employee visits; and the contents of what they view at a given site. The amount of traffic scanned by a given packet sniffer will depend on the location of the computer in the network. If it is located in a relatively secluded area of the network, then the sniffer will be able to scan only a tiny portion of traffic over the network. However, if it is the principal domain server, for instance, the packet sniffer will be able to scan virtually all of the traffic.

The above being said, Mallory still likes packet sniffers, since if successful, he can use them to seize passwords from data packets traversing the network and wreak havoc as described above. One method of thwarting Mallory is to encipher the headers of packets using SSL in browser-based traffic (see Section 5.7).

Ethernet and Promiscuous Mode

Ethernet (as specified in IEEE[9.2] 802.3), is the most commonly employed *Local Area Network* (LAN). Ethernet evolved from a framework called *Alohanet*, named for the *Palo Alto Research Center Aloha Network*, which was developed into Ethernet by XEROX, then further expanded later by DEC, Intel, and XEROX. There exist Ethernet configurations that provide transmission speeds up to 10 billion bits per second, called *Ten-Gigabit Ethernet*, which is specified in IEEE 802.3a. The future of all interconnections of LANs, WANs, and MANs is generally predicted to be via the Ten-Gigabit Ethernet.

Now that we know the basics of Ethernet, we describe the use of packet sniffers in this context. Ethernet was designed to filter out all data traffic not belonging to it. When a packet sniffer is installed in Ethernet hardware, that filter is turned off and the hardware goes into promiscuous mode. Thus, if Alice

---

[9.1] Although we gave a basic definition of a hard drive in Footnote 8.22 on page 324, we will expand it here to get a better idea of how they function. A hard disk is essentially a collection of stacked disks, each storing data electromagnetically recorded in concentric circles, called *tracks*. Two *heads*, one located on each side of a disk, read or write the information on these tracks as the disk spins. The spin speed is anywhere from 4500 to 7200 rpms. Think of the comparison with a phonograph record and its player having a phonograph arm ("head"), to "read" the music.

[9.2] IEEE, pronounced *I-Triple E*, is the Institute of Electrical and Electronics Engineers Incorporated. The AIEE, American Institute of Electrical Engineers, which was founded in 1884, merged with the IRE, Institute of Radio Engineers, in 1963 to form IEEE. The primary function of IEEE, for our interest, is the development of standards for communications security, the most famous of which are the IEEE 802 standards for LANs and WANs. A LAN is a collection of computers and their attendant mechanisms sharing a common communications channel or wireless linkage, and (usually), a shared server. The common server has applications and data storage, which may be accessed by the LAN users who may vary in number from a couple to several thousand. A *WAN* is a *Wide Area Network*, which differs from a LAN in that it is a geographically more dispersed network, which usually includes shared user networks. In size between a LAN and a WAN is a *MAN* or *Metropolitan Area Network*, typically meaning the interconnection of networks in a city into a single large network. A MAN, of course, provides a more efficient connection to a WAN. For more information on IEEE and its standards, visit *http://www.ieee.org/portal/index.jsp*.

and Bob are communicating over an Ethernet channel with a packet sniffer attached, Mallory can read all the traffic between them. Packet sniffers on an Ethernet consist of the following components.

Packet Sniffer Components

1. **Hardware**: In promiscuous mode, every packet is received and read by a *network adapter*, which is a physical device such as a card (and its software driver) that connects a host computer to network traffic, allowing the host to send and receive packets. A network adapter is sometimes called a *network interface*.

2. **Capture Driver**: This type of driver captures the network traffic and stores it to a buffer, for instance. A *driver* in general, is a program that controls a particular device, such as a printer, or disk drive. Either the driver will come with the operating system or have to be loaded when the device is added. Think of a driver as a translator between the device and the programs using the device.

   A *device driver* is a program that controls a specific device such as a printer. Thus, we may (informally) think of a capture driver as a program that controls the capture of information packets for the packet sniffer.

3. **Buffer**: The captured data from the network are stored in a buffer until they can be analyzed.

4. **Protocol Analyzer**: This aspect of the packet sniffer strips off any encoding and analyzes the data (see Section D.6 on page 541).

The antithesis of promiscuous mode is *nonpromiscuous mode* wherein packets are scanned and passed on if those data packets are not theirs. Only the target site device receives and reads the data in this mode.

Now we return to the issue of login security. We have addressed the issue of password selection and checking, remote logins, and attacks that may obtain passwords. We turn to a modern secure method for password storage.

◆ **Security Tokens**

A *security token* is a special device (a physical object usually ranging in size from that of a housekey to that of a credit card), which a user carries for the purpose of authorized access to a network. For example, the device may be embedded in a *key fob*, which has the physical appearance of a key, but has built-in authentication mechanisms consisting of the following:

1. The user's PIN, authenticating, say Alice, as the fob's owner.

2. A login ID, which is displayed after Alice correctly enters her PIN, allowing her to login to the network.

Token Applications

1. A token may be embedded in a *smart card*, which has the physical appearance of a credit card but has the above authorization mechanisms embedded. (We will study smart cards in detail in Section 9.3.) The login ID is not static, and may actually change every few minutes for security reasons. Thus, if a security token is lost, and Mallory finds it, he cannot access the network without Alice's PIN. Furthermore, an additional security measure against the possibility that Mallory might launch a brute-force attack to recover Alice's PIN, is that the device would be disabled after a small number of attempts to enter the PIN, say, three or four. Hence, security tokens provide one of the foremost, modern, practical methods for the storing of secret keys.

2. Since employees of, say, a corporation, need to insert their security token into their computers for network access, the corporate administrators must guard against human laziness. For instance, a user, such as Alice, might decide to leave her office, to get a coffee, say, and not remove the token from her computer, which is a security risk. To guard against this, the employers may require that the token is needed for access to her office, the coffee machine, the filing cabinets, the department office, the rest room, and so on. In this fashion, the token cannot be left unattended, in any reasonable scenario. This makes such a system foolproof, but not idiot-proof. (An adage is that genius knows its limitations, but stupidity is unbounded.)

We will learn about other security options such as biometrics in Section 9.4. For now, we turn to a remote login protocol that is considered to be the industry standard.

◆ The Secure Shell Remote Login Protocol (SSH)

Although there is an older version, SSH1, we will describe only the newer one, SSH2, which corrects failings of the original, including susceptibility to certain attacks. SSH1 and SSH2 are quite different and are actually incompatible under certain configurations. We describe only SSH2 since it is a complete rewriting of the SSH1 protocol, does not use the same networking implementation, and is more secure. We do point out the advantages SSH2 over SSH1 when that benefit is an overwhelming one. For instance, see the automatic mechanism for host authentication on page 337.

Although the protocols in SSH2 described below may have many differing formats, we do not delve into that detail. Instead, suitable references will be provided for the interested reader. We concentrate upon the description of the main protocols and focus upon SSH2 as a development that is on an approach to becoming the new standard for remote login.

For Internet-Drafts documentation[9.3] see [13], [99], [146], [275], and [293]–[296], as well as the elliptic curve, Diffie-Hellman key exchange proposal for SSH transport level protocols in [271], all of which are RFC 2026-compliant, Internet standards process specifications (see [197]).

What is SSH?

Secure Shell or SSH (sometimes called *Secure Socket Shell* — not be be confused with SSL — see Section 5.7), is essentially a Unix-based[9.4] command interface using PKC-oriented, secure remote login protocols. It allows a user to execute commands on a remote computer, as well as securely move files from one host to another. It provides strong authentication and secure communication over an insecure channel. SSH was designed to replace insecure applications such as Telnet and FTP (see page 326).

Basically, How Does SSH work?

The SSH mode of operation is quite simple on the surface. The host computer first authenticates itself to the client, establishing a unilateral server-to-client secure channel. Then a user, Alice, say, on a client computer, employing unilateral public-key and/or password-based protocols, authenticates herself to the server. Once the link is secure, not only can files between hosts be transported, but also other TCP/IP connections may be forwarded over that secure link. All algorithms used to ensure security are negotiated, so if some algorithm is cryptanalyzed, it is a simple matter to eliminate it and switch to another in the cipher suite.

The following is a detailed description of SSH2 (see [99]), the latest version of the protocol.

[9.3] Although Internet-Drafts (working documents for the development of Internet standards) may be distributed by any working group, the IETF is perhaps the most widely known (see Footnote 5.3 on page 219). These documents have a maximum lifespan of six months, after which they are updated or deleted. If a document becomes an RFC (see Footnote 7.2 on page 262), an announcement is made in the *Internet-Drafts Directories*, see *http://ietf.org/1id-abstracts.html*, typically updated daily, where all current Internet-Drafts may be found. The IETF working group for SSH is denoted by *secsh*, see *http://www.ietf.org/html.charters/secsh-charter.html*. If an Internet-Draft is not part of a working group, it is considered to be an individual submission. For instance, the elliptic curve, Diffie-Hellman proposal, [271], cited above, is one such document, whereas all the others cited above are part of *secsh*. IETF is vendor-neutral, maintaining only the standards. The developer of both versions of SSH is SSH Communications Security. It appears, at this juncture, that SSH2 is on its last lap toward becoming an RFC.

[9.4] Unix (pronounced *you-niks*) originated in 1969 at AT&T Bell Labs to provide an interactive time-sharing system. In 1974, Unix attained the status of the first operating system to be written in the C programming language. Being a nonproprietary operating system, it evolved as freeware and eventually became the first standard operating system that could be openly developed by virtually anybody. We may rightfully view both the client-server model and Unix as vital developments in the evolution of the Internet, with a focus toward computing networks and away from independent computers.

◆ SSH Protocol Architecture

We will assume that Alice is the user on the client computer, and she wishes to establish secure communications with the (remote) host computer.

▼ Overview of SSH Protocols

1. **Transport Layer Protocol**: This protocol provides strong host authentication, confidentiality via strong encryption, and integrity protection from the server to the client computer. This layer also thwarts the man-in-the-middle attack. Moreover, it optionally supports compression. Although there are other possible data streams over which this transport layer may run, we assume that it does so over the canonical one, TCP/IP. The other layers of the SSH protocol run on top of the secure tunnel provided by the transport layer.

2. **User Authentication Protocol**: This protocol runs over the transport layer protocol for the purpose of authenticating Alice to the server. The DSA cipher is used for authentication (see page 183). Once this protocol is completed, there is a mutually authenticated secure channel between Alice and the host.

3. **Connection Protocol**: This protocol runs over the encrypted tunnel established above. It multiplexes[9.5] that tunnel into numerous logical channels that may be used for a rich variety of application-support services, including remote program execution, signal propagation, and connection forwarding.

▼ SSH Protocols in Detail

SSH Transport Layer: The purpose of this layer is to ensure secure communication between Alice, as the client user, and the remote server, as the host. Once Alice contacts the server, key data must be exchanged in order to construct the tunnel. With SSH2, it is mandated that DSS be used (see page 183). The host sends its public key, called the *host key* e_S, as identification. In order for Alice to be certain that she is communicating with the correct server, she must have prior knowledge of e_S, for which two trust models are available.

Trust Models for Host Keys: The first is that Alice has a local database available to her at the client machine. This database associates each host name, which Alice enters, with the matching public host key. This requires no PKI infrastructure, which currently is unavailable to the Internet in any case. However, it is clear that maintaining such a database with matching key names may become onerous.

[9.5] *Multiplexing* means the use of a transmission channel to carry two or more signals at the same time.

The second type of trust model structure is via the use of Trent as a CA (see Section 6.2). Alice only knows Trent's root key,[9.6] but can verify the validity of all certified host keys. This trust model eliminates the storage problem of the first model since only Trent's key needs to be stored at that client machine. However, each host key must be certified by Trent before authorization is possible. Moreover, as noted in the discussion of the first trust model, there currently does not exist a comprehensive PKI for the Internet.

SSH2 Advantage

SSH2 eliminates many of the above concerns by automatically maintaining, checking, and updating public host keys. When Alice logs in to a host server for the very first time, that host's public key is stored to a file in Alice's personal directory. Even if that host's ID changes, SSH2 will warn Alice and disable password authentication to prevent attacks. In this fashion, transparency is added to the session. Furthermore, attacks such as a Trojan horse[9.7] are thwarted by the built-in alerts. As well, man-in-the-middle attacks are thwarted by this automatic mechanism (see Footnote 3.7 on page 134).

In any case, once Alice is assured of the validity of e_S, she may initiate a key exchange connection as part of the transport layer construction of the secure tunnel, as follows.

Key Exchange Protocol

It is mandated in [293] and [294] that the Diffie-Hellman key exchange protocol be used to arrive at key agreement. Here is how it is done.

We assume that p is a large safe prime; α is a primitive root modulo p; h is a hash cryptographic hash function; and that identification data has been exchanged in advance such as both Alice's and the server's ID, I_A and I_S, as well as Alice's and the server's protocol versions V_A and V_S, respectively.

1. Alice generates a random number r with $1 < r < p - 1$, then she calculates $c_A \equiv \alpha^r \pmod{p}$, which she sends to the server.

2. The server generates a random number s with $1 < s < p - 1$, and computes each of the following:

   (a) $c_S \equiv \alpha^s \pmod{p}$.

   (b) $K \equiv c_A^s \pmod{p}$.

   (c) $H_S = h(V_A, V_S, I_A, I_S, e_S, c_A, c_S, K)$.

[9.6]A *root key* is a public key for which the matching private key is held by a *root*, which means an end (ultimate) CA, such as Trent say, who signs the certificates of the CAs below him. As root CA, Trent has a self-signed certificate that contains its own public key.

[9.7]A *Trojan Horse* is a program that appears to have a useful purpose, but has a hidden malicious function. Usually such a program exploits authentication mechanisms of a given system. For instance, a *disk defragger* is a class of Trojan Horse that erases a disk rather than (the intended purpose of) reorganizing it, if it were a legitimate defragmenting program. Another class is that of *fake login programs*, which prompt the user for passwords in order to gain access to accounts. A Trojan Horse differs from a virus in that it does not replicate itself. We will learn in depth about about such mechanisms in Section 10.3.

(d) $D_S(H_S)$, the server's digital signature.

Then the server sends $D_S(H_S)$ to Alice.

3. Alice certifies e_S as described in the above discussion preceding the key exchange protocol. Once done, she computes

$$K \equiv c_S^r \pmod{p}$$

and

$$H_S = h(V_A, V_S, I_A, I_S, e_S, c_A, c_S, K).$$

She may then verify the server's signature $D_S(H_S)$. If this is valid, then she accepts the key K as the shared secret session key, which may now be used for encrypting communication between Alice and the server.

Upon completing construction of the secure tunnel via the transport mode described above, it is Alice's turn to authenticate herself to the server.

Authentication

First, the server informs Alice of the various authentication mechanisms supported. She may choose any of these methods. For instance, the server might send Alice a challenge that she signs with her private PKC key, allowing the server to use her public PKC key to authenticate her.

Once the authentication of Alice has occurred, the server will typically log her into the remote computer and provide her with a shell. Thereafter all communications with her remote shell will be automatically encrypted. It should be noted, however, that the SSH shell forbids login to an insecure FTP server, for instance. The remote host is required to posses SSH-enabled software. There is a mechanism, called SFTP, which is an FTP replacement that runs over an SSH tunnel. However, since OpenSSH supports the SSH SFTP protocol, there is no need to use SFTP.[9.8] In other words, simply use SFTP under the SSH shell supported by OpenSSH.

The server can decide which encryption methods it will support, which may be any of 3DES (see page 131), Blowfish (see page 138), Twofish (see page 142), RC4 (see Section 3.7), or CAST128-CBC (see [223]). Alice may choose the order of authentication from the options given by the server.

Given the secure tunnel provided by the transport layer, the authentication methods do not require the level of security that would be required without

---

[9.8]OpenSSH is a version of SSH available over the Internet, supported by the *Open BSD Project*; see *http://www.openbsd.org/*. It contains not only the SSH program, which replaces *rlogin* (remote login) and *telnet*, but also other features such as *SFTP*. Rlogin is a UNIX command allowing a user to login to other UNIX hosts on a network, and interact as if physically present at the remote host. Rlogin is similar to the better known telnet command. However, both are insecure. The OpenSSH suite replaces not only these two UNIX utilities, but also others such as *ssh-add*, *ssh-keygen* and so on, as well as the sftp-server. Sftp is an interactive file transfer program, which operates over an encrypted SSH tunnel, capable of using many features of SSH.

the channel. Once the transport tunnel and the user authentication with key exchange are completed, Alice and the server can create a new channel. This is accomplished as follows.

Connection

When the above protocols are completed, Alice and the server may negotiate the characteristics of each new channel to allow multiplexing the single connection between Alice and the remote host. Each channel is assigned a different number for both ends, Alice and the host, according to [146] and [295]. When Alice wants to open a new channel, she transmits this channel number along with her request. The host stores this data for the purpose of orienting communications to that specific channel, which allows differing sessions to be unaffected and prevents the main SSH connection from being disrupted. This is required since SSH sends different channels over a common secure tunnel. There is a mechanism for these channels, called *flow control*, which ensures the transmission of data in an ordered fashion; for example, the data will not be sent to Alice, say, until she has already been alerted to the fact that a channel is open for the message transfer.

▼ Analysis

SSH, as with IPSec discussed on page 305, thwarts IP spoofing, as well as *IP source routing* where Mallory, a malevolent host, can fake an IP packet on the pretext of coming from a trusted host, and even DNS spoofing, where Mallory falsifies server records. SSH also protects against any attempts by intervening hosts to intercept plaintext passwords or general manipulation of data. However, certain generic implementations of SSH are insecure (see [142]).

SSH2 supports PGP keys as well as the SOCKS firewall (see Section 8.4). However, SSH does not shelter against an attack when either root access has been compromised (see Footnote 9.6 on page 337), or Alice's home directory say, has been accessed by Mallory. In both of these cases there is no security.

Last, we look at how SSH differs from SSL (see Section 5.7). With SSL, authentication is optional, whereas it is mandatory in SSH. A totally anonymous SSL discussed on page 226 is susceptible to the man-in-the-middle attack, whereas, as we saw above, the SSH protocol has built-in mechanisms to thwart such attacks. Also, it is more unwieldy to use the certificate management necessary via a PKI in SSL, whereas the SSH keys are a relatively simple matter to handle. Moreover, SSH has a wide range of client authentication options whereas with SSL only PKC is an option. Last SSH has many more features implicit in its multiplexing via the connection protocol than does SSL at any level. That being said, the PKI certificate management provides SSL with a scalable key management feature that is absent in SSH2.

9.2   Wireless Security

The cell phone industry isn't really interested in providing security against eavesdropping; it's not worth the trade-offs to them. What they are really interested in providing is security against fraud, because that directly affects the companies' bottom line. Voice privacy is just another attractive feature, as long as it does not affect performance or phone size.

Bruce Schneier

— from *Beyond Fear* (see [240, pages 38 and 39])

Why Wireless? Wireless technology has reached the point where it can reach any place on the world, and this success has resulted in its being employed in the computing world on a wider basis through distributed computing over networks. Hence, people worldwide may access and share information on a global scale. The clear advantage of wireless telephony is the removal of the shackle of wired networks. Medical workers do not need to leave a patient's bedside to check paper records for medical history, or other data. A manager of a storage and delivery facility for a large business enterprise can use wireless scanners connected to the main inventory database in order to track current stocks. University students may access course data on wireless terminals across campus. Car rental agencies can facilitate check-ins for their customers using wireless networks. Corporate business meetings, using wireless, may be set up at a moment's notice, and just as easily dismantled. The limits are only those of the reader's imagination.

Of course, with this freedom comes a price, and that price is privacy and security. Whereas a wired LAN (see Footnote 9.2 on page 332), is protected by physical security and potentially additional cryptographic security, WLANs *Wireless Local Area Networks*, sometimes called *Wireless LANs*, use radio waves, not bound by such walls of security. Hence, different cryptographic methods are required since Eve can listen in on a WLAN with her radio receiver. Moreover, and more seriously, it is equally likely that Mallory can use his transmitter to *write* data to a WLAN. Given this ease of access by adversaries, we need serious means to thwart (active) Mallory attacks, as well as (passive) Eve threats.

◆ **WLANs:** We learned about IEEE's development efforts with LANs, MANs, and WANs in Section 9.1 (see Footnote 9.2), via its committee 802. We now discuss its efforts in more detail, especially as it pertains to WLANs. The 802 committee is segmented into the standards upon which it works via their extension numbers. For instance, 802.3 represents task group 3 of the committee. It focuses on development of Ethernet-based wired networks (see page 332). In fact, the term "Ethernet" is often used in place of 802.3. The 802.11 committee develops standards for WLANs, and this is further subdivided into subfamilies, where 802.11 is the original standard, which is publicly ratified. Basically, the three protocols 802.11a, 802.11b, and 802.11g, are protocols that concentrate upon encoding, whereas 802.11c–f, 802.11h–j, and 802.11n are considered to be service improvements and additions, or corrections to earlier specifications. For example, 802.11a is a standard for WLANs operating in the 5 GHz[9.9] radio frequency range and having a data rate of 9–54 Mbps (see Footnote 3.11 on page 160). This is a successor to the widely accepted standard, 802.11b[9.10], which operates in the 2.4-GHz range. The modulation[9.11] scheme chosen for 802.11b is called *Complementary Code Keying* (CCK) allowing for higher data speeds and less transmission errors. CCK is comprised of 64 eight-bit code words employed to encode data for 5.5 and 11 Mbps in the 2.4-GHz band. Unique properties of the code words permit them to be differentiated from one another by a receiver, even if there is a lot of channel noise. CCK only functions with *Direct-Sequence Spread Spectrum* (DSSS), which is a transmission technology wherein a signal to be sent is combined with a higher data rate bit sequence to make it more robust. This sequence is a redundant bit pattern for each bit sent, thus alteration during transmission is minimized; and if there is such an alteration, the original data may be recovered via the redundancy.

Sometimes the task group and standard are referenced interchangeably in the literature. We will occasionally separate the two. For instance, 802.11i is a proposed standard (see page 345), which we discuss at length, so for convenience and to avoid confusion, we use the well-known notation *TGi* to denote the *task group* for the standard implicit in 802.11i; and similarly *TGn* for 802.11n.

▼ **WLAN Standards**: In June of 2003, 802.11g was given official status as the third standard for WLAN encoding. This standard saw wider adoption than that of 802.11a due to the limited range of 802.11a, and the full backward compatibility of 802.11g with 802.11b, the latter of which had already seen a very wide adoption rate, whereas 802.11a is not backward compatible with 802.11b. Furthermore, 802.11g employs data rates of 54 Mbps. 802.11a/b/g are more than adequate for wireless Internet access and for the sharing of small files. However, when big files come into play, they are slower than most methods. For these larger files and other applications requiring higher bandwidth, we need a new standard. The following not-yet-developed standard is the follow-up to 802.11g.

TGn was created in September of 2003 with a goal of substantially increasing

---

[9.9] A GHz is a unit of frequency equal to 10^9 Hertz. A Hertz is the international unit for measuring frequency, equivalent to the older unit of *cycles per second*.

[9.10] The first widespread commercial use of 802.11b was made by Apple Computer under the name *Airport*.[TM]

[9.11] *Modulation* is the varying of some characteristics (for instance, amplitude, frequency, or phase) of an electrical *carrier wave* in order to embed information in it. For instance, *amplitude modulation* (AM) is that in which the amplitude (magnitude) of a carrier signal is varied to encode it with information; whereas *frequency modulation* (FM) is a method of embedding data onto an alternating-current (AC) wave by varying the instantaneous frequency of the wave, but the amplitude remains the same. An illustration from the "Old West" is a stream of smoke (as carrier) from a fire being modulated by waving a blanket to send a "smoke signal". A *modulator* is a device that superimposes data on a wave. It uses an oscillator to create a wave, which it then combines with the data to create the carrier wave.

After the advent of the first wireless transmitters were put into use in the early twentieth century using radiotelegraphy (Morse code), it was modulation that made it possible to broadcast music and voices, which came to be known as radio. With the modern use of data communications, cell phones and the like, the term "wireless" has come back into vogue.

the bandwidth in 802.11g. The development of 802.11n took a step forward in
August 2004, when it passed the proposal posting stage at the IEEE meeting.
Yet the 802.11n specifications may take anywhere from late 2005 to early 2007
to finalize. Essentially, TGn is in the nascent stage of developing its standard.
They are entertaining proposals from various groups, after which they will decide
what to include and exclude. Below we will delineate a couple of the sets of
proposals by certain consortiums and analyze the potential outcomes. The
802.11n standard will have the potential of increasing WLAN speeds to at least
100 Mbps throughput rates,[9.12] but could be as high as 500 Mbps. These high
bandwidths are very important for consumer applications such as HDTV and
streaming video, as well as significant updates for business environments such as
high-density corporate networks. The push for higher bandwidth is also driven
by real and increasing modern-day needs. For instance, the number of mobile
users is increasing, as are the number of business employees with handheld
devices. Moreover, in their private lives, individuals are increasingly connecting
computers to TVs and audio systems in order to move digital sound and video
from one device to another. Lastly, these applications are becoming increasingly
more complex and demanding of higher bandwidth.

▼ Proposals for 802.11n

The WWiSE Proposal

One of the consortiums laying out proposals for the proposed TGn standard
is called *World Wide Spectrum Efficiency* (WWiSE), consisting 12 companies,
including Airgo Networks, Bermai, Broadcom, Conexant, SIMicroelectronics,
and Texas Instruments; with system members: Mitsubishi and Motorola, the
latter of which uses ECC in their wireless phones (see page 190).

1. **Bandwidth**: There should be compulsory employment of already existing
   (and approved) 20-MHz bandwidth channels; and to maximize data rates,
   these channels should employ four MIMO[9.13] antennas. Optionally, higher
   rates may be achieved with 40-MHz channels using two MIMO antennas.

2. **Interoperability**: There should be obligatory modes accommodating interoperability with devices in the 5-GHz and 2.4-GHz bands.

3. **Number of Channels**: There should be 14 channels in the 5-GHz band.

Perhaps equally matched in their power to influence the outcome of TGn's
standardization process is the consortium (originally founded by Agere Sys-

---

[9.12] A *throughput* is the measure of the capacity of a (digital) network. The rate is provided
as a ratio of bits transmitted per unit of time (such as the 100 Mbps, cited above). The more
popular term for throughput is *bandwidth*.

[9.13] *MIMO* stands for Multiple-Inputs, Multiple-Outputs. Essentially, MIMO is a technique
which increases bandwidth on an individual channel by creating more air paths for data
transmission. By employing multiple antennas, both transmitters and receivers, each such
path can carry differing kinds of data at the same frequency. Currently, only a single set of
antennas is employed in each wireless connection. (Note that once a signal is received, it is
interpreted by a *demodulator* that separates the data from the carrier wave, and translates
the information back into its original form.)

tems), *nSynch*, backed by Atheros, Cisco, Intel, Nortel, and Philips; with system members Matsushita, Nokia, Samsung, Sony, and Toshiba.

1. **Bandwidth**: There should be mandatory 10-, 20-, and 40- MHz bandwidth channels. Moreover, there should be mandatory two-antenna MIMO in the 40-MHz channels, and optional four-antenna MIMO in the 20-MHz channels.

2. **Interoperability**: There should be mandatory modes for interoperability with devices in the 5-GHz band, and optionally in the 2.4-GHz band.

3. **Number of Channels**: There should be twenty-four channels in the 5-GHz band.

Analysis: Basically, the general consensus is that the difference between the two aforementioned proposals amounts to 2×40 vs. 4×20. Do we want two MIMO antennas in the 40-MGz channels, or four MIMO antennas in the 20-MGz channels? The former will produce raw throughput of 250 Mbps (although usable bandwidth would be approximately 175 Mbps), but will be more difficult to implement in some areas. The latter will produce raw bandwidth of 216 Mbps (with usable bandwidth of approximately 162 Mbps), but would be easier to implement and face less regulatory barriers. The arguments against the former include that it would reduce the number of available 802.11 channels; while the arguments against the latter include that it would be unnecessarily complicated, expensive, and unsuitable for mobile devices. From a strictly mathematical viewpoint, we see that $2 \times 40 = 80 = 4 \times 20$, so perhaps both proposals are really aiming at the same thing. In the final analysis, a compromise may be the solution.

Whatever the outcome of 802.11n, we will be seeing *Ultra WideBand* (UWB) in the not-too-distant future. This new TGn standard should result in 20 times the speeds we currently have at our disposal.

▼ **Wi-Fi**: The generic term for any type of 802.11 network is *Wi-Fi* or *Wireless Fidelity* for high-frequency WLANs. The *Wi-Fi Alliance* was formed in 1999 as a nonprofit international organization to certify interoperability between WLAN products based on 802.11. Initially, Wi-Fi was used only in reference to the 802.11b standard (ratified in 1999), but Wi-Fi extended the use of the term in a deliberate effort to stem the tide of confusion over WLAN interoperability. However, 802.11b operates with virtually no privacy despite the fact that it supports *Wired Equivalent Privacy* (WEP), which was created with the aim of making a WLAN as secure as a wired network.

▼ **WEP**: The WEP protocol is totally insufficient as a sole means of security for a WLAN. Part of the problem is that the use of RC4 (see Section 3.7), in WEP is flawed partly because they reuse the encryption key.[9.14] Given that

---

[9.14]Do not gather from the above that there is a problem with RC4. It is the implementation of it in WEP, which is problematic. As we have seen previously, the strongest of ciphers is irrelevant if the implementation is done in an improper fashion that renders its use insecure.

it operates only at the physical and data-link protocol layers, it does not offer end-to-end security. In fact, several analyses of WEP have been done over the years and the consensus was that without significant changes, WEP would not provide a sufficient level of security for WLANs, which led to 802.11i. The following is a description of how WEP operates.

In what follows, Alice wishes to send a message m to Bob.

1. **Creating Plaintext**: Alice applies the cyclic redundancy code, CRC-32 checksum (see Appendix D on page 541), which we will denote by ICV, to m to get $ICV(m)$. The plaintext is $P = (m, ICV(m))$.

2. **Creating Ciphertext**: Using RC4, Alice generates an initialization vector v and a secret key k, which we denote by $R_{(k,v)}$. Then the ciphertext $C = P \oplus R_{(k,v)}$ is formed by addition modulo 2, and she sends $\mathcal{C} = (v, C)$ to Bob.

3. **Deciphering and Comparing**: Upon receipt of $\mathcal{C}$, Bob regenerates $R_{(k,v)}$ and forms $P' = C \oplus R_{(k,v)}$. Then he separates P' into m' and c'. Bob computes $ICV(m')$ and compares it to c'. If they match, he knows Alice's message was not altered in transit, since then, $m' = m$.

The problem with the above is that if two messages are enciphered with the same v and the same k, then Mallory can cryptanalyze if he knows one of the two plaintexts. Even if he does not know one of them, he can mount a dictionary attack (see Footnote 5.2 on page 201). Moreover, this is a near triviality for Mallory since the attack is not dependent on the length of k, but rather on the length of v, which WEP mandates to be only 24-bits! Even though the WEP protocol recommends that v, and thus both v and k, be changed after every use, it does not *require* that this be done. Hence, most implementations do default to a reuse of the key pair, which is significantly insecure behaviour. Also, there is no key management protocol in WEP, another security issue.

Cryptanalysis of WEP has been given early and rigorous scrutiny. For instance, see [8], [38], [95], and [274]. The TGi is working on enhanced security for WLANs and is fixing the security in 802.11. These, and other studies, conclude that the following are problems with WEP. We maintain the notation introduced in the above description of WEP.

▼ WEP Design Problems

1. The bitlength of 24 for v, the initialization vector, is insufficient to thwart attacks on confidentiality.

2. The CRC checksum, called the *Integrity Check Value* (ICV), which is employed by WEP for safeguarding integrity, is insecure. It does not thwart attacks where packets can be modified in transit.

3. The mechanism for combining v with k invites attacks where Eve may recover k after the scrutinizing of a mere few million (a relatively small number of) enciphered packets.

4. There is no protection for integrity of the source and target addresses.

Diagram 9.1 WEP Encryption

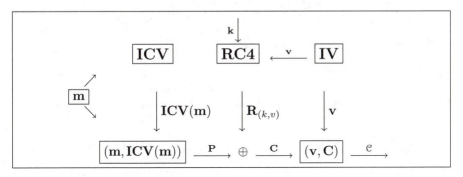

Diagram 9.2 WEP Decryption

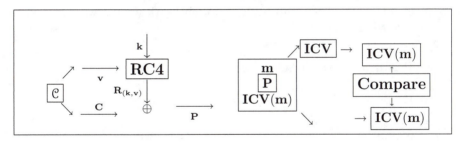

▼ WEP Replacement

The TGi proposed standard, 802.11i, was ratified by IEEE in June of 2004. Before the long-awaited ratification by the TGi, there was a transitional specification called *Wi-Fi Protected Access* (WPA) which was adopted by the Wi-Fi Alliance in November of 2002, largely to respond to the impatience over the long gestation period for the ratification of 802.11i. It was designed to be deployed as a software upgrade to existing WLAN hardware. However, WPA is not an 802.11 standard. Rather, it complements and is based on 802.11i, a strong element of which is the *Robust Security Network* (RSN). Since WPA is essentially a subset of RSN, we look at it first.

Summary of RSN Features

1. AES is employed with a 128-bit key, but supports key lengths up to 256 bits. Although RSN does *not* introduce new hardware, it will be affected by the fact that any RSN-compliant device will be required to support

AES. This, in itself, may mean that no actual upgrading of firmware[9.15] will suffice, that is, the complexity of RSN may mean that it will not be interoperable with anything but the very latest in WLAN hardware. Also, this will mean that certain (AP)s[9.16] will not be RSN-capable (as defined in the IEEE publication, "Wireless LAN Security and IEEE 802.11i" by Chen, Jiang, and Liu — see: *http://wire.cs.nthu.edu.tw/wire1x/*.

2. It uses a 48-bit initialization vector, v.

3. Integrity is achieved via the CBC-MAC in counter mode (see page 262). Since AES replaces RC4, and the counter mode (see page 136) of CBC-MAC is the method of applying it, then this is more suitable for the packet realm than stream data.

4. Since the sequence for v changes as keys change, replay attacks are thwarted.

5. Key management is based upon *Extensible Authentication Protocol* (EAP).

Now we look at the EAP in detail. This protocol is defined in [33] (a document that replaces the one formerly known as RFC 2284bis, and which renders obsolete RFC 2284). RSN employs EAP for authentication of wireless devices to a network, and for provision of dynamic keys as needed. EAP supports numerous authentication schemes.

EAP Authentication Schemes

1. MD5 (see page 255).

2. TLS (see page 219).

3. TTLS, sometimes called *EAP-TLS*, developed by Microsoft. This was accepted as RFC 2716 (see [221]). TTLS, a challenge-response protocol, requires only server-side certificates, and these are used for one-way TLS authentication (network to user). Once a secure channel is established, EAP may be used inside of the TLS tunnel for any other authentication.

4. LEAP, *Lightweight Extensible Authentication Protocol*, developed by Cisco, but they are replacing it (eventually) with the following.

5. PEAP, *Protected Extensible Authentication Protocol*, or Protected EAP, which was developed by Cisco and RSA Security. This is a rival challenge-response protocol to that of TTLS.

[9.15]Firmware typically refers to (permanently stored) software embedded in a hardware device. For instance, firmware may be a program embedded in a ROM-integrated chip. ROM is *Read Only Memory*, memory that may be accessed and read but not altered. The term *Random Access Memory* (RAM) refers to memory space that is basically used to store dynamic data (data that changes during execution of a program).

[9.16]An *access point* is a (base) station that transmits and receives data, and whose function is to both interconnect users on the network as well as interface the WLAN with a wired network. Sometimes, in the WLAN context, an AP is called a *transceiver*.

6. SecurID, developed by RSA Security (see Footnote 9.20 on page 348).

7. SIM, *Subscriber Identity Module*, a card that interfaces with GSM[9.17] technology.

8. AKA, *Authentication and Key Agreement*, is an INTERNET-DRAFT (see [9]), which is based on per-station shared secrets.

EAP was initially intended as an extension of PPP, which is *Point-to-Point Protocol* that provides a mechanism for connecting a computer to the Internet. PPP operates at the data-link layer by transmitting TCP/IP packets to a server, which then places them onto the appropriate Internet site.

EAP is a port-based network access control mechanism that must establish authentication before any port access is allowed. The reason EAP is called "extensible" is that more types of authentication can be introduced in the future, and this may be accomplished without compromising the protocol's specification.

In EAP authentication, a *Master Key* (MK) is produced between Alice and the server. From the MK the authentication server creates the *Pairwise Master Key* (PMK), which binds Alice to the AP for that particular connection, so is given to the AP for that session. The authentication server makes a fresh PMK for every such connection. Other transient keys are created from the PMK, including the *Temporal Key* (TK). TK is the actual device for securing data traffic. When the connection is dropped/terminated, the PMK is discarded.

WPA — The Interim Solution

Due to the key recovery attack on WEP, which became increasingly easier on the Internet, there was a call for an interim solution, out of which came WPA. As noted earlier, WPA is a subset of RSN. It is designed so that only software or firmware upgrades are required to existing WLANs running WEP (by merely running it as a security layer over WEP, namely by running WEP as a sub-component), allowing current WEP hardware to remain unaltered, and with minimal performance degeneration by the fixes it imposes.

Authentication for WPA is essentially done through the above-described EAP process. The mandatory protocols for WPA include RADIUS,[9.18] EAP, and one called 802.1X, whose principal purpose is to control access at a juncture where a client joins a network. Originally designed for wired LANs, 802.1X's objective is to control port access by using the AP as the analogue of a wired

---

[9.17]Originally, *Groupe Spécial Mobile* (GSM) developed in Europe in the early 1990s as a standard for mobile phones. (It is now called, *Global System for Mobile Communications*.) It was the first WLAN architecture to provide user authentication, confidentiality, and key agreement. This is a standard for digital cellular communications, currently used in the 900-MHz and 1800-MHz bands.

[9.18]This is *Remote Authentication Dial In User Service*, defined in [205], which is client-server protocol software allowing remote-access servers to connect with a central server for the purpose of user authentication, for access to whatever system is requested. However, RSN, being a superset of WPA, does not *require* RADIUS for the authentication server that permits more flexibility for implementation.

network switch. This permits the AP to act in the role of a switch since each connection request may be regarded as an *un*authenticated connection until further approval by the authentication server (so upon approval, the AP may "switch on" the connection). Thus, 802.1X may be considered to be a standard for port-based network access control that resides between an authentication protocol and a LAN. Yet, in itself, it is *not* an authentication protocol. That choice (of authentication algorithm, and associated key management) is left to the particular EAP authentication type (one from the list on pages 346–347).

In Diagram 9.3, we see the 802.1X protocol running between the client and the AP for the authentication and key exchange operation. The AP is the link between the client using 802.1X protocol and the RADIUS server running over IP. Thus, the authentication phase is executed previous to the establishment of an IP connection between the client and the network, and *exclusively* 802.1X traffic is permitted and *solely* to the RADIUS server. Once authentication succeeds, the AP switches the client to a network connection.

Diagram 9.3 802.1X Protocol Illustration

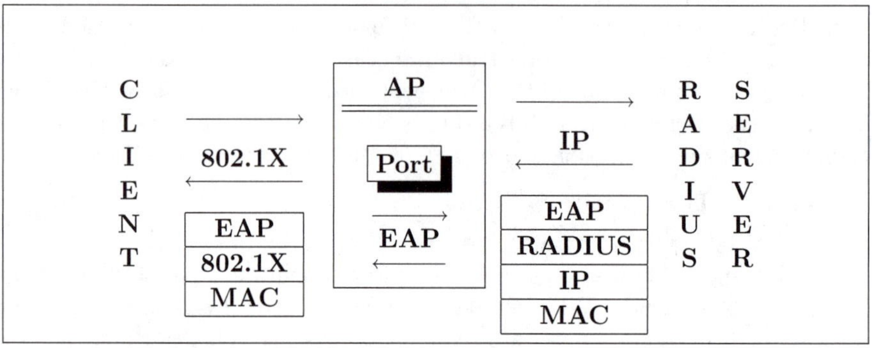

802.1X-EAP Authentication Process

The following is an example of how a common mode of operation for 802.1X would operate with EAP. We assume that Alice is the client (sometimes called the *supplicant*), who wishes to connect to a WLAN. The negotiation takes place among Alice, the AP as intermediary, and the authentication server.

1. Alice requests a connection to a WLAN via the AP.

2. The AP requests ID from Alice, and once received, it forwards this ID to an authentication server, such as *RADIUS*.

3. The authentication server sends a challenge,[9.19] such as a token password scheme,[9.20] for Alice to prove herself, and may send ID to prove itself to

---

[9.19]See pages 197 and 198 for a description of challenge-response protocols.

[9.20]For instance, RSA Security Inc. worked with Microsoft to introduce tokens called SecurID for Microsoft Windows, which provide users with a temporary password every 60 seconds. This "token password" is used with a secret PIN to logon to Windows.

Alice if mutual authentication is being employed; and only *strong* mutual authentication is recommended for WLANs. This message interchange will vary depending on the authentication scheme employed.

4. Alice verifies the server's ID, if mutual authentication is being used, then sends her response to the ID challenge via the AP to the server.

5. The server either accepts or rejects the request by Alice.

6. If her request is accepted, the AP opens a port for her network access.

In the absence of an external authentication server, WPA is capable of executing what is called a *Pre-Shared Key* mode (PSK) to verify ID, for Alice say, both at her client station and the AP. This is accomplished via a password, or some other ID, and she may gain access only if her password matches the AP's password. This password also supplies the material for use by TKIP *Temporal Key Integrity Protocol* (TKIP), which is part of the standard drafted by the TGi, to generate an encryption key for each data packet.

Temporal Key Integrity Protocol

TKIP (also called WEP2), is a collection of algorithms to wrap around WEP in order to patch the security holes, especially the use of static keys in WEP. With WEP, encryption is optional, whereas with TKIP, it is mandatory. Indeed, TKIP replaces WEP with a stronger encryption scheme using computing power in existing wireless devices to execute the required operations. The following are some features of TKIP.

1. **MIC**: This is a *Message Integrity Code*, (MIC)[9.21] employed to thwart forgeries; its code name is *Michael*.[9.22] The 8-byte MIC is placed after the data portion of the TKIP frame and before the 4-byte ICV (Integrity Check Value). (This fixes a WEP problem where Mallory can tamper with the ICV before it is received, even though WEP enciphers it beforehand.) The data, MIC, and ICV portion of the frame are WEP-encrypted. Michael computes a keyed function of the message at the transmission site (as described below), sends the resulting "tag" value together with the message to the receiver. There the tag value is recalculated and compared with the sent value. If the tags match, the message is accepted as authentic. Otherwise it is rejected as a forgery.

[9.21]We know this as a MAC (see page 136), but TGi has already used MAC to mean *Media Access Control*, so we will conform with their acronym here, even though it conflicts directly with the term we used as MIC on page 260, where we used it in reference to an *un*-keyed hash function. With this warning of alternate usage for this discussion *only*, there should be no confusion.

[9.22]MIC was created by Niels Ferguson (see [82]), who is a cryptographic engineer and consultant. His expertise lies in the design of cryptographic algorithms, protocols, and security infrastructures, especially on a large scale. He worked with Bruce Schneier at Counterpane Systems (see page 138), and coauthored a recent book [85] with him.

Also with Michael both the source address and destination address are protected, whereas in WEP there is no such protection. Moreover, Michael essentially enforces the packet sequencing. This is because Michael applies to whole packets, *Media Access Control Service Data Units* (MSDU)s, which includes the MSDU source address (SA), the MSDU destination address (DA), and the MSDU plaintext data.

2. **Packet Sequencing**: To thwart replay attacks, TKIP mandates that the same IV value of 48 bits is never used more than once, and a sequencing mechanism is in place so that there is a discarding of any packet received with an IV value no bigger than the last packet that was received and processed successfully. If the IV were to reach its maximum value, all data traffic would halt.

3. **Per-Packet Key-Mixing Function**: To prevent the recovering of the WEP key (a design problem with WEP, listed as item 3 on page 344), as an automatic feature, a fresh, unique encryption key is generated for each client. Since this is done at periodic intervals, it avoids the insecurities inherent in WEP where the same key may be in use for several weeks.

We now look at the TKIP features in more detail.[9.23]

Message Integrity Code

Background Assumptions: Michael inputs a 64-bit Michael key **MK**, where **MK** is represented as two 32-bit little-Endian words,[9.24] **MK** $= (K_0, K_1)$, and inputs the message m. Michael processes the message by padding it so that its bitlength is congruent to 0 modulo 32. Second, it segments m into a sequence of 32-bit words, $m_1, m_2, \ldots, m_n$. Then it executes the following to compute the tag from the key and the message. First, set $i = 1$, $L = K_0$, $R = K_1$, and let f be a function (that we will not describe explicitly), constructed from shifts, byte swaps, and additions. As usual, $\oplus$ denotes addition modulo 2.

MIC Tag Creation

1. Replace L by $L \oplus M_i$.

2. Replace (L, R) by $f(L, R)$.

3. Replace i by $i + 1$.

4. If $i < n$, go to step 1. Otherwise, output $T = (L, R)$ as the MIC tag.

[9.23]In view of the above discussion, WPA is often written in the form of the following formula: WPA=802.1X+EAP+TKIP+MIC.

[9.24]The term *Endian* refers to the different means of ordering bytes for storage as representation of values. Big Endia means the ordering of bytes in a word such that the most significant digits (or bytes) are positioned on the left. Little Endia refers to the placing of the least significant digits on the left.

Packet Sequencing

The classical method for thwarting replay attacks is to bind a packet sequence number space with a MIC key, and reinitialize the sequence space each time the MIC key is replaced. TKIP does not stray far from the classical paradigm. TKIP employs a 48-bit sequence number, which it binds to the TKIP encryption key (rather than the MIC key). Then, TKIP mixes the sequence number into the key and enciphers the MIC and WEP IV via the following.

Per-Packet Key Mixing Function

As we discussed earlier, in the WEP protocol, the encryption key is vulnerable to attack due to the weak 24-bit IV, among other factors. TKIP fixes this with a mixing function that inputs a 128-bit temporal key **TK**, the 48-bit packet sequence number, **SEQ**, and the transmitter address, **TA**, then outputs a fresh per-packet 128-bit key, called a WEP seed key. The mixing stage is broken down into two phases in order to save on computing time.

The first phase inputs the **TK**, the **TA**, and the first four most significant bytes, **msb**, of **SEQ** to an S-box that outputs an intermediate key **IK**.

In the second phase, **IK** is mixed with the least two significant bytes, **lsb**, of **SEQ** to output the per-packet key, **PPK**. The end result is that a different key is used for each packet that is sent.

Diagram 9.4 Per-Packet Key Mixing

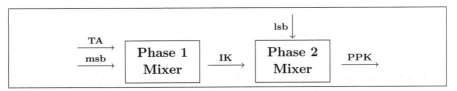

Once TKIP has processed the data and produced the MIC, together with the plaintext MSDU, TKIP appends the MIC to the data field. Then the 802.11 implementation fragments the MSDU into *Media Access Control Protocol Data Packets* (MPDU)s, required for WEP encryption. Once this has been done, each fragment is given a packet sequence number to establish a per-packet encryption key for each such fragment. This is all summarized in Diagram 9.5 on page 352.

Summary

TKIP was meant only for short-term security until the standard RSN became a fact. As a wrapper around WEP it did alleviate some the the problems with the original WEP design, such as removing weak key attacks and thwarting the redirection of packets to unauthorized sites (via Michael's protection of SAs and DAs). However, this comes sometimes at a performance cost, such as the additional key mixing time and rapid rekeying rate, which arises from reuse of WEP packets and IV spaces. Basically, it is a trade-off between security and acceptable performance characteristics. WEP met virtually none of its security goals, and TKIP addressed these problem in the short term. RSN provides the more robust solution.

Diagram 9.5 TKIP Encapsulation

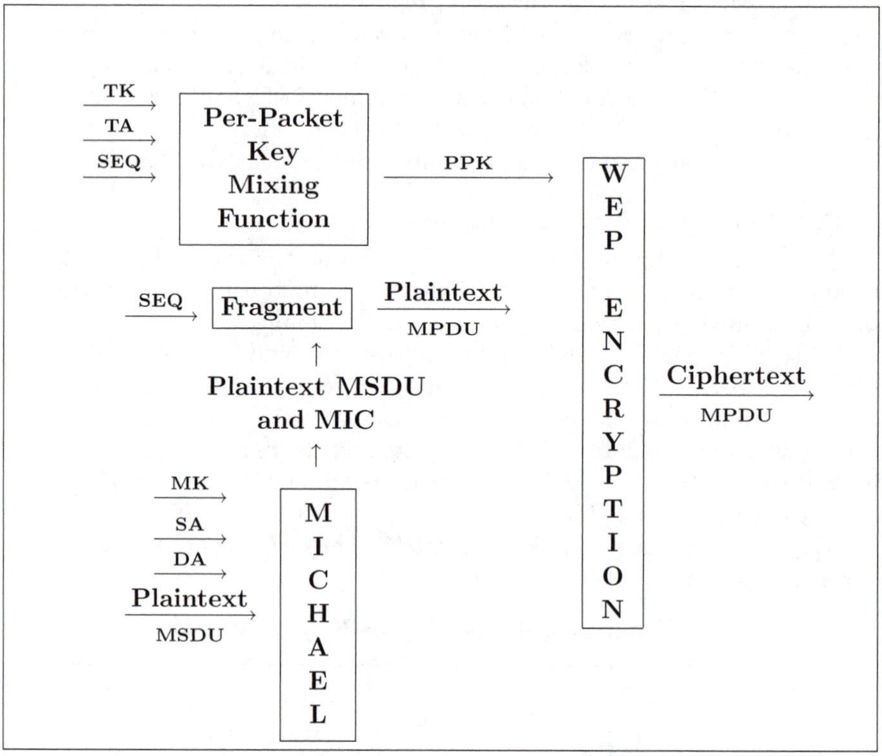

▼ **Conclusions**

TGI's ratification of 802.11i takes the above-described WPA a giant stride forward, and is now often called *WPA2*. Since RSN uses AES, encryption strength is vastly increased. However, some existing hardware cannot simply be upgraded as was possible with the transition to WPA. In other words, some of the older hardware simply has to be replaced, as we mentioned in the context of the RSN summary in part 1 on page 345. However, now that ratification of 802.11i is a fact, we will see the distribution of AES-compliant equipment.

WPA2 advances in another important fashion since it enciphers the entire data frame, whereas WPA encrypts only the payload. That said, coordinating the inter- and backward-compatibility of the system at large is clearly still a challenge to be met. Thus, although the TGi is now disbanded, there is still work to be done. For instance, we await the resolution of the standard from TGn. There also exists the TGr group that works on 802.11r, for *fast hand-off* in those situations where a wireless client is moving, on the same WLAN, from one AP to another. Another is the TGs group working on codifying requirements for standardization of self-configuring mesh networks.

Comparisons

Table 9.1

	WEP	TKIP	RSN
Cipher	RC4	RC4	AES
Key Size	40 or 104 bits	encrypt: 128 authenticate: 64	128 bits
Packet Key	Concatenation	Mixing function	Not Required
Integrity	CRC – 32	Michael	CCM
Replay Protection	None	Use IV	Use IV
Key Management	None	EAP-based	EAP-based

802.11 — Summary

Table 9.2

IEEE	WLAN Applications
802.11	– The *legacy standard* – Provides 1–2 Mbps transmission in 2.4 GHz band, typically using DSSS (see page 341).
802.11a	– Extends 802.11 providing up to 54 Mbps in 5-GHz band – Uses *Orthogonal Frequency Division Multiplexing*, (OFDM), which is a means of sending large quantities of data by radio wave. OFDM operates by separating the signals into smaller ones, sent simultaneously at differing frequencies. – 802.11a is, however, not interoperable with the following
802.11b	– Provides 11 Mbps in the 2.4-GHz band using DSSS – Was ratified in 1999 as the 802.11 standard – Provides functionality comparable to Ethernet
802.11g	– Supplies 20 Mbps in the 24-GHz band, and up to 54 Mbps when operating with 802.11b hardware
802.11i	– The TGi has ratified the standard, upgrading Wi-Fi's short-term solution, WPA, to the security flaws in 802.11a–802.11b
802.11n	– To increase throughput to at least 100 Mbps in the 5-GHz range – Unlike the above, focuses on throughput at the MAC interface, not the physical layer, so throughput will be the highest possible

9.3   Smart Cards

As a human being, one has been endowed with just enough intelligence to be able to see clearly how utterly inadequate that intelligence is when confronted with what exists.

Albert Einstein (1879–1955)
— from a letter to Queen Elizabeth of Belgium, September 19, 1932

The term "smart card" has entered our discussions briefly thus far (see pages 105 and 334, for instance). Now it is time to delve into the details.

What is a Smart Card? Smart cards are made of plastic and are of credit-card-size, having an embedded microprocessor chip with internal memory or merely a memory chip with nonprogramming logic.[9.25]

Types of Smart Cards: Classifications for smart cards is described in the following.

1. **Standard Memory Cards**: These are cards that merely store data. They do not possess data-processing capabilities. Typically, these cards have a *magnetic strip* (so are often called *magnetic strip cards*). These cards store private data, usually employed as credit or debit cards, which require physical contact with a device to read the data on the magnetic strip.

2. **Intelligent Memory Cards**: These cards have a built-in wired logic circuit to access the memory (usually 1 K to 16 K bits) of the card. Sometimes these cards can be configured to restrict access via a password or system key. These cards are often called *protected memory cards*.

3. **Stored-Value Cards**: Sometimes these are called *memory cards with register*. These are cards that have security features hard-wired into the chip at the point of manufacture. Examples of such cards are prepaid phone cards, wherein a terminal inside the pay phone will write a declining balance into the card's memory. The card is discarded when the balance is zero; or if the card has a rechargeable capacity, it can be reset.

4. **Processor Cards**: These cards, perhaps the most deserving of the name *smart card*, contain memory, a processor, and have data-processing capabilities. This is an integrated circuit (IC) card with ISO/IEC 7816 interface.[9.26] If an 8-bit microprocessor had the task of RSA cryptographic calculations, for instance, it could take several minutes. Thus, a

---

[9.25] A *microprocessor* is any integrated circuit (IC) containing the CPU of a small computer. A *CPU* is the *Central Processing Unit*, which controls the operation of a computer, including the execution of arithmetic and logical operations as well as other instructions. In a smart card or microcomputer, the entire CPU is on a single chip. In general a computer *processor* is the logic circuitry that responds to and deals with the instructions that run the computer. However, in the modern day, the term "processor" has been replaced by "CPU".

[9.26] We learned about the ISO on page 218. The IEC is the *International Electrotechnical Commission*, a Switzerland-based organization that sets standards for electronic devices. A committee, JTC1, is joint between ISO and IEC, and its mandate is information technology standardization.

cryptographic coprocessor is typically added to the architecture, thereby reducing cryptographic calculations to a few hundred microseconds.

Types 1–3 are often grouped under the single heading of *memory cards* and type 4 under the heading of *microprocessor chip cards*. Memory cards are, naturally, the least expensive and most common. They contain what is called *Electronically Erasable Programmable Read-Only Memory* (EEPROM) nonvolatile memory.[9.27] For security, the data may be locked in by a PIN of up to eight digits written to a special file on the card.

Chips: There are three kinds of smart card chips as follows.

1. **Memory Chips**: Naturally, the most basic and least expensive are those chips that merely store data and have no processing capabilities. Once created, memory chips cannot be reprogrammed, since they can only hold static data such as personal information that does not require dynamic enciphering capacity. To change the capacity of such a memory card, it would need to be replaced entirely.

2. **Applications Specific Integrated Circuits (ASIC)**: The ASIC chips are hard-wired to keep data and execute a specific processing job. Of course, this processing capacity makes the ASIC chip stronger than the memory chip. Yet, the ASIC chip cannot be reprogrammed, as is the case with the memory chip. However, the ASIC chip does allow for some static encryption, but this is suitable only for low-level security applications.

3. **Microprocessor Chips**: These chips are the most powerful and versatile of the three types. They cannot only do what both the memory and ASIC chips can do, but also they are capable of dynamic encryption, and they can be reprogrammed or updated, unlike the previous two. Processor cards have microprocessor chips that typically come in 8-, 16-, or 32-bit formats. Their data storage may range from 300 to 32,000 bytes.

   Microprocessor-based smart cards have the benefits of (1) a high level of security, having the capacity to execute PKC or SKC protocols, including DES, RSA, and ECC; (2) multiple applications on the same card; and (3) ease of updating existing applications, or the addition of new ones.

   Microprocessor cards have numerous applications: the access medium for GSM (see page 347); for identification; for electronic signatures; for access to restricted areas; to protect data storage; and for e-commerce.

[9.27]Nonvolatile memory means any kind of solid-state memory that does not lose its contents when the computer is turned off. In the case of a memory card, when it is removed from the card reader, the power is cut off, yet the card stores the data. On the other hand *volatile* memory loses its contents when the computer is turned off. Nearly all RAM is volatile, except of course, battery-powered RAM. Included under the heading of nonvolatile memory are not only EEPROM, but also all other forms of ROM such as programmable read-only memory (PROM), erasable programmable read-only memory (EPROM), and flash memory, sometimes called *flash RAM*. The latter type of memory can be erased and reprogrammed. The term "flash" is derived from the fact that in a microchip, a section of memory cells is erased in one solitary act, *in a flash*. Flash memory is employed in PC cards, digital cell phones, printers, and digital cameras, for example.

Card Operating Systems: The microprocessor in a smart card is controlled by a *Card Operating System* (COS), which is a piece of firmware stored in the ROM of the microcontroller IC embedded in the card.[9.28] The COS has the following fundamental tasks.

1. Both establish and control communication between the card and any card-reading device.

2. File management.

3. Memory management.

4. Management of applications including loading and operating.

5. Protect data access.

6. Instruction processing and execution control.

7. Execute and manage cryptographic protocols when communicating with a card-reading device.

Smart Cards and PKI: The structure for smart cards employing PKI is described in RFC 2459 (see [215]). Smart cards may be embedded with functions that generate public and private PKC keys inside the cards, meaning that the private key is not sent to any site outside the card. In other words, the smart card need not export the private key in order to use a given application.

Suppose that Alice interfaces her smart card with her computer for the purpose of using some application, which requires Alice's signature on a document to authenticate her. In order to get the card to communicate with the application, a hash of Alice's document, e-mail for instance, is sent to the card. The card signs the document with her private key (all this taking place inside the card), and the signed document is sent to the application. Hence, her private key is never exposed to the outside, in particular to her computer. Smart cards may employ SSH (see page 334) to authenticate to an application remotely, for instance. In general PKI architecture may support access to a given business enterprise via a local CA or RA for the purpose of certification. Basically, the structures discussed in Section 6.2 may be brought to bear via smart cards and their interaction with various applications.

Contact Vs. Contactless: The communication between a smart card and a card reader or detection device might be direct, namely, physical contact, or *contactless* using radio frequency. Thus, smart cards are further divided into contact and contactless (sometimes called *proximity*) cards. Contactless cards

---

[9.28]Think of a microcontroller as a computer on a chip. A microcontroller is created via the integration of the fundamental components of a microprocessor: RAM; ROM; and digital I/O (input/output) ports into the same chip die. Other features might include: serial I/O, a timer module; analogue to digital converters (ADC); and even serial peripheral drivers. Examples are Motorola's M68HC08 family of 8-bit microcontrollers, and Microchip's PIC17 Family with 16-bit program word.

are embedded with not only a chip, but also an antenna for the purpose of sending a signal to the reading device. Typically, a few centimeters of distance will allow the mechanism to receive the signal and authenticate the card owner for access to that device. Contact cards are usually employed for access to secure areas in a business enterprise, for instance, whereas contactless cards are typically used for mass transit access or for door locks.

Contactless cards use wireless self-powered induction technology, as defined in the standard, ISO/IEC 14443. The latest use for such a card in mass transit is the *Oyster Card* issued in London, England, in January of 2004. The card is rechargeable, secure since, if lost or stolen, it may be cancelled and reissued; and it is valid London-wide including the "Tube", Tramlink, DLR (Docklands Light Railway), and National Rail services across the entire London bus network.

Contactless cards have the benefits of speed of transaction time; convenience; low maintenance (compared to contact cards); and consumer appeal where key fobs, rings, or other devices may be used in place of a plastic card. Many upscale residential areas are looking at replacing locks with contactless smart cards in North America. The fact remains that contacts are the most frequent breakdown points in the electromagnetic system as a result of dirt, and wear on the mechanism. Contactless cards solve these problems and improve performance in the balance, so user acceptance will surely increase.

Last, there are cards which combine certain features, called *combi-cards* or *multifunction* cards. This might involve a combination of password, and biometric such as a fingerprint. Also, there is the possibility of combining both contact and contactless features in one card.

Physical Properties: The actual body of the card is plastic, which may be *polyvinyl chloride* (PVC) or *acrylonitrile butadiene styrene* (ABS). The card itself may contain a signature strip, printed signature, or a cardholder photograph. Of course, the plastic body will be embossed with the proprietary graphics such as with Visa or MasterCard. The size of the card is specified by ISO/IEC 7816-1, namely, $85.6 \times 54 \times 0.76$ mm. This standard includes definitions of resistance to static electricity, electromagnetic radiation and mechanical stress, as well as the location of the card's magnetic strip and embossing area.

The dimension and location of the contacts is specified in ISO/IEC 7816-2. This includes the *module*, which is the smallest part of the card that is capable of accommodating a chip and its contacts. The mechanism for securing the module in place on the card is via encasing them in a resin amalgam, which for security reasons, should be designed so it cannot be removed without destroying the circuitry (see page 361).

There are also cards, called *mini-cards*, which are in size between that of a regular smart card and its module. These are often used for mass transit applications, where the size of the cards mimic the size of the magnetic-strip tickets they replace.

In the following, Diagrams 9.6 and 9.7 give the placement of the electrical contacts in a smart card chip, numbered C1–C8, and describe the function of each.

Diagram 9.6 Smart Card Chip — Electrical Contacts

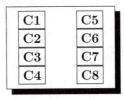

Diagram 9.7 Functions of Electrical Contacts

Position	Abbreviation	Function
C1	VCC	Power Supply Voltage
C2	RST	Reset Microprocessor
C3	CLK	Clock Frequency
C4	RFU	Reserved for Future Use
C5	GND	Ground
C6	VPP	Programming or Write Voltage
C7	I/O	Serial Input/Output Line
C8	RFU	Reserved for Future Use

All the data transmitted to and from a smart card is through the C7 contact point. Once a smart card is inserted into a card reader, for instance, a client-server relationship ensues. The physical transmission is defined in ISO/IEC 7816-3, so any reader must conform to that standard.

Card Origins: The French are responsible for the term "smart card", in development since the 1970s when the French invested a large amount of money into this R&D technology. They originally called these cards *Carte a memoire* or *memory card* in the 1970s. The French government's marketing arm, *Intelimatique*, coined the term *smart card* in 1980. In fact, Roy Bright of Intelimatique (see [45]), was the one who coined the word "smart card" (which is sometimes written as a single word *smartcard*). In 1970, the concept of the smart card was filed in a patent by Kunitaka Arimura of Japan. The patent was restricted to Japan, and to the technical aspects of the smart card idea, namely, to integrate data storage and arithmetic logic on a single silicon chip. Shortly after his patent was filed, the first smart cards were issued in Japan.

Although many credit the first patent for an IC card to the French journalist, Roland Moreno, who filed his patent in in 1974 (see [172]), there was a patent filed in 1968 by the German inventors Jurgen Dethloff and Helmut Grotrupp for the invention of the idea to incorporate an IC into an identification card (see [67]). However, the patent filed by Moreno is considered the first actual "smart card" patent since it was the first to incorporate the two ideas of Arimura and Dethloff-Grotrupp into a single entity, what we consider to be a smart card today. Moreover, Moreno's patent was the first to be broad-based not only in

France, but also in major industrial countries around the globe. By 1977, the first commercial developers of an IC card product were three manufacturers, Bull CP8, SGS Thompson, and Schlumberger. Also, in that year, the French banking system had a smart card payment scheme in place, and by 1978 the first prototype card was produced. In 1979, Motorola introduced the first secure individual chip microcontroller. It was a prototype made in Toulouse, France for Bull CP8, having programmable 1-K memory and microprocessor 6805.

Credit cards contain data including either signature or picture for identification of the person authorized to use it for account access or services. The use of credit cards, on a local scale, actually goes back to the 1920s in the United States, when some oil companies and hotel chains started issuing them to customers for purchases at their enterprises. On a global scale, the first credit card for use at a large multiplicity of businesses was Diners Club Inc., in 1950. Their card employed PVC plastic, which replaced earlier paper-based cards. They were the first to institute charging an annual fee billed to their cardholders. By 1958, American Express entered the stage with its card. The first bank to issue a card was the Bank of America in 1959 with its *BankAmericard* distributed initially in California only, adding other states starting in 1966. In 1976, it was renamed VISA,[9.29] and later MasterCard followed suit. In 1981, MasterCard (formerly called Master Charge), introduced the first gold-card program, and in 1983 it was the first to employ a laser hologram as an antifraud mechanism.

The 1980s saw much field testing of smart cards. The world's first significant IC card test was conducted in France with their testing of serial memory phone cards in 1982. In 1983, the first nationwide smart card scheme was put in place by the French for their public telephone payment system (see [45]). In 1984, the French adopted the Bull CP8 card as their standard for the first version of their bank debit cards Carte Bleue. By 1986, the French also were the first to introduce a smart card scheme in the form of a health card. In 1987, the ISO introduced the first card standards in the form of ISO/IEC 7816-X. The 7816 series of standards today define everything from the physical shape of the card to the format the commands may take when sending to or responses from the card. This includes not only the functionality of the card, but also the very position and shape of the electrical connectors and the protocols defining the power voltages to be applied to them (see Diagrams 9.6 and 9.7).

By the early 1990s the French were involved in field testing of combi-cards. Also in the early 1990s, Germany was involved in memory card distribution on a mass scale. In 1994, they started the distribution of some 80 million serial memory chip citizen health cards. Now, every German citizen has a health smart card. By the mid-1990s, mobile phone use was conducted and paid via smart cards by some three million users. By the late 1990s, the major players in the credit card industry were looking at standards for interoperability. In 1996,

---

[9.29]Internationally, BankAmericard was known by other names before VISA came into being. In Canada, a number of banks, in concert, issued Chargex cards. In the U.K., the BarclayCard was issued by Barclay's Bank. Both of the latter used the blue-white-blue motif familiar to BankAmericard holders. The blue and gold motif on the VISA cards was selected to represent the blue sky and gold-coloured hills of California, where BankAmericard originated.

MasterCard and Visa began developing two types. JavaCard, sponsored by Visa, and *Multi-application Operating System* (MULTOS), sponsored by MasterCard.

Two announcements were made in April of 2004. One was that residents of Lakhpat, Taluka, India, would be the first to be issued a processor card, called the multipurpose National Identification Card to serve as citizenship proof. Another was the fact that many governments were gearing up for a transformation of existing passports to include microprocessor chips embedding biometrics.

Attacks on Smart Cards: There are numerous attacks against smart cards that need to be reviewed so we may better understand the threats and not fall victim to them. Two attacks already discussed are power cryptanalysis and the small RSA enciphering exponent attack (see page 178). These attacks are especially effective against smart cards due to their limited computing power and relatively slow processors, such as the choice of a small enciphering exponent to communicate between the smart card and a larger computer.

Power cryptanalysis (sometimes called *power analysis*) attacks are examples of what are called *side-channel attacks* wherein a cryptanalyst, Mallory, say, has an additional channel of information about the system he is trying to break. Timing analysis of message encryption falls into this category. The reason that side-channel attacks are so effective against smart cards is that Mallory may have full control of the card. Countermeasures for side-channel attacks come from a combination of software implementations and actual hardware.

Countermeasures against timing attacks include the following: (1) blinding signatures (see page 177); (2) avoiding delays (make all operations take the same amount of time); (3) equalization of multiplication and squaring (the time taken to execute multiplication and exponentiation should be set to be very similar); (4) power consumption balancing (operations should be made to appear constant from outside the card, which can be accomplished with dummy gates and the like to even out the power consumption to some constant value); (5) add random noise (enough to stop an attack); and (6) physical shielding.

Magnetic strip cards, having no computing power at all, are subject to what is known as a *skimming attack*. In this case, an illegal card reader can be used to copy the data in the card (once it is swiped through the illegal device) for the purpose of counterfeiting cards and incurring illegal charges. Some criminals have even resorted to planting these devices in legal ATM machines for the purpose of gathering this data. Once the data has been captured, the card owner might be presented with a screen that says there has been a malfunction. In some cases, the criminals engineer the card reader so that it does not interfere with the ATM's function. In this case, the customer will get their cash, when making a withdrawal, say, but their data are still captured for later use by the criminal element. The ATM machines most susceptible to this kind of attack are not usually the ones at banks themselves, but rather at convenience stores, bars, hotel lobbies and the like. Moreover, they are typically the kind of ATM where the card is swiped rather than inserted into the machine directly. Also, skimming may be accomplished by dishonest businesses when your card is taken out of your sight for payment, say at a restaurant, and run through a skimmer.

To thwart skimming attacks, do not use ATMs where something appears to be out of place. Keep all PINS safe and never give them to anyone. Do not let strangers "assist" you at an ATM machine. If your card is not returned after usage in an ATM, immediately contact the institution that issued the card. Treat your cards as if they were cash and do not let them out of your sight.

Returning to IC cards, there are *tampering attacks*, which may be broken down into four subsets: (1) *microprobing*, where the chip itself is accessed, manipulated, and there is direct tampering with the IC; (2) *software attacks*, the exploitation of weaknesses in cryptographic protocols or their implementation via the I/O interface; (3) *eavesdropping*, the monitoring of any electronic radiation produced by the microprocessor's executions; (4) *fault generation*, creating malfunctions in a microprocessor for the purpose of establishing access.

Attacks (2)–(4) are *noninvasive* attacks. On the other hand, microprobing is an *invasive attack* that requires a significant amount of laboratory time, expensive equipment, and expertise. In order to extract the chip, the plastic card is destroyed. Once the chip is removed, it may be mapped, analyzed, and information obtained. One countermeasure for such attacks (already available with some microprocessors), is the embedding of a sensor mesh above the actual chip, so that any tampering would trigger an erasure of nonvolatile memory.

With noninvasive attacks, smart cards are especially vulnerable since their microprocessors are exposed without the safeguards built into larger devices, such as electromagnetic shielding. A microprocessor is basically a collection of a relatively small number of *flipflops* (registers, latches, and SRAM cells),[9.30] which establish its current state, together with a logic design that calculates that state based on a clock cycle and other states. A *register* is a specialized, high-speed storage region of the CPU. No data is capable of being processed before being put into registers. A CPU's power is defined in terms of the number and capacity of registers it possesses. For example, an 8-bit CPU has registers that maintain 8-bit words each, so each command sent to such a CPU is capable of handling 8 bits of information. A *latch* is a digital logic circuit for storing bits. The components of a latch are the data input to it, a clock input, and its output. The term "latch" comes from the function of the clock activity, for example when active, the clock input triggers the data input to be "latched" (stored) and transferred to output when the clock input becomes inactive. The value of the clock output is then set and maintained until the clock input is again activated. This analog effect is one of the vulnerabilities that can be exploited via fault-generation attacks in smart cards, namely, by causing one or more flipflops to take on the incorrect state (see [34]).

Countermeasures to thwart noninvasive attacks include inserting a random-number generator at the clock-cycle level; and embedding a tamper-sensor that will disable the entire microprocessor upon detection of unauthorized activity.

[9.30]*SRAM* is *static* RAM as opposed to the more common *dynamic* RAM or *DRAM*. The term "static" is employed to differentiate it from the conventional form of RAM in that it does not need to be refreshed as does DRAM. Therefore, SRAM is faster and more reliable than DRAM. However, it is more expensive in terms of financial cost, storage space, and power consumption. Thus, DRAM is necessarily volatile memory.

9.4   Biometrics

Biology is the search for the chemistry that work.
                                   R.J.P. Williams (1926–), British chemist
                                   — from a lecture in Oxford, June 1996

◆ **Overview**

The science and technology of quantifying and analyzing biological or behavioral data is what we call *biometrics*. The characteristics to be measured are DNA; ear geometry; eye retina (the nerve endings inside the eyeball that capture and send light to the brain) and irises (the coloured part visible at the front of the eye); facial geometry; fingerprints; hand geometry; and voice frequency. The data to be analyzed is stored in a database for comparison with existing records. Typically, software is used to identify specific *match points*, which are then processed into a value that may be compared with biometric data that is scanned when the owner of a smart card, say, tries to gain access. Biometrics may be used to provide authentication for access to a bank account; to pay for products or services from a business; to pay for telephone charges; and so on. Biometrics can be employed in addition to, or in place of, say, a PIN.

Sensors are used to record the biometric information. Cameras are used for facial, eye, hand, and ear geometry; microphones for voice; chemical laboratories for DNA; and any number of sensors for fingerprints including pressure sensitive, thermal, optic, and capacitive devices.

◆ **Biometrics and Smart Cards**

The idea of embedding biometrics into a smart card together with other personal details has been considered by many governments. For instance, such a card, called Mykad, was mandated in September of 2001 for all Malaysian citizens over the age of twelve. The Mykad deployment started in 1999 when the government awarded the project to an international consortium of technology suppliers. From its official release in September 2001, to April of 2004, nine million cards had been issued with a total of fifteen million expected to be registered by the end of 2005. The Mykad replaces the national ID card and driver's license; it contains medical data; may be used for highway-toll payments; for parking; for public transportation; for ATM transactions; and even e-commerce since it contains PKI infrastructure, including digital signatures. Mykad contains fingerprint biometrics for verification of a given individual.

In 2002, the U.S. Congress mandated a program to issue international visitors "only machine-readable, tamper-resistant visas and other travel and entry documents that use biometric identifiers." The global biometric enrollment program started in September of 2003. By October 26, 2004, all visa-issuing U.S. embassies and consulates will be collecting biometrics for visa applications. Typically, fingerprints of two digits from each individual will be electronically scanned and stored in a database available to the Department of Homeland Security immigration officers for those ports of entry to the United States.

Other countries and agencies using smart card technology in conjunction with biometrics are the following. The U.S. Department of Defense Common Access Card has a photograph together with a fingerprint embedded in its functionality. Spain has a social security card including biometrics in its smart card application. The Netherlands has a system called *Privum* for automated border crossing. Their smart card has a photograph, and iris biometrics. Brunei employs a national ID smart card having a photograph together with fingerprint biometrics. The United Kingdom has the *Asylum Seekers Card*, which is a smart card with a photograph and fingerprint biometrics. It is not long before more countries are added to the list in an effort to secure their borders.

The bottom line for smart cards supported by biometrics is that it raises security levels to very high standards. The reason is that such cards possess the following.

The Three Fundamental Aspects of Authentication

1. Something the user *has* (the smart card, itself)

2. Something the user *knows* (a PIN or password)

3. Something the user *is* (the biometrics)

◆ **Accuracy and Robustness of Biometrics**

Biometric Traits

Biometric traits develop in one of three ways:

1. *Genotypic* (through genetics)

2. *Phenotypic* (through early embryo development)

3. *Behavioural* (through training)

Robust biometrics are those which are not subject to significant changes. Certain biometric traits may vary over time due to aging, growth, injury and later regeneration, wear and tear, and so on. The *least* changeable biometrics are DNA and iris pattern followed by retina, fingerprints, and hand geometry. In terms of *accuracy* (minimal error rates plus clarity and consistency), iris and retina measurements rank *ahead* of DNA, although all three are difficult quantifications to obtain and are costly to process. The reason that DNA trails the other two eye biometrics is that DNA cannot distinguish between monozygotic twins, but the eye biometrics can do so, (and better than the other biometrics). Fingerprints rank roughly fourth on the accuracy scale, but are relatively easy to obtain, and inexpensive to process in comparison to the other three. An iris match against a database can be made 300 times faster than a match to a fingerprint in the same database. Hence, despite the cost differential, the speed and high accuracy of eye biometrics make it vastly superior to the fingerprinting biometric. Once costs descend, this must surely be the medium of choice, if for no other reason than the key factor in selection of an appropriate biometric is its accuracy. At the bottom of the list are face geometry, followed by finger geometry, and voice patterning.

◆ **Verification vs. Identification**

We discussed the use of smart cards and biometrics for *verification* of individuals above, where verification means the following.

Verification

The individual's identity is entered into the system, via a smart card, say, then a biometric feature is scanned. If that scanned trait matches the one previously stored in the card, then verification is successful. This kind of "verification" is also often called "authentication" of the individual.

The notion of verification must be separated from the issue of *identification*, given as follows.

Identification

An individual's recorded biometric feature is compared to *all* the corresponding biometrics in the database. If there is a match, then the individual is identified, and the user's ID may be processed later for verification.

Identification is very useful in fighting crime. For instance, if an individual's fingerprint or DNA, say, is lifted from a crime scene, and a match is made to it after searching a database, this provides crime fighters with evidence to prosecute.

In order for biometrics to be effective, there must be an *enrollment process*, where an individual consents to having a biometric image captured, such as a fingerprint or eye scan, from which the characteristics are extracted. This allows the creation of the user's biometric template, which is stored centrally, in a database, or locally, on a smart card, say. Think of verification as a one-to-one comparison, which confirms that the credential belongs to the individual who is presenting it. The authenticating device need only have access to the individual's enrolled biometric template, which may be stored locally or in a database. Identification, on the other hand, is a one-to-many comparison. It verifies that the given entity exists within a given population and is not enrolled with another ID. Moreover, it will verify that the individual is not on a list of prohibited entities. In this case, the database must contain a set of all entities applying for the access, say, to enter a country, and their biometric templates.

As shown in Diagrams 9.8 and 9.9, the acceptance or rejection will be based upon some threshold value derived from the security policy of the system being accessed.

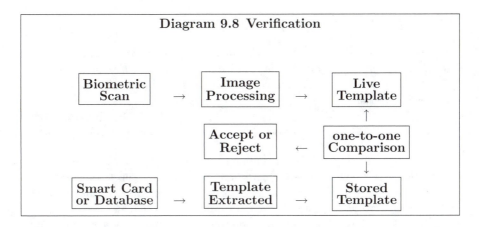

Diagram 9.8 Verification

Biometric Scan → Image Processing → Live Template

Accept or Reject ← one-to-one Comparison

Smart Card or Database → Template Extracted → Stored Template

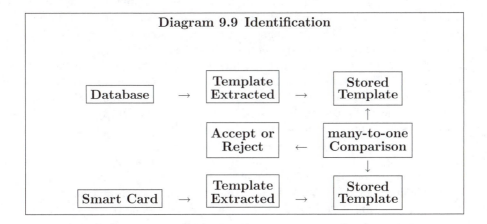

Diagram 9.9 Identification

Database → Template Extracted → Stored Template

Accept or Reject ← many-to-one Comparison

Smart Card → Template Extracted → Stored Template

9.5   Quantum Cryptography

The lesson to be extracted from the latest century of physics is that physical knowledge has greatly expanded and resulted in new and much-improved theories, but that these have been produced largely cumulatively and without a complete break with the past.

Helge Kragh (1944–)
— from *Quantum Generations* (see [140, page 449])

The nobel laureate, Richard Feynman,[9.31] once said: *" I think that it is safe to say that* nobody *understands quantum mechanics."* The science of *Quantum*[9.32] *mechanics* is the branch of physics that accounts for matter at the atomic level. We will not try to explain quantum mechanics beyond this elementary description. However, a cornerstone of quantum cryptography, called the *uncertainty principle*,[9.33] may be relatively easily stated as: One cannot *simultaneously* know both the position and momentum of a given object to arbitrary precision.[9.34] This is usually illustrated as follows. Suppose that we wish to measure the position and momentum of a specific particle. To do this, we must "see" the particle so we must shine light on it. Suppose that light has wavelength λ. To measure the particle's position, λ must be very short, because in order to provide data on position we need wavelengths comparable to the object we want to locate. However, a short wavelength of light transmits a big boost in momentum when it bounces off the particle to provide position data. Thus, the more accurately we measure position, the more uncertainty there is in its momentum. On the other hand, if we want to measure momentum, we use very long wavelengths, which increases uncertainty in its position. Hence, a particle does not have a well-defined *simultaneous* position and momentum. Typically, quantum experiments are done at this subatomic particle level, and this is not in our everyday experience. Yet, there is a means of describing a quantum experiment at a level with which we are all familiar.

[9.31] Richard Phillips Feynman (1918–1988) was born in New York City on May 11, 1918. He was educated in the United States, obtaining a doctorate from Princeton in 1942, wherein his thesis developed a new approach to quantum mechanics. From 1943 to 1945, he worked as a member of the team that developed the first atomic bomb at Los Alamos. In 1965, he was awarded the Nobel Prize in Physics. Despite all his accomplishments, he was not a typical stuffy scientist. To put his life in perspective, we quote from the jacket cover of his book [86]: "In short, here is Feynman's life in all its eccentric glory — a combustible mixture of high intelligence, unlimited curiosity, and raging chutzpah." After eight years of battling abdominal cancer, he succumbed on February 15, 1988, in Los Angeles, at the age of sixty-nine, having taught his students up until two weeks before his death.

[9.32] Quantum theory is a physical theory that holds that certain properties occur only in discrete (as opposed to continuous) amounts, called *quanta*.

[9.33] This is formally known as *Heisenberg's uncertainty principle*, named after Werner Karl Heisenberg (1901–1976) who was awarded the Nobel Prize for Physics in 1932.

[9.34] This does *not* say anything about how precisely a particular object can be known. It *does* say (more generally), that some *pairs* of properties are intimately linked in such a way that they cannot be precisely measured at the same time. Physicists call these pairs *canonically conjugate variables*. For instance, position and momentum is one such pair and another is time and energy. The more precisely one knows the time span when an event occurred, the less precisely one knows the energy involved (and vice versa).

◆ Photons and a Quantum Experiment

We begin by looking at basic properties of light. The particles that constitute light are called *photons*. These photons make up light waves, which are examples of *electromagnetic waves*, meaning that they have an electric field that travels perpendicular to their associated magnetic field. Photons travelling through space vibrate (or oscillate) as they move. This vibration can be horizontal, denoted by →; vertical, denoted by ↑; 45°, denoted by ↗; or 135°, denoted by ↖. The angle of the vibration is known as the *polarization* of the photon. This is a simple type of polarization, called *linear*, meaning that as the photon propagates, the electric field stays in the same plane. This linearity assumption simplifies the situation by allowing only four possible polarizations, rather than the infinitely many possibilities (namely all angles in between).

Now we need to understand a little bit about polarization of light. We are going to look at the effects of a Polaroid filter[9.35] on a light source. We will assume that the axis of the filter is oriented in one of the aforementioned four ways. Quantum theory dictates that if α is the angle that the plane of the electric field of the photon makes with the axis of a Polaroid filter, then there is a probability of $\cos^2 \alpha$ that the photon will emerge with its polarization reset to that of the filter's axis, and a probability of $1 - \cos^2 \alpha$ that it will be absorbed (to be re-emitted later as heat). For example, if the polarizer axis is vertical, then light emitted with random polarization means that if α is only slightly off vertical the photon has a high probability of passing through. If it is 45°, then it has a fifty percent chance of getting through, and this decreases to zero at the horizontally polarized photons. Hence, roughly 50% of the randomly emitted photons get through and as they pass through the vertical filter, they all emerge as ↑ polarizations. Call that polarization filter **V**, and the situation is illustrated in Diagram 9.10.

Diagram 9.10 Polarization with Filter V

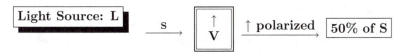

[9.35] Edwin Herbert Land (1909–1991) patented a cellophane-like polarizing filter, the first to polarize light, a process that reduces light glare. In 1932, Land co-founded the Land-Wheelwright Laboratories in Boston. By 1937, Land founded the Polaroid Corporation and began to use his filters in Polaroid sunglasses, and a variety of other applications. However, Land is best known for his invention and marketing of instant photography, called *Polaroid photography*. In 1947 he presented the *Polaroid Land Camera*, which took 1 minute to produce a finished photograph. After his retirement from Polaroid in 1980, he worked with the nonprofit *Rowland Institute of Science* supported by the Rowland Foundation that he founded in 1960. Land stands second only to Thomas Edison in the number of patents issued to him, more than 500. He received a number of awards for his contributions to knowledge about polarized light, photography, and colour perception. Land died in Cambridge, Massachusetts, on March 1, 1991.

Now suppose that we use a polarizer axis that is horizontal, denoted by **H**. Then no light gets past filter **H**, after having passed through filter **V** because all of the photons are polarized ↑, whose angle with **H** is $\alpha = 90°$, and the probability of getting through is $\cos^2 90° = 0$. This process is illustrated in Diagram 9.11.

Diagram 9.11 Polarization with Filters V and H

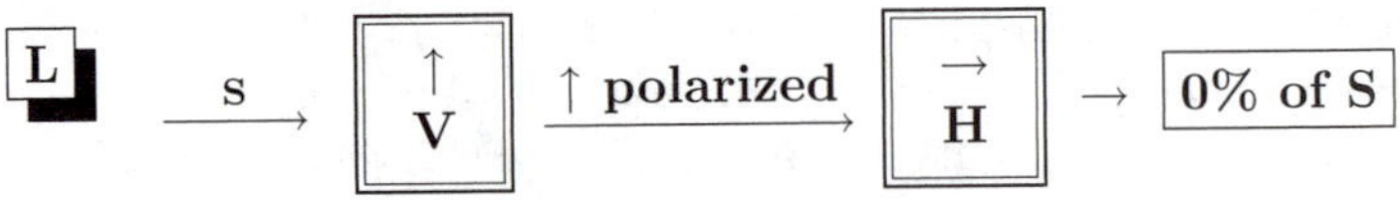

Suppose we now place a filter with polarizer axis 45°, denoted by **F** between **V** and **H**. Then the 50% of those photons that get through **V** now have a 50% chance of getting through **F**, and each of those will be polarized as ↗, so now 25% of the original photons got through. Now we approach **H** and each of the ↗ has a 50% chance of getting through **H**. Hence, once through all three filters, 12.5% of the original photons are emitted. Surprisingly, having put another filter between two that allowed no photons through, now allows 12.5% through.

Diagram 9.12 Polarization with Filters V, F, and H

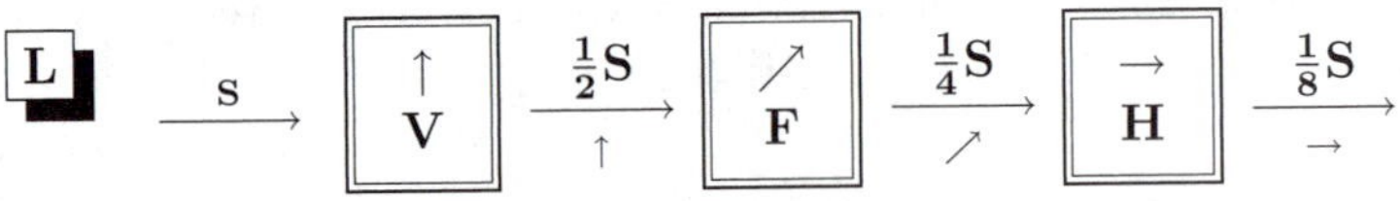

This is the basic principle upon which Polaroid sunglasses work. One can demonstrate this principle, using a pair of Polaroid sunglasses, by taking one lens out and placing it in front of the fixed lens. There will be an orientation that is exactly the same for both lenses, so that the fixed lens has no effect on the loose lens. If the loose lens is now rotated ninety degrees, the effect will be complete blackness. This is because the polarization of the lenses are now perpendicular, so that photons that get through the one lens are blocked by the other. By rotating the loose lens forty-five degrees, one now gets an intermediate stage between complete blackness and no effect. This is because half of the photons that pass through the one lens succeed in getting through the other. Placing a third lens in front of the loose lens with axis perpendicular to the fixed lens, we get about half the light from the first two being filtered through, which is Diagram 9.12 in action.

◆ Quantum Key Generation

Now we turn back to cryptography and show how the above notion of polarization and its effects can be employed to generate cryptograms.

Our goal is for Alice and Bob to communicate in binary using the polarization effects from our earlier discussion. First, we set up two binary schemes based on those polarizations.

Rectilinear Scheme: This scheme will be denoted by $+$, wherein a 1 is represented by $\uparrow$ and a 0 is represented by $\rightarrow$.

Diagonal Scheme: This scheme will be denoted by $\times$, which uses $\nearrow$ for 1 and $\nwarrow$ for 0.

To send a message Alice can randomly switch between these two schemes. For example, she might send a photon string consisting of

$$\rightarrow \rightarrow \uparrow \rightarrow \nearrow \nwarrow \uparrow$$

using the combination of methods: $+ + + + \times \times +$ so the message is:

$$0\ 0\ 1\ 0\ 1\ 0\ 1.$$

▼ Quantum Key-Generation Protocol

In the following, when we say that Alice and Bob "openly" communicate, we mean that they converse over an unsecured telephone line.

1. Alice openly communicates to Bob a string of $n \in \mathbb{N}$ photons with random polarizations in the two schemes, $+$ and $\times$, with the particular polarized photons denoted by $p_1, p_2, \ldots, p_n$. Each polarized photon p_j is associated with one of the schemes $+$ or $\times$, so we denote s_j to denote that scheme under which p_j is polarized, for $j = 1, 2, \ldots, n$. For instance, if $p_1 = \rightarrow$, then $s_1 = +$.

2. Bob has a polarization detector with two settings.

    (a) A $+$ detector that can measure the polarizations $\uparrow$ and $\rightarrow$ with perfect accuracy, but misinterprets $\nearrow$ or $\nwarrow$ as one of $\uparrow$ or $\rightarrow$.

    (b) A $\times$ detector, which can measure $\nearrow$ and $\nwarrow$ with perfect accuracy, but misinterprets $\uparrow$ and $\rightarrow$ as one of the $\times$-polarized ones.

    Both settings cannot be used at the same time due to the uncertainty principle (we cannot simultaneously measure both $+$ and $\times$ polarizations).

    Bob sets the polarization detector at random settings. Sometimes the correct detector (corresponding to Alice's choice) is picked for the incoming photon, and sometimes not. We denote his received photons as $q_1, q_2, \ldots, q_n$, and his corresponding randomly selected schemes as $t_1, t_2, \ldots, t_n$.

3. Alice openly communicates to Bob the s_j for each $j = 1, 2, \ldots, n$, but not p_j. If $s_j = t_j$, then q_j is selected. Otherwise, q_j is discarded. We will label the selected ones as $q_1, q_2, \ldots, q_m$, without loss of generality, for the sake of convenience.

4. Alice openly communicates to Bob her choice of randomly selected $\ell < m$ of the p_j. They compare her ℓ of the p_j with Bob's corresponding q_j If any of these do not match, Alice and Bob know there must be an eavesdropper (see analysis below) and they abort the run. Otherwise, they go to step 5.

5. Alice and Bob discard the ℓ randomly tested $p_j = q_j$, and maintain the remaining $m - \ell$ of them as the secret key.

Analysis: The bitstring corresponding to their agreed-upon secret key is truly random since Alice's initial photon burst was random and Bob's choice of polarization methods was random. Hence, this agreed upon bitstring can be used for a one-time pad.

To see why the above key-generation scheme is the equivalent of a one-time pad, suppose that Mallory has also attempted to measure the initial photon burst from Alice. Then Bob and Mallory are in exactly the same situation since both of them will choose the wrong detector roughly half of the time (but not the same half). The uncertainty principle guarantees that Mallory has no means of duplicating Alice's original settings, so even if Mallory's eavesdropping on the telephone conversation, thereby gaining knowledge of the correct polarization settings, this does not help because Mallory will have measured about half of these incorrectly. Hence, this one-time pad is absolutely unbreakable, since Mallory cannot intercept Alice's message without making errors.

Mallory's presence is detected by the very act of measuring. If Alice sends a $\nearrow$, for instance, and Mallory uses the $+$ detector, then the incoming $\nearrow$ will emerge as one of $\uparrow$ or $\rightarrow$, since this is the only way that photon can get through Mallory's detector. If Bob measured the transformed photon with the $\times$ detector and $\nwarrow$ emerges, then a correct setting of the detector will result in an incorrect reading. In this case Mallory has altered the resulting q_j. Of course, it might also occur that Bob's reading results in the correct $\nearrow$ emerging. Therefore, Mallory has a one in four chance of being detected for each photon checked. Since ℓ of the q_j are checked in step 4, then the probability of detecting Mallory is $1 - (3/4)^{\ell}$. Hence, for arbitrarily large ℓ (and sufficiently large corresponding n), we can make this as close to 1 as we desire.

The above analysis shows that quantum cryptography allows key distribution between two entities (who share no prior keying material) that is provably secure against enemies with unlimited computing power, provided that the entities have access to a conventional channel, aside from the quantum channel.

There was a working prototype for such a similar quantum scheme developed in 1989 by the authors of [20] (having four, rather than two, schemes), but there were some glitches in the prototype, all of which is discussed in the book [291, pages 177 and 178] devoted entirely to quantum computing. Moreover, there have been implementations of the scheme using fiber optical cables over several kilometers (see [125] and [155]). Further advances were announced in Tokyo in March of 2004; NEC Corporation in collaboration with others have succeeded in realizing the world record, a 150-km-long single-photon transmission.

◆ Quantum Computing

The conventional digital computers, with which we are familiar, use bits, 0 and 1, to represent information. Moreover, after each execution, the digital computer has a definite, precisely measurable state; all the bits are either 0 or 1 but not both. A (as yet still hypothetical) *quantum computer* is a quantum analogue of a digital computer, that operates with *quantum bits* involving quantum states. A quantum bit is also called a *qubit*, which may be represented as a (unit) vector in a sphere, where ↑ represents 1, and ↓ represents 0. Any orientations in between such as ╱ and ╲ represents the angle with the vertical axis as a measure of the "0-to-1"-ness of the qubit. Thus, unlike classical bits, a qubit may posses several states at once. The input and output qubits can be linear combinations of basic states, so that the quantum computer functions on all basic states in the linear combinations simultaneously. Hence, a quantum computer is essentially a massively parallel computer. This means that (if one were ever built), it would outperform all classical computers, and would make classical cryptography obsolete. For example RSA could be broken. Even though a quantum computer, per se, does not yet exist, factoring methods for one have been developed. In [252], Shor[9.36] presented a polynomial time quantum algorithm for factoring large integers. He used the quantum property called *interference*, which is the feature of quantum mechanics that dictates that the outcome of a general quantum process is dependent upon all possible histories of that process. Interference makes quantum computers qualitatively more powerful than classical ones, because quantum interference can occur whenever there exists more than one method for obtaining a specific result.

In 1993, the authors of [21] showed how to *teleport* the quantum state of an object, meaning that they presented a scheme for transporting from one location to another *without* passing through the distance between them. They used the notion of *entangled states* (or *entanglement*), which refers to the quantum fact that the properties of a composite system, even when the components are distant and noninteracting, cannot be fully expressed by descriptions of the properties of all the component systems. Entanglement is the feature of quantum mechanics that makes quantum cryptography possible. Hence, with this scenario, quantum cryptography employing quantum computers would involve enciphering qubits using quantum states and teleporting those quantum states from one quantum computer to another without having to pass through an unsecured channel. Although prototypes exist, the construction of a general-purpose quantum computer seems infeasible at this time. Yet, it might be possible to construct a special-purpose quantum factoring machine. After all, Shor has shown us how to use quantum computers to break classical cryptosystems, such as RSA, based on factoring, since his technique reduces the factoring of very large composite numbers to the comparable triviality of multiplying.

[9.36]Peter Shor was born August 14, 1959. He obtained his Ph.D. at MIT, and after a brief stint at the Mathematics Research Center in Berkeley, California, as a postdoctoral fellow, he joined AT&T in 1986. He is currently a mathematician at the AT&T Research Laboratories in Florham Park, New Jersey. His work on quantum factoring earned him the 1998 Nevanlinna Prize at the International Congress of Mathematicians in Berlin.

9.6   Nuclear Test Ban Treaty Compliance

I must create a system, or be enslaved by another man's. I will not reason and compare: my business is to create.
— **William Blake**   (1757–1827), English poet
— from *Jerusalem* (1815 Chapter 1; plate 10, l. 20)

We conclude this chapter with a section on applications of PKC to a modern-day phenomenon given by the title. This was very much a real part of negotiations in the last century when the cold war between the United States and the former Soviet Union were under way to limit nuclear bomb testing.

We begin with a brief biographical description of the man responsible for the idea of using PKC for the application in the title of this section.

Gustavus J. Simmons was born on October 27, 1930 in Ansted, West Virginia. His educational background includes a B.S. in mathematics from New Mexico Highlands University, Las Vegas, in 1955, an M.S. from the University of Oklahoma at Norman in 1958, and a Ph.D. in 1969 from the University of New Mexico, Albuquerque. In 1986, he received both the U.S. Government's E.O. Lawrence Award, and the Department of Energy Weapons Recognition of Excellence Award for "Contributions to the Command and Control of Nuclear Weapons". In May 1991, he was awarded an honorary Doctorate of Technology by the University of Lund, Sweden, which recognized his contributions to the science of communications and to the field of information integrity, in particular.

Simmons spent his working life with Sandia National Laboratories from which he retired in 1993 as a Senior Fellow and the Director for National Security Studies. His work at Sandia mainly centered around integrity and authentication issues surrounding national security, with a special focus on those involving command and control of nuclear weapons. In 1996 he was made an honourary Lifetime Fellow of the Institute of Combinatorics and Its Applications. His many publications were primarily in the areas of combinatorics, graph theory, information theory, cryptography, especially in the application of asymmetric encryption techniques and message authentication. His later work was devoted largely to creating protocols that can be ensured to operate properly, even if some inputs and/or participants in the protocol may not, themselves, be trustworthy.

The following presentation is a simplified version of an idea created by Simmons in the late 1970s and early 1980s, published in a series of papers [253]–[255], as a means for such countries as the United States and the former Soviet Union to ban underground nuclear testing and have a treaty in place to verify compliance using PKC.

In the scenario below, there is no need for secrecy, only guaranteed authentication, called *authentication without secrecy*. What is being sought is authentication *without* covert channels, which means any communication pathway that was neither designed nor intended to transport data. Covert channels, therefore, would only be located and employed by adversaries.

As noted in [253], several SKC schemes were delineated, all with problems concerning authentication that the following PKC scheme solves.

◆ How is the treaty enforced?

Suppose that two countries, A and B, sign a treaty to terminate all underground nuclear weapons testing. Both A and B need to verify that the other is not engaging in underground testing. To do this, country A, say, will put seismic monitors in country B, since one of the most reliable methods of detecting underground tests is the measurement of ground motion from such mechanisms. Also, we need a *monitor*, whom we will call *Monty*, from, say, the United Nations, in country B to relay messages to country A from the sensors. (We assume that both A and B agree on the construction and placement of these devices.) In this scenario, both countries have issues.

Country A wants to ensure that country B does not alter the data, and country B needs to ensure that there is no unauthorized data being transmitted to A. Here is how both of these issues can be resolved.

The seismic device, which we will call HAL, secretly generates primes p and q for an RSA modulus $n = pq$, as well as the enciphering key e. Moreover, we assume that all the security issues discussed on pages 174–179 have been addressed and programmed into HAL, together with a CSRNG (see page 151). Thus, after the random process of generating p, q, and e, the Euclidean algorithm would be used to calculate the decryption key d. Then $n = pq$, and d would be provided to Monty, country A and country B. However, p, q, and e are kept secret within HAL, which is assumed to be deeply buried and tamper-proof. HAL gathers data m and uses e to form the information $c \equiv m^e \pmod{n}$. Both c and m must first pass muster with country B, which verifies that

$$c^d \equiv m \pmod{n} \tag{9.1}$$

so they know that m indeed is the data that corresponds to the encrypted data c. They then forward m and c to country A, who also verify (9.1). Then country A is certain that m could not have been altered. They know this since if B were to choose

$$m_1 \neq m \text{ so that } m_1 \equiv c^d \pmod{n},$$

then this is the same effort as decrypting c, which the RSA conjecture presumes is computationally infeasible (see page 175).

▼ Summary of Treaty features

(1) None of A, B, or Monty can forge messages that would be accepted as authentic.

(2) Since n and d are public, both countries A, and B, as well as Monty may verify the authenticity of messages.

(3) Since e is kept secret from all entities, no unilateral actions are possible by any entity that would be capable of lessening the confidence in the authentication of the message.

(4) No part of the message is concealed from any of the participating enti-
    ties. Hence, the above is an application of authenticity without secrecy
    using PKC, and A and B can try to cheat as much as they wish without
    compromising the system.

▼ **Analysis and Conclusions**

Simmons observed, at the end of [254], that the above mechanism has a direct
analogue for communication between international banks each having branches
in the foreign host country (see Section 5.8). He also gave the opinion that this
methodology is a paradigm for public access systems to important resources or
facilities. It turns out that this opinion was well founded as we have seen from
the multitudinous applications of PKC already demonstrated in this book.

In 1984, Simmons [256] discovered a problem with the above scheme. Although a built-in feature of the scheme is that it does not allow for a covert
channel to be built into the message (since a process is in place for country A
to verify this), HAL could still be used to hide a *subliminal channel* (see page
192). What this means is that a channel can be implanted so that country B
could not detect the use of the covert channel and could not read the hidden
part. In particular, as noted by Simmons [261] in 1993 (with reference to the
Second Strategic Arms Limitation Treaty (SALT II) between the former Soviet
Union and the United States) the subliminal channel could be used to reveal
to the other country which of those silos in the host country were loaded with
missiles and which were empty. What is striking and decisive about this fact is
that the country in possession of this knowledge would be able to successfully
launch a first strike!

In the early 1990s, Simmons [258] and [259] came up with a proposed solution
to the problem (see also [257], [260], [262]). However, in 1996, Desmedt [65]
provided a counterexample to this claim, and demonstrated how several other
protocols in the literature are susceptible to this problem. This was addressed
by Simmons [263] in 1998. The actual details, including the very definition of
subliminal-channel-free protocol is beyond the scope of this text. For details
consult [65] and [66], as well as the aforementioned papers by Simmons.

The aforementioned subliminal channel idea is even mentioned as a steganographic technique in the book on such schemes, [137, page 34].

Chapter 10

Noncryptographic Security Issues

Thought is great and swift and free, the light of the world, and the chief glory of man. (quoted in [245])

Bertrand Russell (1872–1970), British philosopher and mathematician

10.1 Cybercrime

Although (physical) bank robberies are still with us, they are rapidly taking a back seat to a newer crime requiring no direct contact: Cybercrime.

What is Cybercrime? *Cybercrime* (sometimes written *Cyber-Crime*, and also referenced as *e-crime*) is defined, in its most general sense, as any crime involving computer technology and the Internet. According to a 2003 survey conducted, in part, by the U.S. FBI, nearly eighty percent of attacks are launched via the Internet; the most likely source of attackers are either hackers or disgruntled employees (over ninety percent of economic crimes are committed by a company's own employees!); and the cost of computer crime (in that year) was in excess of 200 million dollars (U.S.). One (unfortunate) ubiquitous source of such attackers is the bored (but typically ignorant),[10.1] teenager who employs automated software packages devised by other (more informed) hackers.[10.2] In this case, only the original, inventive, cyber-smart hacker needs to have the computer aptitude to create the software, then distribute it for use over the Internet.

Cyber Terrorism: The extrapolation of the modern-day hacker is the *cyberterrorist*, who may be able to create disaster from outside a given country.

[10.1]Such a computer user with little skill, who employs the software creations of others, is called a *script kiddie*.

[10.2]For now, we will think of "hacker" as a slang term for any entity that attempts to break into a computer system. In Section 10.2, we will look at such entities in more depth.

Imagine what could happen if such entities were able to gain access to the power grid of a given nation. They could shut down or disrupt all transportation, computer networks, power to homes — in short — create a catastrophe.[10.3] If they were successful, how would the home nation prosecute them? These are issues with which governments are now trying to grapple, in order to meet the new challenges of the future. Technology has proven to advance much faster than the means to make it secure before effective countermeasures against system attacks are devised.

Cybercrime and E-commerce: Earlier in the text, we looked at numerous mechanisms for making e-commerce secure. There is good reason for this, of course. As with the scenario presented at the outset of this chapter, digital banks can be robbed, too. Your identity may be stolen (*identity theft*, see page 202), your bank account drained, or you may be conned by a scam[10.4] artist. It happens every day to unsuspecting people. In short, if it can happen in the real world, it can happen in the computer world and then some. Only the venue changes from the direct attack to the digital attack.

Privacy: Privacy invasion is also an issue, whether it be your medical records or other sensitive data that is stored on computer network databases. We have looked at numerous means for securing such information, especially in the previous chapter. There are many reasons for not wanting your personal information to be violated, not the least of which is fraud. Even a simple matter of monitoring your electricity and other utility use could allow a criminal to deduce when you are on vacation and rob your home. Do not assume that any amount of personal data is trivial or useless to the criminal element. If there is a means to use your personal information to extract money it can, and probably will, be exploited. Therefore, do not give out personal data unless you absolutely have to do so to a reputable source. Some culprits deal specifically in gathering personal information via scam "survey", or "opinion getting", for the sole purpose of identity theft. Do not be too trusting of strangers when it comes to divulging information. A telemarketer is not your friend.

A thief need not be near your home to rob you. There are numerous marketing scams that attempt to entice the unsuspecting into parting with their money. Some of the most recent are those from various Third World countries that send you a claim, by e-mail, that they cannot get their money out of the country. If you would just allow them to deposit their money into your account, you would be handsomely compensated. Once you give them your digital bank data, you will find that you are led into a scam (see the discussion of "advance fee fraud" on page 379). Do not, as the adage goes, believe you have access to

---

[10.3]This actually occurred, *without* cyberterrorist intervention, in August of 2003 on the eastern coast of North America, when an overloaded grid caused a domino effect across nearly the entire eastern front.

[10.4]A *scam* is generally viewed as a fraudulent business scheme. Typically, the perpetrators promise significant profit for insignificant risk in order to separate a person from their money or other valuables. Also, a scam may involve one or several individuals, and the scheme is given the appearance of a legitimate enterprise.

something that is too good to be true — that is the point — it **is** too good to be true.

◆ **Types of Cybercrime**: Cybercrime is so new that the very notion of what constitutes a crime is is still open to debate. Certainly, the traditionally viewed crimes, such as espionage, fraud, forgery, larceny, mischief, sabotage, and so forth, are easy to cite as cybercrimes. However, once we are in cyberspace[10.5] then things become more shadowy, and even newer crimes are introduced such as *cyberstalking*, which we will discuss below. However, the United Nations has clarified some of these issues in its document *United Nations Manual on the Prevention and Control of Computer-Related Crime*, (see *http://www.uncjin.org/Documents/EighthCongress.html*), where such activities as computer sabotage, unauthorized access, unauthorized copying, and the like are included. From the first sentence therein: "The burgeoning of the world of information technologies has, however, a negative side: it has opened the door to antisocial and criminal behavior in ways that would never have previously been possible." Indeed, and this is our focus here.

▼ **Computer Espionage**: *Industrial espionage* means obtaining proprietary information from an organization (which might be private enterprise or the government), for the purpose of aiding another company or government, but the term excludes espionage related to national security. A principal motive for industrial espionage is for a business to improve their competitive edge, or for a government to give an advantage to their domestic enterprises. Foreign industrial espionage executed by a government is usually called *economic espionage*. Typically, the greatest threat in this type of espionage is an employee within a given organization or government, who sells the information to another party.

In industrial espionage, stolen data most destructive to the victim are intelligence on pricing, processes used in manufacturing, as well as product development and description. Other types of data stolen are a company's research; its customer lists; as well as data on compensation; costs; negotiating positions; personnel; proposals; sales; and strategic goals. At the least harmful level are activities that are actually legal and are termed separately as *business intelligence*, such as gathering information from a company Web site, examining their patent records, analyzing their corporate publications, and the like in order to deduce the organization's activities.

In economic espionage, the main target is technology-related information. By obtaining intelligence on defence systems, for instance, a country could obtain information to manufacture leading-edge weapons systems without incurring the costs of research and development. They could also sell or trade the information for economic or political reasons. Espionage by foreign governments

---

[10.5]The term *cyberspace* originated with William Gibson, a science fiction (the more recent term being "speculative fiction") author, in his novel *Neuromancer* (see [103]). Today, the term cyberspace is used, typically, to mean the domain of data available via the Internet, and computer networks, as well as the virtual environment created by the Internet. Thus, the term is often employed to describe that new "virtual culture", which is emerging from these electronically interconnected communities.

might include not only economic espionage, but also certain attempts to gain access to unclassified intelligence such as civil defense and emergency readiness; technology employed in manufacturing; trips planned by high-ranking officials, as well as data on satellites; personnel; and payroll; together with files from: policing agencies; investigative departments of government (such as taxation) and security agencies.

▼ **Cyberstalking**: This new crime (also written *cyber-stalking*), spawned by the Internet, refers to the practice of sending harassing messages (which might include threats as serious as threatening another's life), by e-mail. Women are targeted more than men in this regard. Although the majority of cyberstalkers are men and the majority of their victims are women, the number of women cyberstalkers is increasing as are the numbers of adults cyberstalking children; even children cyberstalking children is a new threat. New laws have been enacted in numerous countries to deal with this new criminality.

Cyberstalkers assume they will remain anonymous, so can do as they please with impunity. When they are caught by authorities, most cyberstalkers claim they "didnt mean to do it", or did not "mean it to go so far". Contrary to what one might infer, the vast majority of cases are not from someone known to the victim, rather they are from strangers. Much of the problem stems from the victim putting blind trust in those they meet online. Do not.

To protect yourself online, use your principal e-mail account only for messages to and from those you know and trust. Use some type of free e-mail account for your more frivolous Internet activities. Sites such as *Hotmail* have such accounts for you to access. (MSN Hotmail claims to be the world's largest provider of free, Web-based e-mail.) Even there, use a gender-neutral name for your account, and do not use your real name if it can possibly be avoided. Do not fill out any automatic data features, such as chat-room profiles. If you must go to a chatroom, do *not* engage in lengthy, heated, online arguments with others. The consequences could be more than you are willing to bear. Use filtering programs in your e-mail to get rid of unwanted sources. If you do encounter a cyberstalker, do not respond. That is what they want. Instead, contact their ISP by forwarding their message, and if the harassment persists, call the police.

The dangers in cyberspace are somewhat like drinking alcohol. It lowers one's inhibitions, and allows people to say things they probably would not have said if they were in your physical presence. Again, this stems from the online anonymity factor, as well as the physical separation between entities. One might type something that would provoke a punch in the mouth or a slap in the face if said directly, but the anonymity and physical distance make some individuals think that there will be no consequences to their actions, certainly not an immediate one.

▼ **Fraud**: You might think that as long as your credit cards are "safe" in your wallet or purse, that no criminal can use them. The Internet has changed all that. In point of fact, criminals use cyberspace to target, obtain, and sell credit card data to those who would counterfeit these cards. These counterfeiters use custom-built hardware and software to embed data on plastic cards with magnetic strips. However, this is only one small facet of the total fraud industry. Let us have a look at some of the others.

1 **Accounting Fraud**: By now anyone living on the planet who has access to news media, knows about the Enron accounting scandal, which brought down one of the big-five auditors, Arthur Anderson,[10.6] along with it. In its wake, Worldcom filed for bankruptcy protection in July of 2002, the largest ever corporate insolvency. They engaged in "creative accounting", meaning that they employed unorthodox means of stating assets, income, and liabilities. When used to describe *misrepresentation* of actual income, profits, and so on, creative accounting refers to fraud, for which the aforementioned firms were convicted. Computer data files were "creatively altered" to deceive the shareholders into thinking they were far more profitable than they were in fact. In 2002, this resulted in a major market downturn (see [158] and [266]).

2. **Advance Fee Fraud**: We alluded to this when we discussed privacy issues on page 376. The scam originated in Nigeria, so is often called the *Nigeria scam* and, consequently, sometimes also the *419 scam*, after the pertinent section of the Nigerian criminal code. However, now individuals in many Third World countries actually engage in this type of fraud. Although there are other methods, the most common is now e-mail (this author gets several such letters per week, for instance), which usually says something of the following sort. Some person, Aroujo, say, claiming to be an official, usually in some government department, will suggest that, in order to purchase real estate in the country of the intended victim, say, Alice, a transfer of funds from a Nigerian bank account must have an "overseas agent", Alice. Alternatively, Aroujo might tell Alice he has millions of dollars to "discreetly" transfer abroad. In any case, Aroujo tells Alice that her bank account is required to establish the funds in her country, and for her assistance, she will be richly rewarded. If Alice agrees, then there will be a delay and Aroujo will tell Alice that in order to effect the transfer, she needs to have a Nigerian bank account with a six-figure amount in it. If Alice does this, there will be more delays all the while keeping alive the assurance of the impending transfer. Sometimes more money is requested from Alice to cover such things as bribes for other government officials. In some cases, the most gullible will be invited to Nigeria to meet the government officials and are held there for ransom.

The advance fee fraud is actually a subset of a more general confidence game,[10.7] called the *Spanish prisoner*. This dates to the seventeenth century when con artists would convince the mark that there is a prisoner (of noble birth with vast wealth), in Spain being held under a fictitious

---

[10.6]E-mail records uncovered the document-shredding coverup at the firm.

[10.7]A *confidence game*, or simply a *con* is a fraudulent scheme where a person is persuaded to buy useless property, goods, or services, or to part with their money for a phony scheme. The victim, in this case, is usually called the *mark*. If there is an accomplice to the con artist, the name given is *shill*, a person whose role is to encourage the mark.

identity. He presents himself as the person who has been entrusted to raise money for his release and cannot reveal the prisoner's true identity. If the mark buys this story and money is given, then as with the advance fee scam, there will be delays, and requests for more money, all the while keeping alive the promise of being generously rewarded at the conclusion. However, the conclusion is that the con artist disappears with the loot.

3. **Phishing**: *Password phishing* is the extracting of a password from an individual by pretending to be a legitimate person of authority. Often this scam takes the form of a message from someone pretending to be in authority and asking the victim for their password to "update your account", or "verify billing data", for instance. The term "phishing" was coined in the 1980s by crackers attempting to steal AOL accounts. If the victim gave out the password, the perpetrator would access the account and use it for criminal activity.

   A modern version of phishing involves masquerading online as a victim's bank. The victim might be sent a message saying, for instance, that due to a system error, their account has been deactivated and they have to reenter their banking data to reactivate it. The victim is provided with a link to a Web site that appears to be their bank. Once the data is entered, the criminal has the necessary information to drain the victim's account.

   To protect yourself, simply do not use anything online when presented with such a message. Contact the financial institution directly to confirm that this is legitimate. The online Web site can be very sophisticated and may be difficult, at best, to determine if it is legitimate.

4. **Pump and Dump**: This is a type of financial fraud that involves the artificial inflation of the value of some security or stock, via promotion, for the purpose of selling it at the higher price. Countries typically prohibit this practice under their securities laws. However, the Internet has made it a common and profitable practice. It functions in the following fashion.

   An entity will have a Web site touting their product via press releases, bogus mineral finds, or research claims and the like. If it is a stock, say, then investors will be urged to buy it and when this happens the price escalates, that is, is "pumped up". Then the originators of the scheme sell their stack at its peak — "dump it" — and stop promoting it. Then, of course, the stock drops like a rock and the legitimate investors lose their money. In the 2000 movie *Boiler Room* the scheme is well illustrated.

5. **Miscellaneous Computer Fraud**: Employees may alter computer documents related to their hours of work, or sick leaves taken. This is an example of perhaps the most common computer fraud, called *data diddling*, which is fraud by input manipulation, where an individual having access to data processing at the input level can alter it. Although little computer skill is required to carry this out, detection of data diddling is difficult

to achieve. Similarly, program manipulation is difficult to detect, but requires a computer-savvy individual who alters or inserts new programs, or perhaps subroutines into existing ones. The Trojan Horse is a ubiquitous method for program manipulation (see Footnote 9.7 on page 337), wherein the attacker can write a self-destruct into the program so that evidence of its existence is erased. Yet another type of computer fraud is the practice of transferring (stealing) money from accounts in "thin slices", called *salami slicing*. Salami slicing requires an institution where there are an immense number of transactions, so as not to be noticed, and for which the transactions involve more than two decimal places. Since currencies typically use only two decimals, there is roundoff after those two decimals. A salami-slicing program will round down those decimals and place the excess into an account, usually hidden. This may seem trivial, but it does not take long to accumulate millions by this scheme. Moreover, detection is difficult, and often takes place only after the culprit has left the organization.

▼ **Forgery**: *Computer data forgery* is the alteration, manipulation, or deletion of computer data for the purpose of defrauding or injuring. It may also involve the transmission of such computer data. Computer "data" means any computer-generated text, document, record, or representation, including e-mail, graphics, images, and word-processing documents.

▼ **Identity Theft**: As noted on page 376, when we discussed cybercrime and e-commerce, we have already looked at identity theft. Here we extend the discussion to include methods for thwarting such criminal activity. To protect yourself against identity theft, do not throw away data with information about you unless you destroy it first. For instance, always shred any bank statements, credit card applications or promotions you receive in the mail, credit card receipts, expired credit cards, insurance forms, and any medical statements. Thieves may pick through your garbage or recycling bins to get your personal information. Keep track of your credit card activity at regular intervals, say, once a week. Do not wait for the billing-cycle statement since you may be able to catch any unusual activity early and terminate it before it goes too far. Moreover, if you do not receive your regular billing cycle statement on time, contact the financial institution. Check your address with them since identity thieves often recover bills and change the address when they take over the account. The same holds true for any banking accounts. Keep tabs on activity in order to spot and stop any suspicious activity. If you order cheques from your bank, pick them up. Do not have them sent by mail since that invites intervention by identity thieves. Always cancel inactive credit accounts of any type. If you do not need it, close it. Such accounts are invitations to criminals.

When we talked about privacy issues above, we cautioned against giving out personal information to strangers. This is particularly important in preventing identity theft. Never give out numbers related to your personal identity that is government-related. For instance, if you are Canadian, do not give out your

Social Insurance Number, and if American, do not give out your Social Security Number, unless it is to a reputable source known to you. For those documents not under your control, such as those at your physician's office or your workplace, ask about procedures for disposal. Ensure the methods are secure, and ensure that your information is kept confidential. Use passwords on all financial accounts available to you. Moreover, do not use information such as your mother's maiden name when opening new accounts. Use a password and use one that works (see page 330 for password-selection criteria).

When online, always be suspicious of urgent e-mail requests for your personal financial data. If such is requested, do not click on a link provided. Instead go to the financial institution's WWW page of your own volition and navigate from that site. If you receive an e-mail request for financial or other personal data, do not respond. Instead, enter it at the firm's WWW page. Always call the company directly if in doubt. You are safer to check by phone than by mouse.

▼ **Larceny** *Cyberlarceny* (also written as *cyber-larceny*, and also called *cybertheft*), is the computer-facilitated theft of property. The principal factor differentiating theft of property in the "real world" from cybertheft is that the latter relies upon the electronic maneuvering of data to effect the transfer of property from the legitimate owner to the thief. In other words, cyberlarceny involves the stealing of property via the use of a computer. This may involve electronic siphoning of money (cyberembezzlement); threats employed to force a victim to surrender property (cyberextortion); or electronic communications of falsified data to deceive the victim into parting with property (cyberfraud). Cyberlarceny must be considered to be a part of what one would consider to be *cyberburglary* since real-world break and enter (burglary) is replaced by breaking into a computer system where the property is stored.

▼ **Money Laundering**: This crime refers to the practice of processing illegally gotten gains through electronic channels to make the funds appear to be legitimately obtained. The offenders conceal their true identities and locations, often using cryptographic techniques to do so.

▼ **Sabotage**: *Computer sabotage* means the use of the Internet to interrupt the normal functioning of a computer system by using "malicious code".[10.8] This could involve the use of computer viruses, logic bombs, Trojan horses, worms, and so forth, about which we will learn in Section 10.3.

There are no guarantees, but the more cautious you are, and the better informed you happen to be will give you, at a bare minimum, a sense of greater

---

[10.8]The term *malicious code* has nothing to do with cryptography in this context. Here we are (and will be throughout this chapter) using the term from a programming point of view where an attacker writes a program ("writes code"), with malicious intent. The term "to code" in this context means the act of programming. In particular, *source code* refers to the program written before it is *compiled* by a computer. The computer accomplishes this task via a *compiler*, which is a specific program whose function it is to process statements written in a given programming language and transform them into machine language ("code"). This code is then used by the computer's *processor* (see Footnote 9.25 on page 354). The compiled code used by the processor is called the *object code*.

Figure 10.1: Cybercrime.

This image is via the courtesy of the FBI homepage: *http://www.fbi.gov/*. It accompanied an article on cybercrime, especially related to piracy issues and the damage it does. We will look at these issues more closely in Section 10.4.

control over your own security. Do not leave that security for others.

◆ Cybercrime Law Enforcement

One of the first cybercrime laws to come into existence was the Swedish Data Act of 1973. It was general in scope in that Section 21 of that act incorporated protection from the unauthorized access to *all* types of data. In the 1970s and 1980s many countries began enacting laws to fight cybercrime, but as is usually the case, the criminal was ahead of the law. The United Nations adopted a resolution on cybercrime, which we mentioned on page 377 (see the URL cited there for more information). This was enacted at their eighth *Congress on the Prevention of Crime and Treatment of Offenders*, which was held in Havana, Cuba, in 1990. The United Nations cybercrime manual was published in 1994. However, many countries now see the laws of the 1990s to be woefully inadequate to deal with the new millennium cybercriminal, so many are revising, updating, and expanding those laws. For a case in point, see Footnote 10.28 on page 402.

10.2 Hackers

When you are ignorant of the enemy but know yourself, your chances of winning or losing are equal.

Sun Tzu

(See [279, page 84], as well as page 138 herein.)

◆ **What is a Hacker?**

Hacker is a term with many meanings. The press has co-opted it to mean any malicious attacker of a computer system, but those who consider themselves to be hackers put a different spin on the term. They may see themselves as computer-savvy individuals who are simply dedicated to pushing the limits of a computer system for the fulfillment of the exercise (in this section, we will talk about the pioneers who lived this point of view). Of course, those who consider themselves to be legitimate hackers might likely consider the individual who breaks into a computer system (with malicious intent) to be a *cracker*. However, the victims really do not care about labels, since damage is done to their system.

The term has changed over time. As far back as the 1920s, it meant an amateur who played with radios for the purpose of improving its performance, which may be called *hobby hacking*. The source of *academic hacking* is traced to students at MIT, where a hacker was simply a prankster, and their (technology-based) pranks or practical jokes, were called "hacks" (we will also talk about some of this MIT crowd since they form a highly nontrivial intersection with the aforementioned pioneers). *Network hacking* began with telephone networks and involved attempts to get free phone calls by reproducing certain tones into the telephone receiver. Once computer networks developed, and phone companies went digital, then network hacking took on a new computer-based meaning.

There are many sides to hacking, and this is given weight in the book, *Cyberpunk* [116], where three different stories are told from "the dark side" of hacking to the shy student who wrote a program that brought down a computer network. The *Encyclopedia Britannica* looks at hackers as "irresponsible computerphiles". Perhaps the definition of hacker lies somewhere in the middle of these various illustrations, but we must settle on a definition for our purposes that encompasses much of the new-millennium reality. Therefore, for our purposes a *hacker* will mean an individual who (legally or illegally) gains access to computer systems, or to software, to either make changes, or to inform the system administrators of security flaws. This definition encompasses several acknowledged uses of the term. At the one end, the person who is hired by a firm to discover security flaws and does this by finding weaknesses in a computer system or piece of software, is sometimes called a *white-hat hacker*. On the other end of the spectrum, is the malicious attacker type who breaks into a system illegally to do significant damage for whatever reason from being a disgruntled employee to just doing it for the perverse joy of the act. In this case, the hacker is typically called a *black-hat hacker*. It even encompasses someone in the middle of these two since there are individuals who gain access to systems and make insignificant changes, which is done largely for recognition, or just for entertainment. These are called *grey-hat hackers*. Moreover, we may have nontrivial intersections among the types since it is often the case that white-hat hackers were once black-hat hackers, and that is why they are so good at what they do.

In 1984, Steven Levy, in his book [150], defined the code of conduct of a hacker as the free sharing of information that did not allow for the harming of any data encountered. We now turn to this "ethic" since it, and the people who lived it, were the founders of a culture that was benign in intent and gave us, arguably, some of the best of the modern computer world that we enjoy today. Our "definition" of hacker, given on page 384, is a modern-day interpretation to which we will return once we look at the lives of the pioneers who saw things quite differently.

▼ **Hacker Ethic**: The *hacker ethic* consists of the conviction that information sharing is a positive goal in and of itself, and that it is a social responsibility of hackers to share their expertise by producing freeware and access to computing resources wherever possible. It may also include the notion that breaking into a system for leisure activity is ethically valid provided the hacker commits no criminal offense by altering, deleting, or otherwise tampering with data found. (Perhaps it is the latter that has led to the modern-day notion of the hacker.)

Some hackers extrapolate the above ethic, and maintain that *all* information should be free with any proprietary regulation being unethical. In 1984, a hacker named Richard Stallman (See Figure 10.2, courtesy of Richard Stallman) founded the *GNU* project based on that belief. GNU (pronounced *guh-noo*), is a recursive acronym for *Gnu's Not Unix*. He also founded the nonprofit Free Software Foundation (FSF), which is the main support organization behind the GNU Project (see *www.gnu.org*). Another organization he founded is the League for Programming Freedom.

▼ **Hacker Pioneers**

We are now going to look at those who helped to create the hacker ethic and who lived it. These individuals believed in group effort for improvement, who enjoyed the intellectual challenge, and were dedicated experts and enthusiasts.[10.9]

Stallman: He was born on March 16, 1953 in Manhattan as Richard Matthew Stallman (although he prefers his nickname given by his initials *RMS*). In 1971, as an undergraduate at Harvard University, he became a

---

[10.9]In this context, to "hack" something takes on a more general meaning and may refer to any human enterprise. In other words, in the most general context possible, a hacker is any human being who fully dedicates themselves to their craft whether it be music, poetry, mathematics, physics, architecture, dance, or whatever. If the individual has skill, dedication, and commitment that runs deep, then that is a hacker. In this sense, Richard Feynman was a hacker (see Footnote 9.31 on page 366); Einstein was a hacker; Bach was a hacker; and just about anyone at the upper levels of arts and science is a hacker.

hacker at MIT's Artificial Intelligence[10.10] Lab (AI Lab). After graduating from Harvard in 1974 with a B.A. in physics, he wrote the first extensible Emacs text editor[10.11] in 1975, largely done at that AI Lab. By 1984, he had the GNU project ready to launch, so he resigned from MIT to pursue that project. Although many others left the MIT lab in the early 1980s, Stallman was seemingly the only one with the philosophy that software should be free.

In fact, several of those former MIT hackers established the enterprise called Symbolics, which was a company interested in proprietary software. They even attempted to poach the remaining MIT hackers to work for their company in this endeavor. Stallman actually felt a certain betrayal when he was asked to sign nondisclosure agreements. In 1985, he published the *GNU Manifesto*, which he had first written in 1983. This document outlined his motivation for creating GNU, which he wanted to be Unix-compatible. In fact, by 1991, the final bugs were worked out via Linux, so that now the OS is called *GNU/Linux*.[10.12]

Although Stallman did not complete his Ph.D., he

Figure 10.2: Richard Stallman.

[10.10]*Artificial Intelligence* is intelligence manifested by anything constructed by sentient beings (namely, "self-aware" beings). AI is also commonly called "machine intelligence". Of course, the definition of intelligence itself comes into play here. Since human beings are the only sentient beings we know (so far), then we may accept, as an informal definition here, that AI is any system that can think, and act rationally as do humans (at least most of us).

[10.11]This is a text editor whose source code is freely copyable and redistributable. Moreover, it will run on most machines with differing operating systems (OS)s. Being extensible means that its usage is customizable and programmable to accomplish varying tasks limited only by the users' imagination. "Emacs" actually represents a collection of text editors that evolved from one another in some fashion. Stallman was also the author of GNU Emacs. In the early distribution days of GNU Emacs, a hole was found in the security through which malicious hackers could enter. An interesting real-life tale about one such event is woven in the intriguing book, *The Cuckoo's Egg* (see [273]).

[10.12]Linux is a free Unix-type OS originally created by Linus Torvalds with global assistance from a number of developers.

has since been granted three separate honourary doctorates from various institutions around the world, as well as many awards: the 1991 Association for Computing Machinery's Grace Hopper Award for his work on the original Emacs editor; the 1998 Electronic Frontier Foundation's Pioneer Award together with Linus Torvalds; the 1999 Yuri Rubinski Memorial Award; and the 2001 Takeda Techno-Entrepreneurship Award for Social and Economic Well-Being. Stallman is currently chief GNU of the GNU project and president of the FSF. You may get a flavour of Stallman's ideas by consulting [270].

Gosper and Greenblatt: The hacker ethic, discussed above and epitomized in Levy's book [150], is actually a reflection of the philosophy developed by Stallman and Bill Gosper, who many consider to be Stallman's mentor at the MIT AI Lab. R. William Gosper, Jr. is a computational mathematician. He and Richard Greenblatt, known as the "hacker of hackers", are said to have founded the hacker community. Gosper was associated with the MIT AI Lab in various capacities from 1965 to 1974. He was involved with MACSYMA (see page 167) from 1974 to 1999, which included his being a consultant to Symbolics, discussed above. Richard Greenblatt is a computer scientist who also spent his early days at the MIT AI Lab. He was, in fact, the principal designer of the MIT Lisp machine.[10.13] Some of Greenblatt's exploits are described in Rheingold's book [227]. The director of the MIT AI Lab, Marvin Minsky, Greenblatt's advisor, gave the hackers access to the machines, being impressed with their talent and desire to explore. Many of the MIT students interested in the AI Lab had been exploring the phone switching network and the control systems of the Tech Model Railroad Club (TMRC), near legendary as the "cradle of hackerdom".

▼ **The First Wave**: Gosper and Greenblatt were members of what is commonly considered to be the "first wave" of hackers. This first wave encompassed the pioneer hackers from the age of what we now consider to be the primitive era (those more sentimental might call it the "golden age" of hackers). Other hackers from this time period who are worthy of note are Peter Deutsch, Tom Knight, and Jerry Sussman.

Deutsch: L. Peter Deutsch began early, at the age of twelve, when he was at the MIT AI Lab where he discovered the TX-0, which was a three-million-dollar (U.S.) computer that filled a small room. It was the world's first PC for the MIT hackers who embraced it as such. It was different from computers of its time since it was interactive. This allowed hackers to explore on the machine. He hacked the TX-0 along with his older counterparts. When he was a high-school student in 1963, he created the first interactive Lisp implementation for the PDP-1 computer.[10.14] Although he grew up in Cambridge, Massachusetts, he has lived in the San Francisco Bay Area since 1964. He received his Ph.D.

[10.13]Lisp machines were general-purpose machines running the Lisp language. In the early stages of AI research, AI programs were run using the Lisp language (see Footnote 10.10 on page 386). It was in this context that Greenblatt began the Lisp machine project at MIT in 1974. Although these machines had an interesting developmental history, they went the way of the dinosaur once computers with microprocessors came on the scene. Their advent meant that desktop PCs could run Lisp programs much faster than Lisp machines.

[10.14]The PDP-1 was the first computer in Digital Equipment Corporation's (DEC)'s PDP

from U.C. Berkeley in 1973, but worked at Xerox PARC from 1970 to 1986. In 1986, he began writing the *ghostscript*[10.15] program. From 1986 to 1991 he was employed at ParcPlace Systems. He has been at Aladdin Enterprises since 1991.

Knight: Knight, Greenblatt, and others created the *Incompatible Time Sharing System* (ITS) at the MIT AI Lab in the late 1960s. It was written in Assembly Language and run on DEC's PDP-10s at MIT until 1990 after which it was run at the Stracken Computer Club in Sweden until 1995. The ITS was the first device-independent graphics terminal output; required no password to log on; any user could crash the system, and a message was sent to tell who that was; all users could converse with instant messaging on the terminals of others; and users could watch what was happening on any other user's terminal. These and numerous other truly revolutionary features were incorporated and later used in other operating systems. Even some aspects of GNU/Linux were begun on ITS. Knight was also involved in the development of the Lisp machine at MIT, which came to be the hacker's favorite machine. Currently, Tom Knight is a senior research scientist at the MIT AI Lab.

Sussman Gerald J. Sussman co-invented the *scheme* programming language, which is a descendant of Lisp. He is famed for his book, co-authored with Hal Abelson, *Structure and Interpretation of Computer Programs*, which has become a classic (see [1]). It was used in MIT's introductory computer class for majors, 6.001, developed by Sussman and Abelson, which was taught in scheme. The book [1] has become known as the *Wizard Book*, with legendary characters from it and the 6.001 class emerging, such as "Alyssa P. Hacker", "Louis Reasoner", and "Captain Abstraction".

▼ **The Second Wave**: The "second wave" of hackers were largely based on the west coast, but many such as Stallman, had their beginning at MIT. The west-coast center for this second wave was the Sanford University AI Lab, called

---

(Programmable Data Processor), series produced in 1960. They were famous both for being ground-breaking, and for being influential in creating the hacker culture at MIT.

[10.15]Ghostscript is a PS (PostScript) and PDF (Portable Document Format) translator. This means it is a program inputting a PS or PDF file and generating an alternative-format representation of it as output. Ghostscript is freeware released in two versions: one for commercial use as AFPL (Aladdin Free Public License) ghostscript; and GNU GPL (General Public License) ghostscript. The GNU GPL is a *copyleft free software license*, which is a product of the GNU project from 1988. *Copyleft* is an application of the copy*right* law whose purpose is to force derivative works to be released with a copy*left* license as well. What this means is that any number of individuals may make successive improvements, but those who refuse the terms are not allowed to create derivative works. *Derivative work* is work based upon preexisting work. With respect to software this includes a translation of a computer program into another language or conversion of an existing program onto a new platform. The derivative work is separately copyrightable, but applies only to the *new* work added as well as any amalgamation of the new and old parts of the work. Stallman created the copyleft concept after he supplied a public domain version of his Lisp interpreter to Symbolics, who made improvements to it. When Stallman asked for access to those improvements, Symbolics refused. In 1984, he created a software license to prohibit such behaviour, whence the copyleft idea, to thwart what Stallman called "software hoarding". Think of the purpose of copyleft, with respect to freeware, being to ensure that derivatives of licensed work remain free.

SAIL, under the direction of John McCarthy (see page 167). SAIL closed in 1991, but business research centers such as AT&T and Xerox also had what came to be known as hackers of the second wave due to their expertise. Among this second wave, in addition to Stallman, were Ed Fredkin, Jim Gosling (see Figure 10.3, courtesy of Jim Gosling), Brian Kernighan (see Figure 10.4, courtesy of Brian Kernighan), Brian Reid, and Dennis Ritchie.

Fredkin: Ed Fredkin had no university degrees, yet he learned about computers in the U.S. Air Force in 1956, since he was one of the first to work on the SAGE (Semi-Automatic Ground Environment) computer air defence system. SAGE was established in 1954 by the U.S. Air Force to protect against nuclear bomber attack from the USSR. MIT established the Lincoln Laboratory[10.16] in Lexington, Massachusetts, to produce the SAGE system design. In fact, after leaving the service, Fredkin took a job at the Lincoln Lab, where he earned a reputation as a top-notch, original programmer, so much so that some of his algorithms became standards. To Fredkin hacking meant pride in his abilities at "code crafting." He once commented that nobody could "outcode" him, that is until he met Stewart Nelson, who arrived at MIT in the fall of 1963 as a freshman. Nelson was a brilliant programmer who would work for Fredkin, Gosper, and Greenblatt, the latter three of whom had been hired as full-time hackers at the AI Lab in 1965.

Fredkin is responsible for the founding of more than a dozen institutions, including the famed Information International Incorporated (Triple I), and has served as CEO of several companies including Triple I and RadNet. He has had professorships at MIT and other universities, as well as the directorship of the MIT Laboratory of Computer Science. He is currently Distinguished Service Professor at the Robotics Institute at Carnegie Mellon University in Pittsburgh.

Gosling: While a student at Carnegie Mellon, Gosling wrote a version of Emacs, which he gave to friends and which he negotiated with Unipress Software Inc., of Edison, New Jersey, to sell commercially.

When Unipress tried to sell this version of Emacs, they came in conflict with Stallman's copyleft (see Footnote 10.15 on page 388). After a bitter dispute between Stallman and Unipress, the conflict was ended by rewriting certain aspects.

Gosling's biggest claim to fame is that he is the inventor of the Java programming language. Gosling is now a Fellow at Sun Microsystems Inc., where Java was developed. Java is a computer language specifically designed for writing programs that can be downloaded from the Internet, safely,

Figure 10.3: Jim Gosling.

[10.16]In 1958, MITRE Corporation was formed from the Computer System Division of Lincoln Labs, where the software was developed for SAGE's digital computer system. (See page 97 for a discussion of Horst Feistel's involvement with MITRE.)

specifically without the possibility of transferring a virus or other damaging vehicle to your computer. Java employs miniprograms, called *Applets*, whose functionality includes adding animated images to WWW pages, and generally allows the user to interact with those pages. Java is platform-independent by using *virtual machines* permitting applets to run on any given OS.

Kernighan: In 1970, Brian Wilson Kernighan coined the name *Unix* as a pun on *Multics*, which used enormous system resources and had grown out of control during its development at MIT. The patience of AT&T Bell Labs was waning in terms of the amount of time the development of Multics was taking and basically got tired of waiting for it to be useful, so by 1969, they stopped supporting the system. In that year Unix, the hacker OS of choice, was born.

Kernighan is best known for his book (co-authored with Dennis Ritchie) on the C programming language (see [135]), and for being a co-inventor of the Awk programming language (after the initials of its inventors: Aho, Weinberger, and Kernighan). Awk has a C-like syntax with pattern matching, associative arrays, no declarations, and implicit type casting. The text [135] along with his book [134] on Unix are considered to be classics. He is also known for one of the four major C "indent" styles, called *K&R style*, after Kernighan and Ritchie, also called *kernel style* because the Unix kernel is written in it. The C indent styles all have the goal of making it easier for the reader to visually track the scope of control constructs.

Figure 10.4: Brian Kernighan.

Reid: Brian K. Reid was a hacker who was attracted to computing as a way to avoid the Vietnam-era draft. He is known to have said that due to knowledge he gained breaking into systems when he was young, he was asked to assist in apprehending culprits who broke into other systems. His later education came when he attended Carnegie Mellon as a graduate student. He is perhaps best known for developing *SCRIBE*, a document specification language for the first laser printer and predecessor of several desktop-publishing programs used today. See [97] for more on the life of Reid.

Ritchie: Dennis M. Ritchie, was born on September 9, 1941, in Mount

Vernon, New York. As noted above, he first published the book on the C programming language [135] with Kernighan. His invention of C was for use with the Unix OS, accomplished during his work for AT&T Bell Labs in 1969. In 1983 he was awarded the ACM's A.M. Turing Award (see page 172), along with Ken Thompson, who developed Unix. Moreover, in 1984 he received the IEEE's Pioneer Award (see Footnote 9.2 on page 332).

▼ **The Third Wave**: This group of hackers were the first to be independent of the MIT group. They arose out of northern California, especially the San Francisco Bay area. They were different than the first two waves in that they wanted, and in some cases built, their own computers, and hardware in general. These were the individuals responsible for the PC revolution we know today. Among them are Steve Dompier, Bill Gates, Steve Jobs, and Steve Wozniak. (Jobs and Wozniak are shown in Figure 10.5.)

Dompier: In the mid-1970s, at a meeting of Silicon Valley's Homebrew Computer Club, Steve Dompier showed how he had programmed new MITS Altair 8800 (one of the world's first personal computers) to play the Beatles' *Fool on the Hill* for which he received an unrestrained standing ovation. Even Wozniak (see below) is known to have said that he had no intention of starting a company when he built a computer. It was merely to "show off" to an assembly of that club at one of their meetings. Hence, this club was the catalyst for much ingenious activity and invention for the hacker community. In mid-1975, Louis Solomon, technical editor of *Popular Electronics* magazine, went to Processor Technology to ask about the development of a computer terminal about which he could write an article in his magazine. In July of 1976, an article appeared about a complete computer design, called the *Sol Terminal Computer*, or simply the *Sol-10*, which used the Intel 8080A processor in kit form. "Sol" was, of course, an abbreviation of Solomon's name. Dompier produced the OS for Sol-10, called *CONSOL*. Dompier used the Sol-10 to write a computer game called *Target*, which consisted of a little cannon on the bottom of the screen used to shoot down a series of alien spaceships moving across the top of the screen. Dompier called it a "clever little hack" that he basically gave away since he merely wanted people to have fun with it. Later, when it was presented on the TV show *Tomorrow*, it intrigued the host Tom Snyder so much that he had to be torn away from the game to finish the show. For more on the life of Dompier, see [150].

Gates: In 1975, Bill Gates wrote an interpreter for the programming language BASIC and charged money for it, which was not done at that time. This practice began a new approach to the idea of software development, not seen before since such software development was seen as a hobby and not a business enterprise. However, this was, in and of itself, a betrayal of the hacker ethic. Thus, to many "true" hackers, Bill gates is a traitor for making money from what they deem should be free.

Gates is the co-founder (together with Paul G. Allen) of Microsoft Corporation. Gates is currently Chairman and Chief Software Designer at Microsoft.

Figure 10.5: Steve Jobs (with a blue box) and Steve Wozniak in 1975.

The photo is reproduced with permission of woz.org. The photo was taken by Margaret Wozniak, Steve's mother. Thanks to Alan Luckow of woz.org for the information.

According to *Forbes* magazine, he is the richest person in the world, and Microsoft is the world's largest and best-known software manufacturer. The company was founded in Albuquerque, New Mexico, in 1975 by Gates and Allen to develop and sell BASIC interpreters under the company name *Micro-soft*.

Jobs and Wozniak: Steve Jobs and Steve Wozniak are the co-founders of Apple Computer (1976), with the immensely popular Macintosh PC. Wozniak is also known for his construction of devices, called *blue boxes*, for use in making free telephone calls, called *phone phreaking* today.

Currently, Jobs is CEO of Apple Computer and Pixar, while Wozniak is Founder, Chairman, and CEO of Wheels of Zeus (wOz), which is an organization, based in Los Gatos, California, that has produced a wireless platform designed as a communications tool for both commercial and individual consumption. In January of 2004, it was announced that Motorola plans to employ the wOz platform in aspects of their wireless technology. In fact, Wozniak prefers the nickname *WOZ*. Also, Jobs co-founded Pixar, the animation studio responsible for such academy-award winners as *Finding Nemo* (2003).

▼ Modern-Day Hackers

Now that we have looked at the hacker culture in the sense that it was initially intended, we delve into the modern-day reality of what it has become to many (although the above hacker-culture still lives a separate life in many circles).

We now refer to the definition of hacker that we gave on page 384, where we are looking at gaining access to systems. When the activity is illegal, which is now our focus, the type of modern-day hacker is determined by their goals. The following provides a general delineation of the kinds of intruders we may encounter.

Hacker Types

The hacker with malevolent intent falls into one of three categories.

1. **Covert User**: This is a hacker who somehow gains access to *sysadmin*[10.17] status for the purpose of evading auditing and access controls.

2. **Impostor**: This is a hacker who obtains access control illegally for the purpose of exploiting a legitimate user's account.

3. **Trespasser**: This brand of hacker gains illegitimate access to computer resources (which might include programs and data), with malignant intent. This type also includes those who have legitimate access to such resources, but abuse those privileges.

The tools used by hackers are varied.

Hacker Tools

1. **Exploit**: This is a prepared application that exploits a known weakness in the system. In this case, strong encryption techniques win the day. With no encryption or weak encryption to thwart them, there are tools available to the modern-day hacker that were only a dream a decade ago. For instance, in 1997, a source code was released in *Phrack Magazine* (see [63] and *http://www.phrack.org/show.php?p=50&a=6*), for a TCP hijack tool called *Juggernaut*. In this case, even a script kiddie could hijack a TCP session without understanding how it is done. Not only can the hacker eavesdrop, but also can replace data in packets, and hijack the TCP session by running a daemon (see page 323), which suppresses the legitimate user. That user does not even know the attack is taking place. Rather, to the legitimate user, it will seem that the TCP connection has been dropped (which is not all that unusual, so typically, suspicions will not arise). Hence, as mentioned earlier, strong encryption is necessary to thwart the Juggernaut hijacking and any other such tools. If strong encryption is properly set up, no such tool will ever succeed.

[10.17]This is a system administrator, who is a person with the privileges to administer a computer system and keep it functioning.

2. **Leet**: This is an abbreviation for *elite*, and is jargon employed by hackers to refer to themselves or to the sites they visit for the purpose of sharing pirated software and data.

3. **Rootkit**: This is a basic toolkit for disguising the fact that a computer's security has been compromised. Such kits might include a substitute for system binaries[10.18] rendering it impossible to see activities run by the hacker in the active process tables.[10.19] There are programs such as *rkdet*, which is a small daemon for detecting someone installing a rootkit or running a packet sniffer (see *http://vancouver-webpages.com/rkdet/*).

4. **Social Engineering**: We briefly mentioned social engineering attacks on page 331. Basically, any technique that exploits human weakness or general gullibility can be employed. It consists of using nondigital means to gain digital information from a victim, the most common being masquerading as a bank official to get a person's PIN on the claim that it is needed to fix something concerning the account. Essentially, social engineering attacks involve the obtaining of a person's trust so they will disclose information to the hacker. See [163] for more information.

5. **Vulnerability Scanner**: This is a tool to scan computers for weaknesses. This might include *port scanners*, which check the *open* ports on a computer that are available for access. For example, there is the *nessus* scanner (see *http://www.nessus.org*), which employs modules, so it can be expanded.

   Also, we may include in this category, *brute-force password hacking*, which is software that, given a unix password file or MS Windows registry keys for authentication, goes through a list of common dictionary words to reveal any insecure passwords on the system.

Some of the tools mentioned above were designed for legitimate use to discover security holes in a given system, but as with anything in life, there can be a "good" use and a "bad" use to which they are put.

Defence

There are numerous techniques to protect computers from hacker attacks. We list but a few of the more important ones.

[10.18] A system's *binaries* are the binary, machine-readable forms of programs that have been compiled or assembled, but not to the "source" language forms of programs. In other words, binaries are source code that has been compiled into executable programs. If we are talking about GNU/Linux (see page 386), there are three possibilities: (1) software is distributed as source code only; (2) software package includes both source and binaries; (3) software contains only binary format.

[10.19] A *process* is a program in execution, which may have several states: new; active; waiting; ready; and terminated. An *active process table* is a set of data structures used to represent the process. Thus, if the active process table is compromised, a sysadmin, for instance, cannot see the activities of the hacker.

1. **Audit Records**: A basic tool for detecting intruders is some form of audit record. There may be a *native audit record* that automatically collects data on all user activity. However, it is difficult to be precise about something you are looking to discover, so for that you require *detection-specific audit records*. Audit records may be used quite effectively in conjunction with *statistical anomaly detection*, which analyzes such data over a period of time to determine the profile of the average user. Then an intrusion-detection model may be assembled from this data.

2. **Disable**: One security measure you can take is to disable inactive accounts since these are weak points that hackers love to attack. Also, a sysadmin should not run any servers or daemons that are not needed. Moreover, if such are needed infrequently, then disallow anonymous access, such as anonymous ftp.

3. **Firewalls/Gateways**: We discussed gateways and their associated firewall security at length in Section 8.4. Such measures go a long way toward thwarting hacker attacks.

4. **Intruder-Detection System (IDS)**: There is some overlap with the discussion above since we looked at one form of IDS, namely, statistical anomaly detection used in conjunction with audit records. There are other types as well. If you have a *network-based* IDS, it eavesdrops on network traffic, seeking evidence of an intrusion. A *host-based* IDS might scan for incoming viruses, or altered files, for instance. However, certain IDSs have false negatives (intrusions missed by the IDS), and false positives (false alarms, when an IDS incorrectly concludes there is an intrusion). Anomaly detection, such as the one described in the audit record section, is an IDS that will usually get both false signals. See [119] for instance, which looks in detail at the problems involved with getting an IDS to function in a desired manner. Host and network-based IDSs work in concert. While a network-based IDS is a unit unto itself, so it is attack resistant; a host-based IDS is aware of the state of its computer, so data flow is simplified.

   A recent innovation in the IDS arena is the *honeypot*, which is a decoy system whose function it is to lure the hacker away from sensitive areas in the system. A recent example may be found in [269], which is a honeypot that imitates a complete network. Honeypots work best as a dedicated network-based IDS. Honeypots will not only redirect a hacker from sensitive systems areas, but also collect data about the hacker's activities and possibly keep them online long enough to have a sysadmin take action. This type of IDS is perhaps one of the most effective in existence today, since it not only protects, but also sets a trap for the hacker.

5. **Password Protection**: Your first line of defence should be proper password protection. We have discussed proper choices for passwords on page 330. Moreover, we discussed protocols in Section 5.2 for securely establishing passwords along with authentication, such as SRP-6 (see page 200).

Software such as Nessus, discussed above, was written with the legitimate intention of assisting sysadmins to secure their systems, so although they may be used by hackers, they may be employed by sysadmins to patch the holes in their security schemes as well.

6. **Restrict Access**: You may restrict access to your computer or a sysadmin may restrict access to the server, the latter by specifying a command in a configuration file, and the former using software associated with a firewall, for instance.

7. **Tiger Teams — Sneakers**: These are typically temporary teams formed for the purpose of breaking down the defenses of computer systems, penetrating security, and thus testing security measures in an effort to uncover, and eventually patch, security holes. These may be white-hat hackers, called *sneakers*[10.20] in this case. Hiring reputable people is key here since having anyone poke around inside your computer system is a very risky business. We will not even mention the political aspects, only the technical ones. Testing firewalls is an important exercise since certain weak versions exist and they should be ferreted out. Moreover, this should not be a one-time endeavor. In other words, use a reputable tiger team at regular intervals since environments change quickly in the computer world.

◆ **Conclusions**

There is a thriving community of (nonmalicious) hackers who live the original hacker ethic discussed above, and who deem it to be morally wrong to maliciously hack into a system. They see themselves as the "real hackers" and the modern-day element as the criminalization of the term. The dichotomy is recognized, for example, in Stoll's book, *The Cuckoo's Egg* (see [273]). At the beginning of the book, a colleague is quoted as saying: "Joe's a hacker of the old school. A quick, capable programmer. Not one of those punks that have tarnished the word 'hacker'." (see [273, page 7]); and near the end of the book, Stoll recognizes the "golden age" of computing when he says ([273, page 371]): "I wish that I had lived in a golden age, where ethical behavior was assumed; where technically competent programmers respected the privacy of others; where we didn't need locks on our computers." Thus, as we started this section, we conclude it: there are many meanings of the term "hacker".

Now that we understand the various nuances, we can bring the entire field of vision on this topic into focus.

[10.20]The movie, *Sneakers*, starring Robert Redford is an entertaining look at this phenomenon.

10.3 Viruses and Other Infections

The nature of bad news infects the teller.

William Shakespeare
— from *Antony and Cleopatra*, act 1, scene 2

One of the hacker tools we deliberately omitted from the list in Section 10.2 is that of the *virus construction kit*, since it fits into this section as a convenient segue. Basically the name says it all, and usually these kits come with a GUI (see Footnote 8.28 on page 328), as well as instructions on how to use the kit, so that even script kiddies can use them (unfortunately) with effect, requiring no no knowledge of how they work. Now we will learn about how viruses function.

◆ **Viruses**

A *virus* is a hidden, and typically malicious, program that "infects" your computer, by copying itself into and becoming part of, another program called the *host program*, without which the virus cannot run. The effect varies from the merely annoying to the completely destructive. Viruses might delete files, erase programs, or even your entire hard drive. On the other hand, they may just flash the message "gottcha" without end, for instance. We will learn about the various types of viruses and how they work in this section. We will learn later about other types of infections that do not need such a host program to infect a computer. Moreover, although most viruses are written with a computer in mind, such as those that will only attack a PC, but not a Macintosh, for instance, there are platform-independent viruses (see macro viruses below).

The most common vehicle for infection today is the Internet, and many arrive by e-mail. Downloading files from the Internet or opening an e-mail file may trigger a virus. However, even the exchange of infected disks is a mechanism for spreading infection.

The term "virus" from the biological realm is used here since the computer virus acts in a similar manner to an infectious disease. A biological virus is a string of nucleic acid (DNA or RNA), which may infect a living cell by assuming control of it and instructing it to replicate the virus many times over. Similarly, computer viruses attach themselves, replicate themselves, and spread in a manner akin to a biological one. They may take control of the computer's OS, for instance, and whenever a new piece of software is encountered, it copies itself to that new program thereby infecting it. With the Internet, where you may access resources running on other computers, there is a rich culture for the spread of this kind of infection.

Once a program is infected with a virus program, it becomes the host. The virus program runs secretly when the host program is run, since it stays hidden in the legitimate program, remaining dormant until the infected program is run (or as we will see below, until an infected data file is accessed). A virus may be embedded in an executable program, then once run, the virus code is executed first, then the original program code. The following are the aspects of a computer that a virus attacks.

Virus Targets

Viruses may infect any of the following:

1. **Executable Program Files**: An *executable program* is a set of instructions that can be input to the memory of a computer and executed. In other words, it is a program that may be run as a self-contained procedure, which consists of a main program and, possibly, one or more subprograms. Usually, the name of such a program is all that is required to run it, merely the typing in of the program name and requesting that the computer run it.

2. **File Directories**: A computer's file-directory system keeps track of the location of data files, and without them the computer will not function.

3. **Macros**: Today virus programs can be written so that, for instance, it may attach itself to a macro[10.21] and is launched whenever the macro is run. When we discuss "macro viruses" later, we will see that Microsoft Word (MS-Word), documents are virtually always the target since they contain programs, the macro language, which are automatically executed when one of these "data" files is opened.

4. **System Sector**[10.22]: The system sector refers to special areas on the computer's hard drive containing programs that are executed when the computer is booted. These are not files, but rather small segments of the hard disk that the hardware reads as a single unit. The system sector is required for the normal functioning of the computer, even though they are invisible to normal programs. Sometimes this is called the *boot sector*.[10.23]

How Viruses Work: When an infected program is run, the first action is to invoke the virus program and run it, since this is the first instruction line of the controlling program. The second instruction is for the virus program to check to see if the program it is about to infect has already been infected or not. The mechanism by which this is accomplished is a message, called a *v-marker* or *virus marker*, which the virus program places in the legitimate program.

[10.21] A *macro* is a collection of instructions stored in an executable form, usually written to automate a few steps. Macros may be application-specific, such as a word-processing macro that executes certain steps within that program; or general-purpose, such as a keyboard macro that types in a user's login name when a specific short sequence of keys is pressed on the keyboard.

[10.22] A *sector* is one of the areas (or "pie slices"), into which the disk is segmented. This division of the disk into pie slices is the method of organizing it for access of data to the read-write heads of the disk drive. Moreover, the disk is further divided into concentric circles, so that a given area can be located via the intersection of a given sector and the concentric track passing through it. There are further subdivisions of the tracks into what are called *clusters*, which are the storage units (usually 256 or 512 byte-lengths, which are minimal in terms of allowing the unit to be addressable).

[10.23] *To boot* a computer, also called *booting up*, is the action of loading an OS into the computer's main memory (RAM), see Footnote 9.15 on page 346. On a large computer or a mainframe, booting is sometimes called *initial program load* (IPL). *To reboot* is to reload, or in the case of larger machines, to *re-IPL*.

If the virus program encounters a v-marker, it does not replicate there since it knows that the program is already infected. Then it seeks uninfected executable files (those without v-markers) and infects them. If a virus begins by infecting a program, then each time that program is run, it seeks out uninfected files. Often the virus is embedded in a game, or utility.

Once a virus program determines that there are no more files to infect, it may begin to damage the computer and its data. The virus program may corrupt program or data files so that they either work erratically or not at all. They might destroy all the files on the computer or alter the system files needed to reboot, or any other of a number of damaging actions.

Now we look at the evolution of a given virus from its initial infection to its end goal attacks.

Stages of a Virus

1. **Infection Stage**: The virus infects some area of the computer as discussed earlier. Some viruses then remain dormant until a "trigger" sets it in motion while others go to stage 2 immediately.

2. **Replication Stage**: In this stage, the virus reproduces itself onto other programs using the initial infected program to do so. Then each new infected program will undergo the same replication stage.

3. **Activation Stage**: The virus is triggered to perform its end goal. The trigger may be any number of events from the time of day, the date, or any other event such as the number of times the program is executed.

4. **Execution Stage**: The virus performs its end goal, which may range from erasure of the computer's hard drive to the merely annoying, including simply slowing down the performance of the computer.

Types of Viruses

1. **Boot- (System-) Sector Viruses**: These kinds of viruses infect the *master boot record* (MBR).[10.24] When a computer is rebooted, the virus spreads its infection.

2. **File Viruses**: File-infecting viruses attach themselves to executable program files. Once the program is loaded, the infected program is executed and seeks out uninfected executable files.

[10.24]The *master boot record*, also called the *partition sector*, is the first sector of a computer's hard disk, which indicates the location of the OS and the methodology for finding it. This is necessary for the booting of the OS into the computer's RAM. The MBR is also called the *master partition table* since it contains a table that houses data on each of the hard disk's partitions. The MBR also contains a program whose function it is to read the boot sector record of that partition that contains the OS to be booted into RAM.

3. **Memory-Resident Viruses**: This kind of virus stays in memory after it executes and after its host program is terminated, whereas a *nonmemory-resident* virus only activates when an infected program executes.

4. **Polymorphic Virus**: This is a particularly nasty virus that mutates every time it infects a new program. Therefore, detection of this type of infection is difficult since it leaves no unique trail ("signature") to follow.

5. **Stealth Virus**: This kind of virus is specifically designed to disguise its existence from virus-scanning software. For instance, if a stealth virus has infected the MBR, then its function might be to interrupt a virus-scanning software's request to examine the MBR and then transmitting a (false) copy of the original *uninfected* MBR.

Examples

An example of a virus that is a combination of some of the above is the following.

Multipartite Viruses: These viruses infect in one format type, then transform into another. For instance, one might begin as a boot-system virus, then move to become an attack on executable files.

An example of a memory-resident virus is the following modern-day virus that takes advantage of features found in data-processing software.

Macro Viruses: This type of virus is one of the most recent, and unlike the others, is platform independent. In other words, it will infect those using a Macintosh computer as well as those using Microsoft Windows, for instance. The reason is that these viruses are programs written to attach themselves to macros used in modern-day data-processing systems, such as MS-Word, MS-Excel, and AmiPro. These macro languages fit the three conditions that make them ripe for macro infection, namely, they (1) assign specific macro programs to specific files; (2) copy macro programs from one file to another; (3) pass control to some macro program without the user's explicit permission, that is, they are automatic. The aforementioned word-processing systems were designed to be automatic, and as such, if an infected document is opened, the viral macro will replicate itself into the computer's startup files. From then on, the machine is infected and the macro virus will reside on the computer until eradicated. Any document on the machine that uses the infected application can then become infected. If the machine is on a network, the infection will likely spread to other machines on the network. If a disk with the infection is shared, then the virus will spread to the recipient's machine. Today, macros are deemed to make up two-thirds of all computer viruses according to experts.

The typical agent for spreading macro viruses is via e-mail. The most notorious macro virus was *Melissa*,[10.25] launched in 1999. Melissa was distributed

---

[10.25]In some circles, Melissa is considered to be a worm (see below for our description of worms), since it clogged up systems. However, due to its behaviour as a malicious e-mail attachment and its mechanism for delivery, it is more rightly viewed as a macro virus.

by e-mail and applied to MS-Word documents. Moreover, those recipients who opened the documents found that the first fifty people in their address books also received the virus. This was so effective that on Friday, March 26, 1999, Microsoft Corporation was forced to disable incoming e-mail. Melissa operated by incorporating a message that told the recipient that an important (secret) message was contained in the attachment. Once opened, the infected file was read to the global macro file. Then the virus employed the visual basic language[10.26] to read the first fifty names in the address book, and send them all the virus.[10.27]

Macro viruses are memory-resident since they are active not only when the infected documents are opening or closing, but for the entire time the system is running.

Melissa suggests that e-mail is becoming the medium of choice for attackers and this is indeed the case.

E-Mail Viruses: Malicious software employing e-mail is becoming more common with each passing day. Melissa was just the beginning. More powerful versions of e-mail viruses have emerged wherein the virus is spread to all the e-mail addresses within the address book of the infected host. Thus, the rapid deployment of e-mail viruses is now a major threat.

On Thursday, May 4, 2000, a new e-mail virus called the *"I Love You"* virus, also called the *love bug*, spread itself around the world in a matter of hours. Its name is derived from the fact that it contained a message to check the attached "love letter", which was a file in Visual Basic containing the virus. If the e-mail was deleted without opening the attachment, then the computer was safe. However, if opened, the computer was infected and the virus was distributed via e-mail employing MS-Outlook's address book. This was an advance in the degree of malevolence over Melissa since the latter only sent to the first fifty addresses, whereas the former sent to everyone in the address book. The love bug was much more destructive than Melissa since it copied itself into two vital system directories and added triggers in the Windows registry. This meant that every time an infected computer rebooted, the love bug was executing. It infected data files by overwriting them using Visual Basic, and deleting the original file. Typically files associated with WWW development, and multimedia files were extinguished, such as those of type MP3 (music) and JPG (images). An example, to illustrate the magnitude of the losses, was reported by the Norwegian photo agency Scanpix, which lost over six thousand of its photos, and was able to recover less than twenty-five percent of them. The love bug only affected versions of the Windows and NT operating systems, so Macintosh and Unix platforms were safe. Yet this was enough to cause billions of dollars in

---

[10.26] *Visual Basic* is a graphical programming language introduced by Microsoft in 1990. It is used for developing GUI Windows applications.

[10.27] David Smith, who wrote Melissa, was caught within a week of the virus hitting the Internet. Although he pleaded guilty and was sentenced to ten years in a New Jersey state prison, his sentence was reduced to twenty months when he cooperated in thwarting attacks and aided in the arrest of other hackers.

damages around the globe.[10.28]

In October of 2002, the *Bugbear* virus infected Windows platforms through a hole in the security system in MS-Outlook, MS-Outlook Express, and Internet Explorer. Once a machine was infected, the virus copied all passwords and credit card numbers typed by a user, then it sent the information to numerous e-mail addresses. It was estimated that in the first week it sent roughly 320,000 e-mail messages. In 2003, the virus appeared in a more virulent strain called *Bugbear.B*, which took only one day to cause the damage the previous strain had caused in three days. The reason was that a flaw in MS-Outlook allowed the program to automatically open e-mail attachments. The perpetrator of the Bugbear strains has not been apprehended.

Virus Detection and Prevention

The following steps may be taken to protect and defend yourself from infection by computer viruses.

1. **Check before Use**: Before using any floppy disk or downloaded files, always run a virus-scanner program on them. There are numerous reputable vendors who have relatively inexpensive (or in some cases free), virus-scanning software available. Moreover, updates will be provided as a service by the vendor. As we have seen, the race to beat the attacker is based on knowing what is out there. You should also use the software to do a virus scan after each reboot of your computer.

2. **Create Emergency Disk**: For the worst-case scenario where you get infected and you cannot reboot your machine, the only saviour may be an emergency disk that you have set in advance to use for that scenario. Ensure that the disk is write-protected at the time it is created.

3. **Disable**: Do not allow the enabling of such automatic features as the opening of e-mail attachments, downloading of files, or the like. Disable these features.

4. **Documents (MS)**: Do not open any MS-Word document unless you are certain it is not infected. Remember not to view these as "data files", since they may be infected with a macro virus.

5. **E-Mail**: Be cautious in the extreme about e-mail that you receive, even if you know and trust the sender very well, since anyone may be an unwitting victim. The above-described scenarios should be enough to convince anyone of that. If there is an attachment, especially if it is an executable file, you must verify that it is virus-free. Delete it if there is doubt, or if you believe it to be valid and from a valid source, contact that source *before* opening it. Ask them what is in the file, whether they know if it is

---

[10.28]The author of the love bug, Onel de Guzman, of the Philippines, was never charged with a crime. At the time, there were no laws against cybercrime in the Philippines. Although such laws exist now, he cannot be charged retroactively.

virus-free from having scanned it, say, and why it has been sent to you. Then, and only then, should you attempt to open such a file.

6. **Infection Detected**: If your virus-scanner detects an infection, locate the virus, identify it, and use the software to remove all traces of it. The virus must be removed from all systems in order to restore your computer to health. Remember, it is *detection, identification,* and *removal* in the case of a viral infection. If it is not possible to either identify or remove the virus, then the infected program should be discarded and a new clean backup copy should be reloaded.

7. **Software for Blocking**: Some more sophisticated software exists for the purpose of actually blocking behaviour that is deemed to be malicious. Again, reputable software vendors have numerous such devices available. For instance, there are *Internet filters*, which will screen out any e-mail related to pornography, violence, or other such offensive material as well as potentially malicious e-mail. There are *spam blockers*, to prevent all sorts of irritating e-mail from getting through to you, not just the infected kind. There are *e-mail virus blockers*, which should take care of effectively protecting your computer by identifying and blocking potentially dangerous attachments.

If you are a large corporation concerned with ferreting out holes in your security, for instance, there is *Bugscan*, by HBGary^{TM}, starting at the modest price of $19,500 (U.S.). It will audit code for security gaps, including WWW-based administration and reporting interfaces. However, for the individual with a somewhat smaller bank account, there are numerous scanning devices in the $40 (U.S.) range that work quite nicely, such as the Norton^{TM} AntiVirus package.

Advanced Protection

There exist modern methods that excel in their ability to protect from and eliminate attacks. We look at two of the most common and most effective.

Generic Decryption: Polymorphic viruses may require more sophisticated software. The most modern such device is called a *generic decryption engine* (GDE). Basically a GDE tricks a polymorphic virus into decrypting and revealing itself. If a scanner with GDE is installed, then it makes three assumptions: (1) the body of the polymorphic virus has enciphering to thwart detection; (2) the virus must decrypt before it can execute; and (3) once a polymorphic virus does execute, it must immediately assume control of the computer to decipher the body of the virus, after which the control of the machine is taken over by the completely decrypted virus. The GDE loads each new program file into a self-contained virtual computer that is generated from RAM. It is inside this virtual computer that the program files run as though on a real computer. Therefore, a polymorphic virus can do no damage since it is running in the virtual computer, which is isolated from the real computer. The virtual computer allows the virus

to decrypt after which the virus body is exposed to the GDE scanner, which
can identify the strain via a signature. If there is no virus to expose, the GDE
stops execution and drops the program, proceeding to the next file. Think of
the GDE as a rat and think of the files loaded to it as injections given to the
rat to detect the presence of a virus. If there is no adverse behaviour in the rat,
there is no virus in the injected substance, whereas if there is, then the rat is
observed for symptoms that will identify the virus.

A GDE scanner has five basic components: (1) a processes emulator; (2) a
memory emulator; (3) a system emulator; (4) a virus signature scanner; and (5)
a decision mechanism. The process emulator is an imitation of a CPU, which
reads the instructions in an executable file. This includes software versions of
all registers and other CPU hardware, so the actual processor is unaffected. The
memory emulator imitates the memory of the computer, where the emulated
memory is employed instead of real memory. The system emulator actually
imitates the OS and hardware of a computer. This should also include a virtual
drive that is capable of being read, formatted, and so on. The virus signature
scanner is a module that scans the program code of the loaded file for known
virus signatures. This module interrupts the GDE process to return it to the
scanner for it to look at the code for signatures. The decision as to when to
interrupt is given by the decision-making mechanism, which may be the most
vital part of the GDE since we want to ensure speed. Thus, proper decision
making must be made so that the optimum use of the GDE is ensured. The GDE
innovation seriously reduces the time taken to analyze polymorphic viruses, from
weeks to minutes.

The second type of antivirus device is a comprehensive virus protection
mechanism developed at IBM in the late 1990s. For more data on the original research papers from IBM and related development go to the following site:
http://www.research.ibm.com/antivirus/.[10.29]   In 1999, Symantec entered into
a licensing agreement with IBM to market the idea as antivirus software for
business and personal computing, officially released as a commercial product in
October of 2000.

Digital Immune System (DIS): The idea is, as the title suggests, to
mimic the human immune system in a computer so that a virus is automatically
captured as it enters a system to be analyzed, removed, and ensure that the
system is updated with detection and protection mechanisms (if it is a new
virus). Essentially this builds on the emulation idea described above. The
central goal of the DIS is to drastically reduce the delay time between discovery
of a virus and when a remedy is transmitted to all vulnerable systems. What
we describe here is essentially the version designed by IBM and Symantec.

DIS Closed-Loop Process

We first describe this process, then illustrate the "closed-loop."

[10.29]The idea for a Digital Immune System began with David Chess of IBM in 1991 (see [55]),
then was developed by Kephart and others over a period of years (see [133] for the culmination
of much of that work).

1. **Detection**: A virus is detected at some source point such as a gateway, server, or client machine.

2. **Quarantine**: A sample of the virus is sent to the Digital Immune System central quarantine where it is isolated, and scanned with the latest virus definitions. If it turns out to be a known virus, then the cure can be sent immediately back to the source of infection and no further action is required. Otherwise, central quarantine strips all sensitive data such as MS-word documents (to ensure confidentiality), and the sample is sent to Symantec Security Response. This transmission is accomplished over HTTP on port 80, using SSL, which ensures confidentiality and authentication (see Section 5.7).

3. **Automated Processing**: The DIS automatically analyzes the sample and creates a cure, which is sent back to the administrative console at the source.

4. **Administrative Console**: The new fingerprint is distributed by the administrative console throughout the source network to be added as an update to the current virus definitions.

Diagram 10.1 DIS Closed-Loop Virus Methodology

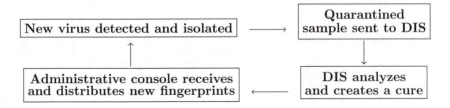

Analysis of DIS: The DIS, arguably, represents the pinnacle of antivirus software currently available. The DIS approach is stronger than other antivirus techniques since it is automated, scalable, and does not require human intervention for decoding viruses and creating signatures. The number of false positives is kept low and supplies end-to-end automation of submission, analysis, and transmission of new fingerprints for virus definition updates. There is relatively little maintenance needed with the DIS system, and costs are minimal given the alternatives. If the administrative console is allowed to streamline the control of the system at the given organizational source, then the maximum benefit will be received, since administrators have control of the level of automation.

There are other kinds of malicious programs requiring a host program, and are not considered to be viruses due to the manner in which they operate. we now look at their morphology.

◆ Logic Bombs

The *logic bomb*, also known as *slag code*, is a much older device than the virus. Like a bomb, it requires a trigger to set it off ("explode"), until which time it remains dormant in a host program. The results are particularly ugly, as would be the effects of a real bomb in a populated area. It may make the entire hard drive unreadable, or it may be more insidious and merely change a byte here and there, avoiding detection until it does irreversible damage. The trigger may be any of a number of vehicles from an elapsed amount of time, a particular date and time (December 31, 1999, at 24:00 hours, for instance), or perhaps the removal of an employee from the payroll file, indicating that he was fired. If he were really clever, the bomb would go off a few months after his termination. In this case, the logic bomb would trigger a piece of malicious code to *slag* (destroy) essential files in the company's system. This use of logic bombs clearly demonstrates the need for audit trails (see page 394), as well as clearly delineated breakdown of individual duties at any organization.

A real-world example comes from Omega Engineering and a (fired) disgruntled employee who turned vicious. A logic bomb slagged all of its research, development, and production programs, including the tape backup. One of Omega's programmer's, Timothy Lloyd, was arrested in 1998 for setting the logic bomb on Omega's network. It exploded and destroyed all their data ten days after he was fired.

A logic bomb may be considered to be a delayed-action virus in terms of effect. They can be eliminated before they explode by using virus-scanning software. If the scanning software is put on auto-protect mode, including e-mail screening, then the probability of catching a logic bomb in time is increased.

◆ Trojan Horse

The name *Trojan horse* comes from the story of Troy (about which you may read on pages 24 and 25). It is piece of malicious code that is inserted into a seemingly benign program. However, it differs from a virus in that it does not replicate itself. For instance, you might download a movie or some music from the Internet and find that it contained a Trojan horse that erases your hard disk. Another popular alternative for downloads that contain Trojan horses are FTP archives (see page 326). Another is peer-to-peer exchanges over an IRC channel.[10.30] You have to be careful since the more you download or exchange, the greater the risk of getting a Trojan horse as part of the deal, since Trojan horses are very common among IRC traders. Do not download from people or sites unless you are 100% certain of them. Even if the peer-to-peer exchange is with a trusted friend, there may be a Trojan horse lurking. In fact, the way most people find out that they have a Trojan horse is that others tell, say, Alice, that they are being infected by her download. Never use auto-download features, since you must check every file first. Moreover, check it out *before* you download it since if you download an executable file that has a Trojan horse and

---

[10.30] *IRC* stands for *Internet Relay Chat*, which was originally designed for people to "chat" in real time. IRC users trade movies, music, games, and software, peer-to-peer sharing.

run it to check it out, then you are already infected. As with the other types of infection discussed above, use a virus scanner, but do not *rely* on it. The fact of the matter is that, even when up to date, it may miss something, especially if the infection is very new.

If you do get infected, then the best eradication is a backup of the entire hard disk, and reinstall the OS and all applications from their original disks. This might become necessary since a typical Trojan horse attack is to destroy the *file allocation table* (FAT) on your hard disk. A FAT is the table that maintains a map of the clusters on the hard disk (see Footnote 10.22 on page 398). Without a FAT or with a damaged FAT, your computer will not operate properly.

An interesting example of the use of a Trojan horse comes from the OpenSSH source (see Footnote 9.8 on page 338). It turns out that in 2002, only the second day after the latest version of OpenSSH was released and ready for download on the Internet, the developers made the somewhat startling discovery that the original package had been exchanged for one with a Trojan horse embedded in it. The checksum (see page 320), was found to have been altered. When installed, the Trojan horse attempted to communicate with another Internet computer to await commands. Fortunately, they caught it early.

Now we look at malicious code that has similarities to a virus, but some differing characteristics that make it a favorite for a network attack.

◆ Worms

A *worm* is (malicious or nonmalicious) code that replicates itself and is self-propagating. Thus, a worm is independent, and designed to thrive in network environments without human intervention. Unlike a virus, it needs no host program. Rather, the computers themselves provide the hosts. The programs running on individual computer hosts are called *segments* of the complete worm. The OS in a given system is not needed to manage the worms since they seek out resources for themselves, finding remote machines and spawning a remote process on that machine. Thus, a worm program is a program that spans machine boundaries as part of a distributed computation. Some worms have a main segment that coordinates the activities of the other segments. Such a worm is sometimes called an *octopus*. Worms that are contained within a single computer are sometimes given the name *host worms*, and those that have many segments on more than one machine are deemed to be *network worms*. A host worm uses the network connections for the sole purpose of copying itself to other machines, whereas the network worm uses the network connections for communication between each of its segments. Those host worms that delete themselves after launching a copy on another host, guaranteeing there is only one version of the worm running on the network at any given time, are sometimes called *rabbits*. It is the network worm that is most common and which will be our focus.

In the 1970s before the Internet was a fact, the first two worms were sent through ARPANET (see page 326), the predecessor of the Internet, as programs called *Creeper* and *Reaper*. First there was Creeper, which used idle processor CPU time in ARPANET to replicate itself on one system and move onto the next. Then Reaper was created to follow the path of Creeper through the network deleting the segments of Creeper as it went. However, these did no damage to the computers they "infected", since they were designed to explore the possibility of making use of idle CPU time. Such nonmalicious worms are called *existential worms*, since their only function is to stay alive and propagate. In 1973, F. Shoch and J.A. Hupp of the Xerox Palo Alo Research Center, developed an existential worm program to move through an Ethernet network. Later, in 1982, these two individuals wrote a paper [251], which contained the first formal definition of the term "worm". Shoch lifted the term from a 1975 science fiction novel, called *Shockwave Rider*, in which the author, John Brunner, conceived of the concept of a worm that takes over a network, and as one of Shockwave Rider's characters puts it: "... now it's so goddamn comprehensive that it cannot be killed. Not short of demolishing the net!" (see [49, page 247]).

As Shoch and Huff found, even the creation of an existential worm opens problems with its control. In the initial stages of development of their worms, they once left one running on a system overnight only to return the next morning to find it had crashed several hosts. Even their attempts to reboot resulted in the worm's crashing the system. Therefore, they had to build a code in the worm that would shut it down when a signal was received through the network. These were problems when the creation of the worms was that based on *benign* intent. When written as malicious code, the consequences proved to be disastrous.

On page 315, we made reference to the *Morris worm*. This was the first true Internet worm. In 1988, a Cornell University graduate student, and son of the chief scientist at NSA's National Security Center, Robert Tappan Morris Jr., wrote a worm program (designated for UNIX systems). Supposedly his intention was that it be an existential network worm. He got it wrong. His program had serious shortcomings in terms of containing the worm. On November 22, 1988, after he released the worm, it propagated itself so many times that it effectively crashed several thousand host machines. It is estimated that as much as ten million dollars (U.S.) was lost in terms of productivity, and this was despite the fact that the worm left no permanent damage once eradicated. Morris was sentenced to three years probation and ordered to pay a fine of ten thousand dollars (U.S.).

Although there have been worms in the last century, the more recent ones in this millennium have been the most devastating in terms of cost. In July of 2001, two variants of the *Code Red* worm were released. It exploited a security weakness in MS-Internet Information Server (MS-IIS). Code Red launched a three-phase attack: scanning, flooding, and sleeping. In the scanning stage, it sought vulnerable machines and ran malicious code on them. In the flooding stage, false IP packets were sent to "flood" machines with useless messaging. At the height of its activity, Code Red infected a couple thousand computers each minute, ultimately contaminating in excess of a third of a million machines, and costing 1.2 billion dollars (U.S.). The final sleep stage was intended to last forever. The culprits who wrote Code Red have not yet been apprehended.

In August of 2003, the *Blaster* worm, also called *Lovesan*, caused mayhem with various Windows servers. Blaster searched for unprotected machines, and

sent itself to those computers. Once it located a vulnerable machine, it sought
out the file *mblast.exe*, retrieved it, then scanned other systems similarly. Blaster
was written to launch a DOS attack (see Footnote 8.8 on page 300), on Microsoft's updated WWW site. Microsoft found a means of thwarting the attack
on their site, but Blaster still infected around a half million computers. Microsoft offered a quarter of a million dollars (U.S.) for information that would
lead to the arrest of Blaster's creators. However, to date, there have been no
arrests. Microsoft has a five-million-dollar reward fund for the apprehension of
the various malicious code authors not yet caught.

On Friday, April 30, 2004, a worm called *Sasser* began spreading over the
Internet. It exploited a vulnerability of MS-Windows Local Security Authority
Subsystem Service (LSASS). Sasser scanned for vulnerable machines, created a
remote connection with them, installed an FTP server and downloaded itself to
the new host. From there it sought out the vulnerable LSASS components on
other machines. Sasser caused the LSASS component of Windows to crash. On
May 7, 2004, German authorities arrested Sven Jashan, an eighteen-year-old
student, who created a total of five separate versions of Sasser. Jaschan is also
responsible for twenty-eight variants of the *Netsky* worm. Key evidence leading
to Jaschan's apprehension was given by a peer group familiar with his activities.
They had approached Microsoft officials in Germany asking about the reward.
Once informed that they would indeed get it, they turned him in, after which
Microsoft paid the quarter million dollar (U.S.) reward to them. This arrest
caused Microsoft officials to have confidence that their reward fund would have
a positive effect on the eventual arrest of the perpetrators of the Blaster and
Code Red worms.[10.31]

Antiworm Countermeasures: At the outset of Section 10.2, we quoted
Sun Tzu on knowing your enemy in battle. This applies equally well today in
the war on malicious code. In fact, we may quote him further: "Know the
enemy and know yourself; in a hundred battles you will never be in peril." (See
[279, page 84].) In the computer word, one must be aware of both internal
and external potential attackers, especially if you are an employer. Disgruntled
employees, as we demonstrated with real-world examples, can be a greater threat
than any external source. We have talked at length about measures against
internal threats such as the use of firewalls (see Section 8.4); monitoring; and
access control. Now we see how to protect against external threats presented
by worms.

Relying solely on firewalls is insufficient. Each server must be protected
as a separate entity. We have already discussed the technological devices such
as IDSs (see page 395); blocking software, including antivirus mechanisms (see
page 403); and access-control software (see page 403). There should also be
human intervention such as Tiger Teams (see page 396); risk analysis; and in-
depth security policies. Using the human and technological devices in concert
can be the most effective of security-management mechanisms.

[10.31]It was reported in August of 2004 in the *Telegram* (Berlin) that teenagers are responsible
for 70% of e-mail viruses.

10.4   Legal Matters and Controversy

When men understand what each other mean, they see, for the most part,
that controversy is either superfluous or hopeless.

John Henry Newman (1801–1890)
English theologian, leader of the Oxford Movement, and Cardinal
— from *Oxford University Sermons* (1843)

◆ **What Is Copyright?**

The term "copyright" refers to the legal right to exclusive distribution, production, and publication for the owner of any original product. Typically, this right is granted by a government to an originator of work, to distribute, produce, publish, or sell the work (subject to certain caveats). It is not always necessary to seek out explicit copyright protection by applying to a government agency, say. For instance, a photographer is automatically granted copyright worldwide, to a photograph via the ownership of the negative (or digital original) of a picture taken.

In our digital world, the issue of "electronic rights" has arisen in addition to the older notion of copyright. In fact, some courts in various countries have ruled that the older notions of copyright did not contemplate electronic databases, so they are not included in the old laws. Several lawsuits pertaining to this issue have tested the new limits of copyright laws, which are now expanding to address these modern bones of contention.

The inception of the *e-book*[10.32] has given publishers pause since some authors have made digital representations of their works available online, putting them in direct conflict with their own publishers. Most lawsuits in this regard have been in favour of the publisher, since new copyright law tends to extend to include the e-book that is based on an existing print book. Many expect that all of the print versions will eventually be converted to digital form. However, copyright laws may interfere with that expectation since the permission of the copyright holder is required to do this. For the older books, the authors may be dead, in which case the copyright passes to the heirs, or the publisher, sometimes as a condition of publication, or even, in the case of many academic publications, to the scholarly journal in which it appeared. In the latter case, this has facilitated the online project called JUSTOR, *Journal Storage*, which is a scholarly journal archive. It is converting back issues of masses of runs of

---

[10.32] An e-book is a digital version of a book, which might be a literal translation of a printed book, created by say, scanning; or it might be created strictly in digital form to be read by an *e-book reader*, which is software that runs on a PC or a hand-held device, for the purpose of downloading and displaying book products. An example is the *Adobe Acrobat e-book reader*, or the *Rocket e-book*. The latter is a hand-held device about the size of a paperback novel, weighing twenty-two ounces, operates on batteries that last about twenty or so hours, and holds about the equivalent of ten novels. It is updatable, has high-resolution liquid crystal display, and can be recharged as well as connected to a PC via a serial connection (see Footnote 10.33 on page 412). If portable e-books are successful, then the Rocket e-book may be the template for their design.

original scholarly material without having to go to each author. In fact, for each journal and all its articles, there is negotiation with a single entity.

With nonscholarly material, there is still the problem of getting permission from the copyright holder, and in the case of older works, this may present problems simply in terms of actually locating that owner. Furthermore, in some countries, such as the United States, laws have been passed to protect highly profitable works by extending the copyright from the life of the author plus fifty years, to the life of the author plus seventy years. This was the result of the 1998 *Sonny Bono Copyright Extension Act* passed by the U.S. Congress. This means that there is a further two-decade-long moratorium on new works entering into the *public domain*, that body of knowledge in general, in which no individual or organization can lay proprietary claim. Thus, in many countries, it will take some time to sort out the copyright issues related to the transferal to digital form.

There are works, other than books, which are intrinsically in digital form, such as music and video. We would be surprised if book publishers wanted to restrict use of photocopy machines away from copyrighted works, or to fight for a royalty tax on such machines to compensate the copyright holders from pirating of printed works. The reason is simple: cost. Even in today's world, if one wants to pirate a book by photocopying it, and binding it, and taking all the time to do this, the cost would be prohibitive. Even if one downloads a digital PDF version of a book and wants to convert it to bound-book format, the costs remain. Yet, with music or movies, which already exist in digital form, and increasingly so, as we will discuss below, then we are not surprised that the audio/video industry seeks a royalty tax on CD, VCR, and DVD players. The problems for the music industry are quantitatively greater than for the movie industry, the reason being that the size of movies prevents most users from downloading, even when available. Moreover, many videos are encrypted, and the network connections required to successfully and quickly download such films are not in the hands of the average user. Actually, in today's world there are bigger problems faced by the music industry that we need to discuss.

◆ Copyright and Piracy Issues

Recent polls taken of teenagers in North America show that, although they are not ignorant of copyright laws when it comes to downloading music from the Internet, more than half admit to simply ignoring those laws. In fact, these polls show that they are more concerned with getting a computer virus than being prosecuted for copyright violations. This is a problem for companies holding those copyrights. How do they get paid for proprietary data if they make it public? Unauthorized copying (pirating) of software, in general, costs the industry billions of dollars. How do manufacturers *prevent* unauthorized copying of their software, called *copy protection*?

There is a copy protection device, called the *dongle*. Basically, it is a mechanism for protecting against unauthorized copying or use of proprietary software. Typically, a dongle will be a hardware device, usually sent with the software package, that plugs into the parallel port[10.33] of a computer. Then the software application sends an encrypted message to the dongle for verification before it will run. Moreover, at selected points during its running, the application will send messages to the dongle. If the dongle does not respond with the program's validation code, the software ceases execution. In other words, the dongle must remain plugged into the computer in order for the software to continue running. Hence, although users may copy the program, they need a new dongle for each copy, so must pay for each one. Modern dongles have a *pass-through* function that allows a printer to be connected, for instance, thereby not tying up the parallel port. Older versions were a nuisance since they did indeed tie up the port and so were not very popular.

Most copyright infringements have been in the area of music downloaded from the Internet. The majority of the recent major court battles and subsequent fights over piracy, copyright violations, and rights of users centered in this arena.

◆ **MP3**: Most digital music downloaded from the Internet is in MP3 format. In order to understand further the copyright/piracy issues, we must have some understanding of this format and how it operates. Prior to the arrival of MP3 on the digital scene, the format for downloading music from the Internet was *WAV* (pronounced "wave", which stood for *waveform* audio), developed by Microsoft. WAV files were monsters that could take vast amounts of storage space for a CD, and take several hours to download it.[10.34]

[10.33] A parallel port (also called an LPT port), is an interface that supports transmission of multiple bits at the same time for connecting an external device, which is usually a printer. Most computers have both a parallel and a *serial* port. A serial port is slower than its counterpart, since it is a general-purpose port in which 1 bit of information is transferred at a time. Thus, it is usually employed for modems, mouses, and some communications devices. Most serial ports use an RS-232C type connector, so are often called RS-232 ports. In the past, serial ports were used to transfer images from digital cameras, but this was painfully slow. It is being replaced by the much faster USB ports on both digital cameras and computers. *USB* means *Universal Serial Bus*, an external bus standard, which supports data transfer rates of 12 Mbps. (Think of a *bus* as a transmission pathway.) USB ports allow one to "daisy-chain" devices, that is connect one device to another. In fact, a single USB port can be used to connect up to 127 peripherals. This means that a new device may be added to a computer without the necessity of adding an adapter card. The port looks like a vertically positioned slim rectangle, typically on the back of a computer or monitor. The widespread use of the USB port really began with the popularity of Apple's iMAC, released in 1998. It is now expected that all serial and parallel ports will eventually be replaced by USB ports. The USB specification was published in 1996 by a consortium of corporations led by Intel Corporation.

[10.34] To understand why, we must learn something about CDs. Digital audio for CDs is obtained by using two-byte numbers to represent samples of the original analog signal. CD *sampling* is akin to frames of a motion picture, but with audio the "samples" are taken at a rate of 44.1 kHz (44,100 times per second), and this is naturally done in stereo (two channels). Therefore, one minute of CD audio contains $60 \times 2 \times 44,100 \times 2 = 10,584,000$ bytes. Since an average song is three minutes long, then we are talking about roughly 30 MB. (Since we get into such large numbers in these discussions, it is convenient to talk in terms of *megabytes* (MB), where $1\ MB = 1,048,576$ bytes or 1024 kilobytes (KB), where $1\ KB = 1024$ bytes.) Supposing that an Internet connection has the (not unreasonable cable-type) rate of about 60 seconds to download 1 MB, then one song will take about 30 minutes, and a complete CD with say a dozen songs on it will take six hours to download!

MP3 is short for MPEG-1, layer 3, where MPEG is the acronym for Motion Picture Experts Group, which operates under the direction of the ISO and IEC (see Footnote 9.26 on page 354), for the development of standards for audio/video compression. In 1988, MPEG met to consider the development of a single codec (compression/decompression) for digital audio. By 1992, the MPEG-1 standard for audio and video coding was created. The new standard produced high-resolution audio compression data in packets vastly smaller than those required for the WAV format. By 1997, an inexpensive software for MP3 was delivered, and this new format is what made it quick and easy to download pirated music stolen from copyrighted CDs. The MP3 compression ratio is about 12:1, meaning that the size of an MP3 clip is one-twelfth the size of an uncompressed audio file. More importantly, MP3 compresses without sacrificing much quality. It does this through a process called *perceptual audio coding scheme* that relies on the property of the human ear to discard the weakest sounds. For MP3, this means that there *is* a loss in quality, but the loss is not perceived by the human ear!

In October of 1998, the MP3 sensation was given a publicly available commercial vehicle, called the *Rio* marketed by Diamond Multimedia, as a portable MP3 player. Initially, users could upload music to certain key sites, such as *MP3.com*. Then anyone could go to that site and download it from the server where it was stored, but the user had to search the Internet for the desired music. Sometimes, IRC (see Footnote 10.30 on page 406) could be employed to chat with others and find an appropriate site. However, a combination of all these functions would have to wait for the idea of a single teenager.

◆ **Napster**: In 1999, eighteen-year-old Shawn Fanning, a Northeastern University student, wrote a program that combined three features:

1. A search engine which was dedicated to finding only MP3 files.

2. File sharing, providing the ability to exchange MP3 files directly, thereby eliminating the need to store them on a server.

3. IRC embedded as part of the program to allow MP3 users to chat online.

The above program became the utility called *Napster*, which was a highschool nickname for Fanning, ostensibly derived from his hair. Napster replaced the server storage mechanism with peer-to-peer sharing. Thus, with Napster, a user would be downloading music from another user's computer, which could be anywhere in the world. The manner in which this was accomplished is explained in what follows.

Suppose that Alice wants to share music over the Internet. In order to use Napster she needs the following:

1. A copy of the Napster utility installed on her computer.

2. A directory on her computer whose privilege she sets to shared mode.

3. An Internet connection set to be on.

Now, if Alice wants to download a specific song, she executes the following.

Alice Downloads via Napster

1. She opens her Napster utility, which verifies the Internet connection, and logs her onto the central server. This Napster server has the purpose of indexing all Napster users, but is not used to hold MP3 files. Once the connection with the Napster server is made, Alice's shared music files also become available to other Napster users, in essence turning Alice's computer into a small server in the Napster network. This is accomplished via Alice's Napster utility informing the central server which files are available on her computer for sharing.

2. She types the information concerning the song she wants, and the Napster utility on her computer *queries* the index server for other Napster users' computers online storing that particular song. For each match found, the Napster server informs the Napster utility on Alice's computer and builds a list in her results window.

3. She chooses a file and selects download. Alice's Napster utility establishes a connection with the host system that hosts her target file, and once accomplished, the file downloads.

4. After the downloading is completed, the host system disconnects with Alice's system. She now has the desired song stored on her shared file for her to access.

As long as Alice is online with the central server, other Napster users may access music files on her computer for download. Fanning's idea was to eliminate the use of a server in the above fashion. Given that Napster grew to literally billions of songs available, there was no server capable of holding them. (At its height, Napster had sixty million users per month.) The central idea in the above is that there is no copyright violation since "friends" may share music with "friends". However, the courts disagreed. The central gap in the logic was the Napster central indexing server. Once the courts ordered Napster to stop using the central database, the Napster network was dealt a lethal blow. In July of 2001, the lawsuit filed by the Recording Industry Association of America (RIAA) was successful. A judge ordered that Napster's servers stop operating. Napster tried to rectify the situation by offering to pay millions to copyright owners but by September 2, 2002, Napster was forced to liquidate its assets under U.S. bankruptcy laws. At a bankruptcy auction in 2002, the Napster name and assets were sold to Roxio, a Silicon Valley software organization.

On October 29, 2003, the Napster division of Roxio made the announcement of their launch of Napster 2.0 where users could purchase a song to download for about a dollar or about ten dollars for a CD, or for a monthly (unlimited-listening) subscription. Although the new Napster is not a peer-to-peer service, it still offers shared playlists, interactive radio, music videos, general access to their "world's largest" music store, and other features. As with the old Napster, a user can browse libraries of others, and participate in the sense of a music community. On March 10, 2004, Napster announced a team effort with IBM in the introduction of their *Super Peer* application for music downloads. It is an open standards-based technology to assist universities, ISPs, and commercial enterprises, as well as their own customers, to safely and rapidly download music legally. On May 20, 2004, Napster announced its move into the British market. Other markets are sure to follow. There are new systems that are similar to the old Napster, but operate in a fashion that makes it difficult for the courts to shut them down. Let us see why by looking at one of them.

◆ **Gnutella**: In early 2000, a subsidiary of AOL,[10.35] called Nullsoft,[10.36] made the first Gnutella program available on its servers as a distributed software project designed to establish a genuine peer-to-peer file-sharing network, without a central server. *Gnutella* (pronounced with a silent *g*), is a word amalgamation of GNU (see page 385), and Nutella, the latter being a registered trademark for a hazlenut spread with a chocolate taste first developed in the 1940s by Pietro Ferrero (currently outselling all peanut butter brands combined all over the world). Ostensibly the name combination, Gnutella, derived from the fact that the two inventors of the initial program, Justin Frankel and Tom Pepper, ate large quantities of the product while working on the project, and they had intended the source code to be released under the GNU GPL. In any case, AOL pulled the plug on Gnutella the day after its release, since they were concerned over the legal issues already plaguing Napster over similar issues at that time. This did not stop the program since it was reverse engineered[10.37] and open source copies were cropping up all over the Internet.

[10.35] *AOL* means *America Online*, which is a corporate ISP, based in Dulles, Virginia, and is owned by Time Warner, the world's largest media organization, headquartered in New York City. Time Warner was created from the merging, on January 10, 1990, of Time Incorporated and Warner Communications. In 1996, Time Warner merged with the Turner Broadcasting System. This ensured that Ted Turner would be the largest shareholder of the group of companies. In 2000, Time Warner merged with AOL to form AOL Time Warner. However, due to the weakness of the performance of the AOL component, the name was changed back to Time Warner on September 17, 2003. AOL's *instant messaging* system works in a fashion similar to that of the original Napster, but of course the purpose is not to pirate music. Instant messaging is essentially a service that allows users to communicate with others in real time through private online chat areas, and alerts them when others are actually online.

[10.36] *Nullsoft* is a play on Microsoft, observing that "null" (meaning "nothing"), is smaller than "micro" (in general terms, meaning "very small"), and is the name of the software company purchased by AOL in 1999. Nullsoft is perhaps best known for its open source installer program, *Nullsoft Scriptable Install System* (NSIS), which is similar to the original Windows installer, while being easier to use and supporting more compression formats. In general, *open source*, in reference to software, means source code that is in the public domain. It may also refer to copyrighted material that is distributed under an open-source license such as the GNU GPL (see Footnote 10.15 on page 388).

[10.37] The process of *reverse engineering* (RE) refers to the taking apart of something, which could be a piece of hardware or software, and analyzing it with the goal of creating a new construct that has the same functions as the original. Of course, reverse engineering collides with copyright and patent laws, since patents, for instance, apply to *what* a product does, not *how* it is implemented to do so.

Bob Downloads Music Using Gnutella

Background: Let us see how Gnutella works so we may achieve some understanding of how it compares and contrasts with the old Napster. There is no central server, which was essentially the legal problem with the old Napster. Each user, such as Bob, has Gnutella client software, which will have built-in preexisting lists of other users IP addresses, and which may use IRC to find suitable addresses from the list when requested. Typically, once the software is installed on a given machine and connected to the Internet, it will try to connect to the IP addresses with which it was shipped until it reaches some preprogrammed quota, and stops, but keeps the addresses not tried.

1. Bob types in the name of the music he is seeking, a song or a CD. The Gnutella software seeks out appropriate machines with that music.

2. The Gnutella software on Bob's computer sends the names of the music Bob wants to the other machines to see if it is located on their local disks. Each one that does have it sends back the file name and IP address of the machine on which it sits.

3. Simultaneously, each of the machines sends out Bob's request to all the machines to which they are connected, and each of these machines repeats the process in turn.

4. There is a limit placed on each request, called a TTH (*Time To Hop*), meaning that the request must go out to say, seven levels deep before it stops its own propagation.

5. Bob selects from the returned files via their IP addresses and his software contacts the sender and negotiates the transfer.

6. Once Bob disconnects, his Gnutella software updates by saving the IP addresses of those users who contacted him.

This is a clever mechanism for employing the peer-to-peer idea without violating any copyright laws since there is no central indexing server, just "friends" sharing files via their computers. However, without that central server, the process can take time. It could be a few minutes for a multilayer deep request to be processed. Furthermore, Bob is giving up time on his own computer to process requests coming in from other users, so he sacrifices bandwidth. Moreover, there is no guarantee he will find what he is looking for even with all the machines involved. Yet the Gnutella idea works every time Bob connects since, as long as there is one user to whom he can connect in the Gnutella network, he is able to query all the interconnected machines. Yet no court is likely to shut this scheme down since there is no central server for which a court could issue the cease-and-desist order. Currently several hundred million copies of the Gnutella software have been downloaded. For more information see *http://draketo.de/inhalt/krude-ideen/gnufu_en.html*.

There are numerous Gnutella client applications, unlike the old Napster. Some of them are *Bearshare* a closed-source, Windows-based platform; *Gnucleus*, a Windows-based, open-source platform written in the C language; *Limewire*, a GPL open-source code in Java, which is multiplatform (but there is a MAC OSX platform based on Limewire, called *Acquisitionx*); and *Poisoned*, an open-source, MAC OSX-based platform. The most common file-sharing applications in use today, which also employ the above clients, are *BitTorrent* and *Kazaalite*; see *http://bittorrent.com/* and *http://www.kazaalite.com/*.

The music industry is responding to the above concerns, especially the MP3 developments, with an initiative called *Secure Digital Music Initiative* (SDMI), which is, an as yet incomplete, standards forum established in 1998 (see *www.sdmi.org*). The goal of SDMI is to provide online, convenient, legal access to digital music, by creating new digital distribution systems, enabling copyright protection for music artists, and promoting development of music-related technologies. Part of the reason that SDMI's goal is, as yet, incomplete is that their SDMI challenge, *Open Letter to the Digital Community*, announced in September 6, 2000, had unexpected outcomes. They invited hackers and cryptanalysts to break their proposed *digital watermarking scheme* for protecting digital music. Their challenge was met in the worst possible way (for them). A group broke the entire system. In other words, not only did they break the proposed scheme, but showed that the basic idea was flawed in that any algorithm based upon it could also be broken. The SDMI project has been inactive since May 18, 2001. However, this hiatus did not invalidate their development of a portable device specification. The idea behind this specification is that future music content would include data that SDMI-compliant players would recognize, and would refuse to play any file containing digital music without these markings. In this fashion, copyrighted material would be protected. However, it is not difficult to see that implementing such a system is going to meet some formidable barriers.

The last topic of this chapter is on controversy surrounding a single computer chip that involved the White House, the NSA, DES, and numerous characters, who attempted to build a wall of security that came crashing down.

◆ The NSA

The National Security Agency (NSA)[10.38] is also regarded as the Central Security Service. It is essentially the U.S. cryptologic citadel, whose function is to organize and direct whatever necessary to safeguard U.S. information systems and to gather foreign intelligence. On November 4, 1952, President Harry S. Truman signed a directive that established the NSA. We have witnessed several cryptologists in the NSA Hall of Honor (see pages 85–88, and 91), as well as images from their National Cryptologic Museum (see page 92–96, and 106), for instance.

[10.38]The acronym is often used for *No Such Agency* and *Never Say Anything* in reference to its cloak of secrecy shrouding the organization.

Figure 10.6: The NSA's 50th Anniversary Exhibit.

This exhibit is in the NSA's National Cryptologic Museum, and is courtesy of the NSA. For more details see *http://www.nsa.gov/museum/museu00021.cfm.*

For more than half a century the NSA has been at the forefront of high-tech development including paving the way for the first large-scale computer, the first solid-state computer, the tape cassette, and modern semiconductor[10.39] technology. The NSA also employs certainly the largest number of mathematicians in the United States, and arguably, in the world. Their cryptologists and mathematicians help the agency to design ciphers and find deficiencies in enemy cryptosystems.

In other words, their job is to make and break codes. In the NSA, this is

---

[10.39]A *semiconductor* is a substance that selectively conducts electricity through the movement of electrons and *holes* (which are electric charge carriers with positive charge, equal in magnitude, but opposite in polarity to the charge of an electron). A semiconductor is intermediate between conductors (like copper that freely conducts electricity), and insulators (such as rubber, which does not conduct an electrical charge). For instance, silicon is a semiconductor, and usually microchips (sometimes called simply *chips* or *integrated circuits* (IC)s), are fabricated on semiconductor substances such as silicon. The substance, *silicon*, present in glass and sand, for instance, is the best known semiconductor material for electronic components. Silicon will conduct electricity in a fashion depending upon the impurities added to it, called *doping*. If the addition of impurities causes the majority of the charge carriers to be negatively charged electrons, then this is called *N-type silicon*. If addition of impurities leads to the majority of charge carriers being positively charged holes, then we have *P-type silicon*. Typically silicon material used in electronics has both N- and P- type material.

Figure 10.7: The NSA's Cryptologic Memorial.

The black granite wall was erected outside the NSA headquarters in Ft. Meade, Maryland, to honour 153 cryptologists who sacrificed their lives for their country; courtesy of the NSA (see *http://www.nsa.gov/gallery/photo/photo00054.jpg*).

sometimes called *Equities*, to designate that both tasks are of equal importance.

Since the NSA is necessarily committed to being on the cutting edge of cryptologic advances, they have a National Cryptologic School. William Friedman was, in fact, the chief cryptologist at the forerunner of the NSA (see page 85), and his influence was probably a precursor to the development of the school.

The above being said, we are interested herein with the controversy surrounding the NSA's involvement in attempting to suppress public access to cryptographic mechanisms. They were involved early on with Feistel's work (see pages 97 and 120), and later with DES (see page 98), as well as the doomed attempt to suppress public-key cryptography (see [169]). However, it is a single chip in which we are interested, since it embodies a whole wealth of stealth that is worth the telling.

◆ The Clipper Chip

The Clipper Chip was designed to be resistant to reverse engineering even against sophisticated and well-funded attackers. It was developed by the defence contractor, Mykotronx of Torrence, California as a secure voice device for AT&T. Mykotronx called the chips MYK-78.

The chip was to be used to protect private voice communications while allowing government agents to employ what was called *key escrow* to access the encrypted communications. The reason was to provide access for law enforcement to be readily able to obtain information in criminal cases from voice communications. The proposal was that the Clipper Chip would hold a master key "in escrow" for release to government agents at a later time as a "secret door" for them to access the encrypted conversations without having to use methods of decryption to get them. This secret door was called *LEAF* or *Law Enforcement Access Field*. To accomplish this, the master key would be in the protected, secure custody of the government. Then the (classified) encryption algorithm on the Clipper Chip, called *Skipjack*, would use the session key to encipher the sounds as they left the transmitting source, and decrypted as they arrived at the received target. There was also to be a unique chip identifier, a unique chip key. LEAF would be generated by Skipjack using the session key and the unique chip key to produce an enciphered session key and the unique chip key. Then the master key would encrypt it all. Below is a stepwise version of how this all works. First we look at Skipjack to see how it is an improvement over DES, and examine what the Clipper Chip contains.

Chip Contents: Each Clipper Chip contains the following.

- The classified SKC, Skipjack, employing 80-bit keys (DES has only 64), and 32 rounds (DES has only 16), supporting DES modes of operation (see Section 3.3).

- An 80-bit family (master key), **F**, common to all Clipper Chips.

- A unique serial number, **N**.

- A unique 80-bit secret key, **U**, for decrypting all messages enciphered with the chip.

The following is a simplified version of the employment of the Skipjack SKC used in conjunction with the Clipper Chip for secure voice communications between Alice and Bob.

Encrypting with the Clipper Chip: Alice and Bob will use the Clipper Chip for an encrypted telephone conversation as follows. We will assume that the conversation is digitized voice.

1. First there is a negotiation of a session key between Alice and Bob that takes place outside the chip, which is embedded in the AT&T security device for each phone. Alice activates her security device to call Bob, and

her device negotiates with Bob's device to create a session key $\mathbf{K}$, using a key exchange such as Diffie-Hellman. Then once $\mathbf{K}$ is established, the Clipper Chip is used to secure the conversation as follows.

2. The telephone security device takes the message stream $\mathbf{M}$ and the key $\mathbf{K}$ as input to the chip. Employing Skipjack encryption, E, output values are

$$E_\mathbf{K}(\mathbf{M}),$$

the encryption of $\mathbf{M}$ using key $\mathbf{K}$; as well as

$$E_\mathbf{F}(E_\mathbf{U}(\mathbf{K}),\mathbf{N}),$$

the LEAF, using master key $\mathbf{F}$ to encipher the concatenation of $\mathbf{N}$ and the value $E_\mathbf{U}(\mathbf{K})$, which is $\mathbf{K}$ encrypted using key $\mathbf{U}$.

3. Both $E_\mathbf{F}(E_\mathbf{U}(\mathbf{K}),\mathbf{N})$ and $E_\mathbf{K}(\mathbf{M})$ are transmitted to Bob's Clipper Chip, which deciphers the voice message via Skipjack decryption, $D_\mathbf{K}$, to get the original message,

$$D_\mathbf{K}(E_\mathbf{K}(\mathbf{M})) = \mathbf{M}.$$

4. The Clipper chips operate in both directions for the conversation in the above fashion now that they are synchronized.

5. The LEAF may be decrypted by law enforcement once a wiretap warrant is ordered in the case of suspected criminal activity.

The following gives one means of having escrow agents, employed as TTPs (see page 182) to program the chips and generate the required keys.

Clipper Chip Programming and Escrow

We will employ Trent and Victor as TTPs to accomplish the task of generating the unique unit keys for each of the Clipper Chips, which are programmed in a secure manner. At least one of Trent or Victor must be independent of any branch of government or law enforcement agency. All Clipper Chips are programmed at Mykotronx in a secure vault. Numerous chips are programmed during a given session as follows. A computer and equipment to program the chips are in the vault.

1. Trent and Victor enter the vault and each enters into the computer a random 80-bit seed, S_T and S_V, respectively.

2. The serial number, $\mathbf{N}$, of a given chip is padded with a fixed block to produce a 64-bit number N_1. Then the following triple encryption is computed using Skipjack,

$$R_1 = E_{S_T}(D_{S_V}(E_{S_T}(N_1))).$$

3. In a similar fashion to step 2, **N** is padded with two more distinct blocks to form two more 64-bit blocks N_2 and N_3, respectively, from which the following are computed:

$$R_2 = E_{S_T}(D_{S_V}(E_{S_T}(N_2))),$$

and

$$R_3 = E_{S_T}(D_{S_V}(E_{S_T}(N_3))),$$

respectively.

4. Then the 192-bit concatenation $R = (R_1, R_2, R_3)$ is formed. The first 80 bits, U_1 is given to Trent, and the second 80 bits, U_2, is given to Victor, with the rest of R being discarded. Then the addition modulo 2 is computed:

$$\mathbf{U} = U_1 \oplus U_2.$$

The value **U** is kept on a separate disk at Mykontronx for later programming.

5. Trent and Victor have now completed a secret-sharing scheme (see Section 5.5), where each of them knows exactly one of the 80-bit parts of the secret key **U**. They leave the vault with their pieces of data, neither of whom may separately generate **U** without the other's participation.

6. Each such value **U**, stored on disk, is used to program the chips. Then all data is discarded from the vault and the computer hard disk is erased to be ready for the next TTPs to enter the vault.

Law Enforcement: If criminal activity is deemed to have occurred in the conversation between Alice and Bob, then law enforcement obtains a wiretap warrant, goes to the service provider of their telephones, and accesses the communications lines. Then the law enforcement agents, Polly and Peter, access the Clipper Chip and perform the following actions.

1. First they use **F** to decrypt LEAF and obtain

$$D_{\mathbf{F}}(E_{\mathbf{F}}(E_{\mathbf{U}}(\mathbf{K}), \mathbf{N})) = (E_{\mathbf{U}}(\mathbf{K}), \mathbf{N}).$$

2. Then Polly and Peter use secure lines to both send the serial number **N**, and authorization documentation to both Trent and Victor.

3. Trent and Victor independently send their respective shared-secret pieces U_1 and U_2 to Polly and Peter, who add them modulo 2 to get **U**.

4. Polly and Peter use **U** to decrypt via

$$D_{\mathbf{U}}(E_{\mathbf{U}}(\mathbf{K})) = \mathbf{K},$$

which is used to obtain $D_{\mathbf{K}}(E_{\mathbf{K}}(\mathbf{M})) = \mathbf{M}.$

Skipjack was a component of what NSA called *Capstone*, a PKC, developed by NSA, that included the DSS (see Section 4.3). Skipjack was started by NSA in 1985, and completed in 1990. The NSA had moved heavily to ensure that the DES keylength was kept relatively small, while developing Skipjack as an improved, albeit classified SKC. Thus, the NSA wanted to have their cake and eat it, too (see pages 98 and 117, for instance). However, the public outcry against what was seen as an extreme invasion of privacy would not allow them much more than a slice of that cake.

The brainchild of the *Clipper Chip* was Clinton Brooks who rose to become assistant deputy director at the NSA. Brooks, who was intent on the success of the Clipper Chip, approached Al Gore after the 1992 election victory of Bill Clinton, even before they entered the White House in December of that year. Being a "techno-geek" himself, Gore was taken with the idea. Gore had a desire to see an *Information Highway* and was intent on making this a reality. Once the Clinton administration took power in 1993, meetings were set in high gear by the NSA with the White House staff. By March 31, 1993, Clinton gave the go-ahead for the Clipper project. However, he had no idea that this move would signal the end, not the beginning, of government control of cryptology.

NIST was required to solicit public input on the Clipper project, and the results were devastating for the pro-Clipper side. Only six percent of those responding were in favour of the proposal. The White House was undeterred. On February 4, 1994, Clinton formally enshrined Clipper as a FIPS standard, known as the *Escrow Encryption Standard* (EES). However, opponents were speaking out against the project in growing numbers with some impressive names doing the talking. Phil Zimmermann (see Section 8.1), posed the important query: Why would anyone want the Clipper Chip when programs like PGP were readily available and free? Why indeed? Whit Diffie (see page 167) appeared at a Senate hearing on the matter and basically focused on the right of the populace to freely communicate and do so in private, without (what had come to be known as) "Big Brother on a Chip" monitoring their every word. Then the flaws of the Clipper Chip idea came to the fore in a way that is even more powerful than the testimony of hundreds of top-notch experts.

Matthew Blaze worked in the cryptography group of AT&T Bell Labs in New Jersey. In early 1994, NSA became interested in Blaze as a potential outside source for testing the Clipper Chip so they invited him to NSA headquarters. He discovered that the certain safeguards built into one of the LEAF fields was a pitiful 16 bits. Using this fact he cracked the scheme in less than an hour. Even though this hole could be filled, the fact that it was there in submitted form, ready to be distributed, created a gaping hole in anyone's trust of the scheme. When the story appeared in *The New York Times*, the writing was on the wall. Public opinion in favour of the Clipper Chip project descended to less than twenty percent. The project died on the operating table, and even its creator, Clinton Brooks, was fed up with the project, and threw in the towel.

The original Clipper plan was retracted, and two new initiatives, called *Clipper 2* and *Clipper 3*, were promoted. In 1996, Clinton promoted the latter, which allowed the use of any encryption technology, but reserved the right of a government to recover any keys exported out of the United States In 1998, Skipjack was declassified along with KEA (see page 223). They had to settle for a small slice of the cake after all.

Other Secret Doors: The idea behind any of the U.S. government key escrow plans, such as the above, was to give law enforcement timely access to plaintext (without the consent or knowledge of the user), in order to solve the problem of criminals enciphering evidence of their nefarious deeds. The public, however, saw the solution as far worse than the perceived problem.

Essentially the security devices, such as those described above for the Clipper Chip, add a subliminal channel (the secret door we talked about earlier), to the users' communications lines (see pages 184, 192, and 374). This notion of building such a channel (or secret door), into hardware at the time of manufacture was being employed elsewhere. On March 18, 1992, the Iranian military counterintelligence service arrested Hans Bueller, who was Crypto AG's[10.40] marketing representative in Tehran. The charges were that he was a spy for Germany and the United States. During his nine-month imprisonment, he was questioned five hours every day. Although Buehler said he was not beaten, he was *told* he would be beaten and, during these interrogations, he was tied to a wooden bench. It turns out that, despite his thirteen years of employment at Crypto AG, Buehler was ignorant of the allegation that the firm was incorporating a secret door in their cryptographic devices (ostensibly, at the behest of the NSA and the BND, *Bundesnacrichtendtendienst*, the German intelligence service). Buehler stated that if he knew anything, they would have gotten it out of him.

In order to sweep the issue under the carpet as quickly as possible, Crypto AG paid a million dollar ransom to Iran to secure Buehler's release in January of 1993. After his return to Switzerland, the firm fired him and demanded that he pay the money back to them. However, not everything went under the carpet since some of the Crypto AG engineers came to Buehler's defence and allegedly, threatened to disclose that the crypto devices had been altered by American and German engineers, who inserted their own cryptosystems.[10.41] The Swiss media was inspired by the Buehler affair, and began to dig into Crypto AG's background. The firm launched a lawsuit in response, but there was an out-of-court settlement days before the trial was to begin. Crypto AG denies all allegations.

[10.40]Crypto AG is a company begun by Boris Hagelin who moved his company to Zug, Switzerland, in 1948, and changed its name as the current incorporated entity in 1959. Previously, it was a company called Aktieboget Cryptograph, based in Sweden, owned by Avrid Damn. After Damn's death in 1927, Hagelin took it over. He became the first millionaire from cryptography due to the royalties earned from the products sold by the firm, such as the M-209 (see page 90).

[10.41]On December 4, 1995, *The Baltimore Sun* reported that the NSA has secretly rigged Crypto AG devices so American intelligence could easily decipher the traffic generated by these machines. The newspaper claimed this information was obtained from former Crypto AG employees whose story was supported by company documents.

Chapter 11

Information Theory and Coding

That was a little more information than I needed to know.
Quentin Tarentino (1963–), American film director
— from the movie *Pulp Fiction* (1994), spoken by Uma Thurman

11.1 Shannon

"Information theory" is a term derived from the seminal work of C.E. Shannon, *The Mathematical Theory of Communication*, published in 1948 (see [249]). In Section 11.2, we will define what "information" means, and see that it is not to be confused with the everyday understanding of the word. We will also see that the Internet would not exist without information theory. Since Shannon is the prime mover and prover of the concepts of perfect secrecy and information theory, we begin with a brief biographical sketch of his life.

Claude Elwood Shannon (see Figure 11.1) was born on April 30, 1916, in Gaylord, Michigan, where he stayed and graduated from the University of Michigan in 1936. He left to do his graduate work at MIT. His supervisor, Vannevar Bush, had Shannon take care of a computing device called the *Differential Analyzer*, which was a concoction of rods and gears that needed manual alignment before a problem could be "input" to the machine. These problems involved finding numerical solutions to ordinary differential equations. It was Shannon's experience with this machine that got him thinking along the lines of replacing the unwieldy mechanical device with electrical circuits. Then he realized that Boolean algebra was similar to the electrical circuit, and from this derived the notion of circuit design according to Boolean algebra to analyze, test, and optimize relay switching circuits. These ideas were expounded in his master's thesis entitled, *A Symbolic Analysis of Relay and Switching Circuits*, for which he earned his master's degree in 1937. His Ph.D., on population genetics, was

Figure 11.1: Claude Shannon.

(Courtesy of Lucent Technologies Inc./Bell Labs.)

granted in 1940.

In 1941, Shannon obtained a job as a research mathematician at the New Jersey AT&T Bell Labs. In 1942, he collaborated with John Riordan to publish a paper whose topic was the number of two-terminal series-parallel networks, which generalized seminal results published by MacMahon in 1892. By 1948 he had published the paper cited at the outset of this section, which essentially founded Information Theory. In this paper, he set forth a linear schematic model of an information system, a revolutionary new idea that described the measurement of information via binary digits. In this paper, he uses the word "bit" for the first time. At that time communication was via nondigital means, the transmission of electromagnetic waves through a wire. What we take for granted today — continuous flow of bits through a wire — was a revolutionary idea at that time. In this paper, he provides a rigorous mathematical definition of information, which was based upon his cryptological work accomplished during World War II. Shannon's assumption was that information sources generate words comprised of a finite number of symbols sent over a channel. He provided a mechanism for analyzing a sequence of error terms in a signal to determine their inherent type, assigning them to the designed type of the control system. He demonstrated how adding extra bits to a signal could correct transmission errors. This and the notions exposed in this paper were used and extended by engineers and mathematicians to provide efficient and error-free transmissions through noisy channels. The development of Information Theory made possible the development of digital systems.

Shannon also worked on AI problems. By 1950, he had written a computer program, which appeared in a publication entitled *Programming a Computer for Playing Chess*. This publication led to the first machine chess game played by the Los Alamos MANIAC machine in 1956. This was also the year in which he published an important paper that demonstrated how a Turing machine can be constructed utilizing only two states. In 1957, he was appointed to the Faculty at MIT, but remained a consultant to AT&T Bell Labs until 1972. He received many awards, among them the Alfred Nobel American Institute of American Engineers Award in 1940, the National Medal of Science in 1966, the Audio Engineering Society Gold Medal in 1985, and the Kyoto prize in that year. In his last few years he suffered from Alzheimer's disease, and was confined to a Massachusetts nursing home. He died at age 84 in Medford, Massachusetts, on February 24, 2001.

Shannon was a genius in his own realm, making contributions that paved the way for our modern digital revolution. Marvin Minsky wrote of him that "For him, the harder the problem might seem, the better the chance to find something new." He certainly found many "new" concepts, without which we could not have the world that we have today. He was an inspiration to generations of mathematicians and computer scientists. One of Shannon's ideas, contained in his paper [249], was a notion related to his formulation of information. We will study this notion in the following section for which the reader will need some familiarity with the basic probability theory presented in Appendix E on page 543.

11.2   Entropy

Heretics are the only bitter remedy against the entropy of human thought.
— see [298, introduction]

Yevgeny Zamyatin (1884–1937), Russian writer

Information Theory is concerned with sending messages via electronic signals in the most efficient and error-free manner. Shannon defined information to mean *a measure of one's freedom of choice when one selects a message*. This "measure" will gain mathematical precision below. The idea is that information refers to the degree of uncertainty that exists in the situation at hand. Therefore, in this (Information Theory) sense of the word, any situation that is totally predictable (namely, whose outcome is certain) has very little information (perhaps none). Thus, redundancy adds little, if any, information. Redundancy, such as in the repeating of a message, helps to eliminate noise in a communications system. Think of "noise" as anything within the communications system that is contrary to the predictability of the outcome of that system.[11.1] The term *entropy* is the degree of randomness (or uncertainty) in a given situation, measured in bits, to which we will give a mathematical flavour below. In other words, entropy is a measure of the amount of information in a given message source. Moreover, in Information Theory, "efficiency" refers to the bits of data per second that can be sent and received, and "accuracy" (error-freeness) is the extent to which transmitted data can be understood (meaning clarity of reception, wherein the message may not have "meaning"). When we put the above into a cryptological situation, where intended plaintext messages *do* have meaning, encryption may be seen as noise added to the cryptosystem. The entropy is the measure of the uncertainty about a message before it leaves the message source. Now we look at all of this from a mathematical viewpoint.

◆ **Properties of Entropy**

Shannon required that entropy must satisfy the following properties.

1. H must be a continuous function of the variables $p_1, p_2, \ldots, p_n$, the probability distribution. In this way, a small change in the probability distribution should not severely alter the uncertainty.

2. When all messages are equally likely, that is, when $p(m_j) = 1/n$ for all $j = 1, 2, \ldots, n$, H should be an increasing function of n. In other words,

$$H\left(n^{-1}, \ldots, n^{-1}\right) \leq H\left((n+1)^{-1}, \ldots, (n+1)^{-1}\right) \text{ for all } n \geq 1,$$

[11.1]In 1948, Shannon was able to precisely determine the maximum data rate achievable over any transmission channel involving noise. Today this is known as *Shannon's theorem*, which says that

$$K = B \cdot \log_2(1 + S/N),$$

where K is the effective limit on the channel's capacity measured in bits per second, B is the bandwidth of the hardware, S is the average signal strength, and N is the average noise strength. S/N is called the *signal-to-noise ratio*. Shannon's theorem places a fundamental limit on the number of bits per second that can be transmitted over a channel. Thus, no amount of engineering innovation will overcome this basic physical law.

which means that there is more choice (uncertainty) when there exist more equally likely outcomes.

3. If the jth outcome is replaced by two successive outcomes, the first with probability cp_j and the second with probability $(1-c)p_j$, for $0 < c < 1$, then the entropy of S should be a weighted sum of the entropies of the two choices. More precisely,

$$H(p_1, \ldots, cp_j, (1-c)p_j, \ldots, p_n) = H(p_1, \ldots, p_j, \ldots, p_n) + p_j H(c, 1-c).$$

This says that the entropy is increased by the uncertainty caused by the choice between the two outcomes, multiplied by p_j.

Based on the above three properties, Shannon established the following definition for entropy.

◆ What Is Entropy?

Entropy, the measure of information (uncertainty), is formalized as follows.

Suppose that we have a set of messages $S = \{m_1, m_2, \ldots, m_n\}$ for some $n \in \mathbb{N}$, and p_j for $1 \le j \le n$ is the probability that message m_j is the message sent, with $\sum_{j=1}^{n} p_j = 1$. Then

$$H(S) = H(p_1, p_2, \ldots, p_n) = \sum_{j=1}^{n} p_j \cdot \log_2(1/p_j) = -\sum_{j=1}^{n} p_j \cdot \log_2(p_j) \quad (11.1)$$

denotes the entropy of S, which is independent of the set of messages since another distinct set $T = \{m_1', m_2', \ldots, m_n'\}$ with the same probability distribution has the same entropy.[11.2] Also, Equation (11.1) defines a mathematical value called *the number of bits per message source*.[11.3]

This definition captures the earlier notion since, for instance, if the outcome is certain to be m_1, namely, $p_1 = 1$ and $p_j = 0$ for all $j > 1$, then $H(S) = 0$, so there is no information in a predictable outcome. In fact, this is the only way that $H(S) = 0$ is possible, that is, if there is an $m_j \in S$ such that $p_j = 1$ and $p_k = 0$ for all $m_k \in S$ with $j \ne k$. Since the goal is to record minimal amounts of bits on a computer to represent information, then there is no point in storing information with zero content such as the above. When $H(S) > 0$, then the following illustrates the measure of certain minimal bits we have to employ in order to record output from S, namely, the average number, given by $H(S)$.

[11.2]Since there is an undefined quantity when $p_j = 0$, we agree by convention that $0 \log_2 0 = 0$.

[11.3]Shannon chose $\log_2$ as a suitable measure of entropy to which he could ascribe the term "bits", which he says in [249], was a word suggested by J.W. Tukey. In that paper, he gave three substantive reasons for the logarithmic measure: (1) It is practically more useful; (2) It is nearer to our intuitive feeling as to the proper measure; and (3) It is mathematically more suitable.

Example 11.1 *If* S *consists of the toss of a coin with equal outcomes, and we encode heads as* 1 *and tails as* 0*, then* $p_1 = p_2 = 1/2$ *with* $S = \{0, 1\}$*, and*

$$H(S) = 1 \log_2(1/2)/2 - 1/\log_2(1/2) = 1 \ bit.$$

In general, if S *has cardinality* n *such that each message has probability* $1/n$*, then* $H(S) = \log_2(n)$*.*

In general, if $n = |S|$, $\log_2(n)$ is *the maximum possible value that* $H(S)$ *may assume*. Moreover, this maximum is assumed if and only if the probability distribution is uniform, for example, when $p_j = 1/n$ for all $j \geq 1$. For instance, we have the following.

Example 11.2 *Suppose we encode* N *equally likely, independent coin tosses as bitstrings of length* N*, then the cardinality of* S *is* $n = 2^N$ *and* $p_j = 2^{-N}$ *for each* j *for which the entropy is given by*

$$H(S) = -\sum_{j=1}^{2^N} 2^{-N} \log_2 2^{-N} = -2^N(2^{-N}(-N)) = N = \log_2(n).$$

Thus, we learn N *bits when we are given a bitstring of length* N*.*

Example 11.2 illustrates that $H(S)$ may be viewed as the measure of the number of bits of information obtained when we know the outcome of S. This example also illustrates the fact that entropy measures the minimal amount of bits required to represent an event on a computer, and then only the *relevant* bits pertaining to the uncertainty.

For the next illustration, we go back to page 208, where we talked about Alice and Bob flipping coins by telephone. Let us look at the entropy of that protocol (event).

Example 11.3 *In this case* $S = \{0, 1\}$ *where* 0 *is the encoding of tails and* 1 *is the encoding of heads. Then as in Example 11.1,* $H(S) = 1$ *bit. In other words, Alice is a source of only* 1 *bit in this coin flipping protocol, irrespective of the size of the integer* x*. The reason is that Bob knows* x *so there is no information there for him. There is, for Bob, uncertainty (information) only in its parity, which is Alice's guess; hence her output:* 1 *bit of entropy.*

We have considered only uniform probability distributions thus far. We now look at other scenarios.

Example 11.4 *Suppose that we flip three coins, which are not fair, where the outcome is the number of tails. Then* $S = \{0, 1, 2, 3\}$*, and we let the probabilities be* $p_0 = 1/8$*,* $p_1 = 3/8$*,* $p_2 = 3/8$*, and* $p_3 = 1/8$*. Thus, the entropy is*

$$H(S) = -\frac{1}{8} \log_2(1/8) - \frac{3}{8} \log_2(3/8) - -\frac{3}{8} \log_2(3/8) - \frac{1}{8} \log_2(1/8) \approx 1.81.$$

We may use this example to illustrate how entropy is related to decision problems (see page 502). In other words, to determine the outcome of the event, one may ask yes or no questions, but how many? One could ask if the number of tails is greater than one. This narrows the field to two possibilities. Then if the answer were yes, for instance, then the next question could be, is the number of tails greater than three? Then we know the outcome after two questions, a number approximately equal to the entropy.

Joint Entropy: When we have two message sources $S = \{s_1, s_2, \ldots, s_n\}$ and $S' = \{s'_1, s'_2, \ldots, s'_{n'}\}$, the *joint entropy* is defined by

$$H(S, S') = -\sum_{s_i \in S} \sum_{t_j \in S'} p_{i,j} \log_2(p_{i,j}),$$

where $p_{i,j}$ is probability that s_i is the outcome of S and s'_j is the outcome of S'. It follows that

$$H(S, S') \leq H(S) + H(S'), \tag{11.2}$$

which says that *the entropy in the pair* (S, S') *is, at most, the information contained in* S *plus the information contained in* S'; and *equality holds precisely when* S *and* S' *are independent.*

To illustrate this new notion, we move from coins to the more "suitable" deck of cards.

Example 11.5 *Suppose that a card is drawn from a standard deck of 52 cards,* $S = \{clubs, diamonds, hearts, spades\} = \{s_1, s_2, s_3, s_4\}$ *with* $p_j = 1/4$ *for each* $j = 1, 2, 3, 4$, *and* $S' = \{ace, 2, 3, 4, \ldots, 10, jack, queen, king\} = \{s'_1, \ldots, s'_{13}\}$, *with* $p_j = 1/13$ *for* $j = 1, 2, \ldots, 13$. *Thus,* S *and* S' *are independent, which means that*

$$p_{i,j} = p_i \cdot p_j = \frac{p_i}{p_j} = \frac{1}{4} \cdot \frac{1}{13} = \frac{1}{52},$$

where p_i *is the probability that* s_i *is the outcome of* S *and* p_j *is probability that* j *is the outcome of* S'. *Then the entropy is given by*

$$H(S, S') = -\sum_{i=1}^{4} \sum_{j=1}^{13} p_{i,j} \log_2(p_{i,j}) = \log_2(52),$$

the maximum entropy possible as discussed above.

Conditional Entropy: Earlier we mentioned the context of cryptography for entropy. We revisit this here in terms of ciphertext and keys, which was one of Shannon's viewpoints when establishing the basics of Information Theory. We may discuss the *conditional entropy* of, say, the key, $k \in \mathcal{K}$, given the ciphertext, $c \in \mathcal{C}$, which is defined as follows,

$$H(\mathcal{K}|\mathcal{C}) = -\sum_{c \in \mathcal{C}} \sum_{k \in \mathcal{K}} p_{c,k} \log_2(p_{k|c}), \tag{11.3}$$

where $p_{c,k}$ is the probability that the outcome of $\mathcal{C}$ is c, and of $\mathcal{K}$ is k, whereas $p_{k|c}$ is the conditional probability that k occurs, given that c occurs.[11.4]

Of course conditional entropy may be invoked with any message source, not just cryptologic. An important property of conditional entropy is the following, which marries the notions of joint and conditional entropy.

The Chain Rule: The joint entropy and the conditional entropy are given by the following *Chain Rule*, where $\mathcal{S}$ is one message source, and $\mathcal{S}'$ is another:

$$H(\mathcal{S}, \mathcal{S}') = H(\mathcal{S}) + H(\mathcal{S}'|\mathcal{S}). \tag{11.4}$$

The Chain Rule tells us that the joint uncertainty of pair $(\mathcal{S}, \mathcal{S}')$ is the uncertainty of $\mathcal{S}$ plus the uncertainty of $\mathcal{S}'$ given that $\mathcal{S}$ is known.

Mutual Information: If $\mathcal{S}$ and $\mathcal{S}'$ are message sources, then their *mutual information* is the uncertainty of $\mathcal{S}'$ reduced when $\mathcal{S}$ is known:

$$I(\mathcal{S}', \mathcal{S}) = H(\mathcal{S}') - H(\mathcal{S}'|\mathcal{S}).$$

Thus, $I(\mathcal{S}', \mathcal{S})$ measures the amount of information learned about $\mathcal{S}'$ that is obtained by learning $\mathcal{S}$. The following material tells us both that mutual information is nonnegative and gives us criteria for when it is zero.

The Role of Conditional Entropy: Perhaps one of the most important facts from Information Theory is the following inequality:

$$H(\mathcal{S}'|\mathcal{S}) \le H(\mathcal{S}'), \tag{11.5}$$

which tells us that the uncertainty about $\mathcal{S}'$ when we know $\mathcal{S}$ is no greater than the uncertainty about $\mathcal{S}'$. As we have seen above, when the events are independent, equality holds (and it can be shown that equality cannot hold otherwise). We may deduce that $\mathcal{S}$ can only yield information about $\mathcal{S}'$, namely, knowing $\mathcal{S}$ cannot increase our certainty about $\mathcal{S}'$. Incidentally, it is clear that Equation (11.5) may be deduced from Equations (11.2) and (11.4).

The Role of Independence: When $\mathcal{S}$ and $\mathcal{S}'$ are independent, the following are equivalent facts.

1. $H(\mathcal{S}, \mathcal{S}') = H(\mathcal{S}) + H(\mathcal{S}')$.

2. $H(\mathcal{S}') = H(\mathcal{S}'|\mathcal{S})$.

3. $H(\mathcal{S}) = H(\mathcal{S}|\mathcal{S}')$.

4. $I(\mathcal{S}', \mathcal{S}) = 0$.

[11.4]From the results in Appendix E, we know that $p_{k|c} = p_{c,k}/p_c$.

11.3   Huffman Codes

A promise made is a debt unpaid, and the trail has its own stern code.
Robert W. Service (1874–1958), Canadian Poet
— From *The Cremation of Sam McGee* (1907)

This section deals with another focus that motivated Shannon. Today we would call it message compression, but he saw it as the problem of finding the optimal method for representing data in the most compact form. This will bring together the relationship between entropy and compression, so the material studied in the previous section will be a valuable tool. We now look at formalizing this notion with an eye to finding an optimal encoding for all possible messages in the message source. First we look at the man behind the idea we present herein.

◆ **Huffman** David A. Huffman (1925–1999) obtained his B.Sc. in electrical engineering from Ohio State University when he was eighteen. He then served as a radar maintenance officer on a destroyer, which served to clear mines in Japanese and Chinese waters following World War II. After returning to Ohio State University where he obtained his M.Sc., he went to MIT for his graduate work. While there, he developed his ideas, under the supervision of Robert Fano, and presented them in a term paper. The ideas were published in 1952. He received the Louis E. Levy Medal from the Franklin Institute for his Ph.D. thesis on sequential switching circuits. Ohio State University also later granted him their Distinguished Alumnus Award. Huffman's ideas on compression eclipsed the ideas put forth by his supervisor and Shannon, which were called *Shannon-Fano codes.*

He served on the faculty at MIT from 1953, but in 1967 he moved to California to take a position at the University of California at Santa Cruz. He founded the Computer Science Department there, and almost single-handedly developed the department, serving as its head from 1970–1973. He retired in 1994, but for a couple of years continued to teach Information Theory and other courses such as signal analysis.

Huffman received many awards, among them the Golden Jubilee Award for Technological Innovation from the IEEE Information Theory Society in 1988; the Computer Pioneer Award from the IEEE Computer Society; and the 1999 Richard W. Hamming Medal from the IEEE in recognition of his outstanding contributions to the general scope of information sciences. He died on October 7, 1999, and is survived by his wife, a son, and two daughters.

Huffman codes can be used in a vast array of compression areas, albeit newer approaches have taken over. Nevertheless, the ideas are seminal in the compression arena, so we will devote some time to understanding them.

◆ **Encoding** If $\mathcal{S}$ is a message source, we may define an injective function $f : \mathcal{S} \mapsto \mathcal{B}$, where $\mathcal{B}$ is the set of all bitstrings of finite length. This is called an *encoding* of messages from $\mathcal{S}$. We denote the bitlength of the image $f(s)$ by

$|f(m)|$. Now we define a *weighted length* of an encoding by

$$L(f) = \sum_{s \in S} p(s)|f(s)|. \tag{11.6}$$

The goal is to find an encoding f that minimizes $L(f)$.

◆ **Huffman Encoding Algorithm** Given a message source $S = \{m_1, m_2, \ldots, m_n\}$ with probability distribution given by p_j for $j = 1, 2, \ldots, n$, we encode as follows.

1. The two elements $m_i \leq m_j$ with lowest probability are encoded, the lowest with a 0 and the largest with a 1. In the event they are equal, m_i is encoded with a 0 and m_j with a 1.

2. The m_i and m_j from step 1 are treated as a single entity $m_{i,j}$, with probability equal to the sum of their individual probabilities, $p_{i,j} = p_i + p_j$.

3. Step 1 and 2 are repeated until there is a single element remaining. Then go to step 4.

4. The binary output for each $s_j \in S$ is obtained by reading backward through the above procedure to m_j in the message source.

Example 11.6 *Let* $S = \{s_1, s_2, s_3, s_4\}$ *with* $p_1 = 0.1$, $p_2 = 0.2$, $p_3 = 0.3$, $p_4 = 0.4$, *we illustrate the above algorithm via the following table.*

s_1	s_2	s_3	s_4
0.1	0.2	0.3	0.4
0	1		
0.3		0.3	0.4
0		1	
0.6			0.4
1			0
1.0			

The encoding is determined by reading backward through the procedure to get $f(s_1) = 100$; $f(s_2) = 101$; $f(s_3) = 11$; $f(s_4) = 0$. *Therefore,* $L(f) = 0.1 \cdot 3 + 0.2 \cdot 3 + 0.3 \cdot 2 + 0.4 \cdot 1 = 1.9$, *and when we compare this with the entropy of* S, *we get*

$$H(S) = 0.1 \cdot \log_2(10) + 0.2 \cdot \log_2(5) + 0.3 \cdot \log_2(10/3) + 0.4 \cdot \log_2(5/2) \approx 1.84644.$$

Hence, the Huffman encoding is approximately equal to the entropy.

Example 11.6 is motivation for the following fact.

Huffman Encoding and Entropy If L is given by Equation (11.6) and the assumptions surrounding it, then

$$H(S) \leq L < H(S) + 1. \tag{11.7}$$

11.4 Information Theory of Cryptosystems

Where is the wisdom we have lost in knowledge?
Where is the knowledge we have lost in information?
T.S. Eliot (1888–1965), Anglo-American poet, critic, and dramatist
— from *The Rock* (1934)

When we defined conditional entropy in Equation (11.3) on page 431, we looked at a cryptological context. It is this interpretation upon which we now concentrate. In fact, the quantity defined in that context for Equation (11.3) is called *key equivocation*.

◆ **Key Equivocation**

The entropy of cryptosystems is a key feature upon which we will focus herein. A cryptosystem may be defined by parameters that include the keyspace $\mathcal{K}$, the message space $\mathcal{M}$, the ciphertext space $\mathcal{C}$ (as well as encryption and decryption transformations), and certain probability distributions given as follows. Each plaintext unit, $m \in \mathcal{M}$, has a certain probability of occurring, and the choice of key $k \in \mathcal{K}$ is assumed to be independent of the choice of m, with probability of a given $k \in \mathcal{K}$ also having a probability distribution from which it follows that

$$H(\mathcal{K}, \mathcal{M}) = H(\mathcal{K}) + H(\mathcal{M})$$

(see part 1 of The Role of Independence on page 432). Also, the possible $c \in \mathcal{C}$ have a probability distribution that depends on the probability distributions for $\mathcal{M}$ and $\mathcal{K}$. Given this setup, the key equivocation satisfies

$$H(\mathcal{K}|\mathcal{C}) = H(\mathcal{K}) + H(\mathcal{M}) - H(\mathcal{C}), \tag{11.8}$$

which is a measure of how much information about the key is revealed by the ciphertext.

Example 11.7 *Let* $\mathcal{M} = \{s_1, s_2, s_3, s_4\}$ *with probabilities,*

$$p_{s_1} = 0.1, \quad p_{s_2} = 0.2, \quad p_{s_3} = 0.3, \quad and \quad p_{s_4} = 0.4;$$

$\mathcal{K} = \{k_1, k_2, k_3\}$ *with probabilities,*

$$p_{k_1} = 0.3, \quad p_{k_2} = 0.3, \quad and \quad p_{k_3} = 0.4;$$

and $\mathcal{C} = \{c_1, c_2, c_3, c_4\}$.
If E_k *is the enciphering transformation for a given* $k \in \mathcal{K}$, *and*

$$E_{k_1}(s_1) = c_1; \quad E_{k_1}(s_2) = c_2; \quad E_{k_1}(s_3) = c_3; \quad E_{k_1}(s_4) = c_4$$

$$E_{k_2}(s_1) = c_2; \quad E_{k_2}(s_2) = c_3; \quad E_{k_2}(s_3) = c_4; \quad E_{k_2}(s_4) = c_1$$

$$E_{k_3}(s_1) = c_3; \quad E_{k_3}(s_2) = c_4; \quad E_{k_3}(s_3) = c_1; \quad E_{k_3}(s_4) = c_2$$

then the probabilities for the ciphertexts, $p_{\mathcal{C}}(c_j)$, for $j = 1, 2, 3, 4$, is derived as follows:

$$p_{c_1} = p_{k_1} \cdot p_{s_1} + p_{k_2} \cdot p_{s_4} + p_{k_3} \cdot p_{s_3} =$$

$$0.3 \cdot 0.1 + 0.3 \cdot 0.4 + 0.4 \cdot 0.3 = 0.27,$$

and similarly

$$p_{c_2} = 0.25, \ p_{c_3} = 0.19, \ and \ p_{c_4} = 0.29.$$

Now we may calculate some conditional probabilities to determine key equivalence. First, we look at individual entropies.

Using the fact cited in Footnote 11.4 on page 432, we calculate the following:

$$p_{k_1|c_1} = \frac{p_{k_1} \cdot p_{s_1}}{p_{c_1}} = \frac{0.3 \cdot 0.1}{0.27} \approx 0.1111,$$

and similarly,

$$p_{k_1|c_2} \approx 0.2400; \quad p_{k_1|c_3} \approx 0.4737; \quad p_{k_1|c_4} \approx 0.4138;$$

$$p_{k_2|c_1} \approx 0.444; \quad p_{k_2|c_2} \approx 0.1200; \quad p_{k_2|c_3} \approx 0.3158; \quad p_{k_2|c_4} \approx 0.3103;$$

$$p_{k_3|c_1} \approx 0.444; \quad p_{k_3|c_2} \approx 0.6400; \quad p_{k_3|c_3} \approx 0.2105; \quad p_{k_3|c_4} \approx 0.2759.$$

Thus, using Equation (11.3) as a formula, we get

$$H(\mathcal{K}|\mathcal{C}) \approx 1.4342.$$

Now we calculate, using Equation (11.1),

$$H(\mathcal{K}) = -0.3 \log_2(0.3) - 0.3 \log_2(0.3) - 0.4 \log_2(0.4) \approx 1.5709,$$

and similarly, $H(\mathcal{M}) \approx 1.8464$, $H(\mathcal{C}) \approx 1.9831$. Therefore,

$$H(\mathcal{K}) + H(\mathcal{M}) - H(\mathcal{C}) \approx 1.4342,$$

which agrees with Equation (11.8), as an illustration.

This tells us that when Eve intercepts encrypted conversation between Bob and Alice, she obtains $H(\mathcal{K}|\mathcal{C})$ information about the key, and this is determined by the right-hand side of Equation (11.8). Moreover, this is, of course, a ciphertext-only attack.

Now define $\mathcal{C}^n$ to denote all n-grams (ciphertexts of length n), and similarly $\mathcal{M}^n$ will denote all n-grams of plaintext, with associated probability distributions, then as with Equation (11.8), we have

$$H(\mathcal{K}|\mathcal{C}^n) = H(\mathcal{K}) + H(\mathcal{M}^n) - H(\mathcal{C}^n), \tag{11.9}$$

Unconditional Security

If the following holds,

$$\lim_{n \mapsto \infty} H(\mathcal{K}|\mathcal{C}^n) \neq 0,$$

then the cryptosystem is said to be *unconditionally secure*.

On the other hand, we have the following.

Breakable — Theoretically

If the following holds,

$$\lim_{n \mapsto \infty} H(\mathcal{K}|\mathcal{C}^n) = 0,$$

then the cryptosystem is said to be *theoretically breakable*.

Associated with the latter case is the following.

Unicity Distance

The shortest length n for which,

$$H(\mathcal{K}|\mathcal{C}^n) \leq 1,$$

is called the *unicity distance*.

Since the unicity distance tells us that there is no more than one bit of uncertainty about the possible key, then it has only two possible values. In other words, any given ciphertext may be decrypted in at most two different ways. A competent cryptanalyst would be able to determine which one. It is sometimes the case where a unicity distance of ∞ is assigned to those cryptosystems that are unconditionally secure, so as to have a unicity distance assigned to all possible cases.

To show how to approximate the unicity distance, we need to explore the following.

◆ Entropy and Redundancy in Languages

The entropy of a cryptosystem is related to the entropy of the underlying language. We now look into this matter and examine languages in general with applications to cryptosystems. Let $\mathcal{L}$ be a given language, such as English. What is the entropy of $\mathcal{L}$? It is given by

$$H(\mathcal{L}) = \lim_{n \to \infty} \frac{H(\mathcal{M}^n)}{n}.$$

This was established by Shannon in [249], and it represents the average amount of information per letter in language text, as well as the degree of uncertainty in determining the next letter given knowledge of a substantial amount of text. It may also be viewed as the average number of bits needed for recording output from $\mathcal{L}$. The *rate of $\mathcal{L}$ for messages of length n* is defined by

$$r_n(\mathcal{L}) = H(\mathcal{L})/n,$$

and the *rate of $\mathcal{L}$* is defined to be

$$r(\mathcal{L}) = \lim_{n \to \infty} r_n(\mathcal{L}).$$

This is the average number of bits of entropy per letter.

The *absolute rate of $\mathcal{L}$* with a k-letter alphabet is given by

$$R(\mathcal{L}) = \log_2(k),$$

which is the maximum number of bits per letter in a string from $\mathcal{L}$. Thus, the *redundancy of $\mathcal{L}$* is defined by

$$D(\mathcal{L}) = R(\mathcal{L}) - r(\mathcal{L}),$$

and the *redundancy rate* is

$$D(\mathcal{L})/R(\mathcal{L}).$$

Redundancy in languages supplies an important cryptanalytic tool in terms of recovering plaintext or keys from ciphertext.

Example 11.8 *Consider $\mathcal{L}$ to be English. Shannon was able to demonstrate that, for English,*

$$1 \le r(\mathcal{L}) \le 1.5,$$

which means that the average information content of the English language is between 1 and 1.5 bits per letter. Moreover, since

$$R(\mathcal{L}) = \log_2(26) \approx 4.7$$

bits per letter, then

$$D(\mathcal{L}) = R(\mathcal{L}) - r(\mathcal{L}) \approx 4.7 - 1.25 \approx 3.5$$

so the redundancy of English is roughly 3.5 bits per letter. Looking at this another way, the redundancy rate is roughly

$$D(\mathcal{L})/R(\mathcal{L}) \approx 3.5/4.7 \approx 75\%.$$

In other words, 3/4 of the English language is redundant. This does not mean that you can discard three quarters of a given English text and still be able to read it. It does mean that there is a Huffman encoding of length n English text that will, for sufficiently large values of n, compress it to about a quarter of what it was (see inequality (11.7)).

Now that we have developed the above, we may return to unicity distance. It can be shown that the it may be approximated by the following.

Unicity Distance Approximation

$$n \approx \frac{\log_2(|\mathcal{K}|)}{D(\mathcal{L})},$$

where $D(\mathcal{L})$ is the redundancy of the underlying plaintext language and $|\mathcal{K}|$ is the cardinality of the keyspace $\mathcal{K}$.

Note that if there is a uniform probability distribution associated with $\mathcal{K}$, then $H(\mathcal{K}) = \log_2(|\mathcal{K}|)$.

Example 11.9 *Let us revisit the Caesar cipher discussed on page 11. Then $|\mathcal{K}| = 26$, and since we are using English, $D(\mathcal{L}) \approx 3.5$. Thus, the unicity distance is given by*

$$log_2(26)/3.5 \approx 1.34$$

letters.

If we take the substitution cipher defined on page 8 on the English alphabet, then

$$\log_2(|\mathcal{K}|) = \log_2(26!) \approx 88.4$$

and since $D(\mathcal{L}) \approx 3.5$, then the unicity distance is

$$n \approx 88.4/3.5 \approx 25,$$

meaning that for ciphertexts of length about 25, there should exist a unique decryption. because the unicity distance is effectively the smallest length of text that has probability near 1 for one of the possible decryptions and probability near 0 for all other possible decryptions.

The above tells us that the unicity distance may be viewed as the average ciphertext length needed for a cryptanalyst to uniquely compute the key, given enough computing time. Often the unicity distance is defined in terms of what are called *spurious keys*, which are those keys that Eve will rule out leaving only "possible keys", assuming she knows that the plaintext is a language such as English, and she is engaged in a ciphertext-only attack. Then the unicity distance is the value of n at which the number of spurious keys has an expected value of 0. That value is exactly the unicity distance we have defined above.

Now we would like to delve into the world of Shannon's perfect secrecy, the land of the one-time pad discussed on page 83.

◆ **Perfect Secrecy**

A cryptosystem is deemed to be *perfect* when the plaintext and ciphertext are mutually independent. In other words, in terms of entropy we have the following.

Perfect Secrecy

A cipher has *perfect secrecy* when $H(\mathcal{M}|\mathcal{C}) = H(\mathcal{M})$.

The above is tantamount to saying that

$$H(\mathcal{C}|\mathcal{M}) = H(\mathcal{C})$$

(see the role of independence on page 432). Basically, in a cryptosystem with perfect secrecy, Eve gains no information from the ciphertext about the plaintext that was not already known previously. Earlier, we talked about uniform distribution of the keyspace with reference to unicity distance. Now we show how this is intertwined with criteria for perfect secrecy.

Theorem 11.1 *If* $\mathbf{C} = \{\mathcal{M}, \mathcal{C}, \mathcal{K}, E\}$ *is a cipher such that*

1. Every key $k \in \mathcal{K}$ has probability $1/|\mathcal{K}|$;

   and

2. For each $m \in \mathcal{M}$ and $c \in \mathcal{C}$, there exists exactly one $k \in \mathcal{K}$ such that $E_k(m) = c$

both hold, then $\mathbf{C}$ *has perfect secrecy.*

Corollary 11.1 (Shannon) *The One-Time Pad has perfect secrecy.*

Shannon's Main Theorem (1949)

For a cryptosystem $\mathbf{C} = \{\mathcal{M}, \mathcal{C}, \mathcal{K}, E\}$, any two of the following imply the third.

1. $H(\mathcal{M}|\mathcal{C}) = H(\mathcal{M})$.

2. $H(\mathcal{K}) = H(\mathcal{M})$.

3. $I(\mathcal{M}, \mathcal{K}) = 0$ (see page 432).

It should be noted that entropy does not take into account the amount of computing time necessary to carry out an action. For instance, it might take the life of the known universe to carry out a computation, but entropy does not cover this. An example is the RSA cipher (see Section 4.2), for which $H(\mathcal{K}) = 0$, since the key may always be determined from public data. The computational time to factor the modulus, which we have seen to be computationally infeasible, is not taken into account by the entropy viewpoint. Hence, the latter (the computational complexity of RSA), is the more accurate assessment of RSA rather than any interpretation via entropy. This is also true of other ciphers based upon the intractability of solving some hard number-theoretic problem. For these cryptosystems, the ciphertext is formed, as with RSA, via some function computed modulo n, and the latter is roughly the size of the key. It follows that the setup of such ciphers forces the cryptanalyst launching a ciphertext-only attack, to try all keys, in other words, a brute-force attack.

11.5 Error-Correcting Codes

The aim of science is not to open the door to infinite wisdom, but to set a limit to infinite error.

Bertolt Brecht (1898–1956), German Dramatist
— from Section 9 of *The Life of Galileo* (1939)

We have already been introduced to the use of *non*cryptographic codes in this chapter. Indeed, Shannon is responsible for the birth of coding theory with his seminal papers, the consequences of which we have studied in the preceding sections. In this book we have been concerned largely with codes from a cryptographic viewpoint. However, at the outset, on page 6, we did promise to look at noncryptographic codes, called *error-correcting codes*, which are methods for detecting and/or correcting errors in the transmission of data.

Noncryptographic error correction is required in virtually anything that works with digitally represented data: satellite communications; telephone communications; fax machines; computers; CD players; and so forth. Coding theory deals with communications over noisy channels. This noise may be caused by human error, lightening, electric impulses, thermal noise, deterioration of machinery, imperfections in the equipment, neighboring channels, and so on. The goal of error-correcting codes is to encode the data, in a fashion (usually involving adding redundancy), so that the original data may be recovered if errors (but not too many of them) have occurred.

A simple example where we replace the original message with an encoding (see pages 433 and 434), which has built-in redundancy, is given as follows. Suppose that we want to send the letter C, which has binary representation 10. If we send a codeword of bitlength 6 as 101010, this is merely the repetition of the original message three times. Thus, if an error is introduced so that the received message is 111010, say, then the receiver can still determine the original message as the most repeated one, namely, 10. Of course, if too many errors are introduced, such as 111111 being the received message, then there is no hope of retrieving the original. Thus, a goal of error-correcting codes is to *minimize the probability* that errors will be introduced.

Once we encode a message, we need a mechanism to decode it after it passes through a noisy channel. For our example, in the above illustration, our decoder would be the recognition of the repetition of 10 twice, and would spit out the corrected message with the three repetitions of it for the intended user. In other words, the decoder recognizes that the nearest codeword is the one with the second 1 replaced by a 0 to make three repetitions of the already twice-repeated 01. If however, it received 111111, it would spit out the same since it has no basis upon which to determine if errors occurred. In this case, we are talking about *binary codewords* of bitlength 6. In terms of the definition given on the aforementioned pages,

$$f(C) = 101010, \text{ with } |f(C)| = 6.$$

This example is depicted in Diagram 11.1.

Example of Binary Encoding/Decoding
Diagram 11.1

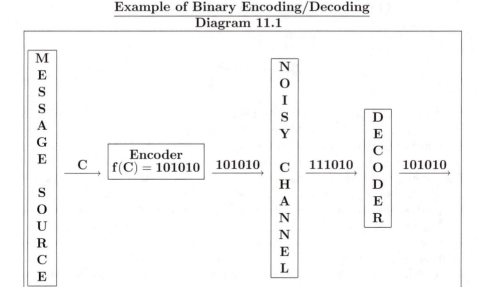

Now we may begin to formalize these notions.

◆ Codewords and Hamming Distance

Let $\mathcal{M}$ be the message space and let $\mathcal{M}^n$ be defined as in Section 11.4. Then a *code of length n* is a nonempty subset C of $\mathcal{M}^n$, and an element of C is called a *codeword*.[11.5] For example, if $\mathcal{M} = \{0, 1\}$, the codes are binary. If $n = 3$, for instance, and we are dealing only with binary repetition codes, then the code is the set,

$$C = \{000, 111\}.$$

In general, if $|\mathcal{M}| = q$, then the codes using the message space $\mathcal{M}$ are called *q-ary codes*. We must place some restrictions on $\mathcal{M}^n$ since a completely arbitrary such set would be unwieldy. Thus, we typically restrict $\mathcal{M}$ to be a finite field $\mathbb{F}_q$ for some prime power $q = p^m$, and the codes are therefore vectors in the vector space $\mathcal{M}^n$.[11.6] With this notion in mind, we now need a concept of distance between the vectors under consideration. This was given by Hamming, see [117] and [118].

The *Hamming distance* $d(u, v)$, for two vectors $u, v \in \mathcal{M}^n$ is defined as the number of components on which they disagree. For instance, if $u = (1, 0, 1, 0, 1, 0)$ and $v = (1, 1, 1, 0, 1, 0)$ from our above example, then $d(u, v) = 1$, since they disagree on only the second component. Similarly, if $u = (3, 4, 2, 1)$ and $v = (3, 2, 2, 0)$ are vectors in $\mathbb{F}_5^4$, then $d(u, v) = 2$ since they differ at the second and fourth components. Hence, the Hamming distance can be employed

---

[11.5]These are often called *block codes* since there exist codes where the codewords do not have fixed length.

[11.6]Such codes are typically called *linear codes*, whose *information rate* or *code rate* is given by $\log_q(M)/n$. Later in this section, we will study linear codes in depth.

as the number of errors to be corrected. The Hamming distance is a function on $\mathbb{F}_q^n$ that satisfies the following.

▼ Properties of the Hamming Function

Each of the following holds:

1. Given $u, v \in \mathbb{F}_q^n$, $d(u, v) = 0$ if and only if $u = v$.

2. For any $u, v \in \mathbb{F}_q^n$, $d(u, v) = d(v, u)$.

3. For any $u, v, w \in \mathbb{F}_q^n$,

$$d(u, v) + d(v, w) \geq d(u, w),$$

called the *triangle inequality*.

If C is a code, then the Hamming distance, $d(C)$, is also defined on C as the minimum distance between any two codewords. In mathematical terms,

$$d(C) = \min\{d(u, v) : u, v \in C, \text{ with } u \neq v\}.$$

For error-correcting codes, this is a valuable number since it provides the smallest number of errors that have to be corrected.

Another fundamental concept in the theory of error detection comes from the geometric arena.

▼ Hamming Sphere

A *Hamming sphere* in $\mathbb{F}_q^n$ of radius r centered at the codeword $u \in \mathbb{F}_q^n$ is the set of all vectors, denoted by $S(u, r)$ given as follows:

$$S(u, r) = \{v \in \mathbb{F}_q^n : d(u, v) \leq r\},$$

which has cardinality given by

$$|S(u, r)| = \sum_{j=0}^{r} \binom{n}{j} (q - 1)^j.$$

Now we need to address the issue of decoding. If we receive a message v sent through a noisy channel, we may decode it to u where u is that value such that $d(u, v)$ is the smallest possible. This process is called the *nearest neighbour decoding*.

To set the stage for the next important result, we need the following definitions. We say that a code C can *detect* up to s errors provided that changing $s+1$ components in a codeword can change it into another codeword, but changing s or fewer components in a codeword cannot alter it to make another codeword. Also, we say that a code C can *correct* up to t errors if changes made to at most t components of a codeword $c \in C$, ensure that the closest codeword remains c. In other words, if c' represents a codeword obtained by altering at most t

components of c, then $d(c, c')$ is a minimum. The following is fundamental in the theory of theory of error-correcting codes.

▼ **Maximal Error-Correction/Detection**

1. A code C can *detect* up to s errors if $d(C) \geq s + 1$.

2. A code C can *correct* up to t errors if $d(C) \geq 2t + 1$.

A consequence of the above is that nearest neighbour decoding may be used to *detect* up to $d - 1$ errors or to *correct* up to $(d - 1)/2$ errors where d is the minimum distance of a code C. This motivates some standard notation used in error-correcting codes.

▼ **Notation for Codes**: A code C of length n, having M codewords, and minimum distance $d = d(C)$, is called an (n, M, d)-code. this notation allows us to formulate more easily the central problems of coding theory.

We also require the following notion that allows us to determine when two codes are essentially the same.

▼ **Equivalent Codes**

Two codes are *equivalent* if a code can be obtained from the other by a finite sequence of operations of the types given in 1 and 2 below:

1. Permute the positions of the code;

2. Permute the symbols appearing in a fixed position of all codewords.

Now we provide a matrix theoretic interpretation of the above. Let a code C be given by the $M \times n$ matrix,

$$\begin{pmatrix} c_{1,1} & c_{1,2} & \cdots & c_{1,n} \\ c_{2,1} & c_{2,2} & \cdots & c_{2,n} \\ \vdots & \vdots & & \vdots \\ c_{M,1} & c_{M,2} & \cdots & c_{M,n} \end{pmatrix},$$

sometimes called the *generator matrix* for C, where each row is a codeword. Then operations of type 1 are merely rearrangements of the columns of the matrix and operations of type 2 are s (relabelling) of entries within a given column. For instance, if we have the binary code given by the matrix,

$$\begin{pmatrix} 0 & 1 & 0 & 1 & 1 & 1 \\ 0 & 0 & 0 & 0 & 1 & 1 \\ 1 & 0 & 0 & 1 & 1 & 1 \\ 0 & 0 & 1 & 1 & 1 & 1 \\ 1 & 0 & 0 & 0 & 1 & 1 \end{pmatrix},$$

then applying the permutation $0 \mapsto 1$ and $1 \mapsto 0$ to column four (an operation of type 2), then interchange columns one and five (an operation of type 1), we get

$$\begin{pmatrix} 1 & 1 & 0 & 0 & 1 & 0 \\ 1 & 0 & 0 & 1 & 1 & 0 \\ 1 & 0 & 0 & 0 & 1 & 1 \\ 1 & 0 & 1 & 0 & 1 & 0 \\ 1 & 0 & 0 & 1 & 1 & 1 \end{pmatrix}.$$

▼ **Central Coding Theory Goals**: An optimal (n, M, d)-code C, is one with small n, large M, and large d. We require a small length n so that we may efficiently transmit the code. We would like to have a large M in order to be able to send a diverse array of messages, and we seek a large d so we can correct as many errors as possible when they occur. Unfortunately, these aims are conflicting, so we usually fix one of the values and optimize the other two. One typical feat is to attempt to minimize d and maximize M for a given length n. For instance, if we have a q-ary (n, M, d)-code where M is a maximum, we denote the maximum value by $A_q(n, d)$. Indeed, the extreme cases are easy to find, namely, $A_q(n, 1) = q^n$ and $A_q(n, n) = q$. In fact, these are special cases of the next result.

The goals for optimizing the values was given mathematical rigour in [264] in 1964. This is thus known as the following.

▼ **The Singleton Bound**: Let C be a q-ary (n, M, d)-code. Then

$$M \leq q^{n-d+1}.$$

Moreover, if $M = q^{n-d+1}$, then C is called a *maximum distance separable* (MDS) code.

Hamming also developed a bound, given as follows.

▼ **The Hamming Bound**: If C is a q-ary (n, M, d)-ocde with $d \geq 2t + 1$, then

$$M \leq \frac{q^n}{\sum_{j=0}^{t} \binom{n}{j}(q-1)^j},$$

where the value on the right is called the *Hamming bound*.

▼ **Perfect Codes**: An (n, M, d)-code with $d = 2t + 1$ for which M equals the Hamming bound is called a *perfect code*. Another characterization of perfect codes is that every vector in F_q^n is a distance of no more than t units from exactly one codeword. In other words, the M spheres of radius t centered on codewords in a perfect t-error correcting code fills the entirety of $\mathbb{F}_q^n$ without overlapping. For instance, the binary code,

$$C = \{(0, 0, \ldots, 0), (1, 1, \ldots, 1)\},$$

of length n where n is odd, is a perfect $(n, 2, n)$-code. Later in this section we will look at some classes of perfect codes including those discovered by Hamming.

There are also lower bounds, the following of which was discovered in the middle of the twentieth century (see [104] and [284]).

▼ **The Gilbert-Varshamov Bound**: Given $n, d \in \mathbb{N}$ with $n \geq d$, there exists a q-ary (n, M, d)-code satisfying,

$$M \geq \frac{q^n}{\sum_{j=0}^{d-1} \binom{n}{j}(q-1)^j}.$$

◆ **Linear Codes**

An $[n, k]$-*linear code* over a field F is a k-dimensional subspace C of F^n. If the minimum distance $d = d(C)$ is given then we call it an $[n, k, d]$-linear code. Note that a a q-ary $[n, k, d]$-code is also a q-ary (n, q^k, d)-code, but not every (n, q^k, d)-code is an $[n, k, d]$-code.[11.7]

If $M_{k \times n}$ is a $k \times n$ matrix whose rows form a basis for the $[n, k]$-code, then $M_{k \times n}$ is called a *generator matrix* for C. It is precisely this mechanism of being able to describe the entire code via a basis of the codewords that make linear codes such a palatable means of error correction/detection.

An example of a binary $[8, 5, 1]$-code is given by the generating matrix,

$$G = \begin{pmatrix} 1 & 0 & 0 & 0 & 0 & 1 & 0 & 1 \\ 0 & 1 & 0 & 0 & 0 & 0 & 0 & 1 \\ 0 & 0 & 1 & 0 & 0 & 0 & 0 & 0 \\ 0 & 0 & 0 & 1 & 0 & 0 & 0 & 1 \\ 0 & 0 & 0 & 0 & 1 & 0 & 1 & 0 \end{pmatrix}.$$

The value of $d = 1$ may be achieved via the notion of *weight*. The weight of a codeword c is the number of nonzero entries in c. The weight of an entire code C is the minimum of the weights of the nonzero codewords in C. Thus, the weight of a given codeword $c \in C$ is $d(\overrightarrow{0}, c)$ where $\overrightarrow{0}$ is the zero vector.

A major advantage to linear codes is the ease with which we can encode. Suppose that G is the generating matrix for an $[n, k]$-code C over a finite field $\mathbb{F}_q$. Then a simple encoding rule for $c \in C$ is the following:

$$c \mapsto cG,$$

the multiplication of the $1 \times k$ vector c by the $k \times n$ matrix G. For instance, if we take the matrix G displayed above and the vector $c = (1, 1, 1, 1, 1) \in C$, then

$$cG = (1, 1, 1, 1, 1, 1, 1, 1).$$

Note that a linear q-ary code cannot be defined unless q is a prime power. This may be considered to a drawback to linear codes. Yet, even this seeming

---

[11.7]The reader may review the notion of vector spaces and dimension in Appendix A, especially Definition A.39 on page 490. It is important for this discussion to recall that a subspace C of dimension k in $\mathbb{F}_q^n$, satisfies the property that every vector of C can be uniquely expressed as a linear combination of the basis vectors $\{v_1, v_2, \ldots, v_k\}$ for C, and that C contains exactly q^k vectors. This explains why every q-ary $[n, k, d]$-code is also a q-ary (n, q^k, d)-code.

disadvantage may be overcome by using a larger message space. For instance, if we want to look at binary 15-ary codes, we need only go to linear codes over $\mathbb{F}_2^{16}$, and omit all codewords containing some fixed value.

Given that there are many equivalent matrices, we need a canonical choice. It can be shown that two $k \times n$ matrices generate equivalent linear $[n, k]$-codes over $\mathbb{F}_q$ if and only if one can be obtained from the other by a sequence of the following operations.

R1. A permutation of the rows.

R2. Multiplication of a row by a nonzero scalar.

R3. Addition of a scalar multiple of one row to another.

C1. A permutation of the columns.

C2. Multiplication of a column by a nonzero scalar.

Furthermore, if G is a generator matrix for an $[n, k]$-code, then operations R1–R3 and C1–C2 can transform G into standard form,

$$[I_k | M_{k,n-k}] = \begin{pmatrix} 1 & 0 & \cdots & 0 & m_{1,k+1} & \cdots & m_{1,n} \\ 0 & 1 & \cdots & 0 & m_{2,k+1} & \cdots & m_{2,n} \\ \vdots & \vdots & & \vdots & \vdots & & \vdots \\ 0 & 0 & \cdots & 1 & m_{k,k+1} & \cdots & m_{k,n} \end{pmatrix},$$

where I_k is the $k \times k$ identity matrix and $M_{k,n-k}$ is a $k \times (n - k)$ matrix. Therefore, $[I_k | M_{k,n-k}]$ has the first k columns to provide the codewords and the remaining $n - k$ columns to add redundancy. Note that the generator matrix G must have rows that are a basis for a k-dimensional subspace of the space of all vectors of length n, namely, our linear code C. Hence, every codeword is uniquely expressible as a linear combination of the rows of G, which must be linearly independent.

The first k bits are called the *information symbols* and the last $n - k$ bits are the *check symbols*. This means, as the above example shows, that in the encoding, the information symbols appear in the clear. Any code that satisfies this property is said to be *systematic*. Moreover, the redundancy in the remaining $n - k$ columns can be employed to do a parity check in the following fashion.

Given a generating matrix $G = [I_k, M_{k,n-k}]$, set $P = [-M_{k,n-k}^t | I_{n-k}]$, where $M_{k,n-k}^t$ is the transpose of $M_{k,n-k}$. Then P is called a *parity-check matrix*. Indeed, any matrix M that, given any $c \in F^n$, satisfies $cM^t = \overrightarrow{0}$ if and only if $c \in C$ is called a parity-check matrix for C.[11.8] Thus, any errors in transmission

---

[11.8]Note that when C is an $[n, k]$-code over $\mathbb{F}_q$ with parity-check matrix P, then $d(C)$ is the smallest number of columns of P that are linearly dependent. It follows that if every subset of $2t$ or fewer columns of P is linearly independent, then C is capable of correcting all errors of weight up to size t. If $q = 2$, this implies that when all possible linear combinations of no more than t columns of P are distinct, then $d(C) \geq 2t + 1$, whence C can correct all errors up to weight t.

may be easily detected via a check that $cP^t = \overrightarrow{0}$ for $c \in C$. For the example of the binary $[8, 5, 1]$-code given on the previous page, we have

$$P = \begin{pmatrix} 1 & 0 & 0 & 0 & 0 & 1 & 0 & 0 \\ 0 & 0 & 0 & 0 & 1 & 0 & 1 & 0 \\ 1 & 1 & 0 & 1 & 0 & 0 & 0 & 1 \end{pmatrix}.$$

We note that for the codeword $(1, 1, 1, 1, 1, 1, 1, 1) = cG$, we have $cGP^t = \overrightarrow{0} = (0, 0, 0)$. Observe, as well, that the subspace of F^n that forms a zero dot product[11.9] with all codewords from C in F^n has dimension $n - k$, and is called the *dual code of C*. This space is typically denoted by $C^\perp$, so a parity-check matrix for C may be defined as a generator matrix for $C^\perp$. In other words, the dual code of a linear $[n, k]$-code C with generating matrix $G = [I_k | M_{k,n-k}]$, is given by

$$C^\perp = \{v \in F^n : v \cdot c = \overrightarrow{0} \text{ for all } c \in C\},$$

which is itself a linear $[n, n - k]$-code with generating matrix,

$$P = [-M_{k,n-k}^t | I_{n-k}].$$

In this way, G may be viewed as a parity-check matrix for $C^\perp$.

Remark 11.1 *There is some linear algebra at work in the above. If $M \in \mathcal{M}_{m \times n}(F)$, then all vectors v such that $vM = \overrightarrow{0}$ is a subspace of F^n, called the left null space of M. This space is often called the* kernel *of M. Our linear codes are constructed via a k-dimensional subspace of F^n, which is manufactured by selecting k linearly independent vectors and taking their span. This is achieved by choosing a $k \times n$ generating matrix G of rank k with entries in F. The set of vectors of the form vG where v runs over all vectors in F^k provides the desired subspace. It is a fact from linear algebra that the left null space space of a matrix $M \in \mathcal{M}_{m \times n}(F)$ of rank r, has dimension $n - r$. Since our generating matrix has rank k, its null space $C^\perp$ has rank $n - k$. (See the discussion in Appendix A of matrix-related data, especially pages 494 and 495.) In any case, the mechanism now exists for checking errors since if $cP^t \neq \overrightarrow{0}$, then we know there is an error. However, the converse need not be true. It may be the case that $cP^t = \overrightarrow{0}$ and there is still an error but the chances are small that this is the case. It has much higher probability that no errors have occurred than that enough errors occurred to create a valid codeword from another. Hence, the parity check is taken as a signal that no errors have arisen.*

If a generating matrix G can be transformed into standard form $[I_k | M_{k,n-k}]$ employing only row operations R1–R3, then the latter will generate exactly the same code as the former. However, if C1–C2 are used, then the latter will generate a code that is equivalent to the former, but not necessarily the same. Furthermore, $[I_k | M_{k,n-k}]$ obtained from G is not unique, since permuting the columns of $M_{k,n-k}$ creates a generator matrix for an equivalent code.

[11.9]Recall that a dot product is the pointwise multiplication of two vectors. For instance, if $c = (c_1, c_2, \ldots, c_n) \in C$ and $x = (x_1, x_2, \ldots, x_n)$, then $cx = (c_1 \cdot x_1, c_2 \cdot x_2, \ldots, c_n \cdot x_n)$ is the dot product, which may be given via the above matrix equations.

▼ **Linear Encoding**: Earlier we stated the ease with which one may encode data by sending a codeword $c \in C$, an $[n, k]$-code over $\mathbb{F}_q$, via $c \mapsto cG$. To summarize, the encoding function $c \mapsto cG$ maps the vector space $\mathbb{F}_q^k$ to a k-dimensional subspace, of $\mathbb{F}_q^n$, namely, the code C. When the generating matrix G is in standard form $[I_k | M_{k,n-k}]$, with $M_{k,n-k} = (m_{i,j})$, then $c = (c_1, c_2, \ldots, c_k)$ is encoded via

$$\mathbf{c} = cG = (c_1, c_2, \ldots, c_k, c_{k+1}, \ldots, c_n),$$

where

$$c_{k+i} = \sum_{j=1}^{k} m_{j,i} c_j \text{ for } 1 \leq i \leq n - k$$

are the *check digits*, and the original c_i for $1 \leq i \leq k$ are the *message digits*. In the example of the $[8, 5, 1]$ binary linear code given above,

$$\mathbf{c} = cG = (c_1, c_2, c_3, c_4, c_5, c_6, c_7, c_8) = (1, 1, 1, 1, 1, 1, 1, 1),$$

with $n = 8$, $k = 5$,

$$M_{k,n-k} = M_{5,3} = \begin{pmatrix} m_{1,1} & m_{1,2} & m_{1,3} \\ m_{2,1} & m_{2,2} & m_{2,3} \\ m_{3,1} & m_{3,2} & m_{3,3} \\ m_{4,1} & m_{4,2} & m_{4,3} \\ m_{5,1} & m_{5,2} & m_{5,3} \end{pmatrix} = \begin{pmatrix} 1 & 0 & 1 \\ 0 & 0 & 1 \\ 0 & 0 & 0 \\ 0 & 0 & 1 \\ 0 & 1 & 0 \end{pmatrix},$$

$c_6 = \sum_{j=1}^{5} m_{j,1} c_j \equiv c_7 = \sum_{j=1}^{5} m_{j,2} c_j \equiv c_8 = \sum_{j=1}^{5} m_{j,3} c_j \equiv 1 \pmod 2$.

Now we add an illustration for linear codes (see diagram 11.2) that supplements Diagram 11.1.

Linear Code Encoding
Diagram 11.2

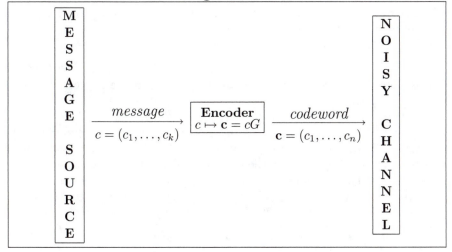

Suppose that the above codeword **c** is sent through the channel and received as $\mathbf{r} = (r_1, \ldots, r_n)$. then we define the *error vector* to be

$$\mathbf{e} = \mathbf{r} - \mathbf{c} = (e_1, \ldots, e_n).$$

Thus, when we want to decode, the decoder must determine from **r** the codeword **c**, if an error occurs, which will be the case when $e_j \neq 0$ for any $j = 1, 2, \ldots, n$. However, it turns out that there is a very elegant means for accomplishing this task as follows. The first thing that we note, as a motivator, is that if P is the parity check matrix for the code, then $\mathbf{c}P^t = \overrightarrow{0}$.

In the following, we need to employ some group theory of a basic kind. The reader needing a refresher on the concept of cosets and related notions should consult Appendix A, especially Definition A.37 on page 488. The following employs the fact that any linear code is an additive subgroup of a suitable $\mathbb{F}_q^n$.

▼ **Syndromes**: If P is a parity-check matrix for a linear $[n, k]$-code C, then for any $\mathbf{x} \in \mathbb{F}_q^n$, the $1 \times (n - k)$ row vector,

$$S(\mathbf{x}) = \mathbf{x}P^t,$$

is called the *syndrome of* **x**.

It is a basic fact that $S(\mathbf{x}) = \overrightarrow{0}$ if and only if $\mathbf{x} \in C$. A pleasant feature of the syndrome is that it depends solely upon the error pattern and not the message itself. The receiver, who detects $\mathbf{e} \neq \overrightarrow{0}$, will know both **r** and **e**, so will be able to determine **c**. To see that the syndrome's dependence is only on the error pattern, consider the following:

$$S(\mathbf{r}) = \mathbf{r}P^t = (\mathbf{c} + \mathbf{e})P^t = \mathbf{c}P^t + \mathbf{e}P^t = \mathbf{e}P^t$$

since $\mathbf{c}P^t = \overrightarrow{0}$.

The syndrome is a valuable piece of information about **e**, but we can do better. There is a limitation since for a given $S(\mathbf{r}) \in \mathbb{F}_q^{n-k}$, the collection of solutions $\mathbf{e}P^t = S(\mathbf{r})$ forms a coset of the code C in $\mathbb{F}_q^n$ of the form, for a fixed value of **e**, given by

$$C + \mathbf{e} = \{\mathbf{c} + \mathbf{e} : \mathbf{c} \in C\}.$$

It follows that two vectors are in the same coset if and only if they have the same syndrome. Hence, there is a one-to-one correspondence between the cosets and the syndromes.

Here we are viewing $\mathbb{F}_q$ as an additive group of q elements and C is a subgroup of the direct product $\mathbb{F}_q \times \cdots \times \mathbb{F}_q$, of which $C + \mathbf{e}$ is a coset.[11.10] Since the cardinality of each coset is q^k, and there are q^{n-k} cosets of C in total, given that there are q^{n-k} syndromes, then the receiver has to distinguish among the q^k possibilities for **e**. We need another concept to simplify the process.

Earlier we spoke about the weight of a binary vector being the number of nonzero elements appearing in it. If we select a vector of minimum weight in

---

[11.10]It is for this reason that linear codes are often called *group codes*.

a given coset, we call this vector a *coset leader*. The coset leader need not be unique, in which case we select one at random and call it the coset leader.

It is possible to form an array of vectors from $\mathbb{F}_q^n$ arranged as the cosets of C in the following fashion.

▼ **Slepian Standard Array**

The following is an algorithm for setting up a $q^{n-k} \times q^k$ matrix for a linear $[n, k]$-code, called a *Slepian standard array*. (See [265].)

1. List the codewords of C beginning with the zero vector $\mathbf{c_1} = \overrightarrow{0}$, as the first row: $\mathbf{c_1}, \mathbf{c_2}, \ldots, \mathbf{c_k}$.

2. Choose any one of the remaining $q^n - q^k$ vectors of minimum weight, $\mathbf{a_1}$ as the first element of the second row, and let

$$\mathbf{a_j} = \mathbf{a_1} + \mathbf{c_j} \text{ for } j = 2, 3, \ldots, q^k$$

be the remaining elements of the second row.

3. Select an element $\mathbf{b_1}$ of minimum weight from the remaining $q^n - 2q^k$ vectors as the first element of the third row, and let

$$\mathbf{b_j} = \mathbf{b_1} + \mathbf{c_j} \text{ for } j = 2, 3, \ldots, q^k$$

be the remaining elements of the third row.

4. Continue in the above fashion until all q^{n-k} rows are filled and every vector of $\mathbb{F}_q^n$ appears exactly once in one of those rows.

Example 11.10 *Let C be the binary $[4, 2]$-code with generating matrix,*

$$G = \begin{pmatrix} 1 & 0 & 1 & 0 \\ 0 & 1 & 1 & 1 \end{pmatrix}.$$

Then $n = 4$, $k = 2$, $q^k = 2^2$, $q^n = 2^4$,

$$C = \{\mathbf{c_1}, \mathbf{c_2}, \mathbf{c_3}, \mathbf{c_4}\} = \{(0,0)G, (0,1)G, (1,0)G, (1,1)G\} =$$

$$\{0000, 0111, 1010, 1101\}.$$

This forms the first row of the Slepian array. The balance are given in the $q^{n-k} \times q^k = 4 \times 4$ matrix,

$$\begin{pmatrix} 0000 & 0111 & 1010 & 1101 \\ 0100 & 0011 & 1110 & 1001 \\ 0010 & 0101 & 1000 & 1111 \\ 0001 & 0110 & 1011 & 1100 \end{pmatrix},$$

wherein the first column consists of the coset leaders, and the rows represent the elements in the cosets of C. (Notice that in the third row, 1000 could also have *qualified as the coset leader. The random choice of* 0010 *as coset leader shows that we obtain the same coset in either case since choice of the latter would just be a permutation of the third-row elements.)*

The parity check matrix is

$$P = \begin{pmatrix} 1 & 1 & 1 & 0 \\ 0 & 1 & 0 & 1 \end{pmatrix},$$

and since vectors in the same coset have the same syndrome, we may look at the received vector $\mathbf{r}$, *calculate its syndrome* $S(\mathbf{r}) = \mathbf{r}P^t$, *then find the position in the row. Since* $\mathbf{r} = \mathbf{c} + \mathbf{a}$ *where* $\mathbf{c}$ *is some codeword and* $\mathbf{a}$ *is the coset leader for the row, then we have a means for decoding* $\mathbf{r}$, *namely,* $\mathbf{c} = \mathbf{r} - \mathbf{a}$. *Hence, the decoding is the vector sitting at the top of the column in which* $\mathbf{r}$ *sits. For instance, if* $\mathbf{r} = (1,0,1,1)$, *then* $S(\mathbf{r}) = (1,1)$, *and this represents the fourth row, with coset leader* $\mathbf{a} = (0,0,0,1)$. *The value at the top of the column in which* $\mathbf{r}$ *sits is* $\mathbf{c} = (1,0,1,0)$. *Indeed,*

$$\mathbf{r} - \mathbf{a} = (1,0,1,1) - (0,0,0,1) = (1,0,1,0) = \mathbf{c}.$$

In fact, since all we use from the array is the coset leader and calculation of the syndrome of the received vector, then these two columns are all that need be stored on a computer, for instance, making this an efficient mechanism for decoding. These two columns make up an array called the syndrome look-up table, *which in our case is given by the following:*

$$\begin{pmatrix} \textbf{Coset Leader} & \textbf{Syndrome} \\ (0,0,0,0) & (0,0) \\ (0,1,0,0) & (1,1) \\ (0,0,1,0) & (1,0) \\ (0,0,0,1) & (0,1) \end{pmatrix},$$

Now we formalize the decoding illustrated in Example 11.10.

▼ Syndrome Decoding

The following algorithm for the decoding of a linear $[n, k]$-code C, needs far fewer iterations than any nearest neighbour decoding scheme, where one searches for the nearest codeword to a received vector. In what follows P is assumed to be the parity-check matrix for C.

1. For a received vector $\mathbf{r}$, calculate the syndrome $S(\mathbf{r}) = \mathbf{r}P^t$.

2. Find $S(\mathbf{r})$ in the second column of the syndrome look-up table, and find its coset leader $\mathbf{a}$ in the first column.

3. Decode via $\mathbf{c} = \mathbf{r} - \mathbf{a}$.

In Example 11.10, if we receive $\mathbf{r} = (1,0,0,0)$, then $S(\mathbf{r}) = (1,0)$, which is in the third row of the second column of the syndrome look-up table, which has coset leader $\mathbf{a} = (0,0,1,0)$. Hence, we decode via

$$\mathbf{c} = \mathbf{r} - \mathbf{a} = (1,0,0,0) - (0,0,1,0) = (1,0,1,0),$$

which we see is the entry at the top of the column of the Slepian array in which $\mathbf{r}$ sits.

The general process of syndrome decoding for a linear $[n,k]$-code C is illustrated as follows.

The notation

$$S(\mathbf{r}) \leftrightarrow \mathbf{a}$$

will denote the decoder's act of calculating the syndrome $S(\mathbf{r})$ and associating it with the coset leader $\mathbf{a}$ for the row in which it sits from the syndrome look-up table. The following illustration (Diagram 11.3) complements Diagram 11.2 on page 449.

Syndrome Decoding
Diagram 11.3

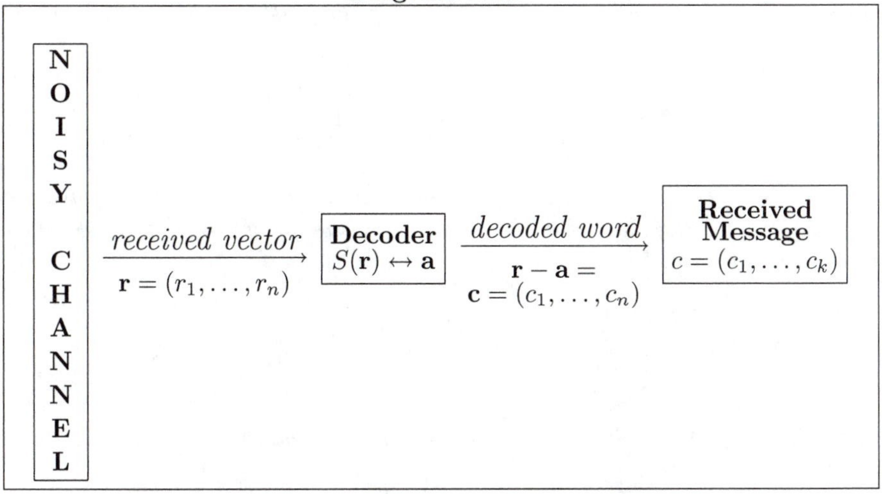

Now we turn to a well-known collection of linear codes. These are important in removing noise, for instance, from long-distance telephone calls. First we need the notion of *single-error-correcting codes*, which are codes capable of correcting all error patterns of weight no bigger than 1. The following are single-error-correcting codes that are easy to employ for encoding and decoding.

◆ The Hamming Codes

Although the codes we discuss herein may be defined over any $\mathbb{F}_q$ (see page 596), we confine our focus to the binary case for ease of presentation and overall simplicity of elucidation.

For an $n \times (2^r - 1)$ matrix, P, whose columns are distinct nonzero vectors in $\mathbb{F}_2^r$, the code C having P as its parity-check matrix is called a *binary Hamming code*, denoted by $\mathrm{Ham}(r, 2)$, which has length $2^r - 1$ and dimension $2^r - r - 1$. Since the columns are nonzero and distinct, then the minimum distance d must be bigger than 2. In fact $d = 3$ since it is possible to find codewords of weight 3 for $r > 1$. Hence, $\mathrm{Ham}(r, 2)$ for $r \geq 2$ is a linear, binary $[2^r - 1, 2^r - r - 1, 3]$ single-error-correcting code, which is also a perfect code. The latter follows from the fact that the Hamming spheres of radius 1 around the codewords exactly fill $\mathbb{F}_2^{2^r-1}$ without any overlapping. Another such class of distinguished linear, binary, perfect codes is the collection of *repetition codes*, which are $[n, 1]$ codes with generator matrix $G = [1, 1, \dots, 1]$, where n is odd. We will investigate yet a third such family, called *Golay codes*, later, and the latter represent the last of the possible such families, namely, the last of the possible linear, binary, perfect codes.

Given the above, $\mathrm{Ham}(r, 2)$ is a binary $(N, 2^k, 3)$-code (equivalently an $[N, k, 3]$-code), where $N = 2^r - 1$, and $k = 2^r - r - 1$, which is a distinguished class of perfect linear codes.

Example 11.11 *First, we observe that the parity check matrix P is given by column j being the binary representation of j. For instance, if $r = 3$, then*

$$P = \begin{pmatrix} 0 & 0 & 0 & 1 & 1 & 1 & 1 \\ 0 & 1 & 1 & 0 & 0 & 1 & 1 \\ 1 & 0 & 1 & 0 & 1 & 0 & 1 \end{pmatrix},$$

where column 1 is the binary representation of 1, column 2 is the binary representation of 2, and so forth. In order to get the generator matrix, we put P into standard form,

$$P = \begin{pmatrix} 0 & 1 & 1 & 1 & 1 & 0 & 0 \\ 1 & 0 & 1 & 1 & 0 & 1 & 0 \\ 1 & 1 & 0 & 1 & 0 & 0 & 1 \end{pmatrix} = [-M_{k,N-k}^t, I_{N-k}],$$

where $N = 2^r - 1 = 7$, $k = 2^r - r - 1 = 4$, then we obtain the generating matrix from it,

$$G = \begin{pmatrix} 1 & 0 & 0 & 0 & 0 & 1 & 1 \\ 0 & 1 & 0 & 0 & 1 & 0 & 1 \\ 0 & 0 & 1 & 0 & 1 & 1 & 0 \\ 0 & 0 & 0 & 1 & 1 & 1 & 1 \end{pmatrix} = [I_k, M_{k,N-k}],$$

which yields the $[7, 4, 3]$ Hamming code. To see that this is a perfect code, let $t = 1$, $n = N$, $k = 2^r - r - 1$, and $q = 2$ in the Hamming bound on page 445 to get

$$M = 2^4 = 2^k = \frac{2^N}{\sum_{j=0}^{t} \binom{N}{j}} = \frac{2^7}{7+1} = 2^{N-r-1},$$

as required.

Now suppose that we want to decode the received vector $\mathbf{r} = (1, 0, 1, 0, 1, 0, 0)$ *using syndrome decoding. Thus, we calculate* $S(\mathbf{r}) = \mathbf{r}P^t = (0, 0, 1)$, *which is the last column of P. Hence, we need only correct the last entry of* $\mathbf{r}$ *to get the correct codeword* $\mathbf{c} = (1, 0, 1, 0, 1, 0, 1)$. *The reason for this is that the error vector can have weight at most 1, so if the syndrome is nonzero, then the correction must be in the position corresponding to the column that the syndrome represents. This simple decoding method is explained in what follows.*

Decoding with Hamming Codes: Decoding with $\text{Ham}(n, 2)$ can be done with ease. Since $\text{Ham}(r, 2)$ is a perfect single-error correcting code, the nonzero coset leaders are the vectors,

$$r_j = (0, 0, \ldots, 1, 0, \ldots, 0),$$

where the only nonzero term is a 1 in exactly the jth position for $j = 1, 2, \ldots, 2^r - 1 = N$ in $\mathbb{F}_2^N$. Thus, calculating the syndrome via the parity-check matrix P, we get

$$S(r_j) = r_j P^t,$$

which is exactly the transpose of the jth column of P. Hence, the syndrome explicitly and directly identifies the error location within the received vector. Therefore, this allows a special, simpler syndrome decoding than in the general case, as follows.

Syndrome Decoding of Hamming Codes

1. For a received vector, $\mathbf{r}$, calculate the syndrome $S(\mathbf{r}) = \mathbf{r}P^t$.

2. If $S(\mathbf{r}) = \overrightarrow{0}$, then $\mathbf{r}$ is the correct codeword.

3. If $S(\mathbf{r}) \neq \overrightarrow{0}$, then $S(\mathbf{r})$ is the transpose of a unique column of P, the jth one, say. To get the correct codeword, add 1 to position j of $\mathbf{r}$.

We illustrated the above at the end of Example 11.11. The Hamming codes are clearly a special case of the general linear codes discussed earlier, since they can correct only one error. They were discovered by Hamming in 1950 (see [117]) and by Golay in 1949 (see [105]–[107]). The latter is responsible for one of the most singular and important codes in the history of error correction.

◆ Golay Codes

In his 1949 paper, Golay presented a binary, linear, perfect $[23, 12, 7]$-code, named the $\mathcal{G}_{23}$-code, that satisfies rather amazing properties. In his search for a perfect code Golay observed that

$$\sum_{j=0}^{3} \binom{23}{j} = 2^{11} = \binom{23}{0} + \binom{23}{1} + \binom{23}{2} + \binom{23}{3} = 2^{23-12},$$

which told him that there might exist a perfect $[23, 12]$ binary code that had the potential to correct up to three errors. In 1949, he made the discovery of $\mathcal{G}_{23}$, which indeed satisfies these properties, and is to this day, the only one known to be capable of correcting any combination of up to three random errors in a vector of length 23 (see Footnote 11.8 on page 447). Later, $\mathcal{G}_{23}$ was extended to what is now known as the $\mathcal{G}_{24}$-code, which is a (nonperfect) $[24, 12, 8]$-code. $\mathcal{G}_{24}$ proved to be extremely useful in satellite transmission. In fact during the years 1979–1981, *Voyager I* and *Voyager II* spacecrafts sent back signals from Jupiter and Saturn that were error-corrected using $\mathcal{G}_{24}$.

It turns out that it is very productive to first define the $\mathcal{G}_{24}$ code and derive the $\mathcal{G}_{23}$ code from it. Since a linear code may be characterized via its generating matrix, we provide the $\mathcal{G}_{24}$ generator matrix first. To construct this 12×24 matrix, we first provide the initial twelve columns that are given by the identity matrix I_{12}, so it has the form $G = [I_{12} | M_{12 \times 12}]$. The matrix $M_{12 \times 12}$ is constructed as follows. The first row of the matrix is given by

$$\mathbf{R_1} = (1, 1, 1, 0, 1, 1, 1, 0, 0, 0, 1, 0) = (r_1, r_2, \ldots, r_{12}),$$

and rows 2 through 11 are given by fixing the first element of this row and cyclically permuting the remaining elements to the right, namely, for $j = 2, 3, \ldots, 11$,

$$\mathbf{R_j} = (1, r_{12-j+2}, r_{12-j+3}, \ldots, r_{12}, r_2, r_3, \ldots, r_{12-j+1}),$$

and row 12 is given by

$$\mathbf{R_{12}} = (0, 1, 1, 1, 1, 1, 1, 1, 1, 1, 1, 1).$$

Hence, we achieve, $G =$

$$\begin{pmatrix}
1 & 0 & 0 & 0 & 0 & 0 & 0 & 0 & 0 & 0 & 0 & 0 & 1 & 1 & 1 & 0 & 1 & 1 & 1 & 0 & 0 & 0 & 1 & 0 \\
0 & 1 & 0 & 0 & 0 & 0 & 0 & 0 & 0 & 0 & 0 & 0 & 1 & 0 & 1 & 1 & 0 & 1 & 1 & 1 & 0 & 0 & 0 & 1 \\
0 & 0 & 1 & 0 & 0 & 0 & 0 & 0 & 0 & 0 & 0 & 0 & 1 & 1 & 0 & 1 & 1 & 0 & 1 & 1 & 1 & 0 & 0 & 0 \\
0 & 0 & 0 & 1 & 0 & 0 & 0 & 0 & 0 & 0 & 0 & 0 & 1 & 0 & 1 & 0 & 1 & 1 & 0 & 1 & 1 & 1 & 0 & 0 \\
0 & 0 & 0 & 0 & 1 & 0 & 0 & 0 & 0 & 0 & 0 & 0 & 1 & 0 & 0 & 1 & 0 & 1 & 1 & 0 & 1 & 1 & 1 & 0 \\
0 & 0 & 0 & 0 & 0 & 1 & 0 & 0 & 0 & 0 & 0 & 0 & 1 & 0 & 0 & 0 & 1 & 0 & 1 & 1 & 0 & 1 & 1 & 1 \\
0 & 0 & 0 & 0 & 0 & 0 & 1 & 0 & 0 & 0 & 0 & 0 & 1 & 1 & 0 & 0 & 0 & 1 & 0 & 1 & 1 & 0 & 1 & 1 \\
0 & 0 & 0 & 0 & 0 & 0 & 0 & 1 & 0 & 0 & 0 & 0 & 1 & 1 & 1 & 0 & 0 & 0 & 1 & 0 & 1 & 1 & 0 & 1 \\
0 & 0 & 0 & 0 & 0 & 0 & 0 & 0 & 1 & 0 & 0 & 0 & 1 & 1 & 1 & 1 & 0 & 0 & 0 & 1 & 0 & 1 & 1 & 0 \\
0 & 0 & 0 & 0 & 0 & 0 & 0 & 0 & 0 & 1 & 0 & 0 & 1 & 0 & 1 & 1 & 1 & 0 & 0 & 0 & 1 & 0 & 1 & 1 \\
0 & 0 & 0 & 0 & 0 & 0 & 0 & 0 & 0 & 0 & 1 & 0 & 1 & 1 & 0 & 1 & 1 & 1 & 0 & 0 & 0 & 1 & 0 & 1 \\
0 & 0 & 0 & 0 & 0 & 0 & 0 & 0 & 0 & 0 & 0 & 1 & 0 & 1 & 1 & 1 & 1 & 1 & 1 & 1 & 1 & 1 & 1 & 1
\end{pmatrix}.$$

Properties of $\mathcal{G}_{24}$

The matrix G is a generator matrix for the $\mathcal{G}_{24}$ linear $[24, 12, 7]$-code, and this code satisfies the following properties.

1. $\mathcal{G}_{24}$ is self dual. In other words, $\mathcal{G}_{24}^{\perp} = \mathcal{G}_{24}$. (See page 448.)

2. If $\mathbf{c}$ is a codeword of $\mathcal{G}_{24}$, then the weight $w(\mathbf{c})$ satisfies that

$$w(\mathbf{c}) \equiv 0 \,(\text{mod } 4),$$

and $w(\mathbf{c}) > 4$. (See page 446.)

The Perfect Golay Code: One can use the $\mathcal{G}_{24}$ Golay code to obtain the $\mathcal{G}_{23}$-code by merely deleting the last entry of each codeword in $\mathcal{G}_{24}$. Thus, $\mathcal{G}_{23}$ is a linear [23.12.7]-code, which unlike the $\mathcal{G}_{24}$-code is perfect, which follows from the Hamming bound on page 445.

There is a mechanism to more naturally introduce the Golay codes, but with more machinery necessary, namely, *cyclic codes*, which we will introduce shortly. For now, we conclude with the comments of a more sophisticated nature.

Advanced Facts Concerning Golay and Related Codes: The following are features of Golay codes, and others studied thus far, for the reader with some deeper mathematical background (or the desire to gain it).

1. The automorphism group (see page 600), of $\mathcal{G}_{24}$ is the Mathieu group M_{24}, one of the so-called *sporadic groups*, first discovered by Mathieu in the late nineteenth century. (For general group theory information, see [277].)

2. The $\mathcal{G}_{24}$-code can be employed to define the *Leech lattice*, which is one of the most efficient sphere-packing mechanisms known today. It was discovered in 1964 by John Leech (see [143]), as the unique lattice with the properties that it is unimodular (namely, it can be generated by the columns of a distinguished 24×24 matrix with determinant 1); the length of any vector in the lattice is an even integer; and the shortest length of any vector in the lattice is 2, meaning that the unit spheres centered at points in the lattice do not overlap. (For general lattice theory information, see [114].)

3. The words of weight 8 in $\mathcal{G}_{24}$ form a $\mathcal{S}(5,8,24)$ *Steiner system*, which is a 24-element set $\mathcal{S}$ together with a set, $\mathcal{T}$, of 8-element subsets of $\mathcal{S}$, with the property that each 5-element subset of $\mathcal{S}$ is contained in exactly one of the subsets in $\mathcal{T}$. One means of constructing this Steiner system is to form the matrix whose rows are the $2^{12} = 4096$ Golay 24-bit codewords. These rows form a group under addition modulo 2. In addition to the row consisting of $\overrightarrow{0}$, there are 759 rows having weight 8; 2576 having weight 12; 759 having weight 16; and one having weight 24. The 759 elements of weight 8 form the aforementioned set $\mathcal{T}$, called *octads*. (For a study of Steiner systems in general, see [59].)

4. There exists the $(11,6)$ ternary Golay code, which is the only known perfect nonbinary code. This $(11,6)$-code over $\mathbb{F}_3$ has minimum distance $d = 5$ and can correct up to two errors. As with the Golay [23, 12] binary code, the $(11,6)$ ternary Golay code can be extended to the $(12,6)$ ternary Golay code, and as with the extended $\mathcal{G}_{24}$ code, the $(12,6)$-code is also unique.

5. A result proved in the mid-1970s shows that the Hamming and Golay codes are the only (nontrivial) perfect codes (up to equivalence). (Here we consider the repetition codes, described on page 454, to be trivial.) In other words, it is known that the only nontrivial q-ary perfect codes are those with parameters given by the Golay and Hamming codes and so are equivalent to one of: (1) the q-ary Hamming codes (see page 596); (2) the

binary Golay code $\mathcal{G}_{23}$; and (3) the ternary $(11,6)$ Golay code, described above. The problem of actually finding all perfect codes having parameters of the Hamming and Golay codes remains difficult and unsolved. What *has* been proved is that if there is a perfect code, then the parameters of these codes must be the same as one of the Golay, Hamming or repetition codes, which is a result is due to van Lint [154] and Tietäväinen [278].

Now we turn to a natural and highly important type of code of which the preceding are examples.

◆ Cyclic Codes

We will be working over some fixed finite field $\mathbb{F}_q$. An (n,k)-code C is called *cyclic* if, whenever

$$\mathbf{c} = (c_1, c_2, \dots, c_n) \in C,$$

so is the cyclic shift,

$$\mathbf{c}' = (c_n, c_1, c_2, \dots, c_{n-1}),$$

For instance, the two Golay codes $\mathcal{G}_j$ for $j = 23, 24$ are both cyclic codes. Moreover, there are always four cyclic (n,k)-codes for any $n > 2$ over $\mathbb{F}_q$, given as follows.

Example 11.12 *Given $n \in \mathbb{N}$, there are always four cyclic codes of length n. They are: (1) the singleton zero vector of length n, $\overrightarrow{0}$, having dimension 0; (2) the collection of constant codewords of length n, $\mathbf{c} = (c_0, c_0, \dots, c_0)$, having dimension 1; (3) any collection of codewords, $\mathbf{c} = (c_1, c_2, \dots, c_n)$, that satisfy the property,*

$$c_1 + c_2 + \cdots + c_n = 0,$$

having dimension $n - 1$; and (4) the entirety of $\mathbb{F}_q^n$, having dimension n. There are also the two Golay codes $\mathcal{G}_j$ for $j = 23, 24$, as well as the Hamming code $\mathrm{Ham}(3,2)$, all studied above.

We will now reindex our reference to the codewords in the cyclic codes so that we look at

$$\mathbf{c} = (c_0, c_1, \dots, c_{n-1}),$$

since we wish to associate this vector with the polynomial,

$$c_0 + c_1 + \cdots + c_{n-1} x^{n-1}.$$

In general, we have the following situation, for which the reader needs familiarity with the notion of polynomial rings and related phenomena in Appendix A, especially pages 484–491.

For a fixed $n \in \mathbb{N}$, we look at

$$R_{n,q} = \mathbb{F}_q[x]/(x^n - 1), \tag{11.10}$$

which are the elements in $\mathbb{F}_q[x]$ modulo $(x^n - 1)$. Hence, we will always be considering polynomials of degree no more than $n - 1$, since any polynomial of degree n or larger may be divided by $x^n - 1$, and we merely take the remainder since we are working modulo the latter. (Note that since $x^n \equiv 1 \,(\mathrm{mod}\; x^n - 1)$, then we need only replace the powers of x accordingly, so division in the usual sense, by $x^n - 1$ is not required.) With this setup, and the notation given in (11.10), general cyclic codes are described as follows.

Polynomials and Cyclic Codes: Suppose that C is a cyclic code of length n over $\mathbb{F}_q$. To each codeword $\mathbf{c} = (c_0, c_1, \ldots, c_{n-1}) \in C$ associate the polynomial,

$$f_c(x) = c_0 + c_1 x + \cdots + c_{n-1}x^{n-1}.$$

Notice that

$$x f_c(x) = c_{n-1} + c_0 x + c_1 x^2 + \cdots + c_{n-2}x^{n-1}$$

corresponds to a single cyclic shift of the vector,

$$(c_0, c_1, \ldots, c_{n-1}).$$

Consequently, multiplying $f_c(x)$ by x^m corresponds to a cyclic shift through m positions to the right.

If we let $g(x)$ be the unique monic polynomial of smallest degree from the set of all such f_c, then let $\mathbf{C}$ denote C embedded in $R_{n,q}$, so

$g(x)$ is called the *generating polynomial* for $\mathbf{C}$.

(Observe that if the polynomial from this set of smallest degree is not monic, we can make it so by dividing through by the coefficient of the highest degree term, which is possible since our coefficients are from a field.)

If we therefore view the embedding $\mathbf{C}$ of C in $R_{n,q}$ via the polynomial identification, we get the following characterizations.

The Generating polynomial and Criteria for Cyclic Codes in $\mathbf{R_{n,q}}$

First, we observe that $\mathbf{C}$ in $R_{n,q}$ as above means that it is cyclic code if and only if it is closed under addition and closed under multiplication from elements of $R_{n,q}$, namely, it is an ideal of $R_{n,q}$. This tells us that any cyclic code can be generated by a polynomial. Hence, the cyclic codes are exactly the ideals of $R_{n,q}$. In other words,

a linear code $\mathbf{C}$ is cyclic if and only if $\mathbf{C}$ is an ideal in $R_{n,q}$.

If $\mathbf{C}$ is a cyclic code in $R_{n,q}$, then there exists a generating polynomial $g(x)$ satisfying each of the following.

1. $\mathbf{C} = \langle g(x) \rangle$, so $g(x)$ is uniquely determined by $\mathbf{C}$, which equals the set of polynomials of the form $f(x)g(x)$, where $\deg(f) \leq n - \deg(g) - 1$.

2. $g(x) \mid (x^n - 1)$.

3. If $x^n - 1 = g(x)p(x)$, then $h(x) \in R_{n,q}$ corresponds to an element of $\mathbf{C}$ if and only if

$$p(x)h(x) \equiv 0 \,(\text{mod } x^n - 1),$$

In other words, a polynomial is in $\mathbf{C}$ if and only if it is a multiple of the generating polynomial. The polynomial $p(x)$ is called the *parity-check polynomial.*

Suppose that the generator polynomial for a cyclic code $\mathbf{C}$ has degree t given by

$$g(x) = c_0 + c_1 + \cdots + c_t x^t,$$

then it follows from part 3 above that every element of $\mathbf{C}$ in $R_{n,q}$ is a polynomial of the form

$$f(x)g(x) \text{ where } \deg(f(x)) \leq n - t - 1,$$

so $f(x)$ may be viewed as a linear combination of the monomials x^j for $j = 0, 1, \ldots, n - t - 1$. In other words, the codewords are linear combinations of the polynomials $x^j g(x)$ for $j = 0, 1, \ldots, n - t - 1$. Hence, $\dim(\mathbf{C}) = n - t$ and a generator matrix for $\mathbf{C}$ is

$$G = \begin{pmatrix} c_0 & c_1 & c_2 & \cdots & c_t & 0 & 0 & \cdots & 0 \\ 0 & c_0 & c_1 & c_2 & \cdots & c_t & 0 & \cdots & 0 \\ 0 & 0 & c_0 & c_1 & c_2 & \cdots & c_t & \cdots & 0 \\ \vdots & \vdots & \vdots & \vdots & \vdots & & \vdots & \vdots & \vdots \\ 0 & 0 & 0 & \cdots & c_0 & c_1 & c_2 & \cdots & c_t \end{pmatrix}.$$

There is also a good reason for the polynomial $p(x)$ in part 3 of the preceding page to be called a parity-check polynomial. Let

$$p(x) = p_0 + p_1 x + \cdots + p_{n-t} x^{n-t},$$

then the $t \times n$ matrix given as follows is the parity-check matrix for C,

$$P = \begin{pmatrix} p_{n-t} & p_{n-t-1} & p_{n-t-2} & \cdots & p_0 & 0 & 0 & \cdots & 0 \\ 0 & p_{n-t} & p_{n-t-1} & p_{n-t-2} & \cdots & p_0 & 0 & \cdots & 0 \\ \vdots & \vdots & \vdots & \vdots & \vdots & & \vdots & \vdots & \vdots \\ 0 & 0 & 0 & \cdots & p_{n-t} & p_{n-t-1} & p_{n-t-2} & \cdots & p_0 \end{pmatrix}.$$

Example 11.13 *Let $g(x) = x + 1 \in R_3$, with $q = 2$, then $t = 1$, $n = 3$, and $c_0 = c_1 = c_t = 1$. Thus,*

$$C = \{(000), (110), (011), (101)\},$$

and C is a subspace of $\mathbb{F}_2^3$ of dimension $n - t = 2$ with basis $\{(110), (011)\}$. Viewed as a code in R_3,

$$\mathbf{C} = \{0, x^2 + x, x + 1, x^2 + 1\},$$

so **C** *is an ideal in* R_3 *of dimension* $n - t = 2$ *since*

$$\mathbf{C} = \langle g(x) \rangle.$$

Also, since

$$g(x)p(x) = (x+1)(x^2 + x + 1) = x^3 - 1 \,(\mathrm{mod}\ 2),$$

then $p(x) = x^2 + x + 1$ *is the parity-check polynomial for* **C**. *Hence, the generator matrix for* C *is given by*

$$G = \begin{pmatrix} 1 & 1 & 0 \\ 0 & 1 & 1 \end{pmatrix},$$

and the parity-check matrix is given by

$$P = \begin{pmatrix} 1 & 1 & 1 \end{pmatrix}.$$

Since the discussion preceding the example tells us that the only possible cyclic codes in R_3 *are those corresponding to polynomial divisors of* $x^3 - 1$, *then the only other possible binary cyclic codes of length 3 can be those associated with the polynomials* $g(x) = 1$ *corresponding to all of* $\mathbb{F}_2^3$; *those corresponding to* $g(x) = x^2 + x + 1$, *namely,* $C = \{(000), (111)\}$; *and the trivial one for* $g(x) = x^3 - 1$ *corresponding to the singleton code* $C = \{(000)\} = \{\overrightarrow{0}\}$. *Since there are no other divisors of* $x^3 - 1$, *this constitutes all possible binary cyclic codes of length 3.*

Although the parity-check matrix allows for *detection* of errors, correcting them for general cyclic codes can be quite arduous. The next set of cyclic codes provide us with more robust decoding, and error-detection features.

◆ BCH Codes

In 1959, a class of codes was discovered by Hocquenghem [124], as well as independently in 1960 by Bose and Ray-Chaudhuri [39] and [40]. The codes were initially named after the latter two authors, but later it was discovered that Hocquenghem had anticipated them, so the codes were renamed and are now accepted as BCH codes.

For a description of the following, the reader must be familiar with primitive roots and the notions surrounding them (see Appendix A, especially pages 480 and 490). First, we need the following.

BCH Bound: Suppose that C is a cyclic $[n, k, d]$-code over $\mathbb{F}_q$ where $q = p^m$ for some prime p and $m \in \mathbb{N}$ with $\gcd(p, n) = 1$. If $\mathbf{C} = \langle g(x) \rangle$ and α is a primitive nth root of unity for which there exist integers r, s such that

$$g(\alpha^r) = g(\alpha^{r+1}) = \cdots = g(\alpha^{r+s}) = 0,$$

then

$$d \geq s + 2.$$

The so-called BCH bound leads us into another fact about cyclic codes, namely, that they can be specified by roots of a generating polynomial considered in a suitable extension field of $\mathbb{F}_q$. When we stipulate that $\mathbf{C} = \langle g(x) \rangle$, we are merely saying that every code polynomial (the elements of $\mathbf{C}$ in $R_{n,q}$), is a multiple of $g(x)$ so they are all equal to 0 at the roots of $g(x)$. Looking at it another way, if $\beta_1, \beta_2, \ldots, \beta_r$ are elements of a finite extension of $\mathbb{F}_q$, and $m_j(x)$ is the minimal polynomial of β_j over $\mathbb{F}_q$ for $j = 1, 2, \ldots, r$, then we may define

$$g(x) = \mathrm{lcm}(m_1(x), \ldots, m_r(x)).$$

If $n \in \mathbb{N}$ such that $\beta^n = 1$ for all $j = 1, 2, \ldots, r$, then $g(x) \mid (x^n - 1)$. Hence, if $\mathbf{C} \subseteq \mathbb{F}_q^n$ is the cyclic code with generator polynomial $g(x)$, then $f(x) \in \mathbf{C}$ if and only if $f(\beta_j) = 0$ for all $j = 1, 2, \ldots, r$. An application of the above discussion is the following.

If $\mathbf{C} = \langle g(x) \rangle$ is the binary cyclic code of length $N = 2^r - 1$, where $g(x)$ is the minimal polynomial over $\mathbb{F}_2$ of a primitive element of F_{2^r}, then $\mathbf{C}$ is equivalent to $\mathrm{Ham}(\mathrm{r}, 2)$.

BCH Defined: Let $b \geq 0$ be an integer and $\alpha \in \mathbb{F}_{q^m}$, a primitive nth root of unity, where $m = \mathrm{ord}_n(q)$. Then a *BCH code* over $\mathbb{F}_q$ of length n and *designed distance* d ($2 \leq d \leq n$) is a cyclic code defined by the roots $\alpha^b, \alpha^{b+1}, \ldots, \alpha^{b+d-2}$ of the generator polynomial. Hence, if $m^{(j)}(x)$ denotes the minimal polynomial of α^j over$\mathbb{F}_q$, then the generator polynomial $g(x)$ of a BCH code is given by

$$g(x) = \mathrm{lcm}(m^{(b)}(x), m^{(b+1)}(x), \ldots, m^{(b+d-2)}(x)),$$

for some nonnegative integer b. When $b = 1$, they are called *narrow-sense BCH codes*. When $n = q^m - 1$, the BCH code is called *primitive*. If $n = q - 1$, then the BCH code is called a *Reed-Solomon code*.

It can be shown that indeed a BCH code of designed distance d has minimum weight at least d. The following illustrates this fact.

Example 11.14 *We maintain the notation from the above discussion. Let $q = 2^3$, $n = 7$, $m = 3$, $t = 4$, $b = 0$, $s = 2$, and $d = 4$. Then the cyclic $[7, 3, 4]$-code, described as follows, illustrates a BCH code. We have the following factorization over $\mathbb{F}_2$:*

$$x^7 - 1 = (x - 1)(x^3 + x^2 + 1)(x^3 + x + 1).$$

If we set $g(x) = x^4 + x^3 + x^2 + 1$, and let α be a primitive 7th root of unity, then the cyclic code,

$$\mathbf{C} = \langle g(x) \rangle,$$

may be described as follows. First, we note that $\mathbb{F}_{2^m} = \mathbb{F}_8 = \mathbb{F}_2(\alpha)$ (see page 487). It can be shown that α^j for $j = 1, 2, 4$ are the roots of $x^3 + x + 1 = 0$ (see page 599). Since $g(1) = g(\alpha) = g(\alpha^2) = 0$, then the BCH bound tells us that $d \geq 4$. Moreover, if we set $m^{(0)}(x) = (x - 1), m^{(1)}(x) = m^{(2)}(x) = x^3 + x + 1$, then

$$g(x) = \mathrm{lcm}(m^{(0)}(x), m^{(1)}(x), m^{(2)}(x)) = \mathrm{lcm}(x - 1, x^3 + x + 1),$$

and since the only remaining roots of $x^7 - 1$ are α^j for $j = 3, 5, 6$, then these are the roots of

$$p(x) = x^3 + x^2 + 1,$$

the parity-check polynomial.

We will not discuss BCH decoding since it is quite complicated in the general case, and simplification to the single error-correcting BCH codes reduces us to what we have already studied, the reason being that there exists an isomorphism between such a code and a Hamming code. For a complete and detailed overview of the general BCH decoding scheme, see [153], for instance. That said, the special case where $n = q - 1$ defined above is worth exploring.

◆ Reed-Solomon Codes

Reed-Solomon codes are a specific kind of BCH code with a vast array of applications in digital communications including digital television; high-speed modems; satellite communications; wireless communications; and storage devices such as barcodes, compact disks, DVDs, and the like. These codes excel at correcting certain types of errors called *burst errors.*[11.11] These are errors that occur close together, as opposed to randomly. As examples: a burst of solar energy can introduce errors in satellite communications; a scratch on a CD can introduce errors in contiguous bits; and electrical interference can be caused when an electric motor starts near a data-carrying cable. Hence, these codes are potent and have great efficacy in the aforementioned applications.

If $n = q - 1$, then $\mathbb{F}_q$ contains a primitive nth root of unity, α. In other words, $\mathbb{F}_q = \mathbb{F}_q(\alpha)$. Select a natural number $d = 2t + i - 1 < n$, where $i, t \in \mathbb{N}$, and set

$$g(x) = (x - \alpha^i)(x - \alpha^{i+1}) \cdots (x - \alpha^{2t+i-1}) = \sum_{j=0}^{2t} c_j x^j \in \mathbb{F}_q[x]. \qquad (11.11)$$

then the code $\mathbf{C} = \langle g(x) \rangle$ is a cyclic $[n, n-2t, t]$-code called a *Reed-Solomon code*, which achieves the maximum possible minimum code distance $d(C) = 2t + 1$ for any linear code with the same encoder input and output block lengths. Often the Reed-Solomon codes, defined above, are denoted by $RS(n, t)$ over $\mathbb{F}_q$ where $q = p^m$, which is a cyclic code of length $n = p^m - 1$, containing q^{n-2t} codewords, having dimension $n - 2t$, with generating polynomial given by (11.11), and minimum distance $2t + 1$.

Example 11.15 *Let us see how $RS(7, 2)$ is constructed. In this case, $q = 2^3$, $n = 7$, $t = 2$, $i = 1$, α, a primitive 7th root of unity, and*

$$g(x) = \prod_{j=1}^{4} (x - \alpha^j) = (x - \alpha)(x - \alpha^2)(x - \alpha^3)(x - \alpha^4) =$$

[11.11]For instance, a Reed-Solomon code may correct an entire byte, even if only one bit is corrupted. This gives these codes a huge advantage over binary codes that might only correct a bit, when several contiguous bits have been corrupted. We look at other codes, with respect to burst errors, in Appendix E (see page 549).

$$\sum_{j=0}^{4} c_j x^j = x^4 + \alpha^3 x^3 + x^2 + \alpha x + \alpha^3.$$

Since the only other root s of $x^7 - 1 = 0$ are $1, \alpha^5, \alpha^6$, then the parity check polynomial for this code is

$$p(x) = (x - 1)(x - \alpha^5)(x - \alpha^6) = x^3 + \alpha^3 x^2 + \alpha^2 x + \alpha^4.$$

The corresponding generating and parity-check matrices are given by

$$G = \begin{pmatrix} \alpha^3 & \alpha & 1 & \alpha^3 & 1 & 0 & 0 \\ 0 & \alpha^3 & \alpha & 1 & \alpha^3 & 1 & 0 \\ 0 & 0 & \alpha^3 & \alpha & 1 & \alpha^3 & 1 \end{pmatrix},$$

and

$$P = \begin{pmatrix} \alpha^4 & \alpha^2 & \alpha^3 & 1 & 0 & 0 & 0 \\ 0 & \alpha^4 & \alpha^2 & \alpha^3 & 1 & 0 & 0 \\ 0 & 0 & \alpha^4 & \alpha^2 & \alpha^3 & 1 & 0 \\ 0 & 0 & 0 & \alpha^4 & \alpha^2 & \alpha^3 & 1 \end{pmatrix}.$$

The dimension of $\mathbf{C} = \langle g(x) \rangle$ is $n - 2t = 3$. Since $\mathbf{C}$ is the embedding of $C \subseteq \mathbb{F}_q$ into $\mathbb{F}_q^7$ via $c \mapsto cG$, then we may illustrate the relationship as follows (see page 446). C has basis,

$$\{(100), (010), (001)\},$$

in $\mathbb{F}_q$ and these correspond to

$$\{1, \alpha, \alpha^2\}$$

via the association,

$$(c_0, c_1, c_2) \leftrightarrow c_0 + c_1 \alpha + c_2 \alpha^2.$$

Notice that, for instance,

$$(100)G = (\alpha^3, \alpha, 1, \alpha^3, 1, 0, 0),$$

$$(010)G = (0, \alpha^3, \alpha, 1, \alpha^3, \alpha, 0),$$

$$(001)G = (0, 0, \alpha^3, \alpha, 1, \alpha^3, 1),$$

which are rows 1–3 of G, respectively, and all codewords of $\mathbf{C}$ are linear combinations of these three rows. for instance,

$$(111)G = (\alpha^3, 1, 0, 0, \alpha^3, \alpha, 1)$$

is the sum of all three rows. Indeed, there are $q^{n-2t} = 8^3 = 512$ codewords in $\mathbf{C}$, which are all linear combinations modulo 8 of the three rows of G.

In Example 11.15, the number of n-tuples is 512 out of a total possible $2^{21} = 2,097,152$, or $1/4096$ of them. This illustrates the fact that if $q = 2^m$, then in the n-tuple space 2^{nm} n-tuples there 2^{km} are codewords in the Reed-Solomon code, where $k = n - 2t$. This is a small proportion of the possible

n-tuples. Hence, when a small proportion, such as this of the total space $\mathbb{F}_q^n$, then $d(C)$ can be allowed to be quite large. Indeed, since $d(C) = 2t + 1$, this helps explains why the maximum such $d(C)$ is achieved. Moreover, although any linear $[n, k, d]$-code has the capacity to correct $\lfloor (d-1)/2 \rfloor$ errors if they lie within the parity-check symbols (see part 2 of "maximum error-correction/detection" on page 444), the RS(n,t) codes can correct *any* $(d - 1)/2 = t$ symbols.

We conclude this section with a look at a generalization of BCH codes, which were discovered in 1970 (see [112]).

◆ **Goppa Codes**

Let $G(x) \in \mathbb{F}_q[x]$ be a polynomial of degree $s \in \mathbb{N}$, and let $(\alpha_0, \alpha_1, \ldots, \alpha_{n-1}) \in \mathbb{F}_{q^m}$, where $m, n \in \mathbb{N}$, and $G(\alpha_j) \neq 0$ for any $j = 0, 1, 2, \ldots, n - 1$. Then a *Goppa code* **C** is defined as follows. An element $\mathbf{c} = (c_0, c_1, \ldots, c_{n-1}) \in \mathbb{F}_q^n$ is in **C** if and only if

$$\sum_{j=0}^{n-1} \frac{c_j}{x - \alpha_j} \equiv 0 \,(\mathrm{mod}\ G(x)),$$

where $G(x)$ is called the *Goppa polynomial*, **C** is a linear $[n, k, d]$-code over $\mathbb{F}_q$ satisfying the properties:

1. $d \geq s + 1$.

2. $k \geq n - ms$.

There exist methods for decoding based upon syndrome calculations, but we do not cover that here. These methods, as well as the decoding methods for Reed-Solomon codes go beyond the scope of the text.

Goppa codes have some of the deepest and most outstanding of results in all of coding theory. Indeed, we may come full circle to the beginning of this chapter by tying together Goppa codes with the theory that Shannon developed in the 1940s from a modern mathematical viewpoint. In 1948, Andre Weil published a monograph (see [288]), that provided what is known as the proof of the Riemann hypothesis for algebraic curves over finite fields, which is well beyond the scope of this book to describe. Yet, we may speak of it insofar as it is intimately linked to coding theory, although in those early years nobody suspected any such connection. Weil's paper was the genesis of an evolution in the area of mathematics called *algebraic geometry*. Then a quarter of a century after the appearance of Weil's work and Shannon's introduction of information theory, Goppa suggested the connection between algebraic curves and codes, which we now call Goppa codes. (See [192] for the advanced theory.)

With this, we have already gone well beyond a mere introduction to coding theory, and its applications. This marks the end of our journey in the main text. Yet, it is our hope that, for the reader, this is merely the beginning of a journey to learn much more about the topics in this book, which is but a brief introduction to such a magnificent edifice of human accomplishment, bringing intellectual innovation and applications to everyday life.

Appendix A: Mathematical Facts

Several of the topics in this appendix are taken directly from the author's books [167]–[170].

A.1   Sets, Relations, and Functions

Definition A.1 (Sets)

A set is a well defined *collection of distinct objects. The terms* set, collection, *and* aggregate *are synonymous. The* objects *in the set are called* elements *or* members. *We write* $x \in S$ *to denote membership of an element* x *in a set* S, *and if* x *is not in* S, *then we write* $x \notin S$.

Set notation is given by putting elements between two braces. For instance, the set $\{1, 2, 3\}$ consists of the elements 1,2, and 3. In general, we may specify a set by properties. For instance, $\{x \in \mathbb{Z} : |x| > 1\}$ specifies those integers that satisfy the property of being bigger than 1 in absolute value. (Recall that $|x| = x$ if $x \geq 0$ and $|x| = -x$ if $x < 0$.)

Definition A.2 (Subsets and Equality)

A set T *is called a* subset *of a set* S, *denoted* $T \subseteq S$, *if every element of* T *is in* S. *On the other hand, if there is an element* $t \in T$ *such that* $t \notin S$, *then we write* $T \nsubseteq S$, *and say that* T *is* not *a subset of* S. *We say that two sets* S *and* T *are equal, denoted* $T = S$ *provided that* $a \in T$ *if and only if* $a \in S$, *namely, both* $T \subseteq S$, *and* $S \subseteq T$. *If* $T \subseteq S$, *but* $T \neq S$, *then we write* $T \subset S$, *and call* T *a proper* subset *of* S. *All sets contain the* empty set, *denoted by* $\varnothing$, *or* $\{\}$, *consisting of no elements. The set of* all *subsets of a given set* S *is called its* power set.

Although the set of all subsets of a given set, namely, the power set, is indeed a set, the set of all sets is not a set. Also, we may get examples of $\varnothing$ by defining sets with vacuous properties. For instance, $\{n \in \mathbb{N} : n < 1\} = \varnothing$, since no natural numbers are less than 1. We can also build sets from existing sets.

Definition A.3 (Complement, Intersection and Union)

The intersection *of two sets* S *and* T *is the set of all elements common to both, denoted* $S \cap T$, *namely,*

$$S \cap T = \{a : a \in S \ and \ a \in T\}.$$

The union *of the two sets consists of all elements that are in either* S *or in* T, *denoted* $S \cup T$, *namely,*

$$S \cup T = \{a : a \in S \ or \ a \in T\}.$$

If $\mathcal{T} \subseteq \mathcal{S}$, *then the complement of* $\mathcal{T}$ *in* $\mathcal{S}$, *denoted* $\mathcal{S} \setminus \mathcal{T}$ *is the set of all those elements of* $\mathcal{S}$ *that are not in* $\mathcal{T}$, *namely,*

$$\mathcal{S} \setminus \mathcal{T} = \{s : s \in \mathcal{S} \text{ and } s \notin \mathcal{T}\}.$$

Two sets $\mathcal{S}$ *and* $\mathcal{T}$ *are called* disjoint *if* $\mathcal{S} \cap \mathcal{T} = \varnothing$.

For instance, if $\mathcal{S} = \{1, 2, 3, 4\}$, and $\mathcal{T} = \{2, 3, 5\}$, then $\mathcal{S} \cap \mathcal{T} = \{2, 3\}$, and $\mathcal{S} \cup \mathcal{T} = \{1, 2, 3, 4, 5\}$. Also, $\mathcal{S} \setminus \{2, 3\} = \{1, 4\}$.

Definition A.4 (Binary Relations and Operations)

Let a, b *be elements of a set* $\mathcal{S}$. *Then we call* (a, b) *an* ordered pair, *where* a *is called the* first component, *and* b *is called the* second component. *If* $\mathcal{T}$ *is another set, then the* Cartesian product *of* $\mathcal{S}$ *with* $\mathcal{T}$, *denoted* $\mathcal{S} \times \mathcal{T}$ *is given by*

$$\mathcal{S} \times \mathcal{T} = \{(s, t) : s \in \mathcal{S}, t \in \mathcal{T}\}.$$

A relation *between* $\mathcal{S}$ *and* $\mathcal{T}$ *is a subset of* $\mathcal{S} \times \mathcal{T}$. *A* binary relation *on* $\mathcal{S}$ *is a subset* R *of* $\mathcal{S} \times \mathcal{S}$. *For* $(a, b) \in R$, *we write* aRb. *A* binary operation *on* $\mathcal{S}$ *is a rule that assigns each element of* $\mathcal{S} \times \mathcal{S}$ *to a unique element of* $\mathcal{S}$ (*see Definition A.5 below*).

Example A.1 *If* $\mathcal{S} = \{\clubsuit, \spadesuit, \text{❂}, \heartsuit\}$, *then*
$$\mathcal{R} = \{(\clubsuit, \clubsuit), (\text{❂}, \clubsuit), (\heartsuit, \text{❂})\}$$
is a binary relation on $\mathcal{S}$, *and so is*
$$\mathcal{R}_1 = \{(\clubsuit, \clubsuit), (\clubsuit, \spadesuit), (\heartsuit, \heartsuit)\}.$$

In Example A.1, $\clubsuit$ does not have a unique second element. There are certain distinguished binary operations that do satisfy the property of uniqueness in this regard.

Definition A.5 (Functions)

A function f (*also called a* mapping *or* map) *from a set* $\mathcal{S}$ *to a set* $\mathcal{T}$ *is a relation on* $\mathcal{S} \times \mathcal{T}$, *denoted by* $f : \mathcal{S} \to \mathcal{T}$, *which assigns each* $\mathcal{S} \in \mathcal{S}$ *a unique* $t \in \mathcal{T}$, *called the* image *of* s *under* f, *denoted by* $f(s) = t$. *The set* $\mathcal{S}$ *is called the* domain *of* f *and* $\mathcal{T}$ *is called the* range *of* f. *If* $\mathcal{S}_1 \subseteq \mathcal{S}$, *then the* image *of* $\mathcal{S}_1$ *under* f, *denoted by* $f(\mathcal{S}_1)$, *is the set* $\{t \in \mathcal{T} : t = f(s) \text{ for some } s \in \mathcal{S}_1\}$. *If* $\mathcal{S} = \mathcal{S}_1$, *then* $f(\mathcal{S})$ *is called the* image *of* f, *denoted by* img$(\mathcal{S})$. *If* $\mathcal{T}_1 \subseteq \mathcal{T}$, *the* inverse image *of* $\mathcal{T}_1$ *under* f, *denoted by* $f^{-1}(\mathcal{T}_1)$, *is the set* $\{s \in \mathcal{S} : f(s) \in \mathcal{T}_1\}$.

A function $f : \mathcal{S} \to \mathcal{T}$ *is called* injective (*also called* one-to-one) *if and only if for each* $s_1, s_2 \in \mathcal{S}$, $f(s_1) = f(s_2)$ *implies that* $s_1 = s_2$. *A function* f *is* surjective (*also called* onto) *if* $f(\mathcal{S}) = \mathcal{T}$, *namely, if for each* $t \in \mathcal{T}$, $t = f(s)$ *for some* $s \in \mathcal{S}$. *A function* f *is called* bijective (*or a* bijection) *if it is both injective and surjective. Two sets are said to be in a* one-to-one correspondence *if there exists a bijection between them. A* composition *of functions* f *and* g *is denoted by* $f \circ g$, *meaning the function defined by* $f \circ g(s) = f(g(s))$.

Each of the following may be verified for a given function $f : \mathcal{S} \to \mathcal{T}$.

(a) If $\mathcal{S}_1 \subseteq \mathcal{S}$, then $\mathcal{S}_1 \subseteq f^{-1}(f(\mathcal{S}_1))$.

(b) If $\mathcal{T}_1 \subseteq \mathcal{T}$, then $f(f^{-1}(\mathcal{T}_1)) \subseteq \mathcal{T}_1$.

(c) The identity map, $1_\mathcal{S} : \mathcal{S} \to \mathcal{S}$, given by $1_\mathcal{S}(s) = s$ for all $s \in \mathcal{S}$, is a bijection.

(d) f is injective if and only if there exists a function $g : \mathcal{T} \to \mathcal{S}$ such that $gf = 1_\mathcal{S}$, and g is called a *left inverse of f*.

(e) f is surjective if and only if there exists a function $h : \mathcal{T} \to \mathcal{S}$ such that $fh = 1_\mathcal{T}$, and h is called a *right inverse for f*.

(f) If f has both a left inverse g and a right inverse h, then $g = h$ is a unique map called the *two-sided inverse of f*.

(g) f is bijective if and only if f has a two-sided inverse.

Notice that in Definition A.4 a binary operation on $\mathcal{S}$ is just a function on $\mathcal{S} \times \mathcal{S}$.

Definition A.6 (Set Partitions)

Let $\mathcal{S}$ be a set , and let $\mathfrak{S} = \{\mathcal{S}_1, \mathcal{S}_2, \ldots\}$ be a set of nonempty subsets of $\mathcal{S}$. Then $\mathfrak{S}$ is called a partition of $\mathcal{S}$ provided both of the following are satisfied.

(a) $\mathcal{S}_j \cap \mathcal{S}_k = \varnothing$ for all $j \neq k$.

(b) $\mathcal{S} = \mathcal{S}_1 \cup \mathcal{S}_2 \cup \cdots \cup \mathcal{S}_j \cdots$, namely, $s \in \mathcal{S}$ if and only if $s \in \mathcal{S}_j$ for some j.

The number of elements in a set is of central importance.

Definition A.7 (Cardinality)

If $\mathcal{S}$ and $\mathcal{T}$ are sets, and there exists a one-to-one mapping from $\mathcal{S}$ to $\mathcal{T}$, then the sets are said to have the same cardinality. A set $\mathcal{S}$ is finite if either it is empty or there is an $n \in \mathbb{N}$ and a bijection $f : \{1, 2, \ldots, n\} \mapsto \mathcal{S}$. The number of elements in a finite set $\mathcal{S}$ is sometimes called its cardinality, or order, denoted by $|\mathcal{S}|$. A set is said to be countably infinite if there is a bijection between the set and $\mathbb{N}$. If there is no such bijection and $\mathcal{S}$ is not finite, then the set is said to be uncountably infinite. Two sets are said to be in one-to-one correspondence if there exists a bijection between them.

Example A.2 If $n \in \mathbb{N}$, then the map $f : \mathbb{N} \mapsto 2\mathbb{N}$ via $f(n) = 2n$ is bijective, so the cardinality of the even natural numbers is the same as that of the natural numbers themselves.

A.2 Basic Arithmetic

◆ Basic Arithmetic

A *natural number* n is one of the so-called counting numbers consisting of the set $\{1, 2, 3, \ldots\}$, denoted by $\mathbb{N}$ (where the ellipsis $\ldots$ means *ad infinitum* or "up to infinity"). The natural numbers, and their negatives, together with 0 is called the set of *integers*, namely, $\{\ldots, -3, -2, -1, 0, 1, 2, 3, \ldots\}$, denoted by $\mathbb{Z}$ (where the ellipsis on the left $\ldots$, denotes from negative infinity and on the right, *ad infinitum*).

Definition A.8 (Primes)
If $p \in \mathbb{N}$ $(p > 1)$ and p has no positive divisors, other than itself and 1, then p is a prime number, or simply a prime. *If $n \in \mathbb{N}, n > 1$, and n is not a prime, then n is* composite.

Once we have primes as the building bricks of the integers, we have a means of representing any given integer.

Definition A.9 (Canonical Prime Factorization)
If $n \in \mathbb{N}, n > 1$, then the factorization $n = \prod_{i=1}^{N} p_i^{a_i}$, where $a_i \in \mathbb{N}$, and $2 \le p_1 < p_2 < \ldots < p_N$, is the canonical prime factorization *of n.*

Moreover, those representations are essentially unique as the following fundamental fact shows.

Theorem A.1 (The Fundamental Theorem of Arithmetic) *Let $n \in \mathbb{N}, n > 1$. If $n = \prod_{i=1}^{r} p_i = \prod_{i=1}^{s} q_i$, where the p_i and q_i are primes, then $r = s$, and the factors are the same if their order is ignored.*

Proof. See [167, Theorem 1.4.2, page 44]. □

If $a, b \in \mathbb{Z}$ with $b \neq 0$, then a/b is a *rational number*. The set of all rational numbers is denoted by $\mathbb{Q}$. Rational numbers have periodic decimal expansions, such as $1/3 = 0.3333\ldots$, but those numbers, such as $\sqrt{2}$, do not have any repeated pattern in their decimal expansions. These numbers are called *irrational numbers*. The collection af all the rational and irrational numbers is called the set of *real numbers*, denoted by $\mathbb{R}$. The collection of only the *positive* reals is denoted by $\mathbb{R}^+$. Also, if $x \in \mathbb{R}$ and $n \in \mathbb{N}$, then $x^n = x \cdot x \cdots x$ multiplied n times is called an *exponentiation* of x.

◆ Divisibility.
If $a, b \in \mathbb{Z}$, then to say that b *divides* a, denoted by $b \mid a$, means that $a = bx$ for a *unique* $x \in \mathbb{Z}$, denoted by $x = a/b$. Note that the existence and uniqueness of x implies that b cannot be 0. We also say that a is *divisible* by b. If b does *not* divide a, then we write $b \nmid a$, and say that a is *not divisible* by b. Note that x is not unique for $a = b = 0$. We say that *division by zero is undefined*.

We may classify integers according to whether they are divisible by 2, as follows.

Definition A.10 (Parity)

If $a \in \mathbb{Z}$, and $a/2 \in \mathbb{Z}$, then we say that a is an even integer. *In other words, an even integer is one that is divisible by 2. If $a/2 \notin \mathbb{Z}$, then we say that a is an odd integer. In other words, an odd integer is one that is not divisible by 2. If two integers are either both even or both odd, then they are said to have the same parity. Otherwise they are said to have opposite or different parity.*

Of particular importance for divisibility is the following algorithm.

Theorem A.2 (The Division Algorithm)

If $a \in \mathbb{N}$ and $b \in \mathbb{Z}$, then there exist unique integers $q, r \in \mathbb{Z}$ with $0 \leq r < a$, and $b = aq + r$.

Proof. See page 599.   □

Now we look more closely at our terminology. To say that b divides a is to say that a is a *multiple* of b, and that b is a *divisor* of a. Also, note that b dividing a is equivalent to the remainder upon dividing a by b being zero. Any divisor $b \neq a$ of a is called a *proper divisor* of a. If we have two integers a and b, then a *common divisor* of a and b is a natural number n, which is a divisor of *both* a and b. There are special kinds of common divisors that are the content of our initial formal definition, first used in Chapter 1 (see page 75).

Definition A.11 (The Greatest Common Divisor)

If $a, b \in \mathbb{Z}$ are not both zero, then the greatest common divisor *or* gcd *of a and b is the natural number g such that $g \mid a$, $g \mid b$, and g is divisible by any common divisor of a and b, denoted by $g = \gcd(a, b)$.*

We have a special term for the case where the gcd is 1.

Definition A.12 (Relative Primality)

If $a, b \in \mathbb{Z}$, and $\gcd(a, b) = 1$, then a and b are said to be relatively prime *or* coprime. *Sometimes the phrase a is prime to b is also used.*

By applying the division algorithm, we get the following according to Euclid.

Theorem A.3 (The Euclidean Algorithm)

Let $a, b \in \mathbb{Z}$ $(a \geq b > 0)$, and set $a = r_{-1}, b = r_0$. By repeatedly applying the division algorithm, we get $r_{j-1} = r_j q_{j+1} + r_{j+1}$ with $0 < r_{j+1} < r_j$ for all $0 \leq j < n$, where n is the least nonnegative number such that $r_{n+1} = 0$, in which case $\gcd(a, b) = r_n$.

Proof. See [167, Theorem 1.3.3, page 37]. □

It is easily seen that any common divisor of $a, b \in \mathbb{Z}$ is also a common divisor of an expression of the form $ax + by$ for $x, y \in \mathbb{Z}$. Such an expression is called a *linear combination* of a and b. The greatest common divisor is a special kind of linear combination, which can be computed using a more general form of Theorem A.3, as follows.

Theorem A.4 (The Extended Euclidean Algorithm)
Let $a, b \in \mathbb{N}$, and let q_i for $i = 1, 2, \ldots, n + 1$ be the quotients obtained from the application of the Euclidean algorithm to find $g = \gcd(a, b)$, where n is the least nonnegative integer such that $r_{n+1} = 0$. If $s_{-1} = 1$, $s_0 = 0$, and

$$s_i = s_{i-2} - q_{n-i+2}s_{i-1},$$

for $i = 1, 2, \ldots, n + 1$, then

$$g = s_{n+1}a + s_n b.$$

Proof. See [167, Theorem 1.3.4, page 38]. □

Corollary A.1 *If $c \mid a$ and $c \mid b$, then $c \mid (ax + by)$ for any $x, y \in \mathbb{Z}$. In particular, the least positive value of $ax + by$ is g.*

We will also need the following notion (see Exercise 4.28 on page 575 in Appendix G, for instance).

◆ **(The Least Common Multiple)**

If $a, b \in \mathbb{Z}$, then the smallest natural number, which is a multiple of both a and b, is the *least common multiple* of a and b, denoted by $\operatorname{lcm}(a, b)$.

◆ **The Sigma Notation**

We can write $n = 1 + 1 + \cdots + 1$ for the sum of n copies of 1. We use the Greek letter upper case *sigma* to denote *summation*. For instance, $\sum_{i=1}^{n} 1 = n$ would be a simpler way of stating the above. Also, instead of writing the sum of the first one hundred natural numbers as $1 + 2 + \cdots + 100$, we may write it as $\sum_{i=1}^{100} i$. In general, if we have numbers $a_m, a_{m+1}, \cdots, a_n$ $(m \leq n)$, we may write their sum as

$$\sum_{i=m}^{n} a_i = a_m + a_{m+1} + \cdots a_n,$$

and by convention,

$$\sum_{i=m}^{n} a_i = 0 \text{ if } m > n.$$

The letter i is the *index of summation* (and any letter may be used here), n is *the upper limit of summation*, m is *the lower limit of summation*, and a_i is a *summand*. In the previous example, $\sum_{i=1}^{n} 1$, there is no i in the summand since we are adding the *same* number n times. The upper limit of summation tells us how many times that is. Similarly, we can write, $\sum_{j=1}^{4} 3 = 3 + 3 + 3 + 3 = 12$. This is the simplest application of the sigma notation. Another example is $\sum_{i=1}^{10} i = 55$.

Theorem A.5 (Properties of the Summation (Sigma) Notation)

Let $h, k, m, n \in \mathbb{Z}$ with $m \leq n$ and $h \leq k$. If R is a ring (see page 483), then:

(a) If $a_i, c \in R$, then $\sum_{i=m}^{n} c a_i = c \sum_{i=m}^{n} a_i$.

(b) If $a_i, b_i \in R$, then $\sum_{i=m}^{n} (a_i + b_i) = \sum_{i=m}^{n} a_i + \sum_{i=m}^{n} b_i$.

(c) If $a_i, b_j \in R$, then

$$\sum_{i=m}^{n} \sum_{j=h}^{k} a_i b_j = \left(\sum_{i=m}^{n} a_i \right) \left(\sum_{j=h}^{k} b_j \right) = \sum_{j=h}^{k} \sum_{i=m}^{n} a_i b_j = \left(\sum_{j=h}^{k} b_j \right) \left(\sum_{i=m}^{n} a_i \right).$$

Proof. See page 599.   □

Of value is the following formula.

Theorem A.6 (A Geometric Formula)

If $a, r \in \mathbb{R}$ $r \neq 1$, $n \in \mathbb{N}$, then

$$\sum_{j=0}^{n} a r^j = \frac{a(r^{n+1} - 1)}{r - 1}.$$

Proof. See [169, Theorem A.30, page 283].   □

A close cousin of the summation symbol is the following.

◆ **The Product Symbol**

The multiplicative analogue of the summation notation is the *product symbol* denoted by Π, upper case Greek *pi*. Given $a_m, a_{m+1}, \ldots, a_n \in R$, where R is a given ring and $m \leq n$, their product is denoted by

$$\prod_{i=m}^{n} a_i = a_m a_{m+1} \cdots a_n,$$

and by convention, $\prod_{i=m}^{n} a_i = 1$ if $m > n$.

The letter i is the *product index*, m is the *lower product limit*, n is the *upper product limit*, and a_i is a *multiplicand* or *factor*.

For instance, $\prod_{i=1}^{7} i = 1 \cdot 2 \cdot 3 \cdot 4 \cdot 5 \cdot 6 \cdot 7 = 5,040$. This is an illustration of the following concept.

Definition A.13 (Factorial Notation!)
   If $n \in \mathbb{N}$, then $n!$ (read "enn factorial") is the product of the first n natural numbers. In other words,

$$n! = \prod_{i=1}^{n} i.$$

We agree, by convention, that $0! = 1$. In other words, multiplication of no factors yields the identity.

The factorial notation gives us the number of distinct ways of arranging n objects. For instance, if you have 10 books on your bookshelf, then you can arrange them in $10! = 3{,}628{,}800$ distinct ways. This motivates the next symbol.

Definition A.14 (Binomial Coefficients)
   If $k, n \in \mathbb{Z}$ with $0 \le k \le n$, then the symbol $\binom{n}{k}$ (read "n choose k") is given by

$$\binom{n}{k} = \frac{n!}{k!(n-k)!}$$

the binomial coefficient.

The binomial coefficient is used in the theory of probability as the number of different combinations of n objects taken k at a time. For instance, the number of ways of choosing two objects from a set of five objects, *without regard for order*, is $\binom{5}{2} = 5!/(2!3!) = 10$ distinct ways (see Appendix E).

Proposition A.1 (Properties of the Binomial Coefficient)
   If $n, k \in \mathbb{Z}$ and $0 \le k \le n$, then

(a) $\binom{n}{n-k} = \binom{n}{k}$.   (**Symmetry Property**)

(b) $\binom{n+1}{k+1} = \binom{n}{k+1} + \binom{n}{k}$.   (**Pascal's Identity**)

(c) $\sum_{i=0}^{n} (-1)^i \binom{n}{i} = 0$.   (**Null Summation Property**)

(d) $\sum_{i=0}^{n} \binom{n}{i} = 2^n$.   (**Full Summation Property**)

   Proof. See [167, Proposition 1.2.1, pages 18 and 19]. □

The full summation property given above can be broken down into two more revealing parts. For this we need a new function as follows.

Definition A.15 (The Greatest Integer Function — The Floor)
   If $x \in \mathbb{R}$, then there is a unique integer $n \in \mathbb{Z}$ such that $n \le x < n + 1$. We say that n is the greatest integer less than or equal to x or the floor of x, denoted by $\lfloor x \rfloor = n$.

For instance, $\lfloor -1/2 \rfloor = -1$, $\lfloor 1/2 \rfloor = 0$, $\lfloor -1.5 \rfloor = -2$, and $\lfloor \sqrt{2} \rfloor = 1$.
Now we are able to revisit the full summation property and break it apart.

Proposition A.2 *If $n \in \mathbb{N}$, then*

(a) $\sum_{i=0}^{\lfloor n/2 \rfloor} \binom{n}{2i} = 2^{n-1}$,

and

(b) $\sum_{j=1}^{\lfloor (n+1)/2 \rfloor} \binom{n}{2j-1} = 2^{n-1}$.

Proof. See [167, Proposition 1.2.2, pages 21 and 22]. $\square$

Now that we have the notion of the floor function, it is valuable to know some of its properties.

Theorem A.7 (Properties of the Greatest Integer Function)

(a) $x - 1 < \lfloor x \rfloor \leq x$.

(b) $\lfloor x + n \rfloor = \lfloor x \rfloor + n$ for any $n \in \mathbb{Z}$.

(c) $\lfloor x \rfloor + \lfloor y \rfloor \leq \lfloor x + y \rfloor \leq \lfloor x \rfloor + \lfloor y \rfloor + 1$.

(d) $\lfloor x \rfloor + \lfloor -x \rfloor = \begin{cases} 0 & \text{if } x \in \mathbb{Z}, \\ -1 & \text{otherwise.} \end{cases}$

Proof. See [167, Theorem 1.2.4, page 22]. $\square$

We not only need the floor function but also the following close cousin. (See Appendix F on pages 555 and 556, for instance.)

Definition A.16 (The Least Integer Function — The Ceiling)
If $x \in \mathbb{R}$, then there is a unique integer $m \in \mathbb{Z}$ such that $x \leq m < x + 1$. We say that m is the least integer greater than or equal to x or the ceiling of x, denoted by $\lceil x \rceil = n$.

Theorem A.8 (Properties of the Least Integer Function)

(a) If $x \in \mathbb{R}$, then $-\lfloor -x \rfloor = \lceil x \rceil$.

(b) If $x \in \mathbb{R}$, then $\lceil x \rceil = \lfloor x \rfloor + 1$ if and only if $x \notin \mathbb{Z}$.

Proof. See [167, Exercises 1.3.1 and 1.3.2, page 40]. $\square$

An important fundamental result involving binomial coefficients that we will need in the text is the following.

Theorem A.9 (The Binomial Theorem)
  Let $x, y \in \mathbb{R}$, *and* $n \in \mathbb{N}$. *Then*

$$(x + y)^n = \sum_{i=0}^{n} \binom{n}{i} x^{n-i} y^i.$$

  Proof. See [167, Theorem 1.2.3, page 19]. $\square$

Note that the full and null summation properties in Proposition A.1 are just special cases of the binomial theorem (with $x = y = 1$ and $x = 1 = -y$, respectively.)

A.3 Modular Arithmetic

Definition A.17 (Congruences)
  If $n \in \mathbb{N}$, *then we say that* a *is* congruent *to* b modulo n *if* $n | (a - b)$, *denoted by*

$$a \equiv b \,(\mathrm{mod}\ n).$$

On the other hand, if $n \nmid (a - b)$, *then we write*

$$a \not\equiv b \,(\mathrm{mod}\ n),$$

and say that a *and* b *are* incongruent *modulo* n, *or that* a *is* not congruent *to* b *modulo* n. *The integer* n *is the* modulus *of the congruence. The set of all integers that are congruent to a given integer* m *modulo* n, *denoted by* $\overline{m}$, *is called the* congruence class *or* residue class *of* m *modulo* n.[A.1]

We have that $a \equiv b \,(\mathrm{mod}\ n)$, if and only if $a = b + nk$ for some $k \in \mathbb{Z}$. Thus, $a \equiv b \,(\mathrm{mod}\ n)$ if and only if $\overline{a} = \overline{b}$ with modulus n. Therefore, it makes sense to have a canonical representative.

Definition A.18 (Least Residues)
  If $n \in \mathbb{N}$, $a \in \mathbb{Z}$, *and* $a = nq + r$ *where* $0 \leq r < n$ *is the remainder when* a *is divided by* n, *given by Theorem A.2, the Division Algorithm, then* r *is called the* least (nonnegative) residue *of* a *modulo* n, *and the set* $\{0, 1, 2, \ldots, n - 1\}$ *is called the set of* least nonnegative residues modulo n.

Example A.3 There are four congruence classes modulo 4, namely,

$$\overline{0} = \{\ldots, -4, 0, 4, \ldots\},$$

$$\overline{1} = \{\ldots, -3, 1, 5, \ldots\},$$

[A.1]Note that since the notation $\overline{m}$ does not specify the modulus n, then the bar notation will always be taken in context.

$$\overline{2} = \{\ldots, -2, 2, 6, \ldots\},$$

and

$$\overline{3} = \{\ldots, -1, 3, 7, \ldots\},$$

since each element of $\mathbb{Z}$ is in exactly one of these sets.

In order to motivate the next notion we let $r \in \mathbb{Z}$, $n \in \mathbb{N}$, and consider the set $\{r, r+1, \ldots, r+n-1\}$. If $r + i \equiv r + j \pmod{n}$ for $0 \le i \le j \le n-1$, then $i \equiv j \pmod{n}$, so by the same argument as above $i = j$. This shows that the $\overline{r+j}$ for $0 \le j \le n-1$ are n distinct congruences classes. Moreover, if $m \in \mathbb{Z}$, then m must be in exactly one of the n congruence classes. In other words, $m \equiv r + j \pmod{n}$ for some nonnegative integer $j < n$. This motivates the following.

Definition A.19 (Complete Residue System)
Suppose that $n \in \mathbb{N}$ is a modulus. A set of integers

$$\mathcal{T} = \{r_1, r_2, \ldots, r_n\}$$

such that every integer is congruent to exactly one element of $\mathcal{T}$ modulo n is called a complete residue system modulo n. *In other words, for any $a \in \mathbb{Z}$, there exists a unique $r_i \in \mathcal{T}$ such that $a \equiv r_i \pmod{n}$. The set $\{0, 1, \ldots, n-1\}$ is a complete residue system, called the* least residue system modulo n.

Example A.4 The least residue system modulo 4 is $\mathcal{T} = \{0, 1, 2, 3\}$. Suppose that we want to calculate the addition of $\overline{3}$ and $\overline{2}$ in $\{\overline{0}, \overline{1}, \overline{2}, \overline{3}\}$. First, we must define what we mean by this addition. Let $\overline{a} \oplus \overline{b} = \overline{a+b}$ where $+$ is the ordinary addition of integers. Since $\overline{3}$ represents all integers of the form $3 + 4k$, $k \in \mathbb{Z}$, and $\overline{2}$ represents all integers of the form $2 + 4\ell$, $\ell \in \mathbb{Z}$,

$$3 + 4k + 2 + 4\ell = 5 + 4(k + \ell) = 1 + 4(1 + k + \ell).$$

Hence, $\overline{3} \oplus \overline{2} = \overline{1} = \overline{3 + 2}$. Similarly, we may define $\overline{a} \otimes \overline{b} = \overline{a \cdot b}$, where $\cdot$ is the ordinary multiplication of integers. The reader may verify that $\overline{2} \otimes \overline{3} = \overline{2} = \overline{2 \cdot 3}$. Notice as well that since $\overline{a - b} = \overline{a + (-b)} = \overline{a} \oplus \overline{-b}$, then $\overline{2} \oplus \overline{-3} = \overline{3} = \overline{2 - 3}$, for instance.

Example A.4 illustrates the basic operations of addition and multiplication in $\{\overline{0}, \overline{1}, \ldots, \overline{n-1}\}$ for any $n \in \mathbb{N}$, namely,

$$\overline{a} \oplus \overline{b} = \overline{a + b} \text{ and } \overline{a} \otimes \overline{b} = \overline{a \cdot b},$$

where $\oplus$ and $\otimes$ are well defined since $+$ and $\cdot$ are well defined. Since it would be cumbersome to use the notations of $\oplus$, and $\otimes$ in general, we maintain the usage of $+$ for $\oplus$ and $\cdot$ for $\otimes$, where we will understand that the the result of the given operation is in the appropriate residue class. The following result formalizes this for us in general.

Theorem A.10 (Modular Arithmetic)

Let $n \in \mathbb{N}$ and suppose that for any $x \in \mathbb{Z}$, $\bar{x}$ denotes the congruence class of x modulo n. Then for any $a, b, c \in \mathbb{Z}$ the following hold.

(a) $\bar{a} \pm \bar{b} = \overline{a \pm b}$. (Modular additive closure)

(b) $\bar{a}\bar{b} = \overline{ab}$. (Modular multiplicative closure)

(c) $\bar{a} + \bar{b} = \bar{b} + \bar{a}$. (Commutativity of modular addition)

(d) $(\bar{a} + \bar{b}) + \bar{c} = \bar{a} + (\bar{b} + \bar{c})$. (Associativity of modular addition)

(e) $\bar{0} + \bar{a} = \bar{a} + \bar{0} = \bar{a}$. (Additive modular identity)

(f) $\bar{a} + \overline{-a} = \overline{-a} + \bar{a} = \bar{0}$. (Additive modular inverse)

(g) $\bar{a}\bar{b} = \bar{b}\bar{a}$. (Commutativity of modular multiplication)

(h) $(\bar{a}\bar{b})c = \bar{a}(\bar{b}\bar{c})$. (Associativity of modular multiplication)

(i) $\bar{1} \cdot \bar{a} = \bar{a} \cdot \bar{1} = \bar{a}$. (Multiplicative modular identity)

(j) $\bar{a}(\bar{b} + \bar{c}) = \bar{a}\bar{b} + \bar{a}\bar{c}$. (Modular Distributivity)

Proof. See [169, Theorem 2.7, page 60]. □

For the notions of rings and fields used in the following, the reader should consult page 483.

Definition A.20 (The Ring $\mathbb{Z}/n\mathbb{Z}$)

For $n \in \mathbb{N}$, the set

$$\mathbb{Z}/n\mathbb{Z} = \{\bar{0}, \bar{1}, \bar{2}, \ldots, \overline{n-1}\}$$

is called the Ring of Integers Modulo n, *where $\bar{m}$ denotes the congruence class of m modulo n.*[A.2]

There is a multiplicative property of $\mathbb{Z}$ that $\mathbb{Z}/n\mathbb{Z}$ does not have, namely, the *Cancellation Law for $\mathbb{Z}$,* which says that if $ac = bc$ where $a, b, c \in \mathbb{R}$, and $c \neq 0$, then $a = b$. This is not the case for $\mathbb{Z}/n\mathbb{Z}$ in general. For instance, $2 \cdot 3 \equiv 2 \cdot 8 \,(\mathrm{mod}\, 10)$, but $3 \not\equiv 8 \,(\mathrm{mod}\, 10)$. In other words, $2 \cdot 3 = 2 \cdot 8$ in $\mathbb{Z}/10\mathbb{Z}$, but $3 \neq 8$ in $\mathbb{Z}/10\mathbb{Z}$. We may ask for conditions on n under which a modular law for cancellation would hold. In other words, for which $n \in \mathbb{N}$ does it hold that:

for any $a, b, c \in \mathbb{Z}/n\mathbb{Z}$ with $a \neq 0$, $ab = ac$ if and only if $b = c$? (A.1)

It can be shown that (A.1) cannot hold if $\gcd(a, n) > 1$, but if $\gcd(a, n) = 1$, then there is a solution $x \in \mathbb{Z}$ to $ax \equiv 1 \,(\mathrm{mod}\, n)$. This motivates the following.

[A.2]Occasionally, when the context is clear and no confusion can arise when talking about elements of $\mathbb{Z}/n\mathbb{Z}$, we will eliminate the *overline bars.*

Definition A.21 (Modular Multiplicative Inverses)
Suppose that $a \in \mathbb{Z}$, *and* $n \in \mathbb{N}$. *A multiplicative inverse of the integer* a *modulo* n *is an integer* x *such that* $ax \equiv 1 \, (\mathrm{mod} \, n)$. *If* x *is the least positive such inverse, then we call it* the least multiplicative inverse of the integer a modulo n, *denoted by* $x = a^{-1}$.

Example A.5 Consider $n = 26$, $a = -7$, and suppose that we want to find the least multiplicative inverse of a modulo n. Since $-7 \cdot 11 \equiv 1 \, (\mathrm{mod} \, 26)$ and no smaller natural number than 11 satisfies this congruence, then $a^{-1} = 11$ modulo 26.

When n is prime $\mathbb{Z}/n\mathbb{Z}$ takes on a new character (see page 483).

Theorem A.11 (The Field $\mathbb{Z}/p\mathbb{Z}$)
If $n \in \mathbb{N}$, *then* $\mathbb{Z}/n\mathbb{Z}$ *is a field if and only if* n *is prime.*

Proof. See page 599. □

We employ the notation F^* to denote the multiplicative group of nonzero elements of a given field F. In particular, when we have a finite field $\mathbb{Z}/p\mathbb{Z} = \mathbb{F}_p$ of p elements for a given prime p, then $(\mathbb{Z}/p\mathbb{Z})^*$ denotes the multiplicative group of nonzero elements of $\mathbb{F}_p$. This is tantamount to saying that $(\mathbb{Z}/p\mathbb{Z})^*$ is the group of units in $\mathbb{F}_p$, and $(\mathbb{Z}/p\mathbb{Z})^*$ is cyclic. Thus, this notation and notion may be generalized as follows. Let $n \in \mathbb{N}$ and let the group of units of $\mathbb{Z}/n\mathbb{Z}$ be denoted by $(\mathbb{Z}/n\mathbb{Z})^*$ (see page 483). Then

$$(\mathbb{Z}/n\mathbb{Z})^* = \{\overline{a} \in \mathbb{Z}/n\mathbb{Z} : 0 \le a < n \text{ and } \gcd(a, n) = 1\}. \qquad \text{(A.2)}$$

Numerous times we will need to solve systems of congruences for which the following result from antiquity is most useful.

Theorem A.12 (Chinese Remainder Theorem)
Let $n_i \in \mathbb{N}$ *for natural numbers* $i \le k \in \mathbb{N}$ *be pairwise relatively prime, set* $n = \prod_{j=1}^{k} n_j$ *and let* $r_i \in \mathbb{Z}$ *for* $i \le k$. *Then the system of* k *simultaneous linear congruences given by*

$$x \equiv r_1 \, (\mathrm{mod} \, n_1),$$

$$x \equiv r_2 \, (\mathrm{mod} \, n_2),$$

$$\vdots$$

$$x \equiv r_k \, (\mathrm{mod} \, n_k),$$

has a unique solution modulo n.

Proof. See [169, Theorem 2.29, page 69]. □

The natural generalization of Fermat's Little Theorem (given below) is the following, which provides the modulus for the RSA enciphering and deciphering exponents, for instance.

Definition A.22 (Euler's ϕ-Function)
For any $n \in \mathbb{N}$ the Euler ϕ-function, *also known as* Euler's Totient $\phi(n)$ *is defined to be the number of $m \in \mathbb{N}$ such that $m < n$ and $\gcd(m, n) = 1$.*

Theorem A.13 (The Arithmetic of the Totient)
If $n = \prod_{j=1}^{k} p_j^{a_j}$ where the p_j are distinct primes, then

$$\phi(n) = \prod_{j=1}^{k} \phi(p_j^{a_j}) = \prod_{j=1}^{k} (p_j^{a_j} - p_j^{a_j-1}) = \prod_{j=1}^{k} (p_j - 1)p_j^{a_j-1} = n \prod_{p_j | n} \left(1 - \frac{1}{p_j}\right).$$

Proof. See [169, Theorem 2.22, page 65]. □

Theorem A.14 (Euler's Generalization of Fermat's Little Theorem)
If $n \in \mathbb{N}$ and $m \in \mathbb{Z}$ such that $\gcd(m, n) = 1$, then $m^{\phi(n)} \equiv 1 \pmod{n}$.

Proof. See [167, Theorem 2.3.1, page 90]. □

Corollary A.2 (Fermat's Little Theorem)
If $a \in \mathbb{Z}$ and p is prime such that $\gcd(a, p) = 1$, then $a^{p-1} \equiv 1 \pmod{p}$.

Example A.6 *Let $n \in \mathbb{N}$. Then the cardinality of $(\mathbb{Z}/n\mathbb{Z})^*$ is $\phi(n)$. Hence, if G is a subgroup of $(\mathbb{Z}/n\mathbb{Z})^*$, $|G| \,\big|\, \phi(n)$.*

The calculus of integer orders and related primitive roots is an underlying fundamental feature of cryptographic problems such as the discrete log problem.

Definition A.23 (Modular Order of an Integer)
Let $m \in \mathbb{Z}$, $n \in \mathbb{N}$ and $\gcd(m, n) = 1$. The order *of m modulo n is the smallest $e \in \mathbb{N}$ such that $m^e \equiv 1 \pmod{n}$, denoted by $e = \text{ord}_n(m)$, and we say that m* belongs to the exponent e modulo n.

Note that the modular order of an integer given in Definition A.23 is the same as the element order in the group $(\mathbb{Z}/n\mathbb{Z})^*$.

Proposition A.3 (Divisibility by the Order of an Integer)
If $m \in \mathbb{Z}$, $d, n \in \mathbb{N}$ such that $\gcd(m, n) = 1$, then $m^d \equiv 1 \pmod{n}$ if and only if $\text{ord}_n(m) \,\big|\, d$. In particular, $\text{ord}_n(m) \,\big|\, \phi(n)$.

Proof. See [169, Proposition 4.3, page 161]. □

Definition A.24 (Primitive Roots)
If $m \in \mathbb{Z}$, $n \in \mathbb{N}$ and
$$\text{ord}_n(m) = \phi(n),$$
then m is called a primitive root modulo n. In other words, m is a primitive root if it belongs to the exponent $\phi(n)$ modulo n.

Theorem A.15 (Primitive Root Theorem)
An integer $n > 1$ has a primitive root if and only if n is of the form $2^a p^b$ where p is an odd prime, $0 \leq a \leq 1$, and $b \geq 0$ or $n = 4$. Also, if m has a primitive root, then it has $\phi(\phi(n))$ of them.

Proof. See [169, Theorem 4.10, page 165]. □

Definition A.25 (Index)
Let $n \in \mathbb{N}$ with primitive root m, and $b \in \mathbb{N}$ with $\gcd(b,n) = 1$. Then for exactly *one* of the values $e \in \{0, 1, \ldots, \phi(n) - 1\}$, $b \equiv m^e \pmod{n}$ holds. This unique value e modulo $\phi(n)$ is the index of b to the base m modulo n, denoted by $\text{ind}_m^n(b)$.

Definition A.25 gives rise to an arithmetic of its own, the *index calculus*. The following are some of the properties.

Theorem A.16 (Index Calculus)
If $n \in \mathbb{N}$ and m is a primitive root modulo n, then for any $c, d \in \mathbb{Z}$ each of the following holds.

(1) $\text{ind}_m^n(cd) \equiv \text{ind}_m^n(c) + \text{ind}_m^n(d) \pmod{\phi(n)}$.

(2) For any $t \in \mathbb{N}$, $\text{ind}_m^n(c^t) \equiv t \cdot \text{ind}_m^n(c) \pmod{\phi(n)}$.

(3) $\text{ind}_m^n(1) = 0$.

(4) $\text{ind}_m^n(m) = 1$.

(5) $\text{ind}_m^n(-1) = \phi(n)/2$ for $n > 2$.

(6) $\text{ind}_m^n(n - c) \equiv \text{ind}_m^n(-c) \equiv \phi(n)/2 + \text{ind}_m^n(c) \pmod{\phi(n)}$.

Proof. See [169, Theorem 4.14, page 166]. □

Proposition A.4 (Primitive Roots and Primality)

(1) If m is a primitive root modulo an odd prime p, then for any prime q dividing $(p - 1)$, $m^{(p-1)/q} \not\equiv 1 \pmod{p}$.

(2) *If $m \in \mathbb{N}$, p is an odd prime, and $m^{(p-1)/q} \not\equiv 1 \,(\mathrm{mod}\ p)$ for all primes $q \mid (p-1)$, then m is a primitive root modulo p.*

Proof. See page 600.   □

Of particular importance is the following notion. If $n \in \mathbb{N}$ and c is an integer, then c is called a *quadratic residue modulo n* if there exists an integer x such that $x^2 \equiv c \,(\mathrm{mod}\ n)$. The *least quadratic residue* of c modulo n is the reduction of c modulo n via Definition A.18. If no such integer exists, then c is called a *quadratic nonresidue modulo n*. The following symbol makes it easier to study quadratic residues.

Definition A.26 (Legendre's Symbol)
If $c \in \mathbb{Z}$ and $p > 2$ is prime, then

$$\left(\frac{c}{p}\right) = \begin{cases} 0 & \text{if } p \mid c, \\ 1 & \text{if } c \text{ is a quadratic residue modulo } p, \\ -1 & \text{otherwise,} \end{cases}$$

and $\left(\frac{c}{p}\right)$ is called the Legendre symbol *of c with respect to p.*

Example A.7 Let p be an odd prime. Then

$$\text{if } c^{(p-1)/2} \equiv 1 \,(\mathrm{mod}\ p), \text{ then } \left(\frac{c}{p}\right) = 1,$$

and

$$\text{if } c^{(p-1)/2} \equiv -1 \,(\mathrm{mod}\ p), \text{ then } \left(\frac{c}{p}\right) = -1.$$

This is called Euler's Criterion for quadratic residuacity.

Theorem A.17 (Properties of the Legendre Symbol)
If $p > 2$ is prime and $b, c \in \mathbb{Z}$, then

(1) $\left(\dfrac{c}{p}\right) \equiv c^{(p-1)/2} \,(\mathrm{mod}\ p)$.

(2) $\left(\dfrac{b}{p}\right)\left(\dfrac{c}{p}\right) = \left(\dfrac{bc}{p}\right)$.

(3) $\left(\dfrac{b}{p}\right) = \left(\dfrac{c}{p}\right)$, *provided $b \equiv c \,(\mathrm{mod}\ p)$.*

Proof. See [169, Theorem 4.28, page 171].   □

Of central importance in the study of the Legendre symbol is the next major result.

Theorem A.18 (The Quadratic Reciprocity Law)
If $p \neq q$ are odd primes, then

$$\left(\frac{p}{q}\right)\left(\frac{q}{p}\right) = (-1)^{\frac{p-1}{2} \cdot \frac{q-1}{2}}.$$

Equivalently,

$$\left(\frac{q}{p}\right) = -\left(\frac{p}{q}\right) \text{ if } p \equiv q \equiv 3 \,(\mathrm{mod}\ 4), \text{ and } \left(\frac{q}{p}\right) = \left(\frac{p}{q}\right) \text{ otherwise.}$$

Proof. See [169, Theorem 4.36, pages 173 and 174]. □

The following generalization of the Legendre symbol will be needed to discuss certain attacks on RSA for instance, (see page 210).

Definition A.27 (The Jacobi Symbol)
Let $n > 1$ be an odd natural number with $n = \prod_{j=1}^{k} p_j^{e_j}$ where $e_j \in \mathbb{N}$ and the p_j are distinct primes. Then the Jacobi symbol of a with respect to n is given by

$$\left(\frac{a}{n}\right) = \prod_{j=1}^{k} \left(\frac{a}{p_j}\right)^{e_j},$$

for any $a \in \mathbb{Z}$, where the symbols on the right are Legendre symbols.

The Jacobi symbol satisfies the following properties.

Theorem A.19 (Properties of the Jacobi Symbol)
Let $m, n \in \mathbb{N}$, with n odd, and $a, b \in \mathbb{Z}$. Then

(1) $\left(\dfrac{ab}{n}\right) = \left(\dfrac{a}{n}\right)\left(\dfrac{b}{n}\right).$

(2) $\left(\dfrac{a}{n}\right) = \left(\dfrac{b}{n}\right)$ *if $a \equiv b \,(\mathrm{mod}\ n)$.*

(3) *If m is odd, then* $\left(\dfrac{a}{mn}\right) = \left(\dfrac{a}{m}\right)\left(\dfrac{a}{n}\right).$

(4) $\left(\dfrac{-1}{n}\right) = (-1)^{(n-1)/2}.$

(5) $\left(\dfrac{2}{n}\right) = (-1)^{(n^2-1)/8}.$

(6) *If $\gcd(a, n) = 1$ where $a \in \mathbb{N}$ is odd, then*

$$\left(\frac{a}{n}\right)\left(\frac{n}{a}\right) = (-1)^{\frac{a-1}{2} \cdot \frac{n-1}{2}},$$

which is the quadratic reciprocity law *for the Jacobi symbol.*

Proof. See [169, Theorem 4.40, pages 175 and 176]. □

A.4 Groups, Fields, Modules, and Rings

Below is listed a set of axioms. Depending on which axioms are satisfied, we are able to determine the structure of the mathematical object we wish to define. After the listing, we describe the various types of such objects. In what follows, $\mathcal{S}$ denotes a set.

(a) For all $\alpha, \beta \in \mathcal{S}$, $\alpha + \beta = \beta + \alpha$. (Commutativity: addition)

(b) For all $\alpha, \beta, \gamma \in \mathcal{S}$, $(\alpha + \beta) + \gamma = \alpha + (\beta + \gamma)$. (Associativity: addition)

(c) There exists a unique $z \in \mathcal{S}$ such that $z + \alpha = \alpha + z = \alpha$. (Additive Identity) (When no confusion can arise, we use the symbol 0 here for the additive identity z, since it mimics the ordinary zero of the integers.)

(d) To each $\alpha \in \mathcal{S}$, there exists a $\alpha^{(0)} \in \mathcal{S}$ such that $\alpha + \alpha^{(0)} = \alpha^{(0)} + \alpha = z$. (Additive Inverse)

(e) For all $\alpha, \beta \in \mathcal{S}$, $\alpha\beta = \beta\alpha$. (Commutativity: multiplication)

(f) For all $\alpha, \beta, \gamma \in \mathcal{S}$, $(\alpha\beta)\gamma = \alpha(\beta\gamma)$. (Associativity: multiplication)

(g) For each $\alpha \in \mathcal{S}$, there exists a unique $i \in \mathcal{S}$ such that $i\alpha = \alpha i = \alpha$. (Multiplicative identity) (Here, as with the additive identity above, we can use the symbol 1 in place of the multiplicative identity i, when no confusion will arise from so doing, since i mimics the function of the multiplicative identity of the integers.)

(h) For all $\alpha, \beta, \gamma \in \mathcal{S}$, $\alpha(\beta + \gamma) = \alpha\beta + \alpha\gamma$. (Distributivity)

(i) For all $\alpha, \beta \in \mathcal{S}$, if $\alpha\beta = z$, then $\alpha = z$ or $\beta = z$. (No zero divisors)

(j) For any $\alpha \in \mathcal{J}$, with $\alpha \neq z$ there exists an element denoted α^{-1} such that $\alpha\alpha^{-1} = i = \alpha^{-1}\alpha$.

Any set which satisfies (a)–(d) is an *additive abelian group*. Any set that satisfies (a)–(d), (f), and (h) is a *ring*. If the ring also satisfies (e), then it is a *commutative ring*. If a commutative ring also satisfies (g), then it is a *commutative ring with identity*. If a ring also satisfies (i), then it is a *ring with no zero divisors*. A commutative ring with identity and no zero divisors is an *integral domain*, namely, those sets that satisfy all of (a)–(i). If a set satisfies all of (a)–(j), it is a *field*. If a set satisfies all of (a)–(j), **except** (e), then it is a *skew field* or *division ring*.

We need the following notion below. A *unit* or *invertible element* u in a commutative ring with identity R is an element for which there exists a multiplicative inverse. In other words, an element $u \in R$ is a unit if there exists an element $u^{-1} \in \mathbb{R}$ such that $uu^{-1} = 1_R$.

The reader should note that the abstract notion of a group is any nonempty set satisfying (b)–(d) of above. Moreover, the operation can be any binary operation (see Definition A.4). Of particular interest in this text is the following notion.

Definition A.28 (Permutation Groups)
    The set of all permutations on the set $\{1, 2, \ldots, n\}$ *is a group* S_n *under composition, with cardinality* $|S_n| = n!$, *called the* permutation group on n symbols.

Another notion that will help the reader understand some more advanced concepts is the following.

Definition A.29 (Modules)
    Suppose that G *is an additive abelian group, and that* R *is a commutative ring with identity* i *that satisfy each of the following axioms:*

(a) *For each* $r \in R$ *and* $g, h \in G$, $r(g + h) = (rg) + (rh)$.

(b) *For each* $r, s \in R$ *and* $g \in G$, $(r + s)g = (rg) + (sg)$.

(c) *For each* $r, s \in R$ *and* $g \in G$, $r(sg) = (rs)g$.

(d) *For each* $g \in G$, $ig = g$.

Then G *is a (two-sided)* module *over* R, *or for our purposes, simply an* R-module.

Definition A.30 (Algebras)
    If R *is a commutative ring with identity, then an* R-algebra *is a ring* A *such that*

(a) A *is an* R-module.

(b) $r(ab) = (ra)b = a(rb)$ *for all* $r \in R$ *and* $a, b \in A$.

Any R-algebra that is (as a ring) a division ring is called a *division algebra*. An algebra over a field K is called a *finite dimensional algebra* over K.

The following notion will be needed in what follows.

Definition A.31 (Characteristic of Rings)
    The characteristic of a ring R *is the smallest* $n \in \mathbb{N}$ *(if there is one) such that* $n \cdot r = 0$ *for all* $r \in R$. *If there is no such* n, *then* R *is said to have* characteristic 0.

✦ Polynomials and Polynomial Rings
    If R is a ring, then a *polynomial* $f(x)$ in an *indeterminant* x with *coefficients* in R is an infinite formal sum,

$$f(x) = \sum_{j=0}^{\infty} a_j x^j = a_0 + a_1 x + \cdots + a_n x^n + \cdots,$$

where the *coefficients* a_j are in R for $j \geq 0$ and $a_j = 0$ for all but a finite number of those values of j. The set of all such polynomials is denoted by $R[x]$. If $a_n \neq 0$, and $a_j = 0$ for $j > n$, then a_n is called the *leading coefficient* of $f(x)$. If the leading coefficient $a_n = 1_R$, in the case where R is a commutative ring with identity 1_R, then $f(x)$ is said to be *monic*.

We may add two polynomials from $R[x]$, $f(x) = \sum_{j=0}^{\infty} a_j x^j$ and $g(x) = \sum_{j=0}^{\infty} b_j x^j$, by

$$f(x) + g(x) = \sum_{j=0}^{\infty} (a_j + b_j) x^j \in R[x],$$

and multiply them by

$$f(x)g(x) = \sum_{j=0}^{\infty} c_j x^j,$$

where

$$c_j = \sum_{i=0}^{j} a_i b_{j-i}.$$

Also, $f(x) = g(x)$ if and only if $a_j = b_j$ for all $j = 0, 1, \ldots$. Under the above operations $R[x]$ is a ring, called the *polynomial ring over R in the indeterminant* x. Furthermore, if R is commutative, then so is $R[x]$, and if R has identity 1_R, then 1_R is the identity for $R[x]$. Notice that with these conventions, we may write $f(x) = \sum_{j=0}^{n} a_j x^j$, for some $n \in \mathbb{N}$, where a_n is the leading coefficient since we have tacitly agreed to "ignore" zero terms.

If $\alpha \in R$, we write $f(\alpha)$ to represent the element $\sum_{j=0}^{n} a_j \alpha^j \in R$, called the *substitution* of α for x. When $f(\alpha) = 0$, then α is called a *root* of $f(x)$. The substitution gives rise to a mapping

$$\overline{f} : R \mapsto R \text{ given by } \overline{f} : \alpha \mapsto f(\alpha),$$

which is determined by $f(x)$. Thus, $\overline{f}$ is called a *polynomial function* over R.

Definition A.32 (Degrees of Polynomials) *If $f(x) \in R[x]$, with $f(x) = \sum_{j=0}^{d} a_j x^j$, and $a_d \neq 0$, then $d \geq 0$ is called the* degree *of $f(x)$ over R, denoted by $\deg_R(f)$. If no such d exists, we write $\deg_R(f) = -\infty$, in which case $f(x)$ is the zero polynomial in $R[x]$. If F is a field of characteristic zero, then*

$$\deg_{\mathbb{Q}}(f) = \deg_F(f)$$

for any $f(x) \in \mathbb{Q}[x]$. If F has characteristic p, and $f(x) \in \mathbb{F}_p[x]$, then

$$\deg_{\mathbb{F}_p}(f) = \deg_F(f).$$

In either case, we write $\deg(f)$ for $\deg_F(f)$, without loss of generality, and call this the degree *of $f(x)$.*

With respect to roots of polynomials, the following is important.

Definition A.33 (Discriminant of Polynomials)

Let $f(x) = a \prod_{j=1}^{n} (x - \alpha_j) \in F[x]$, $\deg(f) = n > 1$, $a \in F$ a field in $\mathbb{C}$, where $\alpha_j \in \mathbb{C}$ are all the roots of $f(x) = 0$ for $j = 1, 2, \ldots, n$. Then the discriminant of f is given by

$$\mathrm{disc}(f) = a^{2n-2} \prod_{1 \le i < j \le n} (\alpha_j - \alpha_i)^2.$$

From Definition A.33, we see that f has a multiple root in $\mathbb{C}$ (namely, for some $i \ne j$ we have $\alpha_i = \alpha_j$, also called a *repeated root*) if and only if $\mathrm{disc}(f) = 0$.

Definition A.34 (Division of Polynomials)

We say that a polynomial $g(x) \in R[x]$ divides $f(x) \in R[x]$, if there exists an $h(x) \in R[x]$ such that $f(x) = g(x)h(x)$. We also say that $g(x)$ is a factor of $f(x)$.

Definition A.35 Irreducible Polynomials over Rings

A polynomial $f(x) \in R[x]$ is called irreducible (over R), if $f(x)$ is not a unit in R and any factorization $f(x) = g(x)h(x)$, with $g(x), h(x) \in R[x]$ satisfies the property that one of $g(x)$ or $h(x)$ is in R, called a constant polynomial. In other words, $f(x)$ cannot be the product of two nonconstant polynomials. If $f(x)$ is not irreducible, then it is said to be reducible.

Remark A.1 Note that it is possible that a reducible polynomial $f(x)$ could be a product of two polynomials of the same degree as that of f. For instance,

$$f(x) = (1 - x) = (2x + 1)(3x + 1) \text{ in } R = \mathbb{Z}/6\mathbb{Z}.$$

The following will be needed in our discussions on secret sharing in Section 5.5, for instance.

Theorem A.20 (The Lagrange Interpolation Formula) Let F be a field, and let a_j for $j = 0, 1, 2, \ldots, n$ be distinct elements of F. If c_j for $j = 0, 1, 2, \ldots, n$ are any elements of F, then

$$f(x) = \sum_{j=0}^{n} \frac{(x - a_0) \cdots (x - a_{j-1})(x - a_{j+1}) \cdots (x - a_n)}{(a_j - a_0) \cdots (a_j - a_{j-1})(a_j - a_{j+1}) \cdots (a_j - a_n)} c_j$$

is the unique polynomial in $F[x]$ such that $f(a_j) = c_j$ for all $j = 0, 1, \ldots, n$.

Now we turn to some facts about fields themselves. For the following example, recall that a *finite field* is a field with a finite number of elements $n \in \mathbb{N}$, denoted by $\mathbb{F}_n$. In general, if K is a finite field, then $K = \mathbb{F}_{p^m}$ for some prime p and $m \in \mathbb{N}$, also called *Galois fields*. The field $\mathbb{F}_p$ is called the *prime subfield* of K. In general, a prime subfield is a field having no proper subfields, so $\mathbb{Q}$ is the prime subfield of any field of characteristic 0 and $\mathbb{Z}/p\mathbb{Z} = \mathbb{F}_p$ is the prime field of any field $K = \mathbb{F}_{p^m}$. Also, we have the following result.

Theorem A.21 (Multiplicative Subgroups of Fields)
If F is any field and F^* is a finite subgroup of the multiplicative subgroup of nonzero elements of F, then F^* is cyclic.[A.3] *In particular, if $F = \mathbb{F}_{p^n}$ is a finite field, then F^* is a finite cyclic group, and a generator of F^* is called a primitive element of F.*

It also follows that if p is prime and $m \in \mathbb{N}$, then there exists a field with p^m elements that is unique up to isomorphism (see [168, Corollary C.19, page 398], for instance). Related to this is the following notion, which we will need, for instance, in Chapter 11 when we discuss coding theory. For $m \in \mathbb{N}$, a *primitive mth root of unity* is a complex number α such that $\alpha^m = 1$, but $\alpha^j \neq 1$ for all natural numbers $j < m$. Primitive roots play a vital role in the proofs of results involving field extensions of various types, especially finite. For instance, it may be shown that $\mathbb{F}_{p^m} = \mathbb{F}_p(\alpha)$ where α is a primitive $(p^m - 1)$th root of unity. (This is related to Galois theory; see [168, Appendix C, pages 393–401] for an overview of this elegant theory.)

✦ **Action on Rings**

Definition A.36 (Morphisms of Rings)
If R and S are two rings and $f : R \to S$ is a function such that $f(ab) = f(a)f(b)$, and $f(a + b) = f(a) + f(b)$ for all $a, b \in R$, then f is called a ring homomorphism. *If, in addition, $f : R \to S$ is an injection as a map of sets, then f is called a* ring monomorphism. *If a ring homomorphism f is a surjection as a map of sets, then f is called a* ring epimorphism. *If a ring homomorphism f is a bijection as a map of sets, then f is called a* ring isomorphism, *and R is said to be* isomorphic to S, *denoted by $R \cong S$. Lastly, $\ker(f) = \{s \in S : f(s) = 0\}$ is called the* kernel *of f. Also, f is injective if and only if $\ker(f) = \{0\}$.*

There is a fundamental result that we will need in the text. In order to describe it, we need the following notion.

[A.3]Recall that a multiplicative abelian group is *cyclic* whenever the group generated by some $g \in G$ *coincides* with G. Note that any group of prime order is cyclic and the product of two cyclic groups of relatively prime order is also a cyclic group. Also, if $\mathcal{S}$ is a nonempty subset of a group G, then the intersection of all subgroups of G containing $\mathcal{S}$ is called the subgroup *generated* by $\mathcal{S}$.

Definition A.37 (Ideals, Cosets, and Quotient Rings) *An ideal I in a commutative ring R with identity is a subring of R satisfying the additional property that $rI \subseteq I$ for all $r \in R$. If I is an ideal in R then a coset of I in R is a set of the form $r + I = \{r + \alpha : \alpha \in I\}$ where $r \in R$. The set $R/I = \{r + I : r \in R\}$ becomes a ring under multiplication and addition of cosets given by*

$$(r + I)(s + I) = rs + I, \text{ and } (r + I) + (s + I) = (r + s) + I$$

for any $r, s \in R$ (and this can be shown to be independent of the representatives r and s). R/I is called the quotient ring of R by I, *or the* factor ring of R by I, *or the* residue class ring modulo I. *The cosets are called the* residue classes modulo I. *A mapping,*

$$f : R \mapsto R/I,$$

which takes elements of R to their coset representatives in R/I is called the natural map *of R to R/I, and it is easily seen to be an epimorphism. The cardinality of R/I is denoted by $|R : I|$.*

Example A.8 Consider the ring of integers modulo $n \in \mathbb{N}$, $\mathbb{Z}/n\mathbb{Z}$ introduced in Definition A.20. Then $n\mathbb{Z}$ is an ideal in $\mathbb{Z}$, and the quotient ring is the residue class ring modulo n.

Remark A.2 *Since rings are also groups, then the above concept of cosets and quotients specializes to groups. In particular, we have the following. Note that an index of a subgroup H in a group G can be defined similarly to the above situation for rings as follows. The* index of H in G, *denoted by $|G : H|$, is the cardinality of the set of distinct right (respectively, left) cosets of H in G. Our principal interest is when this cardinality is finite (so this allows us to access the definition of cardinality given earlier). Then* Lagrange's theorem for groups *says that*

$$|G| = |G : H| \cdot |H|,$$

so if G is a finite group, then $|H| \mid |G|$. In particular, a finite abelian group G has subgroups of all orders dividing $|G|$.

Now we are in a position to state the important result for rings. The reader unfamiliar with the notation "img" of a function should consult Definition A.5 for the description.

Theorem A.22 (Fundamental Isomorphism Theorem for Rings)
   If R and S are commutative rings with identity, and

$$\phi : R \to S$$

is a homomorphism of rings, then

$$\frac{R}{\ker(\phi)} \cong \mathrm{img}(\phi).$$

Example A.9 If $\mathbb{F}_q$ is a finite field where $q = p^n$ (p prime) and $f(x) \in \mathbb{F}_p[x]$ is an irreducible, monic polynomial of degree n (see page 484), then

$$\mathbb{F}_q \cong \frac{\mathbb{F}_p[x]}{(f(x))}.$$

The situation in Example A.9 is related to the following definition and theorem.

Definition A.38 (Maximal and Proper Ideals)

   Let R be a commutative ring with identity. An ideal $I \neq R$ is called maximal *if whenever $I \subseteq J$, where J is an ideal in R, then $I = J$ or $I = R$. (An ideal $I \neq R$ is called a* proper *ideal.)*

Theorem A.23 (Rings Modulo Maximal Ideals).

   If R is a commutative ring with identity, then M is a maximal ideal in R if and only if R/M is a field.

Example A.10 If F is a field and $r \in F$ is a fixed nonzero element, then

$$I = \{f(x) \in F[x] : f(r) = 0\}$$

is a maximal ideal and $F \cong F[x]/I$.

   Another aspect of rings that we will need in the text is the following. If $\mathcal{S} = \{R_j : j = 1, 2, \ldots, n\}$ is a set of rings, then let R be the set of n-tuples $(r_1, r_2, \ldots, r_n)$ with $r_j \in R_j$ for $j = 1, 2, \ldots n$, with the *zero element* of R being the n-tuple, $(0, 0, \ldots, 0)$. Define addition in R by

$$(r_1, r_2, \ldots, r_n) + (r_1', r_2', \ldots, r_n') = (r_1 + r_1', r_2 + r_2', \ldots, r_n + r_n'),$$

for all $r_j, r_j' \in R_j$ with $j = 1, 2, \ldots, n$, and multiplication by

$$(r_1, r_2, \ldots, r_n)(r_1', r_2', \ldots, r_n') = (r_1 r_1', r_2 r_2', \ldots, r_n r_n').$$

This defines a structure on R called the *direct sum* of the rings R_j, $j = 1, 2, \ldots, n$, denoted by

$$\oplus_{j=1}^n R_j = R_1 \oplus \cdots \oplus R_n, \tag{A.3}$$

which is easily seen to be a ring. Similarly, when the R_j are groups, then this is a direct sum of groups, which is again a group.

We conclude this section with some more comments on Example A.9, since we will need these ideas, particularly in Chapter 11 when we discuss error-correcting codes.

If α is a root of $f(x)$ such that

$$\mathbb{F}_p(\alpha) \cong \mathbb{F}_q \cong \frac{\mathbb{F}_p[x]}{(f(x))},$$

then $f(x)$ is uniquely characterized by the conditions that $f(\alpha) = 0$ and $g(\alpha) = 0$ for some $g(x) \in \mathbb{F}_p[x]$ if and only if $f(x)$ divides $g(x)$. In this case, f is called the *minimal polynomial* of α over $\mathbb{F}_p$, and is assumed to be monic. In particular, a polynomial is called *primitive* of degree $n \in \mathbb{N}$ over $\mathbb{F}_q$ if it is a minimal polynomial over $\mathbb{F}_q$ of a primitive element of $\mathbb{F}_{q^n}$.

A.5   Vector Spaces

A *vector space* consists of an additive abelian group V and a field F together with an operation called *scalar multiplication* of each element of V by each element of F on the left, such that for each $r, s \in F$ and each $\alpha, \beta \in V$ the following conditions are satisfied:

1. $r\alpha \in V$.

2. $r(s\alpha) = (rs)\alpha$.

3. $(r + s)\alpha = (r\alpha) + (s\alpha)$.

4. $r(\alpha + \beta) = (r\alpha) + (r\beta)$.

5. $1_F\alpha = \alpha$.

The set of elements of V are called *vectors* and the elements of F are called *scalars*. The generally accepted abuse of language is to say that V is a *vector space over F*. If V_1 is a subset of a vector space V that is a vector space in its own right, then V_1 is called a *subspace of V*.

Definition A.39 (Bases, Dependence, and Finite Generation)
If S is a subset of a vector space V, then the intersection of all subspaces of V containing S is called the subspace generated by S, *or* spanned by S. *If there is a finite set S, and S generates V, then V is said to be* finitely generated. *If $S = \varnothing$, then S generates the zero vector space. If $S = \{m\}$, a singleton set, then the subspace generated by S is said to be the* cyclic subspace generated by m.

A subset S of a vector space V is said to be linearly independent *provided that for distinct $s_1, s_2, \ldots, s_n \in S$, and $r_j \in F$ for $j = 1, 2, \ldots, n$,*

$$\sum_{j=1}^{n} r_j s_j = 0 \text{ implies that } r_j = 0 \text{ for } j = 1, 2, \ldots, n.$$

If S *is not linearly independent, then it is called* linearly dependent. *A linearly independent subset of a vector space that spans V is called a* basis *for V. The number of elements in a basis is called the* dimension *of V. A* hyperplane *H is an $(n-1)$-dimensional subspace of an n-dimensional vector space V.*

Example A.11 For a given prime p, $m, n \in \mathbb{N}$, the finite field $\mathbb{F}_{p^n}$ is an n-dimensional vector space over $\mathbb{F}_{p^m}$ with p^{mn} elements.

A.6 Basic Matrix Theory

If $m, n \in \mathbb{N}$, then an $m \times n$ matrix (read "m by n matrix") is a rectangular array of entries with m rows and n columns. For simplicity, we will assume that the entries come from a field F. If A is such a matrix, and $a_{i,j}$ denotes the entry in the ith row and jth column, then

$$A = (a_{i,j}) = \begin{pmatrix} a_{1,1} & a_{1,2} & \cdots & a_{1,n} \\ a_{2,1} & a_{2,2} & \cdots & a_{2,n} \\ \vdots & \vdots & & \vdots \\ a_{m,1} & a_{m,2} & \cdots & a_{m,n} \end{pmatrix}.$$

Two $m \times n$ matrices $A = (a_{i,j})$, and $B = (b_{i,j})$ are equal if and only if $a_{i,j} = b_{i,j}$ for all i and j. The matrix $(a_{j,i})$ is called the *transpose* of A, denoted by

$$A^t = (a_{j,i}).$$

Addition of two $m \times n$ matrices A and B is done in the natural way.

$$A + B = (a_{i,j}) + (b_{i,j}) = (a_{i,j} + b_{i,j}),$$

and if $r \in F$, then $rA = r(a_{i,j}) = (ra_{i,j})$, called *scalar multiplication*.

Matrix products are defined by the following.

If $A = (a_{i,j})$ is an $m \times n$ matrix and $B = (b_{j,k})$ is an $n \times r$ matrix, then the *product* of A and B is defined as the $m \times r$ matrix:

$$AB = (a_{i,j})(b_{j,k}) = (c_{i,k}),$$

where

$$c_{i,k} = \sum_{\ell=1}^{n} a_{i,\ell} b_{\ell,k}.$$

Multiplication, if defined, is associative, and distributive over addition. If $m = n$, then

$$I_n = \begin{pmatrix} 1_F & 0 & \cdots & 0 \\ 0 & 1_F & \cdots & 0 \\ \vdots & \vdots & \vdots & \vdots \\ 0 & 0 & \cdots & 1_F \end{pmatrix}$$

is called *the $n \times n$ identity matrix*, where 1_F is the identity of F.

Another important aspect of matrices that we will need throughout the text is motivated by the following. Consider the 2×2 matrix with entries from F:

$$A = \begin{pmatrix} a & b \\ c & d \end{pmatrix},$$

then $ad - bc$ is called the *determinant* of A, denoted by $\det(A)$. More generally, we may define the determinant of any $n \times n$ matrix with entries from F for any $n \in \mathbb{N}$. The determinant of any $r \in F$ is just $\det(r) = r$. Thus, we have the definitions for $n = 1, 2$, and we may now give the general definition inductively. The definition of the determinant of a 3×3 matrix,

$$A = \begin{pmatrix} a_{1,1} & a_{1,2} & a_{1,3} \\ a_{2,1} & a_{2,2} & a_{2,3} \\ a_{3,1} & a_{3,2} & a_{3,3} \end{pmatrix}$$

is defined in terms of the above definition of the determinant of a 2×2 matrix, namely, $\det(A)$ is given by

$$a_{1,1} \det \begin{pmatrix} a_{2,2} & a_{2,3} \\ a_{3,2} & a_{3,3} \end{pmatrix} - a_{1,2} \det \begin{pmatrix} a_{2,1} & a_{2,3} \\ a_{3,1} & a_{3,3} \end{pmatrix} + a_{1,3} \det \begin{pmatrix} a_{2,1} & a_{2,2} \\ a_{3,1} & a_{3,2} \end{pmatrix},.$$

Therefore, we may inductively define the determinant of any $n \times n$ matrix in this fashion. Assume that we have defined the determinant of an $n \times n$ matrix. Then we define the determinant of an $(n + 1) \times (n + 1)$ matrix $A = (a_{i,j})$ as follows. First, we let $A_{i,j}$ denote the $n \times n$ matrix obtained from A by deleting the ith row and jth column. Then we define the *minor* of $A_{i,j}$ at position (i, j) to be $\det(A_{i,j})$. The *cofactor* of $A_{i,j}$ is defined to be

$$\text{cof}(A_{i,j}) = (-1)^{i+j} \det(A_{i,j}).$$

We may now define the determinant of A by

$$\det(A) = a_{i,1}\text{cof}(A_{i,1}) + a_{i,2}\text{cof}(A_{i,2}) + \cdots + a_{i,n+1}\text{cof}(A_{i,n+1}).$$

This is called the *expansion of a determinant by cofactors* along the ith row of A. Similarly, we may expand along a column of A.

$$\det(A) = a_{1,j}\text{cof}(A_{1,j}) + a_{2,j}\text{cof}(A_{2,j}) + \cdots + a_{n+1,j}\text{cof}(A_{n+1,j}),$$

called the *cofactor expansion along the jth column of A*. Both expansions can be shown to be equal. Hence, a determinant may be viewed as a function that assigns a real number to an $n \times n$ matrix, and the above gives a method for finding that number.

If A is an $n \times n$ matrix with entries from F, then A is said to be *invertible*, or *nonsingular* if there is a unique matrix denoted by A^{-1} such that

$$AA^{-1} = I_n = A^{-1}A.$$

Here are some properties of invertible matrices.

Theorem A.24 (Properties of Invertible Matrices)
Let R be a commutative ring with identity, $n \in \mathbb{N}$, and A invertible in $\mathcal{M}_{n \times n}(R)$. Then each of the following holds.

(a) $(A^{-1})^{-1} = A$.

(b) $(A^t)^{-1} = (A^{-1})^t$, *where "t" denotes the transpose.*

(c) $(AB)^{-1} = B^{-1}A^{-1}$.

In order to provide a formula for the inverse of a given matrix, we need the following concept.

Definition A.40 (Adjoint)
Let R be a commutative ring with identity. If $A = (a_{i,j}) \in \mathcal{M}_{n \times n}(R)$, then the matrix $A^a = (b_{i,j})$ given by

$$b_{i,j} = (-1)^{i+j} \det(A_{j,i}) = \operatorname{cof}(A_{j,i}) = \left[(-1)^{i+j} \det(A_{i,j}) \right]^t$$

is called the adjoint *of A.*

Some properties of adjoints related to inverses, including a formula for the inverse, are as follows.

Theorem A.25 (Properties of Adjoints)
If R is a commutative ring with identity and $A \in \mathcal{M}_{n \times n}(R)$, then each of the following holds.

(a) $AA^a = \det(A)I_n = A^a A$.

(b) *A is invertible in $\mathcal{M}_{n \times n}(R)$ if and only if $\det(A)$ is a unit in R, in which case $A^{-1} = A^a / \det(A)$.*

For instance, we will need the following in Example 3.2 on page 112.

Example A.12 If $n = 2$, then the inverse of a nonsingular matrix,

$$A = \begin{pmatrix} a & b \\ c & d \end{pmatrix},$$

is given by

$$A^{-1} = \begin{pmatrix} \dfrac{d}{\det(A)} & \dfrac{-b}{\det(A)} \\ \dfrac{-c}{\det(A)} & \dfrac{a}{\det(A)} \end{pmatrix},$$

We will need the following in the text, for example, when we talk about secret-sharing schemes on page 215.

Theorem A.26 (Cramer's Rule)
Let $A = (a_{i,j})$ be the coefficient matrix of the following system of n linear equations in n unknowns:

$$a_{1,1}x_1 + a_{1,2}x_2 + \cdots + a_{1,n}x_n = b_1$$

$$a_{2,1}x_1 + a_{2,2}x_2 + \cdots + a_{2,n}x_n = b_2$$

$$\vdots \qquad \vdots \qquad \vdots \quad \vdots \qquad \vdots$$

$$a_{n,1}x_1 + a_{n,2}x_2 + \cdots + a_{n,n}x_n = b_n,$$

over a field F. If $\det(A) \neq 0$, then the system has a solution given by

$$x_j = \frac{1}{\det(A)} \left(\sum_{i=1}^{n} (-1)^{i+j} b_i \det(A_{i,j}) \right), \quad (1 \leq j \leq n).$$

Of particular importance is a special matrix called the *Vandermonde matrix*, of order $t > 1$, which is given as follows:

$$A = \begin{pmatrix} 1 & x_1 & \cdots & x_1^{t-1} \\ 1 & x_2 & \cdots & x_2^{t-1} \\ \vdots & \vdots & \vdots & \vdots \\ 1 & x_t & \cdots & x_t^{t-1} \end{pmatrix},$$

where

$$\det(A) = \prod_{1 \leq i < k \leq t} (x_k - x_i)$$

is the *Vandermonde determinant*.

A notion that uses vector spaces and matrix theory is the following, which we will use in Appendix C, for instance.

◆ Gaussian Elimination

The term *Gaussian elimination* refers to an efficient algorithm for finding linear dependency relations among vectors in a vector space over a suitable field. Suppose that we have vectors $v_j = (v_{1,j}, v_{2,j}, \ldots, v_{n,j})$ for $j = 1, 2, \ldots, m$ over a field F. We seek field elements $c_1, c_2, \ldots, c_m \in F$ such that

$$\sum_{i=j}^{m} c_j v_j = \overrightarrow{0}, \tag{A.4}$$

where $c_j v_j = (c_j v_{1,j}, c_j v_{2,j}, \ldots, c_j v_{n,j})$, and $\overrightarrow{0} = (0, 0, \ldots, 0)$ is the *zero vector* of length n. Since $\sum_{i=j}^{m} c_j v_j$, the relation given in (A.4) where *not all coefficients* are 0, is a *linear dependency relation* (see Definition A.39 on page 490). Gaussian elimination uses the basic notions of linear algebra to define matrices with the vectors v_j as columns, then performs elementary row operations to put them into a form to determine the dependency relations therefrom, if there are any. The basic point from elementary linear algebra is that *if the number of vectors is greater than the dimension of the vector space over the field, then there must be a dependency relation*. For instance, if $m > n$ in (A.4), then at least one of the $c_j \neq 0$.

In the above, we mentioned the notion of elementary row operations. They are defined as follows.

▼ Elementary Row Operations

1. Interchange two rows of the matrix.

2. Multiply all elements of a row of the matrix by a nonzero scalar.

3. Add to any row of the matrix any other row of it multiplied by a nonzero scalar.

The above operations also hold for columns and may be restated as *elementary column operations* by replacing the word "row" by "column" wherever they occur. It is a fact that any nonzero $M \in \mathcal{M}_{m \times n}(R)$ can be reduced by application of elementary row and column operations to an $m \times n$ matrix of one of the following forms: $(I_m, 0)$, $\begin{pmatrix} I_n & 0 \\ 0 & 0 \end{pmatrix}$, $\begin{pmatrix} I_r \\ 0 \end{pmatrix}$, I_n, where the 0's denote that the matrix has only zero entries in those positions, and the I_j's are identity matrices of the given size for $j = m, n, r$. Since the *rank* of a matrix is equal to the order of the largest nonzero minor, then in the above four cases, the ranks are m, n, r, and n, respectively.

Matrices in the form given above are in what is called *reduced row echelon form*, which is a matrix that has the following properties.

1. All zero rows are in the bottom position(s).

2. Reading left to right, the first nonzero element in a nonzero row is a 1, which is called a *leading* 1.

3. For $j > 2$, if there exists a leading 1 in row j, then it appears to the right of a leading 1 in row $j - 1$.

4. Any column containing a leading 1 has all other elements equal to zero.

When applied to a system of m linear equations in n unknowns, there is an equivalent formulation in which the matrix representation, called the *augmented matrix*, has reduced row echelon form. In practice, the reduction of a system of linear equations to its reduced row echelon form is calculated via the employment of Gaussian elimination, discussed earlier.

A.7 Continued Fractions

For Appendix C, we will need some basic ideas from the theory of continued fractions as follows.

Definition A.41 (Continued Fractions)
If $q_j \in \mathbb{R}$ where $j \in \mathbb{Z}$ is nonnegative and $q_j \in \mathbb{R}^+$ for $j > 0$, then an expression of the form,

$$\alpha = q_0 + \cfrac{1}{q_1 + \cfrac{1}{q_2+}}$$

$$+ \cfrac{1}{q_k + \cfrac{1}{q_{k+1}}}$$

is called a continued fraction. *If $q_k \in \mathbb{Z}$ for all $k \geq 0$, then it is called a* simple continued fraction, *denoted by $\langle q_0; q_1, \ldots, q_k, q_{k+1}, \ldots \rangle$. If there exists a nonnegative integer n such that $q_k = 0$ for all $k \geq n$, then the continued fraction is called* finite. *If no such n exists, then it is called* infinite.

Definition A.42 (Convergents) *Let $n \in \mathbb{N}$ and let α have continued fraction expansion $\langle q_0; q_1, \ldots, q_n, \ldots \rangle$ for $q_j \in \mathbb{R}^+$ when $j > 0$. Then*

$$C_k = \langle q_0; q_1, \ldots, q_k \rangle$$

is the kth convergent *of α for any nonnegative integer k.*

Theorem A.27 (Finite Simple Continued Fractions are Rational)
Let $\alpha \in \mathbb{R}$. Then $\alpha \in \mathbb{Q}$ if and only if α can be written as a finite simple continued fraction.

Proof. See [167, Theorem 5.1.1, page 223]. □

Theorem A.28 (Representation of Convergents)
Let $\alpha = \langle q_0; q_1, \ldots \rangle$ be a continued fraction expansion. Define two sequences for $k \in \mathbb{Z}$ nonnegative:

$$A_{-2} = 0, A_{-1} = 1, A_k = q_k A_{k-1} + A_{k-2},$$

and

$$B_{-2} = 1, B_{-1} = 0, B_k = q_k B_{k-1} + B_{k-2}.$$

Then

$$C_k = A_k/B_k = \frac{q_k A_{k-1} + A_{k-2}}{q_k B_{k-1} + B_{k-2}},$$

is the kth convergent of α for any nonnegative integer k.

Proof. See [167, Theorem 5.1.2, page 224]. □

Theorem A.29 (Irrationals Are Infinite Simple Continued Fractions)
   Let $\alpha \in \mathbb{R}$. *Then* α *is irrational if and only if* α *has a* unique *infinite simple continued fraction expansion,*

$$\alpha = \alpha_0 = \langle q_0; q_1, \ldots \rangle = \lim_{k \to \infty} C_k,$$

where

$$q_{k-1} = \lfloor \alpha_{k-1} \rfloor \text{ with } \alpha_k = 1/(\alpha_{k-1} - q_{k-1}) \text{ and } C_k = A_k/B_k \text{ for } k \in \mathbb{N}.$$

Proof. See [167, Theorem 5.2.1, page 228]. □

Theorem A.30 (Convergents of $\sqrt{D}$)
   Suppose that $D > 0$ *is not a perfect square,* $n \in \mathbb{Z}$, *and* $|n| < \sqrt{D}$. *If* (x, y) *is a* positive solution *of*

$$x^2 - Dy^2 = n,$$

namely, $x, y \in \mathbb{N}$, *then* x/y *is a convergent in the simple continued fraction expansion of* $\sqrt{D}$.

Proof. See [167, Theorem 5.2.5, page 232]. □

Definition A.43 (Periodic Continued Fractions)
   An infinite simple continued fraction $\alpha = \langle q_0; q_1, q_2, \ldots \rangle$ *is called* periodic *if there exists an integer* $k \geq 0$ *and* $\ell \in \mathbb{N}$ *such that* $q_n = q_{n+\ell}$ *for all integers* $n \geq k$. *We use the notation,*

$$\alpha = \langle q_0; q_1, \ldots, q_{k-1}, \overline{q_k, q_{k+1}, \ldots, q_{\ell+k-1}} \rangle,$$

as a convenient abbreviation. The smallest such natural number $\ell = \ell(\alpha)$ *is called the* period length *of* α, *and* $q_0, q_1, \ldots, q_{k-1}$ *is called the* preperiod *of* α. *If* k *is the* least nonnegative *integer such that* $q_n = q_{n+\ell}$ *for all* $n \geq k$, *then*

$$q_k, q_{k+1}, \ldots, q_{k+\ell-1} \text{ is called the fundamental period } of \alpha.$$

If $k = 0$ *is the least such value, then* α *is said to be* purely periodic, *namely,*

$$\alpha = \langle \overline{q_0; q_1, \ldots, q_{\ell-1}} \rangle.$$

Theorem A.31 (Continued Fractions and Recursion)
  Let D be a positive integer that is not a perfect square, and let

$$\alpha_0 = (P_0 + \sqrt{D})/Q_0$$

be a quadratic irrational. Recursively define the following for $k \geq 0$:

$$\alpha_k = (P_k + \sqrt{D})/Q_k, \tag{A.5}$$

$$q_k = \lfloor \alpha_k \rfloor, \tag{A.6}$$

$$P_{k+1} = q_k Q_k - P_k, \tag{A.7}$$

and

$$Q_{k+1} = (D - P_{k+1}^2)/Q_k. \tag{A.8}$$

Then $P_k, Q_k \in \mathbb{Z}$ and $Q_k \neq 0$ for $k \geq 0$, and $\alpha_k = \langle q_k; q_{k+1}, \cdots \rangle$.

  Proof. See [167, Exercise 5.3.6, page 251]. $\square$

Theorem A.32 (Continued Fractions and Quadratic Irrationals)
  Let $\alpha = (P + \sqrt{D})/Q$ be a quadratic irrational and set

$$G_{k-1} = Q_0 A_{k-1} - P_0 B_{k-1} \quad (k \geq -1),$$

where A_{k-1}, B_{k-1} are given in Theorem A.28 on page 496. Then

$$G_{k-1}^2 - B_{k-1}^2 D = (-1)^k Q_k Q_0 \quad (k \geq 1). \tag{A.9}$$

  Proof. See [167, Theorem 5.3.4, page 246]. $\square$

Corollary A.3 *If $\alpha = \sqrt{D}$, then Equation (A.9) becomes*

$$A_{k-1}^2 - B_{k-1}^2 D = (-1)^k Q_k. \tag{A.10}$$

  Proof. See [167, Corollary 5.3.3, page 249]. $\square$

A.8   Elliptic Curves

  We now summarize some basic facts needed for understanding the crypto-graphic schemes described in this text such as the ElGamal cryptosystem on page 190. We begin with the basic definition.

Definition A.44 (Elliptic Curves)
  Let F be a field with characteristics not equal to 2 or 3. If $a, b \in F$ are given such that $4a^3 + 27b^2 \neq 0$ in F, then an elliptic curve E defined over F is given by an equation $y^2 = x^3 + ax + b \in F[x]$. The set of all solutions $(x, y) \in F$ to the equation:

$$y^2 = x^3 + ax + b, \tag{A.11}$$

together with a point $\mathfrak{o}$, called the point at infinity, is denoted by $E(F)$, called the set of F-rational points on E. The value $\Delta(E) = -16(4a^3 + 27b^2)$ is called the discriminant of the elliptic curve E.

▼ Elliptic Curve Facts

We assume that $E(\mathbb{Q})$ is an elliptic curve over $\mathbb{Q}$ given by $y^2 = x^3 + ax + b$ where $a, b \in \mathbb{Z}$, and $\mathfrak{o}$ denotes the point at infinity.

(1) (**Addition of points**): For any two points $P = (x_1, y_1)$ and $Q = (x_2, y_2)$ on E, with $P, Q \neq \mathfrak{o}$ and $P \neq -Q$, define

$$P + Q = (x_3, y_3) = (m^2 - x_1 - x_2, m(x_1 - x_3) - y_1), \qquad \text{(A.12)}$$

where

$$m = \begin{cases} m_1/m_2 = (y_2 - y_1)/(x_2 - x_1) & \text{if } P \neq Q, \\ m_1/m_2 = (3x_1^2 + a)/(2y_1) & \text{if } P = Q, \end{cases} \qquad \text{(A.13)}$$

and

if $P = \mathfrak{o}$, for instance, then $P + Q = Q$ for all points Q on E,

and

if $P = -Q$, then $P + Q = \mathfrak{o}$.

(2) (**Reduction modulo n**): Let $n > 1$ be given and fixed with $\gcd(n, 6) = 1$, and $\gcd(4a^3 + 27b^2, n) = 1$. Then we refer to E reduced modulo n when the coefficients a, b are reduced modulo n, and each point P on E is reduced modulo n in the following fashion. If $P = (r_1/r_2, s_1/s_2)$ where

$$\gcd(r_1, r_2) = \gcd(s_1, s_2) = \gcd(r_2 s_2, n) = 1,$$

then

$$P = (t_1, t_2), \text{ where } t_1 \equiv r_1 r_2^{-1} \pmod{n} \text{ and } t_2 \equiv s_1 s_2^{-1} \pmod{n},$$

with r_2^{-1} and s_2^{-1} being the multiplicative inverses of r_2 and s_2 modulo n, respectively. We denote the reduced curve by $E(\mathbb{Z}/n\mathbb{Z})$, and if n is a prime, then this is a group.

(3) (**Modular group law**): Suppose that P_1, P_2 are points on $E(\mathbb{Q})$ where $P_1 + P_2 \neq \mathfrak{o}$ and the denominators of P_1, P_2 are prime to n. Then $P_1 + P_2$ has coordinates having denominators prime to n if and only if there does not exist a prime $p \mid n$ such that $P_1 + P_2 = \mathfrak{o} \pmod{p}$ on the elliptic curve $E(\mathbb{Z}/p\mathbb{Z})$.

For a more in-depth description of elliptic curve theory as it applies to cryptology, see [169, pages 221–251].

A.9   Complexity

On page 46, we were introduced to an informal definition of the following. An *algorithm* is a well defined computational procedure, which takes a variable input and halts with an output. An algorithm is called *deterministic* if it follows the same sequence of operations each time it is executed with the same input. A *randomized algorithm* is one that makes random decisions at certain points in the execution, so the execution paths may differ each time the algorithm is invoked with the same input.

The amount of time required for the execution of an algorithm on a computer is measured in terms of *bit operations*, which are defined as follows: addition, subtraction, or multiplication of two binary digits; the division of a two-bit integer by a one-bit integer; or the shifting of a binary digit by one position. The number of bit operations necessary to complete the performance of an algorithm is called its *computational complexity* or simply its *complexity*. This method of estimating the amount of time taken to execute a calculation does not take into account such things as memory access or time to execute an instruction. However, these executions are very fast compared with a large number of bit operations, so we can safely ignore them. These comments are made more precise by the introduction of the following notation.

Definition A.45 (Big O Notation) *Suppose that f and g are positive real-valued functions. If there exists a positive real number c such that*

$$f(x) < cg(x) \tag{A.14}$$

for all sufficiently large x, then we write[A.4]

$$f(x) = O(g(x)) \text{ or simply } f = O(g). \tag{A.15}$$

(Alternatively, we may also write $f << g$ to denote $f = O(g)$.)

Big O is *the order of magnitude of the complexity*, an *upper bound* on the number of bit operations required for execution of an algorithm in the *worst-case scenario*, namely, in the case where even the trickiest or the nastiest inputs are given. It is possible that most often, for a given algorithm, even less time will be used, but we must always account for the worst-case scenario.

If the reader is searching for a reason to use complexity as a foundation for cryptography, it is quite simply that it assists us in clarifying assumptions we might make concerning security. Now, the comments made before Definition A.45 may now be put into perspective. The definition of the time taken to perform a given algorithm does not take into consideration time spent *reading and writing* such as memory access, timings of instructions, even the speed or amount of memory of a computer, all of which are negligible in comparison

---

[A.4]Here sufficiently large means that there exists some bound $B \in \mathbb{R}^+$ such that $f(x) < cg(x)$ for all $x > B$. We just may not know explicitly the value of B. Often f is defined on $\mathbb{N}$ rather than $\mathbb{R}$, and occasionally over any subset of $\mathbb{R}$.

with the order of magnitude complexity. The greatest merit of this method for estimating execution time is that it is machine-independent. In other words, it does not rely upon the specifics of a given computer, so the order of magnitude complexity remains the same, irrespective of the computer being used. In the analysis of the complexity of an algorithm, we need not know *exactly* how long it takes (namely, the *exact* number of bit operations required to execute the algorithm), but rather it suffices to compare with other objects, and these comparisons need not be immediate, but rather long term. In other words, what Definition A.45 says is that if f is $O(g)$, then *eventually* $f(x)$ is bounded by *some* constant multiple $cg(x)$ of $g(x)$. We do not know exactly *what* c happens to be or just *how big* x must be before (A.14) occurs. However, for reasons given above, it is enough to account for the efficiency of the given algorithm in the worst-case scenario.

The amount of time taken by a computer to perform a task is (essentially) *proportional* [A.5] to the number of bit operations. In the simplest possible terms, the constant of proportionality, which is the number of nanoseconds[A.6] per bit operation, depends upon the computer being used. This accounts for the machine-independence of the Big O method of estimating complexity since the constant of proportionality is of no consequence in the determination of Big O.

A fundamental *time estimate* in executing an algorithm is *polynomial time* (or simply *polynomial*)[A.7]. In other words, an algorithm is polynomial when its complexity is $O(n^c)$ for some constant $c \in \mathbb{R}^+$, where n is the bitlength of the input to the algorithm, and c is independent of n. (Observe that any polynomial of degree c is $O(n^c)$.) In general, these are the desirable algorithms, since they are the fastest. Therefore, roughly speaking, the polynomial-time algorithms are the *good* or *efficient* algorithms. For instance, the algorithm is constant if $c = 0$; if $c = 1$, it is linear; if $c = 2$, it is quadratic, and so on. Examples of polynomial time algorithms are those for the ordinary arithmetic operations of addition, subtraction, multiplication, and division. On the other hand, those algorithms with complexity $O(c^{f(n)})$ where c is constant and f is a polynomial on $n \in \mathbb{N}$ are *exponential time algorithms* or simply *exponential*. A *subexponential* time algorithm is one for which the complexity for input $n \in \mathbb{N}$ is

$$L_n(r, c) = O(\exp((c + o(1))(\ln n)^r (\ln \ln n)^{1-r})) \qquad (A.16)$$

where $r \in \mathbb{R}$ with $0 < r < 1$, c is a constant, and $o(1)$ denotes a function $f(n)$ such that $\lim_{n \to \infty} f(n) = 0$.[A.8] Subexponential time algorithms are faster than exponential-time algorithms but slower than polynomial-time algorithms. These are, again roughly speaking, the *inefficient* algorithms. Algorithms with

---

[A.5]To say that a is proportional to b means that $a/b = c$, a constant, called the *constant of proportionality*. This relationship is often written as $a \propto b$ in the literature.

[A.6]A nanosecond is $1/10^9$ of a second, that is, a billionth of a second.

[A.7]Recall that a (nonconstant) polynomial is a function of the form $\sum_{i=0}^n a_i x^i$ for $n \in \mathbb{N}$, where the a_i are the coefficients (see page 484).

[A.8]In general, $f(n) = o(g(n))$ means that $\lim_{n \to \infty} f(n)/g(n) = 0$. Thus, $o(1)$ is used to symbolize a function whose limit as n approaches infinity is 0. Also, $\exp(x) = e^x$.

complexity $O(c^{f(n)})$ where c is constant and $f(n)$ is more than constant but less than linear are called *superpolynomial*. It is generally accepted that modern-day cryptanalytic techniques for breaking known ciphers are of superpolynomial-time complexity, but nobody has been able to prove that polynomial-time algorithms for cryptanalyzing ciphers do not exist.

Another notion of time is *expected running time*, which means the *expectation* (in the probability sense) of the runtimes over all the possible inputs, expressed as a function of the input size (see Appendix E). This means that one needs to estimate the probability that a given input occurs, not an easy task in general. For instance, there is expected polynomial time, which we first encounter on page 109 in reference to symmetric-key cryptosystems.

The running time of an algorithm is sometimes difficult to determine. When this is the case, one may be forced to settle for approximations for the running time, called the *asymptotic running time*, which is a measure of how the running time of the algorithm increases as the size of the input increases without bound.

In calculating complexity using the Big O notation, the following properties are essential.

Theorem A.33 (Properties of the Big O Notation)
Suppose that f, g are positive real-valued functions.

(a) If $c \in \mathbb{R}^+$, then $cO(g) = O(g)$.

(b) $O(\max\{f, g\}) = O(f) + O(g)$.

(c) $O(fg) = O(f)O(g)$.

Proof. See [169, Theorem 1.24, page 50].   □

To get some idea of what the various classes of complexity analysis mean in "real-world" terms, let us look at times related to some of these classes. Suppose that the unit of time on the computer at our disposal is a microsecond (a millionth $(1/10^6)$ of a second). Assuming an input of $n = 10^6$ bits, then a constant algorithm (complexity $O(1)$) would take a microsecond to execute, since the number of bit operations is one. A linear algorithm (complexity $O(n)$) would take a second, since the number of bit operations is 10^6. A quadratic algorithm (complexity $O(n^2)$) would take $11.5741 = 10^{12}/(10^6 \cdot 24 \cdot 3600)$ days, since there are 10^{12} bit operations, and a cubic algorithm (complexity $O(n^3)$) would take $31,709 = 10^{18}/(10^6 \cdot 24 \cdot 3600 \cdot 365)$ years, since the number of bit operations is 10^{18}. By the time we get to exponential algorithms, we are looking at times astronomically larger than the age of the known universe. Hence, a problem is called *intractable* if no polynomial-time algorithm could possibly solve it, whereas one that can be solved using a polynomial-time algorithm is called *tractable*. (By a *problem*, we mean a general question to be answered. A *decision problem* is one whose solution is "yes" or "no." A problem may possess *parameters* whose values are left unspecified, and an *instance* of a problem is achieved by specifying values for those parameters.)

Now we need the notion of a *Turing machine*, which is a finite-state machine having an infinite read-write tape, i.e., our theoretical computer has infinite memory and the ability to search for and retrieve any data from memory.

More specifically, a (deterministic, one-tape) Turing machine has an infinitely long magnetic tape (as its unlimited memory) on which instructions can be written and erased. It also has a processor that carries out the instructions: (1) move the tape right, (2) move the tape left, (3) change the state of the register based upon its current value and a value on the tape, and write or erase on the tape. The Turing machine runs until it reaches a desired state causing it to halt. A famous problem in theoretical computer science is to determine when a Turing machine will halt for a given set of input and rules. This is called the *halting problem*. Turing proved that this problem is *undecidable*, meaning that there does not exist any algorithm whatsoever for solving it. The *Church-Turing thesis*, which came out of the 1936 papers of Turing and Church (see page 92), essentially says that the Turing machine as a model of computation is equivalent to any other model for computation. (Here we may think of a "model" naively as a simplified mathematical description of a computer system.) Therefore, Turing machines are realistic models for simulating the running of algorithms, and they provide a powerful computational model. However, a Turing machine is not meant to be a practical design for any actual machine, but is a sufficiently simple model to allow us to prove theorems about its computational capabilities while at the same time being sufficiently complex to include any digital computer irrespective of implementation.

Complexity theory designates a decision problem to be in class **P** if it can be solved in polynomial time, whereas a decision problem is said to be in class **NP** if it can be solved in polynomial time on a *nondeterministic* Turing machine, which is a variant of the normal Turing machine in that it *guesses* solutions to a given problem and checks its guess in polynomial time. Another way to look at the class **NP** is to think of these problems as those for which the *correctness of a guess* at an answer to a question can be proven in polynomial time. Another equivalent way to define the class **NP** is the class of those problems for which a "yes" answer can be verified in polynomial time using some extra information, called a *certificate*.

The class **P** is a subset of the class **NP** since a problem that can be solved in polynomial time on a *deterministic* machine can also be solved, by eliminating the guessing stage, on a nondeterministic Turing machine. It is an open problem in complexity theory to resolve whether **P** = **NP**. However, virtually everyone believes that they are unequal. It is generally held that most modern ciphers can be cryptanalyzed in nondeterministic polynomial time. However, in practice it is the deterministic polynomial-time algorithm that is the end goal of modern-day cryptanalysis. Defining what it means to be a "computationally hard" problem is a *hard problem*. One may say that problems in **P** are *easy*, and those not in **P** are considered to be *hard*. However, there are problems that are regarded as computationally easy, yet are not known to be in **P**. For instance, the Miller-Rabin-Selfridge test is such a problem. It is in the class **RP**, called *randomized* *polynomial time* or *probabilistic polynomial time*. Here,

$$\mathbf{P} \subseteq \mathbf{RP} \subseteq \mathbf{NP}.$$

A practical (but mathematically less satisfying) way to define "hard" problems is to view them as those which have continued to resist solutions after a concerted attack by competent investigators for a long time up to the present.

Another classification in complexity theory is the **NP**-complete problem, which is a problem in the class **NP** that can be proved to be as difficult as any problem in the class. Should an **NP**-complete problem be discovered to have a deterministic polynomial-time algorithm for its solution, this would prove that **NP** $\subseteq$ **P**, so **P** = **NP**. Hence, we are in the position that there is no proof that there are *any* hard problems in this sense, namely, those in **NP** but not in **P**. Nevertheless, this has not prevented the flourishing of research in complexity theory. The classical **NP**-complete problem is the *travelling salesman problem*: A travelling salesman wants to visit $n \in \mathbb{N}$ cities. Given the distances between them, and a bound B, does there exist a tour of all the cities having total length B or less? The next in the hierarchy of complexity classification is **EXPTIME**, problems that can be solved in exponential time.

Thus far, we have been concerned with the *time* it takes for an algorithm to execute, measured (asymptotically, the worst-case scenario) in terms of the number of bit operations required. Another component of complexity is the amount of computer memory (storage required) for the computation of a given algorithm, called the *space requirement*. Time calculation on a Turing machine is measured in terms of the number of steps taken before it enters a halt state, as we have discussed above. The *space* used is defined as the number of tape squares visited by the read-write head (where we think of the tape as having infinitely many squares read, written, or erased by a "read-write head"). Thus, the notion of *polynomial space* takes on meaning, and since the number of steps in a computation is at least as large as the number of tape squares visited, then any problem solvable in polynomial time is solvable in polynomial space. Thus, we define **PSPACE** as those problems that can be solved in polynomial space, but not necessarily in polynomial time. Hence, **PSPACE**-complete problems are those such that if any one of them is in **NP**, then **PSPACE=NP**, and if any one of them is in **P**, then **PSPACE=P**. At the top of the hierarchy of the classification of problems in terms of complexity is **EXPSPACE**, those problems solvable in exponential space, but not necessarily in exponential time. It is known that **P$\neq$EXPTIME** and

$$\mathbf{NP} {\subseteq} \mathbf{PSPACE} {\neq} \mathbf{EXPSPACE}.$$

(There are also the nondeterministic versions, **NPSPACE** and **NEXPSPACE**, which we will not discuss here.) Figure A.1 provides an illustration of the above discussion on the hierarchy of problems in complexity theory.

Figure A.1: Hierarchy of Problems in Complexity Theory.

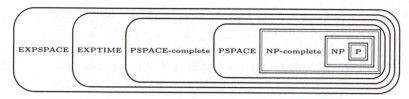

There are other types of complexity such as *circuit complexity*, which looks at the connection between Boolean circuits and Turing machines as a computational model for studying **P** *vis-à-vis* **NP** and affiliated problems. We will not discuss these more advanced themes here.

Roughly speaking, complexity theory can be subdivided into two categories: (a) structural complexity theory, and (b) the design and analysis of algorithms. Essentially, category (a) is concerned with lower bounds, and category (b) deals with upper bounds. Basically, the primary goal of structural complexity theory is to classify problems into classes determined by their intrinsic computational difficulty. In other words, how much computing time (and resources) does it take to solve a given problem? As we have seen, the fundamental question in structural complexity theory remains unanswered, namely, does **P** = **NP**? We have been primarily concerned with the analysis of algorithms, which is of the most practical importance to cryptography.

Appendix B: Pseudorandom Number Generation

In this appendix we look at algorithms for pseudorandomly generating numbers for use in cryptographic schemes (see page 151). In particular, we begin with a FIPS-approved standard for use with DES (see Section 3.2). It is also the basis for the PGPRNG discussed in Section 8.1 (see page 283).

B.1   ANSI X9.17

ANSI X9.17 was introduced in 1985 as the *Financial Institution Key Management* standard, which defined protocols for use by such institutions for encryption key transfer using SKC methods. It was updated in 1995, but had become a FIPS-171 standard in 1992. In 1998, Appendix A of ANSI X9.31 replaced Appendix C of ANSI X9.17, which is the version used by PGP.

X9.17/X9.31 PRNG The goal is to output a string of B 64-bit PRNs, where B is some predetermined bound. We are given the following as input.
(a) a secret, random 64-bit seed, S;

(b) a 64-bit representation, DT, of the current date/time;

   and

(c) a 3-DES key K, which is used for DES encryption, denoted by E_K, which is

$$E_{k_1} \circ D_{k_2} \circ E_{k_1},$$

   a two-key triple encryption, kept secret and used only for the PRNG (see page 131).

Then the algorithm proceeds as follows with a parameter i, and a vector X of length B initialized with entries $X[j] = 0$ for $j = 1, 2, \ldots, B$.

1. Set $I = E_K(DT)$ and set $i = 1$.

2. Compute
$$X_i = E_K(I \oplus S)$$
   and set $X[i] = X_i$.

3. Set $S = E_K(X_i \oplus I)$, and set $i = i + 1$.

4. If $i < B$, go to step 2. Otherwise, output
$$X = (X_1, X_2, \ldots, X_B)$$

   as the sequence of pseudorandom 64-bit numbers.

Any of the X_js may be used as an *IV* for DES modes of operation (see page 135). The PRNs may also be employed for DES keys. On the other hand, PGP employs two PRNGs, the ANSI X9.17/X9.31 generator and a function that measures the entropy from the latency in the user's keystrokes (see Chapter 11). However, PGP is not restricted to 3-DES, since it also has the option of using IDEA or CAST-128 (see [2] and [3] and [159]). In fact, it is a relatively easy task to convert ASNI X9.17/X9.31 to the use of IDEA, for instance.

ANSI X9.17/X9.31 PGP Session Key and IV Generation

We will assume, for convenience, that PGP consists of the four components: (1) the SKC, IDEA; (2) the PKC, RSA (see Section 4.2); (3) the hash MD5 (see page 255); and (4) the PRNG, X9.17/X9.31 in conjunction with user keystroke entropy information. The latter provides *true* random number generation for the purpose of generating RSA pairs, and providing initial and subsequent input to the PRNG. Once a *latency timer*, or *keystroke timer*, anticipates a keystroke from a user, it records the time in 32-bit format and once the keystroke is received, it records the time the key was pressed and the 8-bit value of the keystroke. This time and keystroke data are used to generate a key, which is used to encipher the current value of the random bit buffer. (PGP keeps a 256-byte random bit buffer.) To ensure maximum entropy, the keystrokes should be separated as randomly as possible.

The initial random bitstring from the latency timer is input as a 24-byte seed, called *randseed.bin*, for the X9.17/X9.31 generator. The seed is then *washed* with an IDEA encryption in CFB mode. The IDEA key is an MD5 hash of the plaintext message and a null IV. The outcome of the process is a 16-byte session key and an 8-byte IV, together with a new seed for the next PRNG.

Analysis

PGP's PRNG is a cryptographically solid method for generating temporary SKC keys, which has stood the test of peer review for some time. The seed file, randseed.bin, is kept in a disk file that is reseeded by the generator for each usage. Although randseed.bin should not be revealed, a cryptanalyst would have extreme difficulty in obtaining pertinent data from it, if it were captured, since it is "cryptographically washed" both before and after each use. The long-term RSA key pairs are generated from the "truly random" entropy derived from the keystroke latencies. Thus, the overall strength of the scheme is based on a firm bedrock of iron-clad cryptography.

We now turn to a discussion of other PRNGs. One of the most popular, introduced in 1986 (see [29]), is given in the following. We first need to set the stage with some rather interesting integers.

◆ Blum Integers

For the next algorithm, we need to refer to quadratic residues modulo a given integer n, called a *Blum integer*, which is an integer $n = pq$, where $p \equiv q \equiv 3 \pmod{4}$ are distinct primes. Since Blum integers have such interesting properties, we list some here for the reader as a preparatory introduction.

Properties of Blum Integers

If n is a Blum integer, then each of the following hold, where the symbol $\left(\frac{a}{b}\right)$ is the Jacobi symbol (see Appendix A, especially page 482).

1. $\left(\frac{-1}{p}\right) = \left(\frac{-1}{q}\right) = -1$.

2. If $x \in \mathbb{Z}/n\mathbb{Z}$ and $\left(\frac{x}{n}\right) = 1$, then one of $\pm x$ is a quadratic residue modulo n.

3. If $x^2 \equiv y \pmod{n}$, then the four square roots of y are given by

$$x = \pm \left(upy^{(q+1)/4} + vqy^{(p+1)/4}\right), \text{ and } x = \pm \left(upy^{(q+1)/4} - vqy^{(p+1)/4}\right),$$

where

$$up + vq = 1.$$

4. If $y \equiv x^2 \pmod{n}$, then exactly one (least quadratic residue of a) square root x of y, with $\left(\frac{x}{n}\right) = 1$ satisfies $x \leq n/2$.

Now we are ready for the next generator.

B.2   The Blum-Blum-Shub-(BBS) PRNG

Let $x_0 \in \mathbb{Z}/n\mathbb{Z}$ be a seed quadratic residue where n is a Blum integer. This initializes the BBS-PRNG (also known as the *quadratic residue generator*). The random bit sequence is generated as follows.

1. For $j = 1, 2, \ldots$, compute $x_j \equiv x_{j-1}^2 \pmod{n}$.

2. Let b_j be the least significant bit of x_j.

Then the output pseudorandom bit sequence is $b_1, b_2, \ldots$.

It can be shown that if x_0 is kept secret, then for a cryptanalyst to predict the least significant bits in the above output sequence is computationally equivalent to factoring n (see [110]). (Compare with the RSA conjecture on page 175.)

The BBS-PRNG is considered to be a *cryptographically secure pseudorandom bit generator* (CSPRBG) (see page 151). This takes us into the formal area of *semantic security*, which means that the ciphertext does not reveal any information about the plaintext to a cryptanalyst (whose computational power is polynomially bounded). The reader interested in pursuing this formal theory may consult the pioneering work of Goldwasser and Micali in [109], as well as Blum and Micali in [31], and the much more recent [184].

One drawback to the BBS-PRNG is that it may be very slow in application. To improve speed, one may select the m least significant bits of x_j, and if

$$m \leq \log_2(\log_2(n)),$$

then the scheme is cryptographically secure (see[30]).

Appendix C: Factoring Large Integers

Given the importance of factoring, or rather the difficulty thereof, in the security of RSA and other cryptosystems, it is worth our having a closer look at the issue to which we devote this appendix.

On page 165, we mentioned the integer factoring problem (IFP), but did not delve into its meaning. Now we make this explicit.

◆ **The Integer Factoring Problem — (IFP)**

Given $n \in \mathbb{N}$, find primes p_j for $j = 1, 2, \ldots, r \in \mathbb{N}$ with $p_1 < p_2 < \cdots < p_n$ such that

$$n = \prod_{j=1}^{r} p_j^{e_j}.$$

A simpler problem than the IFP is the notion of *splitting* of $n \in \mathbb{N}$, which means the finding of factors $r, s \in \mathbb{N}$ such that $1 < r \leq s$ such that $n = rs$. Of course, with an RSA modulus, splitting and the IFP are the same thing. Yet, in order to solve the IFP for any integer, one merely splits n, then splits n/r and s if they are both composite, and so on until we have a complete factorization.

First, we look at some older methods that still inspire the methods of today.

C.1 Classical Factorization Methods

Trial Division The oldest method of splitting n is *trial division* by which we mean dividing n by all primes up to $\sqrt{n}$. For $n < 10^8$, or within that neighborhood, this is not an unreasonable method in our computer-savvy world. However, for larger integers, we need more elaborate methods.

Fermat Factoring In 1643, Fermat discovered a factoring scheme based upon the following insight. If $n = rs$ is an odd natural number with $r < \sqrt{n}$, then

$$n = \left(\frac{s+r}{2}\right)^2 - \left(\frac{s-r}{2}\right)^2 = a^2 - b^2. \tag{C.1}$$

Therefore, in order to split n, we need only investigate the values,

$$x = a^2 - n \text{ for } a = \lfloor\sqrt{n}\rfloor + 1, \lfloor\sqrt{n}\rfloor + 2, \ldots, (n-1)/2,$$

until a perfect square is found. This is now called Fermat's *difference-of-squares factoring method*. It has been rediscovered many times and used as a basis for many modern factoring techniques since essentially we are looking at solutions of

$$x^2 \equiv y^2 \pmod{n} \text{ with } x \not\equiv \pm y \pmod{n}, \tag{C.2}$$

and

$$\gcd(x \pm y, n)$$

provides the nontrivial factors.

Although the order of magnitude (see page 500) of Fermat factoring can be shown to be $O(n^{1/2})$, Lehman has shown how to reduce the complexity to $O(n^{1/3})$ when combined with trial division. This is all contained in [144], complete with a computer program. There is also a method, from D.H. Lehmer, for speeding up the Fermat method when all factors are of the form $2k\ell + 1$ (see [48]).

Euler's Factoring Method This method only applies to integers of the form,
$$n = x^2 + ay^2 = z^2 + aw^2,$$
where $x \neq z$ and $y \neq w$. In other words, n can be written in two distinct ways in this special form for a given nonzero value of $a \in \mathbb{Z}$. Then
$$(xw)^2 \equiv (n - ay^2)w^2 \equiv -ay^2 w^2 \equiv (z^2 - n)y^2 \equiv (zy)^2 \pmod{n},$$
from which we may have a factor of n, namely, provided that $xw \not\equiv \pm zy \pmod{n}$. In this case, the (nontrivial) factors of n are given by $\gcd(xw \pm yz, n)$.

The Euler method essentially is predicated on the congruence (C.2), but unlike the Fermat method, not all integers have even one representation in the form $n = x^2 + ay^2$. In fact, the reader who is versed in some algebraic number theory will recognize these forms for n as norms from the quadratic field $\mathbb{Q}(\sqrt{-a})$. It can be shown that Euler's method requires at most $\lfloor \sqrt{n/a} \rfloor$ steps when $a > 0$.

Legendre's Factoring Method This method is a precursor to what we know today as *continued fraction methods for factorization* (see pages 496–498). Legendre reasoned in the following fashion. Instead of looking at congruences of the form (C.2), he looked at those of the form,
$$x^2 \equiv \pm py^2 \pmod{n} \text{ for primes } p, \tag{C.3}$$
since a solution to (C.3) implies that $\pm p$ is a quadratic residue of all prime factors of n. For instance, if the residue is 2, then all prime factors of n are congruent to $\pm 1 \pmod 8$ (see part (5) of Theorem A.19 on page 482). Therefore, he would have halved the search for factors of n. Legendre applied this method for various values of p, thereby essentially constructing a quadratic sieve[C.1] by getting many residues modulo n. This allowed him to eliminate potential prime divisors that sit in various linear sequences, as with the residue 2 example above. He realized that if he could achieve enough of these, he could eliminate primes up to $\sqrt{n}$, thereby effectively developing a test for primality!

The linchpin of Legendre's method is the continued fraction expansion of $\sqrt{n}$ since he was simply finding *small* residues modulo n. Legendre was essentially

---

[C.1]A *sieve* may be regarded as any process whereby we find numbers via searching up to a prescribed bound and eliminate candidates as we proceed until only the desired solution set remains. A (general) *quadratic* sieve is one in which about half of the possible numbers being sieved are removed from consideration, a technique used for hundreds of years as a scheme for eliminating impossible cases from consideration.

building a sieve on the prime factors of n, which did not let him predict, for a given prime p, a different residue to yield a square. This meant that if he found a solution to

$$x^2 \equiv py^2 \pmod{n},$$

he could not predict a solution,

$$w^2 \equiv pz^2 \pmod{n},$$

distinct from the former. If he had been able to do this, he would have been able to combine them as

$$(xw)^2 \equiv (pzy)^2 \pmod{n}$$

and have a factor of n *provided that*

$$xw \not\equiv \pm pzy \pmod{n}$$

since we are back to congruence (C.2).

Gauss invented a method that differed from Legendre's scheme only in the approach to finding small quadratic residues of n; but his approach makes it much more complicated (see [102, Articles 333 and 334, pages 403–406]).

In the 1920s, one individual expanded the idea, described above, of attempting to match the primes to create a square. We now look at his important influence.

Kraitchik's Factoring Method Maurice Kraitchik determined that it would suffice to find a *multiple* of n as a difference of squares in attempting to factor it. For this purpose, he chose a polynomial of the form,

$$kn = ax^2 \pm by^2,$$

for some integer k, which allowed him to gain control over finding two distinct residues at a given prime to form a square, which Legendre could not do. In other words, Kraitchik used quadratic polynomials to get the residues, then multiplied them to get squares (not a square times a small number). Kraitchik developed this method over a period of more than 3 decades, a method later exploited by D.H. Lehmer and R.E. Powers (see [145]). They employed Kraitchik's technique but obtained their residues as Legendre had done. Later this was exploited in the development of an algorithm that systematically extracted the best of the above ideas as follows, which is taken from [169]. First we need to define a couple of terms.

If $B \in \mathbb{N}$, then a number $n \in \mathbb{N}$ is said to be a *B-smooth number* if all primes dividing n are no larger than B, and B is called a *smoothness bound*. A *factor base* is a set of "small" primes that remain the primes under consideration for the algorithm at hand.

C.2   The Continued Fraction Algorithm

Suppose that we wish to factor $n \in \mathbb{N}$ and a smoothness bound B has been selected. Then execute the following steps:

(1) Choose a factor base of primes $\mathcal{F} = \{p_1, p_2, \ldots, p_k\}$ for some $k \in \mathbb{N}$ determined by B and a large upper index value J.[C.2]

(2) Set $Q_0 = 1$, $P_0 = 0$, $A_{-1} = 1$, $A_0 = \lfloor \sqrt{n} \rfloor = q_0 = P_1$. For each natural number $j \leq J$, recursively compute Q_j using the following formulas:

$$Q_j = \frac{n - P_j^2}{Q_{j-1}},$$

$$q_j = \left\lfloor \frac{P_j + \lfloor \sqrt{n} \rfloor}{Q_j} \right\rfloor,$$

$$A_j = q_j A_{j-1} + A_{j-2},$$

$$P_{j+1} = q_j Q_j - P_j,$$

and trial divide Q_j by the primes in $\mathcal{F}$ to determine if Q_j is p_k-smooth. If it is, use its factorization $Q_j = \prod_{i=1}^{k} p_i^{a_{i,j}}$ to form the binary $k+1$-tuple,

$$\mathfrak{v}_j = (v_{0,j}, v_{1,j}, v_{2,j}, \ldots, v_{k,j}),$$

where $v_{0,j}$ is, respectively, 0 or 1 according as j is even or odd, and for $1 \leq i \leq k$, $v_{i,j}$ is, respectively, 0 or 1 according to whether $a_{i,j}$ is even or odd. If Q_j is not p_k-smooth, discard it and return to calculate Q_{j+1}.

(3) For each set $\mathcal{S}$ of the vectors $\mathfrak{v}_j$ constructed in (2), for which it is discovered that

$$\sum_{j \in \mathcal{S}} v_{i,j} \equiv 0 \, (\mathrm{mod} \, 2), \quad 0 \leq i \leq k,$$

we have $x^2 \equiv y^2 \, (\mathrm{mod} \, n)$, where

$$x = \left[\prod_{j \in \mathcal{S}} (-1)^j Q_j \right]^{1/2} \quad \text{and} \quad y \equiv \prod_{j \in \mathcal{S}} A_{j-1} \, (\mathrm{mod} \, n).$$

If $x \not\equiv \pm y \, (\mathrm{mod} \, n)$, then $\gcd(x \pm y, n)$ gives a nontrivial factor of n.

By Corollary A.3 on page 498,

$$A_{j-1}^2 - n B_{j-1}^2 = (-1)^j Q_j,$$

which is the heart of the algorithm. Thus, we have that $n B_{j-1}^2 \equiv A_{j-1}^2 \, (\mathrm{mod} \, p)$, for any prime $p \mid Q_j$, so n is a quadratic residue modulo p. Hence, we only put

---

[C.2] From knowledge about the distribution of smooth integers close to $\sqrt{n}$, the optimal k is known to be one that is chosen to be approximately $\sqrt{\exp(\sqrt{\log(n) \log \log(n)})}$.

primes p in the factor base for which n is a quadratic residue modulo p. The following gives a small illustration of the continued fraction algorithm, called CFRAC by some users.

Example C.1 Let $n = 6109$. Our factor base will be $\mathcal{F} = \{3, 5, 11, 13, 31, 37\}$. Since $\lfloor \sqrt{n} \rfloor = 78$, then we compute the following table (where $J = 3$).

j	P_j	q_j	A_{j-1}	$(-1)^j Q_j$	$\mathfrak{v}_j$
0	0	78	1	1	$(0,0,0,0,0,0,0)$
1	78	6	78	-25	$(1,0,0,0,0,0,0)$
2	72	4	469	37	$(0,0,0,0,0,0,1)$
3	76	17	1954	-9	$(1,0,0,0,0,0,0)$

We have a set $\mathcal{S}$ such that

$$\sum_{j \in \mathcal{S}} v_{i,j} \equiv 0 \,(\mathrm{mod}\ 2) \text{ for each } i = 0, 1, \ldots, 6.$$

This set is $\mathcal{S} = \{1, 3\}$ for which we have $Q_1 = -5^2$, $Q_3 = -3^2$, $A_0 = 78$ and $A_2 = 1954$. We compute $\prod_{j \in \mathcal{S}} A_{j-1} \equiv 5796 \,(\mathrm{mod}\ 6109)$ and since

$$y^2 = \prod_{j \in \mathcal{S}} A_{j-1}^2 \equiv x^2 = \prod_{j \in \mathcal{S}} Q_j = 15^2 \,(\mathrm{mod}\ n),$$

then we check $\gcd(x \pm y, n)$. We compute that both $\gcd(x - y, n) = \gcd(15 - 5796, 6109) = 41$, and $\gcd(x + y, n) = \gcd(15 + 5796, 6109) = 149$. Thus, we have factored $n = 41 \cdot 149$.

The CFRAC algorithm was developed by Brillhart and Morrison in the early 1970s (see [173]). It is widely acclaimed to be the very first efficient *general* factorization algorithm put into use. It is subexponential time (see page 501), which essentially means that if the running time to factor n is n^a, then a slowly decreases as $n \mapsto \infty$.

In 1974, Pollard published a factorization scheme (see [189]), that utilizes Euler's generalization of Fermat's little theorem (see Theorem A.14 on page 479). He reasoned that if $(p - 1) \mid n$ where p is prime, then $p \mid (t^n - 1)$ provided that $p \nmid t$, which follows from Euler's theorem, so p may be found by employing Eulcild's algorithm (see Theorem A.3 on page 470). The following is taken from [169].

C.3 Pollard's $p-1$ Algorithm

Suppose that we wish to factor $n \in \mathbb{N}$, and that a smoothness bound B has been selected. Then we execute the following.

(1) Choose a base $a \in \mathbb{N}$ where $2 \le a < n$ and compute $g = \gcd(a,n)$. If $g > 1$, then we have a factor of n. Otherwise, go to step (2).

(2) For all primes $p \le B$, compute $m = \left\lfloor \frac{\ln(n)}{\ln(p)} \right\rfloor$ and replace a by $a^{p^m} \pmod{n}$ using the repeated squaring method given on page 171. (Note that this iterative procedure ultimately gives $a^{\Pi_{p \le B} p^m}$ modulo n for the base a chosen in (1).)

(3) Compute $g = \gcd(a - 1, n)$. If $g > 1$, then we have a factor of n, and the algorithm is successful. Otherwise, the algorithm fails.

The reasoning behind Pollard's algorithm is given as follows.

Let $\ell = \operatorname{lcm}(p_1^{a_1}, \ldots, p_t^{a_t})$, where $p_j^{a_j}$ runs over all prime powers such that $p_j \le B$. Since $p_j^{a_j} \le n$, then $a_j \ln(p_j) \le \ln(n)$, so $a_j \le \left\lfloor \frac{\ln(n)}{\ln(p_j)} \right\rfloor$. Hence, $\ell \le \prod_{j=1}^{t} p_j^{\lfloor \ln(n)/\ln(p_j) \rfloor}$. Now, if $p \mid n$ is a prime such that $p - 1$ is B-smooth, then $(p-1) \mid \ell$. Therefore, for any $a \in \mathbb{N}$ with $p \nmid a$, $a^\ell \equiv 1 \pmod{p}$, by Fermat's little theorem (A.2). Thus, if $g = \gcd(a^\ell - 1, n)$, then $p \mid g$. If $g = n$, then the algorithm fails. Otherwise, it succeeds.

Example C.2 *Let $n = 13193$, and choose a smoothness bound $B = 13$, then select $a = 2$. We know that a is relatively prime to n so we proceed to step (2). The table shows the outcome of the calculations for step (2).*

p	2	3	5	7	11	13
m	13	8	5	4	3	3
a	6245	1365	1884	3133	5472	396

Then we go to step (3) and check $\gcd(a - 1, n) = \gcd(395, 13193) = 79$. Thus, we have factored $n = 79 \cdot 167$. Observe that $p = 79$ is B-smooth since $p - 1 = 2 \cdot 3 \cdot 13$, but $q = 167$ is not since $q - 1 = 2 \cdot 83$.

The running time for Pollard's $p - 1$ algorithm is $O(B \ln(n)/\ln(B))$ modular multiplications, assuming that $n \in \mathbb{N}$ and there exists a prime $p \mid n$ such that $p - 1$ is B-smooth. This is of course the drawback to this algorithm, namely, that it requires n to have a prime factor p such that $p - 1$ has only "small" prime factors. A generalization of the $p - 1$ method was given by Lenstra using elliptic curves, which we will study later in this appendix. In the Elliptic curve algorithm, we will see that success in factoring depends upon an integer "close" to p having only small prime factors, which is less demanding than the $p - 1$ algorithm and therefore more likely to occur. Another improvement was given by Williams in [289], called the $p + 1$ method of factoring, which is efficient if n has a prime factor p such that $p + 1$ is B-smooth. There have been other refinements and improvements. Pollard also developed another method for factoring in 1975, called the *Monte Carlo factoring method*, also known as the *Pollard rho method* (see [188]). The following is taken from [167].

C.4 Pollard's Rho Method

Given $n \in \mathbb{N}$ composite, and p an (as yet unknown) prime divisor of it, perform the following steps.

(1) Choose an integral polynomial f with $\deg(f) \geq 2$—usually $f(x) = x^2 + 1$ is chosen for simplicity.

(2) Choose a randomly generated integer $x = x_0$, the *seed*, and compute $x_1 = f(x_0)$, $x_2 = f(x_1) \ldots, x_{j+1} = f(x_j)$ for $j = 0, 1, \ldots B$, where the bound B is determined by step (3).

(3) Sieve through all differences $x_i - x_j$ modulo n until it is determined that $x_B \not\equiv x_j \pmod{n}$ but $x_B \equiv x_j \pmod{p}$ for some natural number $B > j \geq 1$. Then $\gcd(x_B - x_j, n)$ is a nontrivial divisor of n.

Example C.3 *If $n = 3161$, and $x_0 = 2$ is the seed with $f(x) = x^2 + 1$, then $x_1 = f(x_0) = 5$, $x_2 = f(x_1) = 26$, $x_3 = f(x_2) = \overline{677}$, $x_4 = f(x_3) = \overline{3146}$, $x_5 = f(x_4) = \overline{226}$, $x_6 = f(x_5) = \overline{501}$, $x_7 = f(x_6) = \overline{1283}$, $x_8 = f(x_7) = \overline{2370}$, $x_9 = f(x_8) = \overline{2965}$, where the bar notation denotes the fact that we have reduced the values to the least residue system modulo n. We find that all $\gcd(x_i - x_j, n) = 1$ for $i \neq j$ until $\gcd(x_9 - x_7, n) = \gcd(1682, 3161) = 29$. In fact, $3161 = 29 \cdot 109$.*

As the number of comparisons of $x_i - x_j$ and n for a gcd gets large, then the method becomes very time consuming. However, there is an observation that cuts down the work considerably. Suppose that $x_i \equiv x_j \pmod{m}$ for some $m \in \mathbb{N}$ and some $j > i > 0$. Then

$$x_{i+1} \equiv f(x_i) \equiv f(x_j) \equiv x_{j+1} \pmod{m},$$

$$x_{i+2} \equiv f(x_{i+1}) \equiv f(x_{j+1}) \equiv x_{j+2} \pmod{m},$$

and, for general $k \in \mathbb{N}$,

$$x_{i+k} \equiv f(x_{i+k-1}) \equiv f(x_{j+k-1}) \equiv x_{j+k} \pmod{m}.$$

In particular, if $k = j - i$, then $x_{2j} \equiv x_j \pmod{m}$. Hence, in our search for a factor of n, we need only look at $x_{2j} - x_j$ for each $j = 1, 2, \ldots$. This modified method has the advantage of checking only one gcd for each j. However, as Example C.3 shows, we would miss the gcd found in there since $5 \neq 2j$ for any j. However, the time saved in general means that we will come across a solution in comparatively less time.

Example C.4 *Referring to Example C.3, and continuing the calculations, we get* $x_{10} = \overline{485}$, $x_{11} = \overline{1312}$, $x_{12} = \overline{1761}$, $x_{13} = \overline{181}$, $x_{14} = \overline{1152}$, $x_{15} = \overline{2646}$, *and* $x_{16} = \overline{2863}$. *We find that* $\gcd(x_{16} - x_8, n) = \gcd(493, 3161) = 29$.

In Example C.4, we made only eight comparisons for $x_{2j} - x_j$ for $j = 1, 2, 3, \ldots, 8$. However, in Example C.3, we made nearly forty of them, since we had to look at $x_j - x_i$ for all $i \neq j$ with $1 \leq i, j \leq 9$.

Now we illustrate the reason behind the name *Pollard rho method*. We take $n = 29$ as the modulus and $x_0 = 2$ as the seed, then we proceed through the Pollard rho method to achieve the following diagram.

Diagram C.1 (Pollard's Rho Method Illustrated)

We take $n = 29$ as the modulus and $x_0 = 2$ as the seed, then we proceed through the Pollard rho method to achieve the following diagram.

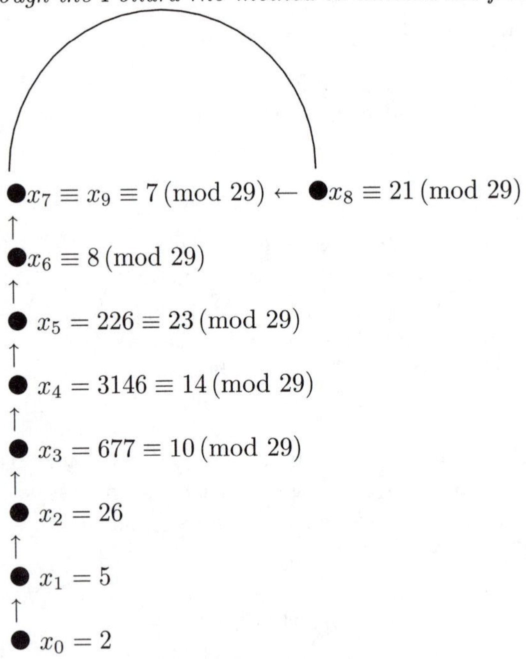

Diagram C.1 shows us that when we reach x_9, then we are in the period that takes us back and forth between the residue system of 7 and that of 21 modulo 29. This is the significance of the left pointing arrow from the position of x_8 back to the position of x_7, which is the same as the reside system of x_9. This completes the circuit. The shape of the symbol is reminiscent of the Greek symbol ρ, *rho*, pronounced *row*. The rho method was made twenty-five percent faster by Brent [43] in 1980. His idea was to stop the algorithm when x_j for $j = 2^i$ occurs, then consider $x_j - x_{2^i} \pmod{p}$ for $3 \cdot 2^{i-1} < j \leq 2^{i+1}$. This has the advantage of revealing the period length after far fewer arithmetic operations than Pollard's original rho method. Adapting this modification, Brent and Pollard found a factor, 1238926361552897, of the eighth Fermat number $F_8 = 2^{2^8} + 1$ in approximately two hours on a mainframe computer (see [44]).

Pollard's two methods above may be invoked when trial division fails to be useful. However, if the methods of Pollard fail to be useful, which they will for *large* prime factors, say, with the number of digits in the high teens, then we need more powerful machinery. The following is one of those.

In the early 1980s, Carl Pomerance was able to fine-tune the parameters in Kraitchik's sieve method (see [190]). The following is taken from [169].

C.5 The Quadratic Sieve (QS)

(1) Choose a *factor base* $\mathcal{F} = \{p_1, p_2, \ldots, p_k\}$, where the p_j are primes for $j = 1, 2, \ldots, k \in \mathbb{N}$.

(2) For each nonnegative integer j, let $t = \pm j$. Compute $y_t = (\lfloor \sqrt{n} \rfloor + t)^2 - n$ until $k + 2$ such values are found that are p_k-smooth. For each such t,

$$y_t = \pm \prod_{i=1}^{k} p_i^{a_{i,t}}, \qquad (C.4)$$

and we form the binary $k + 1$-tuple, $\mathfrak{v}_t = (v_{0,t}, v_{1,t}, v_{2,t}, \ldots, v_{k,t})$, where $v_{i,t}$ is the least nonnegative residue of $a_{i,t}$ modulo 2 for $1 \leq i \leq k$, $v_{0,t} = 0$ if $y_t > 0$, and $v_{0,t} = 1$ if $y_t < 0$.

(3) Obtain a subset $\mathcal{S}$ of the values of t found in step (2) such that for each $i = 0, 1, 2, \ldots, k$,

$$\sum_{t \in \mathcal{S}} v_{i,t} \equiv 0 \, (\mathrm{mod} \, 2). \qquad (C.5)$$

In this case, $x^2 = \prod_{t \in \mathcal{S}} x_t^2 \equiv \prod_{t \in \mathcal{S}} y_t = y^2 \, (\mathrm{mod} \, n)$, where $x_t = \lfloor \sqrt{n} \rfloor + t$, so $\gcd(x \pm y, n)$ provides a nontrivial factor of n if $x \not\equiv \pm y \, (\mathrm{mod} \, n)$.

In step (2), we have that $y_t \equiv x_t^2 \, (\mathrm{mod} \, n)$. Thus, if a prime $p \mid y_t = x_t^2 - n$, we have $x_t^2 \equiv n \, (\mathrm{mod} \, p)$. Thus, we must exclude from the factor base any primes p for which there is no solution $x \in \mathbb{Z}$ to the congruence $x^2 \equiv n \, (\mathrm{mod} \, p)$. In other words, we exclude from the factor base any primes p for which n is *not* a quadratic residue modulo p.

Example C.5 Let $n = 30167$. From Footnote C.2, $k = 11$, so we choose the first eleven primes for which n is a quadratic residue. They comprise our factor base $\mathcal{F} = \{2, 7, 11, 17, 29, 31, 37, 41, 43, 53, 67\}$. We see, by inspection, that a subset $\mathcal{S}$ of the values of t in the table on page 518 (which we computed given

$\lfloor \sqrt{n} \rfloor = 173$) such that $\sum_{t \in \mathcal{S}} v_{i,t} \equiv 0 \,(\mathrm{mod}\ 2)$ for each $i = 0, 1, 2, \ldots, 11$ is $\mathcal{S} = \{0, 18, -23\}$. Thus,

$$\prod_{t \in \mathcal{S}} x_t^2 = 2^2 3^2 5^4 173^2 191^2 \equiv 9062^2 \equiv x^2 \,(\mathrm{mod}\ 30167),$$

and

$$\prod_{t \in \mathcal{S}} y_t = 2^2 7^2 11^2 17^2 41^2 \equiv 16837^2 \equiv y^2 \,(\mathrm{mod}\ 30167),$$

so $y^2 - x^2 \equiv 16837^2 - 9062^2 \equiv 7775 \cdot 25899 \,(\mathrm{mod}\ 30167)$.

By computing both of the values, $\gcd(7775, 30167) = 311 = \gcd(y - x, n)$ and $\gcd(25899, 30167) = 97 = \gcd(x + y, n)$, we get that $n = 30167 = 97 \cdot 311$.

t	x_t	y_t	$\mathfrak{v}_t$
0	173	$-2 \cdot 7 \cdot 17$	$(1, 1, 1, 0, 1, 0, 0, 0, 0, 0, 0, 0)$
-1	172	$-11 \cdot 53$	$(1, 0, 0, 1, 0, 0, 0, 0, 0, 0, 1, 0)$
-5	168	$-29 \cdot 67$	$(1, 0, 0, 0, 0, 1, 0, 0, 0, 0, 0, 1)$
5	178	$37 \cdot 41$	$(0, 0, 0, 0, 0, 0, 0, 1, 1, 0, 0, 0)$
-6	167	$-2 \cdot 17 \cdot 67$	$(1, 1, 0, 0, 1, 0, 0, 0, 0, 0, 0, 1)$
7	180	$7 \cdot 11 \cdot 29$	$(0, 0, 1, 1, 0, 1, 0, 0, 0, 0, 0, 0)$
11	184	$7 \cdot 17 \cdot 31$	$(0, 0, 1, 0, 1, 0, 1, 0, 0, 0, 0, 0)$
14	187	$2 \cdot 7^4$	$(0, 1, 0, 0, 0, 0, 0, 0, 0, 0, 0, 0)$
-15	158	$-11 \cdot 43$	$(1, 0, 0, 1, 0, 0, 0, 0, 0, 0, 1, 0)$
-17	156	$-7^3 \cdot 17$	$(1, 0, 1, 0, 1, 0, 0, 0, 0, 0, 0, 0)$
18	191	$2 \cdot 7 \cdot 11 \cdot 41$	$(0, 1, 1, 1, 0, 0, 0, 0, 1, 0, 0, 0)$
-23	150	$-11 \cdot 17 \cdot 41$	$(1, 0, 0, 1, 1, 0, 0, 0, 1, 0, 0, 0)$
28	201	$2 \cdot 7 \cdot 17 \cdot 43$	$(0, 1, 1, 0, 1, 0, 0, 0, 0, 1, 0, 0)$

Some elementary linear algebra underlies the solution to a factorization problem using the QS as depicted in Example C.5. By ensuring that there are $k + 2$ vectors $\mathfrak{v}_t$ in a $k + 1$-dimensional vector space $\mathbb{F}_2^{k+1}$, we guarantee that there is a linear dependence relation among the $\mathfrak{v}_t$. In other words, we ensure the existence of the set $\mathcal{S}$ in step (3) of the algorithm such that congruence (C.5) holds. There is no guarantee that $x \not\equiv \pm y \,(\mathrm{mod}\ n)$, but there are usually several dependency relations among the $\mathfrak{v}_t$, so there is a high probability that at least one of them will yield an (x, y) pair such that $x \not\equiv \pm y \,(\mathrm{mod}\ n)$. The problem, of course, is that for "large" smoothness bounds B, we need a lot of congruences before we may be able to get these dependency relations.

For k as given in Footnote C.2, the asymptotic running time (see page 502), of the quadratic sieve is $O\left(\exp\left((1 + o(1))\sqrt{\ln(n) \ln(\ln(n))}\right)\right).$

To split $n \in \mathbb{N}$, with the QS, we consider a polynomial,

$$g(x) = (x + \lfloor \sqrt{n} \rfloor)^2 - n,$$

for $x \in (-n^\epsilon, n^\epsilon)$. Then we construct a set of integers $i \in \mathcal{S}$ so that $g(x_i)$ factors over the factor base and

$$\prod_{x_i \in \mathcal{S}} b_i \equiv \prod_{x_i \in \mathcal{S}} g(x_i) \equiv y^2 \,(\text{mod } n).$$

For a single choice of $g(x)$, there is an unreasonable amount of time required to generate a sufficiently large enough set $\mathcal{S}$ over which $g(x)$ will factor. The reason is that for large n, the interval $(-n^\epsilon, n^\epsilon)$ is also large since $g(x) = O(n^{1/2+\epsilon})$ is large as well, and so we will probably not be able to factor most of the $g(x)$ over a small set of primes. The following version, taken from [170], solves this problem by establishing an efficient means of using several polynomials so that the x values may be chosen from smaller intervals rather than one large interval. This means that the average polynomial values are smaller than the average of g and have a higher probability of factoring over small primes than the $g(x)$ values in the ordinary quadratic sieve. This then is a way of running the ordinary quadratic sieve in parallel.

C.6 Multipolynomial Quadratic Sieve (MPQS)

In this algorithm, $n \in \mathbb{N}$ is assumed to be a large composite number. The goal is to split n.

(1) (**Select bounds**): Choose a large smoothness bound B and an $M \in \mathbb{N}$ with $(\sqrt{2n}/M)^{1/4} > B$.

(2) (**Select a factor base**): Choose a set of $L \in \mathbb{N}$ primes as a factor base (see the discussion on page 534) that is fixed for the algorithm:

$$\mathcal{F} = \{p_i : p_i \text{ is prime and } \left(\frac{n}{p_i}\right) = 1 \text{ for } i = 1, 2, \dots, L\},$$

where the symbol is the Legendre symbol. For $p_i \in \mathcal{F}$ with $q_i = p_i^{a_i} < B$, compute solutions t_{q_i} to the congruences,

$$t_{q_i}^2 \equiv n \,(\text{mod } q_i),$$

for $0 < t_{q_i} \le q_i/2$.

(3) (**Create a quadratic polynomial**): Choose $r, k \in \mathbb{N}$ with $1 < k < r$.[C.3] Generate primes $g_1, g_2, \dots, g_r$, which are called *g-primes*, satisfying:

(a) $g_i \approx (\sqrt{2n}/M)^{1/(2k)}$,

(b) $\left(\frac{n}{g_i}\right) = 1$, and

(c) for each $i = 1, 2, \dots, r$, $\gcd(g_i, q) = 1$ for all $q \in \mathcal{F}$.

[C.3] We choose a small values of r for pedagogical purposes but in practice, the MPQS typically uses a value such as $r = 30$, for instance.

For some choice of k of the g-primes where $1 \le i_1 < i_2 < \cdots < i_k \le r$, let

$$a = g_{i_1} g_{i_2} \cdots g_{i_k}.$$

Now, solve for b_i with $i = 1, 2, \ldots, r$ in

$$b_i^2 \equiv n \,(\mathrm{mod}\ g_i^2).$$

Then use the Chinese remainder theorem A.12 on page 478, to solve the system of congruences, for a specific choice of signs:

$$b \equiv \pm b_{i_1} \,(\mathrm{mod}\ g_{i_1}^2); \quad b \equiv \pm b_{i_2} \,(\mathrm{mod}\ g_{i_2}^2); \quad \cdots \quad b \equiv \pm b_{i_k} \,(\mathrm{mod}\ g_{i_k}^2).$$

For this solution b, set $c = (b^2 - n)/a^2$. Then we select

$$W(x) = a^2 x^2 + 2bx + c,$$

where the above generation guarantees that a, b, c satisfy

$$a^2 \approx \sqrt{2n}/M, \quad b^2 - n = a^2 c, \quad |b| < a^2/2. \tag{C.6}$$

(4) (**Test W(x) for divisibility by factor base elements**): If $q_i \mid W(j)$ for some $j \in [-M, M]$, called a *sieve number*, then $q_i \mid (a^2 j + b)^2 - n$, so

$$j \equiv a^{-2}(\pm t_{q_i} - b) \,(\mathrm{mod}\ q_i),$$

since $\gcd(a, q_i) = 1$ from step (3). We compute $a^{-2} \,(\mathrm{mod}\ q_i)$ for all such q_i via

$$a^{-2} \equiv g_{i_1}^{-2} g_{i_2}^{-2} \cdots g_{i_k}^{-2} \,(\mathrm{mod}\ q_i).$$

Thus, for efficiency, with the calculation of g-primes by the methodology in in step (3), we also compute and save, for $i = 1, 2, \ldots, r$, all the numbers $g_i^{-2} \,(\mathrm{mod}\ q_i)$ for each $q_i = p_i^{a_i} < B$ where $p_i \in \mathcal{F}$.

(5) (**Sieving**[C.4]): Define a $(2M + 1)$-tuple:

$$s = (s(-M), s(-M + 1), \ldots, s(j), \ldots, s(M)),$$

which we initialize by setting $s(j) = 0$ for all $j \in [-M, M]$. For each sieve number $j \in [-M, M]$, i.e., those for which some prime power $q_i = p_i^{a_i} \mid W(j)$, reset

$$s(j) = \ln p_i + s(j).$$

(6) (**Selection of factor candidates**): Define the *report threshold* [C.5] to be

$$RT = \ln\left(\frac{1}{2} M \sqrt{n/2}\right).$$

[C.4]In general with the MPQS, the amount of time spent on sieving takes more than 85% of the total computing time.

[C.5]The report threshold is the average of $\ln |W(j)|$ for $j \in [-M, M]$. When $s(j) \ge RT$, $W(j)$ is a good candidate for factoring over the factor base.

Select from step (5) all those values j for which $s(j) \approx RT$, test $W(j)$, and save a, b, j, (and thus tacitly c via the choice in step (3)) only if $W(j)$ factors over $\mathcal{F}$. If the number of $W(j)$ selected is less than $L + 2$, go to step (3). Otherwise, go to step (7).

(7) (**Creation of exponent vector**): Since we have $L + 2$ sieve values j, we form

$$W(j) = (-1)^{b_{j_0}} \prod_{i=1}^{L} p_i^{b_{j_i}}, \text{ and } b_{j_i} \le a_i, \text{ for } j = 1, 2, \ldots, L + 2$$

and associate with $W(j)$ the exponent vector,

$$v_j = (b_{j_0}, b_{j_1}, \ldots, b_{j_L}) \,(\text{mod } 2),$$

so we have a binary $L + 1$-tuple for each $j = 1, 2, \ldots, L + 2$. Since we have $L + 2$ vectors with $L + 1$ coordinates, then there is at least one subset,

$$\mathcal{S} \subseteq \{1, 2, \ldots, L + 2\}^{\text{C.6}},$$

such that

$$\sum_{j \in \mathcal{S}} v_j \equiv 0 \,(\text{mod } 2),$$

so

$$\prod_{j \in \mathcal{S}} W(j) \equiv z^2 \,(\text{mod } n).$$

(8) (**Factor n**): Since $(a^2 x + b)^2 \equiv a^2 W(x) \,(\text{mod } n)$, then

$$X^2 \equiv \prod_{j \in \mathcal{S}} (a^2 j + b)^2 \equiv z^2 \prod_{j \in \mathcal{S}} a^2 \equiv Y^2 \,(\text{mod } n),$$

so if $1 < \gcd(X - Y, n) < n$, then we have a nontrivial factor of n.

One big advantage of the MPQS over the ordinary quadratic sieve is that one can generate many a, b, c values, and switch polynomials when the residues grow too large. In step (3), we see that if we have a fixed k and set of r g-primes, the number of polynomials that may be calculated is

$$2^{k-1} \binom{r}{k} = 2^{k-1} \frac{r!}{(r-k)! k!}.$$

Analysis To see why the conditions in (C.6) hold, we present the following verification. Since (a) holds, then

$$a^2 \approx \prod_{i=1}^{k} \left(\frac{\sqrt{2n}}{M} \right)^{2(1/2k)} = \left(\frac{\sqrt{2n}}{M} \right)^{\sum_{i=1}^{k}(1/k)} = \left(\frac{\sqrt{2n}}{M} \right),$$

[C.6]We can use Gaussian elimination modulo 2 on the matrix whose columns are v_j to find a set $\mathcal{S}$. See Appendix A on page 494.

so the first condition is satisfied in (C.6). Since $b^2 \equiv n \,(\mathrm{mod}\ a^2)$ given the solution to the system of congruences via the Chinese remainder theorem, then $a^2 \mid (b^2 - n)$, so $(b^2 - n)/a^2 = c \in \mathbb{Z}$, which is the second condition. If $b \geq a^2/2$, then replace b by $b - a^2$, and we have $|b| < a^2/2$, which is the last condition.

As mentioned earlier, Lenstra invented a generalization of Pollard's $p-1$ method, which we now present. The following is taken from [170]. The reader needing a reminder of elliptic curve fundamentals may go to page 498 in Appendix A.

C.7   The Elliptic Curve Method (ECM)

In this algorithm, $n \in \mathbb{N}$ is assumed to be composite, prime to 6, and not a perfect power, and $r \in \mathbb{N}$ is a parameter. The goal is to split n.

(1) (**Select and elliptic curve**): Choose a random pair (E, P) where $E = E(\mathbb{Z}/n\mathbb{Z})$ is an elliptic curve:

$$y^2 = x^3 + ax + b \text{ and } P \text{ is a point on } E.$$

Check that $\gcd(n, 4a^3 + 27b^2) = 1$. If not, then we have split n if $1 < g < n$, and we may terminate the algorithm. Otherwise, we select another (E, P) pair.

(2) (**Choosing bounds**): Select $M \in \mathbb{N}$ and bounds $A, B \in \mathbb{N}$ such that the canonical prime factorization for M is $M = \prod_{j=1}^{\ell} p_j^{a_{p_j}}$ for small primes $p_1 < p_2 < \cdots < p_\ell \leq B$ where $a_{p_j} = \lfloor \ln(A)/\ln(p_j) \rfloor$ is the largest exponent such that $p_j^{a_j} \leq A$. Set $j = k = 1$.

(3) (**Calculating multiple points**): Using (A.12) and (A.13) from page 499, compute $p_j P$.

(4) (**Computing the gcd**):

(a) If $p_j P \not\equiv \mathfrak{o} \,(\mathrm{mod}\ n)$, then set $P = p_j P$, and reset k to $k + 1$.
  (i) If $k \leq a_{p_j}$, then go to step (3).
  (ii) If $k > a_{p_j}$, then reset j to $j + 1$, and reset k to $k = 1$. If $j \leq \ell$, then go to step (3). Otherwise go to step (5).

(b) If $p_j P \equiv \mathfrak{o} \,(\mathrm{mod}\ n)$, then compute $\gcd(m_2, n)$ for m_2 in (A.13). If $n > g$, terminate the algorithm, since we have split n. If $g = n$, go to step (5).

(5) (**Selecting a new pair**): Set $r = r - 1$. If $r > 0$, go to step (1). Otherwise, terminate with "failure".

Example C.6 *Let $n = 923$ and select $(E, P) = (y^2 = x^3 + 2x + 9, (0, 3))$. Then* $\gcd(4 \cdot 2^3 + 27 \cdot 9^2, 923) = 1$, *so we choose $B = 4$, based upon (C.7), and let* $A = 3, M = 6 = 2 \cdot 3 = p_1 \cdot p_2$. *Now, using (A.12)–(A.13), with $p_1 = 2$, we calculate*

$$p_1 P = 2(0, 3) \equiv (9^{-1}, -82 \cdot 27^{-1}) \equiv (718, 373) \not\equiv \mathfrak{o} \,(\mathrm{mod}\ n).$$

Thus we set $P = (718, 373)$ and compute

$$p_2 P = 3P \equiv 2P + P \equiv (505, 124) + (718, 373) \equiv \mathfrak{o} \,(\mathrm{mod}\ n).$$

Thus, we have that a denominator in (A.13) is not prime to n. In fact, the calculation of m for $4P + 2P$ yields $m = (124 - 373)/(505 - 718) = 83/71$, and $\gcd(923, 71) = 71$. Indeed, $n = 13 \cdot 71$, and we have split n.

What Example C.6 illustrates is that the failure of the existence of a modular inverse for some m in the calculations may lead to a factor of n. Another way of saying this is that the group law for multiplication actually fails in $\mathbb{Z}/n\mathbb{Z}$ since n is not prime and this allows us to get the factor. Indeed, it is somewhat inaccurate in the ECM algorithm to say that $p_j P \equiv \mathfrak{o} \,(\mathrm{mod}\ n)$, when in fact it is $p_j P \equiv \mathfrak{o} \,(\mathrm{mod}\ p)$ where p is the factor for which we were searching. However, this is legitimate since we were, in a sense, assuming n to be prime and doing the calculations as if it were so, in the *hope* that the calculations would "break down" with an undefined denominator for some value of m in (A.13).

A significant advantage of the ECM is that its running time is highly reliant on the factor, $p \mid n$, found. Hence, one of the most useful means of employing the ECM is for finding "small" prime factors in a number n, which is too large to find *all* its factors. The reasons behind this are as follows. Assuming that p is the smallest prime dividing n, the expected running time of the ECM is known (under certain plausible assumptions) to be

$$O(\exp(\sqrt{(2 + o(1))\ln p(\ln \ln p)}) \cdot \ln^2 n).$$

This may be used in practice to select a smoothness bound B in step (2) of the algorithm as

$$B = \exp(\sqrt{\ln p(\ln \ln p)/2}). \tag{C.7}$$

Since we do not know p in advance, we may nevertheless select (for p) the value $\lfloor \sqrt{n} \rfloor$. In this case, it is estimated that one out of every B iterations will be successful in splitting n.

The worst-case scenario for the ECM is when n is an RSA modulus, in which case we have that the expected running time is

$$O\left(\exp(\sqrt{(2 + o(1))\ln n(\ln \ln n)})\right) = O\left(n^{\sqrt{(2 + o(1))(\ln \ln n)/\ln n}}\right).$$

This being said, it is not surprising that ECM is most successful at splitting *non*-RSA moduli, *usually* finding prime factors of less than 40 decimal digits in large composite numbers.

Now we look at one of the big guns in factoring. This requires some knowledge of the theory of algebraic number fields (see [168] for instance). The following is taken from [170].

C.8 ☞ The General Number Field Sieve

The QS and MPQS studied above is based upon the generation of many smooth quadratic residues of n close to $\sqrt{n}$, for a given composite n. Pollard extrapolated this idea in a manuscript circulated in 1988, where he used cubic integers (those in $\mathbb{Z}[\sqrt[3]{-2}]$) to factor by attempting to generate many smooth cubic residues of n close to $\sqrt[3]{n}$. Later the idea was extended to the fifth degree and used to factor the 9th Fermat number. Then he looked at the more general idea of looking at composite n that are close to being powers in the sense that

$$n = r^t - s \text{ for small} r, |s| \in \mathbb{N} \text{ and larger } t \in \mathbb{N}. \tag{C.8}$$

The special case where $|s| = 1$ are called *Cunningham numbers*, which are the subject of a project unto themselves in the history of factoring, called the *Cunningham project* (see [47]).

The *special number field sieve* dubbed as such by the authors of [50] can factor integers of the form given in (C.8) in expected running time $L_n(1/3.(32/9)^{1/3})$ (see Equation (A.16) on page 501 for a definition of the notation). The general number field sieve (GNFS) that we study below has expected running time (for arbitrary integers n) given by $L_n(1/3, 1.9229)$, making it a fast algorithm for arbitrary integers but the SNFS is faster for the integers of the special type given above.

We now look at the GNFS and point out that despite its sophistication and power, it is still based essentially on Fermat's difference-of-squares method.

The setup and the goal: Given a composite n to split, what we will be setting out to achieve is the following. We select an appropriate monic polynomial $f(x)$, irreducible over $\mathbb{Z}$, where $m \in \mathbb{N}$ with $f(m) \equiv 0 \pmod{n}$, and $\alpha \in \mathbb{C}$ a root of f. This setup allows us to define the natural homomorphism,

$$\phi : \mathbb{Z}[\alpha] \mapsto \mathbb{Z}/n\mathbb{Z}, \text{ via } \phi(\alpha) = m,$$

which ensures that, for any $g(x) \in \mathbb{Z}[x]$, we have $\phi(g(\alpha)) \equiv g(m) \pmod{n}$. Thus, we will seek a set $\mathcal{S}$ of polynomials g over $\mathbb{Z}$ such that both $\prod_{g \in \mathcal{S}} g(\alpha) = \beta^2 \in \mathbb{Z}[\alpha]$, and $\prod_{g \in \mathcal{S}} g(m) = y^2 \in \mathbb{Z}$. Then by setting $\phi(\beta) \equiv x \pmod{n}$ we get

$$x^2 = \phi(\beta)^2 \equiv \phi(\beta^2) \equiv \phi\left(\prod_{g \in \mathcal{S}} g(\alpha)\right) \equiv \prod_{g \in \mathcal{S}} g(m) \equiv y^2 \pmod{n}, \tag{C.9}$$

and we are back to Fermat's method in (C.2) on page 509, where we have a nontrivial factor of n if $x \not\equiv \pm y \pmod{n}$! However, the devil is in the details so here we go.

The Algorithm

We make some initial simplifying assumptions the reasons for which the reader may find in [50]. We assume that a smoothness bound B and the degree d of the polynomial f have been chosen from experimental data.[C.7] Now, we let $m = \lfloor n^{1/d} \rfloor$ and write n in base m,

$$n = m^d + c_{d-1}m^{d-1} + \cdots + c_0, \text{ with } 0 \le c_j \le m - 1 \text{ for } j = 0, 1, \ldots, d. \quad (C.10)$$

Now if we set

$$f(x) = x^d + c_{d-1}x^{d-1} + \cdots + c_0 \in \mathbb{Z}[x],$$

then we have a monic polynomial with $f(m) = n$. However, we wanted f to be irreducible. If it is not, then we have no need of the number field sieve, since then $f(x) = g(x)h(x)$ where g and h have unequal positive degrees, so $g(x)h(x) = f(m) = n$, and we have a nontrivial factor of n. Hence, we may assume that f is irreducible (as are most monic polynomials over $\mathbb{Z}$). Thus, we have our polynomial f, B, and d values, and a number field $F = \mathbb{Q}(\alpha)$ of degree d over $\mathbb{Q}$.

In the following, we have to extend our definition of *smooth* given on page 511. We call $a + b\alpha \in \mathbb{Z}[\alpha]$ *B-smooth* if $|N_F(a + b\alpha)|$ is B-smooth where N_F is the norm map from the field F to $\mathbb{Q}$. Also, for a given prime $p \le B$, set

$$R(p) = \{r \in \mathbb{Z} : 0 \le r \le p - 1 \text{ and } f(r) \equiv 0 \,(\mathrm{mod}\ p)\}.$$

Then whenever (a, b) are coprime, $p \mid N_F(a - b\alpha)$ if and only if $p \mid (a - br)$ for some $r \in R(p)$ with $p \nmid b$. Then r is called the (unique) *signature* of $N(a - b\alpha)$ modulo p. Hence, for each coprime (a, b)-pair, there exist $|R(p)| = \mathfrak{r}$ primes $p \le B$ dividing $N(a - b\alpha)$. We will let these be denoted by $p_1, p_2, \ldots, p_{\mathfrak{r}}$. Then if $a - b\alpha$ is B-smooth, we have

$$N(a - b\alpha) = (-1)^{a(0)} \prod_{i=1}^{\mathfrak{r}} p^{a(p_i)}, \text{ where } a(0) \in \{0, 1\}.$$

Based on this we can now define exponent vectors. Let

$$v(a - b\alpha) = (a(0), a(p_1), a(p_2), \ldots, a(p_{\mathfrak{r}})).$$

However, based on our goal set above, we want not only $a - b\alpha$ to be B-smooth, but also $a - bm$ to be B-smooth. If the latter is the case, then let $q_{\mathfrak{r}+1}, q_{\mathfrak{r}+2}, \ldots, q_{\mathfrak{s}}$ be all the primes less than or equal to B dividing $a - bm$, and write

$$a - bm = (-1)^{b(0)} \prod_{i=\mathfrak{r}+1}^{\mathfrak{s}} q_i^{b(q_i)},$$

[C.7]Heuristic complexity arguments determine the choices to be optimal when $B = L_n(1/3, c)$ for $c = (8/9)^{1/3+\epsilon}$, and $d = ((2/c)^{1/2}[\ln n/(\ln\ln n)]^{1/3}) = \ln(L_n(1/3, (2/c)^{1/2}))$. These choices ensure that $B^d \approx n^{2/d}$. Hence, $n > 2^{d^2}$, which is needed to ensure that n is monic in (C.10), a straightforward exercise to verify.

and define, $v(a - bm) = (b(0), b(q_{\mathfrak{r}+1}), \ldots, b(q_{\mathfrak{s}}))$. Finally set

$$v(a, b) \equiv (v(a - b\alpha), v(a - bm)) \,(\mathrm{mod}\ 2).$$

Hence, $v(a, b)$ is a binary vector of length $\mathfrak{r} + \mathfrak{s} + 2$.

For ease of elucidation, we make the simplifying assumption that if we find a set $\mathcal{S} = \{(a, b) \in \mathbb{Z} \times \mathbb{Z} : \gcd(a, b) = 1\}$ such that $\sum_{(a,b) \in \mathcal{S}} v(a, b)$ is the zero vector modulo 2, then both $\prod_{(a,b) \in \mathcal{S}} (a - bm)$ will be a square in $\mathbb{Z}$ and $\prod_{(a,b) \in \mathcal{S}} (a - b\alpha)$ will be a square in $\mathbb{Z}[\alpha]$ (see [191] for the means of dealing with the obstructions when this is not the case). Hence, all we do is to sieve over coprime integer pairs (a, b) with $0 < b \le B$, $|a| \le B$ until the above is achieved. Then we are in the situation (C.9) and we proceed to factor n.

◆ Summary

In order not to waste time, we must first ensure that the number n that we are trying to factor is indeed composite. One may apply a strong pseudoprime test (see Appendix F). Also, see the deterministic primality test by the authors of [6], which we present in Appendix F.3.

Once convinced of the compositness of n, perform trial division up to 10^4, then apply Pollard's rho method, which finds small factors rather quickly. Trial division is applied first since Pollard's method has a slower running time than trial division for the very small divisors, so it is worthwhile to apply the rho method to a number whose small prime factors have been removed.

Secondly, many methods fail to detect when n is a perfect power, so we must also rule out that possibility. If $n = m^e$, where $m, e \in \mathbb{N}$ and $e > 1$, then we can actually determine m and e as follows. For each prime $p \le \log_2 n$, do a binary search for $r \in \mathbb{N}$ satisfying $n = r^p$, restricting attention to the range $2 \le r \le 2^{\lfloor (\log_2 n)/p \rfloor + 1}$. This calculation may be accomplished in $O((\log_2 n)^3 (\log_2(\log_2(\log_2 n))))$ bit operations. Thus, this is a reasonable pretest for ensuring that we do not have such a power.

Once the above has been completed, use Pollard's $p - 1$ method, followed by Lenstra's ECM. The ECM takes more time than Pollard's rho method for finding relatively small factors, so it may be used to advantage at this juncture.

If all the above fails, the big guns should be brought to bear on the problem. This includes CFRAC, MPQS, and GNFS. Of course, all of this strategy depends on the computing power available, the algorithms at hand, and other factors, but all things being equal and having access to all of the above, then the strategy should ultimately work modulo a suitable waiting period if the heavy machinery mentioned here is employed. The running times given in this appendix provide a good indicator of just how long one should expect to wait as your computer executes its job.

Appendix D: Technical and Advanced Details

In this appendix, we sequester certain technical details about some of the ciphers or other algorithms. Otherwise, the explanation, left in the main text, could present a more onerous task, leaving the novice reader stranded in some cases. Hence, we provide the particulars of these algorithmic fine points for those interested in pursuing the depths of a given cipher or methodology more exhaustively. Moreover, for the reader interested in pursuing material beyond that given in the main text, we provide some advanced algorithms that go well beyond the boundaries of the standard presentations.

D.1 AES

We begin with some specifics concerning the AES cipher introduced in Section 3.5.

◆ The Rijndael S-Box

The means by which Rijndael's invertible S-box, explicitly given below, was constructed consists of composing two functions. First, consider the 3×8 matrix with entries,

$$a_{i,j} = 8i + j - 9 \text{ for } 1 \leq i \leq 32 \text{ and } 1 \leq j \leq 8.$$

Let the map g defined by taking 0 to 0 and for $a_{i,j} \neq 0$,

$$g : a_{i,j} \mapsto a_{i,j}^{-1} = \sum_{k=0}^{7} b_k 2^j,$$

the binary representation of the multiplicative inverse of $a_{i,j}$ in $\mathbb{F}_{2^8}$. If we view the image of g as a column vector,

$$g(a_{i,j}) = \begin{pmatrix} b_7 \\ b_6 \\ b_5 \\ b_4 \\ b_3 \\ b_2 \\ b_1 \\ b_0 \end{pmatrix},$$

then we let f be the Affine function that is applied to the output of g via

$$\begin{pmatrix} 1 & 1 & 1 & 1 & 1 & 0 & 0 & 0 \\ 0 & 1 & 1 & 1 & 1 & 1 & 0 & 0 \\ 0 & 0 & 1 & 1 & 1 & 1 & 1 & 0 \\ 0 & 0 & 0 & 1 & 1 & 1 & 1 & 1 \\ 1 & 0 & 0 & 0 & 1 & 1 & 1 & 1 \\ 1 & 1 & 0 & 0 & 0 & 1 & 1 & 1 \\ 1 & 1 & 1 & 0 & 0 & 0 & 1 & 1 \\ 1 & 1 & 1 & 1 & 0 & 0 & 0 & 1 \end{pmatrix} \begin{pmatrix} b_7 \\ b_6 \\ b_5 \\ b_4 \\ b_3 \\ b_2 \\ b_1 \\ b_0 \end{pmatrix} + \begin{pmatrix} 0 \\ 1 \\ 1 \\ 0 \\ 0 \\ 0 \\ 1 \\ 1 \end{pmatrix} = \begin{pmatrix} s_7 \\ s_6 \\ s_5 \\ s_4 \\ s_3 \\ s_2 \\ s_1 \\ s_0 \end{pmatrix}.$$

Hence,

$$f \circ g(a_{i,j}) = (s_7 s_6 s_5 s_4 s_3 s_2 s_1 s_0)^t$$

is the binary equivalent of the decimal digit appearing in the S-box at position (i,j).

Observe that the column matrix, added on the left of the equality, is binary for the decimal digit 99 (or equivalently, the hexadecimal digit 63).

We may consider the above in terms of polynomials. For instance,

$$a_{11,3} = 8 \cdot 11 + 3 - 9 = 82$$

has representation as the binary polynomial,

$$x^6 + x^4 + x \in \mathbb{F}_{2^8} \cong \mathbb{F}_2[x]/(m(x)),$$

where

$$m(x) = x^8 + x^4 + x^3 + x + 1$$

is the irreducible *Rijndael polynomial* (see Example A.9 on page 489 in Appendix A). The multiplicative inverse of 82 in $\mathbb{F}_{2^8}$ is given by $x^2 + 1$, so

$$(b_7, b_6, b_5, b_4, b_3, b_2, b_1, b_0) = (0, 0, 0, 0, 0, 1, 0, 1).$$

Thus

$$\begin{pmatrix} 1 & 1 & 1 & 1 & 1 & 0 & 0 & 0 \\ 0 & 1 & 1 & 1 & 1 & 1 & 0 & 0 \\ 0 & 0 & 1 & 1 & 1 & 1 & 1 & 0 \\ 0 & 0 & 0 & 1 & 1 & 1 & 1 & 1 \\ 1 & 0 & 0 & 0 & 1 & 1 & 1 & 1 \\ 1 & 1 & 0 & 0 & 0 & 1 & 1 & 1 \\ 1 & 1 & 1 & 0 & 0 & 0 & 1 & 1 \\ 1 & 1 & 1 & 1 & 0 & 0 & 0 & 1 \end{pmatrix} \begin{pmatrix} 0 \\ 0 \\ 0 \\ 0 \\ 0 \\ 1 \\ 0 \\ 1 \end{pmatrix} + \begin{pmatrix} 0 \\ 1 \\ 1 \\ 0 \\ 0 \\ 0 \\ 1 \\ 1 \end{pmatrix} = \begin{pmatrix} 0 \\ 0 \\ 0 \\ 0 \\ 0 \\ 0 \\ 0 \\ 0 \end{pmatrix},$$

and 0 is the decimal entry in position $(11, 3)$ of the S-box:

In summary, all of the values values of $f \circ g$ acting on the $a_{i,j}$ are given by the decimal representations in the following Rijndael S-box.

99	124	119	123	242	107	111	197
48	1	103	43	254	215	171	118
202	130	201	125	250	89	71	240
173	212	162	175	156	164	114	192
183	253	147	38	54	63	247	204
52	165	229	241	113	216	49	21
4	199	35	195	24	150	5	154
7	18	128	226	235	39	178	117
9	131	44	26	27	110	90	160
82	59	214	179	41	227	47	132
83	209	0	237	32	252	177	91
106	203	190	57	74	76	88	207
208	239	170	251	67	77	51	133
69	249	2	127	80	60	59	168
81	163	64	143	146	157	56	245
188	182	218	33	16	255	243	210
205	12	19	236	95	151	68	23
196	167	126	61	100	93	25	115
96	129	79	220	34	42	144	136
70	238	184	20	222	94	11	219
224	50	58	10	73	6	36	92
194	211	172	98	145	149	228	121
231	200	55	109	141	213	78	169
108	86	244	234	101	122	174	8
186	120	37	46	28	166	180	198
232	221	116	31	75	189	139	138
112	62	181	102	72	3	246	14
97	53	87	185	134	193	29	158
225	248	152	17	105	217	142	148
155	30	135	233	206	85	40	223
140	161	137	13	191	230	66	104
65	153	45	15	176	84	187	22

AES Mix Column (MC) Algorithm

In this step of the AES cipher, the columns in the state matrix are treated as polynomials $a(x)$ over $\mathbb{F}_{2^8} \cong \mathbb{F}_2[x]/(m(x))$, where

$$m(x) = x^8 + x^4 + x^3 + x + 1$$

is the irreducible Rijndael polynomial (see construction of the S-Box earlier). Then $a(x)$ is multiplied modulo $M(x) = x^4 + 1$ with a fixed invertible polynomial,

$$c(x) = 3x^3 + x^2 + x + 2,$$

denoted by

$$c(x) \otimes a(x).$$

Here multiplying modulo $x^4 + 1$ means that

$$x^i \pmod{x^4 + 1} = x^{i \pmod 4}.$$

It can be shown that if

$$a_j(x) = a_{3,j}x^3 + a_{2,j}x^2 + a_{1,j}x + a_{0,j}$$

represents column j in the state matrix, then $c(x) \otimes a(x)$ can be represented by the matrix product:

$$CA_j = \begin{pmatrix} 2 & 3 & 1 & 1 \\ 1 & 2 & 3 & 1 \\ 1 & 1 & 2 & 3 \\ 3 & 1 & 1 & 2 \end{pmatrix} \begin{pmatrix} a_{0,j} \\ a_{1,j} \\ a_{2,j} \\ a_{3,j} \end{pmatrix} = \begin{pmatrix} b_0 \\ b_1 \\ b_2 \\ b_3 \end{pmatrix} = B,$$

where the matrix A_j is column j of the state matrix and C is the *circulant* matrix representing $c(x)$. Hence, each column A_j of the state matrix is multiplied in this fashion by C. For instance, if

$$a(x) = x^3 + 1,$$

then

$$c(x) \otimes a(x) = 5x^3 + 4x^2 + 2x + 3,$$

which is given by the matrix product:

$$\begin{pmatrix} 2 & 3 & 1 & 1 \\ 1 & 2 & 3 & 1 \\ 1 & 1 & 2 & 3 \\ 3 & 1 & 1 & 2 \end{pmatrix} \begin{pmatrix} 1 \\ 0 \\ 0 \\ 1 \end{pmatrix} = \begin{pmatrix} 3 \\ 2 \\ 4 \\ 5 \end{pmatrix} = B,$$

D.2   Silver-Pohlig-Hellman

The next algorithm is the Silver-Pohlig-Hellman algorithm for finding discrete logs, which first appeared in 1978 (see [187]). We discussed issues surrounding this algorithm on page 165.

◆ **Silver-Pohlig-Hellman Algorithm for Computing Discrete Logs**

Let α be a generator of $\mathbb{F}_p^*$ and let $\beta \in \mathbb{F}_p^*$, and assume that we have a factorization

$$p - 1 = \prod_{j=1}^{r} p_j^{a_j} \qquad a_j \in \mathbb{N},$$

where the p_j are distinct primes. The technique for computing $e = \log_\alpha \beta$ is to compute e modulo $p_j^{a_j}$ for $j = 1, 2, \ldots, r$, then apply the Chinese remainder theorem (see Theorem A.12 on page 478 in Appendix A). Since we operate on each prime power $p_j^{a_j}$, we replace p_j with q for simplicity in what follows, and simply refer to q^a with the understanding that we are operating on each of the r prime powers in this fashion. To compute e modulo q^a we need to determine e in its base q representation:

$$e = \sum_{i=0}^{a-1} b_i q^i \qquad \text{where } 0 \le b_i \le q - 1 \text{ for } 0 \le i \le a - 1.$$

To find these b_i, we proceed as follows. First, set $\beta_0 = \beta = \alpha^e$, and observe that

$$(p-1)\sum_{k=i}^{a-1} b_k q^{k-i-1} \equiv (p-1)b_i/q \,(\mathrm{mod}\ p-1). \tag{D.1}$$

1. Calculate b_0. By (D.1),

$$\beta_0^{(p-1)/q} \equiv \alpha^{(p-1)b_0/q} \,(\mathrm{mod}\ p), \tag{D.2}$$

   using Fermat's Little Theorem (see Corollary A.2 on page 479). Thus, we compute $\alpha^{(p-1)k/q}\,(\mathrm{mod}\ p)$ until (D.2) occurs, in which case k is b_0.

2. Calculate b_i for $i = 1, 2, \ldots, a-1$. First, recursively define

$$\beta_i = \beta\alpha^{-\sum_{k=0}^{i-1} b_k q^k}.$$

   By (D.1),

$$\beta_i^{(p-1)/q^{i+1}} \equiv \alpha^{(p-1)\sum_{k=i}^{a-1} b_k q^{k-i-1}} \equiv \alpha^{(p-1)b_i/q} \,(\mathrm{mod}\ p), \tag{D.3}$$

   so we compute $\alpha^{(p-1)k/q}$ modulo p for nonzero $k \leq a-1$ until the left and right sides of (D.3) are congruent modulo p, in which case k is b_i.

A small example is in order. This is, of course, not realistic in terms of the degree of difficulty, but for pedagogical purposes, it will suffice, and we will do this often for the same reasons throughout.

Example D.1 *Let $p = 37$. Then $\alpha = 2$ generates $\mathbb{F}_{37}^*$. Given $\beta_0 = \beta = 19$, we want to compute $e = \log_2(19)$ in $\mathbb{F}_{37}^*$. We have*

$$p - 1 = 36 = 2^2 \cdot 3^2 = p_1^{a_1} p_2^{a_2}.$$

All congruences in the balance of this example are assumed to be modulo 37.
    For $p_1 = 2$:

k	0	1
$\alpha^{(p-1)k/p_1}$	1	$2^{18} \equiv 36$

i	0	1
β_i	19	$19 \cdot 2^{-1} \equiv 28$
$\beta_i^{(p-1)/p_1^{i+1}}$	$19^{18} \equiv 36$	$28^9 \equiv 36$
b_i	1	1

Thus, the base 2 representation of $\log_2(19)$ *modulo 4 is*

$$\sum_{i=0}^{a-1} b_i p_1^i = 1 \cdot 2^0 + 1 \cdot 2^1 \equiv 3 \,(\mathrm{mod}\ 4). \tag{D.4}$$

For $p_2 = 3$:

k	0	1	2
$\alpha^{(p-1)k/p_2}$	1	$2^{12} \equiv 26$	$2^{24} \equiv 10$

i	0	1
β_i	19	$19 \cdot 2^{-2} \equiv 14$
$\beta_i^{(p-1)/p_2^{i+1}}$	$19^{12} \equiv 10$	$14^9 \equiv 10$
b_i	2	2

Thus, the base 3 representation of $\log_2(19)$ *modulo 9 is*

$$\sum_{i=0}^{a_2-1} b_i p_2^i = 2 \cdot 3^0 + 2 \cdot 3^1 \equiv 8 \,(\mathrm{mod}\ 9). \tag{D.5}$$

Solving (D.4)–(D.5) *by the Chinese remainder theorem , we get that*

$$e = \log_2(19) = 35 \ in \ \mathbb{F}_{37}^*.$$

If $n = p - 1$, then given a factorization of n, the running time of the Silver-Pohlig-Hellman discrete log algorithm is

$$O\left(\sum_{j=1}^{r} a_j \left(\ln n + \sqrt{p_j}\right)\right)$$

group multiplications. This implies that the Pohlig-Hellman algorithm is only efficient if the prime divisors of $p-1$ are small. This is the reason why we talked about a proper choice of p on page 165 for the intractability of the discrete log problem.

It should also be noted that the above algorithm makes use of what is known as the baby-step giant-step algorithm for computing discrete logs due to the late Dan Shanks, a pioneer in computational number theory. For the sake of completeness, and because it leads to another important method for computing discrete logs, we present it here. The following is taken from [170].

D.3 Baby-Step Giant-Step Algorithm

♦ **Baby-Step Giant-Step Algorithm for Computing Discrete Logs**

Given a generator α of a cyclic group G of order n, and $\beta \in G$, the goal is to compute the discrete logarithm,

$$x \equiv \log_\alpha \beta \,(\text{mod } n).$$

(1) Compute $s = \lfloor \sqrt{n} \rfloor$.

(2) **Baby-Step**: For $j = 0, 1, \ldots, s-1$, compute $(j, \alpha^j \beta)$. Then sort the list by second component in ascending order.

(3) **Giant-Step**: For $i = 1, 2, \ldots, s$ compute (α^{is}, i) and sort by first component in ascending order.

(4) **Search and Compare**: Search the lists in steps (2) and (3) to see if there is an $\alpha^j \beta$ from step (2) and an α^{is} from step (3) such that $\alpha^j \beta = \alpha^{is}$. If so, then compute

$$x \equiv is - j \,(\text{mod } n),$$

which is

$$\log_\alpha \beta \,(\text{mod } n).$$

Example D.2 *Let* $\alpha = 5$, $\beta = 71$, *and* $n = 167$. *We want to determine*

$$x \equiv \log_5(71) \,(\text{mod } 167).$$

First, we calculate $s = \lfloor \sqrt{n} \rfloor = 12$. *The* baby-step *is the computation of*

$$(j, 5^j \cdot 71 \,(\text{mod } 167)) \text{ for } j = 0, 1, \ldots, 11:$$

$(0, 71)$, $(1, 21)$, $(2, 105)$, $(3, 24)$, $(4, 120)$, $(5, 99)$, $(6, 161)$, $(7, 137)$, $(8, 17)$, $(9, 85)$, $(10, 91)$, $(11, 121)$. *Then we sort according to the second element:*

j	8	1	3	0	9	10
$5^j \cdot 71$	17	21	24	71	85	91

j	5	2	4	11	7	6
$5^j \cdot 71$	99	105	120	121	137	161

The giant-step *is the computation of* $(5^{12i} \,(\text{mod } 167), i)$ *for* $i = 1, 2, \ldots, 12$: $(152, 1)$, $(58, 2)$, $(132, 3)$, $(24, 4)$, $(141, 5)$, $(56, 6)$, $(162, 7)$, $(75, 8)$, $(44, 9)$, $(8, 10)$, $(47, 11)$, $(130, 12)$. *Then we order according to the first component:*

15^{12i}	8	24	44	47	56	58
i	10	4	9	11	6	2

15^{12i}	75	130	132	141	152	162
i	8	12	3	5	1	7

Then we search the two lists and find that $\alpha^3\beta \equiv 24 \equiv \alpha^{4\cdot12} \pmod{167}$, so $x = 4 \cdot 12 - 3 = 45$ and indeed

$$\log_5(71) \equiv 45 \pmod{167} \text{ since } 5^{45} \equiv 71 \pmod{167}.$$

The baby-step giant-step method presented above was first used by Shanks in August of 1968 to calculate the class number of an imaginary quadratic field. The running time for the algorithm is $O(\sqrt{n})$ group operations and according to [159, Note 3.67(i), p. 109] is the same as the Silver-Pohlig-Hellman algorithm if n is prime. Moreover, it uses $O(\sqrt{n})$ memory, so this deterministic algorithm has a runtime/memory trade-off. Shanks' method is a kind of square-root method, of which Pollard provided other kinds such as his rho method (see [167, pp. 127–130]). We now look at (arguably) the most potent and efficacious of the methods for computing discrete logs. In its general form, it bears a strong resemblance to some of the most powerful factoring algorithms (such as the *number field sieve* (see [169, Section 5.2, pp. 207–220]), which may be considered to be a variant of the following method). Although the following has a more general formulation for other cyclic groups, we restrict our attention to $\mathbb{F}_p^*$ for the sake of simplicity of presentation. The following is a subexponential time algorithm (see page 501).

D.4   Index-Calculus Algorithm

◆ **The Index-Calculus Algorithm for Computing Discrete Logs**

We solve $\beta \equiv \alpha^x \pmod{p}$ where p is a large prime and α is a primitive root modulo p.

Precomputation stage:

(1) Select a *factor base* (a set of "small primes" that will remain the primes under consideration for the duration of the algorithm): $\mathcal{B} = \{p_1, \ldots, p_B\}$ consisting of the first B primes. (Here the choice for B should be made such that a "considerable number" of the elements of $\mathbb{F}_p^*$ can be expressed as products of powers of elements of $\mathcal{B}$.)

(2) Collect relations by choosing a random nonnegative integer $k \leq p - 2$ and compute the least positive residue of α^k modulo p, if possible, then its canonical prime factorization, $\prod_{j=1}^{B} p_j^{k_j}$ for $k_j \geq 0$. When such relations exist we may take logs and get

$$k \equiv \sum_{j=1}^{B} k_j \log_\alpha(p_j) \pmod{p - 1}. \tag{D.6}$$

Continue to choose (at least) B such k so that we are successful in securing B relations as in (D.6). Here we are trying to solve for $\log_\alpha(p_j)$ for $j = 1, 2, \ldots, B$.

Calculation of discrete logs stage:

(3) For each k in (D.6), determine the value of $\log_\alpha(p_j)$ for $1 \leq j \leq B$ by solving the B (modular) linear equations with unknowns $\log_\alpha(p_j)$.

(4) Select a random nonnegative integer $t \leq p - 2$ and compute $\beta\alpha^t$.

(5) If possible, *factor $\beta\alpha^t$ over $\mathcal{B}$*, namely, write

$$\beta\alpha^t = \prod_{j=1}^{B} p_j^{t_j} \qquad (t_j \geq 0). \tag{D.7}$$

If it is not possible to get (D.7), then go to step (4). If (D.7) is successfully obtained, then

$$\log_\alpha(\beta) + t \equiv \sum_{j=1}^{B} t_j \log_\alpha(p_j) \pmod{p - 1},$$

from which we can calculate $\log_\alpha(\beta)$.

As usual, a small example will suffice to illustrate the algorithm.

Example D.3 *Let $p = 3361$, $\alpha = 22$, and $\mathcal{B} = \{2, 3, 5, 7\}$. We wish to compute $\log_{22}(4)$ in $\mathbb{F}_{3361}^*$ using the index-calculus method. We choose randomly $k = 48, 100, 186, 2986$ and get*

$$22^{48} \equiv 2^5 \cdot 3^2 \pmod{3361}, \qquad 22^{100} \equiv 2^6 \cdot 7 \pmod{3361},$$

$$22^{186} \equiv 2^9 \cdot 5 \pmod{3361}, \qquad 22^{2986} \equiv 2^3 \cdot 3 \cdot 5^2 \pmod{3361}.$$

Thus we get the system of four congruences in four unknowns:

$$48 \equiv 5\log_{22}(2) + 2\log_{22}(3) \pmod{3360},$$

$$100 \equiv 6\log_{22}(2) + \log_{22}(7) \pmod{3360},$$

$$186 \equiv 9\log_{22}(2) + \log_{22}(5) \pmod{3360} \text{ and,}$$

$$2986 \equiv 3\log_{22}(2) + \log_{22}(3) + 2\log_{22}(5) \pmod{3360}.$$

This completes the precomputation stage. Now we use this to compute

$$\log_{22}(2) = 1100; \log_{22}(3) = 2314; \log_{22}(5) = 366; \text{ and } \log_{22}(7) = 220.$$

Suppose that we now select $t = 754$ *at random and compute*

$$\beta \alpha^t = 4 \cdot 22^{754} \equiv 2 \cdot 3^2 \cdot 5 \cdot 7 \,(\text{mod } 3361).$$

Thus, we have

$$\log_{22}(4) + 754 \equiv \log_{22}(2) + 2\log_{22}(3) + \log_{22}(5) + \log_{22}(7) \,(\text{mod } 3360).$$

Hence, $\log_{22}(4) = 2200$, *and we check that indeed*

$$22^{2200} \equiv 4 \,(\text{mod } 3361).$$

D.5   ☞ Brands' Digital Cash Scheme

Now we turn to e-commerce and present the details of Brands' scheme discussed at the end of Section 5.8 on page 232.

● ☞ Brands' Digital Cash Scheme

Setup Stage: The bank performs the following steps:
(1) Choose a large prime p such that $(p-1)/2 = q$ is also prime, and select α to be the square of a primitive root modulo p. Also, we assume that the DLP in $(\mathbb{Z}/p\mathbb{Z})^*$ is intractable.

(2) Choose two random $x_1, x_2 \in (\mathbb{Z}/q\mathbb{Z})^*$, compute $g_1 \equiv \alpha^{x_1} \,(\text{mod } p)$ and $g_2 \equiv \alpha^{x_2} \,(\text{mod } p)$, then discard x_1, x_2. (Note that by (1), $g_1 \equiv g_2 \,(\text{mod } p)$ if and only if $x_1 \equiv x_2 \,(\text{mod } q)$.) Make (α, g_1, g_2) public.

(3) Select a random secret $x \in (\mathbb{Z}/q\mathbb{Z})^*$ and compute

$$h \equiv \alpha^x \,(\text{mod } p), \quad h_1 \equiv g_1^x \,(\text{mod } p), \quad \text{and} \quad h_2 \equiv g_2^x \,(\text{mod } p).$$

Then (h, h_1, h_2) is the bank's public key and x is the bank's private key.

(4) Choose two public cryptographic hash functions,

$$H_1 : ((\mathbb{Z}/p\mathbb{Z})^*)^5 \mapsto (\mathbb{Z}/q\mathbb{Z})^* \quad \text{and} \quad H_2 : ((\mathbb{Z}/p\mathbb{Z})^*)^4 \mapsto (\mathbb{Z}/q\mathbb{Z})^*.$$

(5) The merchant registers identification number M with the bank.

Opening Alice's Account:

(1) Alice generates $e_1, e_2 \in (\mathbb{Z}/q\mathbb{Z})^*$ at random and computes

$$A \equiv g_1^{e_1} g_2^{e_2} \not\equiv 1 \,(\text{mod } p),$$

which she sends to the bank.

(2) The bank stores (A, I_A, N_A) in its database where I_A is a digital data string uniquely identifying Alice and N_A is her account number.

Identification Protocol:[D.1] When Alice wishes to withdraw coins from her account, she must first identify herself to the bank's satisfaction.

(1) Alice generates $f_1, f_2 \in (\mathbb{Z}/q\mathbb{Z})^*$, at random, computes $f \equiv g_1^{f_1} g_2^{f_2} \pmod{p}$, and sends f to the bank.

(2) The bank generates a random $k \in (\mathbb{Z}/q\mathbb{Z})^*$ (the challenge), and sends it to Alice.

(3) Alice computes $\ell_1 \equiv f_1 + ke_1 \pmod{q}$ and $\ell_2 \equiv f_2 + ke_2 \pmod{q}$ (the responses) and sends (ℓ_1, ℓ_2) to the bank.

(4) The bank accepts her response if and only if $fA^k \equiv g_1^{\ell_1} g_2^{\ell_2} \pmod{p}$.[D.2]

(5) If the bank accepts her response in step (4), it sends her an identification number $y_1 = A^x$.

(*By completing step (5), Alice proves that she owns A. She does this by a proof of knowledge of (e_1, e_2).*)

Coin Withdrawal Protocol: For simplicity, we assume that Alice wants to withdraw only one coin, a six-tuple of integers (X, Y, Y_1, Y_2, Y_3, Z), which we will now see how to construct.

(1) The bank chooses a random $w \in (\mathbb{Z}/q\mathbb{Z})^*$, computes $y_2 \equiv \alpha^w \pmod{p}$, $y_3 \equiv A^w \pmod{p}$, and sends (y_2, y_3) to Alice.

(2) Alice selects three random integers $z_1 \in (\mathbb{Z}/q\mathbb{Z})^*$ and $z_2, z_3, \in \mathbb{Z}/q\mathbb{Z}$. She computes the following where all congruences are modulo p:

$$y_1' \equiv A^{z_1}, \quad Y_1 \equiv y_1^{z_1}, \quad Y_2 \equiv y_2^{z_2} \alpha^{z_3} \quad \text{and} \quad Y_3 \equiv y_3^{z_1 z_2} A^{z_1 z_3}.$$

Now she computes $s_1, s_2, t_1, t_2, u_1, u_2 \in (\mathbb{Z}/q\mathbb{Z})^*$ such that

$$e_1 z_1 \equiv s_1 + s_2 \pmod{q}, \quad e_2 z_1 \equiv t_1 + t_2 \pmod{q}, \quad z_1 \equiv u_1 + u_2 \pmod{q}.$$

[D.1]In the Brands scheme this step is often called the *representation problem step*. It turns out that the Brands scheme is built on the Schnorr signature scheme and the representation problem which is given as follows. In a group of prime order G with generators $(g_1, g_2, \ldots, g_s)$ for $s \geq 2$, $g_j \in G$, and a given $h \in G$, find a representation such that $h = \prod_{j=1}^{s} g_j^{b_j}$ for $b_j \geq 0$. The reader will note that this is related to a discrete log problem and so is difficult without knowledge of the b_j.

[D.2]To see that step (4) identifies Alice uniquely, note that since A is unique to Alice and

$$fA^k \equiv g_1^{f_1} g_2^{f_2} (g_1^{e_1} g_2^{e_2})^k \equiv g_1^{f_1 + ke_1} g_2^{f_2 + ke_2} \equiv g_1^{\ell_1} g_2^{\ell_2} \pmod{p},$$

then Alice's identity is indeed verified.

Then she calculates[D.3]

$$X \equiv g_1^{s_1} g_2^{t_1} A^{u_1} \pmod{p} \text{ and } Y \equiv g_1^{s_2} g_2^{t_2} A^{u_2} \pmod{p}.$$

(3) Alice computes a challenge,

$$c_1 = H_1(y_1', Y_1, Y_2, Y_3, X),$$

and blinds it with $c \equiv c_1 z_2^{-1} \pmod{q}$, which she sends to the bank.

(4) The bank sends a response $r \equiv xc + w \pmod{q}$ to Alice, and debits her account. Alice accepts r if and only if [D.4]

$$\alpha^r \equiv h^c y_2 \pmod{p} \text{ and } A^r \equiv y_1^c y_3 \pmod{p}.$$

(5) Alice computes $Z \equiv r z_2 + z_3 \pmod{q}$. Her coin is

$$C = (X, Y, Y_1, Y_2, Y_3, Z),$$

which she can now spend.

(*Essentially* (Y_1, Y_2, Y_3, Z) *is the banks's signature on* (X, Y)*, so we write* $(X, Y, \text{sig}(X, Y))$*) for* C *in what follows for simplicity.*)

Spending Protocol: Alice wishes to purchase some goods from the merchant.

(1) She sends the merchant her coin $(X, Y, \text{sig}(X, Y))$.

(2) The merchant verifies that $XY \neq 1$,[D.5] then sends a challenge,

$$c = H_2(X, Y, M, T_M)$$

to Alice, where T_M is a timestamp with the date and time on it.

[D.3]Note that by this step, $XY \equiv y_1' \pmod{p}$, which is Alice's blinded identity. The reason for this is as follows:

$$XY \equiv g_1^{s_1} g_2^{t_1} g_1^{s_2} g_2^{t_2} (g_1^{e_1} g_2^{e_2})^{u_1} (g_1^{e_1} g_2^{e_2})^{u_2} \equiv g_1^{s_1+s_2} g_2^{t_1+t_2} g_1^{e_1 u_1} g_2^{e_2 u_2} \equiv$$

$$g_1^{e_1 z_1} g_2^{e_2 z_1} (g_1^{e_1} g_2^{e_2})^{u_1+u_2} \equiv (g_1^{e_1} g_2^{e_2})^{z_1} \equiv A^{z_1} \equiv y_1' \pmod{p}.$$

[D.4]These are necessary and sufficient condition for Alice to accept the bank's response because only the bank knows x. Therefore, only the bank can send a response satisfying both

$$\alpha^r \equiv \alpha^{xc+w} \equiv (\alpha^x)^c \alpha^w \equiv h^c y_2 \pmod{p}$$

and

$$A^r \equiv A^{xc+w} \equiv (A^x)^c A^w \equiv m^c y_3 \pmod{p}.$$

[D.5]The merchant must check this since, if Alice is legitimate, then $XY \neq 1$. The reason is that by Footnote D.3, $XY \equiv y_1' \pmod{p}$. Thus, since $y_1' \equiv A^x \pmod{p}$ with $A \not\equiv 1 \pmod{p}$ by step (1) of the protocol for opening Alice's account, and since $x \in (\mathbb{Z}/q\mathbb{Z})^*$, by step (3) of the setup stage, then $x \not\equiv 0 \pmod{q}$, which completes the reasoning.

(3) Alice computes the responses,

$$r_1 = s_1 + s_2 c \,(\mathrm{mod}\ q); \quad r_2 \equiv t_1 + t_2 c \,(\mathrm{mod}\ q); \quad \text{and} \quad r_3 \equiv u_1 + u_2 c \,(\mathrm{mod}\ q)$$

which she sends to the merchant.

(4) The merchant verifies that $g_1^{r_1} g_2^{r_2} A^{r_3} \equiv XY^c \,(\mathrm{mod}\ p)$ holds and if so accepts the payment.[D.6]

(5) The merchant sends $(X, Y, \mathrm{sig}(X,Y), T_M, c, r_1, r_2)$ to the bank.

(6) The bank verifies the signature $\mathrm{sig}(X,Y)$, that no double spending has occurred, and that c and r_1, r_2 are valid challenge response protocols. If all holds true, the bank pays the merchant.

Deposit Protocol:

(1) The merchant sends $(X, Y, \mathrm{sig}(X,Y), T_M, c, r_1, r_2)$ to the bank.

(2) The bank checks that $\mathrm{sig}(X,Y)$ is valid, that the coin has not already been spent, and that the merchant's challenge and Alice's responses r_1, r_2 are valid. If all of this holds true, the bank pays the merchant.

As with the ECash scheme discussed in Section 5.8, Brands' scheme requires the customer to reveal enough information without revealing identity. However, if Alice tries to double-spend, we now show she will be identified and charged with fraud.

If Alice tries to spend the same coin twice, then there will be two distinct challenges c_1 and c_2 to which she will respond with (all congruences being modulo q)

$$r_{c_1}^{(1)} \equiv s_1 + s_2 c_1, \quad r_{c_1}^{(2)} \equiv t_1 + t_2 c_1, \quad r_{c_1}^{(3)} \equiv u_1 + u_2 c_1,$$

and

$$r_{c_2}^{(1)} \equiv s_1 + s_2 c_2, \quad r_{c_2}^{(2)} \equiv t_1 + t_2 c_2, \quad r_{c_2}^{(3)} \equiv u_1 + u_2 c_2,$$

respectively. Hence,

$$r_{c_1}^{(1)} - r_{c_2}^{(1)} \equiv s_2(c_1 - c_2) \text{ and } r_{c_1}^{(2)} - r_{c_2}^{(2)} \equiv t_2(c_1 - c_2),$$

so

$$s_2 \equiv (r_{c_1}^{(1)} - r_{c_2}^{(1)})(c_1 - c_2)^{-1} \text{and } t_2 \equiv (r_{c_1}^{(2)} - r_{c_2}^{(2)})(c_1 - c_2)^{-1}. \tag{D.8}$$

[D.6]This holds for valid responses from Alice since

$$g_1^{r_1} g_2^{r_2} A^{r_3} \equiv g_1^{s_1 + s_2 c} g_2^{t_1 + t_2 c} (g_1^{e_1} g_2^{e_2})^{u_1 + u_2 c} \equiv (g_1^{s_1} g_2^{t_1} A^{u_1})(g_1^{s_2} g_2^{t_2} A^{u_2})^c \equiv XY^c \,(\mathrm{mod}\ p).$$

Similarly, $c_2 r_{c_1}^{(1)} - c_1 - r_{c_2}^{(1)} \equiv s_1(c_2 - c_1)$ and $c_2(r_{c_1}^{(2)} - c_1 r_{c_2}^{(2)}) \equiv t_1(c_2 - c_1)$, so

$$s_1 \equiv (c_2 r_{c_1}^{(1)} - c_1 - r_{c_2}^{(1)})(c_2 - c_1)^{-1}, t_1 \equiv (c_2(r_{c_1}^{(2)} - c_1 r_{c_2}^{(2)}))(c_2 - c_1)^{-1}, \quad (D.9)$$

so from (D.8) and (D.9), the bank can calculate s_1, s_2, t_1, t_2, and thereby

$$e_1 z_1 \equiv s_1 + s_2 \text{ and } e_2 z_1 \equiv t_1 + t_2. \tag{D.10}$$

Lastly, since

$$r_{c_1}^{(3)} - r_{c_2}^{(3)} \equiv u_2(c_1 - c_2) \text{ and } c_2 r_{c_1}^{(3)} - c_1 r_{c_2}^{(3)} \equiv u_1(c_2 - c_1),$$

then

$$u_2 \equiv (r_{c_1}^{(3)} - r_{c_2}^{(3)})(c_1 - c_2)^{-1} \text{ and } u_1 \equiv (c_2 r_{c_1}^{(3)} - c_1 r_{c_2}^{(3)})(c_2 - c_1)^{-1},$$

from which the bank computes $z_1 \equiv u_1 + u_2$. Hence, from (D.10), the bank can compute e_1, e_2, and so $A \equiv g_1^{e_1} g_2^{e_2} \pmod{p}$, which identifies Alice, who is charged with fraud.

If Alice does not try to double-spend and is indeed legitimate, her identity is not revealed. Thus, Brands' scheme provides anonymity to legitimate entities since Alice never has to provide identification, as is the case with paper money. As with the ECash scheme, Brands' scheme also ensures untraceability of legitimate entities. However, as proved above, the bank can identify a double-spender. Brands' scheme possesses authenticity since the scheme is secure against impersonation due to the fact that it is based upon the intractability of the DLP (see page 164).

One of the major advantages of Brands' method is that it does not use any cut-and-choose protocol or secret splitting (see Section 5.5), because the time costs are excessive. Thus, with Brands' scheme, the bank does not have to engage in such protocols. Moreover, since Brands' scheme is based upon the DLP, then the integer factoring problem does not come into play as it does with the use of an RSA modulus, used in the ECash scheme. Now we look at the parameters involved in Brands' method.

Since g_1, g_2 are made public, and $A \equiv g_1^{e_1} g_2^{e_2} \pmod{p}$, then g_1, g_2 must be chosen large enough to make it computationally infeasible for an adversary to compute a representation of Alice's account. Nevertheless, the bank must be able to accommodate all its customers with the pairs (e_1, e_2), so the bank has to ensure that g_1, g_2 are not chosen so large as to prevent this. The exponents e_1, e_2 are in $(\mathbb{Z}/q\mathbb{Z})^*$ and Brands suggests that q should have 140 bits while e_1, e_2 should be around 70 bits. As usual, the system is only as secure as the implementation and security of the private/secret keys.

Although Brands' scheme is relatively complicated mathematically, most of the work is required to preserve both anonymity and to prevent double-spending. Given the above advantages, the consensus is that Brands' scheme is preferable to the ECash scheme in most implementations.

D.6 Radix-64 Encoding

As we saw on pages 277 and 288, both PGP and S/MIME use radix-64 encoding techniques in their execution. Radix-64 is a data encoding scheme (see page 433), consisting of base-64 encoded data with a 24-bit *cyclic redundancy check* (CRC) appended to it, as specified in RFC2440 (see [213]), and see the discussion on page 549. This is necessary to accommodate restrictions in many email systems that only permit the use of blocks consisting of ASCII text. In essence, the radix-64 conversion, also called *ASCII armor*, may be viewed as a wrapper put on the binary message for transmission over nonbinary email channels.

Table D.1 presents the character set of 65 printable characters, one of which, the = sign, is used for padding. However, in order for radix-64 encoded data to travel through mail-handling systems, there are no control characters for such systems to detect when scanned, which results in a text file that is secure against alterations made by email systems. Since one character is used for padding, there are $2^6 = 64$ characters to be employed for representation, so that each character may be used to represent 6 bits of input data. In fact, this is from where "radix-64" is derived since a six-bit number has 64 combinations. We represent the 6-bit input data in their decimal value form for convenience in the table, while the character encodings are represented by upper- and lower-case English alphabet letters, together with the integers 0 through 9, and the symbols +, /, and lastly = for padding.

Radix-64 Conversion
Table D.1

6-*bit Input*	0	1	2	3	4	5	6	7	8	9	10
Encoding	A	B	C	D	E	F	G	H	I	J	K
6-*bit Input*	11	12	13	14	15	16	17	18	19	20	21
Encoding	L	M	N	O	P	Q	R	S	T	U	V
6-*bit Input*	22	23	24	25	26	27	28	29	30	31	32
Encoding	W	X	Y	Z	a	b	c	d	e	f	g
6-*bit Input*	33	34	35	36	37	38	39	40	41	42	43
Encoding	h	i	j	k	l	m	n	o	p	q	r
6-*bit Input*	44	45	46	47	48	49	50	51	52	53	54
Encoding	s	t	u	v	w	x	y	z	0	1	2
6-*bit Input*	55	56	57	58	59	60	61	62	63	PAD	
Encoding	3	4	5	6	7	8	9	+	/	=	

The radix-64 encoding is a mapping denoted by f_{64} acting on 6-bit inputs that are grouped into blocks that are mapped to 32-bit blocks. Each of the four 6-bit input values is mapped to an 8-bit character. In essence, this means that three bytes are mapped to four printable characters. This is illustrated in Diagram D.1.

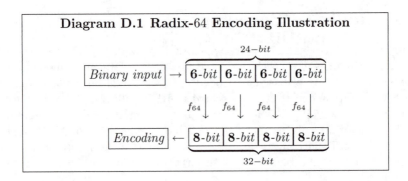

Example D.4 *For instance, suppose that the text for encoding consists of the three bytes* 01010000, 00100001, *and* 10000000, *which are put into four* 6-*bit input values:* 010100, 000010, 000110, *and* 000000, *whose decimal representations are:* 20, 2, 5, *and* 0. *Looking at Table D.1, we get the radix-64 encodings as: UCFA.*

The radix-64 conversion also appends a CRC for the purpose of detecting transmission errors. Essentially this is a *checksum,* meaning a value computed to check the validity of a data transmission, usually by detecting transmission errors. In the case of the armor checksum, a 24-bit CRC is converted to four bytes of radix-64 encoding that is prepended by an = sign to the four-byte code. For the actual mechanism by which this is done, the reader may consult [213].

Appendix E: Probability Theory

We need concepts that are basic to probability at certain points in the text. In particular, to understand entropy in Chapter 11, we must have some understanding of these fundamental concepts. In general, it will assist us throughout the text to be familiar with certain probabilistic tools. For the following, the reader will have to be familiar with the set theory in Appendix A (see Section A.1 on pages 466–468).

E.1 Basic Probability

We have already encountered the symbol that makes it possible to count the number of ways of arranging k objects from a set of n elements. This is given by the binomial coefficient presented in Definition A.14 on page 473 in Appendix A. Now we show this is tied in with the notion of "probability", which we now define.

Suppose we have an *experiment*, S, such as the flipping of a coin, with possible outcomes in a set $\mathcal{S}$. For instance if S is the flipping of a coin, then $\mathcal{S}$ would be the set consisting of $\{heads, tails\}$. Each outcome in $\mathcal{S}$ is assigned a *probability* that is a mapping,

$$p : \mathcal{P}(S) \mapsto \mathcal{S},$$

with real values $0 \leq p_s \leq 1$ for each $s \in S$ and with

$$\sum_{s \in S} p_s = 1,$$

where $\mathcal{P}(S)$ is the power set of S. (In the example of flipping a fair coin, $p_{heads} = p_{tails} = 1/2$.) Furthermore, another property must be satisfied with respect to probabilities of subsets of S. First of all, we must have that $p_S = 1$, called a *certain outcome*, and $p_\phi = 0$, called an *impossible outcome*, where ϕ is the empty set; and if $\{S_j\}_{j=1}^n$ is a collection of pairwise disjoint subsets of S, then

$$p_{\cup_{j=1}^n S_j} = \sum_{j=1}^n p_{S_j}.$$

Now we may look at examples that bring in the binomial coefficient. Suppose that we wish to engage in the experiment of tossing a coin a dozen times, and we want to know the probability that a tail will come up seven times, an outcome we will label t_7. This is related to the number of ways of choosing 7 objects from 12, which is

$$\binom{12}{7} = \frac{12!}{7!5!} = 792,$$

but this is not the probability, which is given by

$$p_{t_7} = \frac{792}{2^{12}} = 792/4096 \approx \frac{1}{5},$$

the ratio of the number of ways of success at getting 7 out of 12 tails divided by the total number of possible outcomes.

In a more general scenario, we could define probability to be a map from a *subset* of the power set satisfying certain closure properties. However, to keep the description simple, we will stick with the power set, which we call the *sample space*, and the subsets of the power set are called *events*, as well as *outcomes*.

Now suppose that we have two experiments S and T with random events $\mathcal{S}$ and $\mathcal{T}$. Then we can put them together and speak about the *joint random events* $\mathcal{U} = (\mathcal{S}, \mathcal{T})$. For instance, suppose that we have a standard deck of 52 cards and S consists of the event *the value of the card*, while $\mathcal{T}$ consists of the event *the suit of the card*. Then $\mathcal{U} = (\mathcal{S}, \mathcal{T})$ represents all the possibilities of the 52 outcomes of choosing a card. If $p_{s,t}$ represents the probability that a card is drawn with value s and suit t, then given a fair deck with $p_s = 1/13$, $p_t = 1/4$, and $p_{s,t} = 1/52$.

In general, we let $p_{s,t}$ denote the probability that $s \in \mathcal{S}$ and $t \in \mathcal{T}$ both occur. It follows that

$$p_s = \sum_{t \in \mathcal{T}} p_{s,t},$$

so the probability of a fixed $s \in \mathcal{S}$ occurring is the sum of all the probabilities of $t \in \mathcal{T}$ occurring along with $s \in \mathcal{S}$ occurring.

Independence

The two random events $s \in \mathcal{S}$ and $t \in \mathcal{T}$, are called *independent* if

$$p_{s,t} = p_s \cdot p_t. \tag{E.1}$$

For instance, in the deck of cards illustration above, the suit and value of the cards are independent events.

Conditional Probability

If we know that event $s \in \mathcal{S}$ has occurred, then the probability that $t \in \mathcal{T}$ will occur *given* that $p(s) > 0$, is defined as follows:

$$p_{t|s} = \frac{p_{s,t}}{p_s}, \tag{E.2}$$

called the *conditional probability of t given s*.

Notice that if we combine (E.1)–(E.2), we get that

s and t are independent events if and only if $p_{t|s} = p_t.$

In other words, the probability that t occurs is unaffected by the probability that s occurs.

In what follows, we assume that s and s' are event subsets of $\mathcal{S}$.

Probability Rules

The Difference Rule

If $s \subseteq s'$, then $p_{s' \smallsetminus s} = p_{s'} - p_s$.

The Sum Rule

$$p_{s \cup s'} = p_s + p'_s - p_{s \cap s'}.$$

The Product Rule

If $p_s > 0$, then $p_{s' \cap s} = p_{s'|s} \cdot p_s$.

Moreover, if s' and s are independent, then $p_{s' \cap s} = p_{s'} \cdot p_s$.

There is a well-known result that is merely the putting together of some of the above facts.

Baye's Theorem

If $s \in \mathcal{S}$ and $t \in \mathcal{T}$ are events such that $p_{\mathcal{S}}(s) > 0$ and $p_t > 0$, then

$$p_{s|t} = \frac{p_s \cdot p_{t|s}}{p_t}.$$

Baye's theorem allows us to formulate the conditional probability of s given t in terms of the conditional probability of t given s. This is a valuable tool in Chapter 11, when we talk about entropy.

Another question of importance in probability is: What is the probability that after n trials of some experiment at least two of the outcomes are the same? For instance, see the birthday attack on page 252.

◆ Probability of Two Outcomes Being the Same

Let S be an experiment that is the choice of an element from S (without removing it from S), with outcomes $\mathcal{S} = \{s_1, s_2, \ldots, s_m\}$ having equal probabilities $p_{s_j} = 1/m$ for all $j = 1, 2, \ldots, m$. Then the probability of two outcomes being the same after n trials is at least

$$1 - e^{-(n-1)n/(2m)}.$$

We see from this fact that if $n > \sqrt{2m \ln 2}$, then the probability that at least two outcomes will be the same is at least 50%. (See page 254 for a comparison with the birthday attack.)

E.2　Randomness, Expectation, and Variance

There is another probabilistic notion that will be used is the notion of "expectation".

A real-valued function $X : S \mapsto \mathbb{R}$ of a set $S = \{s_1, s_2, \dots, s_n\}$ is called a *random variable*. For simplicity, we assume that the random variables take on only finitely many values.

Expectation

If the probabilities of S are given by $p_{s_j} = p_j$ for $j = 1, 2, \dots, n$, then the *expected value* of X is given by

$$\mathcal{E}(X) = p_1 \cdot X(s_1) + p_2 \cdot X(s_2) + \cdots + p_n \cdot X(s_n).$$

Moreover, the *average value*, which will be "close to" $\mathcal{E}(X)$, is given by looking at a large number of independent trials N, say, with outcomes $s_{j_1}, s_{j_2}, \dots, s_{j_N}$, for sufficiently large N:

$$\frac{X(s_{j_1}) + X(s_{j_2}) + \cdots + X(s_{j_N})}{N}.$$

Variance

The *variance* of X is defined by

$$\mathrm{var}(X) = \mathcal{E}((X - \mathcal{E}(X))^2).$$

The square root of the variance is called the *standard deviation* of X. The following are central results on the notion of expectation and variance.

The Expectation Theorem

If X, Y are random variables and $a, b \in \mathbb{R}$ then

$$\mathcal{E}(aX + bY) = a\mathcal{E}(X) + b\mathcal{E}(Y),$$

and if X and Y are independent, then

$$\mathcal{E}(XY) = \mathcal{E}(X) \cdot \mathcal{E}(Y).$$

In order to state the next result, we need a notion related to variance.

Covariance

If X and Y are random variables, the *covariance* is defined by

$$\mathrm{covar}(X, Y) = \mathcal{E}\left[(X - \mathcal{E}(X)) \cdot (Y - \mathcal{E}(Y))\right].$$

The Variance Theorem

If X and Y are random variables, and $a, b \in \mathbb{R}$, then

$$\mathrm{var}(aX + bY) = a^2 \cdot \mathrm{var}(X) + b^2 \cdot \mathrm{var}(Y) + 2ab \cdot \mathrm{covar}(X, Y),$$

and if X and Y are independent, then

$$\mathrm{var}(X + Y) = \mathrm{var}(X) + \mathrm{var}(Y).$$

E.3 Binomial Distribution

An important notion that links the binomial coefficient with the notion of expectation and variance is the following.

The Binomial Distribution Theorem

If $s \in \mathcal{S}$, $p_s = p$, and $B_n(s)$ is the number of occurrences of s in n independent trials, then

1. $B_n(s) \in \{0, 1, \ldots, n\}$, and for any nonnegative integer $k \leq n$, the probability that $B_n(s) = k$ is given by

$$\binom{n}{k} p^k (1-p)^{n-k},$$

2. $\mathcal{E}(B_n(s)) = n \cdot p$,

3. $\mathcal{E}(B_n(s)/n) = p$, where

$$F_n(s) = \frac{B_n(s)}{n}$$

   is a random variable, called the nth *relative frequency of s*.

3. $\mathrm{var}(B_n(s)) = n \cdot p \cdot (1-p)$,

4. $\mathrm{var}(F_n(s)) = p \cdot (1-p)/n$.

See the discussion of attacks on RSA on pages 176 and 177 for an application of the above notions, as well as the references to expectation throughout the text. The above is also of value when discussing such phenomena as the birthday attack (see pages 252–255).

E.4   The Law of Large Numbers

In the theory of probability, there exist numerous versions of what is called the "law of large numbers". What they all essentially say is that if an experiment is performed n times for sufficiently large n, then the difference between the expected and actual values is very small. One way of describing this mathematically is the following.

Law of Large Numbers

If $X_1, X_2, \ldots, X_n$ are independent random variables, and $X = \sum_{j=1}^{n} X_j$, then for any $\varepsilon > 0$,

$$p(|X/n - \mathcal{E}(X/n)| \geq \varepsilon)) \leq \frac{\sum_{j=1}^{n} \mathrm{var}(X_j)}{n^2 \cdot \varepsilon^2},$$

where $p(|X/n - \mathcal{E}(X/n)| \geq \varepsilon)$ is the probability that $|X/n - \mathcal{E}(X/n)| \geq \varepsilon$.

The law of large numbers can be illustrated using the binomial distribution theorem. Let $B_n(s)$ denote the number of heads in n independent coin-flipping trials. Then $p = p_s = 1/2$, $\mathcal{E}(B_n(s)) = n/2$, and $\mathrm{var}(B_n(s)) = n/4$, so if $X = B_n(s)$, then $\mathrm{var}(X) = \sum_{j=1}^{n} X_j = n/4$, and

$$p(|X/n - \mathcal{E}(X/n)| > \varepsilon)) \leq \frac{1}{4n \cdot \varepsilon^2},$$

so as $n \to \infty$, the probability goes to zero. In other words, we may expect that the number of heads will not be far from $n/2$ if the experiment is performed enough times.

E.5   Probability and Error Detection

We conclude this appendix with some data concerning probability and error detection. We are all familiar with power surges and other electro magnetic disruptions. These types of interference may cause transmission errors, that is, the loss, alteration, or insertion of data. Thus, we need mechanisms for detecting when this occurs.

A simple mechanism for error detection is called *parity checking*, which involves the sender's computation of an additional bit, called a *parity bit*, which is attached to each character before sending. The receiver removes the parity bit, and executes the same computation as the sender to verify that the computation agrees with the value of the parity bit. For instance, if *odd* parity is chosen and agreed upon by receiver and sender, then the sender selects a parity bit that will make the total number 1-bits odd. For each character, the receiver computes the parity to ensure it is the same as the sender's computation. If not, an error has been detected. However, if an even number of bits have been altered, then the parity check cannot detect the errors since the total number of 1-bits remains the same. This and other error detection methods, which we will discuss, are subject to some such disadvantage. The need is to reduce the *probability* of the receiver's acceptance of transmission errors.

On pages 320 and 542, we made reference to the notion of a checksum, which helps to detect errors in transmission. The means by which this is accomplished is for the sender to view the data as a sequence of binary integers and compute their sum. This may be a 16- or 32-bit checksum, often built into many networks. Thus, there is ease of computation. However, checksums do not detect all common errors such as simple bit reversal in some packets. For instance, if the last bit in every packet is reversed, then the checksum will remain the same.

One mechanism for error detection that is superior to each of the above is the *cyclic redundancy check* (CRC), which we mentioned on page 541 in Appendix D. The mechanism for computing a CRC is a shift register, which we discussed on page 155, together with addition modulo 2. First, all values in the shift registers are initialized to 0. Then the bits of the message are shifted one at a time until the entire message has been processed into the shift register unit. The receiver uses exactly the same shift register unit to calculate the CRC for the message and to verify its agreement with the CRC transmitted by the sender.

A typical CRC is 16-bit, called CRC-16, where the sender appends an additional sixteen zeros to the message. Then the receiver computes a CRC over the transmitted message together with the transmitted CRC. If there are no errors, the computed value will be zero. As seen in Chapter 11, a mathematical means for representing a message is the use of a binary polynomial. For example,

$$f(x) = 1 + x^5 + x^{12} + x^{16}$$

might be used in CRC-16. Then an n-bit message would be represented by a binary polynomial $d(x)$ of degree $n-1$, and the CRC value corresponding to $d(x)$ is the 16-bit word represented by the polynomial,

$$r(x) = x^{16} \cdot d(x)/f(x).$$

It is a provable fact that CRCs detect more errors than checksums. For instance, errors involving alterations to a small number of bits near one location are called *burst errors*. Such errors are often caused by lightning for instance, so detecting these is an important exercise. It can be shown that CRC-16 can detect all burst errors of bitlength no more than 16, and more than 99% of burst errors of greater bitlength. The downside is that CRCs are more difficult to compute than checksums or parity checks. Yet, a CRC can be implemented with minimal cost, so it remains the error detection mechanism of choice.

Appendix F: Recognizing Primes

F.1   Primality and Compositeness Tests

The methods in Appendix C for factoring might be considered a means for recognizing primes if an attempt to factor n fails. Of course, we need only look at odd integers, so for instance, we could simply trial divide n by all odd integers between 2 and $\sqrt{n}$. However, if n is a 100-digit integer, say, then this would take longer than the life of the universe!

In general, factorization methods are quite time consuming, whereas deciding whether a given n is composite or prime is much more efficient. Part of the reason is that a test for recognizing primes, which are not attempts to factor n, and which determine n to be composite, do not provide the factors of n. Two tests for recognizing primes are given as follows.

(1) The test has a condition for compositeness. If n satisfies the condition, then n *must* be composite. If n fails the condition, it might still be composite (with low probability). In other words, a successful completion of the test — satisfying the condition — always guarantees that n is composite; whereas an unsuccessful completion of the test — failing to satisfy the condition — does *not* prove that n is prime.

(2) The test has a condition for primality. If n satisfies the condition, then n *must must* be prime. If it fails the condition, then n *must* be composite.

We call (1) a *compositeness test* and (2) a *primality test*.

Factorization methods may be used as compositeness tests, but quite expensive ones, as we discussed above. Thus, they may be used only as compositeness tests, and only to find very small factors.

Primality tests, as we have defined them, are sometimes called *primality proofs*, which are typically either complicated to apply or else are applicable only to special numbers such as Fermat numbers, those of the form $2^{2^n} + 1$. In other words, they sacrifice either speed or generality, but always provide a correct answer without failure. We look at one, which employs the converse of Fermat's little theorem (see Corollary A.2 on page 479). The following is attributable to D.H. Lehmer, M. Kraitchik, and others (see page 511 in Appendix C).

◆ **Primality Test via the Converse of Fermat's Little Theorem**

Suppose that $n \in \mathbb{N}$ with $n \geq 3$. Then n is prime if and only if there exists an $m \in \mathbb{N}$ such that $m^{n-1} \equiv 1 \pmod{n}$, but $m^{(n-1)/q} \not\equiv 1 \pmod{n}$ for any prime $q \mid (n-1)$.

A major pitfall with the above primality test is that we must have knowledge of a factorization of $n - 1$, so it works well on special numbers such as Fermat numbers, for instance. However, the above is a general "proof" that n is prime since the test finds an element of order $n - 1$ in $(\mathbb{Z}/n\mathbb{Z})^*$, namely a primitive root modulo n (see page 480). Furthermore, it can be demonstrated that if we have a factorization of $n - 1$ and n is prime, then the above primality test can be employed to prove that n is prime in polynomial time; but if n is composite the algorithm will run without bound, or *diverge*.

Primality proofs sacrifice one of speed or generality. For instance, there is the Lucas-Lehmer test for Mersenne numbers, Pocklington's theorem, Proth's theorem, and Pepin's theorem, all of which the reader may see in detail by consulting [169, Chapter 4, pp. 180–184], for instance.

We will be concerned herein with tests that are used in practice. Usually, these are tests that are simple, generally applicable, and efficient, but unlike the aforementioned tests, they sometimes fail. These are the compositeness tests. Note that in a compositeness test, failure of the *test* means that n does not satisfy the condition for compositeness *and* n is composite. In other words, failure results in a composite number being indicated as a prime, but never is a prime indicated as a composite number. The reason is that the condition is a "proof of compositeness" in the sense that if the condition is satisfied, n is forced to be composite. However, the converse is false. Composite numbers may fail to satisfy the condition. We may again employ the converse of Fermat's little theorem as an illustration.

◆ **Fermat's Little Theorem as a Compositeness Test**

If $n \in \mathbb{N}$, $a \in \mathbb{Z}$ with $\gcd(a, n) = 1$, and

$$a^{n-1} \not\equiv 1 \,(\mathrm{mod}\ n), \tag{F.1}$$

then n is composite.

Note that any n satisfying condition (F.1) *must* be composite by Fermat's little theorem. An application of the above is an interpretation of Lucas' test as a compositeness test by letting n be odd and $a = 2$. See [46] for a recent article by John Brillhart on this famous test by Lucas. He argues that a primality test is an algorithm "whose steps verify the hypothesis of a theorem whose conclusion is "n is prime." which is consistent with our definition.

If n fails condition (F.1), then this is *not* a proof that n is prime. For instance, $2^{340} \equiv 1 \,(\mathrm{mod}\ 341)$, yet $341 = 11 \cdot 31$. Such numbers that fail condition (F.1), and are composite are called *pseudoprimes*. In fact, there are composite numbers for which the choice of the base a is irrelevant in the sense that they will *always* fail the test. For instance, $a^{541} \equiv a \,(\mathrm{mod}\ 541)$ for any $a \in \mathbb{Z}$, yet $561 = 3 \cdot 11 \cdot 17$. This is an example of a *Carmichael number* or *absolute pseudoprime*. Moreover, there are known to be infinitely many Carmichael numbers (see [7]). This shows, in the extreme, that Fermat's little theorem may *not be used as a primality test*. However, the following is a well-known and utilized primality test. The presentation is taken from [170].

F.2   Miller-Selfridge-Rabin

◆ **The Miller[F.1]-Selfridge[F.2]-Rabin [F.3] Primality Test**

Let $n - 1 = 2^t m$ where $m \in \mathbb{N}$ is odd and $t \in \mathbb{N}$. The value n is the input to be tested by executing the following steps, where all modular exponentiations are done using the repeated squaring method described on page 171.

(1) Choose a random integer a with $2 \le a \le n - 2$.

(2) Compute
$$x \equiv a^m \pmod{n}.$$

If
$$x \equiv \pm 1 \pmod{n},$$

then terminate the algorithm with

"n is probably prime".

If $t = 1$, terminate the algorithm with

"n is definitely composite."

Otherwise, set $j = 1$ and go to step (3).

(3) Compute
$$x \equiv a^{2^j m} \pmod{n}.$$

If $x \equiv 1 \pmod{n}$, then terminate the algorithm with

"n is definitely composite."

If $x \equiv -1 \pmod{n}$, terminate the algorithm with

"n is probably prime."

Otherwise set $j = j + 1$ and go to step (4).

[F.1]Gary Miller obtained his Ph.D. in computer science from U.C. Berkeley in 1974. He is currently a professor in computer science at Carnegie-Mellon University. His expertise lies in computer algorithms.

[F.2]This test is most often called the *Miller-Rabin Test* in the literature. However, John Selfridge was using the test in 1974 before Miller first published the result, so we credit Selfridge here with this recognition. John Selfridge was born in Ketchican, Alaska, on February 17, 1927. He received his doctorate from U.C.L.A. in August of 1958, and became a professor at Pennsylvania State University six years later. He is a pioneer in computational number theory.

[F.3]Michael Rabin (1931–) was born in Breslau, Germany (now Wroclaw, Poland), in 1931. In 1956, he obtained his Ph.D. from Princeton University where he later taught. In 1958, he moved to the Hebrew University in Jerusalem. He is known for his seminal work in establishing a rigorous mathematical foundation for finite automata theory. For such achievements, he was co-recipient of the 1976 Turing award, along with Dana S. Scott. He now divides his time between positions at Harvard and the Hebrew University in Jerusalem.

(4) If $j = t - 1$, then go to step (5). Otherwise, go to step (3).

(5) Compute
$$x \equiv a^{2^{t-1}m} \,(\mathrm{mod}\ n).$$
If $x \not\equiv -1 \,(\mathrm{mod}\ n)$, then terminate the algorithm with

"n is definitely composite."

If $x \equiv -1 \,(\mathrm{mod}\ n)$, then terminate the algorithm with

"n is probably prime."

If n is declared to be "probably prime" with base a by the Miller-Selfridge-Rabin test, then

n is said to be a *strong pseudoprime to base a.*

Thus, the above test is often called the *strong pseudoprime test* [F.4] in the literature. The set of all pseudoprimes to base a is denoted by $\mathrm{spsp}(a)$.

Let us look a little closer at the above test to see why it it is possible to declare that "n is definitely composite" in step (3). If $x \equiv 1 \,(\mathrm{mod}\ n)$ in step (3), then for some j with $1 \le j < t - 1$:

$$a^{2^j m} \equiv 1 \,(\mathrm{mod}\ n),\ \text{but}\ a^{2^{j-1}m} \not\equiv \pm 1 \,(\mathrm{mod}\ n).$$

Thus, it can be shown that $\gcd(a^{2^{j-1}m} - 1, n)$ is a nontrivial factor of n. Hence, if the Miller-Selfridge-Rabin test declares in step (3) that "n is definitely composite", then indeed it is. Another way of saying this is that if n is prime, then Miller-Selfridge-Rabin will declare it to be so. However, if n is composite, then it can be shown that the test fails to recognize n as composite with probability at most $(1/4)$. This is why the most we can say is that "n is probably prime". However, if we perform the test r times for r large enough, this probability $(1/4)^r$ can be brought arbitrarily close to zero. Moreover, at least in practice, using the test with a single choice of a base a is usually sufficient.

Also, in step (5), notice that we have not mentioned the possibility that

$$a^{2^{t-1}m} \equiv 1 \,(\mathrm{mod}\ n)$$

specifically. However, if this did occur, then that means that in step (3), we would have determined that

$$a^{2^{t-2}m} \not\equiv \pm 1 \,(\mathrm{mod}\ n),$$

from which it follows that n cannot be prime. Furthermore, by the above method, we can factor n since $\gcd(a^{2^{t-2}m} - 1, n)$ is a nontrivial factor. This

---

[F.4]The term "strong pseudoprime" was introduced by Selfridge in the mid-1970s, but he did not publish this reference. However, it did appear in a paper by Williams [289] in 1978.

final step (4) is required since, if we get to $j = t - 1$, with $x \not\equiv \pm 1 \pmod{n}$ for any $j < t - 1$, then simply invoking step (3) again would dismiss those values of $x \not\equiv \pm 1 \pmod{n}$, and this would not allow us to claim that n is composite in those cases. Hence, it allows for more values of n to be deemed composite, with certainty, than if we merely performed step (3) as with previous values of j.

The Miller-Selfridge-Rabin test is an example of a *Monte Carlo* algorithm meaning a probabilistic algorithm[F.5] that achieves a correct answer more than 50% of the time. More specifically, Miller-Selfridge-Rabin is a Monte Carlo algorithm for compositeness, since it provides a proof that a given input is composite, but only provides some probabilistic evidence of primality. Furthermore, Miller-Selfridge-Rabin is a *yes-biased* Monte Carlo algorithm meaning that a "yes" answer is always correct but a "no" answer may be incorrect. In this case, the answer is to the decision problem:[F.6] "Is n composite?" A yes-biased Monte Carlo algorithm is said to have *error probability* $\alpha \in \mathbb{R}^+$ with $0 \le \alpha < 1$, provided that for any occurrence in which the answer is "yes", the algorithm will give the incorrect answer "no" with probability at most α, where the probability is computed over all possible random choices made by the algorithm for a given input. Therefore, the Miller-Selfridge-Rabin algorithm is a yes-biased Monte Carlo algorithm for the decision problem "Is n composite?" with error probability $\alpha = (1/4)^r$.

There are many related algorithm that we have not discussed here, such as the Solovay-Strassen test, because the Miller-Selfridge-Rabin test is computationally less expensive, easier to implement, and is at least as correct. For information on such tests, the reader may consult [170, pp. 84–86], for instance.

There are several tests that are beyond the scope of this book. Nevertheless, they are worth mentioning as a segue to the next section. Among them is one using Artin symbols, described by Lenstra in [147], which is also given a presentation in [168, Section 4.5, pp. 264–270]. There have also been proofs of the existence of a deterministic polynomial time algorithm for primality testing under the assumption of the Extended Riemann Hypothesis (ERH). The ERH has has not yet been verified, but is widely believed to be true (see [162]). There is also the Goldwasser-Kilian test presented in [108], which is based on elliptic curves. The Goldwasser-Killian test was motivated by a desire to prove that, at least theoretically, it is possible to find a polynomial time primality testing algorithm. Using Schoof's algorithm [243], this can be done, but the procedure is impractical to implement and run. The idea was modified by Adleman and Huang in [5], who presented a randomized algorithm that runs in expected polynomial time on all inputs, as opposed to the restrictions in the Goldwasser-Killian algorithm. There were other advances, but the goal of actually finding an unconditional deterministic polynomial time algorithm for primality testing has only recently been solved, and we present it in the next section. (For a history of primality testing, see [290].)

[F.5]See the discussion of randomized algorithms on page 500 in Appendix A, since probabilistic algorithms use random numbers.

[F.6]See page 502.

F.3 Primes is in P

The following is an unconditional deterministic polynomial-time algorithm for primality testing presented in [6] by M. Agrawal, N. Kayal, and N. Saxena. For notation in what follows, see Appendix A, especially Definition A.22 on page 479 and Definition A.23 on page 479, as well as results on polynomial rings especially as they pertain to finite fields on pages 484–490.

In what follows, $\mathbb{Z}_n$ for a given integer $n > 1$ denotes $\mathbb{Z}/n\mathbb{Z}$, and if $h(X) \in \mathbb{Z}_n[X]$, then the notation,

$$f(X) \equiv g(X) \, (\text{mod } h(X), n)),$$

is used to represent the equation $f(X) = g(X)$ in the quotient ring $\mathbb{Z}_n[X]/(h(X))$. In particular, for suitably chosen r and a, values, we will be looking at equation of the following type:

$$(X + a)^n \equiv X^n + a \, (\text{mod } X^r - 1, n). \tag{F.2}$$

Algorithm F.1 —

Unconditional Deterministic Polynomial-Time Primality Test

Input an integer $n > 1$, and execute the following steps.

1. If $n = a^b$ for some $a \in \mathbb{N}$ and $b > 1$, then terminate with output

   "n is composite".

2. Find the smallest $r \in \mathbb{N}$ such that $\text{ord}_r(n) > 4\log_2^2 n$.

3. If $1 < \gcd(a, n) < n$ for some $a \leq r$, then output

   "n is composite".

4. If $n \leq r$, then output
   "n is prime".

5. Set $a = 1$ and execute the following:

   (i) Compute $Y(a) \equiv (X + a)^n - X^n - a \, (\text{mod } X^r - 1, n)$.

   (ii) If $Y(a) \not\equiv 0 \, (\text{mod } X^r - 1, n)$, output

   "n is composite".

   Otherwise, go to step (iii).

   (iii) If $Y(a) \equiv 0 \, (\text{mod } X^r - 1, n)$, set $a = a + 1$. If $a < \lfloor 2\sqrt{\phi(r)} \cdot \log_2(n) \rfloor$, go to step (i). Otherwise, go to step 6.

6. Output
   "n is prime".

◆ **Analysis**

The reason the authors of [6] considered equations of type (F.2) was that they were able to prove the following.

Polynomial Primality Criterion

If $a \in \mathbb{Z}$, $n \in \mathbb{N}$ with $n > 1$, and $\gcd(a, n) = 1$, then n is prime if and only if

$$(X + a)^n \equiv X^n + a \,(\mathrm{mod}\ n). \qquad (\text{F.3})$$

The satisfaction of polynomial congruence (F.3) is a simple test but the time taken to test the congruence is too expensive. To save time, they looked at the congruence modulo a polynomial, whence congruence (F.2). However, by looking at such congruences, they introduced the possibility that composite numbers might satisfy (F.2), which indeed they do. Yet, the authors were able to (nearly) restore the characterization given in the above polynomial primality criterion by showing that for a suitably chosen r, if (F.2) is satisfied for several values of a, then n must be a prime power. Since the number of a values and the suitably chosen r value are bounded by a polynomial in $\log_2(n)$, they achieved a deterministic polynomial time algorithm for primality testing.

The authors of [6] were able to to establish the following facts about their algorithm. The reader will need the concepts of ceiling and floor functions (see pages 473 and 474 in Appendix A).

Facts Concerning Algorithm F.1

1. The algorithm outputs "n is prime" if and only if n is prime. (Hence, it outputs "n is composite" if and only if n is composite.)

2. There exists and $r \leq \lceil 16 \log_2^5(n) \rceil$ such that $\mathrm{ord}_r(n) > 4 \log_2^2(n)$.

3. The asymptotic time complexity of the algorithm is $O(\log_2^{10.5+\varepsilon}(n))$ for any $\varepsilon > 0$.

4. It is conjectured that the time complexity of the algorithm can be improved to the best-case scenario where $r = O(\log_2^2(n))$, which would mean that the complexity of the algorithm would be

$$O(\log_2^{6+\varepsilon}(n)) \text{ for any } \varepsilon > 0.$$

Two conjectures support the authors' conjecture in part 4 above. They are given as follows.

Artin's Conjecture

If $n \in \mathbb{N}$ is not a perfect square, then the number of primes $q \leq m$ for which $\operatorname{ord}_q(n) = q - 1$ is asymptotically $A(n) \cdot m/\ln(m)$, where $A(n)$ is *Artin's constant* given by

$$A(n) = \prod_{j=1}^{\infty} \left(1 - \frac{1}{p_k(p_k - 1)}\right) = 0.3739558136\ldots,$$

with p_k being the kth prime.

If Artin's conjecture becomes effective for $m = O(\log_2^2(n))$, then it follows that there is an $r = O(\log_2^2(n))$ with the desired properties.

The other conjecture that supports their contention is given as follows.

Sophie Germane's Prime Density Conjecture

The number of primes $q \leq m$ such that $2q+1$[a] is also a prime is asymptotically $2C_2 m/\ln^2(m)$, where C_2 is the *twin prime constant* given by

$$C_2 = \prod_{p \geq 3} \frac{p(p - 2)}{(p - 1)^2} \approx 0.6601611816\ldots.$$

[a]Such primes are called *Sophie Germane primes*.

If the Sophie Germane conjecture holds, then $r = O(\log_2^{2+\varepsilon}(n))$ for any $\varepsilon > 0$ such that $\operatorname{ord}_r(n) \geq 4\log_2^2(n)$. Hence, the algorithm, with this r value, yields a time complexity of $O(\log_2^{6+\varepsilon}(n))$ for any $\varepsilon > 0$.

The authors of [6] leave one more conjecture, the affirmative solution of which would improve the complexity of algorithm F.1 to $O(\log_2^{3+\varepsilon}(n))$ for any $\varepsilon > 0$.

Conjecture F.1 *If r is a prime not dividing $n > 1$ and if*

$$(X - 1)^n = X^n - 1 \,(\operatorname{mod} X^r - 1, n),$$

then either n is prime or $n^2 \equiv 1 \,(\operatorname{mod} r)$.

The result given in Algorithm F.1 is a major breakthrough and the simplicity of the approach is even more noteworthy given the attempts at finding such an algorithm through much more difficult techniques such as those discussed in the previous section. The algorithm uses essentially only elementary properties of polynomial rings over finite fields and a generalization of Fermat's little theorem in that context, quite impressive indeed.

The next section deals with the generation of random primes, another important feature in the recognition of primes.

F.4   Generation of Random Primes

When we talked about SRP on page 200, we discussed safe primes, p, those for which $(p-1)/2$ is also prime. Safe primes are also important in selecting an RSA modulus $n = pq$ since, if p and q are safe primes, then RSA is not vulnerable to $p-1$ and $p+1$ factoring methods discussed in Section C.3 (see page 514). In general, having safe primes in the modulus makes it more difficult to factor. However, finding such primes is also more difficult. We present the following algorithm for so doing, which is taken from [170].

◆ **Algorithm for Generating (Probable) Safe Primes**

Let b be the input bitlength of the required prime. Execute the following steps.

(1) Select a $(b-1)$-bit odd random $n \in \mathbb{N}$ and a smoothness bound B (determined experimentally).

(2) Trial divide n by primes $p \le B$. If n is divisible by any such p, go to step (1). Otherwise, go to step (3).

(3) Use the Miller-Selfridge-Rabin test on page 552 to test n for primality. If it declares that "n is probably prime", then go to step (4). Otherwise, go to step (1).

(4) Compute $2n+1 = q$ and use the Miller-Selfridge-Rabin test on q. If it declares q to be a probable prime, terminate the algorithm with q as a "probable safe prime". Otherwise go to step (1).

There are primes that have even more constraints to ensure security of the RSA modulus. They are given as follows.

Definition F.1 (Strong Primes)

A prime p is called a strong prime *if each of the following hold.*

(1) *$p-1$ has a large prime factor q.*

(2) *$p+1$ has a large prime factor r.*

(3) *$q-1$ has a large prime factor s.*

The following algorithm was initiated in [113].

◆ **Gordon's Algorithm for Generating (Probable) Strong Primes**

(1) Generate two large (probable) primes $r \ne s$ of roughly equal bitlength using the Miller-Selfridge-Rabin test.

(2) Select the first prime in the sequence $\{2js+1\}_{j \in \mathbb{N}}$, and let

$$q = 2js + 1$$

be that prime.

(3) Compute $p_0 \equiv r^{q-1} - q^{r-1} \pmod{rq}$.

(4) Find the first prime in the sequence $\{p_0 + 2iqr\}_{i\in\mathbb{N}}$, and let

$$p = p_0 + 2iqr$$

be that prime, which is a strong prime.

Although it is possible to generate primes that are both safe and strong, the algorithms are not as efficient as Gordon's algorithm. Furthermore, choosing random primes large enough will generally thwart direct factoring attacks. The following, also taken from [170], provides a mechanism for generating large random primes.

◆ Large (Probable) Prime Generation

We let b be the input bitlength of the desired prime and let B be the input smoothness bound (empirically determined). Execute the following steps.

(1) Randomly generate an odd b-bit integer n.

(2) Use trial division to test for divisibility of n by all odd primes no bigger than B. If n is so divisible, go to step (1). Otherwise go to step (3).

(3) Use the Miller-Selfridge-Rabin (MSR) to test n for primality. If it is declared to be a probable prime, then output n as such. Otherwise, go to step (1).

◆ Large (Provable) Prime Generation

Begin with a prime p_1, and execute the following steps until you have a prime of the desired size. Initialize the variable counter $j = 1$.

(1) Randomly generate a small odd integer m and form $n = 2mp_j + 1$.

(2) If $2^{n-1} \not\equiv 1 \pmod{n}$, then go to step (1). Otherwise, go to step (3).

(3) Using the primality test on page 550, with prime bases $2 \leq a \leq 23$, if for any such a,
$$a^{(n-1)/p} \not\equiv 1 \pmod{n}$$
for any prime p dividing $n - 1$, then n is prime. If n is large enough, terminate the algorithm with output n as the provable prime. Otherwise, set $n = p_{j+1}$, $j = j + 1$, and go to step (1). If the test fails go to step (1).

Note that since we have a known factorization of $n-1$ in the above algorithm, and a small value of m to check, then the test is simple and efficient.

F.5 Decision Problem or Primality Test?

On page 551, we discussed Lucas' test as an application of Fermat's compositeness criterion (F.1) (contrapositively speaking). There is, however, a brand of opinion that Lucas' test is really a decision problem (see page 502). Here is the reasoning.

Lucas' test tells us to compute $2^{n-1} \pmod{n}$. If

$$2^{n-1} \not\equiv 1 \pmod{n},$$

then we know that n is not prime and send it off to some factoring routine such as discussed in Appendix C.

If

$$2^{n-1} \equiv 1 \pmod{n},$$

then we send it off for primality testing. Hence, the Lucas test may be viewed as a decision problem on whether to send n for primality testing or factoring. Since decision problems are "yes-no" issues, we phrase the question as

Do we send n for primality testing?

If the answer is yes, we do so, and if the answer is no, we send it to a factoring algorithm.

There is merit to the above argument. Attendant with the above viewpoint is the opinion that assigning probabilities or improving such estimates has nothing to do with primality testing. Thus, this school of thought would not consider the MSR test in Section F.2 to be a test that in any way assists with practical primality testing. Given that MSR does not satisfy our criterion (2) given on page 550, this viewpoint also has some merit. That said, this does not prevent the use of such algorithms in practice.

The aforementioned point of view is presented for mental fodder and general interest in what is often a contentious topic.

Appendix G: Exercises

We do not see things as they are, but as we are.
— from The Talmud, compilations of Jewish civil ceremonial law, dating from the fifth century AD.

G.1   Chapter 1 Exercises

▼ *In Exercises 1.1–1.4, use Table 1.2 for the Caesar cipher on page 11 to decrypt the ciphertext given in each.*

1.1. **ZH EHJLO RXV MRXUQB.**

1.2. **RQ WKH LVODQG RI FUHWH.**

1.3. **ZLWK WKH PLQRDQ FLYLOLCDWLRQ.**

1.4. **D WUXOB PDJQLILFHQW SHRSOH.**

▼ *In Exercises 1.5–1.8, use Table 1.3 on page 11 to decrypt the numeric ciphertext given in each.*

1.5. $20, 18, 4, 12, 14, 3, 20, 11, 0, 17, 0, 17, 8, 19, 7, 12, 4, 19, 8, 2.$

1.6. $11, 8, 13, 4, 0, 17, 0, 17, 0, 8, 18, 0, 12, 24, 18, 19, 4, 17, 24.$

1.7. $4, 0, 18, 19, 4, 17, 8, 18, 11, 0, 13, 3, 8, 18,$
   $17, 8, 2, 7, 8, 13, 7, 8, 18, 19, 14, 17, 24.$

1.8. $3, 4, 2, 8, 15, 7, 4, 17, 19, 7, 4, 15, 7, 0, 8, 18, 19, 14, 18, 3, 8, 18, 10.$

▼ *In Exercises 1.9–1.12, use the Polybius square 1.1 on page 10 to decrypt the numeric ciphertext given in each.*

1.9. $3534315412244543 \ 521143 \ 1123151114 \ 3421 \ 232443 \ 44243215.$

1.10. $253334132543 \ 4215353111131514 \ 44344213231543.$

1.11. $4445423324332 \ 142422244443 \ 24334434 \ 433445331443.$

1.12. $33345224442443 \quad 11 \quad 1211432443 \quad 3421 \quad 32113354$
   $13243523154243.$

1.13. Using the permutation cipher displayed on page 9 decrypt the following ciphertext:

IPEROTEWMURDS.

   (*Hint: See the discussion concerning the finding of inverse permutations on page 122.*)

1.14. Using the description of atbash given on page 23, decipher the following:

WZMRVO   XIBKGZMZOBAVW   RG.

(*Hint: You may visualize the atbash methodology via the following cipher table.*)

Plain	a	b	c	d	e	f	g	h	i	j	k	l	m
Cipher	Z	Y	X	W	V	U	T	S	R	Q	P	O	N

Plain	n	o	p	q	r	s	t	u	v	w	x	y	z
Cipher	M	L	K	J	I	H	G	F	E	D	C	B	A

1.15. Using the above table, decipher the following

GSV   DRHV   NVM   XLFOW   MLG.

1.16. Consider the Alberti disk position as illustrated in Figure 1.22 on page 48. Using this position, decipher *DOLZYIB*. Then assume that the disk is rotated so that the z sits over the C, and decipher *SSIB RGBZRNPV*, keeping in mind that since there is no w on the disk, we use a double v to denote it.

▼ *In Exercises 1.17–1.20, use the Trithemius tableau on page 51 to decrypt each ciphertext.*

1.17. **MPPREQWPILPFWQ.**

1.18. **WFOYXZTWCPPMETTQI.**

This is a quote from St. Francois De Sales (1567–1622), who was Bishop of Geneva.

1.19. **YPXFESTWCCSMXSXQEWMQEQFZCMGQGNLLOSDF.**

This is a quote from Indira Gandhi (1917–1984), *who was prime minister of India until her assassination.*

1.20. **LPQNMTSMEKCPNBWUC.**

This is line 163 of "Lycidas" (1637), *written by the English poet* John Milton (1608–1674).

▼ *In Exercises 1.21–1.24, use the Bellaso polyalphabetic cipher on page 53 employing the keyphrase* quintessential *to decipher each cryptogram.*

1.21. **HZURFFXKYUTCTTCZRFFSFXC.**

This is a quote from Benjamin Franklin (1706–1790) *taken from* "Advice to Young Tradesmen" (1748). *Franklin was an American scientist, diplomat, and publisher.*

1.22. **CIANENMQNFNQEGUUSAYXLGKGBNMTDY**.

This is a quote from Arthur Rimbaud (1854–1891) *taken from* " Une Saison en enfer" (1873). *Rimbaud was a sometimes controversial French poet.*

1.23. **BZIEGNFZNFNQEGUUYBG**

ABMMIBRCSIUCNGNLSZUITEUY.

This is a quote from Sa'Dī (Musharrif-uddīn) (1184–1291) *taken from Chapter 8, maxim 44, of* "Gullistan" [Rose Garden] (circa 1258). *Sa'Dī was a Persian poet.*

1.24. **BZRFOWXONGBXUEIODQRNLWINNQ**.

This is a quote from Lucius Annaeus Seneca (*the Younger*) (circa 4 BC–65 AD) *taken from* "Epistolae Morales". *Seneca was a Spanish-born Roman statesman and philosopher.*

▼ *In Exercises 1.25–1.28, use the Vigenère cipher with the keyphrase given in Example 1.4 on page 57 to decrypt the ciphertext.*

1.25. **FTVSPUVLOJQAGWYKGRBQWMQYIHIIUSLNWSSR**.

This is a quote from Pedro Caldaron De La Barca (1600–1681) *taken from* "La Vida es Sueño" [Life is a Dream]. *He was a Spanish dramatist and poet.*

1.26. **PCVEQWWFWIFPSWDIJZWMTBTQXMMLPW**.

This is a quote from Claude Bernard (1813–1878) *taken from* "Leçons de pathologie expérimentale". *Bernard was a French scientist.*

1.27. **FFRAUNPJKRVXRWZYSYQZULOLOUHKIWBF**.

This is a quote from Virgil (70–19 BC). *Vigil, known formally as* Publius Vergilius Maro, *and was a Roman poet. See the story of Troy on pages 24 and 25 for an explanation of the reason behind the quote.*

1.28. **LFNRPMOSDKTWMNQBZMZG**.

This is a quote from Virgil (see Exercise 1.27). *This quote is the opening line of* "The Aeneid", *written in 19 BC.*

G.2 Chapter 2 Exercises

▼ *In Exercises 2.1–2.4, use the Playfair/Wheatstone cipher described on page 68 to decipher each cryptogram.*

2.1. **ZVKTCSZOFCUHTYUHVNODACUVJDGE**

This is a quote from Menander (circa 342–291 BC) *taken from* "Monosticha". *Menander was a Greek comic dramatist.*

2.2. **FEODQFZJYOYFGPYHNOTOZKAMSNDJEIOT**.

This is a quote from Arthur Schopenhauer (1788-1860) *taken from* "The World as Will and Idea", *published in 1819. Schopenhauer was a German philosopher.*

2.3. **HGGTFOUAOZFUEQKTIEOHTGNHHUZV**.

This, and the completion of it in Exercise 2.4 below, is a quote from John Fitzgerald Kennedy (1917–1963) *taken from one of his speeches. Kennedy was the 35th President of the United States until his assassination. Note that most modern-day presidents used ghost writers. Among Kennedy's writers were Theodore Sorenson, Arthur Schlesinger, Jr. (see Exercise 2.5), and John Kenneth Galbraith (see Exercise 2.6), who wrote most of Kennedy's best-known quotes.*

2.4. **RGGJUGYFOUAOZROEPUNKUYKEWZGU**,

See Exercise 2.3.

2.5. Use the keyphrase given in Example 2.2 on page 75, and employ Kasiski's method to the plaintext given below to show that the keylength is four.

<blockquote>
"The only certainty in an absolute system is

the certainty of absolute abuse."
</blockquote>

(Hint: Look for the repeated groups of ciphertext **UOVRSS** *and* **YR**. *Then compute the* gcd *of their respective distances.)*

The above is a quote from Arthur Schlesinger Jr. (1917–) *taken from an address he gave to the* Indian Council of World Affairs in 1962. *The balance of the quote is:* "Injustice and criminality are inherent in a system of totalitarian dictatorship". *Schlesinger is an American educator and historian (see also Exercise 2.3.)*

▼ *In Exercises 2.6–2.9, use the ADFGVX Field cipher method described on pages 80 and 81 to decrypt the ciphertext in each case, where the key to be used is given in each exercise.*

2.6. In this exercise use the key *SUBNETWORK*.

<blockquote>
**FAAAXDAG DFFFFFFA FXGFFFAV GXGVAFGF
VFVFFAFA XFFFVGDX VFFXFXFX AFGGGVFX
GXGFGXGA AXAFAXXX**
</blockquote>

(Hint: Put each of the above ten rows as individual columns corresponding to the numbering of the letters in the key as done in Example 2.3 on page 81.)

This is a quote from John Kenneth Galbraith (1908–) *taken from Chapter 1 of* "The Affluent Society" (1958). *Galbraith is a Canadian-born American economist. Also see Exercise 2.3.*

2.7. In this exercise use the key *IRONCLAD.*

DXXAXAF VXDFDAA AFAAXVG FFXDFAX
GXAGFVF GAFVFAX FXDXXDA XFXAVGF

This is a quote from Ben Jonson (1573–1637) *taken from* "To the Memory of My Beloved, the Author, Mr. William Shakespeare" (1623). *Jonson was an English actor, poet, and dramatist.*

2.8. In this exercise use the key *FRANCISKEY.*

AXFAA AFDXA FFFXX DFXDV VGAFX
FGFFA VAXXG AGFFX XVXXX XFFXF

This is a quote from Adolph S. Ochs (1858–1935) *that has become the slogan for the New York Times. Ochs was an American newspaper publisher.*

2.9. In this exercise use the key *FLAUBERT.*

XAXXAGFF GFFXXAAF FFVFAXGF DFFAXXXA
AXGFXXXG XAAXXFXX FAGFVFFF AXXXXFVV

This was written by Francis Scott Key (1729–1843) *and is part of* "The Star-Spangled Banner". *Key was an American poet.*

2.10. This exercise refers to Friedman's index of coincidence discussed on pages 85–87.

Calculate the index of coincidence of the following ciphertext using the formula displayed on page 86.

BAETKBESZMZIOMVWSSWYFEUKP
YEBHLNOBIQAMSXAOQFGBDPAE

2.11. Assuming the ciphertext in Exercise 2.10 was produced via the Vigenère cipher with keyword *XANADU*, find the plaintext.

This is a quote from Gustave Flaubert (1821–1880). *Flaubert was a French novelist. Compare with the comment on page 334 in parentheses at the end of part 2 in the discussion of "Token Applications".*

2.12. Show that the probability of choosing identical letters from a text in which there are equal numbers of each letter in English is ≈ 0.0385.

(Hint: The probability of choosing two of any letter is $(1/26)^2$.)

2.13. If you are told that the following ciphertext was created by a modular shift given by $c \equiv m + 7 \pmod{26}$, find the plaintext.

TVKBSHYHYPAOTLAPJPGLHZF

▼ *In Exercises 2.14–2.20, solve the given congruence for* x. Hint: To find the multiplicative inverse of the coefficient a of x in each case, find that integer y such that $ay \equiv 1 \pmod{n}$. For instance, in Exercise 2.15,

$$6 \cdot 6 \equiv 1 \pmod{7},$$

so 6 is the inverse of itself modulo 7. For the reader requiring more background on modular arithmetic, see pages 475–482 in Appendix A.

2.14. $2x + 1 \equiv 3 \,(\mathrm{mod}\ 7)$

2.15. $6x + 5 \equiv 2 \,(\mathrm{mod}\ 7)$

2.16 $17x \equiv 4 \,(\mathrm{mod}\ 26)$

2.17. $23x + 1 \equiv 5 \,(\mathrm{mod}\ 24)$

2.18. $2x - 3 \equiv 6 \,(\mathrm{mod}\ 13)$

2.19. $2x - 5 \equiv 7 \,(\mathrm{mod}\ 9)$

2.20. $5x - 2 \equiv -1 \,(\mathrm{mod}\ 6)$

▼ *In Exercises 2.21 and 2.22, find the values of a and b for which the congruences hold.* Hint: Find the value of a in terms of b from one congruence and plug it into the other.

2.21. $3a + b \equiv 10 \,(\mathrm{mod}\ 29)$ and $a + b \equiv 5 \,(\mathrm{mod}\ 29)$

2.22. $5a + 2b \equiv 1 \,(\mathrm{mod}\ 26)$ and $7a + 9b \equiv 2 \,(\mathrm{mod}\ 26)$

▼ *In Exercises 2.23–2.25, find the plaintext given that the cipehrtext is produced via the congruence $c \equiv 3m + 10 \,(\mathrm{mod}\ 26)$ where the values of m in the plaintext are given in Table 1.3 on page 11.*

2.23. **3, 9, 0, 3, 22, 9, 15, 4, 2, 8, 21, 22, 23, 10, 24, 10, 4, 15,**

   0, 25, 9, 8, 22, 23, 19, 12, 8, 12, 15, 5, 9, 0, 23, 17, 4, 14,

   8, 23, 19, 24, 8, 17, 17, 25, 0, 9, 22, 21, 22, 9, 13, 0, 18, 9, 12

   This is a quote from Martial (Marcus Valerius Martialis) (circa 40-104 AD). *taken from* "Epigrammata". *Martial was an Iberian-born Roman poet.*

2.24. **24, 5, 22, 23, 24, 10, 9, 8, 12, 19, 22, 16, 17, 10, 22,**

   19, 15, 9, 18, 15, 5, 8, 12, 15, 5, 22, 25, 8, 9, 12, 15

   This is a quote from Arthur Ponsonby (1871–1946) *taken from* "Falsehood in Wartime" (1928). *Ponsonby was an English diplomat and writer.*

2.25. **5, 10, 8, 17, 22, 20, 3, 22, 9, 22, 9, 24, 22, 24, 5, 0, 10, 9, 22,**

   10, 13, 0, 18, 15, 15, 0, 3, 22, 9, 8, 17, 0, 18, 9, 17, 8, 21,

   22, 12, 12, 10, 17, 18, 15, 22, 4, 0, 18

   This is a quote from Suetonius (Gaius Suetonius Tranquillus) (circa 70–140 AD) *taken from* [276, p. 320]. Suetonius was a Roman historian (also see page 10).

G.3 Chapter 3 Exercises

3.1. We have had some experience in solving Affine ciphers in the Exercises from Chapter 2. Now show that given any Affine transformation,

$$c \equiv am + b \,(\text{mod } n) \text{ with } \gcd(a, n) = 1,$$

we may achieve the plaintext m. In other words, demonstrate that the inverse function $m \equiv a^{-1}(c - b) \,(\text{mod } n)$ exists and explain why it does.

▼ *In Exercises 3.2–3.5, use the Hill cipher discussed on pages 111 and 112 to decrypt the given ciphertext. Assume that n, r, $\mathcal{A}$, $\mathcal{M}$, $\mathcal{C}$, and $\mathcal{K}$ are those given in Example 3.2 on page 112. The key e will be given in each exercise below. Also, letters corresponding to the numerical equivalents are given in Table 1.3 on page 11.*

3.2. The key is

$$e = \begin{pmatrix} 1 & 5 \\ 2 & 3 \end{pmatrix}$$

and the ciphertext is

HMSPUGPVEREATWUTKEPDFZDMSQIZPYOWSZOWQJ

This is a quote from the American writer, Henry David Thoreau (1817–1862).

3.3. The key is

$$e = \begin{pmatrix} 2 & 3 \\ 1 & 1 \end{pmatrix}$$

and the ciphertext is

ERIKCUJRIGVJSZWNSXVIVZUNPD

This is a quote from the English dramatist, William Wycherley (1640–1716) *taken from Act 3, Scene 3 of* "Love in a Wood" (1671).

3.4. The key is

$$e = \begin{pmatrix} 5 & 6 \\ 2 & 3 \end{pmatrix}$$

and the ciphertext is

BAISNHZJRCYSTJYYWAZSAH

This is a quote from the Aristophenes (circa 450–385 BC) *and is line 1447 of* "The Birds" (414 BC). *Aristophenes was an Athenian poet and dramatist.*

3.5. The key is

$$e = \begin{pmatrix} 7 & 1 \\ 3 & 8 \end{pmatrix}$$

and the ciphertext is

XXVLGWCOCEVHXHOBBKRXKCMJ

This is a quote from Ralph Waldo Emerson (1803–1882). *Emerson was an American poet, Unitarian minister, and philosopher.*

▼ *In Exercises 2.6–2.9, we got some practical experience with the ADFGVX cipher. What is hidden in that cipher is that it is a substitution followed by a transposition. Our experience with the Hill cipher above, as well as Exercise 1.13, provided applications of substitution ciphers. Now we get more experience with transposition/permutation ciphers, discussed in detail on pages 114 and 115. In Exercises 3.6–3.11, use the key given in Example 3.3 on page 114 to decrypt the given ciphertext.*

3.6.   **DEWYSAEGIRRDR**

Hint: Use the inverse permutation given by

$$e^{-1} = \begin{pmatrix} 1 & 2 & 3 & 4 & 5 & 6 & 7 & 8 & 9 & 10 & 11 & 12 & 13 \\ 5 & 9 & 10 & 7 & 12 & 3 & 6 & 13 & 1 & 8 & 11 & 2 & 4 \end{pmatrix}$$

3.7.   **IAESAAGDLLERT**

This and its conclusion in Exercise 3.8 is a quote from the Irish writer, Oscar Wilde (1854–1900). *Wilde's name was actually* Fingall O'Flahertie Wills.

3.8.   **RSNZAGDOREUAE**

3.9.   **FRODITNFCAONA**

This and its completion in Exercises 3.10 and 3.11 is a quote from Louis Agassiz (1807–1873), *taken from a letter in which he refused an offer to give a lecture course. Agassiz, known formally as* Jean Louis Rodolphe Agassiz *was a Swiss-born American naturalist.*

3.10.   **YMTETEATOWISM**

3.11.   **NZGZMMIEAKYNO**

▼ *Exercises 3.12–3.18 are applications of S-DES described in Section 3.2.*

3.12. Apply the initial permutation **IP**, described on page 118 to the input $m = (10101011)$.

3.13. Apply the expansion permutation described on page 119 to the input (1010).

3.14. Given a symmetric key $k = (11101000)$, employ the S-DES key generation method described on pages 119 and 120 to k in order to produce the subkeys k_1 and k_2.

(*Hint: You should get $k_1 = (10000100)$ and $k_2 = (01010110)$.*)

3.15. Apply the S-Boxes $\mathbf{S_0}$ and $\mathbf{S_1}$ to the input (1110).

(*Hint: You should get $\mathbf{S_0}(1110) = (11)$ and $\mathbf{S_1}(1110) = (00)$.*)

3.16. Given $SK = (01010110)$ and $t = (11111011)$, compute $f_{SK}(t)$, the S-DES round function described on page 121.

(*Hint: The end result is $f_{SK}(t) = (11011011)$.*)

3.17. Given input $m = (01110111)$ and key $k = (11101000)$, use the S-DES encryption steps outlined on page 122 to find the ciphertext. Then use the S-DES decryption steps outlined on pages 122 and 123 to verify that your answer is correct.

(*Hint: You have already calculated k_1 and k_2 in Exercise 3.14.*)

3.18. Prove the DES complementation property highlighted on page 127.

(*Hint: Complementation does not affect out modulo 2 addition, namely, $c(x) \oplus c(y) = x \oplus y$.*)

3.19. Verify that the deciphering method for CBC mode described on page 134 actually recovers the plaintext.

3.20. Show that for CFB mode described on page 135, the following decryption method will recover the plaintext: $m_j = c_j \oplus E_k(C_{j-1})$.

3.21. Verify that the CTR random access property, highlighted on page 137, is indeed valid.

3.22. Show that Blowfish decryption, summarized on page 141, is indeed the inverse of the algorithm's encryption technique, summarized on page 140.

3.23. Compare DES and AES from the perspective of their round functions. Also, compare them with respect to the use of confusion and diffusion.

3.24. Given a key for a one-time pad:

$$k = (1110001100100000111100001110001100100000011111000011100)$$

and ciphertext

$$c = (01000111101010000001011001001100001011100001010010110000),$$

find the plaintext string.

3.25. Assuming the bitstring found as plaintext in Exercise 3.24 is to be interpreted as a concatenation of bitstrings of length five, each corresponding
to an English letter whose numerical equivalent in decimal is given by
Table 1.3 on page 11, find the English text equivalent of that bitstring.

▼ *In Exercises 3.26–3.29, assume that $n = 26$ in the Vigenère cipher described
on page 153, and use the given key, as well as the values of r and s given
in each case, to decrypt the ciphertext. Use the numerical equivalents from
Table 1.3 on page 11 to find the English text.*

3.26. Let $r = 3$, $s = 31$ and $k = (3, 7, 9)$.

$$c = \textbf{WONIPAVAAHHLWPXQAXWYDWORVOJWYNG}.$$

This is a quote from Tertullian (circa 160-225 AD), *known formally as
Quintus Septimus Florens Tertullianus. He was a Carthagian-born Latin
church father. The quote is taken from* "Apologeticus" (circa 197 AD).

3.27. Let $r = 5$, $s = 26$, and $k = (11, 13, 17, 19, 20)$.

$$c = \textbf{LALGODRUECQRZLUYRRKFJQVTNS}.$$

This is a quote from Johann Wolfgang Von Goethe (1749–1832), *a German
poet, dramatist, and philosopher. The quote is taken from Act 1, Scene 1
of* "Iphigenie auf Tarris" (1787).

3.28. Let $r = 7$, $s = 34$, and $k = (1, 2, 4, 6, 8, 10, 12)$.

$$c = \textbf{TKPKVMQJUELZSQOFANWGUMNRKDODCGXXII}.$$

This is a quote from Confucius (551–479 BC) *who was a Chinese
sage/philosopher.*

3.29. Let $r = 5$, $s = 15$, and $k = (1, 17, 18, 20, 21)$.

$$c = \textbf{OFLBDOXAHZYTWMN}.$$

This is a quote from Pittacus of Lesbos (650–570 BC), *who was one of
the seven sages of ancient Greece.*

▼ *In Exercises 3.30–3.33, assume $n = 26$ in the autokey Vigenère cipher
described on page 154, and using the given values of r, s, and k in each
case decipher the cryptograms. As usual, employ Table 1.3 for numerical
equivalents.*

3.30. Let $r = 2$, $s = 16$, and $k = 38 = k_1 k_2$.

$$c = \textbf{XVMNXEFXAGYFNWTH}.$$

This is the title of a book by Ralph Nader (1934–), *an American consumer
protectionist.*

3.31. Let $r = 6$, $s = 29$, and $k = (k_1 k_2 k_3 k_4 k_5 k_6) = (237591)$.

$$c = \textbf{ARBHJOLSPGRORXJSIJGACFKWQAVZG}.$$

This is a famous quote attributed to the Duchess of Windsor (Wallis Simpson) (1896–1986), *who was the wife of the former King Edward VIII.*

3.32. Let $r = 4$, $s = 16$, and $k = (k_1 k_2 k_3 k_4) = (7182)$.

$$c = \textbf{SFITYXONEBHANMEP}.$$

This, and its conclusion in Exercise 3.33, is a quote from John Sheffield (First Duke of Buckingham and Normanby) (1648–1721), *who was an English poet and politician. The quote is taken from* "An Essay upon Satire" (1689).

3.33. Let $r = 4$, $s = 17$, and $k = (k_1 k_2 k_3 k_4) = (1234)$.

$$c = \textbf{PTQSHKBKKBHARBTPL}.$$

See Exercise 3.32 for the initial part of this quote.

▼ *In Exercises 3.34–3.44, use the information pertaining to LFSRs as described on pages 156–158. For further reading on LFSRs, the reader may consult* [111] *or* [283].

3.34. Given $\ell = 4$, $(c_1 c_2 c_3 c_4) = (1010)$ and initial state,

$$s_0 = (k_{(3,0)} k_{(2,0)} k_{(1,0)} k_{(0,0)}) = (0111),$$

calculate the period length and each iteration for the LFSR.

3.35. Execute the calculations in Exercise 3.34 using matrix equations as described on pages 157 and 158.

(Hint: Your last calculation should be $CS_6 = S_7$.)

3.36. Show that the maximum number of possible internal states is $2^\ell - 1$.

3.37. Prove that an LFSR must be periodic with period length no larger than $2^\ell - 1$.

Note that if $c_\ell = 0$, then the LFSR is not periodic, but is eventually periodic since it must become periodic after ignoring a finite number of initial terms. In other words, $s_L = s_N$ for some $N \geq L > 0$. When $c_\ell = 0$, the LFSR is called singular, *and is called* nonsingular *otherwise. Hence, we are only considering nonsingular LFSRs since our assumption on page 155 is that $c_\ell = 1$.*

3.38. If an LFSR has period length $2^\ell - 1$, called a *maximum-length* LFSR, show that

$$t(x) = 1 + \sum_{j=1}^{\ell} c_j x^j,$$

called the *tap polynomial*, is irreducible.

(*Hint: See Definition A.35 on page 486 in Appendix A. Assume that $t(x)$ is reducible and argue on the degrees of the individual factors in relation to $t(x)$. Also, note that the individual factors generate LFSRs and the least common multiple of their periods must be at least as big as the period for the LFSR associated with $t(x)$.*)

3.39. Is the converse of Exercise 3.38 true?

(*Hint: Look at Exercise 3.34.*)

3.40. Show that if $t(x)$ is the tap polynomial for a maximum-length LFSR (see Exercise 3.38), then $t(x)$ must divide $x^{2^\ell - 1} - 1$ but $t(x)$ does not divide $x^d - 1$ for any proper divisor d of $2^\ell - 1$. (In this case $t(x)$ is called *primitive*.)

(*Hint: Use matrix theory employing the tap matrix C, defined on page 157, and the matrix theory in Appendix A, especially on page 493, to conclude that $C^{2^\ell - 1} = I$, where I is the identity matrix and all entries are reduced modulo 2. Note that since $t(x)$ must be irreducible by Exercise 3.38, then it divides every polynomial that has the root C of $C^{2^\ell - 1} = I$ in common with it, where $t(x)$ is viewed as the determinant of the matrix $C - Ix$. Thus, $t(x)$ will divide $x^{2^\ell - 1} - 1$ in this case.*)

3.41. Show that if $n(1)$ denotes the number of ones output by a maximum length LFSR, and $n(0)$ denotes the number of zeros output by it, then $n(1) - n(0) = 1$.

(*Hint: The zero sequence cannot be included.*)

3.42. Show that if $t(x)$ is irreducible as the tap polynomial of an LFSR, then the period length is a factor of $2^\ell - 1$.

(*Hint: Use similar techniques to those suggested in the hint for Exercise 3.40.*)

3.43. Show that if $2^\ell - 1$ is prime (called a *Mersenne prime*), then every irreducible polynomial of degree ℓ is the tap polynomial of a maximum-length LFSR.

3.44. Suppose that r is a factor of $2^\ell - 1$ but r is not a factor of $2^d - 1$ for any positive integer $d < \ell$. Show that there is an irreducible polynomial of degree ℓ that is the tap polynomial of an LFSR of period length r.

(*Hint: You can actually show that there are $\phi(r)/\ell$ irreducible polynomials of degree ℓ as tap polynomials of LFSRs of period length r, for each such*)

r. See pages 479 and 480 in Appendix A for a discussion of Euler's ϕ-function.)

G.4 Chapter 4 Exercises

▼ In Exercises 4.1–4.4, assume that the given ciphertext is formed via the permutation given in Example 4.1 on page 163. Find the plaintext by using the inverse permutation and converting to English text from Table 1.3 on page 11.

4.1 $c = (8, 6, 18, 19, 7, 5, 11)$.

4.2 $c = (0, 13, 12, 20, 8, 2, 17)$.

4.3 $c = (0, 18, 2, 8, 18, 2, 11)$.

4.4. $c = (24, 11, 7, 18, 8, 18, 19)$.

4.5. Prove that the DLP presented on page 164 is independent of the generator m of $\mathbb{F}_p^*$.

(Hint: Pick two generators of $\mathbb{F}_p^*$ and show that the log of any element in $\mathbb{F}_p^*$ to one base can be written in terms of the log of that element to the other base. This demonstrates that any procedure for calculating logs to one base can be used to calculate logs to any other base that generates $\mathbb{F}_p^*$. Hence, any such procedure is independent of the choice of base.)

4.6. The Generalized DLP (GDLP) is formulated as follows. Given a finite group G, and elements $g, h \in G$, find an integer e such that $g^e = h$, assuming such an integer exists. Let $e = L_g(h)$. Prove that

$$L_g(h * h') = L_g(h) + L_g(h'),$$

where $h, h' \in G$, $*$ is the group operation, and $g^e * g^f = g^{e+f}$, for integers e, f.

▼ Exercises 4.7–4.10 pertain to the Pohlig-Hellman exponentiation cipher described on page 165. In each case use the data to decrypt the ciphertext and produce the plaintext via Table 1.3 on page 11.

4.7. Let $p = 647$, $e = 67$ and $c = (119, 346, 32, 499, 115, 63, 346, 617)$.

4.8. Let $p = 919$, $e = 47$ and $c = (40, 221, 233, 294, 164, 9, 814)$.

4.9. Let $p = 173$, $e = 99$ and

$$c = (132, 62, 168, 137, 87, 88, 170, 170, 88, 137, 87, 168, 0, 20).$$

4.10. Let $p = 401$, $e = 21$ and $c = (256, 232, 127, 0, 10)$.

▼ *Exercises 4.11–4.15 pertain to the Diffie-Hellman key exchange protocol described on page 166. In Exercises 4.11–4.14, with the given parameters, find the shared secret key in each case.*

4.11. $p = 397$, $\alpha = 5$, $x = 295$, and $y = 301$.

4.12. $p = 643$, $\alpha = 11$, $x = 540$, and $y = 603$.

4.13. $p = 907$, $\alpha = 2$, $x = 101$, and $y = 2$.

4.14. $p = 1181$, $\alpha = 7$, $x = 1000$, and $y = 5$.

4.15. Explain why we choose the generator α, for the Diffie-Hellman protocol, in the range $2 \leq \alpha \leq p - 2$. In other words, why would it be a very bad idea to choose $\alpha = p - 1$?

4.16. Verify the statement made on page 167, namely, if Eve can solve the DLP, she can solve the DHP.

4.17. Suppose that n is an RSA modulus and you know both the enciphering exponent e as well as the deciphering exponent d. Show how this allows you to factor n.

▼ *In Exercises 4.18–4.21, use the repeated squaring method highlighted on page 171 to find the given modular power residue.*

4.18. $5^{72} \pmod{103}$.

4.19. $3^{81} \pmod{303}$.

4.20. $7^{92} \pmod{97}$.

4.21. $2^{51} \pmod{101}$.

4.22. Suppose we alter the modulus in Exercise 4.20 so that we have

$$7^{92} \pmod{105}.$$

Explain how this may be easily done *without* using the repeated squaring method.

(*Hint: Use Euler's theorem A.14 on page 479 once you factor out the gcd of the modulus and the base.*)

▼ *Exercises 4.23–4.26 pertain to the RSA public-key cryptosystem described on page 173. Find the plaintext numerical value of m from the parameters given. You will first have to determine the private key d from the given data via the methodology illustrated in Example 4.5 on page 173. If repeated squaring is not employed (see page 171), then a computer will be required for these calculations. Whenever, we suggest a computer for calculations henceforth, we will be assuming, tacitly, that a software package such as* Maple *or* MATLAB *is available.*

4.23. $(p, q) = (5443, 4327)$, $n = 23551861$, $e = 5$, and $c \equiv 1960142 \,(\mathrm{mod}\ n)$.

   (*Hint: Without repeated squaring, even with a computer, raising the ciphertext to the deciphering exponent d, will take considerable time to execute. Thus, you should first factor d, then raise c to the factors successively until m is found. This will quite considerably reduce the calculation time. The same comment holds for Exercises 4.24–4.26.*)

4.24. $(p, q) = (6113, 7001)$, $n = 42797113$, $e = 11$, and $c \equiv 3430667 \,(\mathrm{mod}\ n)$.

4.25. $(p, q) = (7499, 8237)$, $n = 61769263$, $e = 7$, and $c \equiv 16695987 \,(\mathrm{mod}\ n)$.

4.26. $(p, q) = (8999, 9547)$, $n = 85913453$, $e = 13$, and $c \equiv 63358885 \,(\mathrm{mod}\ n)$.

4.27. Explain why neither $e = 3$ nor $e = 11$ can be employed with the modulus in Exercise 4.26.

4.28. The *Carmichael function*, $\lambda(n)$, is defined as follows. If $n \in \mathbb{N}$, and

$$n = 2^a \cdot p_1^{a_1} \cdot p_2^{a_2} \cdots p_k^{a_k}$$

is its canonical prime factorization, namely, $2 < p_1 < p_2 < \cdots < p_k$, then

$$\lambda(n) = \begin{cases} \phi(n) & \text{if } n = 2^a, \text{ and } 1 \le a \le 2, \\ 2^{a-2} = \phi(n)/2 & \text{if } n = 2^a, a > 2, \\ \mathrm{lcm}(\lambda(2^a), \phi(p_1^{a_1}), \ldots, \phi(p_k^{a_k})) & \text{if } k \ge 1, \end{cases}$$

where ϕ is Euler's totient (see Definition A.22 on page 479), and where lcm is the least common multiple (see page 471).

Suppose that p and q are primes and $n = pq$ is an RSA modulus. Furthermore, assume that $x \in \mathbb{Z}$ with $\gcd(x, n) = 1$ and that e and d are the RSA-enciphering and deciphering exponents, respectively. Show that the following hold.

   1. $x^{\lambda(n)} \equiv 1 \,(\mathrm{mod}\ n)$.

   2. $x^{ed} \equiv x \,(\mathrm{mod}\ n)$.

(*In particular, if p and q are safe primes, then $\lambda(n) = \phi(n)/2$, so the above shows that $\phi(n)/2$ may be used in place of $\phi(n)$ in the RSA cipher. (See page 200 for a definition and application of safe primes.) In fact, even when p and q are not safe primes, $\lambda(n)$ may be employed in the RSA cryptosystem since $\phi(n)$ and $\lambda(n)$ are roughly the same size. The reason for the latter is that the $\gcd(p-1, q-1)$ has an expectation of being small when p and q are chosen arbitrarily.*)

4.29. Exercise 4.28 shows, in particular, that $\phi(n)/2$ may be used instead of $\phi(n)$ in the RSA cipher. Explain why using $\phi(n)/2 + 1$ in place of $\phi(n)$ would be an exceptionally bad idea.

4.30. Show that in Exercise 4.28, n may be any natural number relatively prime to x and part 1 still holds. Also, show that $\lambda(n) \mid \phi(n)$, for any $n \in \mathbb{N}$.

4.31. Exercise 4.29 shows that $\lambda(n)$ is an example of what is called a universal exponent for n, which means an exponent f such that $x^f \equiv 1 \,(\text{mod } n)$ for all integers x relatively prime to n. Show that $\lambda(n)$ is the minimal universal exponent for n.

(*Hint: Use the notion of the order of an integer defined in Definition A.23 on page 479, in conjunction with the Chinese remainder theorem A.12 provided on page 478.*)

4.32. Let n be an RSA modulus and suppose that $m \in \mathbb{Z}$ is a plaintext message unit. Is it necessary that $\gcd(m, n) = 1$ in order to use RSA? If not provide a counterexample. If so prove it.

4.33. Mallory wants to decrypt $c \equiv m^e \,(\text{mod } n)$ sent by Alice to Bob using the RSA cipher. To do so, he intercepts c, masks it by the execution,

$$c' \equiv cx^e \,(\text{mod } n),$$

for a randomly chosen $x \in (\mathbb{Z}/n\mathbb{Z})^*$, and sends c' to Bob. Bob computes $m' \equiv (c')^d \,(\text{mod } n)$ and sends it to Alice. Explain how Mallory can recover m if he intercepts m'.

(*The above is an instance of an* adaptive chosen-ciphertext *attack on RSA. In such attacks, Mallory may send any number of ciphertexts to be decrypted, after which he uses the results to select succeeding ciphertexts. Eventually, Mallory hopes to reveal data about the plaintext, or even the key. Hence, this type of attack may be viewed as an interactive form of the chosen-ciphertext attack (see Footnote 4.3 on page 176). Such attacks may be thwarted by ensuring that the plaintext message, m, has a specified fixed structure, so that if m is disguised as m', then it is unlikely that the latter will maintain that structure. Therefore, if Bob receives a cipher-text that decrypts to a plaintext without that structure, he will discard it as fraudulent. Adaptive chosen-ciphertext attacks can work only when ci-phertexts satisfy what is called* ciphertext malleability, *which means that the ciphertext can be masked in certain ways that have a foreseeable effect on the deciphering. One well-known prevention of such attacks on RSA is OAEP (see page 174).*)

▼ *Exercises 4.34–4.37 refer to the RSA signature scheme developed on pages 181–183. Use the given parameters in each case to first compute the en-cryption key e using the Euclidean algorithm on $\phi(n)$ and d. Then compute*

$c^e \pmod{n}$. If $m \equiv c^e \pmod{n}$, then accept the signature as valid, since $\mathrm{ver}_k(m, c) = 1$. Otherwise, reject the signature, since $\mathrm{ver}_k(m, c) = 0$.

4.34. $n = 17438441$, $\phi(n) = 17430084$, $d = 15845531$, $m = 210314$, and $c = 2673099$.

4.35. $n = 29778839$, $\phi(n) = 29767920$, $d = 17234059$, $m = 186677$, and $c = 17284872$.

4.36. $n = 42486991$, $\phi(n) = 42473952$, $d = 16989581$, $m = 249917$, and $c = 14191108$.

4.37. $n = 42486991$, $\phi(n) = 42473952$, $d = 16989581$, $m = 249917$, and $c = 14191109$.

▼ *In Exercises 4.38–4.41, use the description of the DSA given on pages 183 and 184, applied to the given parameters in each case, to verify that Bob should accept Alice's digital signature. For simplicity we use very small parameters, as with the above applications of other algorithms, and in this case we assume that $h(m) = m$, to further simplify the calculations. Furthermore, the primes p and q are not selected with the values suggested in the description of DSS, rather are artificially small for pedagogical purposes.*

4.38. $p = 1549$, $q = 43$, $\alpha = 104$, $\beta = 252$, $m = 21$, $\gamma = 29$, $\sigma = 7$.

4.39. $p = 2699$, $q = 71$, $\alpha = 896$, $\beta = 1850$, $m = 21$, $\gamma = 11$, $\sigma = 33$.

4.40. $p = 3359$, $q = 73$, $\alpha = 2451$, $\beta = 1185$, $m = 45$, $\gamma = 43$, $\sigma = 48$.

4.39. $p = 9439$, $q = 13$, $\alpha = 4139$, $\beta = 2471$, $m = 4$, $\gamma = 5$, $\sigma = 8$.

▼ *In Exercises 4.42–4.45, employ the Elgamal cryptosystem described on pages 185 and 186 to recover the plaintext m from the ciphertext c via the parameters given by the prime p and Bob's private key a in each case.*

4.42. $p = 2099$, $a = 17$, $c = (\alpha^b, m\alpha^{ab}) = (1700, 304)$.

4.43. $p = 3313$, $a = 7$, $c = (\alpha^b, m\alpha^{ab}) = (1697, 770)$.

4.44. $p = 4657$, $a = 19$, $c = (\alpha^b, m\alpha^{ab}) = (1640, 4556)$.

4.43. $p = 7177$, $a = 35$, $c = (\alpha^b, m\alpha^{ab}) = (1416, 7104)$.

▼ *Exercises 4.44–4.47 pertain to the ElGamal signature scheme delineated on pages 187 and 188. For the given parameters, determine if Bob should accept the signature as valid.*

4.44. for $p = 463$, $\alpha = 3$, $y = 454$, $\beta = 243$, and $\gamma = 153$, so Alice sends, $m = 96$ and $\mathrm{sig}_k(m, r) = (\beta, \gamma) = (243, 153)$.

4.45. for $p = 2689$, $\alpha = 19$, $y = 2221$, $\beta = 954$, and $\gamma = 2154$, so Alice sends, $m = 96$ and $\mathrm{sig}_k(m, r) = (\beta, \gamma) = (954, 2154)$.

4.46. for $p = 4657$, $\alpha = 15$, $y = 3484$, $\beta = 284$, and $\gamma = 2503$, so Alice sends, $m = 1111$ and $\mathrm{sig}_k(m, r) = (\beta, \gamma) = (284, 1865)$.

4.47. for $p = 8761$, $\alpha = 23$, $y = 5807$, $\beta = 2973$, and $\gamma = 2678$, so Alice sends, $m = 2069$ and $\mathrm{sig}_k(m, r) = (\beta, \gamma) = (2973, 2678)$.

▼ *In Exercises 4.48 and 4.49, use the elliptic curve given in Example 4.9 on page 190 to decipher the given cryptogram with the private key a, provided in each case.*

4.48. $c = ((11, 1), (1, 3))$, and $a = 2$.

4.49. $c = ((11, 1), (11, 12))$, and $a = 3$.

▼ *Exercises 4.50–4.56 look at the cryptographic applications of Dickson polynomials, defined below.*

4.50. The *Dickson polynomial* [G.1] of the *first kind* of degree $n \in \mathbb{N}$ in the indeterminant x with parameter $a \in R$, where R is a commutative ring with identity, is defined by

$$D_n(x, a) = \sum_{j=0}^{\lfloor n/2 \rfloor} \frac{n}{n-j} \binom{n-i}{i} (-a)^j x^{n-2j}.$$

Prove that $\frac{n}{n-j}\binom{n-j}{j}$ is an integer. Then establish that, when $R = \mathbb{F}_q$, a finite field, we may write the polynomials in the form,

$$D_n(x, a) = \left(\left(x + \sqrt{x^2 - 4a} \right) / 2 \right)^n + \left(\left(x - \sqrt{x^2 - 4a} \right) / 2 \right)^n.$$

Moreover, if $A = (x + \sqrt{x^2 - 4a})/2$, then $A^2 - Ax + a = 0$, and

$$D_n \left(A + \frac{a}{A}, a \right) = A^n + \left(\frac{a}{A} \right)^n. \tag{G.1}$$

(*Hint: You may use* Waring's Formula, *given as follows:*

$$A^n + B^n = \sum_{j=0}^{\lfloor n/2 \rfloor} (-1)^j \frac{n}{n-j} \binom{n-j}{j} (AB)^j (A+B)^{n-2j},$$

[G.1] In 1896, L.E. Dickson had these polynomials as part of his doctoral thesis at the University of Chicago. In a paper [244], published in 1973, I. Schur put Dickson's name to these polynomials in his honour, and related these Dickson polynomials to the well-known Chebyshev polynomials. From the 1970s to the present, the theory of Dickson polynomials has flourished. Much of this development is due to W.B. Nöbauer and his followers (see [179] and [180], for instance). In particular, applications to cryptography have come to the fore, and it is this in which we are interested, albeit many other applications of these polynomials abound.

and you may use the fact that if $f(x) = x^2 - cx + d \in F[x]$ for any field F, then c is the sum of the roots of $f(x)$ and d is the product of the roots. Note, as well, that $(x + \sqrt{x^2 - 4a})/2$ lies in the extension field $\mathbb{F}_{q^2}$. Also, Equation (G.1) is called the functional equation of $D_n(x, a)$.)

☆ 4.51. A *permutation polynomial* on a finite field $\mathbb{F}_q$ is a polynomial $f(x) \in \mathbb{F}_q[x]$ such that f permutes the elements of $\mathbb{F}_q$. In other words, f is a one-to-one and onto mapping of $\mathbb{F}_q$ to itself.

Prove that for $a \in \mathbb{F}_q^*$, $D_n(x, a)$ is a permutation polynomial on $\mathbb{F}_q$ if and only if $\gcd(n, q^2 - 1) = 1$.

(*Hint: Prove first that x^n permutes $\mathbb{F}_q$ if and only if $\gcd(n, q - 1) = 1$. Then use Exercise 4.50.*)

4.52. Let $k = \prod_{j=1}^t p_j^{a_j}$ be the canonical prime factorization of $k \in \mathbb{N}$. A polynomial is a *permutation polynomial* modulo k, or permutation polynomial of $\mathbb{Z}/k\mathbb{Z}$, if f is a one-to-one and onto function of $\mathbb{Z}/k\mathbb{Z}$. With somewhat more difficulty than the above, it can be shown that $D_n(x, a)$ is a permutation polynomial modulo k if and only if $\gcd(n, \nu(k)) = 1$, where

$$\nu(k) = \mathrm{lcm}_j(p_j^{a_j - 1}(p_j^2 - 1)).$$

(Exercise 4.51 is provided as an indication of that process for the simpler case.) Once we have a permutation polynomial, we can form its inverse and this provides a basis for developing a cryptosystem.

The following public-key cipher is called a *Dickson cryptosystem*.

Let Bob's public enciphering exponent be $e_B \in \mathbb{Z}/k\mathbb{Z}$ with $\gcd(e_B, \nu(k)) = 1$. Alice enciphers a message m to Bob via computation of

$$D_{e_B}(m, a) \equiv c \,(\mathrm{mod}\ k)$$

where $a = 1$ or $a = -1$ in $\mathbb{Z}/k/Z$. To decrypt, Bob uses his private key d_B obtained via the linear congruence,

$$e_B d_B \equiv 1 \,(\mathrm{mod}\ \nu(k)),$$

and computes $D_{d_B}(c, a) \,(\mathrm{mod}\ k)$. Prove that the latter actually recovers m. Moreover, show that if $k = pq$ is an RSA modulus, and

$$\mu(k) = (p^2 - 1)(q^2 - 1)$$

is used in place of $\nu(k)$, then this is similar to the RSA cipher, and for $a = 0$ it *is* the RSA cipher.

4.53. With reference to Exercises 4.50–4.52, Dickson polynomials may also be used as a basis for digital signatures. Suppose that Alice wishes to sign a message m, and k is given to be a product of large primes. If (e_A, d_A) is Alice's public/private key pair, assume she computes

$$D_{d_A}(m, a) \equiv s \,(\mathrm{mod}\ k),$$

and sends the signature s to Bob.

1. Show how Bob can form a computation to recover m, and uniquely verify that this is Alice's signature.

2. How can Alice sign a message using Dickson polynomials and Bob's public key to achieve the same effect?

4.54. The Dickson polynomial schemes devised above can be employed to form a type of Diffie-Hellman key exchange as follows. Let $\alpha \in \mathbb{F}_q$ be chosen such that $\alpha = \gamma^{q-1} + \gamma^{-(q-1)}$ where γ is a primitive element of $\mathbb{F}_{q^2}$. Randomly chosen positive integers a and b, chosen by Alice and Bob, respectively, are kept private, whereas q, α, $D_a(\alpha, 1)$, and $D_b(\alpha, 1)$ are made public. The following is a *Dickson key exchange*.

1. Alice gets Bob's public data, $D_b(\alpha, 1)$ and computes $D_a(D_b(\alpha, 1)) = D_{ab}(\alpha, 1)$.

2. Bob gets Alice's data $D_a(\alpha, 1)$ and computes $D_b(D_a(\alpha, 1), 1) = D_{ba}(\alpha, 1)$.

.

Show that the shared key $D_{ab}(\alpha, 1) = D_{ba}(\alpha, 1)$ in this key exchange scheme depends on the DLP as does the Diffie-Hellman exchange.

4.55. The three-pass protocol (sometimes called *Shamir's three-pass scheme*) discussed on pages 198 and 199, can be generalized to Dickson polynomials developed in Exercises 4.50–4.54.

Assume that Alice wishes to send a message $m \in \mathbb{F}_p$ (p a prime), to Bob. To do so, the following steps are executed.

1. Alice selects an integer a with $\gcd(a, p^2 - 1) = 1$, where a is kept secret. She sends $D_a(m, 1) \equiv c \pmod{p}$ to Bob.

2. Bob picks an integer b such that $\gcd(b, p^2 - 1) = 1$, where he keeps b secret. Then he sends $D_b(c, 1) \equiv c' \pmod{p}$ to Alice.

3. Alice computes a' such that

$$aa' \equiv 1 \pmod{p^2 - 1}$$

and sends

$$D_{a'}(c', 1) \equiv d' \pmod{p}$$

to Bob.

4. Bob recovers $m \pmod{p}$ by computing b' satisfying

$$bb' \equiv 1 \pmod{p^2 - 1}$$

and

$$D_{b'}(D_{a'}(D_b(D_a(m, 1), 1), 1), 1) \equiv m \pmod{p}.$$

Describe the property satisfied by the Dickson polynomial that allows Bob to recover m in step 4, and verify that step 4 indeed does recover m as suggested.

4.56. Dickson polynomials may also be used for multiple encryptions. The notation and concepts used in this exercise were developed in Exercise 4.52.

Suppose that $D_n^{(s)}(x, a)$ denotes the composition of $D_n(x, a)$ $s \in \mathbb{N}$ times. If c is a ciphertext, show that there are integers r, s such that

$$D_n^{(r)}(c, a) \equiv D_n^{(s)}(c, a) \equiv c \,(\text{mod } k).$$

This implies the existence of an integer t such that

$$D_n^{(t)}(c, a) \equiv c \,(\text{mod } k) \qquad\qquad (\text{G.2})$$

Show how this may be used to recover the plaintext without factoring the modulus. (It can be shown that the fixed point given by (G.2) may be used as an attack to factor k (see [174] and [181])). For more information on Dickson polynomials and their applications, see a book devoted entirely to this topic ([152]).

G.5   Chapter 5 Exercises

5.1. This problem refers to the explanation of oblivious transfer covered in Section 5.1 on pages 191–194.

Suppose that Alice and Bob are secret agents for two different countries and Bob wants to buy a secret from Alice. Moreover, Bob wants to buy a secret without Alice knowing which one.

Suppose that Alice has a list of ℓ secrets $s_1, s_2, \ldots, s_\ell$, all expressed as bitstrings of equal bitlength. She also possesses a one-way function f for which only she possesses f^{-1}. She lets Bob know what each of the secrets represent by giving him a list of questions $q_1, q_2, \ldots, q_\ell$ to which the s_j are the answers. The oblivious transfer protocol is described as follows.

1. Bob selects q_k from the list of questions to which he wants to buy the secret s_k. He chooses random numbers $r_1, r_2, \ldots, r_\ell$ from the domain of f. Then he computes

$$c_j = \begin{cases} r_j & \text{if } j \neq k, \\ f(r_j) & \text{if } j = k, \end{cases}$$

and sends $(c_1, c_2, \ldots, c_\ell)$ to Alice.

2. Alice computes $m_j = f^{-1}(r_j)$, and sends the values $x_j = m_j \oplus s_j$ (where $\oplus$ is the mod 2 addition of the bitstrings), for $j = 1, 2, \ldots, \ell$, to Bob.

3. Bob knows $m_k = f^{-1}(f(r_k)) = r_k$, so Bob can compute $m_k \oplus x_k = s_k$.

Show that Bob cannot compute s_j for $j \neq k$. Explain how Bob can cheat Alice by altering his computation of c_j in step 1.

5.2. Explain how Kerberos (see pages 196 and 197), is vulnerable with respect to (1) host security; (2) Carol's password (encryption keys); and (3) offline attacks on Carol's ticket.

5.3. Show how the three-way authentication protocol described on page 198 would be vulnerable to the man-in-the middle attack (see Footnote 3.7 on page 134), if the DSS is not employed. In other words, if signatures are not employed, show how Mallory can impersonate Alice and successfully convince Bob that he is talking to her.

▼ *In Exercises 5.4–5.7, use the Fiege-Fiat Shamir identification protocol presented on pages 202 and 203 to show that Bob should accept Alice's proof, given the parameters in each case.*

5.4. Let $p = 523 \cdot 1637 = 856151$, $s_A = 5$, $a = 2$, and assume that in round 1, Alice selects $m = 651$, and Bob chooses $c = 0$, while in round 2, Alice picks $m = 1516$ and Bob selects $c = 0$.

5.5. Let $p = 613 \cdot 2281 = 1398253$, $s_A = 7$, $a = 2$, and assume that in round 1, Alice selects $m = 3291$, and Bob chooses $c = 1$, while in round 2, Alice picks $m = 1923$ and Bob selects $c = 1$.

5.6. Let $p = 739 \cdot 2557 = 1889623$, $s_A = 25$, $a = 3$, and assume that in round 1, Alice selects $m = 3681$, and Bob chooses $c = 1$; in round 2, Alice picks $m = 111$ and Bob selects $c = 0$; and in the third round Alice picks $m = 38888$ and Bob chooses $c = 1$.

5.7. Let $p = 857 \cdot 3323 = 2847811$, $s_A = 49$, $a = 3$, and assume that in round 1, Alice selects $m = 333$, and Bob chooses $c = 1$; in round 2, Alice picks $m = 723$ and Bob selects $c = 1$; and in the third round Alice picks $m = 111111$ and Bob chooses $c = 1$.

▼ *In Exercises 5.8–5.11, employ the Schnorr identification protocol delineated on page 205 to show that Bob should accept Alice's identity. As usual, the parameters are artificially small to make computation reasonable. In other words, we are not choosing $p \geq 2^{1024}$, or $q > 2^t$ with $t \geq 40$, for instance. We will assume that all certificates and signature verifications have taken place. All that is required is the calculation of the commitment, the response, and the verification as outlined in steps 1–4 of the protocol given on the aforementioned page.*

5.8. Let $p = 1249$, $q = 13$, $e = 12$, $k = 2$, $r = 55$.

5.9. Let $p = 2251$, $q = 5$, $e = 4$, $k = 3$, $r = 23$.

5.10. Let $p = 6619$, $q = 1103$, $e = 110$, $k = 234$, $r = 233$.

5.11. Let $p = 9377$, $q = 293$, $e = 11$, $k = 43$, $r = 199$.

▼ *Exercises 5.12–5.15 pertain to the protocol for coin flipping by telephone using discrete logs that we studied on page 209. With the parameters given, decide whether the outcome of the coin toss is heads or tails.*

5.12. Let $p = 3011$, $\alpha = 2$, $\beta = 7$, $x = 9$, and $y \equiv 1 \,(\text{mod } p)$, which Alice guesses is a function of α.

5.13. Let $p = 4129$, $\alpha = 13$, $\beta = 14$, $x = 5$, and $y \equiv 3812 \,(\text{mod } p)$, which Alice guesses is a function of β.

5.14. Let $p = 4561$, $\alpha = 11$, $\beta = 13$, $x = 7$, and $y \equiv 2840 \,(\text{mod } p)$, which Alice guesses is a function of β.

5.15. Let $p = 7481$, $\alpha = 6$, $\beta = 7$, $x = 13$, and $y \equiv 5128 \,(\text{mod } p)$, which Alice guesses is a function of α.

▼ *In Exercises 5.16–5.19, we are going to play poker by telephone in a fashion similar to that described on page 210. However, instead of five cards dealt, we assume that only one card is dealt by Alice to Bob. In each exercise, the values of the cards are given specific numerical values. You will need computing power. The hint to Exercise 4.23 applies here.*

5.16. Let $n = 26904167$, $(e_A, d_A) = (5, 21515021)$; $(e_B, d_B) = (11, 9779555)$, $c_j \equiv 5^j \,(\text{mod } n)$ for $j = 2, 3, 4, \ldots, 14$ where $c_j = j$ for $j = 2, 3, \ldots, 10$, whereas c_j for $j = 11, 12, 13, 14$ represent the Jack, Queen, King, and Ace, respectively. Bob is dealt $h_j \equiv 14473275 \,(\text{mod } n)$. Determine the card dealt to Bob.

(*Hint: Compute* $c_j \equiv h^{d_B} \,(\text{mod } n)$.)

5.17. In Exercise 5.16, assume that $g_j \equiv 22526833 \,(\text{mod } n)$. Determine the card that was initially enciphered by Alice.

(*Hint: Compute* $c_j \equiv g_j^{d_A \cdot d_B} \,(\text{mod } n)$.)

5.18. Given $f_j \equiv 24498353 \,(\text{mod } n)$ with the parameters in Exercise 5.16, determine the card enciphered by Alice.

(*Hint: Calculate* $c_j \equiv f_j^{d_A} \,(\text{mod } n)$.)

5.19. With reference to the analysis of Jacobi symbols concerning poker playing by telephone, given on pages 210 and 211, determine which of the values -1 or $+1$ is more advantageous to Alice given the parameters in Exercise 5.16. How may we alter the numbering of the cards to eliminate this advantage?

5.20. Blum integers and their properties are examined in Appendix B on pages 507 and 508. There is a coin flipping scheme based on Blum integers that we now describe.

1. Bob generates a Blum integer n, a random $x \in \mathbb{N}$ relatively prime to n, and computes $x_0 \equiv x^2 \pmod{n}$ and $x_1 \equiv x_0^2 \pmod{n}$. He sends n and x_1 to Alice.

2. Alice guesses the parity of x and sends the guess to Bob.

3. Bob sends x and x_0 to Alice.

4. Alice checks that both $x_0 \equiv x^2 \pmod{n}$ and $x_1 \equiv x_0^2 \pmod{n}$. Thus, Alice can determine if the guess is correct.

Explain how Bob can cheat freely if this is *not* a Blum integer, where $p \equiv q \equiv 1 \pmod{n}$ for primes p and q.

▼ *Exercises 5.21–5.24 contain parameters for Shamir's threshold scheme described on pages 212–214. Use the parameters to determine the message $m = c_0$. Again, a computer may be necessary for some of the calculations.*

5.21. Let $(t, w) = (2, 2)$, $p = 1009$, $(x_1, m_1) = (1, 172)$, $(x_2, m_2) = (2, 244)$.

5.22. Let $(t, w) = (3, 3)$, $p = 3271$, $(x_1, m_1) = (1, 1234)$, $(x_2, m_2) = (2, 1578)$, and $(x_3, m_3) = (3, 2144)$.

5.23. Let $(t, w) = (4, 4)$, $p = 1433$, $(x_1, m_1) = (3, 372)$, $(x_2, m_2) = (5, 859)$, $(x_3, m_3) = (7, 50)$, and $(x_4, m_4) = (11, 720)$.

5.24. Let $(t, w) = (4, 4)$, $p = 6367$, $(x_1, m_1) = (7, 3401)$, $(x_2, m_2) = (11, 2822)$, $(x_3, m_3) = (12, 4239)$, and $(x_4, m_4) = (13, 1821)$.

▼ *In Exercises 5.25.–5.28, use the description of Blakely's secret-sharing vector scheme given on pages 214 and 215, to determine the message m for each of the set of parameters given. Some calculations may require a computer.*

5.25. Let $t = 2$, $p = 3359$, $n_1^{(1)} = 358$; $n_1^{(2)} = 953$; and $(c_1, c_2,) = (1001, 1111)$.

(*Hint: Form the equation given by (5.4) on page 215, and solve the equation:*

$$X \equiv A^{-1}C \pmod{p}.$$

The same hint holds for Exercises 5.26–5.28.)

5.26. Let $t = 3$, $p = 2551$, $(c_1, c_2, c_3) = (109, 526, 2128)$, and

$$(n_1^{(1)}, n_2^{(1)}) = (7, 9); \quad (n_1^{(2)}, n_2^{(2)}) = (27, 361); \quad (n_1^{(3)}, n_2^{(3)}) = (100, 2).$$

5.27. Let $t = 4$, $p = 757$, $(c_1, c_2, c_3, c_4) = (26, 399, 711, 192)$, and

$$(n_1^{(1)}, n_2^{(1)}, n_3^{(1)}) = (3, 21, 31); \quad (n_1^{(2)}, n_2^{(2)}, n_3^{(2)}) = (5, 26, 10);$$
$$(n_1^{(3)}, n_2^{(3)}, n_3^{(3)}) = (7, 71, 5); \quad (n_1^{(4)}, n_2^{(4)}, n_3^{(4)}) = (11, 20, 1).$$

5.28. Let $t = 5$, $p = 9001$, $(c_1, c_2, c_3, c_4, c_5) = (2706, 3743, 95, 3239, 6475)$, and,

$$(n_1^{(1)}, n_2^{(1)}, n_3^{(1)}, n_4^{(1)}) = (217, 1, 3549, 2);$$

$$(n_1^{(2)}, n_2^{(2)}, n_3^{(2)}, n_4^{(2)}) = (900, 25, 867, 27);$$

$$(n_1^{(3)}, n_2^{(3)}, n_3^{(3)}, n_4^{(3)}) = (1002, 111, 257, 29);$$

$$(n_1^{(4)}, n_2^{(4)}, n_3^{(4)}, n_4^{(4)}) = (1, 261, 69, 96);$$

$$(n_1^{(5)}, n_2^{(5)}, n_3^{(5)}, n_4^{(5)}) = (2, 900, 8999, 21).$$

5.29. Construct a multiauthority election protocol that extends the notions described on pages 216 and 217 by allowing the independent choice of *more than one* generator of $\mathbb{Z}/q\mathbb{Z}$.

(*Hint: Allow each voter to encrypt their vote with respect to all the generators such that exactly one of the encipherings reveals the actual vote, by an interactive proof of knowledge, say.*)

5.30. Explain how SSL/TLS described in Section 5.7 on pages 218–226, is susceptible to Eve's doing a *traffic analysis* on the communications. In other words, explain how Eve may use her observations concerning the number of messages being sent to and from a specific Internet address to extract information, even if she does not know the particular content of those messages.

(*Hint: Assume you are a lawyer (just for this one time), and you want to know about the legal activities of competing law firms in terms of the volume of their activity. (Also, see Footnote 8.11 on page 309.)*)

5.31. Outline the mechanism for the use of different coin denominations via differing RSA exponents mentioned on page 231 of Section 5.8 on digital cash.

G.6   Chapter 6 Exercises

▼ *Exercises 6.1–6.8 refer to the Blom key predistribution protocol presented on pages 234–236. Assuming, as we did in Example 6.1 on page 235, that there is a network of only three users, use the parameters given to determine the keys k_{AB}, k_{AC}, and k_{BC}.*

6.1. $p = 1297$, $(r_1, r_2, r_3) = (12, 79, 721)$, $(u_A, u_B, u_C) = (92, 219, 691)$.

6.2. $p = 2843$, $(r_1, r_2, r_3) = (29, 289, 378)$, $(u_A, u_B, u_C) = (919, 1001, 2004)$.

6.3. $p = 3253$, $(r_1, r_2, r_3) = (38, 391, 499)$, $(u_A, u_B, u_C) = (111, 1111, 2000)$.

6.4. $p = 4799$, $(r_1, r_2, r_3) = (49, 492, 585)$, $(u_A, u_B, u_C) = (479, 1078, 3003)$.

6.5. $p = 5009$, $(r_1, r_2, r_3) = (59, 581, 609)$, $(u_A, u_B, u_C) = (3232, 1137, 1194)$.

6.6. $p = 7321$, $(r_1, r_2, r_3) = (78, 777, 832)$, $(u_A, u_B, u_C) = (2590, 2495, 1979)$.

6.7. $p = 8293$, $(r_1, r_2, r_3) = (87, 888, 929)$, $(u_A, u_B, u_C) = (186, 4047, 152)$.

6.8. $p = 9497$, $(r_1, r_2, r_3) = (96, 999, 9000)$, $(u_A, u_B, u_C) = (2152, 1601, 6133)$.

6.9. Explain what would be necessary to embed a PKI into Kerberos (see Section 6.2).

6.10. Explore the ramifications of eliminating the dual signature in the SET protocol described in Section 6.3 in favour of a standard signature.

6.11. How do SSL/TLS and SET differ with respect to suitability of e-commerce applications?

G.7   Chapter 7 Exercises

7.1 This exercise pertains to the birthday attack and related issues described on pages 252–255. Suppose that we want to solve the DLP, namely, given a large prime p, a generator m of $\mathbb{F}_p^*$ and an element $c \in \mathbb{F}_p^*$, we want to find e such that $c \equiv m^e \pmod{p}$. How does the following aid in solving the problem using the birthday attack?

Alice compiles two lists, A and B of length $\approx \lfloor \sqrt{p} \rfloor$, satisfying the two properties:

1. List A consists of all numbers $m^x \pmod{p}$ for approximately $\lfloor \sqrt{p} \rfloor$ randomly selected values of x.

2. List B contains the values $cm^{-y} \pmod{p}$ for approximately $\lfloor \sqrt{p} \rfloor$ randomly chosen values of y.

7.2 Suppose that there are 150 students in a class. What is the probability that at least two of them have the same birthday?

7.3. What is the minimum number of people who should be in a room to ensure a probability of 99% that at least two of them have the same birthday?

7.4. What is the probability that we have a collision from a randomly chosen pair of 16-bit numbers, given a random hash function applied to them?

(*Hint: See the formula for $P_2(n, m)$ on page 253 and set $n = 2^{16}$, $m = 1$.*)

7.5. Suppose that you choose a pair of 16-bit numbers as in Exercise 7.4, but if you do not get a collision, you keep trying until you do. If you do this n times, how big must n be in order to have a 50% chance of success?

7.6. Suppose that you have a sorted list, A, of 16-bit values m_j and a sorted list, B, of their hash values $h(m_j)$ for $j = 1, 2, \ldots 2^8$. You compare the lists A and B pairwise, namely,

$$(m_j, h(m_j)) \quad \text{against} \quad (m_k, h(m_k)),$$

in order to seek out collisions, where $h(m_j) = h(m_k)$. How many comparisons must you make to guarantee a success rate of at least 50%?

7.7. What is the maximum number of iterations to produce a collision with SHA-1? (See pages 255–258.)

7.8. RIPEMD-160, described on page 259, uses little-Endian architecture (see Footnote 9.24 on page 350). Explain why it must be ensured that the message digest is independent from the underlying architecture.

7.9. Is A CBC-MAC a one-way transformation? (See page 261.)

7.10. Explain the security problems with an HMAC if it were possible to invert the hash function H (see pages 263–264). In general, explain why hash function must be practically noninvertible.

7.11. If it were possible to have a "perfect" hash function, then we would have a powerful imaginary function called a *random oracle*. In other words, if we could prove that hash functions behave like truly random functions, we would have the existence of a random oracle. A random oracle has output that is not only uniform, but also deterministic and efficient. This idealized model for a hash function is often called the *random oracle model*, which was introduced in [14]. In the random oracle model, a hash function is randomly selected, but we are not given an explicit description of how to compute the hash values. We are allowed only to query a so-called *oracle* who has access to the function. Think of this as equivalent to looking up a value of $h(m)$ for a given m in some table. Thus, if we are given m then h will output the same value $h(m)$ every time the oracle is queried with that value. Moreover, if the oracle is not queried with a specific value, then the output is a random value that has a uniform probability of being chosen throughout the possible values in the range of h.

Based on the above description, suppose that Alice sends messages m to Bob but does so as $h(m)$ where h is a random oracle. Suppose that Mallory has access to an oracle of his own that allows him to submit ciphertext (c, m), which responds with *true* if $h(m) = c$ and *false* otherwise. Is this system secure against Mallory's attack?

7.12. Explain how Mallory could successfully impersonate Alice to Bob, in the X.509 three-way authentication protocol described on pages 269 and 270 if timestamps are not employed.

(*Hint: See Exercise 5.3.*)

G.8   Chapter 8 Exercises

8.1. When discussing PGP in Section 8.1, we assumed that triple DES was used. There are others options such as CAST-128 and IDEA (see page 275). Explain why (single) DES would not be suitable for use with PGP.

8.2. We assumed, in our description of PGP (see Exercise 8.1), that CFB mode is employed. Why would PGP use CFB over say CBC, which is more common in usage?

8.3. Explain how PKI can ensure a greater degree of trust in the use of PGP, especially with respect to ensuring we are in possession of the actual owner's public key.

8.4. Without a PKI, why is the web of trust discussed in detail on pages 280–286, insufficient to guarantee that, for instance, Alice really knows Bob's public key?

8.5. Explain how MIME helps your Internet browser recognize a text file, assuming that the remote web server has not already identified it for your browser.

(*Hint: See page 290.*)

8.6. If you had to list only two primary goals of IPSec, detailed in Section 8.3, what would they be?

8.7. Explain why it is desirable to have encryption before authentication in SA bundling (see page 309).

8.8. Make an argument for employing authentication before encryption in SA bundling.

(*Hint: see pages 266 and 267.*)

8.9. On page 305, we illustrated configurations for end-to-end security using SA tunnels. What configuration would constitute end-to-end authentication and encryption without nesting as illustrated in Diagrams 8.24 and 8.25 on page 310?

8.10. Compare transport mode and tunnel mode SAs for AH and ESP with authentication. (See pages 302–312.)

8.11. Speculate as to how HMAC might be used with RIPEMD-160 (see page 259), within ESP and AH in IPSec (see Section 8.3). For actual technical details, see RFC 2857 [222].

G.9 Chapter 9 Exercises

9.1. Suppose we have a large codebook that consists of short words with corresponding 6-digit numbers. Devise a means of choosing passphrases from this list based upon the roll of dice.

9.2. Assume that you have a password p and each time you logon to a WWW site the host uses a one-way function f to calculate $f(p)$, and compares this with a stored value. Devise a means using only this function, that will prompt you to change your password after, say, 100 logins.

9.3. The SSH protocol presented on pages 334–339, has a significant additional feature called *port-forwarding*, which means that either Alice or the server can bind a socket to a collection of specified ports. In practice, what this means is that when Alice, say, connects to one of these ports, the call is relayed to the other end of this particular SSH call, from which another call is made to some other predetermined port. Effectively, this is an SSH built-in tunneling mechanism (see page 337 for details). Discuss the pros and cons of such tunnels.

(*Hint: Consider setting up such tunnels to avoid firewalls.*)

9.4. Explain how a basic wireless system operates. (See Section 9.2.)

9.5. Explain the relationship between frequency of a signal and the size of an antenna. In particular, why is it possible for cell phones to have such small antennas?

(*Hint: The higher the frequency, the smaller the wavelength, so the closer the antenna the same size is to the wavelength it receives, the better the reception. In practice, an antenna is some fraction of the wavelength, such as half, say.*)

9.6. A cell phone has internal memory that contains the wireless phone number, a system identification number or SID, which identifies the phone to the system to which it subscribes, and an electronic serial number identifying that specific phone as a measure against fraud. When a cell phone is turned on, it compares its SID with any overhead signals and when a match occurs, the phone knows it is operating within the subscriber network. How would a cell phone operate when it is away from its subscriber network and needs *roaming mode*?

9.7. Cell phone networks are made up of cells each of which is associated with a base station. These cells overlap, so when one is travelling, there needs to be a *hand-off* (see page 352), from one cell to another. How would a mobile switching center perform this hand-off?

(*Hint: A mobile switching station continuously monitors the power level of a given cell phone, together with the power level of the base station. Speculate how the switching station would react when a cell is getting close to the edge of the current base station's cell.*)

9.8. Digital cell phones not only digitize speech, but also compress it. Why is this necessary?

9.9. There is a standard for wireless networks that competes with IEEE 802.11 (see pages 340–353), called *Bluetooth*,[G.2] which is a network that does not require a server or other central access source. In other words, the devices on a Bluetooth network can find each other independently and can communicate directly with each other, i.e., a peer-to-peer network (see page 220). Since Bluetooth allows not only telephones, but also computers, personal desktop assistants (PDA)s, and even TVs and audio equipment to communicate with one another, it may be used in concert with 802.11.

Bluetooth devices have microchips embedded in them containing software, called a *link controller*, allowing one Bluetooth device to recognize another, by continually sending out signals in search of other such devices in its range. The software also contains profiles encoded into each device so that one can determine if it is appropriate to form a connection (cell phones and TVs need not communicate, for instance). Once connection is established between two or more Bluetooth devices, this is a network called a *piconet*. Bluetooth devices that are placed too close to one another may have their radio frequencies interfere with one another. Speculate as to how Bluetooth fixes this problem (via some monitoring technique).

9.10. If a cell phone has a *microbrowser* installed, it can interface with the WWW to display a Web page using a protocol called *Wireless Access Protocol* (WAP), which employs its own language, the *Wireless Markup Language* (WML). To accomplish this, the cell phone must send a request that is routed through a *landline* (wire-based network), which is sent to a Web server where the page is located. Then the page is sent to to what is called a WAP gateway. Assuming the WWW page is in HTML, speculate on how the WAP gateway would deal with the page in order for it to be displayed on the cell phone once it is relayed back.

9.11. A step above cell phones are satellite phones, which employ GSM technology, so the phone must have a SIM installed (see page 347). A signal is sent from the phone to a satellite, which receives the call. Assuming there are numerous satellites in the system, how would this system of satellites work in a fashion similar to a cell phone network?

9.12. Exploration satellites, such as the Voyager series, send signals back to earth that contain pictures and other data. The transmitter on some of these satellites can be a mere 23 watts. Given an antenna of fourteen feet on board, and a signal in the 8-GHz band, sent to NASA satellites that are 100 feet in diameter, explain how the signal can reach earth.

(*Hint: See Exercise 9.5.*)

[G.2]The name is a (rough) translation from the Danish word *Blatand*, the surname of the king of Denmark from 940 to 985 AD. King Harald Blatand united Denmark and Norway. Thus, it was deemed appropriate to name a uniting communications feature after this man.

9.13. Suppose that a smart card (see Section 9.3), uses the RSA cipher with public encryption exponent $e = 3$. Next assume that m is Alice's credit card number and she buys merchandise from three shops whose public moduli are n_1, n_2, and n_3, respectively. Thus, each shop computes

$$m^3 \ (\text{mod } n_j) \text{ for } j = 1, 2, 3,$$

respectively. If Mallory has been observing these transactions, how can he recover m?

(*Hint: See page 178.*)

9.14. Cite some problems that might occur with the use of voice as a biometric identifier. (See Section 9.4.)

9.15. Compare fingerprint and iris scanning as biometric identifiers from the perspective of which is more accurate and least open to replication. (See Section 9.4.)

9.16. Given the quantum schemes described on page 369, translate the following into binary integers:

$$\uparrow\nwarrow\searrow\rightarrow\rightarrow; \quad \nearrow\nwarrow\uparrow\rightarrow\rightarrow; \quad \nwarrow\nwarrow\rightarrow\rightarrow\rightarrow; \quad \nearrow\nearrow\rightarrow\uparrow;$$

$$\uparrow\rightarrow\rightarrow\uparrow\uparrow; \quad \nearrow\nwarrow\swarrow\nearrow\nwarrow; \quad \uparrow\uparrow\rightarrow\rightarrow .$$

9.17. Translate the binary integers found in Exercise 9.16 into decimal digits and convert to plaintext via Table 1.3 on page 11.

9.18. Suppose that in the nuclear test ban treaty compliance, presented in Section 9.6, Monty selects p, q, e. Then he downloads $n = pq$ and e into HAL, and gives n and d to Hostvania. Demonstrate how Monty can now produce undetectable forgeries. Moreover, in this scenario show how Hostvania can break the treaty and blame Monty.

G.10   Chapter 10 Exercises

10.1. Name ten different priority needs that must be addressed in cybercrime (see Section sec:crime).

(*Hint: Begin with the need for "public awareness" at the top of your list.*)

10.2. Name five advantages to information sharing between policing agencies in various countries when it comes to cybercrime.

10.3. Name five steps that might be used in the gathering of digital data in a cybercrime.

(*Hint: Start with: "Evaluate the target".*)

10.4. Name three methodologies for a malicious hacker to break into a host from the Internet. (See Section 10.2.)

10.5. Compare and contrast worms and viruses, and the mechanisms to protect against them. (See Section 10.3.)

10.6. Review the Clipper Chip enciphering, deciphering, and law enforcement. Decide on advantages and disadvantages of the Skipjack scheme, discussed on pages 420–424.

G.11   Chapter 11 Exercises

11.1. Establish identity (11.4) on page 432.

11.2. Verify Inequality (11.5) on page 432.

11.3. Prove the equivalence of items 1–4 in the "Role of Independence" at the bottom of page 432.

▼ *In Exercises 11.4–11.7, calculate the entropy for the set* $S = \{s_1, s_2, s_3, s_4\}$, *where the probabilities for each* s_j *is given in the exercises via* p_j *for* $j = 1, 2, 3, 4$. *See pages 433–434.*

11.4. $(p_1, p_2, p_3, p_4) = (0.5, 0.25, 0.125, 0.125)$.

11.5. $(p_1, p_2, p_3, p_4) = (0.4, 0.4, 0.1, 0.1)$.

11.6. $(p_1, p_2, p_3, p_4) = (0.3, 0, 3, 0.1, 0.1)$.

11.7. $(p_1, p_2, p_3, p_4) = (0.7, 0.1, 0.1, 0.1)$.

11.8. Calculate the Huffman codes for each of the situations in Exercises 11.1–11.4.

▼ *In Exercises 11.9–11.12, use the data given to calculate each of* $H(\mathcal{K})$, $H(\mathcal{M})$, $H(\mathcal{C})$, *and* $H(\mathcal{K}|\mathcal{C})$. *Then compare the latter with the former three. Assume that* $\mathcal{M} = \{s_1, s_2, s_3, s_4\}$ *and* $\mathcal{C} = \{c_1, c_2, c_3, c_4\}$. *See pages 435–436.*

11.9. $p_{s_1} = 0.2$, $p_{s_2} = 0.3$, $p_{S-3} = 0.4$, $p_{s_4} = 0.1$, $\mathcal{K} = \{k_1, k_2\}$ with $p_{k_1} = 0.4$, $p_{k_2} = 0.6$, $E_{k_1}(s_1) = c_1$, $E_{k_1}(s_2) = c_2$, $E_{k_1}(s_3) = c_3$, $E_{k_1}(s_4) = c_4$, $E_{k_2}(s_1) = c_3$, $E_{k_2}(s_2) = c_2$, $E_{k_2}(s_3) = c_1$, and $E_{k_2}(s_4) = c_4$.

11.10. $p_{s_1} = 0.1$, $p_{s_2} = 0.1$, $p_{s_3} = 0.5$, $p_{s_4} = 0.3$, $\mathcal{K} = \{k_1, k_2, k_3\}$ with $p_{k_1} = 0.5$, $p_{k_2} = 0.2$, $p_{k_3} = 0.3$, $E_{k_1}(s_1) = c_1$, $E_{k_1}(s_2) = c_2$, $E_{k_1}(s_3) = c_3$, $E_{k_1}(s_4) = c_4$, $E_{k_2}(s_1) = c_2$, $E_{k_2}(s_2) = c_3$, $E_{k_2}(s_3) = c_4$, $E_{k_2}(s_4) = c_1$, $E_{k_3}(s_1) = c_3$, $E_{k_3}(s_2) = c_4$, $E_{k_3}(s_3) = c_2$, and $E_{k_3}(s_4) = c_1$.

11.11. $p_{s_1} = 0.2$, $p_{s_2} = 0.2$, $p_{s_3} = 0.2$, $p_{s_4} = 0.4$, $\mathcal{K} = \{k_1, k_2, k_3\}$ with $p_{k_1} = 0.3$, $p_{k_2} = 0.3$, $p_{k_3} = 0.4$, $E_{k_1}(s_1) = c_1$, $E_{k_1}(s_2) = c_2$, $E_{k_1}(s_3) = c_3$, $E_{k_1}(s_4) = c_4$, $E_{k_2}(s_1) = c_4$, $E_{k_2}(s_2) = c_3$, $E_{k_2}(s_3) = c_2$, $E_{k_2}(s_4) = c_1$, $E_{k_3}(s_1) = c_2$, $E_{k_3}(s_2) = c_3$, $E_{k_3}(s_3) = c_4$, and $E_{k_3}(s_4) = c_1$.

11.12. $p_{s_1} = 0.3$, $p_{s_2} = 0.3$, $p_{s_3} = 0.3$, $p_{s_4} = 0.1$, $\mathcal{K} = \{k_1, k_2, k_3\}$ with $p_{k_1} = 0.1$, $p_{k_2} = 0.1$, $p_{k_3} = 0.8$, $E_{k_1}(s_1) = c_1$, $E_{k_1}(s_2) = c_2$, $E_{k_1}(s_3) = c_3$, $E_{k_1}(s_4) = c_4$, $E_{k_2}(s_1) = c_2$, $E_{k_2}(s_2) = c_3$, $E_{k_2}(s_3) = c_4$, $E_{k_2}(s_4) = c_1$, $E_{k_3}(s_1) = c_1$, $E_{k_3}(s_2) = c_3$, $E_{k_3}(s_3) = c_2$, and $E_{k_3}(s_4) = c_4$.

11.13. Prove that if $\mathcal{M}$ is the message space and $\mathcal{C}$ is the ciphertext space for a one-time pad, then $H(\mathcal{M}|\mathcal{C}) = H(\mathcal{M})$. (See pages 439 and 440.)

11.14. Calculate the unicity distance of a block cipher with 56-bit keys and 64-bit blocks of plaintext, assuming use of the English language. (See page 439.)

11.15. In Section 11.5, we looked at error-correcting codes. This book has an ISBN number,[G.3] meaning an *International Standard Book Number*, given by, $1 - 58488 - 470 - 3$. The first digit, 1 refers to the country, area, or language group, in this case the fact that the book is published in the English language. The second group 58488 identifies the publisher. The third group 470 identifies this particular book, and the last digit 3 is a checksum digit. In the case of an ISBN, what this means is that the last digit is chosen such that the following occurs. Suppose that the ten digits in the ISBN are $d_1, d_2, \ldots, d_{10}$. Then we must have that the weighted sum satisfies

$$\sum_{j=1}^{10} j d_j \equiv 0 \,(\mathrm{mod}\ 11).$$

Note that the first nine digits d_j for $j = 1, 2, \ldots, 9$ are in $\{0, 1, 2, \ldots, 9\}$, but $d_{10} \in \{0, 1, 2, \ldots, 10\}$, but in the ISBN number a 10 will be represented as the Roman numeral X. In the case of this book,

$$1 \cdot 1 + 2 \cdot 5 + 3 \cdot 8 + 4 \cdot 4 + 5 \cdot 8 + 6 \cdot 8 + 7 \cdot 4 + 8 \cdot 7 + 9 \cdot 0 + 10 \cdot 3 \equiv 0 \,(\mathrm{mod}\ 11),$$

as required. Are the following valid ISBN numbers?

  1. $0 - 4523 - 2345 - 4$.

  2. $0 - 13 - 061817 - 9$.

  3. $1 - 23 - 098733 - X$.

  4. $2 - 432 - 23459 - 6$.

[G.3]The current ISBN system is reaching the end of its viability. On January 1, 2007, it will be replaced by a 13-digit ISBN. Although the 10-digit ISBN system, designed for printed books in the late 1960s, has the capacity to assign a billion numbers, the internal structure of the ISBN restricts the capacity of the system. The new 13-digit ISBN system will be better suited to integrate with current bar-code technology. See *http://www.isbn-international.org/en/revision.html*.

If an incorrect ISBN number has been transmitted, what is the best way to correct it?

11.16. Prove that the Hamming distance defined on page 443 is indeed a function as asserted. Then establish that the properties, displayed on page 443, of the Hamming function hold.

11.17. Establish the two conditions 1 and 2, for the Hamming function's ability to detect/correct codes, given on page 444.

11.18. Verify the statement made on page 444 to the effect that nearest neighbour decoding may be used to either detect up to $d-1$ errors or to correct up to $(d-1)/2$ errors.

11.19. Prove the facts, stated on page 445, that $A_q(n,1) = q^n$ and $A_q(n,n) = q$.

(Hint: To show the latter it suffices to show that you can always find at least one (n, M, d)-code. To do this, let C be a collection of vectors $(a, a, a, \ldots, a, a_0, \ldots, a_0)$ where there are d repetitions of a, and $n - d$ repetitions of a_0, where $a_0 \in \mathcal{M}$ fixed and $a \in \mathcal{M}$ arbitrary. Conclude that there are q such vectors, with distance $d(c, c') = d$ for $c \neq c'$.)

11.20. Prove that any (n, M, d)-code over $\mathbb{F}_q$ is equivalent to an (n, M, d)-code which has a row of zeros in some matrix representation of it (see page 444).

11.21. Prove that the Singleton bound given on page 445 holds. Conclude that the code rate for such a code is at most $(n - d + 1)/n$. See Footnote 11.6 on page 442.

11.22. Let $w(c)$ denote the number of 1s appearing in a binary code c, called its *weight* (see page 446). Show that if $c, c' \in \mathbb{F}_2^n$, then $d(c, c') = w(c + c') \leq w(c) + w(c')$. Indeed, show that $d(c, c') = w(c) + w(c') - w(c \cap c')$, where $c \cap c'$ is the codeword consisting of 1s in precisely the places j where c and c' both have a 1 in position j.

11.23. Show that if d is even, then there exists a binary $(n - 1, M, d - 1)$-code if and only if a binary (n, M, d)-code exists.

(Hint: Use Exercise 11.22. In particular, when you assume that C is a binary $(n - 1, M, d - 1)$-code, proceed as follows. For each codeword $c = (c_1, \ldots, c_{n-1}) \in C$, define $c' = (c_1, c_2, \ldots, c_{n-1}, c_n)$, where

$$c_n = \sum_{j=1}^{n-1} c_j \,(\mathrm{mod}\ 2).$$

The set C' of new codewords can be shown to be an (n, M, d)-code. This construction of C' from C is often called adding an overall parity check *to C.)*

11.24. Prove that an MDS code, defined on page 445, possesses the largest possible value of d for a given M and n.

11.25. Verify that the cardinality of the Hamming sphere displayed on page 443 is indeed valid.

(*Hint: Calculate* $|\{v \in S(u,r) : d(u,v) = m\}|$ *for each* $m \in \mathbb{N}$.)

11.26. Prove that the Hamming bound displayed on page 445 holds.

(*Hint: Place a Hamming sphere around each codeword so that it has radius t. Then count the number of vectors in these spheres and multiply by the number of codewords.*)

11.27. Establish that the Gilbert-Varshamov bound given on page 446 holds. Conclude that this is also a lower bound for $A_q(n,d)$.

(*Hint: Use an iterative process where you begin with some fixed vector and remove all vectors within a Hamming sphere radius of $d - 1$ from it. Then from the remaining vectors choose another and do the same thing, continuing in this fashion until no vectors remain. Then employ the cardinality of the Hamming sphere established in Exercise 11.25 to conclude that ultimately the bound is achieved.*)

(*Alternative hint: There is an alternative proof for those who are comfortable with the notion of cosets, at this juncture, and have solved Exercise 11.22. In this case, assume that the code has fewer than the number of elements in the bound. Then there is a coset where all the words have weight at least d. However, the union of a linear code with one of its cosets can be shown to be a linear code. Once confirmed this establishes the result.*)

11.28. Prove that $d(C) = \min\{w(c) : c \in C \text{ where } c \neq \overrightarrow{0}\}$ for a linear code C. (See Exercise 11.22 and the discussion on page 446.)

11.29. Prove that two matrices generate equivalent linear $[n,k]$-codes if one matrix can be obtained from the other by operations R1–R3 and C1–C2 described on page 447.

(*Hint: The rows of a generator matrix are linearly independent. Show that the row operations preserve this independence and that the column operations create a generator matrix for an equivalent code.*)

11.30 Prove that if G is the generator matrix for a linear $[n,k]$-code, then operations R1–R3, and C1–C2 transform G into standard form $[I_k|M_{k,n-k}]$, as described on page 447.

11.31. Prove that the matrix P defined on page 447 satisfies that $cP^t = \overrightarrow{0}$ for all $c \in C$.

11.32. Prove that $C^\perp$ defined on page 448 is indeed a linear $[n, n-k]$-code with generating matrix $P = [-M_{k,n-k}^t|I_{n-k}]$, and that G is a parity check matrix for $C^\perp$ as claimed on the aforementioned page.

11.33. Prove that if G is a generator matrix for an $[n, .k]$-code, then it can be reduced to standard form $[I_k | M_{k,n-k}]$ employing *only* row operations R1–R3 if and only if the first k columns of G are linearly independent.

11.34. A linear code C is *self-dual* if $C^\perp = C$ (see page 448). Prove that if C is a self-dual linear $[n, k, d]$-code, then n is even.

11.35. With reference to Exercise 11.34, prove that $(C^\perp)^\perp = C$ for any linear $[n, k]$-code C.

▼ *Exercises 11.36–11.40 pertain to the discussion on page 450 concerning cosets of linear codes.*

11.36. Establish that two vectors are in the same coset if and only if they have the same syndrome. Conclude that there is a one-to-one correspondence between cosets and syndromes.

11.37. Prove that if $\mathbf{e} + C$ is a coset of C and $\mathbf{f} \in \mathbf{e} + C$, then $\mathbf{e} + C = \mathbf{f} + C$.

11.38. Prove that every coset $C + \mathbf{e}$ has exactly q^k vectors.

11.39. Verify that every vector of $\mathbb{F}_q^n$ is in some coset of C.

11.40. Show that two cosets are either identical or are disjoint.

(*Hint: Use Exercise 11.37.*)

11.41. Prove that if C is a linear $[n, k]$-code over $\mathbb{F}_q$ with parity check matrix P, then the minimum distance of C is d if and only if $d - 1$ columns of P are linearly independent, but d columns are linearly dependent.

11.42. If q is a prime power and $N = (q^n - 1)/(q - 1)$ for a given $n \in \mathbb{N}$, prove that there exists a $[N, N - n, 3]$-code.

(*Hint: Consider the sets $\mathcal{S}_v = \{\lambda v : \lambda \in \mathbb{F}_q, \lambda \neq 0\}$ for each nonzero $v \in \mathbb{F}_q^n$. Then $|\mathcal{S}_v| = q - 1$ for each such v and there are N such sets. Select one vector from each $\mathcal{S}_v$, no two of which are linearly dependent. Now use Exercise 11.41 to conclude.*)

11.43. The codes constructed in Exercise 11.42 are called *q-ary Hamming codes*. Prove that these codes are perfect single-error-correcting codes. (See page 445.)

11.44. Show that the repeating $[n, 1]$ linear code with generator matrix $G = [1, 1, \ldots, 1]$ with n odd is a perfect code where the Hamming spheres of radius $(n - 1)/2$ completely fill $\mathbb{F}_q^n$ without overlapping.

11.45. Prove the allegation stated in Footnote 11.8 on page 447.

11.46. In the Example on page 454 of the $[7, 4, 3]$ Hamming code, construct the 8×16 Slepian array. Then extract the syndrome lookup table from it, and use syndrome decoding to find the original message from the following received vectors.

  1. $(0, 1, 1, 0, 0, 1, 0)$.
  2. $(0, 1, 1, 0, 0, 0, 1)$.
  3. $(0, 0, 1, 1, 1, 0, 1)$.
  4. $(0, 1, 1, 0, 1, 0, 0)$.
  5. $(1, 0, 0, 1, 0, 1, 0)$.
  6. $(1, 0, 0, 1, 0, 0, 1)$.
  7. $(0, 0, 0, 1, 1, 0, 0)$.
  8. $(1, 0, 1, 0, 1, 0, 1)$.

☆ 11.47. Prove that the matrix G given on page 456 is the generator matrix for
the Golay code $\mathcal{G}_{24}$, and that properties 1 and 2, listed therein, hold for
this code.

(*Hint: If you have a desire and ability at programming and a computer to
execute your algorithm, you can list all $2^{12} = 4096$ codewords and verify
that $d(\mathcal{G}_{24}) = 8$ directly. Otherwise, first show that the code is self-dual.
To do this, employ the fact that rows 2 through 11 of the matrix $M_{12\times12}$
are cyclic permutations of row 1 as presented on the aforementioned page.
This reduces the work in establishing that any given row of G forms a
zero dot product with all the rows of G. Now you can easily establish that
$[M_{12\times12}|I_{12}]$ is a generator matrix for $\mathcal{G}_{24}$ since $\mathcal{G}_{24}^{\perp}$ has generator matrix
$[M_{12\times12}^{t}|I_{12}]$ and $M_{12\times12} = M_{12\times12}^{t}$. To show that every codeword of $\mathcal{G}_{24}$
has weight divisible by 4, observe that the weight of the intersection is the
dot product, which is even since the code is self dual. Since all rows have
weight divisible by 4, it follows from Exercise 11.22 that the weight of the
sum of any two codewords is divisible by 4. This can be employed to verify
that the weight of any linear combination of rows of G has weight divisible
by 4. To show that there exists no codeword of weight 4, look at all the
possibilities for the weight of such a codeword by breaking it into left and
right components. The outcome will be that $\overrightarrow{0}$ is the only such word. That
G is the generating matrix for $\mathcal{G}_{24}$ now follows from these facts.*)

11.48. Prove that $\mathcal{G}_{23}$ is a linear $[23, 12, 7]$-code.

11.49. Prove that the Golay code $\mathcal{G}_{23}$ is cyclic.

11.50. Prove that the ternary $(11, 6)$ Golay code is perfect.

11.51. Prove that $\mathrm{Ham}(r, 2)$ is a cyclic code for any $r \geq 2$.

11.52. Show that the polynomial $g(x)$ chosen with minimal degree for cyclic
codes described on pages 459 and 460 satisfies properties 1–3 listed therein.

11.53. Show that the matrix G displayed on page 460 is the generator matrix
for the cyclic code C as claimed.

11.54. Prove that P given on page 460 is the parity-check matrix for the cyclic
code C, as claimed.

11.55. Find all binary cyclic codes of length 5.

11.56. Prove that a generator matrix for the binary code Ham(3, 2) is given by

$$G = \begin{pmatrix} 1 & 1 & 0 & 1 & 0 & 0 & 0 \\ 0 & 1 & 1 & 0 & 1 & 0 & 0 \\ 0 & 0 & 1 & 1 & 0 & 1 & 0 \\ 0 & 0 & 0 & 1 & 1 & 0 & 1 \end{pmatrix}$$

corresponding to the polynomial $x^3 + x + 1$ in R_3, and that this matrix may be obtained from the matrix given in Example 11.11 via the operation R1–R3 and C1–C2 on pages 447.

11.57. Suppose that $\mathbf{C}$ is a cyclic $[n, k]$-code with parity-check polynomial $p(x) = p_0 + p_1 x + \cdots + p_{n-k}^{n-k}$. Prove that $\mathbf{C}^\perp$ is a cyclic code generated by the polynomial,

$$p_{n-k} + p_{n-k-1} x + \cdots + p_0 x^{n-k}.$$

☆ **11.58.** Consider the binary cyclic code generated by the polynomial,

$$g(x) = 1 + x^2 + x^4 + x^5 + x^6 + x^{10} + x^{11}.$$

Show that this is a perfect [23, 12, 7]-code that is equivalent to the Golay code $\mathcal{G}_{23}$.

11.59. Show that the ternary code in R_{11} generated by $g(x) = x^5 + x^4 - x^3 + x^2 - 1$ is a 12, 6, 5]-code, which is perfect. This is equivalent to the ternary Golay code discussed on page 457.

11.60. Prove that the BCH bound, $d \geq s + 2$, stated on page 461, actually holds under the conditions given.

(*Hint: Prove by contradiction. Assume there is a code with weight w no bigger than $s + 2$, and select a polynomial $m(x)$ with the code elements as coefficients. Then $g(x) \mid m(x)$. Select the nonzero coefficients $\{c_{i_j}\}_{1 \leq i_j \leq w}$ of $m(x)$ and form a matrix that represents the equation $m(x) = \sum_{j=1}^{w} c_{i_j} x^{i_j} = 0$ at the roots α^{r+k} for $r = 0, 1, 2, \ldots, w - 1$. Then consider the determinant of the coefficient matrix in light of the Vandermonde determinant given on page 494. Once shown that the determinant is nonzero, this forces the coefficients to be zero, a contradiction.*)

11.61. Prove the statement given on page 462: the binary, cyclic $[N, N - r]$-code for $N = 2^r - 1$ (having generator polynomial the minimal polynomial of a primitive element of $\mathbb{F}_{2^r}$ over $\mathbb{F}_2$) is equivalent to Ham(r, 2). (See page 490.)

11.62. Suppose that $p(x)$ is a primitive polynomial (see page 490) over $\mathbb{F}_2$ of degree r. Prove that the cyclic code $\mathbf{C} = \langle p(x) \rangle$ is Ham(r, 2).

(*Hint: Use Exercise 11.61.*)

11.63. Prove that a BCH code of distance d has weight $w \geq d$. (See page 462.)

11.64. Let α be a primitive pth root of unity where p is a prime (see page 487). Prove that
$$\alpha^{p-1} + \alpha^{p-2} + \cdots + \alpha + 1 = 0.$$

11.65. Use Exercise 11.63 to prove that if α is a primitive 7th root of unity, then α, α^2, and α^2 are roots of $x^3 + x + 1 = 0$ over $\mathbb{F}_2$, as claimed in Example 11.14 on page 462.

11.66. Prove that the polynomials $g(x)$ and $p(x)$ given in Example 11.15 on page 463 have the additive structure corresponding to the multiplicative one given. In other words, verify the coefficients of the various powers of x. Also, show that in $\mathbb{F}_q$, the codewords $(110), (011), (111), (101)$ correspond to $\alpha^3, \alpha^4, \alpha^5$, and α^6, respectively.

(*Hint: Use Exercises 11.64 and 11.65.*)

11.67. Establish the properties of Goppa codes listed as 1 and 2 on page 465.

G.12   Appendices Exercises

App.1. Prove the division algorithm presented as Theorem A.2 on page 470.

App.2. Establish the properties of the sigma notation given as Theorem A.5 on page 472.

App.3. Prove Theorem A.11 on page 478.

▼ *Exercises App.4–App.12 concern Euler's function defined on page 479.*

App.4. Prove that $\phi(n) \equiv 2 \,(\mathrm{mod}\ 4)$ if and only if $n = 4$, $n = p^a$ or $n = 2p^a$, where $p \equiv 3 \,(\mathrm{mod}\ 4)$ is prime and $a \in \mathbb{N}$.

App.5. Prove that if $n \in \mathbb{N}$ is even, then
$$\sum_{d|n} (-1)^{n/d} \phi(d) = 0.$$
What is this sum if n is odd?

App.6. Prove that if $n \in \mathbb{N}$ is composite and $\phi(n) \mid (n-1)$, then n is square-free.

App.7. Prove that $\phi(d) \mid \phi(n)$ for all divisors $d \in \mathbb{N}$ of $n \in \mathbb{N}$.

App.8. Prove that for any $a \in \mathbb{Z}$, $a^m \equiv a^{m-\phi(m)} \,(\mathrm{mod}\ m)$ for all $m \in \mathbb{N}$.

App.9. Prove that $\phi(a^n - 1) \equiv 0 \,(\mathrm{mod}\ n)$.

App.10. Prove that $\phi(n) \leq n - \sqrt{n}$ for all composite $n \in \mathbb{N}$.

App.11. Prove that there are infinitely many $n \in \mathbb{N}$ such that $\phi(n) > \phi(n+1)$.

App.12. Suppose that $b, n \in \mathbb{N}$, p is a prime not dividing a, and $g = \gcd(\phi(p^b), n)$. Prove that

$$a^{\phi(p^b)/g} \equiv 1 \,(\text{mod } p^b),$$

if and only if there exists an integer x such that

$$a \equiv x^n \,(\text{mod } p^b).$$

◆ *Exercises App.13–App.16 pertain to indices and related material. See pages 479 and 480.*

App.13. Prove Proposition A.4 on page 480.

App.14. Prove that if c is an integer and p is an odd prime, not dividing c, then there exists an integer x such that

$$c \equiv x^2 \,(\text{mod } p)$$

if and only if $\text{ind}_a^p(c)$ is even for any primitive root a modulo p.

App.15. Assume that p is an odd prime, and $\text{ord}_p(c)$ is odd. Prove that $c^x \equiv -1$ (mod p) has no solution $x \in \mathbb{N}$.

App.16. Given that $m, n \in \mathbb{N}$ are relatively prime. Prove that a is a primitive root modulo mn if and only if a is a primitive root modulo both m and n.

App.17. Let $n \in \mathbb{N}$ be odd. Prove that the Jacobi symbol, $(\frac{m}{n}) = 1$, for all natural numbers $m < n$ with $\gcd(m, n) = 1$ if and only if $n = a^2$ for some $a \in \mathbb{N}$. (See page 482.)

App.18. Given $a \in \mathbb{Z}$. Prove that $x^2 \equiv a \,(\text{mod } p)$ has a solution $x \in \mathbb{Z}$ for all primes p if and only if $a = b^2$ for some $b \in \mathbb{Z}$.

App.19. Prove that $2x^2 - 219y^2 = -1$ is not solvable for any integers x, y.

App.20. Prove that the congruence $2x^2 - 219y^2 \equiv -1 \,(\text{mod } n)$ has solutions $x, y \in \mathbb{Z}$ for all $n \in \mathbb{N}$.

App.21. Prove that a finite integral domain is a field. (See page 483.)

App.22. Suppose that R is a ring, and $\alpha : R \mapsto R$ is an isomorphism (see pages 487–489). Then α is called an *automorphism* of R. Prove that the set of all automorphisms of a group forms a group itself, under composition. This group is typically denoted by $Aut(R)$.

App.23. If K is a field extension of F and $\alpha \in Aut(K)$, such that $\alpha(f) = f$ for all $f \in F$, then α is called an F-automorphism of K. Prove that the group of all F-automorphism of K is a subgroup of $Aut(F)$. This group is denoted by $\mathcal{G}al(K/F)$, called the *Galois group* of K over F.

App.24. With reference to Exercise App.23, determine $\mathcal{G}al(\mathbb{C}/\mathbb{R})$. (For a detailed view of Galois theory and its ramifications, see [168].)

◆ *Exercises App.25–App.29 pertain to continued fractions. See pages 496–498.*

App.25. Suppose that the period length ℓ of the simple continued fraction expansion of $\sqrt{D}$ is even. Prove that all (positive) solutions of the Pell equation $x^2 - Dy^2 = 1$ for $D \in \mathbb{N}$ not a perfect square are given by $x = A_{k\ell-1}$ and $y = B_{k\ell-1}$ for $k \in \mathbb{N}$.

(*Hint: See Corollary A.3 on page 498.*)

App.25. Prove that in the situation given in Exercise App.24, there are no solutions to the Pell equation $x^2 - Dy^2 = -1$.

App.26. Suppose that the period length ℓ of the simple continued fraction expansion of $\sqrt{D}$ for nonsquare $D \in \mathbb{N}$ is odd. Prove that all positive solutions of $x^2 - Dy^2 = 1$ are given by $x = A_{2k\ell-1}$, $y = B_{2k\ell-1}$ for $k \in \mathbb{N}$; and all solutions of $x^2 - Dy^2 = -1$ are given by $x = A_{(2k-1)\ell-1}$, $y = B_{(2k-1)\ell-1}$ for $k \in \mathbb{N}$.

App.27. With reference to Theorem A.32 on page 498, prove that

$$G_{k-1} = P_k B_{k-1} + Q_k B_{k-2},$$

for any nonnegative integer k.

App.28. If ℓ is the period length of the simple continued fraction expansion of $\sqrt{D}$, show that $Q_j = Q_{\ell-j}$ for $0 \le j \le \ell$, and $P_{\ell-j} = P_{j+1}$ for $0 \le j \le \ell - 1$.

App.29. If $D = pq$ where $p \equiv q \equiv 3 \,(\mathrm{mod}\ 4)$ are primes with $p < q$, and $\sqrt{D} = \langle q_0; \overline{q_1, \ldots, q_\ell} \rangle$, prove that ℓ is even. Also, verify that the following Legendre symbol identity holds:[G.4]

$$\left(\frac{p}{q}\right) = (-1)^{\ell/2}.$$

App.30. Let p be a prime and define and elliptic curve $E(\mathbb{F}_p)$ over $\mathbb{F}_p$ by $y^2 = x^3 + ax + b$ for integers a, b. Prove that the number of points on E counting the point at infinity is given by the following Legendre symbol formula.

$$p + 1 + \sum_{x \in \mathbb{F}_p} \left(\frac{x^3 + ax + b}{p}\right).$$

(See pages 498 and 499.)

[G.4]This idea and related issues have been substantially generalized by this author in [171].

App.31. Verify the four properties of Blum Integers given on page 508.

♦ *Use the continued fraction algorithm, described on pages 512 and 513 to factor the integer n given in Exercises App.32–App.36.*

App.32. $n = 6457$.

App.33. $n = 75433$.

App.34. $n = 387181$.

App.35. $n = 98759$.

App.36. $n = 689863$.

App.37. Use Pollard's $p - 1$ algorithm to factor the integers in Exercises App.32–App.36. (See pages 514 and 515.)

App.38. Use Pollard's Rho-Method to factor the integers in Exercises App.32–App.36. (See pages 515–517.)

App.39. Use the QS method to factor the integers in Exercises App.32–App.36. (See pages 517–519.)

App.40. Use the MPQS method to factor the integers in Exercises App.32–App.36. (See pages 519–522.)

App.41. Use the ECM method to factor the integers in Exercises App.32–App.36. (See pages 522–524.)

App.42. Compare the factoring methods used in Exercises App.37–App.41.

♦ *In Exercises App.43–App.46, use the Silver-Pohlig-Hellman algorithm presented on pages 530–532, to find the value of $\log_\alpha \beta$ from the given parameters.*

App.43. $p = 73$, $\alpha = 5$, $\beta = 8$.

App.44. $p = 1637$, $\alpha = 2$, $\beta = 15$.

App.45. $p = 2689$, $\alpha = 19$, $\beta = 27$.

App.46. $p = 2999$, $\alpha = 17$, $\beta = 38$.

App.47. Use the baby-step giant-step algorithm described on pages 533 and 534, to calculate the values in Exercises App.43–App.46.

App.48. Use the index-calculus method, delineated on pages 534–536, to calculate the values in Exercises App.43–App.46.

♦ *In Exercises App.49–App.52, use radix-64 encoding via the description on pages 541 and 542, to encode the given three-byte segments.*

App.49. 01010010, 10000011, 10101011.

App.50. 10101101, 11111100, 11010100.

App.51. 11110000, 00001111, 01010111.

App.52. 11111111, 10000000, 01111110.

◆ *Using the probability theory basics that we learned in Appendix E (pages 543–549) to solve Exercises App.53–App.56.*

App.53. If you have two fair dice, thrown one at a time, find the probability that the sum is 8. What is the probability that the first die is less than the second die? What is the probability that the second die is a 4, given that the first is a 3?

App.54. Given two cards dealt, one after the other, from a fair 52-card deck, what is the probability that they are the same suit? What is the probability they are the same value?

App.55. Suppose that the probability that an event occurs is $p = 0.1\%$. Find the probability that this event occurs no more than twice in a series of 1000 independent trials.

App.56. Assume that Alice buys a lottery ticket for \$2, and her possible winnings are \$1,000, \$10,000, and \$100,000 with respective probabilities of 0.3%, 0.005%, and 0.001%. Determine Alice's possible winnings.

◆ *Use Fermat's compositeness test on page 551 to determine show that the values in Exercises App.57–App.60 are composite*

App.57. $n = 296977$.

App.58. $n = 36977$.

App.59. $n = 45671$.

App.60. $n = 77571$.

App.61. Use the Miller-Selfridge-Rabin test in Section F.2 on page 552, to determine the (probable) status of the values: $n \in \{561, 1729, 14081, 296987\}$.

App.62. Employ the algorithms for probable prime generation in Section F.4 on pages 558 and 559, to produce a list of a half dozen probable primes.

Bibliography

[1] H. Ableson and G.J. Sussman, **Structure and Interpretation of Computer Programs**, MIT Press, Second Edition (1996). (*Cited on page 388.*)

[2] C. Adams, *The CAST-128 Encryption Algorithm*, Internet Request for Comments 2144, (May 1977). (*Cited on pages 275, 507.*)

[3] C. Adams, *Constructing symmetric ciphers using the CAST design procedure*, Designs, Codes, and Cryptography **12**, November (1997), 71–104. (*Cited on pages 275, 507.*)

[4] C. Adams and S. Lloyd, **Understanding Public-Key Infrastructure**, New Riders Publishing, Indianapolis, IN (1999). (*Cited on page 239.*)

[5] L. M. Adleman and M.-D. Huang, *Primality testing and two dimensional Abelian varieties over finite fields*, LNM, Springer-Verlag, Berlin **1512** (1992). (*Cited on page 554.*)

[6] M. Agrawal, N. Kayal, and N. Saxena, *Primes is in P*, preprint. (*Cited on pages 526 and 555–557.*)

[7] W.R. Alford, A. Granville, and C. Pomerance, *There are infinitely many Carmichael numbers*, Ann. Math. **140** (1994), 703–722. (*Cited on page 551.*)

[8] W. Arbaugh, *An inductive chosen plaintext attack against WEP/WEP2*, IEEE doc 802.11-02/30, May (2001) (see: *http://grouper.ieee.org/groups/802/11/*). (*Cited on page 344.*)

[9] J. Arkko and H. Haverinen, *EAP AKA authentication*, INTERNET-DRAFT, *draft-arkko-pppext-eap-aka-11.txt*, October (2003). (*Cited on page 347.*)

[10] D. Atkins, M. Graff, A.K. Lenstra, and P.C. Leyland, *The magic words are SQUEAMISH OSSIFRAGE* in **Advances in Cryptology**, ASIACRYPT '94, Springer-Verlag, Berlin, LNCS **917** (1995), 263–277. (*Cited on page 175.*)

[11] Marcus Aurelius, *Meditations*, from *Marcus Aurelius and His Times, Translations from Paganism to Christianity*, W.J. Black, Inc., Roslyn, NY (1973). (*Cited on page 116.*)

[12] T. Balister, **The Phaistos Disk**, Copyright T. Balister, Production: Ebner, Ulm, Germany, English translation: M. Sheer, first edition, ISBN # 3-9806168-0-0. (*Cited on page 2.*)

[13] M. Bellare and T. Kohno, *SSH transport layer encryption modes*, INTERNET-DRAFT, *draft-ietf-secsh-newmodes-01.txt*, October (2003). (*Cited on page 335.*)

[14] M. Bellare and P. Rogaway, *Random oracles are practical: A paradigm for designing efficient protocols*, in First ACM Conference on Computer and Communications Security, ACM Press, New York (1993), 62–73. (*Cited on page 587.*)

[15] M. Bellare and P. Rogaway, *Optimal asymmetric encryption* in *Advances in Cryptology*, EUROCRYPT '94, Springer-Verlag, Berlin, LNCS **950** (1994), 92–111. (*Cited on page 174.*)

[16] S.M. Bellovin, *Security problems in the TCP/IP protocol suite*, Computer Communications Review **19** (1989), 32–48. (*Cited on page 315.*)

[17] S.M. Bellovin and M. Merritt, *Encrypted key exchange: Password-based protocols secure against dictionary attacks*, in Proceedings of the 1992 IEEE Computer Society Conference on Research in Security and Privacy (1992), 72–84. (*Cited on page 199.*)

[18] S.M. Bellovin and M. Merritt, *Cryptographic protocol for secure communications*, U.S. Patent 5241599, August 31 (1993). (*Cited on page 199.*)

[19] S.M. Bellovin and M. Merritt, *Augmented encrypted key exchange: A password-based protocol secure against dictionary attacks and password file compromise*, Technical report, AT&T Bell Labs (1994). (*Cited on page 199.*)

[20] C. Bennett and G. Brassard, *The dawn of a new era for quantum cryptography: The experimental prototype is working!*, SIGACT News **20**, Fall (1989), 78–82. (*Cited on page 370.*)

[21] C. Bennett, G. Brassard, C. Crépeau, R. Josza, A. Peres, and W. Wooters, *Teleporting an unknown quantum state via dual classical and Einstein-Podolsky-Rosen channels*, Physical Review Letters **70** (1993), 1895–1899. (*Cited on page 371.*)

[22] E. Biham and L.R. Knudsen, *Cryptanalysis of the ANSI X9.52 CBCM mode*, J. Cryptol. (2002), 47–59. (*Cited on page 132.*)

[23] E. Biham and A. Shamir, *Differential cryptanalysis of the full 16-round DES*, Advances in Cryptology — Crypto '92, Springer-Verlag (1993), 487–496. (*Cited on page 127.*)

[24] E. Biham and A. Shamir, **Differential Cryptanalysis of the Data Encryption Standard**, Springer-Verlag, New York (1993). (*Cited on page 127.*)

[25] G.R. Blakely, *Safeguarding cryptographic keys*, Proceedings National Computer Conf., American Federation of Information Processing Societies **48** (1979), 242–268. (*Cited on page 214.*)

[26] D. Bleichenbacher, *Generating ElGamal signatures without knowing the secret key*, in **Advances in Cryptology** — EUROCRYPT '96, Springer-Verlag, Berlin, LNCS **1070** (1996), 10–18. (*Cited on page 189.*)

[27] R. Blom, *An optimal class of symmetric key generation systems*, EURO-CRYPT '84, Springer-Verlag, Berlin, LNCS **209** (1985), 335–338. (*Cited on pages 234, 236.*)

[28] M. Blum, *Coin flipping by telephone: A protocol for solving impossible problems*, in Proceedings of the Twenty-Fourth IEEE Computer Conference, IEEE Press (1982), 133–137. (*Cited on page 208.*)

[29] L. Blum, M. Blum, and M. Shub, *A simple unpredictable pseudo-random number generator*, Siam J. Comput. **15** (1986), 364–383. (*Cited on page 507.*)

[30] M. Blum and S. Goldwasser, *An* efficient *probabilistic public-key encryption scheme which hides all partial information*, in **Advances in Cryptology** — CRYPTO '84, Springer-Verlag, Berlin, LNCS **196** (1985), 289–299. (*Cited on page 508.*)

[31] M. Blum and S. Micali, *How to generate cryptographically strong sequences of pseudo-random bits*, in Proceedings of the Twenty-third IEEE Symposium on Foundations of Computer Science (1982), 112–117. (*Cited on page 508.*)

[32] C. Blundo, A. De Santis, A. Herzberg, S. Kutten, U. Vaccaro, and M. Yung, *Perfectly-secure key distribution for dynamic conferences*, in **Advances in Cryptology**, CRYPTO '92, Springer-Verlag, Berlin, LNCS **740** (1993), 471–486. (*Cited on page 234.*)

[33] L. Blunk, J. Vollbrecht, B. Aboba, J. Carlson, H. Levkowetz, *Extensible authentication protocol (EAP)*, INTERNET-DRAFT, *draft-ietf-eap-rfc2284bis-09.txt*, February 15 (2004). (*Cited on page 346.*)

[34] D. Boneh, R.A. DeMillo, and R.J. Lipton, *On the importance of checking cryptographic protocols for faults*, in **Advances in Cryptology**, EURO-CRYPT '97, Springer-Verlag, Berlin, LNCS **1233** (1997), 37–51. (*Cited on page 361.*)

[35] D. Boneh and G. Durfee, *Cryptanalysis of RSA with private key d less than $N^{0.292}$*, IEEE Transactions on Information Theory **46** (2000), 1339–1349. (*Cited on page 179.*)

[36] D. Boneh, G. Durfee, and Y. Frankel, *An attack on RSA given a fraction of the private key bits*, in **Advances in Cryptology**, ASIACRYPT '98, Springer-Verlag, Berlin, LNCS **1514** (1998), 25–34. (*Cited on page 179.*)

[37] D. Boneh, A. Joux, and P.Q. Nguyen, *Why textbook ElGamal and RSA encryption are insecure*, in **Advances in Cryptology**, ASIACRYPT 2000 (Kyoto), Springer-Verlag, Berlin, LNCS **1976** (2000), 30–43. (*Cited on page 174.*)

[38] N. Borisov, I. Goldberg, and D. Wagner, *Intercepting mobile communications: The insecurity of 802.11*, in Proceedings of the International Conference on Mobile Computing and Networking, July (2001), 180–189. (*Cited on page 344.*)

[39] R.C. Bose and D.K. Ray-Chaudhuri, *On a class of error correcting binary code groups*, Information and Control **3** (1960), 68–79. (*Cited on page 461.*)

[40] R.C. Bose and D.K. Ray-Chaudhuri, *Further results on error correcting binary group codes*, Information and Control **3** (1960), 279–290. (*Cited on page 461.*)

[41] A. Bosselaers, H. Dobbertin, and B. Preneel, *The RIPEMD-160 cryptographic hash function*, Dr. Dobb's Journal, January (1997). (*Cited on page 259.*)

[42] A. Bosselaers and B. Preneel, eds., *Integrity primitives for secure information systems: Final report of the RACE integrity primitives evaluation RIPE-RACE 1040*, Springer-Verlag, Berlin, LNCS **1007**, New York (1995). (*Cited on page 259.*)

[43] R.P. Brent, *An improved Monte Carlo factorization algorithm*, Nordisk Tidskrift för informationsbehandling (BIT) **20** (1980), 176–184. (*Cited on page 516.*)

[44] R.P. Brent and J.M. Pollard, *Factorization of the eighth Fermat number*, Math. Comp. **36** (1981), 627–630. (*Cited on page 517.*)

[45] R. Bright, **Smart Card Principles, Practice, Applications**, LS Horward Books, Chinchester (1988). (*Cited on pages 358–359.*)

[46] J. Brillhart, *Commentary on Lucas' Test*, Fields Inst. Comm. **41** (2004), 103–109. (*Cited on page 551.*)

[47] J. Brillhart, D.H. Lehmer, J.L. Selfridge, B. Tuckerman, and S.S. Wagstaff, Jr., **Factorizations of $b^n \pm 1, b = 2, 3, 5, 6, 7, 10, 11, 12$ up to High Powers**, Contemp. Math. **22**, American Mathematical Society, Providence, RI (1983). (*Cited on page 524.*)

[48] J. Brillhart and J.L. Selfridge, *Some factorizations of* $2^n \pm 1$ *and related results*, Math. Comp. **21** (1967), 87–96. (*Cited on page 510.*)

[49] J. Brunner, **The Shockwave Rider**, Ballentine Books, New York (1975). (*Cited on page 408.*)

[50] J.P. Buhler, H.W. Lenstra Jr., and C. Pomerance, *Factoring integers with the number field sieve*, in **The Development of the Number Field Sieve**, A.K. Lenstra and H.W. Lenstra Jr. eds., LNM, Springer-Verlag, Berlin **1554** (1993), 50–94. (*Cited on pages 524–525.*)

[51] K.W. Campbell and M.J. Weiner, *Proof that DES is not a group* in **Advances in Cryptology** — CRYPTO '92 Proceedings, Springer-Verlag, Berlin, LNCS **740** (1993), 518–526. (*Cited on page 132.*)

[52] J. Chadwick, **The Decipherment of Linear B**, Cambridge University Press, Cambridge, U.K. and New York, USA, Second Ed. (1990). (*Cited on page 31.*)

[53] D. Chaum, *Blind signatures for untraceable payments*, in **Advances in Cryptology**, CRYPTO '82, Plenum Press, New York (1983), 199–203. (*Cited on page 232.*)

[54] D. Chaum, *Online cash checks*, in **Advances in Cryptology**, EURO-CRYPT '89, Springer-Verlag, Berlin, LNCS **434** (1990), 288–293. (*Cited on page 232.*)

[55] D.M. Chess, *Virus verification and removal tools and techniques*, Virus Bulletin, November (1991). (*Cited on page 404.*)

[56] B. Chorr and R.L. Rivest, *A knapsack type public key cryptosystem based on arithmetic in finite fields* in **Advances in Cryptology** — CRYPTO '84, Springer-Verlag, Berlin, LNCS **196** (1985), 54–65. (*Cited on page 105.*)

[57] B. Chorr and R. L. Rivest, *A knapsack type public key cryptosystem based on arithmetic in finite fields* in IEEE Transactions on Information Theory, **34** (1988), 901–909. (*Cited on page 105.*)

[58] C.C. Cocks, *A note on "non-secret" encryption*, GCHQ/CESG publication, November 20 (1973), 1 page. (*Cited on page 103.*)

[59] C.J. Colbourn and J.H. Dinitz, eds., **CRC Handbook of Combinatorial Designs**, CRC Press, Boca Raton, FL (1996). (*Cited on page 457.*)

[60] D. Coppersmith, *The data encryption standard (DES) and its strength against attacks*, IBM Journal of Research and Development, **38**, May (1994), 243–250. (*Cited on pages 117–118.*)

[61] D. Coppersmith, *Small solutions to polynomial equations, and low exponent RSA vulnerabilities*, J. Cryptol. **10** (1997), 233–260. (*Cited on pages 178–179.*)

[62] R. Cramer, R. Gennero, and B. Schoenmakers, *A secure and optimally efficient multi-authority election scheme*, in **Advances in Cryptology**, EUROCRYPT '97, Springer-Verlag, Berlin, LNCS **1233** (1997), 103–118. (*Cited on page 216.*)

[63] Daemon9: Juggernaut, *Phrack Magazine* **50**, April (1997). (*Cited on page 393.*)

[64] C. Davies and R. Ganesan, *BApasswd: A new proactive password checker*, Proceedings Sixteenth National Computer Security Conference, September (1993). (*Cited on page 330.*)

[65] Y. Desmedt, *Simmons' protocol is not free of subliminal channels*, in 9th Foundations Workshop, IEEE Computer Society, Kenmare, Ireland (1996), 170–175. (*Cited on page 374.*)

[66] Y. Desmedt and M. Yung, *Minimal cryptosystems and defining subliminal-freeness*, in Proceedings of the IEEE International Symposium on Information Theory (1994), 347. (*Cited on page 374.*)

[67] J. Dethloff, *Intellectual property rights and smart card patents*, Smart Card Europe, London, December 12 (1995). (*Cited on page 358.*)

[68] T. Dierks and C. Allen, *The TLS protocol, version 1.0*, Internet Request for Comments, 2246 (January 1999). (*Cited on 220.*)

[69] W. Diffie and M.E. Hellman, *New directions in cryptography*, IEEE Transactions on Information Theory **22** (1976), 644–654. (*Cited on pages 98, 100–101.*)

[70] W. Diffie and M.E. Hellman, *Exhaustive cryptanalysis of the NBS data encryption standard*, Computer, June (1977). (*Cited on page 130.*)

[71] W. Diffie, P.C. Van Oorschot, and M.J. Weiner, *Authentication and authenticated key exchanges*, Designs, Codes, and Cryptography **2** (1992), 107–125. (*Cited on page 233.*)

[72] H. Dobbertin, A. Bosselaers, and B. Preneel, *RIPEMD-160: A strengthened version of RIPEMD*, in Proceedings, Third International Workshop on Fast Software Encryption, Springer-Verlag, New York (1996). (*Cited on page 259.*)

[73] M. Dworkin, *Recommendation for block cipher modes of operation — Methods and techniques*, National Institute of Standards and Technology, December (2001). (*Cited on page 136.*)

[74] T. ElGamal, *A public key cryptosystem and signature scheme based on discrete logarithms*, IEEE Transactions on Information Theory **31** (1985), 469–472. (*Cited on pages 185, 187.*)

[75] T. ElGamal, *A public key cryptosystem and a signature scheme based on discrete logarithms*, in **Advances in Cryptology**, CRYPTO '84, Springer-Verlag, Berlin, LNCS **196** (1985), 10–18. (*Cited on page 187.*)

[76] J.H. Ellis, *The possibility of secure non-secret digital encryption*, GCHQ–CESG publication, January (1970). (*Cited on page 103.*)

[77] J.H. Ellis, *The history of non-secret encryption*, GCHQ–CESG publication (1987). (*Cited on page 103.*)

[78] Sir A.J. Evans, **The Palace of Minos: A Comparative Account of the Successive Stages of the Early Cretan Civilization as Illustrated by the Discoveries at Knossos**, Macmillan, London, (1921–1935), (4 volumes plus index, bound into 7 books). (*Cited on page 31.*)

[79] U. Feige, A. Fiat, and A. Shamir, *Zero-knowledge proofs of identity*, Proceedings nineteenth Annu. ACM Symp. Theor. Comput. (1987), 210–217. (*Cited on page 202.*)

[80] U. Feige, A. Fiat, and A. Shamir, *Zero-knowledge proofs of identity*, J. Cryptol. **1** (1988), 77–94. (*Cited on page 202.*)

[81] H. Feistel, *Cryptography and computer privacy*, Sci. Am., May (1973). (*Cited on page 142.*)

[82] N. Ferguson, *Michael: An improved MIC for 802.11 WEP*, IEEE doc 02-020r0, January 17 (2002) (see: *http://grouper.ieee.org/groups/802/11/*). (*Cited on page 349.*)

[83] N. Ferguson, J. Kelsey, S. Lucks, B. Schneier, M. Stay, D. Wagner, and D. Whiting, *Improved cryptanalysis of Rijndael*, in *Fast software encryption*, Seventh International Workshop, FSE 2000, **1978** of LNCS, Springer-Verlag, Berlin (2000), 213–230. (*Cited on page 150.*)

[84] N. Ferguson, R. Schroeppel, and D. Whiting, *A simple algebraic representation of Rijndael*, in **Selected Areas in Cryptography**, S. Vaudenay and A.M. Youssef, ed., SAC, 2001, **2259** of LNCS, Springer-Verlag, Berlin (2001), 103–111. (*Cited on page 150.*)

[85] N. Ferguson and B. Schneier, **Practical Cryptography**, Wiley, Indianapolis, IN (2003). (*Cited on pages 142, 151, 349.*)

[86] R.P. Feynman, **Surely You're Joking Mr. Feynman**, W.W. Norton and Co., New York (1985). (*Cited on page 366.*)

[87] A. Fiat and A. Shamir, *How to prove yourself: Practical solution to identification and signature problems*, in **Advances in Cryptology**, CRYPTO '86, Springer-Verlag, Berlin, LNCS **263** (1987), 186–194. (*Cited on page 202.*)

[88] FIPS 180-1, *Secure Hash Standard* April 17, 1995. (*Cited on page 255.*)

[89] FIPS PUB 180-2, *Secure Hash Standard* (SHS) August 26, 2002. (*Cited on page 255.*)

[90] FIPS PUB 185, *Escrowed Encryption Standard* (EES), February 9, 1994. (*Cited on page 223.*)

[91] FIPS 186, *Digital signature standard*, Federal Information Processing Standards Publication 186, U.S. Department of Commerce/N.I.S.T. National Technical Information Service, Springfield, VA (1994). (*Cited on pages 183, 223.*)

[92] FIPS 186-2, *Digital signature standard*, February (2002). (*Cited on page 183.*)

[93] FIPS 197, *Specification for the Advanced encryption standard (AES)*, November 26 (2001). (*Cited on page 143.*)

[94] FIPS 46-3, *Data encryption standard (DES), defines and specifies the use of DES and triple DES*, November (1999). (*Cited on page 131.*)

[95] S. Fluhrer, I. Mantin, and A. Shamir, *A weakness in the key schedule for RC4*, in Proceedings of the Fourth Annual Workshop on Selected Areas of Cryptography (2001). (*Cited on page 344.*)

[96] S. Fortune and M. Merritt, *Poker protocols* in **Advances in Cryptology**, CRYPTO '84, Springer-Verlag, Berlin, LNCS **196** (1985), 454–466. (*Cited on page 210.*)

[97] K.A. Frenkel, *Brian Reid, A Graphics Tale of a Hacker Tracker*, Comm. ACM **30** (1987), 820-823. (*Cited on page 390.*)

[98] W.F. Friedman and E. Friedman, **Shakespearean Ciphers Examined**, Syndics of the Cambridge University Press, London (1958). (*Cited on page 85.*)

[99] J. Galbraith, T. Ylonen, and S. Lehtinen, *SSH file transfer protocol*, INTERNET-DRAFT, *draft-ietf-secsh-filexfer-05.txt*, January (2004). (*Cited on page 335.*)

[100] M. Gardiner, *A new kind of cipher that would take millions of years to break*, Sci. Am. **237**, August (1977), 120–124. (*Cited on pages 175, 271.*)

[101] M.R. Garey and D.S. Johnson, **Computers and Intractability**, Freeman, New York, Twenty-second printing (2000). (*Cited on page 169.*)

[102] C.F. Gauss, **Disquisitiones Arithmeticae** (English Edition), Springer-Verlag, Berlin (1985). (*Cited on page 511.*)

[103] W. Gibson, *Neuromancer*, Ace Books (Reissue Edition) (2003). (*Cited on page 377.*)

[104] E.N. Gilbert, *A comparison of signalling alphabets*, Bell Systems Technical Journal **31** (1952), 504–522. (*Cited on page 446.*)

[105] M.J.E. Golay, *Notes on digital coding*, Proc. IEEE **37** (1949), 657. (*Cited on page 455.*)

[106] M.J.E. Golay, *Binary coding*, Trans. IRE PGIT **4** (1954), 23–28. (*Cited on page 455.*)

[107] M.J.E. Golay, *Notes on the penny-weighing problem, lossless symbol coding with nonprimes, etc.*, IEEE Transactions on Information Theory **4** (1958), 103–109. (*Cited on page 455.*)

[108] S. Goldwasser and J. Kilian, *Almost all primes can be quickly certified*, Proceedings Eighteenth Annu. ACM Symp. Theor. Comput. (1986), 316–329. (*Cited on page 554.*)

[109] S. Goldwasser and S. Micali, *Probabilistic encryption*, Journal of Computer and System Sciences (1984), 270–299. (*Cited on page 508.*)

[110] S. Goldwasser, S. Micali, and P. Tong, *Why and how to establish a private code on a public network*, in Proceedings of the Twenty-third IEEE Symposium on Foundations of Computer Science (1982), 134–144. (*Cited on page 508.*)

[111] S. Golomb, **Shift Register Sequences**, Revised edition, Aegean Park Press (1982). (*Cited on pages 158 and 571.*)

[112] V.D. Goppa, *A new class of linear error-correcting codes*, Problems of Information Transmission **6** (1970), 207–212. (*Cited on page 465.*)

[113] J. Gordon, *Strong primes are easy to find*, in **Advances in Cryptology**, EUROCRYPT '84, Springer-Verlag, Berlin, LNCS **209** (1985), 216–223. (*Cited on page 558.*)

[114] G. Grätzer, **Lattice Theory: First Concepts and Distributive Lattices**, W.H. Freeman, San Francisco (1971). (*Cited on page 457.*)

[115] R. Graves, **The White Goddess — A Historical Grammar of Poetic Myth**, Carcanet Press Ltd. (1999). (*Cited on page 15.*)

[116] K. Hafner, **Cyberpunks: Outlaws and Hackers on the Cyber Frontier — revised**, Simon and Schuster (1995). (*Cited on pages 315, 384.*)

[117] R.W. Hamming, *Error detecting and error correcting codes*, Bell Systems Technical Journal **29** (1950), 147–160. (*Cited on pages 442 and 455.*)

[118] R.W. Hamming, **Coding and Information Theory**, Prentice-Hall, Englewood Cliffs, NJ (1980). (*Cited on page 442.*)

[119] M. Handley, C. Kreibich, and V. Paxson, *Network intrusion detection: Evasion, traffic normalization, and end-to-end semantics*, Proceedings of the USENIX Security Symposium (2001), 115–131. (*Cited on page 395.*)

[120] J. Hastad, *Solving simultaneous modular equations of low degree*, Siam J. Comput. **17** (1988), 336–341. (*Cited on page 178.*)

[121] S. Hawking, **A Brief History of Time**, Bantam, New York (1988). (*Cited on page 159.*)

[122] H.M. Heys, *A tutorial on linear and differential cryptanalysis*, Technical Report CORR 2001-17, Department of Combinatorics and Optimization, University of Waterloo, Waterloo, Canada (2001). (*Cited on page 132.*)

[123] L. S. Hill, *Cryptography in an algebraic alphabet*, Amer. Math. Monthly, **36**, 306–312 (1929). (*Cited on page 87.*)

[124] A. Hocquenghem, *Codes correcteurs d'erreurs*, Chiffres **2** (1959), 147–156. (*Cited on page 461.*)

[125] R.J. Hughes, G.G. Luther, G.L. Morgan, C.G. Peterson, and C. Simmons, *Quantum cryptography over underground optical fibers*, in Advances in Cryptology — Crypto '96, LNCS **1109**, Springer-Verlag, Berlin (1996), 329-342. (*Cited on page 370.*)

[126] ISO/IEC 9594-8, *Information technology — Open Systems Interconnection — The Directory: Authentication framework*, International Organization for Standardization, Geneva, Switzerland (1995). (*Cited on page 268.*)

[127] ITU-T Recommendation X.500, *The directory — Overview of concepts and models*, ITU, Geneva, Switzerland (1997). (*Cited on page 240.*)

[128] ITU-T Recommendation X.509, *Information technology — Open systems interconnection — The directory: Authentication framework* (June 1997). (*Cited on page 238.*)

[129] D.P. Jablon, *Strong password-only authenticated key exchange*, Computer Communications Review, ACM SIGCOMM **26** (1996), 5–26. (*Cited on page 199.*)

[130] R.R. Juenman, *Analysis of certain aspects of output feedback mode* in Advances in Cryptology — Crypto '82, Plenum Press (1996), 99–128. (*Cited on page 136.*)

[131] D. Kahn, **The Codebreakers**, Macmillan, New York (1967). (*Cited on pages 4, 79.*)

[132] L. Keliher, H. Meijer, and S. Tavares, *Improving the upper bound on the maximum average linear probability for Rijndael*, in **Selected Areas in Cryptography**, S. Vaudenay and A.M. Youssef, ed., SAC, 2001, LNCS **2259**, Springer-Verlag, Berlin (2001), 112–128. (*Cited on page 150.*)

[133] J.O. Kephart, G.B. Sorkin, M. Swimmer, and S.R. White, *Blueprint for a computer immune system*, Proceedings of the 1997 International Virus Bulletin Conference, San Francisco, October (1997). (*Cited on page 404.*)

[134] B.W. Kernighan and R. Pike, **The UNIX Programming Environment**, Prentice Hall, Computer Books, Upper Saddle River, NJ (1984). (*Cited on page 390.*)

[135] B.W. Kernighan and D. Ritchie, **C Programming Language**, Prentice Hall PTR, Upper Saddle River, NJ, Second Edition (1988). (*Cited on pages 390,391.*)

[136] J. Kilian and P. Rogaway, *How to protect DES against exhaustive key search*, in Advances in Cryptology - Crypto '96, Springer-Verlag, Berlin (1996), 252-267. (*Cited on page 132.*)

[137] G. Kipper, **Investigator's Guide to Steganography**, Auerbach Pub., Boca Raton, FL (2004). (*Cited on pages 192, 374.*)

[138] D.E. Knuth, **The Art of Computer Programming**, Volume **2**, **Seminumerical Algorithms**, Third Edition, Addison-Wesley, Reading MA and Paris (1998). (*Cited on page 83.*)

[139] P. Kocher, *Timing attacks on implementations of Diffie-Hellman, RSA, DSS, and other systems*, in **Advances in Cryptology**, CRYPTO '96, Springer-Verlag, Berlin, LNCS **1109** (1996), 104–113. (*Cited on page 176.*)

[140] H. Kragh, **Quantum Generations**, Princeton University Press, NJ (1999). (*Cited on page 366.*)

[141] E. Kranakis, **Primality and Cryptography**, Wiley, New York (1986). (*Cited on page 212.*)

[142] H. Krawczyk, *The order of encryption and authentication for protecting communications (or: How secure is SSL?)*, in **Advances in Cryptology** — CRYPTO 2001, Springer-Verlag, Berlin, LNCS **2139** (2001), 310–331. (*Cited on pages 225, 267, 339.*)

[143] J. Leech, *Some sphere packings in higher space*, Canad. J. Math. **16** (1964), 657–682. (*Cited on page 457.*)

[144] R.S. Lehman, *Factoring large integers*, Math. Comp. **28** (1974), 637–646. (*Cited on page 510.*)

[145] D.H. Lehmer and R.E. Powers, *On factoring large numbers*, Bull. Amer. Math. Soc. **37** (1931), 770–736. (*Cited on page 511.*)

[146] S. Lehtinen, *SSH protocol assigned numbers*, INTERNET-DRAFT, *draft-ietf-secsh-assignednumbers-05.txt*, October (2003). (*Cited on pages 335, 339.*)

[147] H.W. Lenstra Jr., *Primality testing with Artin symbols*, in **Number Theory Related to Fermat's Last Theorem**, N. Koblitz, ed., **Progress in Math. 26**, Birkhäuser, Boston (1982), 341–347. (*Cited on page 554.*)

[148] A.K. Lenstra and M.S. Manasse, *Factoring by electronic mail*, in **Advances in Cryptology**, EUROCRYPT '89, Springer-Verlag, Berlin, LNCS **434** (1990), 355–371. (*Cited on page 175.*)

[149] A.K. Lenstra and E.R. Verheul, *Selecting cryptographic key sizes*, J. Cryptol. **14** (2001), 255–293. (*Cited on page 178.*)

[150] S. Levy, **Hackers: Heros of the Computer Revolution**, Anchor (1984) (updated January 2001 by Penguin USA). (*Cited on pages 385, 387, 391.*)

[151] S. Levy, **Crypto**, Penguin Books, New York (2001). (*Cited on pages 97, 120, 165, 168, 187.*)

[152] R. Lidl, G.L. Mullen, and G. Turnwald, **Dickson Polynomials**, Longman Science and Technical (1993). (*Cited on page 581.*)

[153] R. Lidl and H. Niederreiter, **Finite Fields**, Encyclopedia of Mathematics and its Applications, **20**, Addison-Wesley, Reading, MA (1983). (*Cited on page 463.*)

[154] J.H. van Lint, *A survey of perfect codes*, Rocky Mountain J. Math. **5** (1975), 199–224. (*Cited on page 458.*)

[155] C. Marand and P. Townsend, *Quantum key distribution over distances as long as 30 km*, Optic Letters **20** (1995), 1695–1697. (*Cited on page 370.*)

[156] M. Matsui, *Linear cryptanalysis method for the DES cipher* in **Advances in Cryptology** — EUROCRYPT '93, Springer-Verlag, Berlin, LNCS **765** (1994), 386–397. (*Cited on page 132.*)

[157] M. Matsui, *The first experimental cryptanalysis of the Data Encryption Standard* in **Advances in Cryptology** — CRYPTO '94, Springer-Verlag, Berlin, LNCS **839** (1994), 1–11. (*Cited on page 132.*)

[158] B. McLean and P. Elkind, **The Smartest Guys in the Room: The Amazing Rise and Scandalous Fall of Enron**, Portfolio (2003). (*Cited on page 379.*)

[159] A.J. Menezes, P.C. van Oorschot, and S.A. Vanstone, **Handbook of Applied Cryptography**, CRC Press, Boca Raton, FL (1997). (*Cited on pages 252, 275, 507, 534.*)

[160] R.C. Merkle and M. E. Hellman, *Hiding information and signatures in trapdoor knapsacks*, IEEE Trans. Inform. Theory, **24** (1978), 525–530. (*Cited on page 104.*)

[161] R. Merkle and M. Hellman, *On the security of multiple encryption*, Communications of he ACM **24** (1981), 465–467. (*Cited on page 130.*)

[162] G.L. Miller, *Riemann's hypothesis and tests for primality*, J. Comput. Sys. Sci. **13** (1976), 300–317. (*Cited on page 554.*)

[163] K.D. Mitnick, W.L. Simon, and S. Wozniak, **The Art of Deception: Controlling the Human Element of Security**, Wiley, New York (2002). (*Cited on page 394.*)

[164] R.A. Mollin, ed., **Number Theory and Applications**, NATO ASI **C265**, Kluwer Academic Publishers, Dordrecht, The Netherlands (1989). (*Cited on page xi.*)

[165] R.A. Mollin, ed., **Number Theory**, Proceedings First Conf. of the Canadian Number Theory Association, Walter de Gruyter, Berlin (1990). (*Cited on page xi.*)

[166] R.A. Mollin, **Quadratics**, CRC Press, Boca Raton, FL (1996). (*Cited on page xi.*)

[167] R.A. Mollin, **Fundamental Number Theory with Applications**, CRC Press, Boca Raton, FL (1998). (*Cited on pages xi, 466, 469, 471, 473–474, 479, 496–498, 515, 534.*)

[168] R.A. Mollin, **Algebraic Number Theory**, Chapman Hall/CRC Press, Boca Raton, FL (1999). (*Cited on pages xi, 466, 487, 524, 554, 601.*)

[169] R.A. Mollin, **An Introduction to Cryptography**, Chapman Hall/CRC Press, Boca Raton, FL (2000). (*Cited on pages xi, 115, 145, 158, 190, 202, 419, 466, 477, 479–482, 499, 502, 511, 513, 517, 534, 551.*)

[170] R.A. Mollin, **RSA and Public-Key Cryptography**, Chapman Hall/CRC Press, Boca Raton, FL (2003). (*Cited on pages xi, 44, 170, 179, 197, 199, 207, 254, 466, 519, 522, 524, 532, 551, 554, 558–559.*)

[171] R.A. Mollin, *A continued fraction approach to the Diophantine equation $ax^2 - by^2 = \pm 1$*, JP Jour. Algebra, Number Theory and Appl. **41** (2004), 159–207. (*Cited on page 601.*)

[172] R. Moreno, and P. Le Clech, *IPR and smart card patents — France*, (Innovatron), Smart Card Europe, London, December 12 (1995). (*Cited on page 358.*)

[173] M.A. Morrison and J. Brillhart, *A method of factoring and the factorization of F_7*, Math. Comp. **29** (1975), 183–205. (*Cited on page 513.*)

[174] W.B. Müller and and R. Nörbauer, *Cryptanalysis of the Dickson-scheme* in Advances in Cryptology, EUROCRYPT '85 (Linz 1985), Springer-Verlag, Berlin, LNCS **219** (1986), 50–61. (*Cited on page 581.*)

[175] S. Murphy, *The cryptanalysis of FEAL-4 with 20 chosen plaintexts*, J. Cryptol. **2** (1990), 145-154. (*Cited on page 127.*)

[176] M. Myers, R. Ankeny, A. Malpani, S. Galperin, and C. Adams, *X.509 Internet Public Key Infrastructure On-Line Certificate Status protocol — OCSP*, Internet Request for Comments 2560 (June 1999). (*Cited on page 240.*)

[177] R.M. Needham and M.D. Schroeder, *Using encryption for authentication in large networks of computers*, Commun. ACM **21** (1978), 993–999. (*Cited on page 197.*)

[178] R.M. Needham and M.D. Schroeder, *Authentication revisited*, Operating Syst. Rev. **21** (1987), 7. (*Cited on page 197.*)

[179] W.B. Nöbauer, *Über Gruppen von Dickson-Polynomfunktionen und einige damit zusammenhängende zahlentheoretische Fragen*, Monatsh. Math. **77** (1973), 330–344. (*Cited on page 578.*)

[180] W.B. Nöbauer, *Über die Fixpunkte der Dickson-Permutationen*, Österreich. Akad. Wiss. Math.-Natur. Kl. Sitzungsber. II **193** (1984), 115–133. (*Cited on page 578.*)

[181] W.B. Nöbauer, *Rédei-Funktionen für Zweierpotenzen*, Period. Math. Hungar. **17** (1986), 37–44. (*Cited on page 581.*)

[182] T. Okamoto and K. Ohta, *Universal electronic cash*, in **Advances in Cryptology**, CRYPTO '91, Springer-Verlag, Berlin, LNCS **576** (1992), 324–337. (*Cited on page 231.*)

[183] P. van Oorschot, *A comparison of practical public-key cryptosystems based on integer factorization and discrete logarithms*, in **Contemporary Cryptography: The Science of Information Integrity**, G. Simmons, ed., IEEE Press, Piscatoway, NJ (1992), 289–322. (*Cited on page 165.*)

[184] D.H. Phan and D. Pointchevel, *About the security of ciphers* (*Semantic security and pseudo-random permutations*), in Proceedings of the Eleventh Annual Workshop on Selected Areas of Cryptography, Waterloo, Canada (H.Handschuh and A. Hasan eds.) (2004). (*Cited on page 508.*)

[185] PKCS1, *Public key cryptography standard no. 1 version 2.0*, RSA Labs. (*Cited on page 174.*)

[186] E.A. Poe, **Edgar Allan Poe Selected Works**, Random House, Toronto, New York (1990), 357–381. (*Cited on page 37.*)

[187] S.C. Pohlig and M.E. Hellman, *An improved algorithm for computing logarithms in GF(p) and its cryptographic significance*, IEEE Transactions on Information Theory **24** (1978), 106–111. (*Cited on page 530.*)

[188] J.M. Pollard, *Theorems on factorization and primality testing*, Proc. Cambr. Philos. Soc. **76** (1974), 521–528. (*Cited on page 515.*)

[189] J.M. Pollard, *An algorithm for testing the primality of any integer*, Bull. London Math. Soc. **3** (1971), 337–340. (*Cited on page 513.*)

[190] C. Pomerance, *The quadratic sieve factoring algorithm* in in **Advances in Cryptology**, EUROCRYPT '84, Springer-Verlag, Berlin, LNCS **209** (1985), 169–182. (*Cited on page 517.*)

[191] C. Pomerance, *The number field sieve*, Proc. Symp. Appl. Math. **48** (1994), 465–480. (*Cited on page 526.*)

[192] O. Pretzel, **Codes and Algebraic Curves**, Oxford Lecture Series in Math. and Apps., Clarendon Press, Oxford (1998). (*Cited on page 465.*)

[193] RFC 1602,* *The Internet standards process — Revision 2*, March (1994). (*Cited on page 326.*)

[194] RFC 1928, *SOCKS protocol version 5*, March (1996). (*Cited on page 319.*)

[195] RFC 1929, *Username/password authentication for SOCKS V5*, March (1996). (*Cited on page 320.*)

[196] RFC 1961, *GSS-API authentication method for SOCKS Version 5*, June (1996). (*Cited on page 320.*)

[197] RFC 2026, *The Internet standards process — Revision 3*, October (1996). (*Cited on page 335.*)

[198] RFC 2045, *Multipurpose Internet Mail Extensions (MIME) Part One: Format of Internet Message Bodies*, November (1996). (*Cited on page 290.*)

[199] RFC 2046, *Multipurpose Internet Mail Extensions (MIME) Part Two: Media Types*, November (1996). (*Cited on page 290.*)

[200] RFC 2047, *MIME (Multipurpose Internet Mail Extensions) Part Three: Message Header Extensions for Non-ASCII Text*, November (1996). (*Cited on page 290.*)

[201] RFC 2048, *Multipurpose Internet Mail Extensions (MIME) Part Four: Registration Procedures*, November (1996). (*Cited on page 290.*)

*All RFC documents may be downloaded from the "Active RFC Index", by specifying the number, at *http://www.faqs.org/rfcs/rfc-activeT.html*.

[202] RFC 2049, *Multipurpose Internet Mail Extensions (MIME) Part Five: Conformance Criteria and Examples,* November (1996). (*Cited on page 290.*)

[203] RFC 2109, *RFC 2109 — HTTP state management mechanism,* February (1997). (*Cited on page 324.*)

[204] RFC 2119, *Key words for use in RFCs to indicate requirement levels, BCP 14,* March (1997). (*Cited on pages 288, 303.*)

[205] RFC 2138, *Remote Authentication Dial In User Service (RADIUS),* April (1997). (*Cited on page 347.*)

[206] RFC 2313, B. Kaliski, *PKCS #1: RSA encryption version 1.5,* March (1998). (*Cited on page 181.*)

[207] RFC 2402, *IP authentication header,* November (1998). (*Cited on page 303.*)

[208] RFC 2403, *The Use of HMAC-MD5-96 within ESP and AH,* November (1998). (*Cited on page 306.*)

[209] RFC 2404, *The use of HMAC-SHA-1-96 within ESP and AH,* November (1998). (*Cited on page 306.*)

[210] RFC 2405, *The ESP DES-CBC cipher algorithm with explicit IV,* November (1998). (*Cited on page 308.*)

[211] RFC 2406. *IP Encapsulating Security Payload (ESP),* November (1998). (*Cited on page 303.*)

[212] RFC 2409. *The Internet Key Exchange (IKE),* November (1998). (*Cited on page 297.*)

[213] RFC 2440, *OpenPGP Message Format,* November (1998). (*Cited on pages 541, 542.*)

[214] RFC 2451, *The ESP CBC-Mode cipher algorithms,* November (1998). (*Cited on page 307.*)

[215] RFC 2459, *Internet X.509 public key infrastructure certificate and CRL profile,* January (1999). (*Cited on page 356.*)

[216] RFC 2630, *Cryptographic message syntax, defines a cryptographic algorithm independent format for signed and encrypted data,* June (1999). (*Cited on pages 287, 289.*)

[217] RFC 2631, *Diffie-Hellman key exchange method, defines a variant of the Diffie-Hellman cryptographic algorithm as a mandatory key agreement for S/MIMEV3,* June (1999). (*Cited on page 287.*)

[218] RFC 2632, *S/MIME Version 3 certificate handling, describes how an S/MIME V3 client uses public key infrastructure (PKI) to establish that a public key is valid,* June (1999). (*Cited on page 287.*)

[219] RFC 2633, S/MIME Version 3 message specification, describes the protocol for adding cryptographic signature and encryption services to MIME data, June (1999). (*Cited on page 287.*)

[220] RFC 2634, *Enhanced security services for S/MIME, describes four optional security service extensions: signed receipts; security labels; secure mailing lists; and signing certificates,* June (1999). (*Cited on page 287.*)

[221] RFC 2716, *PPP EAP TLS authentication protocol,* October (1999). (*Cited on page 346.*)

[222] RFC 2857, *The Use of HMAC-RIPEMD-160-96 within ESP and AH,* June (2000). (*Cited on page 588.*)

[223] RFC 2984, *Use of the CAST-128 Encryption Algorithm in CMS,* October (2000). (*Cited on page 338.*)

[224] RFC 3566, *The AES-XCBC-MAC-96 algorithm and its use with IPsec,* September (2003). (*Cited on page 306.*)

[225] RFC 3602, *The AES-CBC cipher algorithm and its use with IPsec,* September (2003). (*Cited on page 308.*)

[226] RFC 3610, *Counter with CBC-MAC (CCM),* September (2003). (*Cited on page 262.*)

[227] H. Rheingold, **Tools for Thought: The History and Future of Mind-Expanding Technology**, Prentice-Hall (1986); Second revised edition by MIT Press (2000). (*Cited on page 387.*)

[228] R.L. Rivest, *The MD4 Message Digest Algorithm,* in **Advances in Cryptology**, CRYPTO '90, Springer-Verlag, Berlin, LNCS **537** (1990), 303–311. (*Cited on page 255.*)

[229] R.L. Rivest, *The MD5 Message-Digest Algorithm,* RFC 1321, April (1992). (*Cited on page 255.*)

[230] R.L. Rivest, A. Shamir, and L. Adleman, *A method for obtaining digital signatures and public-key cryptosystems,* Communications of the A.C.M. **21** (1978), 120–126. (*Cited on page 101.*)

[231] P. Rogaway, *The security of DESX,* CryptoBytes **2** (Summer 1996). (*Cited on page 132.*)

[232] E. Schaefer, *A simplified data encryption standard algorithm,* Cryptologia, January (1996). (*Cited on pages 115, 123.*)

[233] H. Schliemann, **Troy and Its Remains**, (1875), reprinted by Ayer Pub. Co., Manchester, NH (1968). (*Cited on page 31.*)

[234] H. Schliemann, **Ilios: The City and Country of the Trojans** (1880), reprinted by Ayer Pub. Co., Manchester, NH (1968). (*Cited on page 31.*)

[235] B. Schneier, *Description of a new variable-length key, 64-bit block cipher* (*Blowfish*), Proceedings Workshop on Fast Software Encryption, Springer-Verlag, New York, December (1993). (*Cited on page 138.*)

[236] B. Schneier, *The Blowfish encryption algorithm*, Dr. Dobb's Journal, April (1994). (*Cited on page 138.*)

[237] B. Schneier, *The Blowfish encryption algorithm — One year later*, Dr. Dobb's Journal, September (1995). (*Cited on page 138.*)

[238] B. Schneier, **Applied Cryptography**, Second Edition, Wiley, New York (1995). (*Cited on page 138.*)

[239] B. Schneier, **Secrets and Lies**, Wiley, New York, (2000). (*Cited on page 138.*)

[240] B. Schneier, **Beyond Fear**, Wiley, New York (2003). (*Cited on pages 138, 340.*)

[241] B. Schneier, J. Kelsey, D. Whiting, D. Wagner, C. Hall, and N. Ferguson, **The Twofish Encryption Algorithm: A 128-bit Block Cipher**, Wiley, New York (1999). (*Cited on page 142.*)

[242] C.P. Schnorr, *Efficient signature by smart cards*, J. Cryptol. **4** (1991), 161–174. (*Cited on page 205.*)

[243] R. Schoof, *Elliptic curves over finite fields and the computation of square roots mod p*, Math. Comp. **44** (1985), 483–494. (*Cited on page 554.*)

[244] I. Schur, *Arithmetisches über die Tschebyscheffschen Polynome*, Ges. Abhandlungen, **III**, Springer-Verlag, Berlin-New York (1973), 422–453. (*Cited on page 578.*)

[245] G. Seldes, **Great Thoughts**, Ballentine Books, New York (1996). (*Cited on page 375.*)

[246] A. Shamir, *How to share a secret*, Commun. ACM **22** (1979), 612–613. (*Cited on page 212.*)

[247] A. Shamir, *A polynomial-time algorithm for breaking the basic Merkle-Hellman cryptosystem* in **Advances in Cryptology** — CRYPTO '82 Proceedings, Plenum Press, New York (1983), 279–288. (*Cited on page 104.*)

[248] A. Shamir, *A polynomial-time algorithm for breaking the basic Merkle-Hellman cryptosystem*, IEEE Trans. Inform. Theory, **30** (1984), 699–704. (*Cited on page 104.*)

[249] C.E. Shannon, *A mathematical theory of communication*, Bell Systems Technical Journal **27** (1948), 379–423, 623–656. (*Cited on pages 425–427, 429, 437.*)

[250] C.E. Shannon, *Communication theory of secrecy systems*, Bell Systems Technical Journal **28** (1949), 656–715. (*Cited on pages 128, 129.*)

[251] J.F. Shoch and J.A. Hupp, *The 'worm' programs: Early experience with a distributed computation*, Commun. ACM **25** (1982), 172–180. (*Cited on page 408.*)

[252] P. Shor, *Algorithms for quantum computation: discrete logarithms and factoring*, Proceedings 35-th IEEE Annual Symposium on Foundations of Computer Science (1994), 124–134. (*Cited on page 371.*)

[253] G.J. Simmons, *Message authentication without secrecy: A secure communications problem uniquely solvable by asymmetric techniques*, in IEEE Electronics and Aerospace Systems Convention, EASCON '79 Record, Arlington, VA (1979), 661–662. (*Cited on pages 372, 373.*)

[254] G.J. Simmons, *Message authentication without secrecy*, in **Secure Communications and Asymmetric Cryptosystems**, G.J. Simmons, ed., AAAS Selected Symposia Series **69**, Westview Press, Boulder, Colorado (1982), 105–139. (*Cited on pages 372, 374.*)

[255] G.J. Simmons, *Verification of treaty compliance-revisited*, in Proceedings of the 1983 IEEE Symposium on Security and Privacy, IEEE Computer Society Press, Oakland, CA (1983), 25–27. (*Cited on page 372.*)

[256] G.J. Simmons, *The prisoner's problem and the subliminal channel*, in **Advances in Cryptology**, CRYPTO '83, Plenum Press, New York (1984), 51–67. (*Cited on page 374.*)

[257] G.J. Simmons, *The subliminal channels in the U.S. digital signature algorithm (DSA)*, in Proceedings of the Third Symposium on: State and Progress of Research in Cryptography, Rome, Italy (1993), 35–54. (*Cited on page 374.*)

[258] G.J. Simmons, *An introduction to the mathematics of trust in security protocols*, in Proceedings: Computer Security Foundations Workshop VI, IEEE Computer Society Press, Franconia, NH (1993), 121–127. (*Cited on page 374.*)

[259] G.J. Simmons, *Cryptanalysis and protocol failures*, Commun. ACM **37** (1994), 56–65. (*Cited on page 374.*)

[260] G.J. Simmons, *Subliminal communication is easy using the DSA*, in **Advances in Cryptology**, EUROCRYPT '93, Springer-Verlag, Berlin, LNCS **765** (1994), 218–232. (*Cited on page 374.*)

[261] G.J. Simmons, *Subliminal channels: Past and present*, Euro. Trans. Telecommun. **5** (1994), 459–473. (*Cited on page 374.*)

[262] G.J. Simmons, *Protocols that ensure fairness*, in **Codes and Ciphers**, P.G. Farrell, ed., Royal Agricultural College, Cirencester, 1993 (1995), 13–15. (*Cited on page 374.*)

[263] G.J. Simmons, *Results concerning bandwidth of subliminal channels*, IEEE J. Selected Areas Commun. **16** (1998), 463–473. (*Cited on page 374.*)

[264] R.C. Singleton *Maximum distance q-nary codes*, IEEE Transactions on Information Theory, **10** (1964), 116–118. (*Cited on page 445.*)

[265] D. Slepian, *Some further theory of group codes*, Bell Systems Technical Journal **39** (1960), 1219–1252. (*Cited on page 451.*)

[266] R. Smith and J.R. Emshwiller, *24 Days: How Two Wall Street Journal Reporters Uncovered the Lies that Destroyed Faith in Corporate America or Infectious Greed*, HarperCollins (2003). (*Cited on page 379.*)

[267] E. Spafford, *Observing reusable password choices*, Proceedings UNIX Security Symposium III, September (1992). (*Cited on page 330.*)

[268] E. Spafford, *OPUS: Preventing weak password choices*, Computers and Security, **3** (1992). (*Cited on page 330.*)

[269] L. Spitzner, **Honeypots: Tracking Hackers**, Addison Wesley, Boston (2002). (*Cited on page 395.*)

[270] R. Stallman, **Free Software, Free Society: Selected Essays of Richard M. Stallman**, Free Software Foundation (2002). (*Cited on page 387.*)

[271] D. Stebila, *Elliptic-curve Diffie-Hellman key exchange for the SSH transport level protocol*, INTERNET-DRAFT, *draft-stebila-secsh-ecdh-00*, November (2003). (*Cited on page 335.*)

[272] M. Steiner, G. Tsudik, and M. Waidner, *Refinement and extension of encrypted key exchange*, ACM Operating Systems Review **29**, July (1995). (*Cited on page 200.*)

[273] C. Stoll, **The Cuckoo's Egg**, Pocket Books, New York (2000). (*Cited on pages 326, 386, 396.*)

[274] A. Stubblefield, J. Ioannidis, and A. Rubin, *Using the Fluhrer, Mantin, and Shamir attack to break WEP*, in Proceedings of the 2002 Network and Distributed Systems Security Symposium (2002), 17–22. (*Cited on page 344.*)

[275] S. Suehring, *SCP/SFTP/SSH URI Format*, INTERNET-DRAFT, *draft-ietf-secsh-scp-sftp-ssh-uri-01.txt*, October (2003). (*Cited on page 335.*)

[276] C. Suetonius Tranquillus, **The Lives of the Twelve Caesars**, Corner House, Williamstown, MA (1978). (*Cited on pages 10 and 566.*)

[277] M. Suzuki, **Group Theory I**, Springer-Verlag, Berlin (1982). (*Cited on page 457.*)

[278] A. Tietäväinen, *On the nonexistence of perfect codes over finite fields*, SIAM J. Appl. Math. **24** (1973), 88–96. (*Cited on page 458.*)

[279] Sun Tzu, **The Art of War**, translated and illustrated by S.B. Griffith, Oxford Paperbacks, Oxford University Press, London (1971). (*Cited on pages 138, 384, 409.*)

[280] M. Weiner, *Cryptanalysis of short RSA secret exponents*, IEEE Transactions on Information Theory **36** (1990), 553–558. (*Cited on page 179.*)

[281] M.J. Williamson, *Non-secret encryption using a finite field*, GCHQ–CESG publication, January 21 (1974). (*Cited on pages 103, 104.*)

[282] Unknown author, *Final report on project C43*, Bell Telephone Laboratory, October (1944), p. 23. (*Cited on page 103.*)

[283] J. van der Lubbe, **Basic Methods in Cryptography**, Cambridge University Press (1998). (*Cited on pages 158 and 571.*)

[284] R.R. Varshamov, *Estimate of the number of signals in error correcting codes*, Dokl. Akad. Nauk SSSR **117** (1957), 739–741. (*Cited on page 446.*)

[285] S. Vaudenay, *Cryptanalysis of the Chor-Rivest cryptosystem*, J. Cryptol. **14** (2001), 87–100. (*Cited on page 105.*)

[286] V. Voydock and S. Kent, *Security mechanisms in high-level network protocols*, Computing Surveys, June (1983). (*Cited on page 136.*)

[287] S.S. Wagstaff, **Cryptanalysis of Number Theoretic Ciphers**, Chapman and Hall/CRC Press, Boca Raton, FL (2003). (*Cited on pages 158, 184.*)

[288] A. Weil, **Sur les Courbes Algebriques et les Variétés qui s'en Deduisent**, Hermann, Paris, (1948). (*Cited on page 465.*)

[289] H.C. Williams, *A $p + 1$ method of factoring*, Math. Comp. **39** (1982), 225–234. (*Cited on pages 515, 553.*)

[290] H. C. Williams, **Édouard Lucas and Primality Testing**, Canadian Mathematical Society Series of Monographs and Advanced Texts, Vol. **22**, Wiley-Interscience, New York and Toronto, (1998). (*Cited on page 554.*)

[291] C.P. Williams and S.H. Clearwater, **Explorations in Quantum Computing**, Springer-Verlag, New York (1998). (*Cited on page 370.*)

[292] M.J. Williamson, *Thoughts on cheaper non-secret encryption*, GCHQ–CESG publication, August 10 (1976). (*Cited on pages 103, 104.*)

[293] T. Ylonen, *SSH protocol architecture*, INTERNET-DRAFT, *draft-ietf-secsh-architecture-15.txt*, October (2003). (*Cited on pages 335, 337.*)

[294] T. Ylonen, *SSH transport layer protocol*, INTERNET-DRAFT, *draft-ietf-secsh-transport-17.txt*, October (2003). (*Cited on pages 335, 337.*)

[295] T. Ylonen, *SSH connection protocol*, INTERNET-DRAFT, *draft-ietf-secsh-connect-18.txt*, October (2003). (*Cited on pages 335, 339.*)

[296] T. Ylonen, T.J. Rinne, and S. Lehtinen, *Secure shell authentication agent protocol*, INTERNET-DRAFT, *draft-ietf-secsh-agent-02.txt*, January (2004). (*Cited on page 335.*)

[297] G. Yuval, *How to swindle Rabin*, Cryptologia **3** (1979), 187–190. (*Cited on page 254.*)

[298] Y. Zamyatin, **The Dragon and Other Stories**, Penguin, London (1975). (*Cited on page 428.*)

[299] J. Ziv and A. Lempel, *A universal algorithm for sequential data compression*, IEEE Transactions on Information Theory, May (1977). (*Cited on page 277.*)

Index

Index

X